Rocky Mountain Section
of the
Geological Society of America

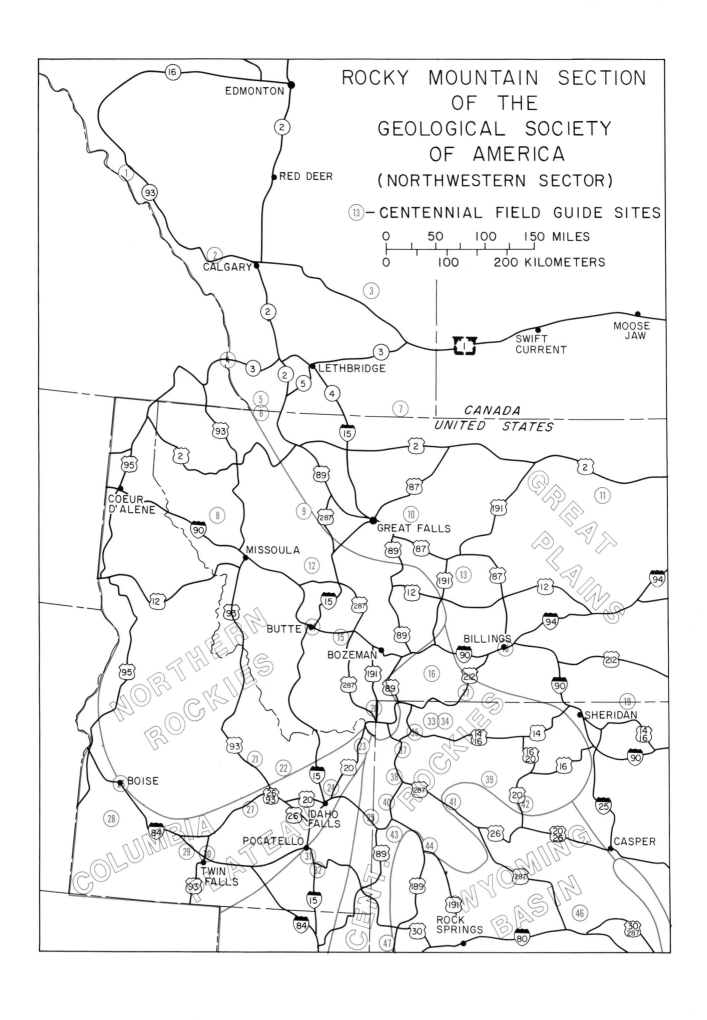

ROCKY MOUNTAIN SECTION
OF THE
GEOLOGICAL SOCIETY
OF AMERICA
(NORTHWESTERN SECTOR)

⑬ —CENTENNIAL FIELD GUIDE SITES

0   50   100   150 MILES
0   100   200 KILOMETERS

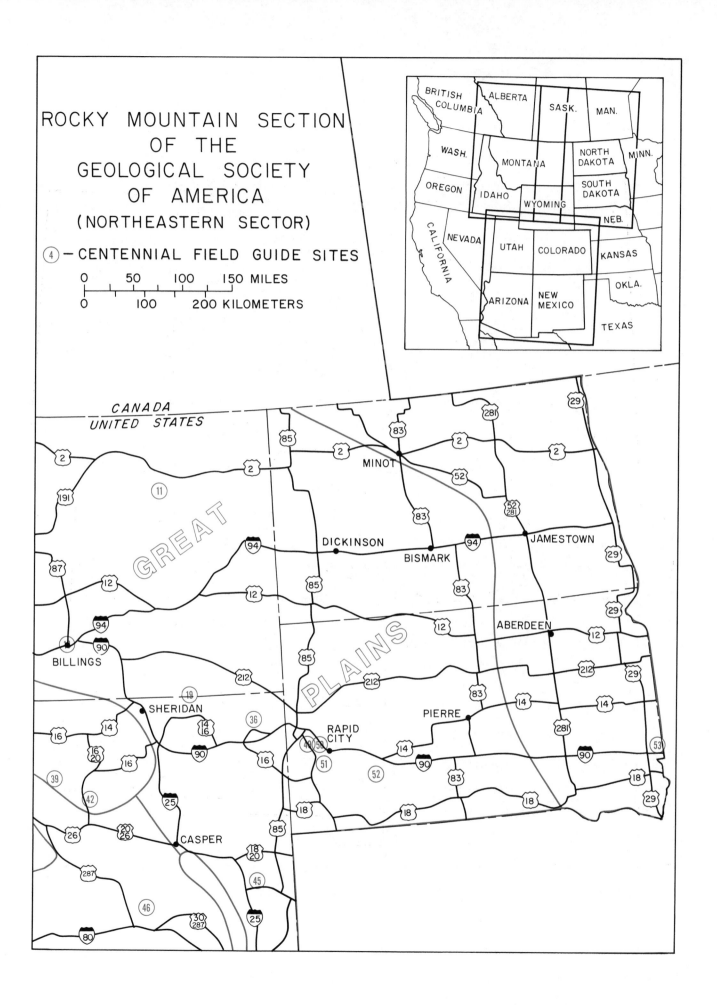

ROCKY MOUNTAIN SECTION
OF THE
GEOLOGICAL SOCIETY
OF AMERICA
(NORTHEASTERN SECTOR)

④ – CENTENNIAL FIELD GUIDE SITES

0   50   100   150 MILES

0   100   200 KILOMETERS

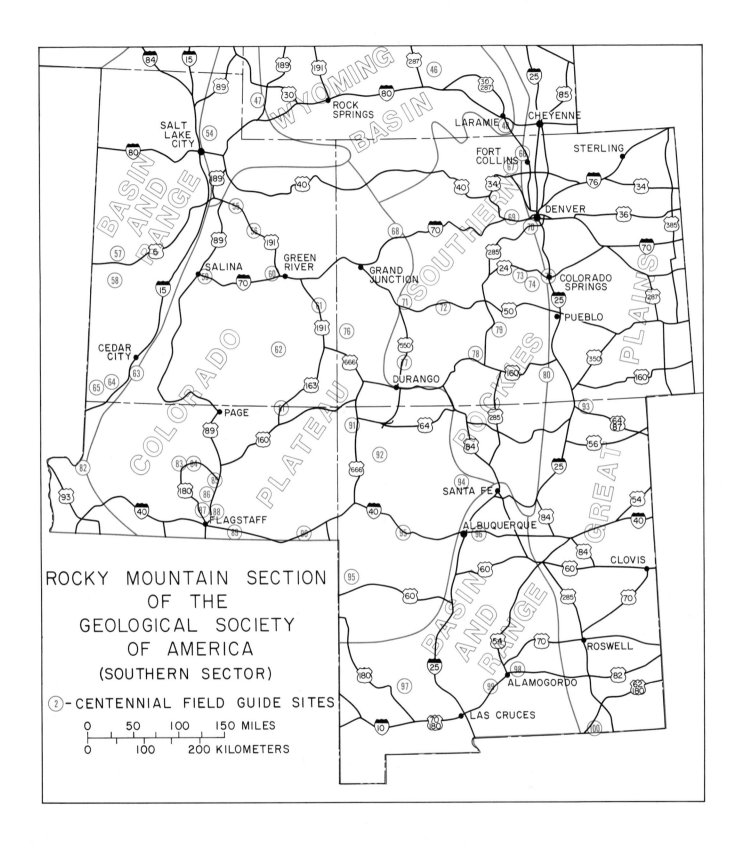

ROCKY MOUNTAIN SECTION
OF THE
GEOLOGICAL SOCIETY
OF AMERICA
(SOUTHERN SECTOR)

②-CENTENNIAL FIELD GUIDE SITES

0   50   100   150 MILES

0   100   200 KILOMETERS

Centennial Field Guide Volume 2

Rocky Mountain Section
of the
Geological Society of America

Edited by

Stanley S. Beus
Department of Geology
Northern Arizona University
Flagstaff, Arizona 86001

1987

Acknowledgment

Publication of this volume, one of the Centennial Field Guide Volumes of *The Decade of North American Geology Project* series, has been made possible by members and friends of the Geological Society of America, corporations, and government agencies through contributions to the Decade of North American Geology fund of the Geological Society of America Foundation.

Following is a list of individuals, corporations, and government agencies giving and/or pledging more than $50,000 in support of the DNAG Project:

ARCO Exploration Company
Chevron Corporation
Cities Service Company
Conoco, Inc.
Diamond Shamrock Exploration
    Corporation
Exxon Production Research Company
Getty Oil Company
Gulf Oil Exploration and Production
    Company
Paul V. Hoovler
Kennecott Minerals Company
Kerr McGee Corporation
Marathon Oil Company
McMoRan Oil and Gas Company
Mobil Oil Corporation
Pennzoil Exploration and Production
    Company

Phillips Petroleum Company
Shell Oil Company
Caswell Silver
Sohio Petroleum Corporation
Standard Oil Company of Indiana
Sun Exploration and Production Company
Superior Oil Company
Tenneco Oil Company
Texaco, Inc.
Union Oil Company of California
Union Pacific Corporation and
    its operating companies:
        Champlin Petroleum Company
        Missouri Pacific Railroad Companies
        Rocky Mountain Energy Company
        Union Pacific Railroad Companies
        Upland Industries Corporation
U.S. Department of Energy

Published by the Geological Society of America, Inc.
3300 Penrose Place, P.O. Box 9140, Boulder, Colorado 80301

Printed in U.S.A.

Cover photo: The Maroon Bells, Pitkin County, Colorado; predominantly Pennsylvanian red clastics of the Maroon Formation. Photo by Lee Gladish.

Library of Congress Cataloging-in-Publication Data
(Revised for vol. 2)

Centennial field guide.

    "Prepared under the auspices of the regional
Sections of the Geological Society of America as a
part of the Decade of North American Geology (DNAG)
Project"—V. 6, pref.
    Vol.   : maps on lining papers.
    Includes bibliographies and index.
    Contents: —v.2 Rocky Mountain Section of
the Geological Society of America / edited by Stanley
S. Beus—   —v. 6. Southeastern Section of the
Geological Society of America / edited by Thornton L.
Neathery.
    1. Geology—United States—Guide-books.   2. Geology—
Canada—Guide-books.   3. United States—Description
and travel—1981-   —Guide-books.   4. Canada—
Description and travel—1981-   —Guide-books.
I. Geological Society of America.
QE77.C46   557.3   86-11986
ISBN 0-8137-5406-2 (v. 6)

Contents

Contents

Contents

TOPICAL CROSS REFERENCES

Topic	Alberta	Montana	Idaho	Wyoming	South Dakota	Utah	Colorado	Arizona	New Mexico
Geomorphology									
Mass Wasting	4	20	25	39					
Glacial	1,5,6,7	8,17	31	35,38,44					
Lacustrine			28,30,32			55			95
Fluvial	3,4		29,30,32	35	52	55,64	71	81,82	
Eolian				46					99
Stratigraphy									
Cenozoic		11,19	26	41	52	55,56,59 63,65	71,73,75	81,82	92
Mesozoic	3	11,18		41,42,47	53	56,59-63 65	68,70,71 75,76	81,90	92
Paleozoic		13	22	41,42,45 48	50	57,58, 60-63,65	70,75,76 77	81,83	96,98 100
Precambrian		15,17	31	42	53	65	67,71,75 77	84	
Sedimentary structures	3	13		48					
Tectonic structures	2	9,15,17 20	21,22 25,31	33,34,36 38,40-43 47	49,50	59,60 61,63,65	66,67,69 70,71,75 76,77,79	81,82,85	97
Volcanic rocks			23,24,27 28,31	35,36,37		64	71,73,78	86,87,88	91,93,94 95,97
Plutonic rocks		10,12,14 16,17		33,34,36	51,53	54	67,69,72 74,78,80		91,96
Metamorphic rocks		12,17	31		49,53	54	67		
Springs,karst and geysers			26,29	37,45					
Mineralization		12,14,16					69,72,76		
Meteorite impact								89	
Paleontology	3	11			52	57,58	73	90	
Physiographic Provinces									
Columbia Plateau			23,24, 26-30						
Northern Rocky Mountains	1,2,4,5	8,9,12,14 15,20	21,22						
Central Rocky Mountains		16,17	25	33-35, 37-41 43,44,47		54			
Southern Rocky Mountains							66,67,69 70,72-75 77-80		
Colorado Plateau						55,65,59 60-63	68,71,76	81,83-90	91,92 94,95
Basin and Range			31,32			57,58,64 65		82	96-99
Wyoming Basins				41,46,48					
Great Plains	3,5,7	10,11,13 18,19		36,45	49-52				93,100
Central Lowlands					53				

Preface

This volume is one of a six-volume set of Centennial Field Guides prepared under the auspices of the regional Sections of the Society as a part of the Decade of North American Geology (DNAG) Project. The intent of this volume is to highlight, for the geologic traveler and for students and professional geologists interested in major geologic features of regional significance, 100 of the best and most accessible geologic localities in the area of the Rocky Mountain Section. The leadership provided by the editor, Stanley S. Beus, and the support provided to him by the Rocky Mountain Section of the Geological Society of America and the Department of Geology of Northern Arizona University are greatly appreciated.

Drafting services were offered by the DNAG Project to those authors of field guide texts who did not have access to drafting facilities. Particular thanks are given here to Ms. Karen Canfield of Louisville, Colorado, who prepared final drafted copy of many figures from copy provided by the authors.

In addition to Centennial Field Guides, the DNAG Project includes a 29-volume set of syntheses that constitute *The Geology of North America,* and 8 wall maps at a scale of 1:5,000,000 that summarize the geology, tectonics, magnetic and gravity anomaly patterns, regional stress fields, thermal aspects, seismicity, and neotectonics of North America and its surroundings. Together, the synthesis volumes and maps are the first coordinated effort to integrate all available knowledge about the geology and geophysics of a crustal plate on a regional scale. They are supplemented, as a part of the DNAG project, by 23 Continent–Ocean Transects providing strip maps and both geologic and tectonic cross sections strategically sited around the margins of the continent, and by several related topical volumes.

The products of the DNAG Project have been prepared as a part of the celebration of the Centennial of the Geological Society of America. They present the state of knowledge of the geology and geophysics of North America in the 1980s, and they point the way toward work to be done in the decades ahead.

Allison R. Palmer
Centennial Science Program Coordinator

Foreword

The Rocky Mountain Section of the Geological Society of America encompasses the intermountain states of Montana, Idaho, Wyoming, Utah, Colorado, and New Mexico as well as the northern third of Arizona, the Dakotas, and the Canadian provinces of Alberta and Saskatchewan. This region includes much of the mountain crest of the continent, plus adjacent parts of the Basin and Range, plateaus to the west and south, and the high plains to the east. A number of important geologic concepts were either first recognized or greatly amplified here by the pioneer geologists of the last century. G. K. Gilbert, a member of an early survey party of this region in the nineteenth century, was the first to recognize laccoliths and extensive block faulted mountains. J. Wesley Powell was the first to traverse the canyons of the Colorado River by boat and first defined the concept of antecedent streams.

Some areas in this region have received intense geologic scrutiny, particularly those near important mining sites or fossil fuel localities and near major universities. Other parts of the region are sufficiently remote to have been only slightly studied. There remain many challenging and stimulating geologic problems to be pursued.

The original editorial committee for this field guide included Donald L. Smith, Montana State University; Kenneth Kolm, Colorado School of Mines; J. Keith Rigby, Brigham Young University in Utah; and myself, as editor. Midway through the project, Don Smith was tragically killed in an accident. His place on the committee was filled by Stephan G. Custer, Montana State University. The selection of the sites to be included, the obtaining of reviews, and the preliminary editing were done by all five members of the committee. However, I accept responsibility for the final editing and any errors that may have been missed. A great debt of appreciation is owed to the other four members of the editorial committee for their generous service in organizing this field guide. They have also each written one or more site descriptions for the book.

The charge given the editorial committee by the Management Board of the Section in 1983 was to select the 100 geologic sites that best display significant and instructive aspects of Rocky Mountain Section regional geology and/or unique examples of basic geologic concepts. Site nominations were solicited by a mailed invitation to all GSA members of the section and also by announcements at annual section meetings and correspondence with colleagues in industry, educational, and state and federal institutions throughout the Section. The editorial committee then invited others familiar with the local geology of the various regions to assist in selecting the 100 most appropriate sites from the more than 300 originally nominated. An attempt, only partially successful, was made to achieve a regional and topical balance of sites. In some cases, site descriptions from areas not originally nominated were actively solicited. Even so, the northeastern region of the section is not as well represented as the rest, owing in part to lack of nominations and the withdrawl of some sites originally selected.

The topical distribution of sites is more balanced and covers a wide spectrum of features. It was agreed by the editorial committee, as initially suggested by GSA guidelines, to avoid sites that were of interest chiefly for mineral or fossil collections. However, several sites are included wherein fossils of biostratigraphic significance or minerals occur as unique examples of mineralization processes. One site of an impact crater having both earthly and extraterrestrial significance is included.

The geographic and topical distribution of sites is summarized in Table 1. Geographical arrangement is both by state or province and physiographic province. A number of sites include strata from more than one erathem or major features of more than one type; they have multiple entries in the table.

The limitation to 100 localities in such a vast and varied region necessitated omission of many significant sites. No doubt some will find a favorite site not included. It is hoped that the site descriptions contained here will provide a useful introduction for students and professionals to the regional geology. It is recognized that many of the localities described here are treated in local or regional guidebooks or other older literature. However, many of these references are either out of print or not conveniently available, particularly to those travelling through this region for the first time.

The site descriptions are arranged and numbered in groups by state or province beginning in the north, and generally from north to south within a state or province. Each site description includes an index map of the locality, access information, significance of the site, and description of features to be observed. The reference list at the end of each description was purposely limited to the most appropriate citations. The references within these citations provide a more complete bibliography. At some sites all the geologic features are available at a single location, but in many a series of separate stations, or observation viewpoints, is included in the site description. Most sites are located along or near a road, but some require major traverses on foot. Several localities wherein two or more closely related or closely spaced features occur are treated as double sites, as at Yellowstone, Grand Canyon, the Black Hills, and Heart Mountain, Wyoming. Many of the sites are on federal or state lands and are thus readily open to the public. Sites on private land may require permission to visit as indicated in individual site descriptions. At some sites, collecting or extensive hammering on the rocks may be permissible. At others, particularly on private land, such activities may be neither necessary nor appropriate and are expressly prohibited in national, provincial, or state parks or monuments.

Appreciation is extended to the 137 authors and co-authors who contributed to the field guide for their generally enthusiastic responses and patience during the several years required for completion. I am especially grateful to the unnamed colleagues from throughout the continent who provided critical and helpful reviews. Appreciation is also extended to those who nominated sites that could not be included because of space limitations. Perhaps at some future date there will be a second volume containing the second 100 significant localities in the section.

The Rocky Mountain Section of GSA provided financial assistance in mailing and other editorial expenses associated with this field guide. The Geology Department at Northern Arizona University assisted in mailing and secretarial chores. Assistance in editing and in preparation of the index map was provided by the following students: John Anthony of Colorado School of Mines, Lee Murray of Montana State University, and Hilary Mayes of Northern Arizona University.

Stanley S. Beus
January, 1987

1

The Athabasca Glacier, Alberta, Canada

Gerald Osborn, *Department of Geology and Geophysics, University of Calgary, Calgary, Alberta T2N 1N4, Canada*

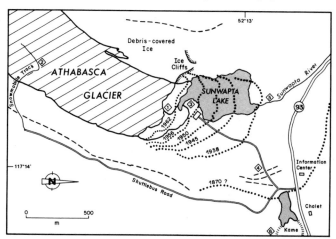

Figure 1. The proglacial area of the Athabasca Glacier in 1979; modified from Inland Waters Directorate, 1979. Diamonds enclose stop numbers. Dashed lines: moraine crests. Dotted lines: former positions of toe at dates shown. Date of outermost toe position is estimated at 1870 by Environment Canada, but Luckman (1982) suggests date is 1840–50.

LOCATION AND ACCESSIBILITY

The Athabasca Glacier, a major tourist attraction in the Canadian Rockies, lies close to Alberta 93 (the Icefields Parkway) in Jasper National Park, 100 mi (160 km) northwest of Banff and 60 mi (95 km) southeast of Jasper. The Icefields Chalet and Parks Canada Information Center (Fig. 1) are accessible by bus from Banff and Jasper. Interpretive exhibits are on display at the center. From the highway a paved road leads to a parking lot within 0.3 mi (0.5 km) of the glacier terminus (Stop 1, Fig. 1). The proglacial area is easily explored on foot, and the ice-front sometimes can be traversed with ordinary boots. However, steep slopes, crevasses (sometimes hidden by snow), and meltwater pools occur in the terminal area; extended travel without partners, rope, crampons, and ice axe is not recommended. A less-steep, less-dissected part of the glacier surface is accessible via the snowmobile track that descends the lateral moraine on the southeast side of the glacier (Stop 2, Fig. 1); the head of this road can be reached by shuttle bus from the Information Center. Foot travel is generally safe on the track itself but may not be off the track.

SIGNIFICANCE

Apart from constituting spectacular scenery, the Athabasca Glacier and its surroundings provide excellent examples of glacial and glacial-geomorphological features. By virtue of its accessibility, the glacier is one of the most studied in the world, and much information has been published on its geometry and flow. Glacial deposits and erosional forms are abundant and obvious in the

Figure 2. Part of the Columbia Icefield as seen from Mt. Andromeda. On the skyline is Mt. Columbia, the highest point in the vicinity of the icefield. 1: Upper icefall, marking the head of the Athabasca Glacier (which descends to lower right). 2: Snowline (lower margin of previous winter's snow). 3: Splaying crevasses.

area. Of further interest is the relatively well-known recent history of the glacier; advances have been dated using tree rings, and glacier retreat over the last century is well documented.

THE GLACIER

The Athabasca is an outlet glacier of the Columbia Icefield, an ice cap straddling the continental divide (Fig. 2). The icefield, roughly 116 mi² (300 km²) in area, occupies a gently rolling plateau ringed by peaks, the highest of which is Mt. Columbia (12,295 ft; 3,748 m). The thickest parts of the ice cap, measured with radar techniques, are up to 1,200 ft (365 m) thick. Thick-

Figure 3. The Athabasca Glacier descending from the Columbia Icefield. Obvious are the prominent lateral moraines on either side of the glacier and the black supraglacial debris along the northwestern (right-hand) margin of the glacier. Photograph taken in 1985. Mt. Andromeda is the vantage point of Figure 2.

nesses on the order of 320 ft (100 m) along edges of the ice cap can be seen from the highway. Ice on the plateau discharges through several outlet valley glaciers, one of which is the Athabasca (Fig. 3). Flowing from the ice cap, the Athabasca Glacier drops over a series of three bedrock steps, each marked by a short icefall. Below the lowest icefall, the 0.6-mi (1-km) wide valley glacier extends for about 2 mi (3.5 km).

Thickness of the glacier has been studied with seismic reflection, gravity methods, and boreholes; references can be found in Kucera (1981). A profile is shown in Fig. 4. In 1963, centerline ice thicknesses were 300 ft (92 m) above the middle icefall and 1,050 ft (320 m) about 0.6 mi (1 km) below the lower icefall.

Measurements of flow of the ice have been made in several studies. Using stakes along the longitudinal centerline, Savage and Paterson (1963) measured surface velocities ranging from 240 ft/yr (74 m/yr) at the foot of the lower icefall to 50 ft/yr (15 m/yr) near the terminus. They concluded that ice would travel from the lower icefall to the 1960 terminus in about 150 years. More recently, Kucera (1981) measured an average flow rate of 68 ft/yr (20.8 m/yr) near the terminus, with short-term summer rates as high as 0.2 in/hr (0.5 cm/hr). Kucera attributes faster summer flowage to meltwater lubrication. Savage and Paterson (1963) measured velocity distributions at depth using borehole instruments; a typical profile is shown in Figure 5. The proportion of surface velocity due to basal sliding varied considerably in the glacier, ranging from about 10% in the borehole shown in Figure 4 to 75% elsewhere.

Several surface phenomena can be seen on the Athabasca Glacier, some from a distance and some by walking on the ice. These include the snowline, crevasses, foliation, thrust faults, meltwater streams, and moulins.

The snowline, or firn limit, is the lower margin of the previous winter's snow cover. During spring and summer the snowline recedes upglacier, normally to or above the upper icefall by the end of the summer (Fig. 2). The late-summer snowline is the approximate position of the equilibrium line, which separates the accumulation area on the Columbia Icefield from the ablation area encompassing the entire Athabasca Glacier.

A variety of crevasse types occurs on the glacier. Splaying crevasses, common near the head of the glacier (Fig. 2) as well as at the terminus, parallel the glacier's longitudinal axis at their upstream ends but splay out downglacier. Transverse crevasses occur in the icefalls and also where the ice thins over a bedrock bulge at the terminus (Stop 1). Marginal crevasses, which are oriented upglacier at a 45° angle from the ice margin, characterize the middle part of the glacier near Stop 2. Kucera (1981), who

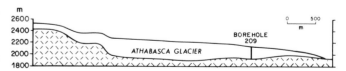

Figure 4. Longitudinal profile of the Athabasca Glacier in 1959–60. The two steps in the left-hand portion of the profile are the middle and lower icefalls. Modified from Kanasewich, 1963.

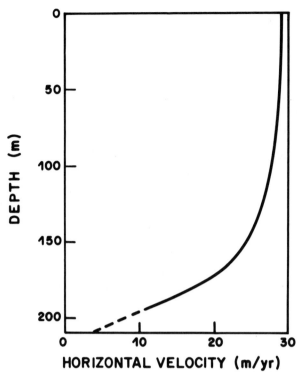

Figure 5. Vertical velocity distribution of the Athabasca Glacier at Borehole 209. Modified from Savage and Paterson (1963) and Paterson (1969).

Figure 6. Closed fractures (possibly thrust faults) and associated dirt cones and dirt patches near the toe of the Saskatchewan Glacier. Similar fractures are sometimes apparent on the Athabasca Glacier.

measured several of the largest crevasses on the glacier, found none deeper than 105 ft (32 m).

Layered structures commonly visible in glaciers are foliation (somewhat analogous to metamorphic foliation in rocks) and annual layering (analogous to stratification in sedimentary rocks). Foliation is well developed around the toe of the Athabasca Glacer, where it consists of thin, alternating layers of white bubbly ice and bluish clear ice. Along the margins of the glacier, the foliation dips steeply toward the longitudinal centerline; at the terminus the foliation arcs around the snout in a spoon-shaped pattern. A walk on the ice (Stop 2) will usually reveal several examples of foliation deformed by small folds and faults. Annual layers are not easily visible on the glacier except in some crevasses. Such layers, however, can usually be seen from Stop 2, in ice cliffs in the small glaciers southeast of the Athabasca Glacier.

Closed fractures parallel or subparallel to foliation in the ice are apparent near the terminus in some years (Stop 1). They may be lined with dirt or studded with small black dirt cones. Examples from the nearby Saskatchewan Glacier are shown in Figure 6. These fractures are probably thrust faults (commonly designated as "shear planes") that bring subglacial sediment up through the ice to the surface.

Surface meltwater streams, and the microrelief produced by them, are extremely common in middle reaches of the glacier in summer (Stop 2). Surface water is somewhat less common near the toe because many rivulets drop into crevasses before reaching the terminus. Among the obvious features associated with the surface runoff are well-developed incised meanders and deep vertical shafts called moulins. The latter form as water running down cracks gradually melts away ice.

Hydrology and Sediment Yield. Most surface meltwater drops through crevasses and moulins to the base of the ice before reaching the terminus. This water, and any added by basal melting, travels in subglacial streams that emerge from tunnels at the terminus. These streams flow into Sunwapta Lake near the north corner of the terminus; the lake waters meanwhile discharge to form the Sunwapta River (Figs. 1, 3). Meltwater streams on, under, and in front of the glacier show diurnal variations in discharge related to air temperature fluctuations. Occasional large floods apparently unrelated to weather conditions are interpreted to be jökulhlaups (sudden releases of water confined by obstructions in the glacier).

Sunwapta Lake, occupying a basin at least partly formed by glacial excavation, formed after 1938 and reached its maximum size in 1966 (Baranowski and Henoch, 1978). It is presently decreasing in size due to greater input than output of sediment. Meltwater streams are building deltas into the lake; the parking lot near the toe of the glacier is situated on one such delta (Stop 3).

Glacial Erosional Forms. Large-scale glacial erosional forms are obvious in the vicinity of the Columbia Icefield, and cirques and aretes in particular are well displayed above the margins of the Athabasca Glacier. Smaller forms occur in the proglacial area wherever bedrock crops out. For example, polished and striated bedrock is common along the short path from the parking lot to the glacier toe (Stop 1), and rock flour produced by contemporary glacial abrasion colors the streams along the path. Miniature roche moutonnées (bedrock ridges, smooth and streamlined on the upglacier end and steep and jagged on the downglacier end) can be found between Sunwapta Lake and the glacier toe.

Glacial Depositional Forms. Glacially transported debris can be seen on, under, and marginal to the glacier. Along the toe a compact, sticky, mainly fine-grained subglacial ("lodgement") till extends from underneath the ice into the proglacial area (Stop 1). Superglacial debris, consisting largely of loose, coarse, angular, nonstriated clasts, is produced by intermittent rockfall from cliffs above the glacier. Along the northwestern margin of the glacier, such debris covers a black, dirty-ice cliff southwest of Sunwapta Lake (Figs. 1, 3). Because the debris partially insulates underlying ice from ablation, the debris-covered ice stands higher above the lake than does the clean ice.

The prominent lateral moraines flanking the glacier (Figs. 1, 3) were built during the recent advance of the glacier, discussed below. Because these moraines are ice cemented and possibly ice cored, their thawing causes unstable masses of till to slump along the steep flanks facing the glacier. The terminal moraine continuous with the lateral moraine is not well developed, and parts of it have been destroyed by construction activity. However, the glacier is known to have extended slightly beyond the highway to the north of the Information Center during its recent advance. At that time the glacier merged with the lower end of the Dome Glacier to the west.

During subsequent retreat the glacier has left a series of end moraines. The larger recessional moraines (about 10 to 20 ft; 3 to 6 m high) were produced during temporary stillstands of the glacier toe (Kucera, 1981). Three such moraines are shown in Figure 1 (Stop 4). Smaller "annual moraines" (2.3 to 6.5 ft; 0.7 to 2 m high) can be seen along the road southwest of Stop 4. Their origin has been described by Kucera (1981). During the summer the glacier toe retreats several feet, but during the succeeding winter the rate of ice flow exceeds the rate of melting, and the toe advances slightly. During the winter advance, till is pushed into a gap between the ice and proglacial snow. The ridge of till is revealed upon melting of the snow.

Other types of deposits associated with glacial environments can be seen. One is the outwash in the floodplain of the Sunwapta River (Stop 5). The braided pattern and relatively rapid channel shifts are characteristic of rivers carrying heavy bed loads. A second type is the kame deposit constituting the hill across the highway from the chalet (Stop 6). Exposures on the east flank of the hill show stratified glaciofluvial sediment probably deposited in a small ice-marginal lake impounded by the glacier when the latter was more extensive (Luckman and Osborn, 1979). The deposit predates the recent advance of the glacier.

RECENT GLACIAL HISTORY

A minor glacial advance (or advances) occurring in the last few centuries is known in many parts of the world. This advance, sometimes popularly termed the Little Ice Age, is called the Cavell Advance in the Canadian Rockies (Luckman and Osborn, 1979). The main advances in the region of Banff, Jasper, and Yoho parks appear to have occurred from the late 17th to the early 18th century, in the early to mid 19th century, and from the late 19th century to the early 20th century (Luckman and Osborn, 1979). Heusser (1956) dated the maximum extent of the Athabasca Glacier at A.D. 1714, based on tree-ring studies of a tree tilted by the glacier near its former confluence with the Dome Glacier. Following a retreat, the glacier readvanced to almost the same point in about 1840, according to Heusser. Recent work by Luckman and colleagues (Luckman, 1982) suggests that the mid–19th century advance was slightly more extensive than the earlier one. Retreat began again by 1870, but the fronts of the Athabasca and Dome glaciers were still coalesced when the earliest white explorers described the area around the turn of the century.

Retreat has continued over the last century; some recessional positions are shown in Figure 1. The recent volumetric decrease of ice in the Athabasca Glacier was calculated by Kite and Reid (1977). They estimated that by 1971 the glacier below the second icefall had lost one-third of its 1870 volume. If the average rate of retreat of the toe from 1870 to 1971 (43 ft/yr; 13 m/yr; Kite and Reid, 1977) were to continue, the toe would reach the lower icefall in 250 to 300 years.

REFERENCES

Baranowski, S., and Henoch, W.E.S., 1978, Glacier and landform features in the Columbia Icefield area, Banff and Jasper National Parks, Alberta, Canada: Report for Parks Canada, 131 p.

Heusser, C. J., 1956, Post glacial environments in the Canadian Rocky Mountains: Ecological Monographs, v. 26, p. 263–302.

Inland Waters Directorate, 1979, Athabasca Glacier: Environment Canada Inland Waters Directorate Glacier Map Series No. 8, Sheet No. 6.

Kanasewich, E. R., 1963, Gravity measurements on the Athabasca Glacier, Alberta, Canada: Journal of Glaciology, v. 4, p. 617–631.

Kite, G. W., and Reid, I. A., 1977, Volumetric change of the Athabasca Glacier over the last 100 years: Journal of Hydrology, v. 32, p. 279–294.

Kucera, R. E., 1981, Exploring the Columbia Icefield: Canmore, Alberta, High Country Publishers, 64 p.

Luckman, B. H., 1982, The Little Ice Age and oxygen isotope studies in the middle Canadian Rockies: Unpublished report to Parks Canada, 31 p.

Luckman, B. H., and Osborn, G. D., 1979, Holocene glacier fluctuations in the middle Canadian Rocky Mountains: Quaternary Research, v. 11, p. 52–77.

Paterson, W.S.B., 1969, The physics of glaciers: London, Pergamon Press, 250 p.

Savage, J. C., and Paterson, W.S.B., 1963, Borehole measurements in the Athabasca Glacier: Journal of Geophysical Research, v. 68, p. 4521–4536.

The McConnell thrust at Mount Yamnuska, Alberta, Canada

P. E. Gretener, Department of Geology and Geophysics, University of Calgary, Calgary, Alberta T2N 1N4, Canada

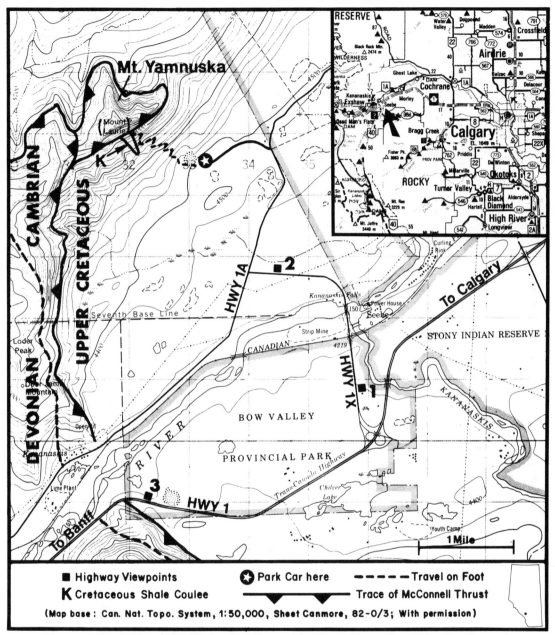

Figure 1. Topographic and simplified geological map of the Yamnuska area. Consult for viewpoints and access. Insert gives location in respect to the nearest major city, Calgary, Alberta.

LOCATION

McConnell thrust can be reached by driving west from Calgary for 49 mi (78 km, official mileage from the city's center) along the Trans-Canada (T-C) Highway 1 to the intersection with T-C 1X leading north to the old Calgary-Banff highway (T-C 1A; Fig. 1). Those in a hurry to get to Banff may go past the intersection to viewpoint no. 3 for a distant view of the McConnell thrust, which rises from the valley floor up a ramp into the saddle on the west side of Mt. Yamnuska (Fig. 2). A small splay fault occurs in the saddle, and the eastward continuation of the thrust along the base of the Yamnuska wall is essentially horizontal in the direction of transport. For a closer view, turn north onto T-C 1X and stop at the entrance to the campground; better yet, proceed to the vicinity of the junction with T-C 1A and pull over onto the shoulder for a view as in Figure 3. In all cases, field glasses will be

Figure 2. Highway viewpoint no. 3 is located to the left of the gravel pit in the lower left-hand corner of the picture. On the far left the well-bedded Devonian Fairholme Formation is visible. The main wall is the Cambrian Eldon Formation. The McConnell thrust rises from the valley to the saddle just southwest of Mt. Yamnuska and is horizontal under the mountain. The fault closely hugs the base of the cliff, with a small splay fault cutting up through the saddle west (left) of Yamnuska.

helpful. Those ardent souls who wish to touch the thrust plane should turn east on T-C 1A and drive for 1.3 mi (2.1 km) to the dirt road leading northwest into the quarry at the foot of Mt. Yamnuska. They may park here and then proceed on foot to the upper east end of the marked shale coulee (Fig. 4)—a steep climb of about 1,600 ft (480 m), which should take about 1¼ hours.

SIGNIFICANCE

The McConnell thrust of the southern Canadian Rocky Mountains is one of the world's major thrusts. It has a lateral extent of about 250 mi (400 km), and its maximum total displacement is on the order of 25 mi (40 km) (Dahlstrom, 1969). Over much of this range, it forms the boundary between the Foothills and the Front Ranges, as is the case at Mt. Yamnuska. The Yamnuska locality is unique because the thrust plane is actually exposed over a distance of about 500 ft (150 m)—within viewing and walking distance of a major Highway (T-C 1). Anyone working in thrust-faulted terranes appreciates the rarity

Figure 3. Mt. Yamnuska from highway viewpoint no. 2. The fault follows the base of the cliff, which consists of middle Cambrian Eldon carbonates. Lower grassy slopes are upper Cretaceous Belly River Formation, a sequence of shales and sandstones. The Cretaceous shale coulee is visible near the center of the photograph (arrow).

Figure 4. Close-up of the shale coulee where the fault is exposed at the upper termination over a distance of about 500 ft (150 m). Note the fold in the Belly River in the lower left of the coulee. A late normal fault displaces the thrust plate at the centre of the head of the coulee. Displacement is about 10 ft (3 m).

of such a coincidence. Details of the location are found on the geological map by Price (1970).

SITE INFORMATION

The McConnell thrust was recorded by McConnell in 1887 and named in his honor by Clark in 1949. Detailed mapping in the Southern Canadian Rocky Mountains, supported by extensive geophysical surveys and drilling on the part of the oil industry, has produced a coherent picture of the structure of the Front Ranges (Bally and others, 1966; Price and Mountjoy, 1970).

West of Calgary, the McConnell thrust separates the Foothills and the Front Ranges provinces (Price and Mountjoy, 1970, Fig. 2-1). Thus it occupies a position equivalent to the one held by the better-known Lewis thrust just north and south of the international border.

At Yamnuska (Figs. 3 and 4), the McConnell thrust places Middle Cambrian Eldon limestone, which forms vertical cliffs, over Upper Cretaceous Belly River clastic rocks that form the grassy and wooded slope below. Figure 2 shows the McConnell fault rising at about 20° over a ramp from the Bow Valley floor (4,400 ft; 1,340 m) to the saddle west of Mt. Yamnuska, where there is a small splay fault, a backlimb thrust of Douglas (1950). It flattens out along the wall of Yamnuska to an essentially horizontal attitude at an elevation of about 6,100 ft (1,860 m). The thrust is horizontal only in the direction of transport but dips into the valley on the north side of Yamnuska (Fig. 1; Price 1970). Along strike the McConnell thrust near the Bow Valley has a distinctly wavy attitude.

Yamnuska is not far from the area of maximum displace-

Figure 5. At the head of the coulee the fault is fully exposed. Note the sharp contact.

ment of the McConnell thrust. In fact a half-window formed by the underlying Panther River Culmination just 30 mi (48 km) northwest of Yamnuska reveals Cambrian over Cretaceous for a minimum distance of 10 mi (16 km). The surprising element about the thrust surface is its sharp nature and the lack of gouge formation (Fig. 5) in view of the large transport distance. Stratigraphic displacement at Yamnuska is given as 13,000 ft (4,000 m) by Clark (1949). This is a conservative estimate; modern cross-sections such as by Bally and others (1966) suggest a stratigraphic displacement in excess of 20,000 ft (6,100 m).

Minor features that can be observed along the fault plane include basal tongues (Gretener, 1977, Fig. 3). These apparent intrusions into the upper plate by material from the footwall appear to be a common feature for large thrusts. Offset of the thrust plane by late normal faults has created what climbers refer to as "the cave" (Fig. 4, top center of shale coulee; also Gretener, 1977, Fig. 11).

The severe tectonic distortions observed throughout the Eldon Formation remain enigmatic. Near the base of the thrust, secondary faults and folding are both common, and a closer examination of the seemingly massive Yamnuska wall reveals that the Eldon limestones are strongly deformed right up to the

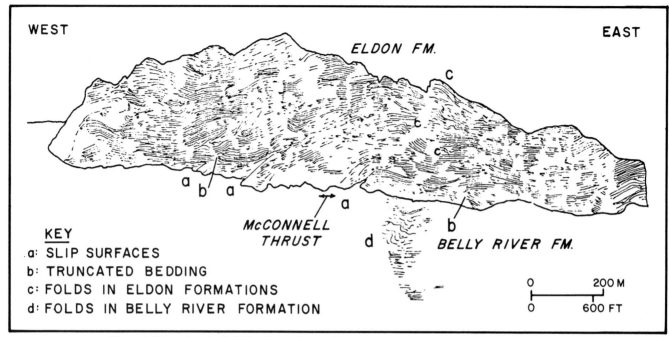

Figure 6. Distortions in the Eldon Formation on Mt. Yamnuska, as recorded by Venter (1973). Note that deformation ranges to the mountain top and is not restricted to the vicinity of the fault. The wall is 1,000 ft (300 m) high.

top of the mountain, 1,000 ft (300 m) above the thrust plane (Gretener, 1977, Fig. 10). Venter (1973) traced the major elements of these deformations, as shown in Figure 6. As one follows the Eldon down the ramp (Fig. 2), it is obvious that this is equally true for the Eldon wall to the right of Loder Peak. The two light-gray bands visible in the wall just left of the center of Figure 2 are a give-away for the severe tectonic deformation the Eldon has undergone. Above the Eldon is the well-bedded Devonian Fairholme Formation, seen at the very left in Figure 2. These beds, separated from the Eldon by a major erosional unconformity, are completely unaffected. How these relations can be interpreted in the light of current theories of thrust mechanics is uncertain.

In contrast, the overridden Cretaceous Belly River Formation seems little deformed. This nonmarine formation contains numerous sandstone channels forming discontinuous ledges on the grassy lower slopes of Yamnuska. All these shelves dip uniformly 5 to 10° SW. The indicated structural tranquility may, however, be more apparent than actual, as is shown by the substantial fold exposed on the lower left side of the shale coulee in Figure 4. This suspicion is confirmed by Spratt (personal communication, 1985), who notes that "although the deformation in the Eldon is important and unexpected since it is much more competent than the Belly River, the Belly River deformation is not minor; it is just harder to see because of the cover of vegetation and scree. On the east side of Yamnuska and at End Mountain to the north, there is substantial folding in the Belly River. Although some fold axes trend northwesterly, the majority I have measured trend northeasterly, like the 'waves' in the overlying Eldon." Yet this observation cannot detract from the fact that the distortions in the Cretaceous are mild when compared with the drastic structural deformation suffered by the massive Eldon Formation as the basal unit of the McConnell thrust plate in the Bow Valley.

REFERENCES CITED

Bally, A. W., Gordy, P. L., and Stewart, G. A., 1966, Structure, seismic data, and organic evolution of the Southern Canadian Rocky Mountains: Bulletin of Canadian Petroleum Geology, v. 14, no. 3, p. 337–381. (Also Special Reprinting, Canadian Society of Petroleum Geology, 1970, 1982.)

Clark, L. M., 1949, Geology of Rocky Mountain Front Ranges near Bow River, Alberta: Bulletin of the American Association of Petroleum Geologists, v. 33, no. 4, p. 614–633.

Dahlstrom, C.D.A., 1969, Balanced cross sections: Canadian Journal of Earth Sciences, v. 6, no. 4, p. 743–757.

Douglas, R.J.W., 1950, Callum Creek, Langford Creek, and Gap map-areas, Alberta: Geological Survey of Canada Memoir 255, 124 p.

Gretener, P. E., 1977, On the character of thrust faults with particular reference to the Basal Tongues: Bulletin of Canadian Petroleum Geology, v. 25, no. 1, p. 110–122.

Price, R. A., 1970, Geology of the Canmore sheet (east half): Geological Survey of Canada Map 1265A, scale 1:50,000.

Price, R. A., and Mountjoy, E. W., 1970, Geologic structure of the Canadian Rocky Mountains between Bow and Athabaska Rivers; A Progress Report: Geological Association of Canada Special Paper 6, p. 7–25.

Venter, R. H., 1973, McConnell Thrust and associated structures at Mt. Yamnuska, Alberta, Canada [M.S. thesis]: Calgary, University of Calgary, 118 p.

3

Upper Cretaceous coastal plain sediments at Dinosaur Provincial Park, southeast Alberta

Emlyn H. Koster, *Alberta Geological Survey, Alberta Research Council, 4445 Calgary Trail South, Edmonton, Alberta T6H 5R7, Canada*
Philip J. Currie, *Tyrrell Museum of Palaeontology, P.O. Box 7500, Drumheller, Alberta TOJ OYO, Canada*

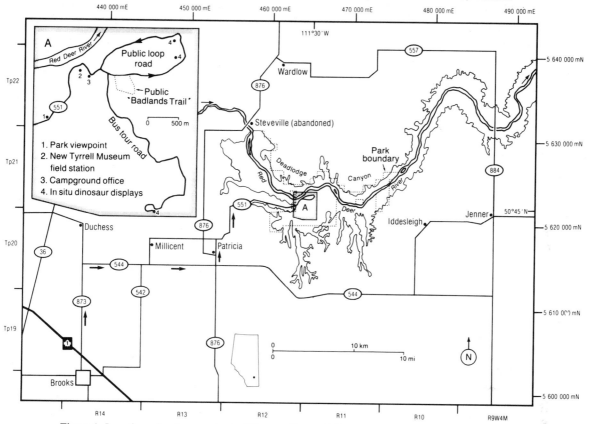

Figure 1. Location of and access to the Dinosaur Provincial Park area showing generalized extent of badlands. Both the 6-mile township-range lines of the Dominion Land Survey System and the 10-km coordinates in Zone 12 of the Universal Transverse Mercator Grid are shown. Map and air photo coverage is listed in Koster (1984). Insert A shows the layout and points of interest in the area surrounding the present Park headquarters.

LOCATION AND ACCESSIBILITY

Dinosaur Provincial Park, a UNESCO World Heritage Site 112 mi (180 km) east of Calgary, Alberta, occupies part of a large tract of badlands along the Red Deer River valley in the semi-arid southeast Alberta plains (Fig. 1). Access is from the Trans-Canada Highway at Brooks (where all major services are available), northeastward 30 mi (48 km) on Alberta Highways 873, 554, and 551. The Park is fully operative from May 15 to September 15; camping facilities are limited with a small retail food outlet. Access to outcrop requires dry weather, and precautions should be taken for extreme heat, rough terrain and occasional rattlesnakes. Provincial regulations prohibit the extraction of any fossil material.

In terms of visitor access, the badlands are divided into two

areas (Fig. 1). First, Dinosaur Park encloses 23 mi^2 (60 km^2) of crown land flanking the river for 16 mi (25 km) downstream from the Steveville bridge on Alberta Highway 876. It is sub-divided into a small public area surrounding the campground and a Natural Preserve into which bus tours and hikes are regularly conducted by Park officials. Inquiries about further access to the Natural Preserve should be directed to Park officials (403/378-4587). Plans are underway to relocate the Park headquarters approximately 1.2 mi (2 km) upstream and to expand visitor services. Secondly, the extensive privately owned badlands that surround the Park continue eastward to the bridge crossing on Alberta Highway 884, north of Jenner. With a rugged vehicle and local landowner permission, the prairie edge is reached by a network of dirt tracks leading inward from the encircling highways.

9

Subaerial	Freshwater
Herbivorous dinosaurs	Mesoreptiles
Hadrosaurids - 8	Turtles - 11
Ceratopsids - 7	Crocodiles - 4
Nodosaurids - 2	Champsosaurs - 1
Pachycephalosaurids - 4	
Ankylosaurids - 2	Fish - 14
Thescelosaurids - 1	
Carnivorous dinosaurs	Salamanders - 7
	Molluscs - 4
Dromaeosaurids - 4	
Tyrannosaurids - 3	Frogs - 3
Ornithomimids - 3	
Caenagnathids - 2	
Elmisaurids - 1	Marine
Troodontids - 1	
Mammals - 20	Fish - 2
Lizards - 11	
	Plesiosaurs - 2
Flying reptiles - 2	
	?Unreworked plankton
Birds - 1	

*Numbers indicate current numbers of species.

SIGNIFICANCE

In the Park area, the Red Deer River valley originated during deglaciation as a deep, joint-aligned spillway system through near-horizontal Upper Cretaceous strata. Except where the bedrock surface is cut by buried valleys, the capping veneer of Pleistocene deposits rarely exceeds 10 ft (3 m). Subsequent erosion of the spillway walls has led to the modern terrain along Deadlodge Canyon—the largest tract of badlands in western Canada.

This major geomorphic feature provides excellent exposure of the upper 200–330 ft (60–100 m) of the late Campanian Judith River Formation (McLean, 1971, 1977). It represents a coastal plain succession richly fossiliferous with vertebrates (Table 1; Dodson, 1983). Vertebrate quarry sites became protected by establishment of the Provincial Park in 1955. The UNESCO World Heritage Site status since 1979 signifies the unrivalled abundance and variety of articulated vertebrate remains.

East of 472500mE (Fig. 1) above 2,412–2,428 ft (735–740 m) a.s.l., the basal 82 ft (25 m) of the overlying marine Bearpaw Formation also crops out. This sequence of concretionary fissile shales contains the ammonites *Placenticeras meeki* and *Hoploscaphites* and the ornate bivalve *Cymella montanensis,* as well as shark teeth, fish scales, and mosasaur and plesiosaur remains. The Judith River–Bearpaw boundary represents a continental-marine transition separating the last two cycles of foreland sedimentation along the Western Interior. The actual transition is conspicuously marked by a coaly argillaceous zone, 16–50 ft (5–15 m) thick, that locally includes several feet of herringbone cross-laminated, greenish sandstone containing *Ophiomorpha nodosa, O. bornensis,* and bored oyster valves.

BACKGROUND INFORMATION

The Park area outcrop of the Judith River Formation is treated here as one 'locality'. This unconventional approach reflects the unusually large extent of outcrop, the varying ease of accessibility to different parts of the badlands, the fairly representative nature of any one part, and the plans to change the layout of publicly accessible badlands within the Park. This section describes recurring sedimentological features, catalogs four taphonomic states for vertebrate remains (Koster, 1987), and briefly interprets the depositional framework.

The optimal itinerary of a brief 1–2 day visit with easy logistics should include (1) the viewpoint at the Park entrance on Alberta Highway 551, (2) at least one scheduled tour into the Natural Preserve, (3) a north-south traverse across an area of three-dimensional exposure outside the Park (e.g., Little Sandhill Creek coulée immediately south of Alberta Highway 551 near the Park entrance), and (4) an overview of one or two extensive two-dimensional exposures commonly found on south- or west-facing walls of coulées tributary to Deadlodge Canyon. The first two components are introductory, the third provides for close examination of the sedimentary and taphonomic features outlined below, and the last enables appreciation of larger scale aspects of stratigraphic architecture.

PALEOGEOGRAPHIC SETTING

The aggrading coastal plain supported a rich ecosystem of vertebrates (Table 1) and was well vegetated under a warm, equable climate with seasonally variable precipitation (Koster, 1984). Perennial drainage from newly uplifted parts of the Cordillera flowed across the foreland basin to the encroaching shoreline of the Western Interior Seaway. In theory, the combined effect of tidal circulation (Parrish and others, 1984) and a rising base-level would have rendered the lower reaches of the low-gradient drainage susceptible to an estuarine influence.

STRATIGRAPHIC ARCHITECTURE

Vertebrate-bearing 'coarse members,' commonly 13–20 ft (4–6 m) thick, form most of the outcropping Judith River strata. These variable channel deposits occur between a three-dimensional complex of vertically accreted, inter-channel 'fine members' (Fig. 2). Dodson (1971) found that paleochannel units made up about 70% of his measured sequences, although park wide the vertical ratio between coarse/fine members varies between 0.4 and 2.9. Higher ratios apply to the central and eastern areas and reflect the common occurrence of multistory channel units.

'FINE MEMBERS'

Intervals of interchannel sediment (Fig. 2) have conformable bases and erosional tops and mostly consist of bentonitic olive-gray mudstones. Thicknesses locally reach 60 ft (18 m) in western outcrops, although eastward where 'coarse members' are commonly multistory 7–20 ft (2–6 m) is more typical. Trenching generally reveals a massive structure to the mudstones with conformable or pedogenic changes between levels of different texture, color, carbon content, degree of rooting, and abun-

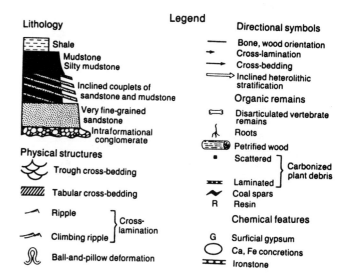

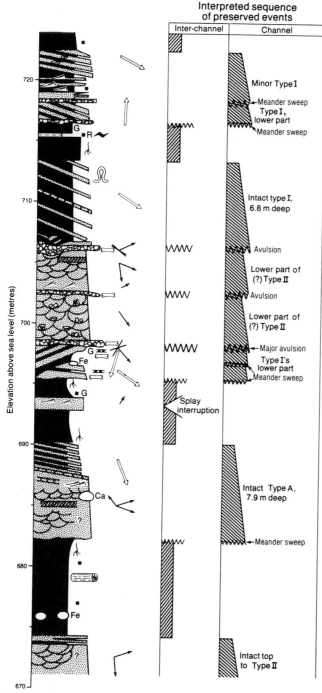

Figure 2. A 172-ft (52-m) section of Judith River Formation from 684550mE, 5621430mN, just outside the southeast limit to the Park, showing the typical alternation of channel (single- and multi-story) and interchannel facies.

dance of macrofloral debris. However, X-ray radiographs have revealed convolute lamination, possible gas escape structures, and bioturbation due probably to animal trampling. Mudstones are rich in palynomorphs, although macroflora is limited mostly to waterworn and split logs, and vertebrate remains are extremely

rare (Dodson, 1971). Present-day weathering is responsible for surficial concentrations of gypsum crystals that commonly lie strewn over sulfur-rich intervals.

Relatively thick 'fine members' are locally interrupted by horizons of ironstone concretions, splay sands locally capped by diverse ichnofauna, *Unio*-bearing iron-cemented sandstones, or volcanic ash layers.

'COARSE MEMBERS'

End-Member Facies. Variability in paleochannel units (Fig. 2) resolves into a mixture of two end-member facies assemblages representing extremes in the nature of channel-fill processes. These are inclined heterolithic stratification (IHS) due to low-energy, rhythmic accretion of point bars in meandering reaches (Figs. 3A and B) and cross-stratified sand (CSS) sequences due to high-energy, episodic aggradation in wider, nonmigratory reaches (Figs. 3C and D).

IHS units have an average dip of 7° and a bimodal pattern of vectors indicating a slightly translational meander geometry on an ESE paleoslope (Fig. 4A). 'Heterolithic' refers to sand-mud couplets on a dm-scale, in which the relative abundance of traction-deposited layers varies from 15 to 90%. Above a planar erosional base, an overall upward-fining is common. Couplets typically consist of a slightly scoured base to a very fine-grained sand with ripple cross-lamination (vector mean of 094°) and an upper conformable unit of brown or olive-gray bentonitic mudstone. Successive couplets are either parallel with uniform dip or occur as large discordant packages bounded by higher-order erosional surfaces. Horizontal partings are commonly smothered with plant remains, but their degree of preferred orientation and carbonization is variable. Other conspicuous features of some IHS sequences are repeated development of ledge-forming ironstones and syndepositional deformation of upper levels into ball-and-pillow structures. In mud-dominant couplets, ripple bedforms are extensively mud draped, giving rise to flaser or wavy bedding. Upper mud layers to couplets, some of which contain evenly spaced silt laminae, yield a well-preserved, abundant as-

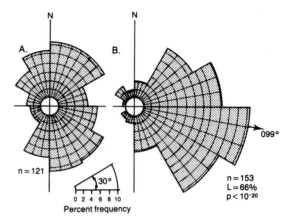

Figure 4. Rose diagrams of lateral accretion (A) and axes of trough cross-beds (B). L and p denote the vector magnitude and the probability that the observed distribution stems from a randomly directed population.

semblage of palynomorphs as well as occasional damaged specimens of dinoflagellates and acritarchs. Trace fossils are rare, mostly limited to *Skolithos* and escape structures.

At the other extreme of paleochannel facies assemblages, CSS sequences are characterized by extensive cosets of large-scale cross-bedding, little vertical organization, frequent signs of reactivation, minimal occurrence of mud-size sediment apart from intraclasts, and no direct sign of lateral accretion. Vertebrate remains abound in these "clean sandstones" (Dodson, 1971), particularly at lower levels. In order of abundance, preserved structures include trough cross-bed sets up to 5 ft (1.5 m) deep and 62 ft (19 m) wide, smaller cosets of planar cross-beds, thin intervals of ripple or climbing-ripple cross-laminations, tabular units of horizontal lamination, and rare single sets of large-scale planar cross-beds over areas approaching 1600 ft^2 (150 m^2). Current structures are best examined in concretionary intervals or across dissected slopes with low light conditions. Lower foreset to bottomset slopes of some cross-bed sets contain bundled drapes of various types (see fig. 5 in Dodson, 1971). Trough axes yield a reliable vector mean of 099° (Fig. 4B)—a direction that bisects the meander migration pattern deduced from IHS units (Fig. 4A). Other types of cross-beds show greater spread about a similar mean direction. The low-relief base of single-story CSS sequences is commonly strewn with mudstone intraclasts, abraded skeletal fragments, and—occasionally—articulated remains. Once aggradation was underway, relief of the sand bed varied from scour holes in front of bedform crests up to 6 ft (1.8 m) deep to relatively flat reactivation surfaces. Intact upper boundaries to cross-bed sets occasionally bear unornamented tube-like burrows.

Form and Variability. IHS and CSS sequences, as described in the previous section, are the primary architectural elements in a spectrum of paleochannel sequence types. Representing a trend of increasing energy and apparently diminishing sinuosity, the spectrum consists of five main types as follows: Type I: composed entirely of IHS (Fig. 3A, B); Type A: IHS above CSS, with the conformable contact between point bar and

Figure 3. A: A 21-ft (6.3-m) thick IHS unit with a basal elevation of ca. 2,211 ft (670 m) a.s.l. at 656550mE, 5622140mE (see person at lower left for scale). Above a flat erosional base on interchannel mudstone, the unit accreted at between 5–10° toward 159–201°. B: Close-up of one couplet in A with details as described in text. C: Lower-to-middle part (ca. 15.75 ft; 4.8 m) of a CSS-dominated unit at the southeast corner of the *Centrosaurus* bonebed (text gives location). Trough cross-beds, with bottomsets commonly outlined by drapes of plant-rich mud or hematite stain, characterize the sequence above an uneven, erosional base. This prominent landmark, known locally as the 'Citadel,' has a summit elevation of 2,257 ft (684 m) a.s.l. D. A well-cemented, creek-bed outcrop at 641500mE, 5622010mN, elevation 2,135 ft (647 m) a.s.l., showing part of a coset of trough cross-beds with a local east-northeast flow into the face (pick is 32 in; 80 cm long).

Figure 5. Examples of the intermediate types of paleochannel sequences as discussed in the text. Note sharp, erosional base in B and conformable tops in A and B. A: Type A at 456000mE, 5629500mN, basal elevation 2,237 ft (678 m) a.s.l. Sequence is 9.3 ft (3 m) thick with lateral accretion at 11° toward 207° (person at center). B: Type B at 459920mE, 5621350mN, basal elevation 2,221 ft (673 m) a.s.l. Sequence is 35 ft (10.5 m) thick with lateral accretion at 8° toward 217° (person at basal contact, lower left). C: Close-up of Type C facies at 460815mE, 5617940mN, at ca. 2390 ft (724 m) a.s.l. Midlevel of a sequence laterally accreting to northeast, oblique to this face (view is ca. 8 ft; 2.5 m high).

coarser thalweg facies either intercalated or relatively abrupt at 3-10 ft (1–3 m) above the base (Fig. 5A)—locally, single cosets of trough cross-beds show a landward direction of paleoflow; Type B: 'streaky' occurrence of IHS at mid-upper levels in a CSS-dominated sequence (Fig. 5B); Type C: inclined intervals of densely-packed intraclastic breccia composed of reworked IHS and fragments of siderite concretions, alternating with CSS (Fig. 5C); and Type II: composed entirely of CSS (Fig. 3C, D).

The above list covers observations from vertical stratigraphic sections. Complete channel sequences, when viewed in a north-south transverse orientation, show either simple or composite form. Simple sequences are infilled by one end-member vertical sequence type. In Type I cases, the IHS starts at a cut bank and ends at an abandoned, 115–165-ft (35–50 m) wide cross section of the channel bend. In Type II cases, the CSS is contained within a broad bowl-shaped cross section for which the total width/thickness ratio tends to increase down paleoslope from ca. 40 to 70. In composite channel fills, the more common transitions involve I → A, I → C, B → II and C → II, indicating that the lateral accretion process was prone to various reduced forms of identity along dip.

Amalgamation. Erosion surfaces below each channel episode in a multistory 'coarse member' (Fig. 2) represent time gaps equivalent to at least the upper part of the preceding episode and possibly also an overbank 'fine member' that had been vertically accreted above it. Amalgamation is most conspicuous in stacked point bar sequences in which sets of IHS typically have opposing dip directions. Amalgamation between CSS-dominated episodes is more subtle because of their cleaner mud-free nature, unidirectional paleocurrent trends, and apparent vertical aggradation. However, an extensive lag horizon consisting of skeletal debris, mudstone intraclasts, and ironstone concretions provides unequivocal evidence. Large abraded/aligned bone fragments and

displaced blocks of bank sediment indicate shear stresses far greater than those required to entrain the enclosing sand. Amalgamation was therefore associated with high-energy, degradational events.

Vertebrate Fauna. The Tyrrell Museum of Palaeontology in Drumheller, Alberta, which maintains a field station in Dinosaur Provincial Park, works on four types of preservation.

(1) Articulated skeletal remains represent the richest and most diverse dinosaur fauna anywhere (Dodson, 1983). To date, more than 300 specimens have been recovered. Quarry sites are assigned numbers and staked, and records are kept on all specimens encountered. After flotation in flow deeper than the animal's girth (up to 5 ft; 1.5 m) from an upstream mortality site, carcass remains most commonly became stranded in the basal levels of non-amalgamated, CSS-dominated sequences. Skin impressions, where present, indicate that burial preceded complete decomposition of soft-parts.

(2) Bonebeds are areal concentrations (10^3–10^5 ft^2; 10^2–10^4 m^2) of disarticulated skeletal remains and represent a tremendous, virtually untapped resource for paleoecological studies. The majority contain components of numerous genera and are associated with channel-base lags, particularly in multistory sequences. This suggests that intermixed bone and intraclastic material were filtered out from reworked intervals in the multistory sequence, combined with some horizontal sorting. Although 'monogeneric' bonebeds are rare, the extensive *Centrosaurus* Bonebed (Currie, 1982) at 465950mE, 5622350mN, is being intensively worked. This category is apparently associated with IHS—bearing sequences and may represent mass drowning.

(3) 'Microsites' are important in understanding composition of the local fauna (Dodson, 1985) and constitute the primary source of information on Late Cretaceous amphibians, lizards, birds, and mammals. Mostly associated with channel-base lags,

they are concentrations of bones and teeth from small vertebrates as well as the teeth from dinosaurs and other larger vertebrates (Currie, 1985). Hydraulically equivalent diameters range from coarse sand to small pebbles, and their worn surfaces and even texture suggest that multiple sorting events preceded their final burial. Modern runoff often concentrates the material of microsites even further, making it possible to collect several thousand specimens in a single day.

(4) Single bones, often abraded and/or fragmentary, tend to occur at the lower levels of sand-dominated channel sequences. Long bones are typically aligned transverse to the local paleoflow direction. Disarticulation of the parent skeleton long preceded the burial of its single bone derivatives and generally took place far upstream of environments represented by the sequence at Dinosaur Provincial Park. Isolated bone occurrences have the potential for major significance and in recent years have produced pterosaurs and many new species of dinosaurs.

INTERPRETATION

Four features of the coastal plain succession are considered critical to interpretation—origin of IHS, variability in paleochannel processes, vertebrate taphonomy, and the nature of 'fine members.' Each is briefly discussed below in the context of an estuarine setting. As used here, an estuary refers to the near-coastal, widening reach of a river down which alluvial processes are increasingly modified by tidal circulation and saltwater incursion (cf. Jouanneau and Latouche, 1981; Dorjes and Howard, 1975). Commonly, a transitional series of zones exists between strictly fluvial conditions upstream of the fall line and the open sea. Zonal boundaries oscillate on time scales relating to the lunar tidal cycle, seasonal river regime, and periodic onshore storm surges, whereas the entire estuarine system will shift in response to transgression or regression.

IHS is viewed here as a distinctive style of lateral accretion in tidally influenced point bars. Sand-mud couplets relate to rhythmic additions of traction-deposited sand and fallout of suspended fines. Complete coverage of the point bar slope by undesiccated and unrooted mud drapes shows that bank-full velocity was minimal and subaerial exposure was briefly, if ever, attained. Along the upper sinuous reaches of modern estuaries, sluggish river flow alternates with slackwater, the latter induced both by rising tide and diminished slope of the water surface. When river discharge is high there is less opportunity for backwater and mud deposition. An estuarine setting also provides a coherent framework for the variability in paleochannel types, with Type I as the upstream end-member in the 'tidally influenced river' zone. Type II sequences are considered to have formed within the lower estuary under fluctuating, relatively high-energy flow (cf. Lithofacies I of Terwindt, 1971). It is envisaged that Types A and B formed at intermediate locations in the 'stratified estuary' zone. Type C represents a major local change in channel process and form, whereby IHS-forming point bars were suddenly rendered unstable by extreme flow conditions.

Clearly, both avulsion and amalgamation of paleochannels played a major role in concentrating vertebrate remains. Whereas buried carcasses indicate a short, single episode of movement from a nearby site of death, disarticulated bone fragments commonly show signs of prolonged wear. Given that fluvial transport of fossil material is effectively halted within the estuarine zone (Frey and others, 1975), only a small portion of the preserved population may have inhabited the immediately surrounding interchannel environments.

Because modern estuaries are flanked by various floral communities (e.g., Dorjes and Howard, 1975), the paleochannels probably inherited a mixed sample of estuarine flora (cf. Jarzen, 1982). The low frequency of splay events in overbank areas is a consequence of the low-energy nature of bank-full flow in tidally influenced channels.

REFERENCES

Currie, P. J., 1982, Hunting dinosaurs in Alberta's huge bonebed: Canadian Geographic, v. 10, no. 4, p. 34–39.
——, 1985, Small theropods of Dinosaur Provincial Park, Alberta: Geological Society of America Abstracts with Programs, v. 17, no. 4, p. 215.
Dodson, P., 1971, Sedimentology and taphonomy of the Oldman Formation (Campanian), Dinosaur Provincial Park, Alberta (Canada): Palaeogeography, Palaeoclimatology, Palaeoecology, v. 10, p. 21–74.
——, 1983, A faunal review of the Judith River (Oldman) Formation, Dinosaur Provincial Park, Alberta: The Mosasaur (Delaware Valley Paleontological Society), v. 1, p. 89–118.
——, 1985, Studies of dinosaur paleoecology by repeated microfaunal sampling in Campanian sediments, southern Alberta: Geological Society of America Abstracts with Programs, v. 17, no. 4, p. 216.
Dorjes, J., and Howard, J. D., 1975, Fluvial-marine transition indicators in an estuarine environment, Ogeechee River–Ossabaw Sound: Senckenbergiana Maritima, v. 7, p. 137–179.
Frey, R. W., Voorhies, M. R., and Howard, J. D., 1975, Fossil and recent skeletal remains in Georgia estuaries: Senckenbergiana Maritima, v. 7, p. 257–295.
Jarzen, D. M., 1982, Palynology of Dinosaur Provincial Park (Campanian), Alberta: National Museum of Natural Sciences, Syllogeus No. 38, 69 p.
Jouanneau, J. M., and Latouche, C., 1981, The Gironde Estuary, *in* Fuchtbauer,

H., Lisitzyn, A. P., Milliman, J. D., and Seibold, E., eds., Contributions to Sedimentology 10: Stuttgart, E. Schweizerbart'sche Verlagsbuchhandlung (Nagele u. Obermiller), 115 p.
Koster, E. H., 1984, Sedimentology of a foreland coastal plain; Upper Cretaceous Judith River Formation at Dinosaur Provincial Park: Canadian Society of Petroleum Geologists, Field Trip Guidebook, 115 p.
——, 1987, Vertebrate taphonomy applied to the analysis of ancient fluvial systems, *in* Ethridge, F. G., and Flores, R. M., Recent Developments in Fluvial Sedimentology: Society of Economic Paleontologists and Mineralogists Special Publication 39, p. 159–168.
McLean, J. R., 1971, Stratigraphy of the Upper Cretaceous Judith River Formation in the Canadian Great Plains: Saskatchewan Research Council, Geology Division, Report No. 11, 96 p.
——, 1977, Lithostratigraphic nomenclature of the Upper Cretaceous Judith River Formation in southern Alberta; Philosophy and practice: Bulletin of Canadian Petroleum Geology, v. 25, p. 1105–1114.
Parrish, J. T., Gaynor, G. C., and Swift, D.J.P., 1984, Circulation in the Cretaceous Western Interior Seaway of North America; A review, *in* Stott, D. F., and Glass, D. J., eds., The Mesozoic of Middle North America: Canadian Society of Petroleum Geologists Memoir 9, p. 221–231.
Terwindt, J.H.J., 1971, Litho-facies of inshore estuarine and tidal-inlet deposits: Geologie en Mijnbouw, v. 50, p. 515–526.

The Frank Slide, southwestern Alberta

D. M. Cruden, *Departments of Civil Engineering and Geology, University of Alberta, Edmonton, Alberta T6G 2G7, Canada*
C. B. Beaty, *Department of Geography, University of Lethbridge, Lethbridge, Alberta T1K 3M4, Canada*

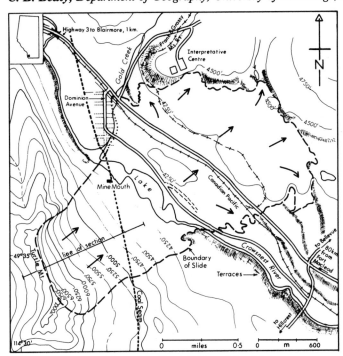

Figure 1. Map of the Frank Slide and its vicinity.

Figure 2. Oblique aerial view from the northeast, photograph by C. Beaty, September, 1976.

LOCATION

The Frank Slide is at the east end of the Municipality of Crowsnest Pass, southwestern Alberta, lying within the valley of the Crowsnest River in the Front Ranges of the Canadian Rockies. Alberta 3 crosses the slide and, immediately north of the slide debris at the bridge over Gold Creek, an access road leaves the highway for a new Orientation Center, 1.1 mi (1.7 km) to the northeast (Fig. 1). The center, an excellent starting point for a tour of the slide, provides a superb view of Turtle Mountain, the slide scar on its eastern flank, and the immense pile of shattered limestone on the valley floor (Fig. 2).

SIGNIFICANCE

For many years, the Frank Slide was the only well-described, historic example in the English language literature of what Varnes (1978) has called a rockslide-avalanche. The importance of the Frank Slide has been reinforced by nearly every North American textbook of physical geology or rock mechanics. Recent investigations have increased our knowledge of the geological factors contributing to the occurrence of the slide, but an understanding of the immediate trigger that set it in motion and a satisfactory explanation of its movement across the valley floor have proved elusive.

Stop 1. Frank Slide Interpretative Center. The Frank

Slide occurred at 4:10 on the morning of April 29, 1903. An enormous wedge of rock moved down the east face of Turtle Mountain, shot across the entrance to the Frank Mine of the Canadian American Coal and Coke Company and the Crowsnest River, obliterated the eastern end of the town of Frank, and buried the main road and Canadian Pacific Railway through the Crowsnest Pass. Rising ground to the east absorbed the momentum of much of the debris, but huge boulders of limestone covered a spur railway line then under construction in the lower valley of Gold Creek. In about 100 seconds, 1.2 mi^2 (3 km^2) had been submerged by an average of 46 ft (14 m) of rubble. About 1 billion ft^3 (30 million m^3) of rock had moved. Dust thrown up from the debris caused violent electrical discharges during the slide and settled like a pall over the valley. At least 70 people were killed.

The high bedrock bench on which the Orientation Center is located is one of the few localities, all elevated, where the debris has a sharply defined edge of boulders. Below the bench, to the left and east, the deposit has two edges. The outer one is steep but not much more than 3 ft (1 m) in height. Inwards is a low area with small, frequent hummocks about 3 ft (1 m) high and littered with scattered boulders. The dense cover of brush and trees indicates a substantial proportion of fines in the debris. This is the "splash area," up to 330 ft (100 m) wide, hypothesized to have formed as the more fluid fines at the base of the debris ran ahead of the coarser material, which was left behind to form a second scarp up to 25 ft (8 m) high (Hungr, 1981, p. 335).

So the steep lateral margins of the slide described by McConnell and Brock (1904, p. 9), and taken by Shreve (1968, p. 38) to be evidence of debris transport on an air cushion are

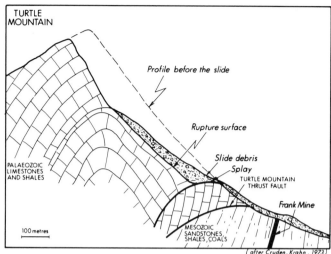

Figure 3. Cross section along the line shown in Figure 1.

probably caused by this marginal high ground catching only the uppermost layers of the moving material. W. Pearce, an engineer who interviewed survivors and eyewitnesses only a few days after the event, commented: "There was apparently no very great rush of wind with the slide except immediately in front of it . . . On the edge of the slide, some very peculiar movements are discernible. Thus a portion (of the debris) would strike a bank or ridge and would curve off to one side or another until its course would be deflected perhaps 120° from its original direction. This and the action which this deflection took in the way of throwing mud, gravel and sticks . . . led many to believe that a large amount of this slide had been hurled through the air" (Pearce, 1903, p. 490). However, McConnell and Brock (1904, p. 7) stated that survivors of the event described the noise of the slide as "resembling that of steam escaping under high pressure."

Returning to Alberta 3 from the Orientation Center and turning south for 1,000 ft (300 m) brings a closer view of the structure of Turtle Mountain.

Stop 2. The Frank Slide Highway Sign on the South Side of Alberta 3. The Canadian Department of the Interior, in a report completed within a month of the disaster, attributed "the primary cause" of the slide to "the form and structure of Turtle Mountain" (McConnell and Brock, 1904, p. 17). Interpretation of the bedrock structure has changed since that report, but the basic conclusion has not. From this site, the structure of Turtle Mountain can conveniently be reviewed.

This structure is summarized in Figure 3, a vertical section across the north-south trending ridge. The beds of limestone that form the ridge have been folded into a great arch, the Turtle Mountain anticline, and thrust eastwards along the Turtle Mountain thrust fault during the Rocky Mountain orogeny. They now overlie vertical to overturned, younger, coal-bearing rocks. In the area of the slide, the east limb of the anticline moved out along a splay of the fault upslope from the coal mine (Cruden and Krahn, 1977).

Looking south from the highway sign, some of the structure shown in Figure 3 may be seen in the eastern flank of the mountain where it slopes precipitously from South Peak, the prominent summit to the left. The crest of the Turtle Mountain anticline is about 160 ft (50 m) below South Peak. Limestone beds dipping eastwards, parallel to the slope and partially exposed by removal of the slide mass, can be followed almost to a distinct break in slope below the trace of the Turtle Mountain thrust. Immediately above the break in slope, the limestones are overturned and dipping westwards. The splay of the Turtle Mountain thrust, which formed the toe of the rupture surface of the slide, is not exposed beyond the south margin of the slide. The splay can be seen at the north margin, better viewed from Stop 3.

It is apparent that the bulk of the slide mass, mainly Palaeozoic carbonates of the Livingstone Formation, moved on steeply eastward-dipping bedding planes, rather than on joint planes and across the bedding as McConnell and Brock (1904, p. 12) suggested. Joints probably controlled the form of the scarp and crown of the slide.

Reconstruction of the Canadian Pacific Railway line required a deep cut visible to the south through nearly the full depth of the slide. Assuming the base of the debris is within 6.6 ft (2 m) of the new track, the maximum thickness here is about 66 ft (20 m), 20 ft (6 m) more than the estimated overall average for the slide. The section shows a distinct upward coarsening of the rubble.

Hungr (1981) has distinguished "base" material identifiable in at least the lowest 16 ft (5 m) of the cut. This sandy gravel is made up dominantly of angular limestone fragments and contains only a few boulders. Occasional rounded pebbles, probably derived from till or Crowsnest River alluvium, and traces of dark, organically-stained sand have been found. Grain size distributions of the "base" material are similar to those of debris in the "splash" areas, which also contains rounded pebbles. The implication is strong that the slide eroded surficial materials as it traversed the valley floor and hence was in contact with the preexistent surface along at least part of its path.

The coarsest debris is at the top of the cut, dominated by boulders that form the surface of the slide. Such reverse grading has been observed in some debris flows and may be a consequence of dispersive pressures in the flows.

To reach Stop 3, continue eastwards on Alberta 3 for 1.2 mi (2 km) and turn right on the road to Hillcrest. Immediately after the level crossing of the Canadian Pacific Railway, 0.3 mi (0.5 km) down the Hillcrest Road, and before the bridge over the Crowsnest River, take the right turn signposted "7 Avenue." Follow this old highway for 1.2 mi (2 km) past abandoned lime kilns to the right at 0.5 mi (0.8 km) and into the slide debris.

Stop 3. Crest of Old Road through the Slide Debris. From here, we can see the abandoned approach to the Frank Mine, on the west bank of the Crowsnest River just above river level, at the northern edge of the slide (Fig. 4). Mining was from a horizontal adit into the near-vertical seam and drew down coal from large rooms extending upwards to within 30 ft (10 m) of the

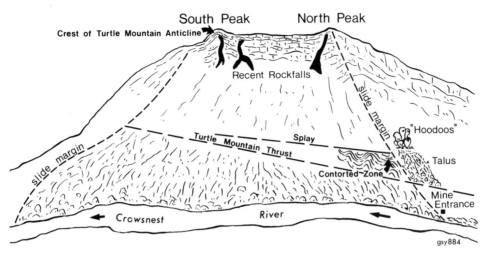

Figure 4. Sketch of the slide scarp from the old road through the debris.

ground surface. The slide debris moved rapidly over the limits of the mine but apparently damaged only the surface and near-surface workings.

Men working inside the mine during the slide recalled to McConnell and Brock (1904, p. 7) that although some coal was dislodged, no really significant movements of the mine workings occurred. Discovering the mine entrance to be blocked by debris, they dug their way to the surface through an up-raise. Production was resumed shortly after the slide and continued until 1918. The subsidence visible on the south margin of the slide clearly post-dates deposition of the debris. The last episode of mining extended the underground workings more than 1.2 mi (2 km) beyond the south margin of the slide.

There was considerable controversy at the time about the role of mining in triggering the movements (Pearce, 1903). A more recent analysis suggested that the movements resulting from mining were not likely to have been significant factors in the slide. Reasonable assumptions about rock-mass strength in the slope before mining now indicate that it was unstable even then (Cruden and Krahn, 1977).

At the north edge of the slide, the belt of minor folds forming the "contorted beds" noted by McConnell and Brock (1904, p. 15) can be clearly seen (Fig. 4). Its upper margin is the splay of the Turtle Mountain Fault, which can be followed northwards into a talus beyond the north margin of the slide. Immediately above the talus is the "hoodoo area" of McConnell and Brock (1904, p. 15). The "hoodoos" are isolated blocks of the muddy carbonates of the Banff Formation, which dip eastwards and stratigraphically underlie the limestones of the Livingstone Formation. The peculiar pattern of weathering is caused by intense deformation around the fault and in the hinge zone of the Turtle Mountain anticline. The crest of the anticline in the Banff Formation is about 150 ft (50 m) higher. Clearly, the anticlinal crest is plunging northwards, and the east limb of the anticline is not exposed to the north. The mass of light-grey–weathering limestone protruding from the talus is overturned and westerly

dipping Livingstone Formation, deformed in the thrust slice above the Turtle Mountain fault, which also plunges northwards.

Continuing north along the old highway past a memorial to the slide victims and over Gold Creek brings us, after 0.75 mi (1.2 km) onto Dominion Avenue, former "main street" of the original town of Frank.

Stop 4. Northeast End of the Old Town of Frank. The antique fire hydrant immediately west of the road here may have been illustrated in McConnell and Brock's report (1904, Plate 5). Basements of buildings along Dominion Avenue are the only other obvious remains of the old townsite. At the recommendation of the Commission of 1911 (Daly and others, 1912), most of the buildings were moved to the present townsite north-east of the highway. The Commission believed that the geological structure of Turtle Mountain was essentially similar in sections through the South and North peaks and north of the North Peak. Because slopes are steeper north of the North Peak, slide risk was assumed to be greater there.

Following reports of continuing slide danger, the Provincial Government offered residents threatened by potential movements from the South Peak of Turtle Mountain half the cost of moving their homes to new sites away from the southern margin of the slide. The Government also relocated Alberta 3 to its present position to the east of the railroad.

A monitoring program begun in 1933 has established that rock falls from the crown of the slide continue. A recent upgrading of the monitoring network should make large-volume movements from the South Peak more perceptible. Monitoring techniques at Turtle Mountain are the subject of research collaboration among the University of Alberta, the University of Calgary, and Alberta Environment (Cruden, 1985).

Some of the natural processes that contributed to the slide—the destruction of cohesion along bedding planes by limestone solution and by freeze and thaw weathering, for instance—have probably been accelerated by the movement-induced fracturing of rock in the crown of the slide. Other processes, such as the removal of lateral support for the rock mass by slush avalanches down the steeply inclined gullies on the east face of the mountain, are probably less important because the slide itself reduced the mass requiring support.

The major source of concern today is the possibility of a sudden closure of the 1903–1918 workings of the abandoned Frank mine. The 6 ft (2 m) or so of horizontal movement that might take place could be transferred up the slope and result in some loosening of rock around the South Peak, triggering further movements. The fire burning in the mine in 1918 has probably been extinguished.

Continuing north-westwards, Dominion Avenue crosses the Canadian Pacific Railway tracks and rejoins Alberta 3 660 ft (200 m) north of the Gold Creek Bridge.

REVIEW

At least two aspects of the Frank Slide remain incompletely understood. First, the trigger that set the slide in motion has not been identified. There is general agreement that mining at the base of Turtle Mountain was not the direct cause of the slide, the literal trigger, although removal of a large volume of coal must have led to a further weakening of an already unstable structure. A small earthquake, undetected by the sparse seismic network then in operation, might have provided the trigger. But there were no reports of earthquake activity in the area on the morning of April 29, 1903.

A more likely cause could have been the local weather. As noted by McConnell and Brock (1904, p. 14), "The night of the slide was excessively cold. . . .Those outside say the temperature was down to zero (°F). The day before and the preceeding days had been very hot, so that the fissures in the mountain must must have been filled with water . . ." Since the month of March 1903

had been wet in the southern Alberta Rockies, a probably source of the water was copious melt from a heavy snowpack on Turtle Mountain.

Given appropriate meteorological conditions, deep freezing might have produced sufficient stresses within the fissures to have wedged or pried loose a large prism of rock. Indeed, McConnell and Brock concluded (1904, p. 17): "The heavy frost on the morning of the slide . . . appears to have been the force which severed the last thread and precipitated the unbalanced mass."

Second, the mechanism of movement that enabled the slide to spread so widely has been the subject of considerable discussion and speculation. McConnell and Brock (1904, p. 17) wrote: "The motion of the slide was complex in detail, but as a whole resembled that of a viscous fluid." They also stated (p. 8) that a shelf of rock in the basin of the slide apparently ". . . hurled most of the material over the coal mine at the base of the mountain into the river bed, or beyond."

Until the advent of Shreve's air-layer lubrication hypothesis (1968), many geologists and engineers evidently favored flow as a consequent of fluidization, with compressed air, steam, dust, or water and mud from the Crowsnest River mentioned as possible fluidizing agents. A recent discussion of possible acoustic fluidization as an explanation of the behaviour of long-runout landslides included the Frank Slide (Melosh, 1983). The air-layer lubrication mechanism has been rigorously criticized by Hsü, (1975), but it at least has the virtue of providing a possible explanation for two of the slide's more conspicuous characteristics: the impressive lateral extent of the debris and the absence of large masses of shattered rock at the immediate foot of Turtle Mountain.

As is the case with the search for a trigger, the currently available field evidence and the limited and often contradictory contemporary accounts of the disaster allow for multiple interpretations of the behavior of the slide debris.

REFERENCES CITED

Cruden, D. M., 1985, Rock slope movements in the Canadian Cordillera: Canadian Geotechnical Journal, v. 22, p. 528–540.

Cruden, D. M., and Krahn, J., 1977, Rockslides and avalanches, *in* Voight, B., ed., Frank slide, Alberta, Canada: Amsterdam, Elsevier, 833 p.

Daly, R. A., Miller, W. G., and Rice, G. S., 1912, Report of the Commission appointed to investigate Turtle Mountain, Frank, Alberta: Geological Survey of Canada Memoir 27, 34 p.

Hsü, K. J., 1975, Catastrophic debris streams (Sturzstroms) generated by rock-falls: Geological Society of America Bulletin, v. 86, p. 129–140.

Hungr, O., 1981, Dynamics of rock avalanches and other types of slope movements [Ph.D. thesis]: Edmonton, University of Alberta, 506 p.

McConnell, R. G., and Brock, R. W., 1904, Report on the great landslide at Frank, Alberta, Canada: Canadian Department of the Interior Annual Report, 1902–1903, pt. 8, 17 p.

Melosh, H. J., 1983, Acoustic fluidization: American Scientist, v. 71, p. 158–165.

Pearce, W., 1903, The great rockslide at Frank, Alberta: Engineering News, v. 49, p. 490–492.

Shreve, R. L., 1968, The Blackhawk landslide: Geological Society of America Special Paper 108, 47 p.

Varnes, D. J., 1978, Slope movement types and processes, *in* Landslides: Analysis and Control Transportation Research Board, National Academy of Sciences Special Report 176, p. 11–33.

Zone of interaction between Laurentide and Rocky Mountain glaciers east of Waterton-Glacier Park, northwestern Montana and southwestern Alberta

Eric T. Karlstrom, Department of Geography, University of Wyoming, Laramie, Wyoming 82071

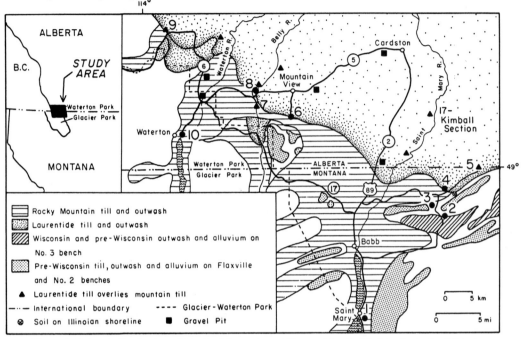

Figure 1. Inferred distribution of Rocky Mountain and Laurentide glacial deposits in the Waterton-Glacier Park area and location of Flaxville (Number 1), Number 2, and Number 3 Bench remnants; gravel pits; sections in which Laurentide drift overlies Rocky Mountain drift; and Stops 1 through 10.

LOCATION AND ACCESSIBILITY

The evidence for interaction between Laurentide and Rocky Mountain glaciers east of Waterton-Glacier Park can best be studied by examining a number of sites in the zone of glacier overlap and possible confluence. The 10 stops described in this field guide are located in the Saint Mary, Belly, and Waterton drainages in Montana and Alberta (Fig. 1). With the exceptions of Stops 1 and 10, which occur on the Blackfeet Indian Reservation and in Waterton Park, respectively, all stops occur on public roads through private land. Landowner's permission is not required for the excursion as described but should be obtained before collecting samples or extensive off-route exploration. The trip can be completed in a long half day with two-wheel drive vehicle. Those with less time may limit their trip to Stops 1 through 5 on the U.S. side or the approximately equivalent Stops 6 through 10 in Canada. The U.S. portion of the trip is covered by the Cutbank 1:250,000 topographic sheet and the Saint Mary, Babb, Duck Lake, Wetzel, Hall Coulee, Emigrant Gap, and Pike Lake 1:24,000 sheets. The Canadian portion is covered by the Lethbridge 1:250,000 sheet and the Cardston and Waterton Lakes (82 H/3 and H/4) 1:50,000 sheets.

SIGNIFICANCE

This site provides a unique opportunity to examine the field evidence indicating the nature of interaction between Laurentide continental and Rocky Mountain piedmont glaciers in their southwesternmost zone of overlap. In numerous locales (e.g., Stops 5, 6, 7, and Kimball Section, Fig. 1), Laurentide drift overlies mountain till and, in some places, an intervening soil. This suggests that mountain piedmont glaciers had reached their maximum extent and partly or completely retreated prior to the advance of the corresponding Laurentide ice sheet. However, converging lateral moraines and/or interbedding of drift sheets at some locales (e.g., Stops 3, 4, 6, and 9) have been interpreted to mean that the glaciers may have temporarily coalesced (Wagner, 1966; Mickelson and others, 1983). Distribution of shoreline deposits, ice-rafted erratics, and spillways indicates the approximate elevations of proglacial lakes temporarily impounded by Laurentide glaciers. Reconstructing the timing of glacier interactions in this area has proven problematic and controversial due to lack of absolute dating of deposits, repeated mixing of sediments in extremely unstable proglacial environments, and varying interpretations of the geomorphic evidence by different workers.

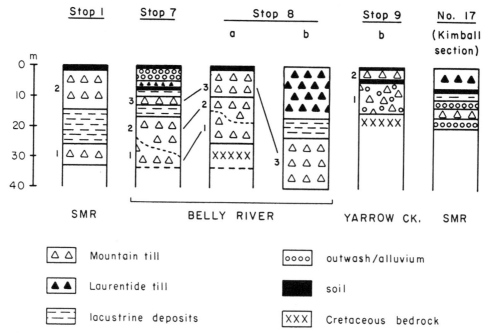

Figure 2. Stratigraphy of tills, lacustrine deposits, outwash/alluvium, and soils in sections exposed by stream cuts and road cuts in the Waterton-Glacier Park area.

DESCRIPTION

Rocky Mountain and Laurentide glacial deposits can generally be distinguished on the basis of lithology and texture. Rocky Mountain till is relatively coarse-textured, gravelly sandy loam that contains mostly Precambrian Belt Supergroup (Purcell) sedimentary rock clasts from the mountains. These tills typically include about 57 to 69% red and green argillite, 6 to 14% quartzite, 10 to 20% limestone and dolomite, 1 to 4% diorite and basalt, and in areas east of the mountain front, about 4 to 10% Cretaceous sandstone and shale. Laurentide drift tends to have a clayey texture and fewer, smaller, and more rounded clasts (2 to 5%). Clast lithologies include about 3 to 10% Precambrian granite and gneiss from the Canadian shield, 12% Cretaceous sandstone and shale, 0 to 36% limestone and dolomite, 20 to 32% quartzite, and about 21 to 40% argillite. However, it should be cautioned that deposits in this zone of oscillating ice fronts have probably been subject to repeated mixing by glacial, glaciofluvial, glaciolacustrine, and colluvial processes. Therefore, deposits must be examined carefully in order to determine provenance, mode of deposition, and relative age.

Four main terminal and/or recessional positions of the Laurentide glacier have been mapped east of Waterton Park. These are, from oldest to youngest, the outer continental, Kimball, Glenwoodville, and Lethbridge moraines of Horberg (1954), or the approximately equivalent Waterton I through IV positions of Stalker and Harrison (1977). Likewise, Richmond and others (1965) map four to five terminal positions of mountain glaciers in the Saint Mary Valley. Although some of these authors have

assigned local names to deposits (Stalker, 1983), and others have applied Rocky Mountain terms (Richmond and others, 1965), classical North American Midwestern terminology is used here informally because of the obvious importance of the Laurentide ice sheet in the region.

Ages of glacial deposits have been estimated by geomorphic relations and relative degree of soil formation (Alden, 1932; Horberg, 1954; Richmond and others, 1965; Karlstrom, 1981) and by correlation with radiocarbon and paleomagnetically dated deposits elsewhere (Barendregt, 1984). However, chronometric data are not yet conclusive, so the dating of the last major overlap of ice sheets remains the subject of a minor but persistent border dispute. In general, the current "American school" (Mickelson and others, 1983) holds that the event occurred during the late Wisconsin stage (approximately 23,000 to 10,000 years B.P.). The "Canadian school" (Rutter, 1980, following original interpretations of Alden, 1932), however, places the event in the early Wisconsin stage (about 100,000 to 53,000 years B.P.). According to the latter interpretation, the late Wisconsin Laurentide glacier (Waterton IV) terminated near Lethbridge, Alberta, whereas the corresponding mountain glaciers advanced only to the mountain front. Chronometric resolution of this problem will be critical for reconstructing the history of the "ice-free corridor" through which human beings supposedly migrated about 13,000 years ago.

Stop 1. The route begins about 0.4 mi (0.64 km) south of the town of Saint Mary, Montana (Fig. 1; SW¼,NE¼,S3, T34N,R14W. Saint Mary 7½-minute Quadrangle), where a road cut in a late Wisconsin moraine exposes an approximately 108-ft

(33-m) thick section that includes two mountain tills separated by lacustrine deposits (Fig. 2). Here the 23-ft (7-m) thick lower till unit is light-gray, compact, gravelly sandy loam that is weakly cemented by carbonates and includes about 25% angular to subrounded Belt rocks from the mountains. Glacial striations are abundant, particularly on green argillite clasts. The approximately 40 ft (12 m) of overlying gray lacustrine clays, separated by light-gray silt bands, probably represent an interstadial or proglacial lake phase that separated the two advances. The uppermost 45 ft (14 m) of the section is another weakly cemented, light-gray mountain till that includes about 10% Belt rocks and is capped by a weakly developed soil under aspen and pine forest. This belt of lateral moraines can be traced about 9 mi (14 km) north to near Babb, where terminal moraines and kame terraces have been breached by the Saint Mary River (Fig. 1).

A series of older Wisconsin and/or Illinoian lateral moraines flanks the western edge of Saint Mary Ridge up to 1,200 ft (366 m) above Saint Mary Lake (Fig. 1). These moraines represent a much larger piedmont glacier (or glaciers) that included tributary glaciers from Divide, Red Eagle, Boulder, and Swiftcurrent creeks and extended at least 25 mi (40 km) east-northeast from the mountain front onto the plains. Soils developed on these moraines are considerably thicker and better developed than those on late Wisconsin moraines at Stop 1. The post–late Wisconsin soils examined are weakly developed Cryochepts that include 9- to 20-in (23- to 50-cm) thick, pale-brown to dark yellowish-brown cambic B horizons but lack argillic horizons. By contrast, soils on older (early Wisconsin and/or Illinoian) till exposed in road cuts some 0.7 to 4.2 mi (1.1 to 6.7 km) south of Stop 1 on U.S. 89 are moderately to strongly developed Cryoboralfs. They occur in weathered till with numerous decomposed clasts and include 23- to 47-in (58- to 120-cm) thick, yellowish-brown argillic horizons with up to 22% more clay than underlying C horizons.

Now drive 6.7 mi (10.7 km) north of Saint Mary on U.S. 89, paralleling late Wisconsin lateral moraines, then turn right (east) onto the Duck Lake Road immediately before U.S. 89 crosses the Saint Mary River. About 2.6 to 5.0 mi (4.2 to 8.0 km) past the turnoff, pass through a belt of prominent, rock-cored, lateral moraines (of Wisconsin and/or Illinoian age), then turn left (north) at the T junction onto a well-graded dirt road and drive 3.4 mi (5.4 km) across several levels of outwash and alluvium included in Alden's (1932) Number 3 Bench (late Pleistocene). Turn right past an airplane hangar onto an old, unused dirt road and drive 0.3 mi (0.5 km) to a borrow pit (now a dump) on the southern valley wall overlooking the North Fork of the Milk River (Fig. 1; NE¼,NW¼,S5,T36N,R12W, Hall Coulee 7½-minute Quadrangle).

Stop 2. Pre-Wisconsin alluvium and outwash gravels exposed in this gravel pit lie above the level of mapped moraines and predate the cutting of Hall Coulee and the North Fork of the Milk River immediately to the north. Cryoturbation features, including vertically oriented clasts, festoons, and—in other locations—ice-wedge casts, are common in these gravels (Karl-

strom, 1986). In addition, Number 3 Bench gravels are capped by a strongly developed paleosol typical of pre-Wisconsin soils elsewhere in the region. This Argiboroll (or truncated Paleustoll?) consists of a truncated, up to 51-in (130-cm) thick, reddish-yellow argillic horizon that contains some 28% clay and is capped by a modern soil developed in 16 in (40 cm) of Wisconsin-age loess. Return to the graded dirt road, continue north, and descend into the North Fork of the Milk River drainage, observing slumped Cretaceous bedrock and outwash from mountain glaciers along valley-side slopes. About 2.4 mi (3.8 km) past Stop 2, turn left and drive 0.15 mi (0.24 km) to another gravel pit excavated in mountain gravel on the eastern edge of Hall Coulee, an outwash channel and spillway (Fig. 1; center of SW¼,S29, T37N,R21W, Hall Coulee 7½-minute Quadrangle.

Stop 3. This stop overlooks the part of Willow Creek drainage mantled by deposits of the Saint Mary piedmont glaciers and the Saint Mary lobe of the Laurentide glacier. Lateral moraines of the continental ice sheet overlap those of the mountain glacier at elevations between 4,400 and 4,700 ft (1,341 and 1,423 m) on each side of the valley, suggesting either that the Laurentide glacier advanced into the area after the piedmont glacier had retreated or that the glaciers may have actually coalesced here briefly. This gravel pit, at 4,680 ft (1,426 m), exposes relatively unweathered mountain outwash or kame terrace deposits capped by a moderately developed Cryoboroll or Argiboroll soil, which locally includes an 8-in (20-cm) thick argillic (Bt) horizon. About 200 yd (183 m) to the northwest, granite boulders at the surface were probably ice-rafted across a proglacial lake.

The flat Number 2 Bench remnant at about 4,900 ft (1,492 m) elevation to the east is an erosion surface (pediment) bevelled across Cretaceous bedrock and mantled with rounded alluvial gravels (early Pleistocene?) from the mountains (Alden, 1932). This escarpment has held in proglacial lakes at least twice. Moraines, kames, shoreline deposits, and ice-rafted erratics of the most recent interaction event (early Wisconsin?) can be traced up to 4,700 ft (1,433 m). Older shoreline deposits, consisting predominantly of discoidally shaped mountain gravels, occur higher on the flanks of the Number 2 Bench interfluve at elevations between 4,820 and 4,880 ft (1,469 and 1,487 m). The strongly developed paleosol developed in this beach shingle deposit includes a 30-in (76-cm) thick, reddish-yellow argillic horizon and is similar to the pre-Wisconsin soil at Stop 2. In addition, numerous large (up to 3-ft or 1-m diameter) ice-rafted granite and gneiss erratics flank a series of broad and generally shallow spillways along the crest of the Number 2 Bench at about 4,730 to 4,880 ft (1,442 to 1,487 m), indicating that pre-Wisconsin proglacial lakes overtopped and spilled across the interfluve to the southeast.

Return to the dirt road, turn north, and descend through mixed mountain and Laurentide drift, then turn right (east) after crossing the Saint Mary Canal. Drive 0.6 mi (1 km) to a north-south trending Cretaceous bedrock ridge that marks the approximate maximum position of the Wisconsin Laurentide glacier in this drainage (Fig. 1; center of NW¼,S21,T37N,R12W, Hall Coulee 7½-minute Quadrangle).

Figure 3. At Stop 5, Laurentide till and outwash (LT and LO) overlie Rocky Mountain till (MT) in section excavated in ice-pressed feature near the Weathered ranch.

Stop 4. Continental outwash gravels, shoreline deposits, and large angular granite and gneiss boulders can be observed by hiking about 0.1 to 0.25 mi (0.2 to 0.4 km) up the bedrock ridge to the left (north). Numerous small lakes occur to the northeast on clayey Laurentide till and lacustrine deposits. Organic horizons in alluvial surface near the Martin ranch, about 2.6 mi (4.2 km) northwest of this stop, yield dates of 42,170 ± 2,140 (A-1998) and greater than 38,000 (A-1997) C-14 years B.P. (Karlstrom, 1981). Although the dates may be suspect because of possible contamination by locally derived Cretaceous coal, geomorphic relations and soils nonetheless suggest that the dated alluvium could be pre–late Wisconsin in age.

Now drive 12.6 mi (20.2 km) east along the canal road, crossing Laurentide till and lake deposits. Note the piles of large continental boulders to the left and the Laurentide lateral moraine that flanks the Number 2 Bench escarpment at about 4,600 to 4,700 ft (1,402 to 1,433 m) to the right. Turn left (north) on the well-graded Emigrant Gap road, drive 0.5 mi (0.8 km), then turn left (west) and drive 3.2 mi (5.1 km) to the Weathered ranch. Turn left (southeast) and drive 0.2 mi (0.32 km), then turn left again before reaching a gate, and drive 0.1 mi (0.2 km) to a gravel pit area (Fig. 1; SW¼,NE¼,S3,T37N,R12W, Hall Coulee 7½-minute Quadrangle).

Stop 5. Laurentide till and outwash overlie mountain till in a 23-ft-high (7-m) section excavated in a small, rounded, ice-pressed feature (Fig. 3). Mountain till and outwash, exposed in the lower 18 ft (5.5 m) of the section, are relatively coarse textured, pale brown to light gray, and include about 30% angular to subrounded mountain rocks up to 20 in (50 cm) in diameter. This unit is overlain by about 3 ft (1 m) of light-gray to pale-yellow,

highly compact and fractured, clayey Laurentide till that includes 1 to 3% small, rounded gravels up to 4 in (10 cm) in diameter. No soil or other evidence of a subaerial weathering interval separates the two drift sheets. The section is capped with a thin layer of Laurentide outwash. Deformation of drift sheets in this ice-pressed feature probably occurred under the stagnating continental ice sheet. Lacustrine silts, loess, and soils, exposed in other excavated sections some 66 ft (20 m) west of this section, probably record the presence of a kettle lake depression formed during the last ice retreat from the area.

Stratigraphic relations at this site indicate that mountain glaciers advanced beyond this point, then retreated prior to the advance of the corresponding Laurentide glacier. The presence of mountain outwash in Emigrant Gap about 3.5 mi (5.6 km) east-southeast of this location suggests that proglacial streams from mountain glaciers were diverted to the southeast through the spillway at Emigrant Gap by the Laurentide glacier.

Now return to Stop 4 and continue west along the Saint Mary Canal road. Cross the bridge over the Saint Mary River and return to U.S. 89. Turn right (north) and drive about 4.8 mi (7.7 km) to the Carway International Customs station, then drive another 16.8 mi (26.9 km) across loess-covered Laurentide drift on Alberta 2 to Cardston. Turn left (west) onto Alberta 5 at Cardston and drive about 16.2 mi (26.0 km) to a dirt road 0.2 mi (0.3 km) west of Mountain View. Turn left (south) and drive 2.9 mi (4.6 km) to the north end of Beaverdam Lake (Fig. 1; 2100 ft or 640 m north of the SE corner of T2,R28,W.4th mer., UTM-54434 N., 7484 E., Waterton Lakes 82 H/4 Quadrangle).

Stop 6. This site marks the end position of one of the Laurentide advances into the region (the Kimball moraine of

Horberg, 1954). Wave-cut bluffs along the north shore of Beaverdam Lake expose both Laurentide and Rocky Mountain drift. Interbedding of tills, without soils or other evidence of significant time separation, could represent coalescence of ice bodies and/or ice-thrusting of sediments along the glacier margin. Laurentide till, which occurs to the west of the road end, generally consists of light-gray, dense, mottled, highly fractured clay loam with coarse strong prismatic structure. It contains 2 to 3% rounded cobbles up to 10 in (25 cm) in diameter. About 2% of these are shield crystalline rocks and Cretaceous sandstone and coal. The soil (Cryoboroll) developed in this unit includes a 4- to 10-in (10- to 25-cm) thick A horizon, a 15- to 38-in (37- to 97-cm) thick, brown to light yellowish-brown Bw horizon, and a 16- to 33-in (40- to 83-cm) thick calcic horizon with lime coatings along seams and ped faces.

Mountain till occurs along the bluff east of the road end and is generally coarser textured, less compact, brown material that contains about 1 to 50% rocks of predominantly mountain provenance. Approximately 1,380 ft (420 m) east of the road end, about 6.5 ft (2 m) of mountain till overlies about 6.5 ft (2 m) of Laurentide till, suggesting either: (1) contact of the two ice sheets, (2) slumping of unstable proglacial deposits, or (3) ice-thrusting or shoving along the Laurentide glacier margin. The latter interpretation is reinforced by apparent thrust-faulting of till materials and inclusion of Cretaceous bedrock blocks within drift units.

Prominent, sharp-crested moraines in a series south of the lake are separated by deep gullies thought to represent spillways that drained glacial Lake Belly River when it was impounded by the Laurentide ice sheet (Horberg, 1954). These moraines are overlain by continental crystalline boulders and were mapped as "outer continental" end moraines by Horberg (1954). However, inspection of road cuts and soil pits indicates that the underlying drift consists almost entirely of mountain rocks. Hence, they are probably mountain lateral moraines veneered with ice-rafted continental boulders.

Return to Alberta 5, then turn left (west), drive 4.5 mi (7.2 km), and turn left again onto a dirt road that leads 1.3 mi (2 km) to a gravel pit and stream cut along the east side of the Belly River (Fig. 1, NE¼,SE¼,S8,T2,R28,W.4th mer., UTM-5445 N., 7418 E., Waterton Lakes 82 H/4 Quadrangle).

Stop 7. The approximately 70 ft (21 m) of exposed sediment include three superposed mountain tills, intercalated lacustrine deposits, and a thin lens of Laurentide till that overlies a dark organic and clay-rich soil (Fig. 2). The soil and lacustrine deposits underlying Laurentide drift in this section suggest that mountain glaciers retreated, proglacial lakes developed, and the surface was subject to soil formation prior to the advance of the Laurentide glacier into the region. A better-developed soil underlies Laurentide drift in the Kimball section along the Saint Mary River (Fig. 1).

From top to bottom, this section includes: (1) a modern soil developed in about 2.5 ft (0.75 m) of loess, (2) about 20 ft (6 m) of outwash from the mountains, (3) a thin (10 in; 25 cm) lens of brown continental till with coarse, strong prismatic structure that

locally overlies an organic-rich soil, (4) about 5 to 15 ft (1.5 to 4.6 m) of light yellowish-brown to dark-gray stratified and deformed lake deposits, (5) about 5 ft (1.5 m) of dark-brown to buff mountain till with about 40% relatively unweathered rocks, (6) about 8 ft (2.4 m) of dark-gray lacustrine deposits, and (7) about 25 ft (7.6 m) of oxidized, weathered mountain till that includes about 30 to 40% partly decomposed mountain rocks. This lower unit is inset into an eighth layer, a partly indurated pinkish till and outwash unit that includes about 60% mountain rocks and forms most of the section some 100 yd (91 m) downstream.

Other sections in the area that expose multiple superposed tills have been described by Horberg (1954), Stalker (1983), and Stalker and Harrison (1977). Horberg recognizes three Laurentide drift sheets and three correlative mountain tills in the Waterton region, whereas Stalker distinguishes three to nine Laurentide drift sheets overlying a mountain till at the base in sections exposed along the Oldman and Belly Rivers near Lethbridge, Alberta.

Now return to Highway 5 and take the first small dirt road to the right, which leads to another stream cut along the Belly River (Fig. 1; NW¼,NE¼,S17,T2,R28,W4th mer., UTM-54474 N., 7415 E., Waterton Lakes 82 H/4 Quadrangle).

Stop 8. This stream cut below a terrace along the east side and some 70 ft (21 m) above the Belly River also exposes three superposed mountain tills above Cretaceous bedrock (Fig. 2). The uppermost till unit consists of about 30 ft (10 m) of weakly cemented, buff mountain till that includes 10 to 20 percent rounded mountain rocks up to 1.5 ft (0.5 m) diameter. This unit overlies about 5 to 20 ft (1.5 to 6 m) of gray till with about 50 to 60 percent mountain rocks; it in turn overlies a basal pinkish till and outwash unit with about 50 to 60 percent mountain rocks. Up to 20 ft (6 m) of buff lake silts capped by a modern soil (Mollisol) overlie the upper till unit in road cut exposures to the east of this section.

About 0.5 mi (0.8 km) downstream from this site and immediately behind a prominent, discontinuous Laurentide end moraine, stream cuts expose about 50 to 60 ft (15 to 18 m) of clayey, brown Laurentide till overlying some 20 ft (6 m) of stratified lake deposits and 50 to 60 ft (15 to 18 m) of buff mountain till (Fig. 2). Gravel pits on the opposite (north) side of the Belly River expose continental outwash.

The elevation of Laurentide end moraines drops from about 4,600 to 4,700 ft (1,402 to 1,480 m) in the Willow Creek and Yarrow Creek drainages to about 4,200 ft (1,280 m) in the Belly and Waterton drainages. This could indicate that the Waterton lobe of the Laurentide glacier was blocked by the stagnating or retreating margin of the Waterton piedmont glacier near this point. More probably it indicates that the Laurentide glacier margin terminated here because glacier ice calved into a proglacial lake.

Return to Alberta 5, continue west to the Waterton Park entrance station, then turn north onto Alberta 6 and drive across a zone of eskers, kames, and drumlins deposited by the compound Waterton mountain piedmont glacier. The road parallels Indian Springs Ridge, a complex of early Wisconsin lateral moraines which flanks Lakeview Ridge. One mile (1.6 km) past Dungarvan Creek, turn left and drive 3.7 mi (5.9 km) to Stop 9 (Fig. 1; NE¼,S23,T3,R30,W.4th mer., UTM-54574 N., 7226 E., Waterton Lakes 82 H/4 Quadrangle).

Stop 9. According to Wagner (1966), low-angle road cuts here expose interbedded Rocky Mountain and Laurentide tills that indicate contemporaneous deposition of moraines and confluence of glaciers. In addition, relatively sharp-crested lateral moraines from the mountains converge with low, rounded, hummocky moraines of the Laurentide glacier in this area. Note the numerous kettle lake depressions developed on clayey Laurentide drift to the northeast. Regional geomorphic relations suggest that the Laurentide glacier flowed to the southeast along the mountain front in this area during the last overlap event (A. Mac S. Stalker, personal communication, 1985). Some 0.85 mi (1.4 km) to the west of this stop, a stream cut on the east side of Yarrow Creek south of the bridge exposes Horberg's (1954) Drywood Soil (mid-Wisconsin or possibly Sangamon?), which occurs between two mountain drift sheets (Fig. 2).

Return to the Waterton Park entrance station, enter the park, cross the Blakiston Creek delta (which separates Middle and Lower Waterton lakes), and drive to the Prince of Wales Hotel (NE¼,S23,T1,R30,W4th mer., UTM-54378 N., 7267 E., Waterton Lakes 82 H/4 Quadrangle).

Stop 10. A resistant bedrock ridge crosses the valley at this point, separating the Upper and Middle Waterton lakes. Stalker and Harrison (1977) place the terminus of the late Wisconsin (Waterton IV) glacier here. Horberg (1954), however, suggests that the late Wisconsin Waterton glacier extended at least 10 mi (16 km) northeast of this point. Upper Waterton Lake occupies a U-shaped glacial trough carved by successive mountain glaciers that drained an area of about 200 mi^2 (320 km^2). The largest of these glaciers, however, were probably augmented by Cordilleran ice that spilled across the continental divide at Kootenai Pass (5,700 ft; 1,737 m) and Brown Pass (6,255 ft; 1,907 m). Stalker and Harrison (1977) note three sharp breaks in slope along the walls of Waterton Valley, which they correlate with four main glacial events of successively lesser magnitude. During the first, or Illinoian "Great Glaciation," a sea of ice filled the valley to a height of some 8,000 ft (2,438 m), leaving only the highest peaks protruding as nunataks. The Albertan till of this glacier is recognized some 48 mi (77 km) northeast of the park at Lethbridge, Alberta (Stalker, 1983). Three subsequent and less extensive glaciations further deepened Waterton Valley by an estimated 800 ft (244 m).

REFERENCES

Alden, W. C., 1932, Physiography and glacial geology of eastern Montana and adjacent areas: U.S. Geological Survey Professional Paper 174, 133 p.

Barendregt, R. W., 1984, Correlation of Quaternary chronologies using paleomagnetism; Examples from southern Alberta and Saskatchewan, *in* Mahaney, W. C., ed., Correlation of Quaternary chronologies, York Symposium: Norwich, England, Geo Books, p. 59–71.

Horberg, L., 1954, Rocky Mountain and continental Pleistocene deposits in the Waterton region, Alberta, Canada: Geological Society of America Bulletin, v. 65, p. 1093–1150.

Karlstrom, E. T., 1981, Late Cenozoic soils of the Glacier and Waterton Parks area, northwestern Montana and southwestern Alberta, and paleoclimatic implications [Ph.D. thesis]: Calgary, Alberta, Department of Geography, University of Calgary, 358 p.

——— , 1986, Probable ice-wedge casts east of Glacier Park, northwestern Montana: American Quaternary Association Program and Abstracts of the Ninth Biennial Meeting, University of Illinois, p. 140.

Mickelson, D. M., Clayton, L., Fullerton, D. S., and Borns, H. W., Jr., 1983, The late Wisconsin glacial record of the Laurentide ice sheet in the United States, *in* Porter, S. C., ed., Late Quaternary environments of the United States: Minneapolis, University of Minnesota Press, p. 3–37.

Richmond, G. M., Mudge, M. R., Lemke, R. W., and Fryxell, R., 1965, Relation of alpine glaciation to the Continental and Cordilleran ice sheets, Pt. E, *in* Guidebook for Field Conference E, Northern and Middle Rocky Mountains, International Association for Quaternary Research: Lincoln, Nebraska Academy of Science, p. 53–68.

Rutter, N. W., 1980, Late Pleistocene history of the western Canadian ice-free corridor, *in* Rutter, N. W., and Schweger, C. E., eds., Special AMQUA issue; The ice-free corridor and peopling of the new world: Canadian Journal of Anthropology, v. 1, no. 1, p. 1–8.

Stalker, A. MacS., 1983, Quaternary stratigraphy in southern Alberta report III; The Cameron ranch section: Geological Survey of Canada Paper 83-10, 20 p.

Stalker, A. MacS., and Harrison, J. E., 1977, Quaternary glaciation of the Waterton–Castle River region of Alberta: Bulletin of Canadian Society of Petroleum Geologists, v. 25, no. 4, p. 882–906.

Wagner, W. P., 1966, Correlation of Rocky Mountain and Laurentide glacial chronologies in southwest Alberta, Canada [Ph.D. thesis]: Ann Arbor, Department of Geology, University of Michigan.

Multiple soils in pre-Wisconsin drift on Mokowan Butte, southwestern Alberta

Eric T. Karlstrom, Department of Geography, University of Wyoming, Laramie, Wyoming 82071

LOCATION AND ACCESSIBILITY

The stratigraphy and morphology of superposed pre-Wisconsin soils on Flaxville surface remnants (Alden, 1932) east of Waterton-Glacier Park can best be studied at Mokowan Butte (Fig. 1). The site is located on private land about 1.2 mi (1.9 km) east of the Waterton Park boundary and 3.6 mi (5.8 km) north of the International Boundary and is reached by driving about 21 mi (33.6 km) west of Cardston, Alberta, on a dirt road. The trip can be completed in a half day with a robust two-wheel or four-wheel drive vehicle. Landowners' permission is not required.

SIGNIFICANCE

A number of Flaxville bench remnants (Alden, 1932) form interfluves about 600 to 1,600 ft (183 to 488 m) above modern streams in the foothills east of Waterton-Glacier Park (Fig. 1). These Miocene-Pliocene erosion surfaces (Alden, 1932) are beveled across folded and faulted Cretaceous sandstone and shale and mantled by about 50 to 250+ ft (15 to 75+ m) of diamicton and alluvial gravels derived from the nearby mountains. This material is interpreted as pre-Wisconsin (Nebraskan or Kansas) glacial till and outwash (Alden and Stebinger, 1913; Alden, 1932; Richmond, 1957) and termed "Kennedy drift" (Horberg, 1956). Within the "Kennedy drift" is the most complete sequence of superposed pre-Wisconsin soils yet recognized in the Rocky Mountains. These soils record at least five relatively warm and moist periods of subaerial weathering during the Pleistocene and probably late Tertiary.

DESCRIPTION

"Kennedy drift" typically consists of 40 to 50% Precambrian Belt Supergroup (Purcell) sedimentary rocks in a sandy loam matrix. Lithologies include about 65% green and red argillite, 20% limestone and dolomite, 7% quartzite, 4% sandstone, and 4% diorite and basalt. Striations that occur throughout the section, particularly on green argillite clasts, suggest a glacial origin for the material. Clast compositions and fabrics indicate that the sediments were deposited by mountain glaciers and streams that flowed east-northeast onto the plains prior to the establishment of modern drainage (Alden, 1932). Volcanic ash within "Flaxville gravels" in eastern Montana yields fission-track dates of about 7 Ma (Colton and others, 1983). Hence, if the Flaxville surfaces are correlative across the state, the lower units within the "Kennedy drift" could date back to the late Miocene and represent either preglacial or late Tertiary glacial deposition. Cryoturbation features, including ice-wedge casts, observed in the Cypress Hills

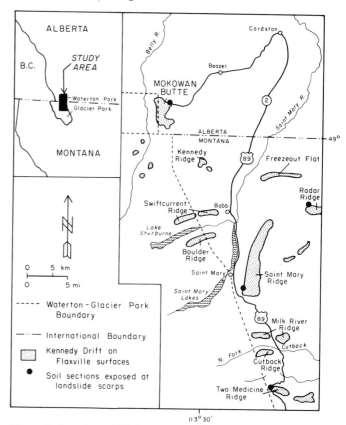

Figure 1. Location of Mokowan Butte and other Flaxville surfaces mantled by "Kennedy drift" or outwash gravels east of Waterton-Glacier Park.

gravels (Oligocene) in southeastern Alberta (Jungerius, 1966) and in "Kennedy drift" and alluvial gravels on Flaxville benches east of Glacier Park (Karlstrom, 1986) suggest that these flat-topped benches developed, in part, as a result of cryoplanation processes.

Stratigraphy and morphology of pre-Wisconsin soils may be examined at a number of accessible landslide scarps that occur along Flaxville ridges (Fig. 1). Some of these have been described (Horberg, 1956; Richmond, 1957; Karlstrom, 1981). Horberg concluded that the 19-ft (5.7-m) thick, reddish B horizon that caps Two Medicine Ridge is similar to Red-Yellow Podzolic or Reddish Prairie soils (Paleustalfs, Paleudults, or Paleudolls; Soil Survey Staff, 1975) of the southeastern and southcentral U.S., and that this polygenetic soil formed mainly during warm, perhaps moist, Yarmouth and Sangamon interglacial climates. Richmond (1957) recognized three buried soils within "Kennedy drift" on Saint Mary and Swiftcurrent ridges and correlated these soils with the Aftonian, Yarmouthian, and Sangamonian interglacial stages

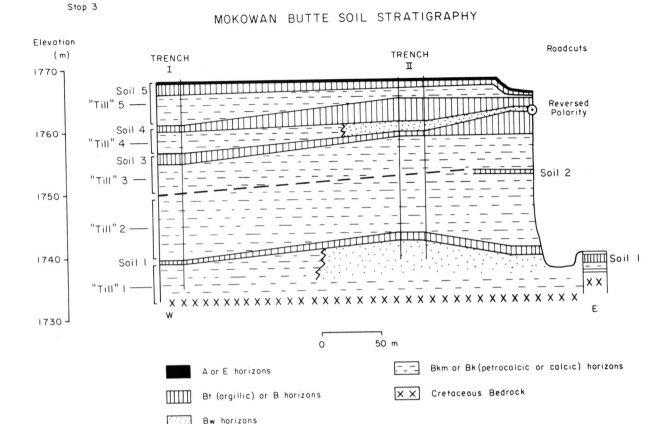

Figure 2. Soil and till stratigraphy at Trenches I and II and road cut sections at Mokowan Butte.

of the Midcontinent. Underlying tills, then, are correlated with the Nebraskan, Kansas, and Illinoian tills of the Midcontinent and the presumed equivalent Washakie Point, Cedar Ridge, and Sacagawea Ridge tills of the Wind River Range, Wyoming. Karlstrom (1981) recognized five soils in five "tills" on Mokowan Butte and Saint Mary Ridge; two soils in three "tills" on Two Medicine Ridge; one soil in one "till" on Milk River, Cutbank, Boulder, and Cloudy ridges; and one truncated soil in the cryoturbated alluvial gravels that cap Freezeout Flat and Radar Ridge (Fig. 1).

The best and most complete sequence of superposed pre-Wisconsin soils yet observed in the region occurs at the northeastern edge of Mokowan Butte (Karlstrom, 1987). This site (Fig. 1; NE¼,NW¼,S24,T1,R28,W.4th mer., UTM-543635 N., 74529 E., Waterton Lakes 82 H/4 Quadrangle) is reached by turning west off Alberta 2 onto a well-graded dirt road south of a Gulf station at the south end of Cardston, Alberta. Turn left again in 1.3 mi (2.1 km) at a Yield sign and follow the road along Lee Creek to Beazer. Continue about 4 mi (6.4 km), then bear left and drive 0.5 mi (0.8 km) to a gate. Turn right (west) after the gate and continue about 4 mi (6.4 km) to the top of Mokowan Butte. Please close gates behind you.

Mokowan Butte is the northernmost of the Flaxville surface remnants in the area and stands about 1,150 ft (350 m) above the Belly River and 1,380 ft (420 m) above the plains to the north. "Kennedy drift" on this surface is 115+ ft (35+ m) thick and mostly indurated by $CaCO_3$ reprecipitated from leached upper horizons. With the exception of Soil 2, the superposed, very strongly developed soils on Mokowan Butte (Fig. 2) are much thicker, redder, and more weathered than soils developed in mountain and Laurentide glacial deposits of Wisconsin and possibly Illinoian age that mantle the Waterton River drainages (Table 1).

Stratigraphy and morphology of soils at this site can be studied in road cut exposures and two trenches excavated on a south-facing landslide scarp (Fig. 2). The top three soils (Soils 5, 4, and 3) are best exposed in Trench I, excavated in 1985. This site is reached by hiking past graded road cuts to the end of the road on the northeastern edge of Mokowan Butte at "Pole Haven," then walking about 900 ft (274 m) west-southwest along the southern edge of the escarpment to the steepest segment of the landslide scarp. Stiff mountaineering boots and an ice ax or shovel facilitate climbing on this steep (35 to 44 degrees) section.

The upper soil (Soil 5) includes a 15-in (37-cm) thick, pale brown E horizon and a 68-in (173-cm) thick, red to yellowish-red argillic (Bt) horizon over a 15-ft (465-cm) thick, white petro-

TABLE 1. PROPERTIES* OF PALEOSOLS ON MOKOWAN BUTTE AND A REPRESENTATIVE POST-WISCONSIN SOIL

Soil[1]	Horizon	Thickness (cm)	Soil Color	pH	Clay (%)	CaCO$_3$ (%)	Clay Minerals[2]	HPI[3]	CAI[4]
Mokowan Butte Soils									
5	E	37	10 YR 6/3 dry	4.7	30	0		125	3448
(TI)	Bt	173	5 YR 4/6 moist	4.8	34	0	3S, 2I, 2K		
	Bkm	465	10 YR 8/2 dry	8.0	14	46	3I, 2K, 2C		
4	BC	36	5 YR 5/4 moist	7.5	24	0		54	2654
(TI)	Bt	71	5 YR 5/6 moist	5.5	36	0	3I, 2C, 2H, 1S		
	Bkm	386	7.5 YR 8/2 dry	8.2	6	40	3I, 2C, 1K, TS		
3	Bt	158	5 YR 5/6 moist	7.8	30	0	3S, 2I, 2H, 1C, 1C-S	83	3960
(TI)	Bkm	400+	7.5 YR 8/2 dry	8.1	5	35	3S, 3I, 2C		
2	Bt	40	5 YR 5/6 moist	7.8	36	0	3S, 1I, 1K, TC	28	960
(RC)	Bkm	70+	2.5 YR 8/2 dry	8.1	8	56			
1	Bt	120	5 YR 5/6 moist	7.5	48	0	2S, 2I, 1K, 1C-V, TC, TV	59	3350
	Bk	115	7.5 YR 5/6 moist	8.1	10	18			
Post-Wisconsin Soil in Waterton Valley									
	A	7	10 YR 4/3 moist	6.7	8	0	2I, 1K, 1C	19	70
	E	11	10 YR 6/3 moist	6.5	12	0			
	Bt	14	10 YR 5/6 moist	7.1	18	0	2S, 2I, 1K, 1C		
	Bw1	26	10 YR 5/4 moist	7.4	6	0			
	Bw2	85+	10 YR 5/3 moist	7.9	13	0	3S, 2I, 1K, 1C		

* Properties are averaged when more than one horizon is included in horizon designation.

[1] T = trench; RC = roadcut.

[2] S = smectite; I = illite; K = kaolinite; C = chorite; H = halloysite; V = vermiculite; 3 = dominant; 2 = moderate; 1 = minor; T = trace amounts.

[3] HPI = Harden profile index (Harden, 1982).

[4] CAI = clay accumulation index (Levine and Ciolkosz, 1983).

calcic (Bkm) horizon or calcrete (Table 1). The argillic horizon is acidic (pH = 3.8 to 6.0), is leached of primary CaCO$_3$ and carbonate rocks, and includes about 70% decomposed clasts. It has thick, continuous clay films on ped faces, up to 35% more clay than the underlying Bkm horizons, and a fine/total clay ratio of up to 0.423. This soil thins to the east where it is exposed in Trench II and in road cuts (Fig. 2), perhaps as a result of differential stripping by colluvial and/or cryoturbation processes.

Soils 4 and 3 are similar to Soil 5 in color, texture, and weathering characteristics. Again, soil thicknesses vary in the three different sections (Fig. 2). Nonetheless, both soils include approximately 3- to 16-ft (1- to 5-m) thick, yellowish-red argillic horizons that contain 26 to 38% clay, 1.4 to 1.8% free iron, and 55 to 66% weathered clasts. Argillic horizons are virtually plugged with illuvial clay, and microfeatures include ferriargillans, papules, and prominent large mangans. Bulk density values of 1.87 to 1.97 reflect the lack of pore space due to intense lessivage and probably compaction of materials by overriding glaciers. Clay minerals in argillic horizons of the upper three soils include smectite, kaolinite, illite, weathered chlorite, mixed-layer chlorite-smectite, chlorite-vermiculite, and halloysite (Table 1). Illite, chlorite, and some kaolinite are inherited from Belt Supergroup rocks and predominate in the upper two calcrete "tillites," whereas crystalline smectite (probably inherited from Cretaceous bedrock), illite, and chlorite predominate in the lower two petrocalcic horizons.

Although classification of these soils presents problems, the upper three soils are morphologically, chemically, and mineralogically most similar to Mediterranean soils and Alfisols (Soil Survey Staff, 1975) and possibly Luvisols (Canadian Soil Survey Committee, 1978). Based on weathering properties and depth of carbonate leaching, these soils probably represent long periods (0.5 to 1.0+ m.y.) of soil formation, and formed, at least in part, under interglacial climates that were warmer and moister than today's. They are tentatively considered late Pliocene (Soil 3), early to middle Pleistocene (Soil 4), and middle to late Pleistocene (Soil 5) in age, but could be even older. Paleomagnetic analysis of eight samples taken from sands and silts between Soils 3 and 4 at a road cut exposure (Fig. 2) indicates that each sample has strong reversed polarity with a normal overprint. It is considered most probable that parent materials ("Tills 3 and 4") for Soils 3 and 4 were deposited during the Matuyama Reversed Epoch between 2.4 and 0.7 Ma.

The stratigraphic contexts of Soils 2 and 1 are less well defined than those of the upper three soils. Soil 2 is a thin, truncated, and discontinuous, but strongly developed Paleustalf (Soil Survey Staff, 1975), which was originally sampled from a road cut that has subsequently been graded. This soil consists of a 16-in (40-cm) thick, yellowish-red argillic horizon over a petrocalcic horizon. Soil properties (Table 1) suggest that the soil formed during a relatively warm, semi-arid climate. Soil 1 underlies carbonate-cemented diamictite (tillite?) in Trenches I and II but is exhumed and exposed at the surface above Cretaceous bedrock in a road cut on the north side of the road about 600 ft (183 m) east of and 330 ft (100 m) below the ridge crest at "Pole Haven." This soil is a very strongly developed Eutroboralf (Soil Survey Staff, 1975) or Luvisol (Canadian Soil Survey Committee, 1978), which includes a 2-ft (120-cm) thick, leached argillic horizon with up to 38% more clay than the underlying Bk horizon

(Table 1). The lower B and Bk horizons include many ghost stones and weathered clasts. Based on its stratigraphic position, this may be a composite or welded soil that formed during the earliest (late Miocene to Pliocene?) and latest (middle to late Pleistocene?) intervals of soil formation recorded in the sequence (represented by Soils 1 and 5, respectively).

In conclusion, the five superposed soils on Mokowan Butte permit subdivision of "Kennedy drift" into at least five diamicts or till units. Soil properties suggest that late Tertiary to Pleistocene soil-forming environments were periodically wetter and warmer than today's. At present, the Waterton-Glacier Park area climate is continental with strong seasonal influences from Pacific Maritime systems. Based on pre-Wisconsin soil properties, the influence of Pacific Maritime and/or Gulf of Mexico air masses may have been stronger during the past. In addition, warmer soil-forming conditions might be explained by warmer late Tertiary and Pleistocene interglacial climates and/or by the possibility that some of these soils formed at lower elevations prior to significant uplift of the region during the late Miocene and Pliocene.

REFERENCES

Alden, W. C., 1932, Physiography and glacial geology of eastern Montana and adjacent areas: U.S. Geological Survey Professional Paper 174, 113 p.

Alden, W. C., and Stebinger, E., 1913, Pre-Wisconsin glacial drift in the region of Glacier National Park, Montana: Geological Society of America Bulletin, v. 24, p. 529–572.

Canadian Soil Survey Committee, 1978, The Canadian system of soil classification: Research Branch Canada Department of Agriculture Publication 1646, 164 p.

Colton, R. B., Naeser, N. D., and Wilcox, R. E., 1983, Seven million-year-old volcanic ash on Missouri–Yellowstone River drainage divide near Circle, Montana: Geological Society of America Abstracts with Programs, v. 15, no. 5, p. 414.

Harden, J. W., 1982, A quantitative index of soil development from field descriptions; examples from a chronosequence in central California: Geoderma, v. 28, p. 1–28.

Horberg, L., 1956, A deep profile of weathering on pre-Wisconsin drift in Glacier Park, Montana: Journal of Geology, v. 64, p. 201–218.

Jungerius, P. D., 1966, Age and origin of the Cypress Hills Plateau surface in

Alberta: Geographical Bulletin, v. 8, no. 4, p. 307–318.

Karlstrom, E. T., 1981, Late Cenozoic soils of the Glacier and Waterton Parks area, northwestern Montana and southwestern Alberta, and paleoclimatic implications [Ph.D. thesis]: Calgary, Alberta, Department of Geography, University of Calgary, 358 p.

—— , 1986, Probable ice-wedge casts east of Glacier Park, northwestern Montana: American Quaternary Association Program and Abstracts of the Nineth Biennial Meeting, University of Illinois, Champaigne-Urbana, p. 140.

—— , 1987, Stratigraphy and genesis of five superposed pre-Wisconsin soils in Kennedy drift on Mokowan Butte, southwestern Alberta: Canadian Journal of Earth Science (in press).

Levine, E. R. and Ciolkosz, E. J., 1983, Soil development in till of various ages in northeastern Pennsylvania: Quaternary Research, v. 19, p. 85–99.

Richmond, G. M., 1957, Three pre-Wisconsin glacial stages in the Rocky Mountain region: Geological Society of America Bulletin, v. 68, p. 239–262.

Soil Survey Staff, 1975, Soil taxonomy: A basic system of soil classification for making and interpreting soil surveys: Soil Conservation Service, U.S. Department of Agriculture Handbook no. 436.

The Milk River Canyon, Alberta: Superposition and piping in a semi-arid environment

C. B. Beaty and R. W. Barendregt, Department of Geography, University of Lethbridge, Lethbridge, Alberta T1K 3M4, Canada

LOCATION AND ACCESSIBILITY

The Milk River Canyon is located in southeastern Alberta some 62 mi (100 km) east of the town of Coutts (Fig. 1). Access to a valley overlook by passenger car on paved and unpaved county roads is relatively easy and requires no permissions to cross private land, but the easternmost unpaved section of road (beyond Black Butte, Fig. 2) is to be avoided if wet. The canyon itself is not visible until the overlook on its rim is reached. The map (Fig. 2) shows readily recognizable points on the route from Coutts and indicates distances between conspicuous landmarks. As is customary, barbed wire gates, if encountered, should be left as found: open if open, closed if closed. There are *no* services available east of Coutts, so gas tanks should be comfortably filled.

SIGNIFICANCE

As well as providing excellent exposures of the Upper Cretaceous Foremost and Oldman formations, the Milk River Canyon contains a textbook example of a so-called "underfit" stream. The river today occupies a wide, deep valley that for a time carried virtually all surface runoff out of southern Alberta and into Montana. The pre-glacial Milk River drained northeast into an ancestral South Saskatchewan system and was superposed by glacial deflection across a low divide extending from the Sweetgrass to the Cypress hills (see Fig. 1). The walls of the canyon are at present retreating, primarily by piping, and provide a spectacular demonstration of that particular geomorphic process. The surficial glacial features include till sections with evidence of at least three continental advances and a series of giant bedrock grooves along the south side of the river.

GEOLOGY AND GEOMORPHOLOGY

The bedrock geology of the Milk River Canyon region is comparatively simple. The valley is cut in essentially horizontal Upper Cretaceous sediments, undisturbed by significant tectonic activity (Russell and Landes, 1940). Small igneous intrusions (presumably outliers of the Tertiary Sweetgrass Hills of Montana) crop out at Black Butte (passed on road to overlook, Fig. 2) and as a number of widely-spaced dikes along the canyon. Topographically, the intrusions are minor elements of the landscape.

Two Upper Cretaceous formations—the Foremost and Oldman—make up most of the walls of the Milk River Canyon. A third formation—the Pakowki—lies below the Foremost on parts of the valley floor but is covered in most places by alluvial and colluvial debris (for a general description and map of bedrock geology see Irish, 1968).

The Foremost and Oldman formations are non-marine, and both contain significant amounts of bentonite. The Foremost con-

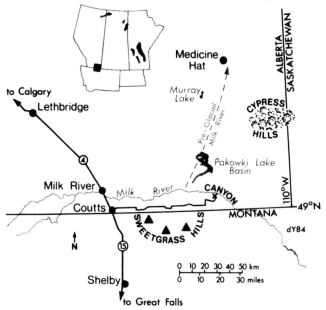

Figure 1. Regional map showing location of Milk River Canyon and associated features.

sists primarily of green, grey, and dark grey carbonaceous shales, grey and green siltstones, and grey and pale brown sandstones. Ironstone beds are common. Some of the carbonaceous shales grade into coal seams, and the boundary between the Foremost and the overlying Oldman is marked by a conspicuous band of lignite visible in the valley for many mi (km). Estimated thickness of this formation along the river is 260 to 330 ft (80 to 100 m).

Strata of the Oldman Formation consist of green, grey, and light grey shales and silty shales interbedded with grey and light grey sandstones. This formation was in the past referred to as the "Pale Beds," in reference to the soft pastel colors typical of these sediments. Hard calcareous sandstone beds and thin ironstone bands are found throughout the succession. Crossbedded lenticular sandstone units occur in the upper part of the formation, and the mineable Lethbridge Coal Member marks its approximate top. Mean thickness of the Oldman in the canyon is 560 ft (170 m). Fragmentary reptilian fossils, and shells—principally *Ostrea*—are present in several horizons.

In general, the two formations making up the walls of the Milk River Canyon are characterized by poorly consolidated, easily erodable sediments. The climate is semi-arid; the natural vegetation is sparse—mostly native bunch grasses—and the soils lack profile development over much of the area. Differences in permeability in surficial materials and bedrock units are marked, with bentonitic clays, tills, lacustrine deposits, and finer alluvium virtually impenetrable when wet. It is the presence of bentonite, in particular, that is responsible for the dramatic development of

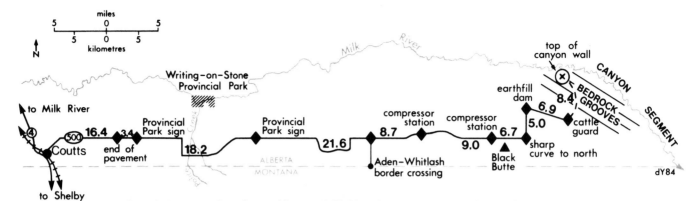

Figure 2. Route map from Coutts, Alberta to Milk River Canyon overlook. Beginning of route at Coutts is signed Alberta Secondary 500. Distances between indicated points are in kilometers. Travel beyond Black Butte is *not* advised if road is wet. Total distance from Coutts to overlook: 64.8 mi (104.3 km).

piping, which is probably the dominant geomorphic agent in canyon wall recession.

The indicated route from Coutts brings the observer to the south rim of the upstream end of the deepest part of the valley. At this point the canyon is 360 to 390 ft (110 to 120 m) deep and 0.75 mi (1.2 km) wide; 6 mi (10 km) downstream it attains a width of 0.9 mi (1.5 km) and a maximum depth of more than 490 ft (150 m) (Fig. 3). Along most of the length of the canyon, access to its floor by foot is possible, although care should be taken not to step accidentally into the open top of a vertical pipe; also, rattlesnakes are present during the warmer months of the year.

The size of the Milk River Canyon is attributable to a number of interrelated factors. Subsurface investigations have revealed the existence of a much larger pre-glacial Milk River system draining toward the northeast (Fig. 1), separated from an ancestral tributary net of the pre-glacial Missouri River by a topographic high extending NE-SW between the Sweetgrass Hills of Montana and the Cypress Hills upland of Alberta-Saskatchewan (Williams and Dyer, 1930; Bretz, 1943). The present course of the Milk River is, in effect, superimposed on this pre-glacial divide, with perceptibly lower ground to the north and south of the canyon. The modern valley was formed when runoff was deflected across the divide by glacial ice.

The contemporary location of the river is thus considerably south of its pre-glacial channel, and in its southeasterly direction of flow, the stream is tributary to the Missouri drainage system. Its pre-glacial course was through the basins of Pakowki and Murray lakes to the north (Fig. 1), where it connected with an ancestral South Saskatchewan–Hudson Bay drainage system. As noted, its present location straddles the divide between the Sweetgrass and Cypress hills. This anomalous position resulted from stagnation of receding Laurentide ice (probably Illinoian) across the divide, leading to the formation of an ice-marginal glacial spillway that conducted water to the southeast.

For an undetermined period of time, then, the Milk River system must have carried all or at least most of the glacial melt-waters as well as runoff resulting from precipitation in southern

Alberta. Until the margin of the retreating Laurentide ice cleared the western and northern flanks of the Cypress Hills, drainage was forced to follow this route. With the opening of an alternative course north of the Cypress Hills, most of the surface runoff in southern Alberta was diverted to the more logical path *down* the regional slope, and discharge of the Milk River must have decreased dramatically. The modern river is but an insignificant shadow of its former self, a classic example of an underfit stream.

At least three advances of Laurentide ice moved through the region, and the Pleistocene stratigraphy includes several till units separated in places by sands and gravels and in other places by lacustrine silts and clays (Westgate, 1968). A series of bedrock grooves along the south side of the river (traversed by the last few miles [km] of the road to the viewpoint) strongly suggests NW to SE movement of ice (Fig. 2). The grooves occur as depressions 3 to 5 mi (5 to 8 km) long, 65 to 130 ft (20 to 40 m) wide, and 7 to 20 ft (2 to 6 m) deep. Although best seen on air photos, they are readily identifiable in the field. Tributary coulees here have mainly formed either parallel or perpendicular to the grooved surface, imparting a rectangular pattern to the drainage.

The Milk River Canyon last served as a spillway during deglaciation following the late Wisconsin, approximately 8,000 to 10,000 years ago. Since that time, the valley has been widened primarily by badland erosion and piping, which have produced parallel retreat of its walls. Investigations in nearby badlands indicate an average slope recession of 0.2 in (5 mm) per year (Barendregt and Ongley, 1979). Slope retreat at this mean rate would account for post-glacial canyon widening of at least 260 ft (80 m), yielding approximately 1.0×10^8 tons of sediment to the Milk River. However, earlier measurements by Barendregt and Ongley (1977) showed that a much larger volume of debris is supplied to the river by piping. They estimated that 1.11×10^5 tons of sediment reached the valley floor through pipe networks in a single summer. Piping and surface erosion associated with badland development may thus have contributed as much as 2.5 $\times 10^9$ tons of sediment to the river during post-glacial time. This would represent an average overall canyon widening of 0.3 mi (0.5 km), or roughly one-third of the present width, and emphasizes the importance of these two geomorphic processes in the area.

Piping along the Milk River is largely restricted to the canyon segment, and many of the pipes are neither visible nor accessible. In the main, they are long, narrow underground channels connecting surficial collapse pits and vertical shafts. The size and morphology of the pipes vary greatly, but diameters may exceed several ft (m). Pipes have been observed at two or more levels, with the lower set pirating water from upper, abandoned systems. Piping occurs throughout the entire height of the canyon wall. All stages of development may be observed, from small, discontinuous rills to large, continuous tunnels and extensively collapsed systems that leave a rugged, dissected canyon wall.

The tunnels parallel the canyon wall slope and frequently have gradients of up to 30°. Vertical shafts leading to the tunnels tend to be incised in the coarser, more porous and weakly cemented sandstones and siltstones of the lower Oldman Formation. The vertical shafts often function as sinkholes, conducting surface runoff into connecting pipes via small rills cut into the shaft walls. Where the exit of a pipe system is not a gully or valley bottom, it usually occurs in or just above carbonaceous shale or bentonitic clay.

Differences between the bedrock canyon wall pipes with their steep gradients and those of lesser inclination near the floor of the valley are due mainly to the nature of the materials in which the latter have developed. The lower deposits consist of alluvial and colluvial material that is lensed and slopes toward the river. This contains a much higher percentage of calcium carbonate, derived from leaching of glacial deposits above and growth and decay of the short-grass prairie vegetation below. This cementing agent increases the lifespan of the lower pipes before they collapse.

Several conditions seem essential for piping to occur (Parker, 1963). They are, in decreasing order of importance: (1) a high percentage of swelling clays (montmorillonite in the bentonite in this case) and pronounced dry periods to desiccate the clay; (2) high-intensity rainfall to provide enough water to saturate a part of the bedrock or surficial deposits; (3) a relatively permeable zone above the local base level of erosion; (4) a steep hydraulic gradient; and (5) outlets above the base level of erosion (usually appearing first as seepage zones).

It is presumed that swelling and cracking of montmorillonite-containing surficial materials are essential to the initiation of piping. Heavy rainfall fills the desiccation cracks with water. Upon wetting and swelling, the clays become dispersed, slippery, and non-cohesive. If throughflow is possible, the clay- and silt-size particles are easily detached and transported in suspension by a process of eluviation similar to that found in soils. The coarser grains provide a skeletal matrix or packing network through which the water and finer particles can move. If throughflow is not immediately possible, a large hydraulic head develops and is maintained until passageways become available. Once the water starts to move, mechanical erosion occurs very rapidly. Joints and fractures permit water to penetrate deeply into bedrock. Excep-

Figure 3. Aerial view of a portion of the Milk River Canyon looking west. The line passing from upper left to lower right on the photo and marked by a land-use change from cultivated to non-cultivated is the Canada-U.S.A. border. West Butte of Sweetgrass Hills, Montana, visible on skyline.

tionally heavy rain may cause surface cracks to be closed by swelling. However, by this time, subsurface channels, too large to be closed, have come into existence. The system self-perpetuates, as pipes are of essentially unlimited permeability, continuously increasing flow and corrasion as the size of the network increases.

Although not required for the development of pipes, impermeable layers probably control the areal extension of the process by impeding vertical drainage and encouraging lateral flow. Eventually, frost action and subsurface water will open passageways through the impervious layers, causing the greater head of water above them to drain suddenly, thus destroying their effect. Along steep bedrock slopes, pipe outlets are often found just above an impermeable layer. However, in the gullies, tributary pipe outlets occur at floor level, which represents the local base level of erosion regardless of structural or lithologic conditions.

The amount of sediment transported to the river through piping networks is impressive. The load of a subterranean stream is large because of the additional sediment supplied by continued collapse of roof and walls. The piping network produces a uniquely rugged landscape, with a memorable display of disappearing stream courses, dry valleys, "sinkholes," blind and hanging valleys, waterfalls, natural bridges, residual hills, and caves. Piping is thus indicated to be the main mechanism of valley-wall recession and badland development in the Milk River Canyon area.

Inspection of the valley from the rim quickly leads to an appreciation of the large volumes of sediment generated along the canyon walls and deposited on its floor. Fresh alluvial deposits fan out toward the present river course, covering old meander scrolls and point bars. Closer investigation reveals three terrace-like surfaces that have probably resulted from continued lateral cutting by the Milk River since Wisconsin time. Only remnants of these surfaces remain, because the orderly downstream migration of river meanders has at one time or another led to erosion of most valley floor sediments. Since the present Milk River is an "underfit" stream, it is unable immediately to remove all of the material supplied to it by the disproportionately much larger valley. Storage of sediments is thus inevitable, and removal must await the migration of river meanders through the accumulating debris.

Undercutting of banks on the outside bends of meanders has produced vertical exposures, in places 30 ft (10 m) high, which show colluvial slope wash interfingering with fine-grained alluvium laid down in calm backwater areas and oxbows. Occasional flood deposits associated with overbank flow are also seen. Since these sections do not reveal any bedrock, it is here assumed that the Milk River is not downcutting at present. The river appears to be "graded"; the rate of sediment removal is roughly in equilibrium with the rate of sediment supply.

The overall longitudinal profile of the river through the canyon portion has a gradient 1.5 times that of the non-canyon reaches. This increase in gradient is likely related to the river's oblique crossing of the Sweetgrass-Cypress Hills pre-glacial divide in this area. The river has markedly increased its sinuosity over the steeper gradient to maintain equilibrium and, consequently, the condition of grade.

The present floodplain of the Milk River consists of cross-bedded coarse sands invaded during periods of low discharge by an annual growth of willow, poplar, sweet clover, and horsetail. Accumulations of wind-blown sand derived from exposed bars are found on the leeward bank; in a few places these mantle the terrace surfaces for a considerable distance. The modern river carries a high sediment load (hence its name) composed of sands, silts, and clays quarried from the older floodplain deposits. Coarser materials are infrequently encountered along the canyon length, occurring as narrow bands varying in particle size from granules to boulders up to 8 in (20 cm) in diameter. This larger debris is derived from till clasts and bedrock concretions brought to the river by its tributaries.

SUMMARY

The geologically recent history of the Milk River Canyon and an understanding of its origin and significance are moderately well in hand today. Diversion of the ancestral Milk River by Laurentide ice across a pre-glacial divide accounts in large part for the size of the present valley, particularly its *depth,* and valley-wall recession by piping has contributed significantly to its *width.* It is a reasonable expectation that effective piping will continue into the foreseeable future since all of the necessary conditions are present in the area.

Nevertheless, there are a number of unresolved issues and unanswered questions about the canyon and its surrounding region. For example, field workers in Alberta and adjacent Montana do not agree on the extent of the Late Wisconsin glaciation. Correlation across the international boundary has not definitively been achieved and probably must await determination of more absolute dates than are currently available. The true nature of the terrace-like surfaces on the floor of the valley has yet to be elucidated, and an unequivocal explanation for the steeper gradient of the river through its canyon segment remains to be discovered. There are, in short, problems aplenty remaining for the student of glacial geology in this part of the northwestern Great Plains.

REFERENCES CITED

Barendregt, R. W., and Ongley, E. D., 1977, Piping in the Milk River Canyon, southeastern Alberta; A contemporary dryland geomorphic process, *in* Erosion and solid matter transport in inland waters—symposium: Paris, France, Proceedings of the Paris symposium, July, 1977, International Association of Hydrological Sciences Publication no. 122, p. 233–243.
——— , 1979, Slope recession in the Onefour Badlands, Alberta, Canada; An initial appraisal of contrasted moisture regimes: Canadian Journal of Earth Sciences, v. 16, p. 224–229.
Bretz, J. H., 1943, Keewatin end moraines in Alberta, Canada: Geological Society of America Bulletin, v. 54, p. 31–52.
Irish, E.J.W., 1968, Geology—Foremost: Geological Survey of Canada Map 22-1967.
Parker, G. G., 1963, Piping, a geomorphic agent in landform development of the drylands: Berkeley, California, International Association of Scientific Hydrology Publication 65, p. 103–113.
Russell, L. S., and Landes, R. W., 1940, Geology of the southern Alberta plains: Geological Survey of Canada Memoir 221, 223 p.
Westgate, J. A., 1968, Surficial geology of the Foremost-Cypress Hills area, Alberta: Research Council of Alberta Bulletin 22, 122 p.
Williams, M. Y., and Dyer, W. S., 1930, Geology of southern Alberta and southwestern Saskatchewan: Geological Survey of Canada Memoir 163, 160 p.

8

Multiple catastrophic drainage of Glacial Lake Missoula, Montana

David D. Alt, Department of Geology, University of Montana, Missoula, Montana 59812

LOCATION AND ACCESS

The lake-drainage site is in Sanders County, Montana, approximately 65 mi (105 km) northwest of Missoula, and about the same distance southwest of Kalispell (Fig. 1).

Paved, two-lane highways and easily passable side roads provide interconnecting access to most parts of the area at all seasons. Montana 382 between Perma and its junction with Montana 28 passes through Markle Pass and across the giant ripples of Camas Prairie. An unpaved track leads from that road to Wills Creek Pass and the Schmitz Lakes. Montana 28 between Plains and Hot Springs crosses the debris fill in the valley of Boyer Creek, and passes Rainbow Lake and through Duck Pond Pass above its head. An unpaved county road connects Rainbow Lake and Camas Prairie.

Ownership is a mixture of private ranch land, public and private forest land, and Flathead-Salish tribal lands, none of which are posted. Access is generally open to all parts of the area except that people planning to spend significant amounts of time on the tribal lands around Rainbow Lake should buy a recreation permit.

The varve site comprises large roadcuts through stream and glacial lake deposits on and near I-90 in the vicinity of the Nine-mile Exit, about 19 mi (31 km) west of Missoula, in Missoula County (Fig. 1).

SIGNIFICANCE

The lake-drainage site contains a complex of spectacular erosional and depositional features that illustrate the effects of grossly torrential discharge. In a sense, the area is the world's most extravagant natural flume experiment. The evidence of great discharge also demonstrates that Glacial Lake Missoula drained catastropically, and was the source of the Spokane floods.

One large roadcut in the varve site contains a record of at least 36 fillings of Glacial Lake Missoula, presumably during Pinedale(?) time, which suggest that drainage occurred many times. Another roadcut exposes a similar, but unstudied record of deposition in a higher terrace that probably formed during Bull Lake(?) time and may indicate an older history of lake formation and drainage.

SITE INFORMATION

Glacial Lake Missoula and the Spokane Flood. In the first significant study of Glacial Lake Missoula, Pardee (1910) provided a general description of the lake, and the ice dam that impounded the Clark Fork drainage near the present site of Pend Oreille Lake in northern Idaho. He did not then address the question of how the lake drained. A radical view of catastrophic

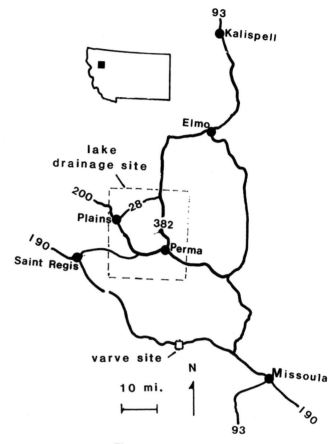

Figure 1. Location map.

lake drainage was proposed by Bretz (1923) who inferred that the Channeled Scabland of eastern Washington had been very rapidly eroded by a catastrophic "Spokane flood." Bretz (1925) vigorously defended this radical contention before a thoroughly skeptical audience. Flint (1938) best exemplifies the intellectually conservative faction that attempted to attribute erosion of the Channeled Scabland to normal noncatastrophic processes.

Bretz' greatest problem in defending his interpretation was that of demonstrating a plausible source for the Spokane flood. Pardee (1942) provided the source. He showed that Glacial Lake Missoula had indeed drained catastrophically when the ice dam failed, and was therefore the source of the flood. The primary site contains the most striking evidence of sudden drainage of Glacial Lake Missoula: three major flood spillways with their associated complexes of erosional and depositional features. All carried water spilling over the divide from the Little Bitterroot Valley south toward the Flathead River.

Erosion of the passes. When Glacial Lake Missoula

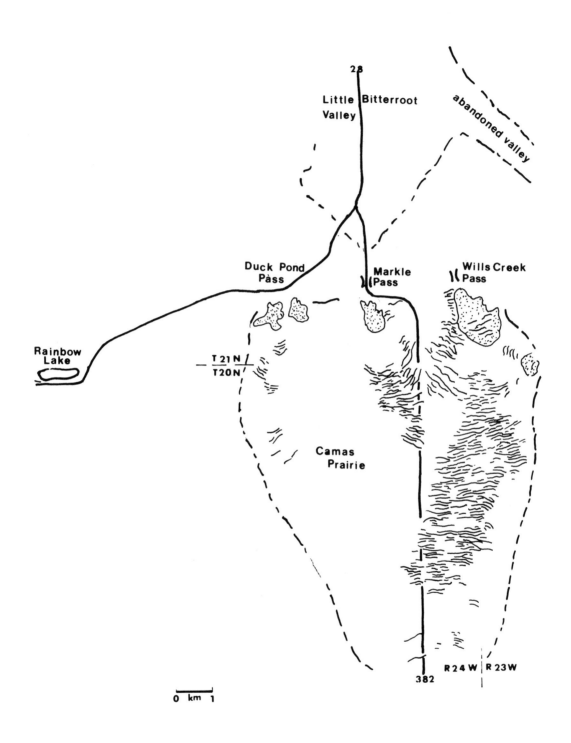

Figure 2. Crested traces of giant ripples in Camas Prairie. Stippled areas below the passes outline large deltaic bars.

drained, Wills Creek Pass, at an elevation of 3,320 ft (1,012 m), and Markle Pass, at an elevation of 3,420 ft (1,042 m), were the major spillways between the Little Bitterroot Valley and Camas Prairie (Fig. 2). Duck Pond Pass injected large volumes of water into the valley of Boyer Creek west of Camas Prairie. All contain basins eroded into sound bedrock, presumably by cavitation plucking.

Markle Pass contains five bedrock basins. One deep pit beside the highway holds a shallow pond during wet seasons. Wills Creek Pass contains about 30 bedrock basins, several of which hold the small but perennial Schmitz Lakes. Flow through Duck Pond Pass at the head of Boyer Creek gouged a deep trough that now holds marshes near the pass, and Rainbow Lake a few kilometers below.

Bedrock in all the passes is Precambrian quartzite and argillite metamorphosed in the greenschist facies, extremely resistant rock. Bretz (1925, 1969), Bretz and others (1956), and Baker (1973, 1978) all argued that the Spokane floods eroded the Channeled Scabland by plucking the columnarly jointed basalt. It seems reasonable to extend that argument to the eroded bedrock in the spillways of Glacial Lake Missoula. Such extensive plucking suggests that discharge through the passes was almost certainly in upper regime flow.

After the lake level dropped below the elevation of Wills Creek Pass, the water still in the Little Bitterroot Valley drained through the abandoned valley of the Flathead River.

Deposition below the passes. The north end of Camas Prairie contains a pair of enormous dumps of sediment immediately beneath Markle and Wills Creek passes. Each grades downslope into a train of giant ripples composed of angular debris dominantly in the cobble-size range. Vantage points near Markle or Wills Creek Pass clearly show one train of giant ripples west of the road below Markle Pass and another much larger set east of the road below Wills Creek Pass (Fig. 2). The two ripple trains meet and interfere approximately along the line of Montana 282.

Pardee (1942) showed that the largest giant ripples on the steep slope below Markle and Wills Creek passes attain amplitudes as great as 35 ft (11 m) from crest to trough and wave lengths in excess of 300 ft (91 m). They diminish as the slope flattens, and pass into a large area of nearly flat terrain on the floor of Camas Prairie.

Farther south, downslope from the area of flat valley floor, much smaller ripples composed mostly of silt and sand continue for several kilometers, their grain size becoming progressively finer down the valley. Lister (1981) showed that average height of those ripples decreases down slope at a rate of 3.7 ft per mi (0.7 m per km), their average chord length at about 79 ft per mi 15 m per km). They become imperceptible near the southern end of Camas Prairie.

Lister (1981) found through study of their form and internal structure that at least some of the high ripples near the head of Camas Prairie are antidunes. That observation, coupled with the plucking in the passes, suggests that they formed under water in

upper regime flow. Lower ripples farther down the valley are dunes that evidently formed beneath water in lower regime flow. It therefore seems likely that the flat valley floor between the two trains of giant ripples corresponds to a zone of transition between upper and lower regime flow.

Another much less conspicuous train of giant ripples exists in the woods below Rainbow Lake on the top of an enormous deposit of angular debris that almost completely fills most of the valley of Boyer Creek. That feature is best appreciated by driving north from Plains on Montana 28. After following a normal stream valley for several miles, the road abruptly angles up the steep front slope of an enormous debris deposit. Large roadcuts disclose crude foreset bedding in angular debris that includes many blocks as much as 6.6 ft (2 m) on a side.

Repetitious drainage of Glacial Lake Missoula. Early workers generally assumed some modest number of Spokane floods, fewer than 10. Chambers (1971) and Alt and Chambers (1970) counted alternating varved and cross-bedded sequences exposed in a large roadcut through Glacial Lake Missoula deposits about 19 mi (31 km) west of Missoula, near the Ninemile Exit from I-90. They attributed the cross-bedded sequences to stream deposition, and thus concluded that Glacial Lake Missoula filled at least 36 times—the number of varved sequences. Waitt (1980) later counted at least 40 separate flood sequences on the Columbia Plateau. The discrepancy is not important because it is not necessary to assume that all fillings of the lake reached elevations that would flood far into western Montana.

Chambers (1971) found through varve counts that the earliest filling of Glacial Lake Missoula lasted for 58 years, and that each later filling lasted fewer years than the one before. The latest filling recorded in their measured exposure lasted just 9 years. It follows that successive fillings were to progressively lesser depth, and that not all, perhaps as few as a third, were deep enough to flood the passes between the Little Bitterroot Valley and Camas Prairie.

Deposits so far studied appear to record only the Pinedale(?) filling of Glacial Lake Missoula. Varved glacial lake deposits in widely scattered high terraces are clearly older than the Pinedale(?) sediments, and appear to record earlier, presumably Bull Lake(?), fillings of Glacial Lake Missoula that await investigation. Roadcuts beside the old highway near the Ninemile Exit from I-90 expose such sediments in a high and considerably dissected terrace. The extent to which pre-Pinedale(?) drainages contributed to development of erosional and depositional features associated with drainage of Glacial Lake Missoula features in the site remains quite unknown.

REFERENCES CITED

Alt, D., and Chambers, R. L., 1970, Repetition of the Spokane flood [abs.], *in* American Quaternary Association Meeting, 1st, Yellowstone Park and Bozeman, Montana, 1970, Abstracts: Bozeman, Montana, Montana State University, p. 1.

Baker, V. R., 1973, Paleohydrology and sedimentology of Lake Missoula flooding in eastern Washington: Geological Society of America Special Paper 144, 79 p.

——, 1978, Paleohydraulics and hydrodynamics of scabland floods, *in* Baker, V. R., and Nummendal, D., eds., The channeled scabland: Washington, D.C., National Aeronautics and Space Administration, p. 59–79.

Bretz, J. H., 1923, The Channeled Scabland of the Columbia Plateau: Journal of Geology, v. 31, p. 617–649.

——, 1925, The Spokane flood beyond the Channeled Scabland: Journal of Geology, v. 33, p. 97–115.

——, 1969, The Lake Missoula floods and the Channeled Scabland: Journal of Geology, v. 77, p. 505–543.

Bretz, J., Smith, H.T.U., and Neff, G. E., 1956, Channeled Scabland of Washington; New data and interpretations: Geological Society of America Bulletin, v. 67, p. 962–1049.

Chambers, R. L., 1971, Sedimentation in Glacial Lake Missoula [M.S. thesis]: Missoula, University of Montana, 100 p.

Flint, R. F., 1938, Origin of the Cheney-Palouse scabland tract, Washington: Geological Society of America Bulletin, v. 49, p. 461–524.

Lister, J. C., 1981, The sedimentology of Camas Prairie basin and its significance to the Lake Missoula flood [M.S. thesis]: Missoula, University of Montana, 66 p.

Pardee, J. T., 1910, The Glacial Lake Missoula: Journal of Geology, v. 18, p. 376–386.

——, 1942, Unusual currents in Glacial Lake Missoula, Montana: Geological Society of America Bulletin, v. 53, p. 1569–1600.

Waitt, R. B., Jr., 1980, About forty last Glacial Lake Missoula jökulhlaups through southern Washington: Journal of Geology, v. 88, p. 653–679.

Structural geology of the Sawtooth Range at Sun River Canyon, Montana Disturbed Belt, Montana

David R. Lageson, Department of Earth Sciences, Montana State University, Bozeman, Montana 59717

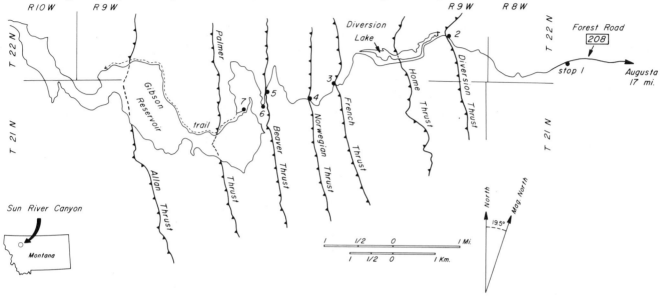

Figure 1. Map of the lower Sun River Canyon showing stop localities (1 to 6) along Forest Service Road 208, major thrust faults, and Gibson Reservoir and Diversion Lake.

LOCATION

The Sawtooth Range is located in northwest Montana in parts of Pondera, Teton, and Lewis and Clark counties. The range is part of the Rocky Mountain fold and thrust belt, or Disturbed Belt, and lies between Glacier National Park to the north and the Helena salient to the south. In an analogy with the Canadian Rocky Mountains, the Sawtooth Range would represent the Front Ranges of the Rocky Mountain thrust belt (Mudge, 1972; Bally and others, 1966).

The Sun River and its tributaries form the major drainage system in the southern part of the Sawtooth Range. Two dams have been constructed on the Sun River creating Gibson Reservoir and Diversion Lake (Fig. 1).

The entrance to the Sun River Canyon is approximately 20 mi (32 km) northwest of Augusta, Montana. Turn west on Montana 435 (U.S. Forest Service Road 208) at the intersection of highways U.S. 287 and Montana 434-435 in Augusta. Montana 435 is a year-round gravel road that ends at Gibson Dam (Fig. 1). Spectacular exposures of imbricate thrust faults and related structures are found in the lower part of the canyon downstream from Gibson Dam.

SIGNIFICANCE

The Sawtooth Range is one of the best exposed examples of imbricate thrust faulting in the foreland fold and thrust belt of the

western United States. It is an outstanding field classroom for the demonstration of stratal shortening and the various structural features associated with thrust belt terranes. The range is composed of approximately 20 major imbricate thrust faults and many smaller faults and folds that form a north-northwest–trending belt 25 mi (40 km) wide and 80 mi (129 km) long. The Sun River Canyon forms an outstanding cross section through the southeastern margin of this belt, offering excellent exposures of thrust contacts along the road and panoramic views of stacked thrust sheets.

In particular, the Sun River Canyon traverse demonstrates the dependence of structural style on relative competence of units in the stratigraphic section. Large thrust sheets are formed by massive Paleozoic carbonates that deform internally by concentric folding, whereas closely spaced imbricate thrust faults and pseudo-similar folding characterize the more ductile Mesozoic strata. Mudge (1972) provides a detailed stratigraphic section for the Sun River Canyon.

GEOLOGY

The Sawtooth Range is composed of an anastomosing array of imbricate thrust faults that flatten and merge downward into a regional décollement or sole detachment fault in Cambrian shales. This regional décollement dips west and steps down-

Figure 2. Photo looking west at Diversion Thrust Fault (lower cliff) and the Front Ranges of the Disturbed Belt. The two massive cliffs are Mississippian carbonate rocks that have been thrust over Cretaceous rocks at the base of each cliff.

section to the west above a largely autochthonous basement complex, eventually incorporating Proterozoic Belt Supergroup strata into the hanging wall of the Lewis and other major thrust faults west of the Sawtooth Range (Mudge and Earhart, 1980). The deformational style of foreland fold and thrust terranes has been extensively reviewed in the literature (e.g., Dahlstrom, 1970; Mudge, 1972, 1982) and will not be discussed here.

The entire Sawtooth Range verges eastward in the direction of tectonic transport. Individual thrust faults and thrust sheets dip west, giving the "sawtooth" appearance to the range. In the eastern Sun River Canyon, the Mississippian Castle Reef Dolomite and Allan Mountain Limestone have been thrust over the Cretaceous Kootenai Formation on five major thrust faults (Mudge, 1965, 1966, 1972). From east to west (as one drives into the canyon), these are: (1) Diversion thrust (Fig. 2), (2) Home thrust, (3) French thrust, (4) Norwegian thrust, and (5) Beaver thrust (site of Gibson Dam). In addition, smaller thrust faults within larger thrust sheets repeat parts of the Jurassic and Cretaceous sections west of the Diversion thrust.

East of the Diversion thrust, in the "foothills" of the Sawtooth Range, over 25 small imbricate thrust faults repeat parts of the Cretaceous section from the Kootenai to the Two Medicine Formation (Mudge, 1965). Although most of these faults are too subtle to observe while driving on the road, in places one can pick out sandstone ledges repeated within the Blackleaf Formation. These imbricate thrusts are grouped into two "imbricate zones," the Stecker imbricate zone to the west and the Neal imbricate zone to the east (Mudge, 1965). It is likely that these imbricate zones are map expressions of small duplex fault zones, with the roof thrust located on the west side of each imbricate zone. Furthermore, the faults within these zones probably join at depth with a single thrust fault that displaces Paleozoic rocks.

Evidence for timing of deformation in the Disturbed Belt has been presented by Hoffman and others (1976), based on K-Ar dates of Cretaceous bentonites. They report an age range from 72 to 56 Ma, which reflects burial metamorphism due to tectonic loading. However, this age range does not necessarily indicate either the beginning or the end of thrusting, but rather it represents the time range during which maximum temperatures were reached, which may correspond to thrust emplacement. In general, Mudge (1982) has concluded that most, if not all deformation in the Disturbed Belt occurred during Paleocene time.

The imbricate style of the Sawtooth Range contrasts dramatically with the broad, open style of deformation exhibited by the Lewis thrust at Glacier National Park to the north. The reader is referred to Raup and others (1983) for an excellent roadside geology review of the Lewis thrust fault and Glacier National Park along Going-To-The-Sun Road.

ROADSIDE STOPS

The following localities are on the road (Montana 435, U.S. Forest Service Road 208) between Augusta, Montana, and the trail head at Gibson Dam in the Sun River Canyon (Fig. 1). Total distance covered is 26 mi (42 km), and all mileages are measured from the intersection of U.S. 287 and Montana 434-435 in Augusta. Refer to Mudge (1965, 1966) for detailed structural and stratigraphic information.

Stop 1 is 17 mi (27 km) from Augusta, Montana, on U.S. Forest Service Road 208 (Montana 435). South of the road (9:00) is a well-exposed, overturned, thrust-faulted anticline in the Cretaceous Blackleaf Formation (see Mudge, 1972, p. B15). The fore- and back-limbs of this asymmetric anticline have both been thrust faulted. This is an excellent example of the style of deformation in Cretaceous rocks of the foothills of the thrust belt.

Stop 2 is 19.6 mi (31.5 km) from Augusta, 2.6 mi (4.2 km) from the previous stop. Here, the trace of Diversion thrust crosses the road and forms the entrance to Sun River Canyon. The Mississippian Allan Mountain Limestone and Castle Reef Dolomite have been thrust over the Cretaceous Blackleaf Formation, forming the first exposures of Paleozoic carbonates and marking the east boundary of the Sawtooth Range. Diversion Lake Dam is just ahead on the right.

Stop 3 is 21.8 mi (35 km) from Augusta, 2.2 mi (3.5 km) west of Stop 2. The roadcut on the south side of the road provides an excellent exposure of the French thrust fault, where the Mississippian Castle Reef Dolomite has been displaced over black shales of the Cretaceous Blackleaf Formation. This is a unique opportunity to actually observe, at close range, the features associated with a thrust fault contact, which are usually covered and inferred from stratigraphic relations. Note the slickensides on the fault contact and the intensely sheared Cretaceous shales, whereas Mississippian carbonate rocks in the hanging wall are relatively undeformed (Mudge, 1972, p. B22).

Stop 4 is 22.1 mi (35.5 km) from Augusta, 0.3 mi (0.5 km) west of the previous stop. This is the trace of the Norwegian thrust fault, one of three closely spaced thrust faults that crop out in this area. In general, this fault is marked by the Mississippian Allan Mountain Limestone overlying the Cretaceous Kootenai Formation. Turn right 0.3 mi (0.5 km) past trace of Norwegian thrust, cross the bridge over the Sun River, and continue west on the main road. Gibson Dam is to left.

Stop 5 is 23.7 mi (38 km) west of Augusta, 1.6 mi (2.6 km) past Stop 4. The Beaver thrust is exposed on the steep hillside to the right (3:00), where the Mississippian Allan Mountain Limestone has been thrust over redbeds of the Cretaceous Kootenai Formation.

Stop 6 is 24.1 mi (38.8 km) west of Augusta, 0.4 mi (0.64 km) from the previous stop. Stop at the Gibson Dam overlook at the crest of the ridge, which is formed by the resistant Mississippian Castle Reef Dolomite. The panoramic view to the east takes in the lower Sun River Canyon and the major imbricate thrust faults that comprise the eastern Front Ranges of the Disturbed Belt. To the south, the Beaver and Norwegian thrust sheets form spectacular dipslopes on the Mississippian Castle Reef Dolomite.

Stop 7 is 26 mi (42 km) west of Augusta, 1.9 mi (3 km) past the previous stop, and is at the end of the road. This is the boat access to Gibson Reservoir and the trail head for U.S. Forest Service Trail 201. A short hike along this trail will offer good exposures of Devonian and Cambrian strata along the Palmer thrust fault (Fig. 1); those more adventurous will find outcrops of Proterozoic Belt strata and outstanding scenery in the Bob Marshall Wilderness several miles farther west.

REFERENCES CITED

Bally, A. W., Gordy, P. L., and Stewart, G. A., 1966, Structure, seismic data, and orogenic evolution of the southern Canadian Rocky Mountains: Bulletin of Canadian Petroleum Geology, v. 14, no. 3, p. 337–381.

Dahlstrom, C.D.A., 1970, Structural geology in the eastern margin of the Canadian Rocky Mountains: Bulletin of Canadian Petroleum Geology, v. 18, no. 3, p. 332–406.

Hoffman, J., Hower, J., and Aronson, J. L., 1976, Radiometric dating of time of thrusting in the Disturbed Belt of Montana: Geology, v. 4, p. 16–20.

Mudge, M. R., 1965, Bedrock geologic map of the Sawtooth Ridge Quadrangle, Teton and Lewis and Clark counties, Montana: U.S. Geological Survey Geologic Quadrangle Map, GQ-381, scale 1:62,500.

——, 1966, Geologic map of the Patricks Basin Quadrangle, Teton and Lewis and Clark counties, Montana: U.S. Geological Survey Geologic Quadrangle Map, GQ-453, scale, 1:62,500.

——, 1972, Structural geology of the Sun River Canyon and adjacent areas, northwestern Montana: U.S. Geological Survey Professional Paper 663-B, 52 p.

——, 1982, A resume of the structural geology of the Northern Disturbed Belt, northwestern Montana, in Powers, R. B., ed., Geological studies of the Cordilleran thrust belt: Rocky Mountain Association of Geologists, p. 91–122.

Mudge, M. R., and Earhart, R. L., 1980, The Lewis thrust fault and related structures in the Disturbed Belt, northwestern Montana: U.S. Geological Survey Professional Paper 1174, 18 p.

Raup, O. B., Earhart, R. L., Whipple, J. W., and Carrara, P. E., 1983, Geology along Going-To-The-Sun Road, Glacier National Park, Montana: West Glacier, Montana, Glacier Natural History Association, 62 p.

Shonkin Sag laccolith; A differentiated alkaline intrusion, Highwood Mountains, Montana

W. P. Nash, *Department of Geology and Geophysics, University of Utah, Salt Lake City, Utah 84112-1183*
Lawrence M. Monson, *Fort Peck Tribal Minerals Resource, P.O. Box 595, Poplar, Montana 59255*

LOCATION AND ACCESS

The Shonkin Sag laccolith is on the northeast flank of the Highwood Mountains in Sec. 22–27,T.21N.,R.11E., in Choteau County, Montana. The laccolith lies entirely within the Geraldine, Montana, 1:24,000 Quadrangle. It is approximately 44 mi (71 km) east of Great Falls and 54 mi (87 km) northwest of Lewiston. The nearest town is Geraldine, about 3 mi (5 km) to the north (Fig. 1). The best access is through Geraldine, which lies on Montana 80, 27 mi (43 km) south of Fort Benton or 38 mi (61 km) north of Stanford. The shortest route from Great Falls is by way of U.S. 87 to Geyser, a distance of 47 mi (76 km). From Geyser, a gravel road can be used to reach the Buck Ranch access, 27.3 mi (43.9 km) to the north. This route contains many excellent outcrops and views of other alkalic igneous bodies, including Square Butte and Round Butte.

Outcrops of the Shonkin Sag laccolith can be seen on the Buck Ranch, which is entered 3.7 mi (6.0 km) south of Geraldine (Fig. 2). Follow the ranch access 1.5 mi (2.4 km) to the ranch buildings. The western outcrop is adjacent to this road. Trails east of the ranch yard lead to the top of the laccolith and permit easy examination of an entire cross section by traversing down the canyon draining to the southwest. This cross section can be viewed from the Flat Creek Road, which intersects Montana 80, 0.5 mi (0.8 km) north of the small town of Square Butte (Fig. 1). Because of the underlying Eagle Sandstone cliffs, access to the igneous rocks is extremely difficult from this road. Permission to examine any of the rocks should be obtained from Mr. Dexter Buck, Geraldine, Montana 59446; telephone (406) 737-4349.

SIGNIFICANCE

Shonkin Sag is but one of several laccoliths in the Highwood Mountains. The most spectacular is Square Butte laccolith, more than 1,640 ft (500 m) thick, which forms a prominent landmark visible from as far as 62 mi (100 km). All of the laccoliths are generally similar in petrography, but Shonkin Sag laccolith is well exposed, has easy access, and has a rich investigative history. For almost a century Shonkin Sag laccolith has contributed to the development of models, often controversial, for the differentiation of mafic alkaline magmas. The laccolith is small, circular, and ideally suited for study. It provides a superb cross section of a laccolith with upper and lower contacts well exposed. Margins of the laccolith provide excellent exposures of the interfingering of magma and host rock that reveal features associated with the forcible injection of magma. Differentiation of alkaline magma may be observed, with rock types evolving from a chilled margin of mafic phonolite to shonkinite, syenite, and alkaline syenite.

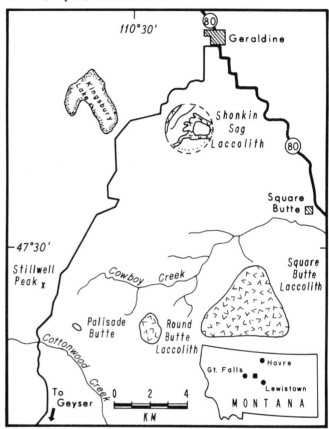

Figure 1. Location map for Shonkin Sag laccolith.

GEOLOGIC SETTING

Introduction. Shonkin Sag laccolith is adjacent to the Highwood Mountains, in the northern part of Larsen's "Montana Petrographic Province" (1940) which stretches from Yellowstone National Park to Canada and includes a wide variety of alkaline volcanic and plutonic rocks. The Highwood Mountains and adjacent intrusions cover almost 300 mi^2 (1000 km^2) and have been dated at 50–53 Ma (Marvin and others, 1980).

Following early work on the Highwood Mountains (Pirsson, 1905), teams from Harvard and Yale described the geology of the Highwoods, including Shonkin Sag, in great detail (Hurlbut, 1939; Larsen and others, 1941; Barksdale, 1937). A general geologic map of the Highwood Mountains is given by Larsen and others (1941, p. 1733).

Shonkin Sag laccolith lies within nearly horizontal Cretaceous sediments of the Eagle Sandstone Formation. The enclosing sediments are best exposed on the east flank of the

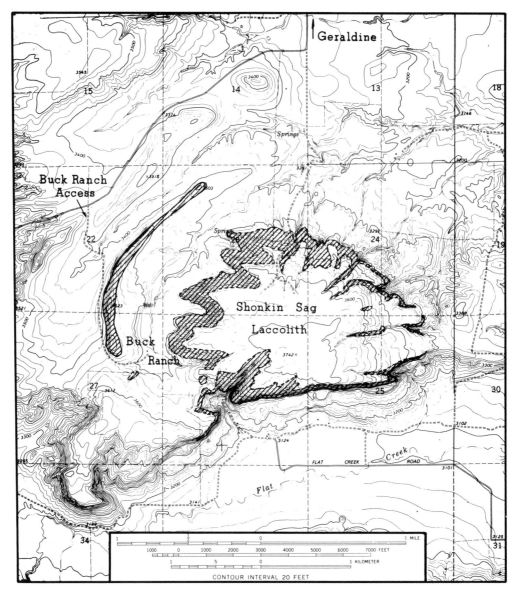

Figure 2. Map showing access and generalized outcrop pattern of Shonkin Sag laccolith. (Base map is from the Geraldine, Montana, 1:24,000 quadrangle).

laccolith, and large-scale layering within the laccolith can be observed from this location as well. Toward the eastern end of the cliffs formed by the laccolith, the intrusion can be seen to thin and interfinger with the host sediments. Pollard and others (1975) provide a thorough description of the geometry of the laccolith as well as an analysis of the intrusion mechanism. Detailed geologic maps of the laccolith are provided by Barksdale (1937) and Hurlbut (1939, plate 3).

Petrology. Figure 3 is a diagrammatic cross section of the laccolith showing the major rock types present. At the base of the laccolith the underlying sandstone has been thermally altered into a light blue, flinty rock. Thermal alteration is commonly only a few inches thick, ranging occasionally up to a thickness of 3 ft (1 m). The chilled margin of the intrusion is a dense porphyritic

rock containing phenocrysts of diopsidic-augite (salite), pseudoleucite, olivine, phlogopitic biotite, and apatite. The fine-grained groundmass consists of pseudoleucite, olivine, salite, sanidine, biotite, titanomagnetite, apatite, zeolites, and minor carbonate (Table 1). The rock of the chilled margin has been known by several names, including leucite basalt (Weed and Pirsson, 1895), pseudoleucite-augite vogesite (Barksdale, 1937), and mafic phonolite (Hurlbut, 1939).

Within about 16 ft (5 m) from the contact, the mafic phonolite grades upward into the lower shonkinite through an increase in grain size and the development of a granitic texture, while pseudoleucites lose their euhedral outlines. The lower shonkinite ranges in thickness from 10 to 164 ft (3 to 50 m), depending upon proximity to the margin of the laccolith. The shonkinite has the

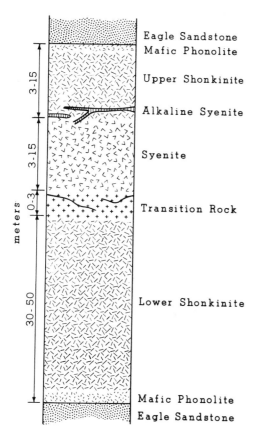

Figure 3. Diagrammatic cross-section of Shonkin Sag laccolith (after Hurlbut, 1939).

TABLE 1. CHEMICAL ANALYSES AND MODES OF REPRESENTATIVE LITHOLOGIES FROM SHONKIN SAG LACOLITH*

wt %	1 Mafic phonolite**	2 Lower shonkinite	3 Syenite	4 Soda Syenite
SiO_2	47.00	47.87	49.44	52.47
TiO_2	0.80	0.75	0.77	0.12
Al_2O_3	12.91	11.92	18.10	22.71
Fe_2O_3	1.30	3.25	3.40	1.96
FeO	7.20	5.18	3.60	1.55
MnO	0.15	0.15	0.10	0.10
MgO	7.75	8.72	3.20	0.10
CaO	9.70	9.92	5.48	1.22
Na_2O	1.85	2.95	4.25	8.25
K_2O	6.45	5.15	7.25	6.60
P_2O_5	0.91	1.11	0.71	0.08
H_2O+	1.74	1.71	2.76	4.33
H_2O-	0.40	0.24	0.36	0.50
CO_2	1.01	0.01	----	----
S	0.06	0.06	0.03	----
Total	100.00	99.63	100.18	100.09

ppm	1	2	3	4
Ba	6410	5940	6850	1550
Sr	2195	1465	1595	570
Rb	200	140	245	135
La	80	55	60	145
Ce	155	135	80	210
Zr	190	160	180	615
Nb	15	10	20	50
Cu	120	100	135	40
Co	90	85	60	15
Ni	65	90	20	5
Cr	170	225	40	10

Volume %	1	2	3	4
Clinopyroxene	19.5	27.1	13.3	6.2
Olivine	5.8	3.9	tr	---
Biotite	0.5	17.5	6.7	1.6
Pseudoleucite	5.6	---	---	---
Sanidine	---	30.3	54.1	63.0
Opaques	---	1.4	2.7	0.3
Apatite	tr	1.8	0.9	0.3
Sphene	---	---	tr	tr
Amphibole	---	---	---	0.1
Garnet	---	---	---	0.1
Other (Zeolite and carbonate)	--	18.0	22.3	28.4
Groundmass	68.6	---	---	---

*Analyses from Nash and Wilkinson (1979, 1971).

**Chilled lower margin.

same mineral assemblage as the chilled margin. Nepheline has also been reported in shonkinite (Hurlbut, 1939), although it is rare and usually represented by zeolite alteration products. Clinopyroxene often has narrow rims of aegerine-augite, and olivine is commonly surrounded by biotite. The type shonkinite occurs at Square Butte laccolith.

Shonkinite grades upward over a short interval into a transition rock that separates the lower shonkinite from the main syenite. Its coarse grain size is distinct, and it contains abundant star-shaped clusters of radiating augite crystals commonly 4 in (10 cm) in length. Other than the textural difference, the transition rock is mineralogically similar to shonkinite. Barksdale (1937) has described the transition rock in considerable detail.

The contact between the transition rock and syenite is also distinct and takes place over a few inches. The syenite is light gray, weathering to chalky white. The term syenite is misleading, for the rock is chemically equivalent to nepheline syenite, and the abundant zeolites almost certainly are alteration products of nepheline, which has been found on rare occasions. Syenite reaches a maximum thickness of 49 ft (15 m) near the center of the laccolith. The syenite differs from shonkinite only in the proportions of its mineral constituents and contains euhedral and sub-

hedral grains of diopsidic augite, biotite, rare olivine, titanomagnetite, and apatite surrounded by sanidine and zeolites.

The upper shonkinite overlies the syenite, and the transition between the two is often abrupt. The upper shonkinite also has a maximum thickness of about 49 ft (15 m). It grades vertically into an upper chilled margin about 7 ft (2 m) thick, which is overlain by the sedimentary cover. Within the upper shonkinite, and at times separating it from syenite, are thin sills of distinct alkaline syenite, termed "upper transition rock" by Pirsson (1905), "aegerite syenite" by Hurlbut (1937), and "soda syenite" by Nash and Wilkinson (1970). This variety is intrusive to the upper shonkinite and is distinguished from the main syenite by the absence of olivine. Compared to the main body of syenite, these rocks possess distinctly higher Na_2O/K_2O ratios (Table 1) and greater amounts of acmitic pyroxene. On the basis of chemistry and mineralogy, Nash and Wilkinson (1970) concluded that the most evolved rock of the laccolith is represented by a thin, irregular, leucocratic vein approximately 4 in (10 cm) thick, which cuts coarse-grained syenite. In addition to acmitic pyroxene, the soda syenites contain euhedral iron-rich biotite elongated parallel to the b crystallographic axis, sanidine, zeolites, magnetite, apatite, sphene, and traces of arfvedsonite, melanite, and rare britholite.

Mineralogy and Chemistry. The laccolith has several noteworthy mineralogical features. Detailed analytical data are provided in Nash and Wilkinson (1970) and Nash (1972). Clinopyroxene displays the well-documented chemical evolution from diopsidic augite through aegerine-augite to acmite characteristic of differentiating alkaline magmas. Micas range in chemical composition from $annite_{24}$ ($phlogopite_{76}$) in the chilled margin to $annite_{100}$ in soda syenite. Olivine changes much less in composition from Fo_{80} to Fo_{60} and is absent in soda syenite. Arfvedsonite occurs in soda syenite, often intergrown with acmite. Melanite garnet is present only in highly evolved soda syenites; it contains fluorine, and under the microscope it exhibits low birefrigence and sector twinning. Apatite shows systematic variation in minor element chemistry and morphology throughout the laccolith; aspect ratios range from 3:1 for phenocrysts in the chilled margin to 300:1 in soda syenite. Britholite, Ce_3Ca_2-$(SiO_4)_3OH$, occurs as rare grains in the highly evolved soda syenite. It resembles apatite in appearance and is readily obtained from apatite by the coupled substitution of $REE^{3+} + Si^{4+}$ for $Ca^{2+} + P^{5+}$. Pseudoleucite is confined primarily to the upper and lower chilled margins of the laccolith. It is usually altered as shown by abundant calcite and zeolites, but on occasion nepheline is preserved in pseudoleucite. Sanidines in the chilled margin are Or_{92} and contain up to 2.3% BaO.

Chemical analyses of each lithology are given in Table 1. The rocks of Shonkin Sag laccolith reflect the distinctive potassic nature of the Highwood petrographic province: K_2O predominates over Na_2O except in the small volume of the most highly differentiated rocks. The general differentiation pattern is from potassic mafic magma toward the nepheline-syenite (phonolite) minimum in petrogeny's residua system. With respect to trace elements, the intrusion is enriched in Ba and Sr overall. Transition metals such as Co, Cr, and Ni are enriched in shonkinite, whereas incompatible elements such as Zr and Nb are enriched in syenite varieties.

Shonkin Sag laccolith has inspired lively and continuing debate about its petrogenesis and, in particular, its mechanism of differentiation. In 1905, Pirsson advocated differentiation in place involving convection to produce a shonkinite shell and a residual syenite core. Osborne and Roberts (1931) advocated a crystal fractionation mechanism producing differentiation via crystal set-

tling. Reynolds (1937) suggested that the syenite formed from shonkinite by assimilation of host sandstones. Barkesdale (1937) argued that the laccolith was the result of three separate intrusions: The first formed a laccolith of shonkinite; the second, syenite in composition, was intruded to form the core; and a third minor intrusion emplaced pegmatite between shonkinite and syenite. Hurlbut (1939), returning to the theme of Osborne and Roberts, concluded that the intrusion was the result of differentiation of a single magma body in place via settling of mafic minerals and the floatation of leucite. Nash and Wilkinson (1970, 1971) did not discuss differentiation mechanisms; however, they concluded that the intrusion evolved from shonkinite through syenite to soda syenite in temperature and time, with temperatures of individual lithologies overlapping in part. Kendrick and Edmond (1981) suggested differentiation of syenite from shonkinite via liquid immiscibility; this interpretation is based on textural features at Square Butte laccolith. However, their analytical data do not argue persuasively for liquid immiscibility, and differentiation via crystal-liquid equilibria may be a viable alternative (Nash, 1982). The issue remains controversial, and at this writing awaits a more thorough quantitative evaluation.

REFERENCES

Barksdale, J. D., 1937, The Shonkin Sag laccolith: American Journal of Science, v. 33, p. 321–359.

Hurlbut, C. S., Jr., 1939, Igneous rocks of the Highwood Mountains, Montana; Pt. I, The laccoliths: Geological Society of America Bulletin, v. 50, p. 1043–1112.

Kenrick, G. C., and Edmond, C. L., 1981, Magma immiscibility in the Shonkin Sag and Square Butte laccoliths: Geology, v. 9, p. 615–619.

Larsen, E. S., 1940, Petrographic province of central Montana: Geological Society of America Bulletin, v. 51, p. 887–948.

Larsen, E. S., and others, 1941, Igneous rocks of the Highwood Mountains, Montana, Pts. II through VII: Geological Society of America Bulletin, v. 52, p. 1733–1868.

Marvin, R. F., Hearn, B. C., Jr., Mehnert, H. H., Naeser, C. W., Zartman, R. E., and Lindsey, D. A., 1980, Late Cretaceous-Paleocene-Eocene igneous activity in north-central Montana: Isochron/West, no. 29, p. 5–25.

Nash, W. P., 1972, Apatite chemistry and phorphorus fugacity in a differentiated igneous intrusion: American Mineralogist, v. 57, p. 877–886.

——, 1982, Comment on 'Magma immiscibility in Shonkin Sag and Square Butte laccoliths': Geology, v. 10, p. 444–445.

Nash, W. P., and Wilkinson, J.F.G., 1970, Shonkin Sag laccolith, Montana; I. Mafic minerals and estimates of temperature, pressure, oxygen fugacity, and silica activity: Contributions to Mineralogy and Petrology, v. 25, p. 241–259.

——, 1971, Shonkin Sag laccolith, Montana; II. Bulk rock geochemistry: Contributions to Mineralogy and Petrology, v. 33, p. 162–170.

Osborne, F. F., and Roberts, E. J., 1931, Differentiation in the Shonkin Sag laccolith, Montana: American Journal of Science, v. 22, p. 331–353.

Pirsson, L. V., 1905, Petrography and geology of the igneous rocks of the Highwood Mountains, Montana: U.S. Geological Survey Bulletin 237, 208 p.

Pollard, D. D., Muller, O. H., and Dockstader, D. R., 1975, The form and growth of fingered sheet intrusions: Geological Society of America Bulletin, v. 86, p. 351–363.

Reynolds, D. L., 1937, The Shonkin Sag laccolith: American Journal of Science, Fifth Series, v. 34, p. 314–315.

Weed, W. H., and Pirsson, L. V., 1895, Highwood Mountains of Montana: Geological Society of America Bulletin, v. 6, p. 389–422.

Upper Cretaceous–Paleocene sequence, Bug Creek area, northeastern Montana

Robert E. Sloan, Department of Geology and Geophysics, University of Minnesota, Minneapolis, Minnesota 55455

LOCATION AND ACCESSIBILITY

The area described here is in McCone County, Montana, approximately 100 mi (160 km) northwest of Glendive and 40 mi (64 km) south of Glasgow, on the east side of the Big Dry Arm of the Fort Peck Reservoir. It is located on the Jordan 1:250,000-scale map, the Fort Peck Lake East 1:100,000-scale map, and the Bug Creek 1:24,000-scale topographic map. The Bug Creek area (Fig. 1) is reached by proceeding south on Montana 24 for 24 mi (39 km) from the Fort Peck Dam or north 25 mi (40 km) from the intersection of Montana 200 with Montana 24. At this point a dirt road heads due west for 2.5 mi (4 km), then swings north at the rim of a mesa, and then swings south to the crossing of Bug Creek. The dirt road is passable by any vehicle when dry; four-wheel drive is required when the road is wet. The dirt access road is about 2.5 mi (4 km) south of the crossing of Montana 24 over the South Fork of Rock Creek and 5.5 mi (8 km) south of the road sign to Rock Creek State Park.

More than 2 mi (3 km) west of Montana 24, the land is part of the Charles M. Russell National Wildlife Range. Much of the other land is controlled by the U.S. Bureau of Land Management (BLM); some is privately owned. Collection of vertebrate fossils on federal land is by permit only. No permission is required to examine exposures on federal or state land; good manners (if nothing else) requires that gates be closed behind you. The nearest gasoline and food are 56 mi (90 km) southeast in Circle or in Fort Peck. Camping sites are available in Rock Creek State Park 8 mi (12.8 km) from Bug Creek as the crow flies, 15 mi (24 km) by dirt road. Very strong winds during fortnightly cold front passage make the use of tents with multiple guy ropes advisable: These winds chew up pup tents, wall tents, umbrella tents, and tubing poles!

If you wish to fly in by light plane, the Nelson Ranch on the Rock Creek Park road has a 2,400-ft (730-m) gravel strip.

SIGNIFICANCE

Study of the Late Cretaceous–Paleocene sequence in the Bug Creek Area and the rest of the Fort Peck Fossil Field (Fig. 1) has contributed greatly to our understanding of the paleoecology and stratigraphy of dinosaur extinction and the primary radiation of Tertiary placental mammals. No section of terrestrial sediments across the K/T boundary has been studied in as many ways as this one. Some 133 species of spores and pollen have been identified, and 93 species of Cretaceous vertebrates are known (including 30 species of Cretaceous mammals and 19 genera of dinosaurs) as are 24 species of Paleocene mammals.

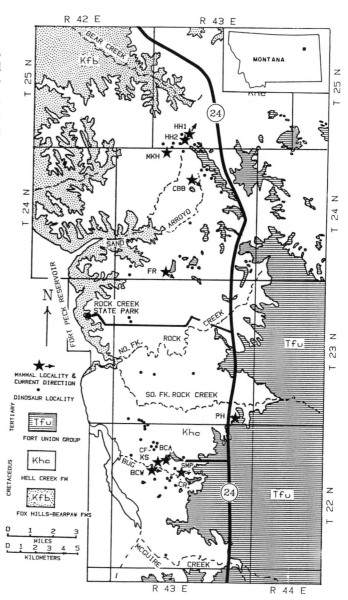

Figure 1. Geologic map of the Fort Peck Fossil Field, McCone County, Montana.

The oldest ungulate and the oldest primate have been found in these latest Cretaceous rocks. The fossil leaves and wood have been described. Here or in similar deposits in adjacent Garfield County to the west, the K/T boundary has been dated, magneto-stratigraphy has been described, the iridium layer has been identified, and shocked grains of St. Peter–type sand have been found in the iridium layer. The stratigraphy and sedimentary

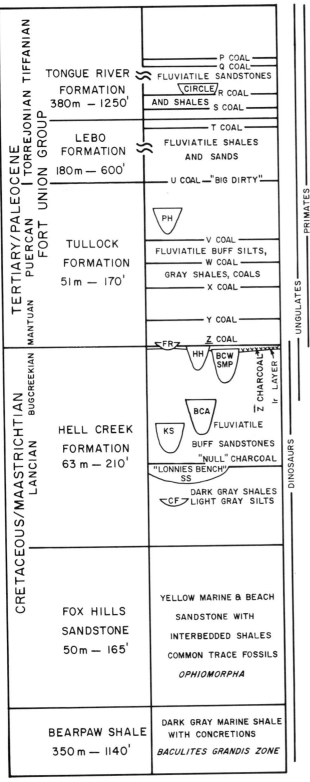

facies have been mapped and studied in detail. The extremely rapid rates of evolution of the placental and multituberculate mammals that have been demonstrated here in the latest Cretaceous and early Paleocene have a bearing on the duration of the Paleocene land mammal ages. Coal bed tracing from here to western North Dakota into the marine Cannonball Formation permits a direct correlation between land mammal ages and zones and the planktonic foraminiferal zones. This has been and will continue to be a major locality at which to study evidence for various hypotheses about the events at the K/T boundary.

GEOLOGY AND PALEONTOLOGY

The general structure of the bedrock in the Fort Peck Fossil Field is a simple homocline dipping 85 ft/mi (25 m/km) to the southeast. The general geology was well described by Collier and Knechtel (1939). The oldest bedrock exposed around the Fort Peck Reservoir is the upper part of the upper Campanian–lower Maastrichtian Bearpaw Shale, zones *Baculites reesidei* to *B. grandis*. The Bearpaw Shale is overlain by the beach and littoral sands of the Fox Hills Sandstone, some 50 to 100 ft (15 to 30 m) in thickness. Oysters and the trace fossil *Ophiomorpha* are common in the Fox Hills sandstone (Fig. 2).

Conformably above the Fox Hills is the Hell Creek Formation, which locally varies in thickness from 400 ft (122 m) in central Garfield County to about 180 ft (55 m) in the Fort Peck Fossil Field. The Hell Creek, deposited during the Maastrichtian regression and eustatic lowering of sea level at the end of the Cretaceous, represents a variety of coastal plain and fluvial sediments including channel sandstones, crevasse splay sands, levees, floodplain shales and silts, oxbow deposits, and rare thin coaly streaks from swamps and forest fires. The channel sandstones make up 35% of the formation; floodplain bentonitic swelling shales make up most of the balance.

Nineteen genera of dinosaurs have been described from the Hell Creek; major specimens are in museums in New York, Michigan, Wisconsin, Minnesota, Montana, and California. Most of the articulated dinosaurs are from the floodplain shales. The most common dinosaurs are *Triceratops* and *Anatosaurus*. Small dinosaurs occur as loose bones and teeth in the channel sands with the rest of the vertebrates. Bug Creek was given its name by the local ranchers in 1906, when Barnum Brown collected from near the mouth of the creek the skeleton of *Tyrannosaurus* now mounted at the American Museum of Natural History in New York. The ranchers decided it was the biggest "bug" anyone had seen, so the creek became "Bug Creek"! The major collections of vertebrates come from the bases of the channel sands where isolated bones and teeth of small vertebrates and teeth and claws of the dinosaurs are found as hydraulic concentrates in clay pebble conglomerates. Collection is usually by underwater screening, but the large red harvester ants make things easy for the paleontologist. Their large anthills have yielded mammal teeth for almost a century.

The overlying Tullock Formation of the Fort Union Group

Figure 2. Stratigraphic column of Late Cretaceous and Paleocene rocks of McCone County, Montana, showing horizons of major fossiliferous river channels and stratigraphic ranges of dinosaurs, ungulates, and primates. Note that the contact between the Hell Creek and Tullock formations is at the upper Z Coal (Z̲); the K/T boundary is at the lower Z Coal (Z̄), within the upper Hell Creek Formation.

Figure 3. Oblique air photo of the valley of Bug Creek looking north 30 degrees east, altitude 1,000 ft (300 m) above ground. Buttes are all capped by channel sandstones, which produces a topographic reversal. Major localities and the Z Coal are marked with arrows. The Z Coal zone above the BCW/SMP channel is here merged into one coal bed. The contrast between the coal-rich Tullock with few sandstones and the coal-poor Hell Creek with more sandstones is obvious in the photo. (Photo, Don Beckman, 1963).

is about 180 ft (55 m) thick and is exposed widely near Montana 24. It was deposited on the same coastal plain during the early Danian transgression. In contrast to the drab-colored Hell Creek, sands make up only 15% of the formation in the Tullock channel and coals are much more common, sheetlike, and extensive, often crossing levees and low paleodivides; the floodplain sediments are often yellowish. Collier and Knechtel (1939) gave the coals letter-names in inverse alphabetical order: Z, the lowest, is at the base of the Tullock, and U, the "Big Dirty," is at the boundary with the overlying 300 ft (91 m) of Lebo Shale, which has only one major coal, the T bed. Bed S occurs near the base of the Tongue River Sandstone, and P occurs as the highest coal bed in the county, still

in that formation. The Tullock represents the Mantuan and Puer-can early Paleocene mammal ages and about one-tenth of the thickness of the whole Paleocene. The Lebo and lower Tongue River Formations locally represent the middle Paleocene Torrej-onian mammal age, about two-tenths of the Paleocene, and the rest of the Tongue River and the Sentinel Butte formations repre-sent the balance of the Paleocene, the Tiffanian, and Clarkforkian mammal ages. This uneven distribution of ages and thicknesses came about because when the mammal ages were first defined, approximately equal amounts of evolutionary change were as-signed to the three ages. Unfortunately, the rate of evolution of the mammals was most rapid during and immediately after dinosaur extinction and then declined exponentially. This has caused a number of interesting misinterpretations. When Tiffanian mammals were found stratigraphically close to dinosaur-bearing rocks, often an unneeded unconformity was introduced into the geologic history. The mid-Tiffanian, about coal bed R, represents the stratigraphically highest Cannonball Formation and Cannonball Sea equivalents and the top of the Danian. In general, the coals represent the highest stands of the Cannonball Sea, as was shown by Roland Brown (1962). Coal bed U, the "Big Dirty," represents the highest eustatic Danian sea level and maximum expansion of the sea, with the coast at the North Dakota–Montana border.

The Z Coal is actually a zone of one to seven coals, which separate and merge as a result of differential compaction and deposition of the sediments. The iridium layer is usually located within the lowest of the coals in the Z Coal zone. The K/T pollen break also falls within the zone. In Garfield County, a bentonite in the Z Coal has been dated as 64 to 66 Ma.

The rate of sedimentation during the deposition of the Hell Creek Formation was about 35 bubnoffs (meters/million years), while during the deposition of the Fort Union Group it was about 40 b.

The terrestrial biostratigraphy of the K/T boundary is not as simple as the catastrophists would have one believe. A freshwater sting ray, two multituberculate mammals, 5 opossums, and the 19 dinosaurs go extinct—but not all at the same time. Not much else became extinct of the remaining 66 vertebrates (Estes and Berberian, 1970).

The K/T boundary and the iridium layer are about midway in magnetic anomaly 29r, but the final extinction process actually began at the base of 29r about 300,000 years earlier. Dinosaur specimen abundance between 90 and 120 ft (27 and 37 m) below the lower Z Coal is 5 dinosaurs per square kilometer, whereas in the top 30 ft (9 m) of the Hell Creek, abundance is down to 1.4 dinosaur specimens per square kilometer. Intermediate levels are intermediate in abundance. Although 19 genera of dinosaurs are known from the lower Hell Creek, the number of dinosaurs in successively higher stream channels in the upper 90 ft (27 m) of the formation drops from 13 to 12 to 11 to 8! The number of dinosaur teeth per ton of stream sediment washed for fossils drops from an average of 170 in the lower channels to an average of 25 in the upper channels. Precisely at the beginning of this reduction, four new mammal species whose ancestors were previously present only in Asia appear at the base of 29r. Their descendants then rapidly diversified.

The first and lowest of these uppermost Cretaceous channels is the famous Bug Creek Anthills (BCA) channel (Fig. 3), the richest Mesozoic mammal locality in the world. Some 60 tons (54 t) of this locality have been washed for fossils. The number of fossils collected is in the hundreds of thousands. The first hoofed mammal or ungulate, *Protungulatum*, first occurs at BCA and is ancestral to all later ungulates, including whales and seacows. The successive channels stratigraphically above BCA include the

Bug Creek West/Scmenge Point channel (BCW/SMP) with 3 ungulates, Harbicht Hill (HH) with 5 ungulates, and Ferguson Ranch (FR) with 8 ungulates. The top of FR occurs 5 ft (1.5 m) above the iridium layer. All of these channels contain unreworked dinosaur teeth. There do appear to be a few Paleocene dinosaurs! The oldest primate occurs in HH. These localities document the fastest rate of evolution in the fossil record!

Although the new Paleocene pollen types come in just above the iridium layer, many of the Cretaceous types were becoming extinct long before. There are 114 pollen types in the lower part of the Hell Creek but only 84 in the top 40 ft (12 m) of the formation. Vegetation in the Hell Creek Formation was a humid subtropical rain forest with palms, figs, breadfruit, and other tropical angiosperms. Vegetation in the Tullock Formation was dominated instead by the conifers (*Metasequoia*) and by cypress and ferns.

Inferred reasons for dinosaur extinction, besides the asteroid impact (which surely did not help, although it is hardly the sole reason), include the global drop in temperature during the last 15 million years of the Cretaceous, the eustatic sea-level drop that would make terrestrial climate more seasonal, the resulting major restriction of vegetation toward the tropics, and competition from the only other group of terrestrial herbivores, the multituberculate mammals and ungulates, both of which were rapidly spreading and diversifying.

The top of the first hill southeast of the crossing of the south fork of Rock Creek by Montana 24 is Purgatory Hill, composed of rocks deposited about 1 million years after the iridium layer and the K/T boundary. Ten tons (9 t) of channel sand were freighted off the top of this hill and washed for fossils, producing 600 useful mammal teeth representing more than 24 species. This hill was the discovery site for the earliest Primate, *Purgatorius*; it is now an undeveloped state park. The very top of this hill is a lag of Quaternary stony till, presumably from the earliest (Nebraskan ?) glaciation; this is as far south as the till occurs. The entire 800 ft (245 m) relief of these badlands then is Quaternary in age!

SELECTED REFERENCES

Brown, R. W., 1962, Paleocene flora of the Rocky Mountains and Great Plains: U.S. Geological Survey Professional Paper 375, 119 p.

Collier, A. J., and Knechtel, M. M., 1939, The coal resources of McCone County, Montana: U.S. Geological Survey Bulletin 905, 88 p.

Estes, R., and Berberian, C., 1970, Paleoecology of a Late Cretaceous vertebrate community from Montana: Breviora, Museum of Comparative Zoology, no. 343, p. 1–36.

Sloan, R. E., 1970, Cretaceous and Paleocene terrestrial communities of western North America: Proceedings of the North American Paleontological Convention, p. 427–453.

Sloan, R. E., 1970, Rigby, J. K., Jr., Van Valen, L. M., and Gabriel, D., 1986, Gradual dinosaur extinction and simultaneous ungulate radiation in the Hell Creek Formation: Science, v. 232, p. 629–633.

Van Valen, L., and Sloan, R. E., 1977, Ecology and the extinction of dinosaurs: Evolutionary Theory, v. 2, p. 37–64.

Marysville stock and contact aureole, western Montana

Robert E. Derkey and Pamela Dunlap Derkey, Montana Bureau of Mines and Geology, Montana College of Mineral Science and Technology, Butte, Montana 59701

LOCATION AND ACCESSIBILITY

The Marysville stock is about 17 mi (27 km) northwest of Helena, Montana, in Lewis and Clark County (Fig. 1). To reach the area from Helena, go 7 mi (11 km) north on I-15 to exit 200 and turn west (left) off the exit ramp onto Montana 279 (Lincoln Road). At 8.5 mi (14 km) from I-15, turn southwest (left) onto a gravel road marked by signs for Marysville and the Belmont Ski Area. The old mining town of Marysville is about 6 mi (10 km) from this turnoff. Most of the area is readily accessible by passenger car although some areas are accessible only by four-wheel drive or foot (Fig. 2). The mines are on private land. Permission should be obtained from owners with posted land or locked gates.

SIGNIFICANCE

The Marysville stock and its contact aureole (Fig. 2) is the site where the processes of magmatic stoping, contact metamorphism, and contact metasomatism were recognized and described by Joseph Barrell (1907). Discovery of mineralization associated with the Marysville stock led to the organization of the Marysville mining district in the late 1800s. Before the turn of the century, mines in the Marysville district produced more than $30 million in gold and silver (Weed, 1903, p. 88), when gold was $20.67 per ounce. The value of the gold at today's current price of more than $300 per ounce would be about $500 million. This higher gold price has caused renewed interest in the district and has resulted in reopening of the Drumlummon and Cruse-Belmont mines (points 5 and 18, Fig. 2) for exploration.

Other key references to the geology of the Marysville district include Knopf (1950) and Blackwell and others (1975). The report of Blackwell and others, on the Marysville Geothermal Project, is used for this summary but is not readily available. Rice (1977) discusses contact metamorphism of the Helena Formation in the southeast part of the contact aureole, and Chadwick (1981) presents an overview of igneous activity in southwestern Montana.

IGNEOUS HISTORY

The sequence of igneous events at Marysville proposed by Blackwell and others (1975) is emplacement of: (1) Precambrian(?)-age microdiorite sills, (2) the 78 ± 4 Ma Marysville stock, (3) the 47.8 ± 2 Ma Bald Butte stock and hornblende diorite dikes southwest of the Marysville stock, and (4) the 37 to 40 Ma Empire Creek stock. The Bald Butte and Empire Creek stocks are not exposed, but their presence has been confirmed by drilling for molybdenum mineralization at Bald Butte (point 11,

Fig. 2) and geothermal resources at Empire Creek (point 14, Fig. 2).

Knopf (1950) believed that the Marysville stock (78 ± 4 Ma) was an outlier of the Boulder batholith, which is exposed about 6 mi (10 km) to the south, because the two bodies are similar in composition and age. The Bald Butte stock (47.8 ± 2 Ma) and the Empire Creek stock (37 to 40 Ma) are coeval, respectively, with the Lowland Creek Volcanics exposed between Butte and Helena and rhyolitic volcanic rocks in the Helena and Avon areas (Chadwick, 1981). All three stocks intensely metamorphosed strata of the Empire and Helena formations of the Middle Proterozoic Belt Supergroup.

STRATIGRAPHY

Empire Formation. The Empire Formation, the eastern correlative of the St. Regis Formation, is the oldest rock unit exposed in the Marysville district. Unmetamorphosed Empire Formation, outside of the Marysville district, consists of pale-green quartzite, siltite, and argillite in wavy, nonparallel-laminated, fining-upward couplets (James Whipple, USGS, personal communication, 1984). The contact between the Empire Formation and the overlying Helena Formation is a transition zone about 130 ft (40 m) thick (Blackwell and others, 1975). The Empire Formation is about 1,000 ft (300 m) thick; however, only the upper 600 ft (180 m) are exposed in the Marysville district.

Helena Formation. The Helena Formation is the eastern facies of the middle Belt carbonate, of which the Wallace Formation is the western facies. The Helena consists of two interbedded lithologies, a gray limestone and a dark brown to black siliceous dolomite (Blackwell and others, 1975). The Helena Formation overlies the Empire Formation and is about 4,000 ft (1,200 m) thick in the Marysville district (Barrell, 1907).

MARYSVILLE STOCK

The Marysville granodiorite stock crops out over a 3 mi^2 (8 km^2) area (Fig. 2). A typical mode for the stock (Barrell, 1907) is:

Quartz	22.2%	Hornblende	5.5%
Orthoclase	15.6%	Apatite	0.2%
Plagioclase	47.5%	Sphene	0.1%
Biotite	7.2%	Magnetite	1.7%

Plagioclase is typically zoned and ranges from An_{15} to An_{40}. Medium to coarse equigranular texture is dominant, but finer grain sizes occur near chilled margins and toward the center of

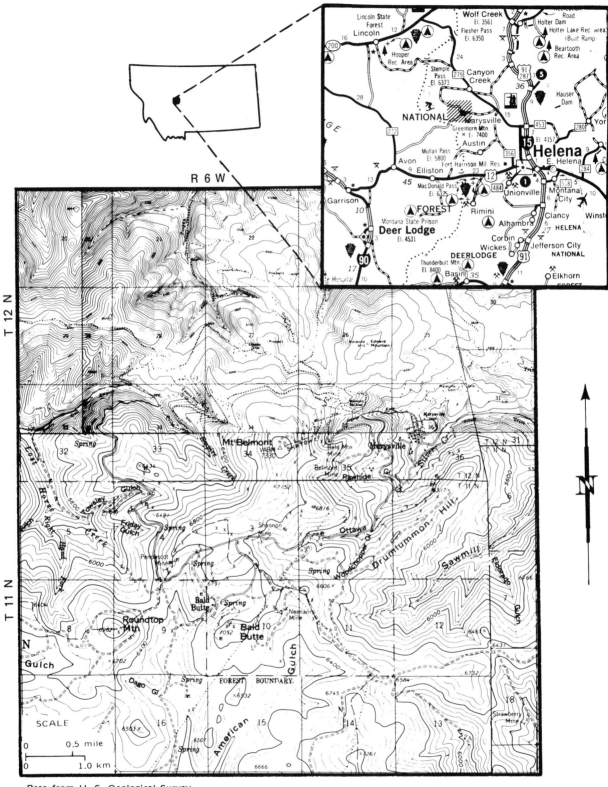

Base from U. S. Geological Survey
Canyon Creek 7.5' (1968), Granite Butte 7.5'
(1968) and Elliston 15' (1959) quadrangles.

Figure 1. Location and topographic maps for the Marysville area.

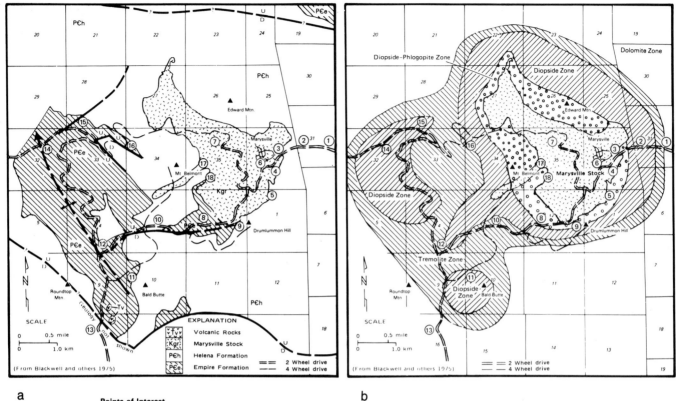

a

Points of Interest

1—Helena Formation (dolomite zone) exposed in roadcuts. Dredge tailings in Silver Creek extend downstream for several miles east of Marysville.

2—Helena Formation (diopside zone) exposed in roadcuts.

3—Contact of Marysville stock with Helena Formation.

4—Drumlummon stamp mill. All that remains is the stone foundation on hillside south of Silver Creek.

5—Drumlummon mine, largest mine in the district.

6—Marysville.

7—Belmont ski area. There are several roadcuts in the Marysville stock between Marysville and the ski area.

8—Ottawa Gulch. Cordierite porphyroblasts in the Empire Formation (diopside-phlogopite zone) occur in outcrops north of the road.

b

9—Drumlummon Hill. Scapolite-diopside hornfels (Helena Formation, diopside zone) is exposed in small prospect pits near the summit.

10—Shannon mine.

11—Bald Butte mine and former town site. The unexposed Bald Butte stock underlies this area.

12—Penobscot mine.

13—Bald Butte stamp mill.

14—Marysville geothermal test well. The unexposed Empire Creek stock underlies this area.

15—Empire mill and former town site.

16—Empire mine.

17—Bald Mountain mine.

18—Cruse-Belmont mine.

Figure 2. Geologic and metamorphic zone maps of the Marysville stock and contact aureole. Numbers in circles indicate points of interest. (a) Geology, land net, principal roads, and points of interest. (b) Marysville stock, metamorphic zones, land net, principal roads, and points of interest.

the stock. Good exposures of the stock occur in roadcuts between Marysville and the Belmont ski area (points 6 and 7, Fig. 2).

The irregularly-shaped stock dips gently to the southwest and northwest and steeply to the southeast and northeast. These contact relationships were determined from surface outcrop pattern, three-dimensional observations of the contact in mines, and from the variable width of the contact metamorphic aureole (Barrell, 1907, p. 20–21). The contact relationships were confirmed by a ground magnetometer survey (Blackwell and others, 1975).

Careful mapping and correlation of contact-topographic relations also led Barrell (1907) to develop the theory of magmatic stoping for emplacement of the Marysville stock. Recent mapping (Blackwell and others, 1975) showed that some contact metamorphosed rocks mapped as Empire Formation by Barrell

(1907), actually are Helena Formation. This reinterpretation indicates that magmatic stoping proceeded higher into the stratigraphic section and was a much more dominant process than the combined doming and stoping envisioned by Barrell.

CONTACT METAMORPHISM AND METASOMATISM

The contact metamorphic aureole around the Marysville stock ranges from 0.3 to 3 mi (0.5 to 5 km) in width, being narrowest along the northern, eastern, and southern margins of the area (Fig. 2). The wide aureole along the western and southwestern margins is, in part, related to the concealed Bald Butte and Empire Creek stocks (Blackwell and others, 1975).

Within the aureole, the Empire Formation was metamorphosed to a dark cordierite hornfels and the Helena Formation to a banded light-colored diopside and tremolite hornfels.

Four zones of progressive contact metamorphism are distinguished within the aureole (Blackwell and others, 1975) and are shown on Figure 2. From outermost to innermost, the zones are dolomite, tremolite, diopside, and diopside-phlogopite. The dolomite zone is characterized by a dolomite-quartz assemblage. The appearance of talc in rocks of the Helena Formation marks the inner boundary of the dolomite zone. Dolomite zone rocks (Helena Formation) are exposed in roadcuts on the north side of Silver Creek (point 1, Fig. 2) and appear to be unmetamorphosed in hand sample. The tremolite zone is characterized by the assemblage tremolite-calcite-quartz. Rocks of the diopside zone contain the assemblage diopside-tremolite ± plagioclase, calcite, and scapolite. Good exposures of diopside zone rocks (Helena Formation) occur in roadcuts on the north side of Silver Creek (point 2, Fig. 2). Here, beds of tremolite hornfels are interbedded with those of diopside hornfels. Scapolite-bearing diopside hornfels (Helena Formation) is exposed in small prospect pits and as float near the summit of Drumlummon Hill (point 9, Fig. 2). The innermost diopside-phlogopite zone is characterized by a diopside-phlogopite assemblage for calcite-free calc-hornfels and by cordierite for pelitic rocks (Blackwell and others, 1975). Cordierite hornfels (Empire Formation) crops out on the north side of the upper part of Ottawa Gulch (point 8, Fig. 2).

Rice (1977), however, recognized five metamorphic zones separated by four distinct isograds for rocks of the Helena Formation on the east and southeast flanks of the Marysville stock. His inner phlogopite-bearing zone on the east side of the stock is broader than the diopside-phlogopite zone of Blackwell and others (1975).

Metasomatism in the Marysville district occurred predominantly in the inner half of the contact metamorphic aureole and resulted in the formation of scheelite (Knopf, 1950, p. 842),

hornblende, diopside, garnet, epidote, quartz, and possibly some occurrences of biotite and pyrite (Barrell, 1907, p. 144). Barrell (1907) discerned that, for rocks affected both by contact metamorphism and metasomatism, metasomatic alteration followed fractures in contact metamorphosed rocks and, thus, occurred later than contact metamorphism.

MINERALIZATION

Gold and silver mineralization in the Marysville district occurs principally in quartz-calcite, steeply dipping fissure veins that, generally, are confined to the margin of the Marysville stock and the surrounding hornfels. Both wall rock and veins are brecciated, indicating that the veins were emplaced along faults that were active before, as well as after, mineralization. Within the veins, gold occurs in shoots and is finely divided and not usually visible. Vein quartz has a lamellar habit and is believed to be pseudomorphic after calcite.

Three types of veins were noted: (1) molybdenite-pyrite-fluorite veins with only a trace of gold; (2) gold with pyrite, chalcopyrite, sphalerite, and galena; and (3) gold with tetrahedrite and chalcopyrite. Type 2 veins were mined at the Bald Butte mine (point 11, Fig. 2) and type 3 veins were mined at the Drumlummon mine (point 5, Fig. 2). With the exception of the Drumlummon mine, producing veins had an average strike length of 1,000 ft (300 m), a thickness less than 6 ft (2 m), and were mined down dip generally not more than 500 ft (150) (Knopf, 1913, p. 64). The largest producer in the district, the Drumlummon mine, was mined to a depth of 1,200 ft (360 m) with the vein walls varying in width from 2 to 20 ft (0.6 to 6 m). Evidence of former mining activity includes mine dumps, dredge tailings, the foundation and wood frame of an old gold stamp mill, and a building that housed an ore concentrating mill (Fig. 2).

REFERENCES CITED

Barrell, J., 1907, Geology of the Marysville mining district, Montana: U.S. Geological Survey Professional Paper 57, 178 p.

Blackwell, D. D., Holdaway, M. J., Morgan, P., Petefish, D., Rape, T., Steel, J. L., Thorstenson, D. C., and Waibel, A. F., 1975, Results and analysis of exploration and deep drilling at Marysville geothermal area, in McSpadden, W. R., project manager, The Marysville geothermal project: Richland, Washington, Battelle, Pacific Northwest Laboratories, report no. 23111-01410, section E, p. E.1–E.116.

Chadwick, R. A., 1981, Chronology and structural setting of volcanism in southwestern and central Montana, in Tucker, T. E., ed., Guidebook to southwest Montana: Montana Geological Society, 1981 Field Conference

and Symposium, p. 301–309.

Knopf, A., 1913, Ore deposits of the Helena mining region, Montana: U.S. Geological Survey Bulletin 527, 143 p.

—— , 1950, The Marysville granodiorite stock, Montana: American Mineralogist, v. 35, p. 834–844.

Rice, J. M., 1977, Progressive metamorphism of impure dolomitic limestone in the Marysville aureole, Montana: American Journal of Science, v. 277, p. 1–24.

Weed, W. H., 1903, Gold mines of the Marysville district, Montana: U.S. Geological Survey Bulletin 213, p. 88–89.

13

Mississippian Waulsortian bioherms in the Big Snowy Mountains, Montana

Donald L. Smith (Deceased) and Stephan G. Custer, Department of Earth Sciences, Montana State University, Bozeman, Montana 59717

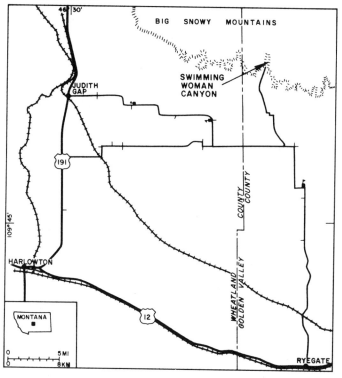

Figure 1. Location and route map.

LOCATION AND ACCESSIBILITY

Mississippian-age Waulsortian bioherms in the Lodgepole Formation are well exposed and accessible for field study in Swimming Woman Canyon, Big Snowy Mountains, Central Montana. Swimming Woman Canyon is east of Judith Gap, northeast of Harlowtown, and north of Rygate, Montana (Fig. 1). The site can be reached by automobile via gravel roads from U.S. 191 or from U.S. 12. Three routes to the site are shown. The route selected will depend on the direction the visitor is traveling. Travelers using the Judith Gap route may be tempted to continue east after driving 15 mi (24 km). Part of the road to the east of this point has been abandoned.

Mileages to turns along the route can be scaled from Figure 1, which is adapted from the Roundup 1 × 2° Army Map Service (AMS) sheet published by the U.S. Geological Survey. The only other available access map is the Lewis and Clark National Forest (Jefferson Division), Montana, Forest Visitor's Map (planimetric) at a scale of 1:126,720. No topographic map at a scale larger than 1:250,000 is currently available. The Irene 3 NE 1:24,000

Topographic Quadrangle covers the field area and is in preparation. Travelers may want to keep track of mileage using their automobile odometer while off the main highway. The distances from the various starting points to Swimming Woman Canyon Road are: 25 mi (40.2 km) from Judith Gap, 24.2 mi (38.9 km) from mile marker 12 along U.S. 191, and 23.7 mi (38.1 km) from Rygate.

At the intersection of Judith Gap Road and Swimming Woman Canyon Road, turn north and travel 8.6 mi (13.8 km) north to the Lewis and Clark National Forest Boundary. Please leave all gates you encounter as you find them after you pass; open if open, closed if closed. From the Forest Boundary, proceed approximately 1.1 mi (1.8 km) north to the bioherms (Sec.9,T.-11N.,R.19E.). You will drive across Swimming Woman Creek three times (no bridges). This road is not recommended during Spring runoff (May and early June) or during the snow season (November to April). There is one fork in the road; bear left (west). You should be able to see the bioherms to the southeast as you pass from the narrow erosion-resistant part of Swimming Woman Canyon into the open part of the canyon where the stream has access to more erodable rock. Once you have seen the bioherms at a distance, return to the creek crossing and gate approximately 0.1 mi (0.2 km) back down the canyon, unless you prefer wading a cold mountain stream to get to the outcrop.

The best places to examine the rock require good hiking boots and a steep climb. For those who would like to examine the bioherms with less effort, some of the facies are more poorly exposed along the road that follows Swimming Woman Creek.

SIGNIFICANCE

While framework-supported reefs in the modern ecologic sense are absent in Mississippian Rocks, massive mud mounds have been recognized at several locations throughout the world (Wilson, 1975). Such mounds differ from the more familiar modern mounds in that they are composed of approximately 60 percent lime mud, with the remaining material composed of crinoid and bryozoan skeletal debris rather than growth-position skeletons of frame-building organisms. Despite the lack of obvious framework organisms, the mounds may have had a depositional relief of as much as 328 ft (100 m) with respect to the surrounding horizontally deposited, thin-bedded, dark, argillaceous, carbonate rocks. This site provides an excellent opportunity to study the lateral and vertical facies relationships at well-exposed outcrops that have been described in the literature (Cotter, 1965, 1966; Smith, 1972, 1977, 1982). This opportunity takes on additional significance in light of the fact that such Waulsortian bioherms can form important hydrocarbon reservoirs (Ahr and Ross, 1982).

TABLE 1. STRATIGRAPHIC COLUMN OF EXPOSED ROCKS WITH MILEAGES (KM) NORTH FROM THE
LEWIS AND CLARK FOREST BOUNDARY*

Mile	(km)	Geologic Age	Unit	Thickness (approximate) (ft)	(m)	Description
0.1	(0.2)	K	Kootenai Fm	300	(98)	Glittering white quartz salt-and pepper ss and cherty ss intbd with red sltst and mdst; cgl base.
		K	Morrison Fm	225	(74)	Red, yellow, and gray sltst and mdstn; lenses of brown ls and ss.
		J	Ellis Gp (Swift, Rierdon, Piper? Fms)	200	(66)	Ostracod-bearing glauconitic salt-and-pepper ss; brown ls.
		P	Quadrant Fm	100	(33)	White to pink quartz ss.
0.3	(0.5)	P	Amsden Gp (Devils Pocket, Alaska Bench, Tyler Fms)	530	(174)	Intbd gray dolst, quartz ss and red sltstn; gray ls intbd with red sltst and sh; red sh and sltst intbd with cgl, ss, and gray ls.
0.5	(0.8)	M	Big Snowy Gp (Heath, Otter, Kibbey Fms)	700	(230)	Black marine sh; gray, purple, green, and black sh with nodular limestone; red-yellow quartz ss intbd with red sh.
0.7	(1.1)	M	Charles Fm	400	(131)	Massive light brown fine-crystalline ls with chert nodules; light brown to yellow-brown ls breccia zones.
0.8	(1.3)	M	Mission Canyon Ls	500	(164)	Massive medium-crystalline dolomitic light brown fossiliferous ls.
0.9	(1.4)	M	Lodgepole Ls	640	(210)	Thin-bedded light brown to gray fossiliferous ls; lower part biohermal; basal black sh.
1.2	(1.9)	D	Three Forks Sh	13	(4)	Light tan to gray dolomitic sh.
		D	Jefferson Dl	150	(49)	Light tan medium-crystalline fossiliferous dlst.
1.6	(3.0)	e	Pilgrim Ls	616	(202)	Light gray intraclastic ls with intbd green sh.
		e	Park Fm	160	(52)	Green micaceous sh.
		e	Meagher Fm	50	(16)	Gray ls intbd with green sh.
		e	Woolsey Sh	180	(59)	Green to gray micaceous sh.
		e	Flathead Ss	110	(36)	Conglomeratic rippled quartz ss.
		pe	Belt Supergroup	300	(98)	Dark gray limey argillite.

*Modified from Reeves (1930), Douglass (1952), Strickland and others (1956), Wurden and others (1956), and Maughan and Roberts (1967).

SITE DESCRIPTION

General Setting. Unlike the adjacent isolated mountains of the central Montana plains, the Big Snowy Mountains are composed entirely of sedimentary rocks and are not known to be cored by intrusive igneous rocks. Examination of any generalized geologic map (Geologic Map of the United States, Geologic Map of Montana) reveals that the Big Snowy Mountains are a large asymmetric anticline with steep dips on the south side and gentle dips on the north side. A complete section from the Cretaceous Kootenai Formation to the Proterozoic Belt Supergroup is exposed in Swimming Woman Canyon (Table 1). Younger Cretaceous rocks are present, but are generally covered by Tertiary alluvium deposited by coalescing alluvial fans. Although the valley form suggests glaciation, no glaciers are known to have existed in this region (Reeves, 1930; Douglass, 1952).

Stratigraphic Setting. The bioherms are in the thin-bedded Lodgepole Formation, which is underlain by the Devonian Jefferson Dolomite and is overlain by the massive limestone of the Mission Canyon Formation (Fig. 2). Together, the Lodgepole, Mission Canyon, and Charles formations comprise the Madison Group. The Lodgepole Formation is composed of three

members. The lowermost is the Cottonwood Canyon Member, which contains black, thin-bedded shale and dolomitic shale deposited as the Madison Sea advanced into the Big Snowy Trough in central Montana. The overlying Paine Member contains thin-bedded argillaceous limestones and the bioherms. This unit is interpreted to have formed in relatively quiet, deep water far from shore. Overlying the Paine Member is the Woodhurst Member, which contains oolitic and bioclastic limestones rather than the micritic limestone of the Paine Member. The Woodhurst deposits are cyclic (Smith, 1972; Fig. 2). Each cycle is composed of a basal nonresistant, fine-grained, bioclastic pelletal grainstone, which is overlain by a more resistant, coarser-grained, cross-bedded, bioclastic, oolitic grainstone. This unit is interpreted to represent a shallower-water facies that prograded across the Mississippian seaway.

Bioherm Facies. The Bioherms have been described in detail elsewhere (Cotter, 1965, 1966; Smith, 1972, 1977, 1982). A comparison of the facies model and a keyed photograph of the bioherms in Swimming Woman Canyon should help the interested field geologist find the important rocks (Figs. 3 and 4). There are three bioherms (I, II, III) at Swimming Woman Canyon; they are 230 ft (70 m) thick, up to 2,300 ft (700 m) long,

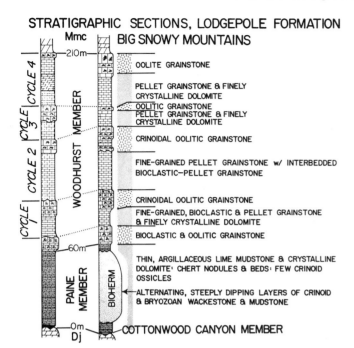

Figure 2. Stratigraphic section for the Lodgepole Formation. Left section is from Rock Creek (Sec.7,T.12N.,R.18E.). Right section is from Swimming Woman Canyon (Sec.9,T.11N.,R.19E.). Dj is the Devonian Jefferson Dolomite; Mmc is the Mission Canyon Formation (from Smith, 1972, p. 31).

Figure 3. View of the Lodgepole bioherms looking southeast. B refers to bioherm facies, F refers to flank facies, D refers to dark mudstone facies, and S refers to shallow-water facies. These letters are also used on the facies model in Figure 4 as are the roman numerals to identify the bioherms (from Smith, 1977, p. 196).

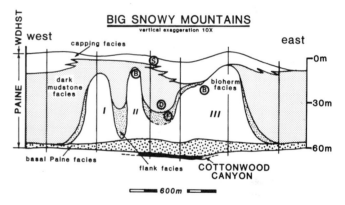

Figure 4. Facies model of the bioherm area. There are three separate biohermal masses in the canyon designated I, II, and III by Cotter (1965). Letters correspond to those in Figure 3. Vertical lines refer to measured sections. Figures 3 and 4 are mirror images of one another because Figure 4 is drawn facing north but the photograph in Figure 3 was taken facing south (modified from Smith, 1982, p. 56).

and between 330 and 660 ft (100 and 200 m) apart. The bioherms rest on mottled and burrowed glauconitic wackestones, packstones, and grainstones, which comprise the basal part of the Paine Member of the Lodgepole Formation. The basal unit contains a diverse fauna including crinoids, bryozoans, brachiopods, ostracods, and corals deposited offshore, but in water shallow enough to occasionally be affected by storm waves. The storm-agitated zones give way to deeper, more quiet-water facies associated with the bioherms. The contact with the overlying rocks is largely covered but appears to be abrupt and conformable.

There are four facies associated with the bioherms above this basal unit. The bioherm facies (B in Figs. 3 and 4) is composed of layers of lime mudstone and bioclastic wackestone, packstone, and grainstone. Lime silt and mud constitutes 60 percent of the mounds. Bedding planes are inclined as much as 35° to the surrounding dark mudstones, can be traced up to 165 ft (50 m) vertically, and show small-scale soft-sediment-deformation structures. Articulated crinoids, and unbroken fenestrate bryozoans, are the dominant fossil component with less abundant brachiopods, coelenterates and cephalopods, trilobites, and foraminifera. Some dikelike structures filled with coarse-grained crinoidal packstone and grainstone cut the bioherm layers.

The origin of the distinctive mound shape of the bioherm facies is problematic. Hypotheses include growth by unknown organisms, sediment trapping by algae, sediment baffling by bryozoans and crinoids, accumulation of sediment in the lee of

thickets of organisms, and shaping by subsea currents (Wilson, 1975). Some researchers believe the mounds grew to wave base and then expanded to the sides by lateral accretion.

The flank facies (F in Figs. 3 and 4) is composed of crinoid ossicles in a lime-mud matrix. Bedding planes display articulated crinoid stems with no preferred orientation. This poorly exposed facies covers parts of both the sides and tops of the bioherms and interfingers with the dark mudstone facies. This relationship implies that the flank facies was formed after the mound began to rise above the surrounding dark mudstone facies. The flank facies may have formed as the bioherms reached storm wave base. If so, the water must have been at least as deep as the height of the

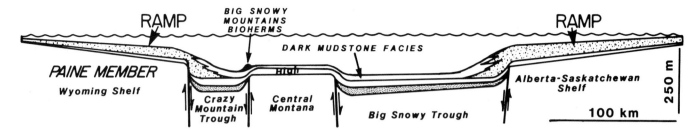

Figure 5. Cross section showing the tectonic setting of the bioherms (modified from Smith, 1982, p. 61).

bioherm above the dark mudstone facies plus the distance from storm-wave base to the surface (164 ft + 66 ft = 230 ft; 50 m + 20 m = 70 m).

The dark mudstone facies (D in Figs. 3 and 4) surrounds the bioherms. This facies is composed of thin-bedded, horizontally laminated, argillaceous lime mudstone. Occasional articulated crinoid fragments can be found in the rock as well as unbroken fenestrate bryozoan fronds and isolated small thin-shelled brachiopods and feeding traces such as *Zoophycus, Cosmoraphe,* and *Sclarituba.* The mudstones are interpreted to have been deposited in moderately deep water well below wave base. The bioherms probably rose above these deposits so that at any given level the bioherm and dark mudstone facies may have slightly different but unresolvable ages.

The bioherms are overlain by shallow-water facies (S in Figs. 3 and 4) of the Paine Member. This facies is composed of medium- to thick–bedded, horizontal and cross-stratified bioclastic and pelletal mudstone, wackestone, and grainstone. Fossils include crinoids, fenestrate and ramose bryozoans, rugose coral, and very large *Syringopora* colonies in growth position. This unit represents the end of quiet water deposition and the end of bioherm development.

Diagenesis. Cotter (1966) described the diagenetic features present in the facies associated with the bioherms. Two types of

sparry calcite filled cavities shortly after deposition. First, radiaxial fibrous cloudy calcite precipitated, and then clear equant crystals formed. The cavities include brachiopod valves and *Stromatactis* structures. During this period, chalcedonic chert was diagenetically mobilized and deposited in patches that were often fossiliferous. The chert extends into the fibrous sparry calcite but not the younger equant sparry calcite. Sucrosic brown dolomite is present in a zone from 0 to 50 ft (0 to 15 m) thick in the upper part of the three bioherms. Dolomite crystals range from 0.002 to 0.157 in (0.05 to 4 mm) in diameter. The larger crystals replace calcite crystals in crinoid fragments. Dolomite also occurs in association with equant sparry calcite in the centers of cavities. The dolomite may have been formed during a period of subaerial exposure as the water depth dropped at the end of the period of bioherm formation. Subaerial exposure is suggested by red, smectite clay mixed with calcite that geopetally fills the lower parts of cavities in a zone in the upper part of the bioherm.

Tectonic Setting. Waulsortian mounds in the Big Snowy Mountains are thought to have developed along fault-generated breaks in slope on the unstable shelf along the margin of the Central Montana High (Fig. 5; Smith, 1982). Periods of subsidence or eustatic sea level rise may have intermittently terminated bioherm growth and allowed the bioherms to be enveloped and eventually partially buried by dark lime mudstone.

REFERENCES CITED

Ahr, W. M., and Ross, S. L., 1982, Chappel (Mississippian) bioherm reservoirs in the Hardeman Basin, Texas: Transactions of the Gulf Coast Association of Geological Societies, v. 32, p. 185–193.

Cotter, E. J., 1965, Waulsortian-type carbonate banks in the Mississippian Lodgepole Formation of central Montana: Journal of Geology, v. 73, p. 881–888.

—— , 1966, Limestone diagenesis and dolomitization in Mississippian carbonate banks in Montana: Journal of Sedimentary Petrology, v. 36, p. 764–774.

Douglass, M. R., 1952, Geology and geomorphology of the south-central Big Snowy Mountains, Montana [M.S. thesis]: Lawrence, University of Kansas, 105 p.

Maughan, E. K., and Roberts, A. E., 1967, Big Snowy and Amsden Groups and the Mississippian-Pennsylvanian Boundary in Montana: United States Geological Survey Professional Paper 554B, 27 p.

Reeves, F., 1930, Geology of the Big Snowy Mountains, Montana: U.S. Geological Survey Professional Paper 165D, p. 135–149.

Smith, D. L., 1972, Depositional cycles of the Lodgepole Formation (Mississippian) in Central Montana, *in* 21st Annual Field Conference of the Montana Geological Society: Billings, Montana, Montana Geological Society, p. 29–35.

—— , 1977, Transition from deep to shallow water carbonates, Paine Member, Lodgepole Formation, central Montana, *in* Enos, P., and Cook, H. E., eds., Deep-water carbonate environments: Society of Economic Paleontologists and Mineralogists Special Publication 25, p. 187–201.

—— , 1982, Waulsortian bioherms in the Paine Member of the Lodgepole Limestone (Kinderhookian) of Montana, U.S.A., *in* Bolton, K., Lane, R. H., and LeMone, D. V., eds., Proceedings of the Symposium on the Paleoenvironmental Setting and Distribution of the Waulsortian Facies: El Paso, Texas, El Paso Geological Society and the University of Texas at El Paso, p. 51–64.

Strickland, J. W., Foster, D. I., Wilson, J. McM., and Knupke, J., 1956, Measured section: Swimming Woman Canyon, Big Snowy Mountains: Billings (Montana) Geological Society Seventh Annual Field Conference Guidebook, p. 30.

Wilson, J. L., 1975, The lower Carboniferous Waulsortian facies, *in* Wilson, J. L., ed., Carbonate facies in geologic history: New York, Springer Verlag, p. 148–168.

Wurden, F. H., Edwards, W. D., and Foster, D., 1956, Side Trip to Swimming Woman Canyon: Billings (Montana) Geological Society Seventh Annual Field Conference Guidebook, p. 29.

Geology of the Butte mining district, Montana

Lester G. Zeihen, Richard B. Berg, and Henry G. McClernan, *Montana Bureau of Mines and Geology, Montana College of Mineral Science and Technology, Butte, Montana 59701*

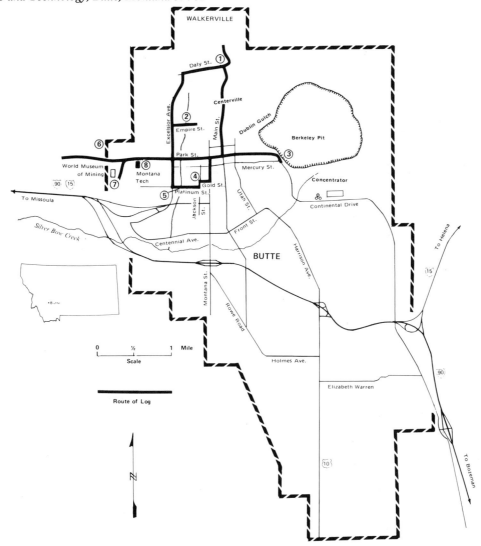

Figure 1. Map of Butte, Montana, showing field trip stops.

LOCATION AND ACCESSIBILITY

The Butte mining district is in Silver Bow County in southwestern Montana at the intersection of I-15 and I-90. The district proper lies north and west of these highways, and most of it is within the business and residential areas of Butte. Because city streets provide access to all of the stops, they are easily accessible by automobile or on foot as shown in Figure 1. The sites described lie on private land and direct access is not practical to most of them; however, because of their large size they can be easily viewed from the public right of way. A pair of binoculars is useful in viewing some of these features. Unfortunately, the best exposures of veins are in the underground workings no longer accessible or in open pits to which access is prohibited.

SIGNIFICANCE

The Butte mining district is one of the major mining districts of the world with continuous production from both underground and open pit mines for 119 years, from 1864 until June 30, 1983. In 1986, plans were underway for the resumption of mining in this district. The fascinating history of this mining camp has been the subject of numerous books (see Glasscock, 1935; Malone, 1981).

There have been many firsts in Butte both in mining methods and also in the development of the techniques of mine mapping and the detailed recording of geological data. Methods initiated by Reno H. Sales are practiced throughout the world by the many geologists who received their training at Butte. Detailed

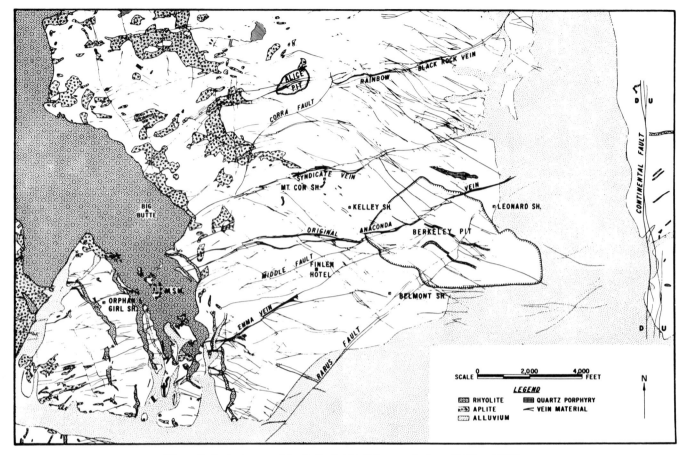

Figure 2. Geologic map of the Butte district (Meyer and others, 1968, p. 1380).

research on the mineralogy and genesis of the veins as well as accompanying hydrothermal alteration continue (Miller, 1973a; Brimhall and Ghiorso, 1983; and Brimhall and others, 1985).

GEOLOGY

Although many articles have been written on the geology of the Butte district, the following description of this district is mainly from two sources (Meyer and others, 1968; Miller, 1973a).

The host rock for the veins of the Butte district is the Butte Quartz Monzonite, the largest of the plutons of the Boulder batholith. Plutons of the Boulder batholith were emplaced in the interval between 68 and 78 Ma (Robinson and others, 1968, p. 566). Quartz and orthoclase compose nearly 45 percent of typical Butte Quartz Monzonite, and plagioclase nearly 40 percent. Biotite and hornblende together make up between 15 and 20 percent of the rock. Aplite and pegmatite dikes and irregular masses occur sporadically throughout this pluton and can be examined in roadcuts along I-90 a few miles east of Butte. Aplite and pegmatite are more difficult to identify in the hydrothermally altered rock next to the veins, but relict aplitic textures persist even in the pervasively altered sericitic zone.

In terms of form and composition, the Butte deposit consists of two distinct vein types, pre–main-stage veins and main-stage veins. Pre–main-stage veins form a low-grade, dome-shaped disseminated copper-molybdenum zone with its upper limit just above the 2,800 level (2,800 ft; 850 m below the surface) in the central part of the district. Apparently this deep mineralized zone extends eastward to the Continental fault. Relative upward movement on the east side of this high-angle fault (Fig. 2) has exposed the zone of disseminated mineralization at the surface where it has been mined in the East Berkeley and Continental pits. Pre–main-stage veins are typically small, extending for only a few tens of feet at most and are generally less than several inches thick, most less than 1 in (2.5 cm) thick. They fill fractures and do not show evidence of replacement of the host rock. Mineralogically there are several varieties of pre–main-stage veins. The early dark micaceous (EDM) veins are characterized by a dark alteration envelope consisting of secondary alkali feldspar, quartz, sericite, secondary biotite, disseminated chalcopyrite, and minor andalusite. These EDM veins consist of quartz or quartz-molybdenite and may contain chalcopyrite and pyrite either in the vein or in the alteration envelope. Some of the quartz-molybdenite veins are not surrounded by an alteration envelope, whereas others are accompanied by an alkali feldspar-anhydrite alteration assemblage. Dikes and irregular bodies of quartz porphpyry cut some of the pre–main-stage veins and are cut by some of these veins, indicating that the quartz porphyry was intruded during the interval of pre–main-stage mineralization.

Although recent production from the East Berkeley and Continental pits has been from porphyry-type copper-molybdenum ore of the pre–main-stage veins, more than 80 percent of the district's total production has been from underground mining of large main-stage veins that cut through the pre–main-stage dome of mineralization. Some of these large veins have been mined to depths of more than 5,000 ft (1,520 m) and extend laterally more than 1 mi (1.6 km) beyond known limits of the dome. Most of the large veins also extend to the surface. Mineralogically, main-stage veins and their sericitic and argillic alteration envelopes are much more complicated than the pre–main-stage veins. Many of the elements concentrated in the main-stage veins do not appear to be present in the earlier veins. The source of the copper and zinc in the large main-stage veins is still under discussion by geologists familiar with the district. Brimhall and Ghiorso (1983) have suggested that metals were concentrated into the main-stage veins from the disseminated pre–main-stage veins by continuing hydrothermal activity.

Sales (1914) recognized two major main-stage vein systems in the district, the Anaconda system of steeply dipping, approximately east-west–trending veins and the younger northwest-trending Blue fault veins, which displace Anaconda veins with left lateral offset. Detailed mapping of underground workings since Sales' original work has substantiated the distinction between these two vein systems. The veins of the Anaconda system have been the most productive of the district, with average reported widths of 20 to 30 ft (6 to 9 m) and local widths of up to 100 ft (30 m). Near the eastern edge of the district where the Berkeley pit is situated, Anaconda veins split into a series of smaller veins, which generally strike northwest and are all developed on the same side of an individual Anaconda vein, resembling the splaying out of a horse's tail. The horsetail zones have produced a substantial amount of ore because these veins are abundant and closely spaced. Major ore minerals of the Anaconda veins in the central part of the district are chalcocite and enargite with lesser bornite, covellite, digenite, and djurleite in a quartz-pyrite gangue. In addition to these minerals, veins of the horsetail zones contain the rare mineral colusite named after the Colusa mine.

The Blue veins, which are smaller than veins of the Anaconda system, were emplaced along northwest-striking faults that offset the Anaconda veins. They range from 5 to 20 ft (1.5 to 6 m) wide and up to 2,400 ft (730 m) long, but because they tend to be more closely spaced than veins of the Anaconda system they have contributed significantly to the production of high-grade ore. A third system of veins, the Steward, is of much less importance than either the Anaconda or Blue veins. Steward veins are most prominent in the eastern part of the district where they were emplaced along the steeply dipping, east-north-east–trending Steward faults.

Wall rock alteration surrounding the main-stage veins is more extensive than that surrounding the pre–main-stage veins. The alteration sequence consists of the sericitic zone closest to the vein and separated from unaltered quartz monzonite host rock by the argillic zone. The argillic zone is subdivided into the outer propyllitic zone, intermediate montmorillonitic zone, and the inner kaolinitic zone. The inner part of the sericitic zone next to the vein is designated the advanced argillic zone and contains kaolinite, dickite, pyrophyllite, and topaz. A K-Ar date on sericite from the sericitic zone accompanying a main-stage vein gives an age of 57.5 ± 1.8 Ma (Meyer and others, 1968, p. 1386).

The Rarus and Middle faults (Fig. 2) both strike northeast and displace veins of the district, but are themselves not mineralized. Farther to the east, displacement along the north-south–trending Continental fault has downdropped the veins in the main part of the district relative to the veins mined in the East Berkeley and Continental pits east of the Continental fault.

The Butte district is well known for its zonal distribution of metals first described by Sales (1914, p. 58–61). The central copper zone is characterized by chalcocite-bornite-covellite-enargite veins with quartz-pyrite gangue and minor silver. The intermediate zone is distinguished from the central zone by the occurrence of chalcopyrite, sphalerite, and galena in addition to those copper minerals found in the central copper zone. The manganese minerals rhodochrosite and rhodonite also occur in minor amounts in the intermediate zone. Bornite, chalcopyrite, and tennantite are more abundant in this zone than in the central copper zone. The outermost or peripheral zone, in contrast to the inner two zones, is noted for the production of manganese, zinc, lead, and silver, but not copper. Rhodochrosite is the only economic manganese mineral. Sphalerite and galena also occur in the peripheral zone. In addition to native silver, the silver minerals acanthite, stephanite, proustite, pyrargyrite, pearceite, and polybasite occur in this zone.

A blanket of supergene mineralization, mainly sooty chalcocite, has been an important source of ore in the Berkeley pit. This blanket had an average thickness of 176 ft (54 m) and an average grade of 0.75% copper (McClave, 1973, p. K-1). Oxidation of pyrite with subsequent formation of sulfuric acid resulted in leaching of copper sulfide from the upper part of the deposits and its reprecipitation at a greater depth to form the supergene blanket.

A postore "rhyolite" dike, actually ranging in composition from rhyolite to quartz latite, with an east-west strike and nearly vertical dip, intersects veins across the district. This dike, which is cut by steeply dipping north-south "rhyolite" dikes, is exposed at the surface only on the east side of the district; in the rest of the district it is only encountered in workings at depths 2,000 to 3,000 ft (610 to 914 m) below the surface. K-Ar dates on biotite from one of the north-south dikes indicate an age of about 43 Ma (Woakes, 1959, p. 19).

FIELD STOPS

Stop 1. Alice pit. Go north on Montana Street to Park

Figure 3. Aerial view of the Berkeley pit looking west. The viewing stand is at the left edge of the photo above the Rarus fault. Veins identified on the photo can be recognized by the white or pale green discoloration on the northwest face of the pit. These veins are in the Blue system (northwest-trending system). Since the discontinuance of dewatering of the underground workings in April 1982, the water level in all of the workings rose with water now partially filling the Berkeley pit. Photo by Hugh Dresser with geology by George Burns. The feature indicated by Q. P. is a "quartz-porphyry" dike.

Street and then right two blocks to Main Street. From Park Street, go north on Main Street 1.2 mi (1.9 km) to Daly Street, Walkerville. Offset one-half block east (right) on Daly and then continue 0.2 mi (0.3 km) north to boardwalk at the east end of the Alice pit. Although this is inaccessible, there is a good view of dark brown to black oxidized manganese and iron-oxide–stained quartz veins of the Alice–Rainbow–Black Rock veins system and associated alteration. These veins are part of the Anaconda system. The ore from this mine in the peripheral zone contained native silver, acanthite, rhodochrosite, rhodonite, and sphalerite. By continuing north from the Alice pit for approximately a mile, one can examine typical outcrops of unaltered Butte Quartz Monzonite, the host rock for the veins in the Butte district.

Stop 2. Syndicate pit. From the Alice pit go back to Daly Street and then west on Daly Street to Excelsior Avenue (Daly Street curves to the left into Excelsior Avenue). Continue south on Excelsior Avenue to Empire Street (0.7 mi; 1.1 km from the west end of Daly Street) and then go two blocks east on Empire Street to the Syndicate pit on the north side of the street. This pit is inaccessible. View the oxidized, dark-brown–stained quartz veins of the Syndicate vein system and altered wall rock, all part of the Anaconda system of veins. Fault slickensides indicating strike-slip movement are prominent. This mine in the intermediate zone contains the ore minerals bornite, chalcocite, chalcopyrite, tennantite, acanthite, stromeyerite, and sphalerite.

Stop 3. Berkeley pit viewing stand. Return to Excelsior Avenue and go south to Park Street, then east on Park Street through the business district to the Berkeley pit viewing stand on the left side of the street. Excellent view of the Anaconda vein and veins of the Blue vein system (Fig. 3). Because of the large size of the pit the ore mineralogy includes that of the central, intermediate, and even the peripheral zones. Much of the ore grade depended on the presence of the supergene ore minerals chalcocite and covellite. A self-activated tape gives a brief historical description of the district and methods of mining formerly employed in the Berkeley pit. From this vantage point the benches of the East Berkeley pit are visible across the valley near the base of East Ridge. The Clyde E. Weed concentrator southeast of the Berkeley pit was used in the concentration of ore from the Butte mines before it was shipped to the smelter at Anaconda.

Stop 4. Emma vein of the Anaconda system. Return to the intersection of Park Street and Excelsior Avenue and go south on Excelsior Avenue to Platinum Street, then east on Platinum Street over the hill to Jackson Street. Turn left on Jackson Street and go one block to the north. The Emma vein is exposed on the west side of Jackson Street where Gold Street runs into Jackson Street. Accessible black manganese–iron-oxide–stained and silicified exposure of the Emma vein. The vein is 20 ft (6 m) wide with sericitic alteration on the hanging wall. The Emma vein was mined from underground workings east of this exposure. Near the eastern end of this vein it was reported to reach a width of 100 ft (30 m) of almost pure rhodochrosite. In addition to rhodochrosite, this vein of the peripheral zone contains minor sphalerite and chalcopyrite.

Stop 5. Emma vein of the Anaconda system. Return to Platinum Street and go west to parking lot at the professional building at 800 West Platinum, a few hundred feet east of Excelsior Avenue. Accessible exposures are on the east side of the parking lot. The dark brown, manganese–iron-oxide–stained, siliceous composite vein 17 ft (5 m) wide is in the peripheral zone.

Stop 6. Orphan Boy and associated veins of the Anaconda system. Go north on Excelsior Avenue to Park Street and then west on Park Street 0.3 mi (0.5 km) west of the turnoff to the World Museum of Mining. Surface workings about 300 ft (91 m) north of the continuation of Park Street expose the oxidized portion of this vein. Although this vein contains rhodochrosite, acanthite, native silver, and sphalerite at depth, at the surface it consists of manganese and iron oxides in quartz. These veins are in the peripheral zone.

Stop 7. World Museum of Mining. The grounds of this museum surround the head frame of the Orphan Girl mine (closed in 1958) southeast from Stop 6. Exhibits feature mining equipment, methods, and a look at life around the turn of the century in the Butte district.

Stop 8. Montana Tech Mineral Museum. Return to the campus of the Montana College of Mineral Science and Technology and turn right at the sign for Mineral Museum parking. Displays include a very fine collection of Butte minerals.

REFERENCES

Brimhall, G. H., Jr., and Ghiorso, M. S., 1983, Origin and ore-forming consequences of the advanced argillic alteration process in hypogene environments by magmatic gas contamination of meteoric fluids: Economic Geology, v. 78, p. 79–90.

Brimhall, G. H., Alpers, C. N., and Cunningham, A. B., 1985, Analysis of supergene ore-forming processes and ground-water solute using mass balance principles: Economic Geology, v. 80, p. 1227–1256.

Glasscock, C. B., 1935, The war of the copper kings: New York, Grosset and Dunlap, 314 p.

Malone, M. P., 1981, The battle for Butte; Mining and politics on the northern frontier, 1864–1906: Seattle, University of Washington Press, 281 p.

McClave, M. A., 1973, Control and distribution of supergene enrichment in the Berkeley Pit, Butte District, Montana, *in* Miller, R. N., ed., Society of Economic Geologists, Butte Field Meeting, Guidebook: Butte, Montana, Anaconda Company, p. K1–K4.

Meyer, C., Shea, E. P., Goddard, C. C., Jr., and staff, 1968, Ore deposits at Butte, Montana, *in* Ridge, J. D., ed., Ore deposits of the United States, 1933–1967, Graton-Sales Volume: American Institute of Mining, Metallurgical, and Petroleum Engineers, v. 2, p. 1373–1416.

Miller, R. N., ed., 1973a, Guidebook for the Butte Field Meeting, Society of Economic Geologists August 18–21, 1973: Butte, Montana, Anaconda Company, 102 p. (reprinted by the Montana Bureau of Mines and Geology, 1983).

—— , 1973b, Production history of the Butte district and geological function, past and present, *in* Miller, R. N., ed., Society of Economic Geologists, Butte Field Meeting, Guidebook: Butte, Montana, Anaconda Company, p. F1–F10.

Robinson, G. C., Klepper, M. R., and Obradovich, J. D., 1968, Overlapping plutonism, volcanism, and tectonism in the Boulder batholith region, western Montana, *in* Coats, R. R., Hay, R. L., and Anderson, C. A., eds., Studies in volcanology: Geological Society of America Memoir 116, p. 557–576.

Sales, R. H., 1914, Ore deposits at Butte, Montana: Transactions of the American Institute of Mining Engineers, v. 46, p. 3–109.

Woakes, M. E., 1959, Potassium-argon dating of mineralization at Butte, Montana [M.S. thesis]: Berkeley, University of California, 42 p.

The Jefferson River Canyon area, southwestern Montana

Christopher Schmidt, *Department of Geology, Western Michigan University, Kalamazoo, Michigan 49008*
Richard Aram, *Phillips Petroleum Company, 8055 E. Tufts Aveunue, Denver, Colorado 80237*
David Hawley, *(deceased), Department of Geology, Hamilton College, Clinton, New York 13323*

LOCATION AND ACCESSIBILITY

The Jefferson River Canyon area is on an important structural boundary between the Cordilleran thrust belt and the Rocky Mountain foreland (Fig. 1). The features described here reflect the tectonic history of that boundary from late Precambrian through late Cenozoic time. The area is in Jefferson and Madison counties. Most of the features described may be observed along U.S. 10 through the canyon and along the access road to the Lewis and Clark Caverns State Park (Fig. 2). Stops of particular interest discussed in the text are numbered and indexed in Figure 2 with mileages indicated from LaHood Park (bar, restaurant, and sometime gas station and motel) and/or from the intersection of U.S. 10 and the state park access road. The area is located in the center of the Jefferson Island 15-minute Quadrangle.

Many of the features that can be seen across the Jefferson River from U.S. 10 (south side of river) are currently accessible via the abandoned track bed of the Chicago, Milwaukee, St. Paul, and Pacific Railroad. The track bed is easily traversed by most vehicles and may be accessed on the west from County Road 359 at Jefferson Island and on the east from Montana 287 at Sappington. The access route at Jefferson Island is privately owned and the owner (Vernon Shaw of Cardwell) should be contacted for permission.

SIGNIFICANCE

The Jefferson River Canyon area displays three long-separated episodes of tectonic activity. These occurred in late Precambrian, Late Cretaceous, and late Cenozoic time. The late Precambrian Willow Creek fault is not exposed, but is suggested by thick, arkosic conglomerates of the LaHood Formation (Middle Proterozoic–Precambrian Y). These rocks indicate the presence of east-west, fault-bounded highlands that shed clastic sediments to the north into an east-trending embayment of the Belt basin. Several different facies of the LaHood Formation are exposed at highway level in the canyon.

This fault-controlled shoreline profoundly influenced the structural configuration of the region in Late Cretaceous and Paleocene time. The dominant regional structural trend is NNW-SSE, but this grain is interrupted by a major east-west zone of faulting (the southwest Montana transverse zone; Fig. 1) that coincides with the Precambrian Willow Creek fault. Faults of the transverse zone are oblique-slip faults and combine both right slip and reverse movement. The zone also marks the boundary of two major tectonic styles: "thin-skinned," Sevier style thrusting on the north in the fold and thrust belt; and "thick-skinned,"

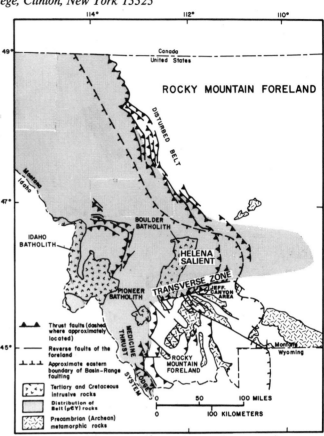

Figure 1. Tectonic map of western Montana showing location of the Jefferson Canyon area (modified from Schmidt and Hendrix, 1981).

Laramide style reverse faulting on the south that involved the Precambrian (Archean) basement rocks of the Rocky Mountain foreland (Fig. 1).

The third episode of faulting trends perpendicular to the earlier east-west faults. Basin and range style normal faulting separated the Jefferson basin to the west from the Jefferson Canyon–London Hills area to the east. At least 1,500 ft (450 m) of offset has occurred on the major north-trending fault (Starretts Ditch fault; Fig. 2) since mid-Pliocene time. The uplift of the London Hills forced the ancestral Jefferson River to abandon its former high position and downcut to its present level. This entrenchment created both the Jefferson River Canyon and the Lewis and Clark Caverns.

LATE PRECAMBRIAN HISTORY

The LaHood Formation at the base of the Belt Supergroup provides strong evidence of late Precambrian faulting along the

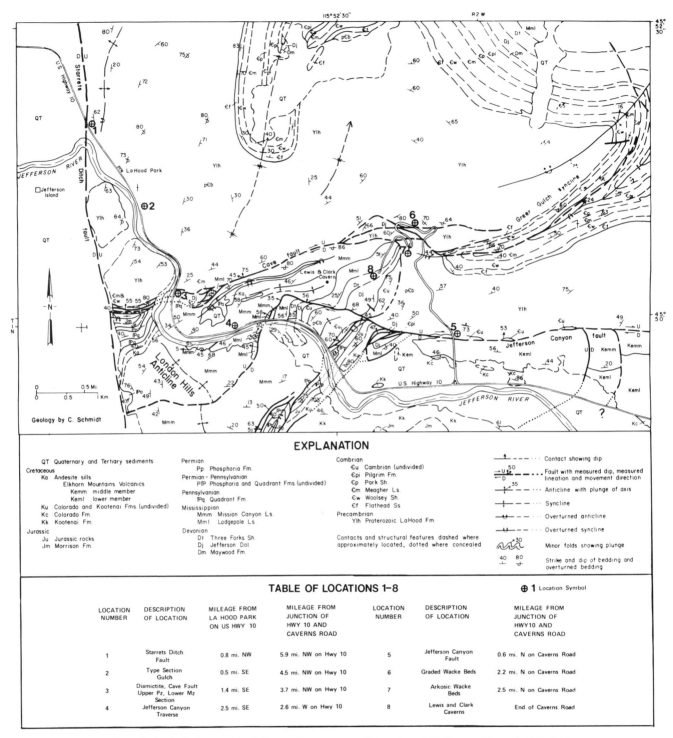

Figure 2. Geological map of the Jefferson Canyon area with locations 1–6 (discussed in text). (Modified from Schmidt and O'Neill, 1982.)

southern margin of the Belt basin. The LaHood rocks are coarse, arkosic conglomerates derived from eroded Archean basement rocks. The LaHood is absent south of the faults of the southwest Montana transverse zone and generally becomes finer grained northward away from the zone. These relationships and the fact that the LaHood rocks are everywhere allochthonous suggest the

presence of a fault-controlled shoreline in Proterozoic time that controlled the trend and position of the later faults of the transverse zone. The Proterozoic fault, named the Willow Creek fault by Robinson (1963), is inferred to have bounded a rugged Archean source terrane, and downward movement north of the fault provided the depositional basin.

The LaHood Formation was defined by McMannis (1963) as "all dominantly coarse Belt strata along the southern margin of the central Montana embayment of the Belt geosyncline," modifying Alexander's original description (1955). McMannis applied the name "LaHood" where Belt strata contain less than 75% fine-grained rocks and more than 25% coarse arkosic debris. The eastern limit of this unique lithosome is in the Bridger Range, north of Bozeman, and the western limit is 80 mi (130 km) distant in the western Highland Mountains, south of Butte.

The LaHood is at least 8,000 ft (2,400 m) thick in some parts of the Jefferson Canyon area. However, the base of the LaHood is nowhere clearly seen. McMannis (1963) interpreted two small areas of Archean metamorphic rock near, and seemingly structurally below, the oldest LaHood beds as lying unconformably beneath the LaHood. Schmidt and Hendrix (1981) suggested that these elongate narrow exposures of Archean metamorphic rocks are remnants of the foot wall of the Proterozoic Willow Creek fault that were elevated to their present position by thrusting.

The sedimentary character of the LaHood varies greatly from one area of exposure to another. The dominant characteristic is graded bedding, two types of which are present: (1) very thick crudely graded beds, and (2) thinner, well-graded beds. The thicker beds may be tens of feet (m) thick without apparent bedding. Locally they contain pebbly zones that begin abruptly and grade finer upward. Cobbles as large as 1 ft (0.3 m) in diameter are seen "floating" in finer grained sediment and the finer component is generally a poorly sorted arkosic to lithic wacke. Sequences of this type have been called fluxoturbidites (transitional between turbidites and slides deposited near source on gentle bottom slopes) and are included in one or another of the subfacies of facies A of Walker and Mutti (1973). The thinner beds are uniformly graded and are 1 in to 8 ft (2.5 cm to 2.4 m) thick or more, interbedded with laminated siltstone and shale, and carbonate beds in some places. It is common for these beds to have thin, plane-laminated sandy zones, and small-scale current gross-laminated zones above the graded beds (Bouma intervals a, b, and c; facies C of Walker and Mutti [1973].

Rubble beds are locally interbedded with thick sections of graded beds. These zones contain unsorted pebbles, boulders, and blocks embedded in a wacke matrix and are characteristic of facies A-1 of Walker and Mutti (1973). They range from a few feet to about 200 ft (1 to 60 m) thick and locally contain clasts of pegmatite as long as 30 by 100 ft (9 by 30 m). The rubble beds are interpreted as olistostromes, formed when massive slides developed off a steep shoreline.

LATE CRETACEOUS AND PALEOCENE STRUCTURE

Most of the structures of the Jefferson Canyon area are part of a zone of east-northeast–trending folds and faults that bound the southern margin of the Helena structural salient of the Montana disturbed belt (Fig. 1). This Late Cretaceous–Early Tertiary boundary has been termed the southwest Montana transverse zone (Schmidt and O'Neill, 1982) because it trends transverse to the general north-to-northwesterly trend of the fold and thrust belt in Montana. The fault zone is about 6 mi (10 km) wide and 75 mi (120 km) long. It extends from the southwestern Highland Mountains near Melrose, Montana, eastward to the Bridger Range north of Bozeman. It is one of the longest zones of transverse folds and faults in the Cordilleran fold and thrust belt. It has been interpreted by Schmidt and Garihan (1986) as a large lateral or transverse ramp zone in which décollement thrusts at or near the base of the Belt Supergroup rocks in the Helena salient on the north were forced to ramp laterally up the north-facing step (Willow Creek fault) of Archean basement rocks, which presumably formed the southern boundary of the Proterozoic Belt basin. The faults of transverse zone are demonstrably oblique slip with roughly equal components of reverse slip and dextral slip (Schmidt and Hendrix, 1981). The principal faults of the zone in the Jefferson Canyon area are the Cave fault and the Jefferson Canyon fault (Fig. 2). Belt (LaHood) rocks have been translated on these faults against rocks as young as the Late Cretaceous Elkhorn Mountains Volcanics, and at one location (3 mi [4.8 km] east of park access road), a large slice of Archean amphibolite forms part of the hanging wall of the Jefferson Canyon fault.

The fault pattern in the transverse zone is noticeably affected by the interaction of thrust belt structures with structures of the Rocky Mountain foreland. The foreland structure is comprised of a regularly spaced 4 to 6 mi (7 to 10 km) set of northwest-trending faults and associated folds that involve Archean crystalline rocks. Initial fault movement occurred in Middle Proterozoic time and may have influenced the depositional patterns in the Belt rocks at the southern margin of the Belt basin (Schmidt and Garihan, 1986). Movement on these faults in Late Cretaceous–early Tertiary time was oblique (left-reverse) and occurred just prior to thrusting. This partly accounts for the sinuosity of the transverse zone, which developed as the thrust belt structures impinged in the foreland structures.

LATE TERTIARY NORMAL FAULTING AND GEOMORPHOLOGY

The modern basin and range topography of southwest Montana formed primarily in the Neogene and partially masks the Precambrian and Late Cretaceous features. These normal faults generally trend north-south or northwest-southeast and, in the Jefferson Canyon area, were not strongly influenced by the east-west fault zones previously discussed. The faulting that created the London Hills triggered the erosion that carved both the Jefferson River Canyon and Lewis and Clark Caverns.

The Sevier-Laramide uplifts were eroded and buried (Fig. 3A, B), and several thousand ft (m) of fluvial clastics, volcanic ash, and minor lake sediments blanketed the region from latest Eocene until the mid-Pliocene (Bozeman Group). Since mid-Pliocene time the region has been broken up into many smaller basins and intervening mountain ranges by normal faulting.

C. Schmidt and Others

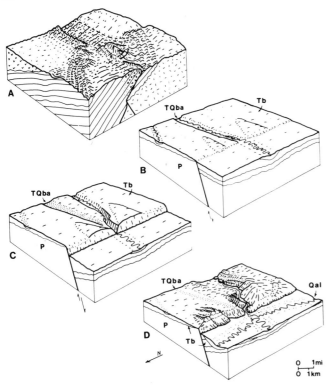

Figure 3. Cenozoic evolution of the Jefferson Canyon area: (A) interpreted Paleocene landscape; (B) Possible minor faulting on Starrets Ditch fault in Oligocene time prevents deposition of mid-Tertiary sediments on London Hills; (C) Continued movement on Starrets Ditch fault in Pliocene and Pleistocene time initiates Jefferson River Canyon; (D) modern landscape. P, pre-Cenozoic rocks; Tb, Tertiary Bozeman sediments, TQba, Ballard gravels; Qal, late Quaternary alluvium (modified from Aram, 1981).

The Starretts Ditch fault separated the Jefferson Canyon–London Hills area from the Jefferson Basin to the west with at least 1,500 ft (450 m) of offset. Sediments of the late Miocene–mid-Pliocene Six Mile Creek Formation dip up to 45° into the fault (Fig. 3B, C); dating the major movement along the fault as post Six Mile Creek.

The Jefferson River Canyon formed primarily by antecedence as the Starretts Ditch fault raised the London Hills (Fig. 4). Sediments of the ancestral Jefferson River can be found up to 1,500 ft (450 m) above the modern river level. A distinctive lithology of orthoquartzite and andesite clasts allow mapping of the ancestral Jefferson River deposits, called the Ballard gravels, for more than 10 mi (16 km) east from the fault. The clasts range in size from boulders on the west nearest the fault to coarse sand and fine gravel to the east.

The Ballard gravels were deposited during the early stages of movement along the Starretts Ditch fault (Fig. 3B). They rest unconformably on Precambrian and lower Paleozoic rocks nearest the fault on the west, and unconformably on the Oligocene Renova Formation to the east. This subcrop relationship shows that the London Hills had already begun to rise along the Starretts Ditch fault prior to deposition of the gravels. The gradient along the base of the Ballard gravels is approximately 10 times steeper

than the modern Jefferson River gradient, suggesting that the Ballard gravels have been tilted significantly by subsequent uplift of the London Hills. The Ballard gravels are tentatively dated as late Pliocene or earliest Pleistocene.

As the normal faulting and canyon cutting progressed, older rocks and structures were exhumed. The increasingly steeper stream gradients encouraged the flow of groundwater through the Mississippian Mission Canyon Limestone, forming the Lewis and Clark Caverns. The caverns formed along fractures associated with a minor anticline. Most of the known cave passages are solutionally enlarged bedding plane faults formed by bedding plane slip during the folding, and the largest known rooms of the cave formed along the hinge of the fold, where the limestone was most fractured. Large trunk passages extended west, discharging the cave water into the Jefferson Canyon.

Subsequent deepening of the canyon cut off the supply of water to the cave, changing it greatly. The caverns, formerly filled with phreatic water, slowly drained, and its ceilings collapsed as the bouyant force of the water was lost. The resulting boulder rubble reaches a depth of at least 300 ft (90 m) on the upper level trunk passage, and completely hides the trunk passage on the lower level of the cave. The circulation patterns of the cave winds provide evidence that the lower trunk is still partly open. Lewis and Clark Caverns now lie 1,400 ft (420 m) above the Jefferson River.

FIELD LOCATIONS (Fig. 2)

The important aspects of the geologic history of the Jefferson Canyon area in the preceding discussion are amplified in the following description of specific locations (stops).

Location 1 (Fig. 2), The Starretts Ditch Fault. The Starretts Ditch fault is a late Cenozoic basin-range feature that juxtaposes Belt (LaHood) and Paleozoic rocks against Tertiary strata of the Bozeman Group. On the north side of the highway the westernmost exposures of the LaHood Formation in the canyon are on the footwall of the fault. Thick, graded wacke beds interbedded with laminated silty shale dip westward at 60°–65° along the west edge of the hills close to the mine dumps and lie on the west flank of an overturned anticline. The beds at the fold hinge a short distance east of here are probably the oldest LaHood rocks in the vicinity of LaHood Park.

The fault trends north-south and generally cuts across the more easterly trending Sevier-Laramide structures seen in subsequent stops. Movement along the fault in Neogene time was responsible for the entrenchment of the Jefferson River into its canyon, the formation of basin on the west, and the preservation of the Tertiary sediments in that basin.

Location 2 (Fig. 2), "Type Section Gulch." (Park on broad shoulder on east side of highway.) The type section of the LaHood Formation, first described by Alexander (1955) begins 0.4 mi (0.6 km) up this gulch. The bluff at the north side of the entrance to the gulch has an assortment of beds typical of the LaHood in this area. A rubble bed 30 ft (9 m) thick is composed of various types of metamorphic clasts as large as 4 ft (1.2 m) embedded in a sandy-gravelly matrix. This bed is offset by a left-lateral strike-slip fault that dips steeply northeastward. Upsection to the east is a sequence of graded beds of feldspathic wacke from about 4 in to 5 ft (10 cm to 1.5 m) thick, and more rubble beds. Most of the coarse beds are separated by laminated silty shale zones <1 in to about 3 ft (<2.5 cm to 0.9 m) thick. Farther to the east no stratification may be apparent through 10 ft (3 m) or more of gravelly wacke. To the east 0.5 mi (0.8 km), about 900 ft (270 m) up on the rocky bluffs, is the trace of a massive rubble bed. The cream-colored pegmatite blocks, 50 to 100 ft (15 to 30 m) across, can be seen from several places along this highway.

Hawley estimates that the LaHood is at least 8,000 ft (2,400 m) in this general area and that the section is nearly continuous between location 1 and the overlying Cambrian Flathead Sandstone in the syncline northeast of location 2 (Fig. 2). Pre-Flathead dips between 17° and 51° south have been measured here indicating that thousands of ft (m) of very coarse LaHood beds were eroded on the unconformity between here and the I-90 road cuts 2.5 mi (4 km) to the north.

Location 3 (Fig. 2), The Cave Fault. Location 3 is a broad grassy parking area on the east side of U.S. 10 that runs north-south in this portion of the canyon. It can easily accommodate a large group. The Cave fault and an excellent section of strata from the Mississippian Mission Canyon Limestone to the Jurassic Ellis Group may be seen on the opposite (west) side of the river (Fig. 4). The fault was named and first described by Alexander (1955) and is an important fault of the southwest Montana transverse zone. It overlaps the Jefferson Canyon fault to the east (Fig. 2) and may have been originally connected with the Jefferson Canyon fault.

On the west side of the river the principal splay of the Cave fault places Cambrian rocks (Meagher Formation) against Permian rocks (Phosphoria Formation). The upper Paleozoic and lower Mesozoic section on the footwall (south side of the fault) is folded into a tight, overturned syncline (Fig. 4). The LaHood, Wolsey, and Meagher formations on the hanging wall are overturned. The Flathead Sandstone is missing. The fault plane dips about 75°N. The footwall syncline plunges about 40° northwesterly, so progressively older rocks are exposed on the footwall as the fault is traced toward the northeast. Just behind the parking area on the east side of the highway the LaHood Formation is against the Devonian Jefferson Dolomite and the Mississippian Lodgepole Limestone. Maximum dip separation of the fault is between 3,000 and 6,000 ft (900 and 1,800 m) where the La-Hood is against the Mission Canyon Limestone. Separation decreases rapidly just north of the Lewis and Clark Caverns, where the fault splays into smaller segments and is eventually lost in the overturned north limb of the Greer Gulch syncline (Fig. 2). This rapid loss of separation eastward from the caverns suggests that displacement was transferred to the Jefferson Canyon fault about 0.6 mi (1 km) to the south. The progressive increase of displacement along the Jefferson Canyon fault can be easily seen along the traverse that begins at location 4.

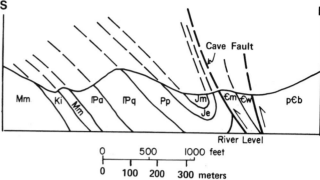

Ki	Andesite sill	Je	Ellis Group
Mm	Mission Canyon Limestone	Jm	Morrison Formation
ℙa	Amsden Formation	€m	Meagher Limestone
ℙq	Quadrant Formation	€w	Wolsey Shale
Pp	Phosphoria Formation	p€b	Belt (LaHood Formation)

Figure 4. Sketch cross section and photo of south side of Jefferson River looking west from location 3 (modified from Schmidt, 1976).

Measurements of the minor folds on the footwall of the Cave fault and of slickensides on the fault surface at two locations clearly suggest that the Cave fault is oblique slip with nearly equal components of dextral and reverse slip. Net slip is 3,900 to 7,800 ft (1,200 to 2,400 m).

The curving pattern of the Jefferson Canyon here suggests that the course of the river was established on less-resistant sediments of the Tertiary Bozeman Group and that the river was entrenched with this meandering path during uplift of the London Hills east of the Starretts Ditch fault.

Location 4 (Fig. 2), Jefferson Canyon Traverse. In order to observe the structural relationships along the canyon here, a west to east foot traverse of approximately 1 mi (1.6 km) is recommended beginning at location 4. Please note that this is a narrow, dangerous road. Location 4 is the parking area of the abandoned Limespur quarry, where Mission Canyon Limestone was quarried in the early 1900s for flux material in Butte smelters. The following description applies to a west to east traverse beginning here.

At the starting point near the quarry the massive bedding of the Mission Canyon Limestone dips 30°–50°NW. The contact between the Mission Canyon Limestone and the underlying, more thinly bedded Lodgepole Limestone is easily recognized in the cliff face along the highway. A short distance eastward from this contact is the first exposure of the Jefferson Canyon fault.

Here the fault dips 58°W and follows bedding in the Lodgepole, which is thrust over the more massive and pervasively fractured lower portion of the Mission Canyon. Stratigraphic throw here is only about 200 ft (61 m). The fault is also exposed across the river on the south side of the canyon. Numerous minor folds are developed in the Lodgepole on the hanging wall and several may be seen from the highway. The folds verge east and trend parallel to the fault suggesting primarily dip slip movement. This is confirmed by down dip slickensides on the fault plane at one location. Along the highway the contact between the Mission Canyon and the Lodgepole may be observed again on the footwall a very short distance east of the fault.

Just east of the last roadcut of Lodgepole, the Jefferson Canyon fault may be seen to bend sharply eastward in the cliffs several hundred ft (m) above the highway. The fault dips 35°W west of this abrupt bend and 56°N after it bends eastward. The hanging wall rocks are the steep cliffs of Mission Canyon Limestone. On the footwall an entire vertical section from Lodgepole on the west to LaHood Formation on the east strikes north into the east-west trace of the fault. Near the fault the section is bent into a more easterly strike, which probably reflects a significant dextral movement component. Portions of the stratigraphic section are absent because of tectonic thinning, but the conspicuous absence of the Flathead Sandstone above the highway is due to non-deposition, and the Wolsey Shale rests disconformably on the LaHood Formation.

For nearly 1 mi (1.6 km) along the east trace of the fault, younger rocks are thrust on older rocks and stratigraphic throw increases very rapidly eastward between here and the state park access road (Fig. 2). The younger against older relationship is a function of the strong dextral component of slip along this fault segment.

Approximately midway along the traverse, as the highway curves gradually southeastward, there is a roadcut through rocks of the LaHood Formation, the only occurrence of Belt rocks on the footwall side of the Cave–Jefferson Canyon fault system. The rocks lie in the core of a tight anticline whose eastern limb is thrusted with Cambrian carbonates (Meagher Limestone) over highly folded Pennsylvanian rocks (Amsden Formation). Most of the minor folds here are within the Amsden and overlying Quadrant formations. An inverted sequence of upper Paleozoic and Mesozoic rocks continues to the gulch at the end of the traverse. The last rocks to be observed before the highway turns eastwardly are gastropod-rich carbonates at the top of the Cretaceous Kootenai Formation and black shales of the Cretaceous Colorado Formation. The section here is replete with minor faults that bend eastwardly as they join the Jefferson Canyon fault.

Location 5 (Fig. 2), The Jefferson Canyon Fault. The

Lewis and Clark State Park access road cuts through the Jefferson Canyon fault here. The LaHood Formation and a sliver of Cambrian rock (mostly carbonate in the Wolsey Shale) are against mudflows of the Upper Cretaceous Elkhorn Mountains Volcanics, the youngest rocks in the region to be involved in thrusting. The fault here dips 50° to 60°N. The minimum strike

separation is about 11,000 ft (3,400 m) and dip separation is considerably greater. Orientation of 115 slickensides along the fault from this point to a point several mi (km) to the east indicates oblique slip with the reverse (dip-slip) component somewhat greater than the dextral component. Net slip is on the order of 6 mi (9.6 km).

Location 6 (Fig. 2), Lewis and Clark Caverns. The

cavern's geology has been summarized by Aram (1981).

The visitors center at the top of the caverns road provides a good overview of many of the geologic features of the area. Looking west towards the cave, the highest cliffs to the right or north are the basal beds of the Misson Canyon Limestone. The grassy slopes below are the argillaceous Lodgepole Limestone, and the deep valley directly ahead is the Devonian Three Forks Shale. The small peak to the left is capped by the Devonian Jefferson Dolomite. The beds strike N55°–70°E along most of the trial to the cave until the trail climbs upsection at the switchbacks, where the beds turn to N20°W. This abrupt change of strike marks the hinge of the Colter anticline, which plunges steeply to the northwest. Lewis and Clark Caverns formed along the fractures of this anticline. This location gives a striking view of the proximity of the cave to the Jefferson Canyon.

REFERENCES CITED

Alexander, R. G., 1955, Geology of the Whitetail area, Montana: Yellowstone-Bighorn Research Project Contribution 195, 111 p.

Aram, R. B., 1981, Geologic history of Lewis and Clark Caverns, Montana, *in* Tucker, T. E., ed., Field conference and symposium guidebook to southwest Montana: Montana Geological Society, p. 285–300.

McMannis, W. J., 1963, LaHood Formation; A coarse facies of the Belt Series in southwestern Montana: Geological Society of America Bulletin, v. 74, p. 407–436.

Robinson, G. D., 1963, Geology of the Three Forks Quadrangle, Montana: U.S. Geological Survey Professional Paper 370, 143 p.

Schmidt, C. J., 1976, Structural development of the Lewis and Clark Cavern State Park area, southwest Montana: Montana Bureau of Mines and Geology, Special Publication 73, p. 141–150.

Schmidt, C., and Garihan, J., 1986, Middle Proterozoic and Laramide tectonic activity along the southern margin of the Belt Basin, *in* Winston, D., and Robberts, S., eds., A guide to the belt: Montana Bureau of Mines and Geology Special Publication 94 (in press).

Schmidt, C. J., and Hendrix, T. E., 1981, Tectonic controls for thrust belt and Rocky Mountain foreland structures in the northern Tobacco Root Mountains–Jefferson Canyon area, southwestern Montana, *in* Tucker, T. E., ed., Field conference and symposium guidebook to southwest Montana: Montana Geological Society, p. 167–180.

Schmidt, C., and O'Neill, J. M., 1982, Structural evolution of the southwest; Montana traverse zone, *in* Powers, R. W., ed., Geological studies of the western overthrust belt: Rocky Mountain Association of Geologists, p. 193–218.

Walker, R. G., and Mutti, E., 1973, Turbidite facies and facies associations, *in* Lecture notes for short course; Turbidites and deep water sedimentation: Pacific Section, Society of Economic Paleontologists and Mineralogists, 157 p.

The Stillwater Complex, southern Montana; A layered mafic intrusion

W. W. Atkinson, Jr., Department of Geological Sciences, University of Colorado, Boulder, Colorado 80309
Robert E. Derkey, Montana Bureau of Mines and Geology, Montana College of Mineral Science and Technology, Butte, Montana 59701

LOCATION AND ACCESSIBILITY

The Stillwater Complex is about 75 mi (120 km) southwest of Billings, Montana, and northeast of Yellowstone National Park (Fig. 1). Exposures of the NE-dipping complex extend over a 30-mi-long (48 km) strike along the north margin of the Beartooth Mountains. The Benbow and Mountain View areas southwest of Absarokee, Montana (Fig. 1), are readily accessible by passenger car. The Benbow area offers easy access to a variety of rocks from the ultramafic series and chromite deposits but only limited exposures of features from the banded series. The Mountain View area offers easy access to most of the banded series and the platinum deposits.

The Benbow and Mountain View areas are reached by exiting I-90 at Columbus, Montana (Fig. 1), onto Montana 78 and proceeding south to Absarokee. County Highway 419 to both areas begins 2.5 mi (4.0 km) south of Absarokee. It is a narrow blacktop road that passes through Fishtail, where a small motel and restaurant are located. The Benbow turnoff at Dean, a collection of several houses, is 13.9 mi (22.4 km) from Montana 78. Turn left onto the gravel road marked "Benbow mine" if you wish to tour the Benbow area (Table 1). Continue on County Highway 419 if you wish to tour the Mountain View area (Table 2). The Nye Post Office and Carters Camp are located 20.4 mi (32.8 km) from the Montana 78 junction. Continuing ahead 5.4 mi (8.7 km) will bring you to the platinum mining operations at the end of the blacktop road. Platinum is mined from a mineralized unit called the J-M reef in the banded series of the complex. The road log to the Mountain View area (Table 2) begins at the hairpin curve to the right near the end of the blacktop and the platinum mining operations. The Beartooth Guest Ranch is 2.5 mi (4.0 km) up river on Woodbine Road, which begins 0.1 mi (0.2 km) before reaching the platinum mining operation. Many of the researchers working on the complex stay here. We suggest you call ahead should you plan to stay overnight.

Tables 1 and 2 are condensed road logs for the Benbow and Mountain View areas respectively. Both areas are readily accessible by passenger car; however, the road is relatively steep. A large part of the Benbow and Mountain View areas is on private land. The road to the Benbow chromite mine is open to the public. If you use extreme caution around open mine workings and respect the private property, the area should continue to be accessible for geologists to examine and sample. To tour the Mountain View area, check first at the platinum mining operations office. Please do not oversample or destroy classic igneous features, which include limited exposures of igneous layering in road cuts. Loose

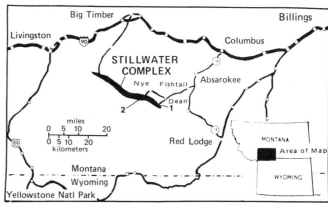

Figure 1. Location map of the Stillwater Complex with respect to major highways in the area. Number 1: Benbow area. Number 2: Mountain View area.

material is available for collecting without destroying the classic outcrops of the complex.

SIGNIFICANCE

Layered complexes provide the most visible evidence of processes of magmatic differentiation; thus, they occupy a central place in the study of igneous petrology. Such complexes are not only scientifically rewarding, they are host to several types of mineral deposits, including copper-nickel, chromium, and platinum-group elements. The Stillwater Complex is one of the world's great layered mafic intrusions, distinguished not so much by its size as by the fact that it is tilted on its side, and erosion has exposed the layering to ready access.

The Stillwater volume, edited by Czamanske and Zientek (1985), is a comprehensive summary of the Stillwater Complex that presents review articles on various aspects of the geology, field guides to selected areas, and a bibliography of Stillwater articles. The present article is based largely on that summary.

GENERAL GEOLOGY

The Stillwater Complex (Fig. 2) forms a part of the Precambrian terrane of the Beartooth Mountains, which are surrounded and partly covered by Paleozoic and Mesozoic sedimentary and Tertiary volcanic rocks. Metasedimentary rocks of the range are intruded by Precambrian dioritic to granitic igneous rocks and by the Stillwater Complex. Complexly folded, Early Archean rocks in the footwall of the Stillwater Complex are magnesium-, iron-, nickel-, and chromium-rich clastic and

TABLE 1. CONDENSED ROAD LOG FOR THE BENBOW AREA (from Lipin and others, 1985)

Mi (km)

0.0 (0.0). Begin at Dean (Fig. 1), turn left onto the Benbow road, U.S. Forest Service road 1414.

4.3 (6.9). Road to left goes to Benbow millsite. Continue straight ahead.

5.7–8.7 (9.1–13.9). Spectacular exposures of lower Mesozoic and Paleozoic sedimentary units.

8.7 (13.9). Cambrian unconformably overlying the Stillwater Complex.

8.9 (14.2). Plagioclase cumulates, lower anorthosite zone.

9.0 (14.4). Contact between lower anorthosite and lower gabbro zone.

9.2 (14.7). Lower gabbro zone.

9.5 (15.2). Fault, Cambrian in contact with complex.

9.6 (15.4). Grading and layering, lower gabbro zone.

9.7 (15.5). Fault, Cambrian in contact with complex.

10.0 (16.0). Cambrian unconformable on complex.

10.8 (17.3). Bronzitite zone, ultramafic series.

10.9 (17.4). Follow road to right.

12.0 (19.2). Stop, Benbow chromite mine. G chromitite crops out and can be safely examined and sampled about 200 ft (60 m) southwest of the headframe. Turn around and return back down the road.

12.3 (19.7). Stop at the fork in the road to the right. A short walk on this road passes through exposures in the periodite zone including thin chromitite seams and pegmatoid. Please preserve these exposures in the road cut and do not hammer on them.

TABLE 2. CONDENSED ROAD LOG OF THE MOUNTAIN VIEW AREA (from Page and others, 1985b).

Mi (km)

0.0 (0.0). End of blacktop, hairpin curve to the right at the platinum mining operations.

2.7 (4.3). Road to the West Fork of the Stillwater River. Take a sharp turn to the left instead.

3.4 (5.4). Locked gate restricts access to the ultramafic series and the Mountain View chromite mine and the basal series and copper-nickel mineralization in the Verdigris Creek area. Refer to Page and others (1985b) for detail of this area. If you wish to see features in this area, we suggest you obtain permission from the platinum mining operations office. Turn around to continue trip through exposures of the banded series along the road over which you have just traveled.

4.1 (6.6). West Fork Stillwater River Road. Turn right.

4.5 (7.2). Inch-scale layering in the lower banded series.

4.8 (7.7). Fine-scale layering in the banded series.

4.9 (7.8). Fork in the road to the right. Park your car here and hike up the road behind the locked gate to exposures of the J-M reef in the lower banded series. Remember, this is on private land. Check with the platinum mining office for permission to examine this exposure. They will give you information on the exact location of this outcrop.

5.1 (8.2). Syndepositional structures exposed in road cuts in the lower banded series.

6.3 (10.1). Numerous features of igneous layering that resemble sedimentary structures in the many outcrops of the lower banded series in the vicinity of Minneapolis adit. This is one of the main entries to the platinum mine.

chemical sediments (banded iron formation and blue quartzite) in which sedimentary features are preserved locally. The rocks range from pyroxene hornfels near the contact to biotite hornfels away from the contact; they consist of assemblages of orthopyroxene, cordierite, quartz, plagioclase, biotite, and anthophyllite. Mineral assemblages and compositions indicate that the Stillwater Complex was emplaced at a depth of about 7.5 mi (12 km) into rocks whose temperature was about 400°C.

Various stratigraphic schemes have been proposed for the layering of the Stillwater Complex, but the classification in Czamanske and Zientek (1985) is adopted here to show the distribution of the units in the Complex (Fig. 3). The units can be divided into three ranks: series, zones, and subzones. This nomenclature is widely used by Stillwater geologists and is adapted from Wager and others (1960). Rock names are based principally on "cumulus" mineral grains, which apparently accumulated by settling or flotation. The cumulus minerals are subhedral to euhedral and form a skeletal framework like that of a sandstone. The interstices may be filled by various types of postcumulus growth, such as

randomly oriented crystals; adcumulus overgrowths on the cumulus crystals; or by poikilitic crystals, called oikocrysts, enclosing the cumulus crystals. Variation in postcumulus mineral proportions may take place laterally within a single layer, so that the use of cumulus nomenclature is essential to identification of individual stratigraphic layers.

Most of the Stillwater Complex consists of four dominant minerals: olivine, orthopyroxene (called "bronzite" in the field), plagioclase, and augite. These minerals are abbreviated o, b, p, a respectively; the abbreviation for chromite is c. The "cumulate" is abbreviated with an uppercase C to distinguish it from chromite. Thus, examples of rocks in the Stillwater Complex are abbreviated oC for olivine cumulate, obC for olivine-bronzite cumulate, paC for plagioclase-augite cumulate, and so on. In addition, traditional names based on mineral proportions are also used to some extent, such as harzburgite (olivine and orthopyroxene), norite (plagioclase + orthopyroxene), troctolite (plagioclase + olivine), anorthosite, pyroxenite, dunite, and peridotite.

The basal series, the lowermost division of the Stillwater

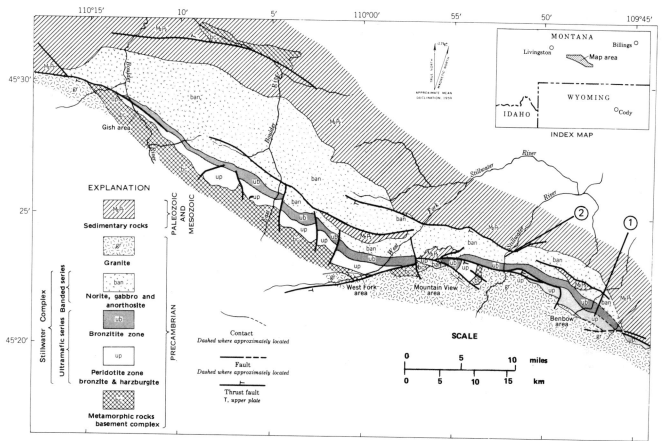

Figure 2. Geologic map of the Stillwater Complex showing relative locations of the areas suggested for touring. Number 1: Benbow area. Number 2: Mountain View area. (Adapted from Jackson, 1961)

Complex, is divided into two zones (Zientek and others, 1985). The lower one, the basal norite zone, consists of numerous intrusions varying in grain size and cumulate minerals. Rocks such as bronzite, olivine, bronzite-plagioclase, bronzite-olivine, and bronzite-augite cumulates are observed. Noncumulate chilled rocks are also present. The intrusions are crowded with hornfels xenoliths to the extent that the zone might be regarded as an igneous breccia. The thickness of the unit varies from 0 to 330 ft (0 to 100 m). The upper unit of the basal series, the basal bronzite cumulate, consists of a fairly homogeneous bronzite cumulate, which extends over the length of the Stillwater Complex, varying in thickness from 100 to 650 ft (30 to 200 m).

Sills and dikes beneath the Stillwater Complex are not included in the stratigraphic units. Several different compositions have been identified. Two of these, mafic norite and high-Mg gabbronorite, have compositions that suggest they formed from the parent magma of the Stillwater Complex.

The Ultramafic series consists of cumulus olivine, bronzite, and chromite (Jackson, 1961), averages about 3,600 ft (1,100 m) thick, and is divided into a lower Peridotite zone and an upper Bronzitite zone (Zientek and others, 1985). The Peridotite zone is characterized by cyclic units beginning with olivine-chromite cumulate (ocC) and passing into olivine-bronzite cumulate (obC) and finally into bronzite cumulate (bC). Olivine gradually disappears upward then reappears abruptly at the base of the next cycle. In most cases, this ideal sequence does not occur, and one or two of the upper units is absent. In traditional terminology, the olivine cumulates are either dunite or poikilitic harzburgite. The olivine cumulates contain less than 2 modal percent chromite, but individual chromitite seams contain up to 80 modal percent chromite. Chromitite seams are identified by letters, with A for the lowest. The number of cyclic units varies along strike from as few as 8 to as many as 21. Correlation is somewhat uncertain due to limited exposures. The upper unit of the series, the Bronzitite zone, consists of a bronzite cumulate with sparse interstitial plagioclase and augite. Its thickness averages about 1,150 ft (350 m). The rock shows lamination defined by subtle variation in grain size in thin section but in the field appears to be nearly uniform and massive. The upper contact of the ultramafic series is sharp, marked by the sudden appearance of cumulus plagioclase.

The banded series (maximum thickness is 14,850 ft; 4,500 m) is marked by the appearance of plagioclase as a cumulus mineral and usually contains plagioclase throughout, either alone or in combination with one or more of cumulate bronzite, augite, and olivine. Cumulus pigeonite, inverted to orthopyroxene, also occurs near the top of the Complex (Zientek and others, 1985). The makeup of the series is diverse, consisting of layers ranging from a few millimeters to tens of feet thick. Although some thick units have been mapped over the length of the Stillwater Complex, lateral discontinuity and variation of mineral proportions within individual layers are quite pronounced. The series has been studied in detail along certain sections, but knowledge is far from complete. Subdivisions of the banded series are the lower banded, the middle banded, and the upper banded series.

The lower banded series begins with norite zone I (Fig. 3, using the terminology of Todd and others, 1982), a plagioclase-bronzite cumulate (pbC) consisting of numerous individual layers of variable mineral proportion. The top of norite zone I is marked by the appearance of cumulate augite at the base of the gabbro-norite zone I. This unit consists of plagioclase, bronzite, and augite cumulates. The next unit is troctolite-anorthosite zone I, a complex succession of olivine, plagioclase, bronzite, and augite cumulates that contains the J-M reef, a mineralized unit containing economic quantities of platinum-group elements. Overlying this zone is norite zone II, a plagioclase-bronzite cumulate (pbC) and gabbro zone II, a thick monotonous 2-pyroxene gabbro, plagioclase, bronzite, augite cumulate (pbaC). This is overlain by troctolite-anorthosite zone II, a thin, complex zone of plagioclase, olivine, bronzite, and augite cumulates, similar to troctolite-anorthosite zone I with mineralized units containing subeconomic concentrations of platinum-group elements.

The middle banded series begins with a thick anorthosite zone mappable along the entire length of the Stillwater Complex. It contains 10–20 percent augite-inverted pigeonite oikocrysts 4 in (10 cm) across and shows an apparent absence of mineral fabric and layering. The next three zones, troctolite-gabbro zone III, anorthosite-troctolite zone IV, and troctolite-gabbro zone IV, are a complex sequence of plagioclase, bronzite, augite, and olivine cumulates. They are overlain by another thick continuous anorthosite zone, anorthosite zone V.

The upper banded series begins with troctolite-gabbro Zone V, consisting of plagioclase-olivine cumulate (poC) and plagioclase-bronzite-augite cumulate (pbaC). It is overlain by gabbro Zone VI, a plagioclase-bronzite-augite cumulate (pbaC). The uppermost exposed unit in the Stillwater Complex is the pigeonite-gabbro zone VI, consisting of cumulate plagioclase, augite, and inverted pigeonite with accessory postcumulus magnetite.

ORIGIN

Recent isotopic studies shed some light on the origin of the Stillwater Complex (Lambert and others, 1985). Radiometric dating by Sm-Nd and U-Pb methods indicates an age of about 2,700 Ma. Studies of Sr, Nd, Pb, and Hf isotopic content suggest either that the parent Stillwater magma or magmas came from a mantle source region enriched in Rb/Sr, Nd/Sm, U/Pb, light rare earths, and Lu/Hf or, alternatively, that the magma was somewhat contaminated by older continental crust. The oxygen

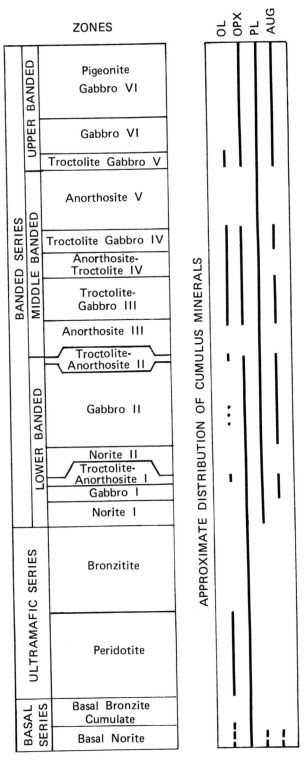

Figure 3. Subdivisions of the Stillwater Complex stratigraphy using the terminology of Todd and others (1982). Right side of figure shows approximate distribution of cumulus minerals in the complex. (Adapted from Zientek and others, 1985)

isotopes are typical of a primary magmatic origin (5.5 to 8.0 per mil), showing no influence of meteoric water. The sulfur isotopes (−3.8 to +3.0 per mil) indicate a mantle source, with no contribution from sedimentary sources such as the wall rocks.

STRUCTURE

The general structure of the Stillwater Complex shows a WNW-trend of layering, dipping steeply (about 65°) to the north (Page and Zientek, 1985). The wall rocks on the south have similar trends, although they were isoclinally folded prior to intrusion. Locally, dips are more gentle—as little as 20° at Chrome Mountain. Part of the tilting, perhaps 20 to 30°, took place in late Precambrian time; the present attitude (dips range from 20° to overturned 60°) is due to further deformation in the Laramide orogeny. Paleozoic and Mesozoic rocks have been turned up vertically along the northern and eastern flanks of the Beartooth Mountains in response to southward and northward thrusting, which has locally juxtaposed Precambrian rocks of the complex against Paleozoic rocks. Four types of faults have been noted: Longitudinal ramp and thrust faults strike N 40–80° W and dip NE, south-dipping longitudinal steep reverse or flat thrusts strike E-W and dip 30–70°S, with measured offsets as much as 2,970 ft (900 m), vertical E-W trending faults show left-lateral wrench movement; and steeply dipping transverse faults strike N30°E to N30°W, offsetting layers up to 1,980 ft (600 m) in apparently both right- and left-lateral senses. Although such faults have locally disrupted the complex, the body as a whole is still largely intact, preserving the relations between layers.

MINERAL DEPOSITS

The Stillwater Complex is host to three types of potentially economic mineral deposits: copper-nickel, chromite, and platinum-group elements (Page and others, 1985a). The basal series hosts Cu-Ni deposits. Sulfide minerals occur in high concentrations, are locally massive, and appear to have solidified from a sulfide liquid. Pyrrhotite is the principal mineral, with small amounts of pentlandite and chalcopyrite. The largest known Cu-Ni deposits occur in the Mountain View and Benbow areas and have an average grade of 0.25 percent Ni and 0.25 percent Cu. Low metal prices and environmental concerns have prevented exploitation of this resource.

Chromite bands occur throughout the length of the complex in the peridotite zone (Page and others, 1985a). During World War II, the U.S. government considered the complex as an alternative source of chromite in case foreign supplies were cut off. Exploration and development took place in the Mountain View, Benbow, and Gish areas. More than one million tons of concentrates with a grade of 39 percent Cr_2O_3 were eventually stockpiled. However, Stillwater chromite cannot currently compete with foreign sources because of its somewhat lower Cr_2O_3 content and its high Fe/Al ratio.

Platinum-group elements have attracted considerable attention since Johns-Manville Corporation geologists discovered a mineralized layer resembling the Merensky reef of the Bushveld Complex, a major source of the world's platinum. The unit, known as the J-M reef, consists of troctolite, pegmatoidal pyroxenite, and anorthosite. The platinum-group elements are associated with pyrrhotite, pentlandite, and chalcopyrite and occur as unusual sulfides, tellurides, arsenides, and metallic alloys. The grade at the West Fork adit is 5 g/ton Pt, 17.3 g/ton Pd.

In 1978, Anaconda Minerals Company began exploration of its property on the J-M reef in the Mountain View area. This work led to formation of the Stillwater Mining Company in 1982, a partnership between Johns Manville, Chevron Resources, and Anaconda companies. During the summer of 1985, Lac Minerals purchased the Anaconda holdings in the platinum venture. Exploration and development continue.

SUMMARY

The origin of the complex is a subject of controversy. The bulk composition led most workers in the 1960s to conclude that the complex crystallized from a single basaltic magma (Jackson, 1961), but recent theories diverge from that point. Recent evidence shows that new magma may have been injected with each ultramafic cycle in the banded series. Simple crystal settling is responsible for formation of the ultramafic series, as envisioned by Jackson, but the mechanisms for deposition of the crystals in the banded series cannot be so simple. The banded series has a chaotic sequence, and the regularity found in the ultramafic series is not present. The density of plagioclase is very close to that of the parent magma, so that it is unclear whether the crystals floated or sank. Other problems remain to be answered, such as the formation of thick, nearly monomineralic anorthosite layers. The platinum-group element–rich units may have been produced by injection of new magma that was responsible for deposition of olivine cumulates associated with platinum-group element occurrences in the banded series. For further details, the reader must consult Czmanske and Zientek (1985), which includes an extensive list of references for the complex.

REFERENCES

Czamanske, G. K., and Zientek, M. L., 1985, The Stillwater Complex, Montana; Geology and guide: Montana Bureau of Mines and Geology Special Publication 92, 396 p.

Jackson, E. D., 1961, Primary textures and mineral associations in the ultramafic zone of the Stillwater Complex, Montana: U.S. Geological Survey Profes-sional Paper 358, 106 p. (Reprinted 1984, Montana Bureau of Mines and Geology Reprint 4)

Lambert, D. D., Unruh, D. M., and Simmons, E. C., 1985, Isotopic investigations of the Stillwater Complex; A review, *in* Czamanske, G. K., and Zientek, M. L., The Sillwater Complex, Montana: Geology and guide: Montana Bureau of Mines and Geology Special Publication 92, p. 46–54.

Lipin, B. C., and 6 others, 1985, Guide to the Benbow area, *in* Czamanske, G. K., and Zientek, M. L., The Sillwater Complex, Montana; Geology and guide: Montana Bureau of Mines and Geology Special Publication 92, p. 125–146.

Page, N. J., and Zientek, M. L., 1985, Geologic and structural setting of the Stillwater Complex, *in* Czamanske, G. K., and Zientek, M. L., The Sillwater Complex, Montana; Geology and guide: Montana Bureau of Mines and Geology Special Publication 92, p. 1–8.

Page, N. J., and 7 others, 1985a, Exploration and mining history of the Stillwater Complex and adjacent rocks, *in* Czamanske, G. K., and Zientek, M. L., The Sillwater Complex, Montana; Geology and guide: Montana Bureau of Mines and Geology Special Publication 92, p. 77–92.

Page, N. J., and 9 others, 1985b, Geology of the Stillwater Complex exposed in the Mountain View area and on the west side of the Stillwater Canyon, *in* Czamanske, G. K., and Zientek, M. L., The Sillwater Complex, Montana; Geology and guide: Montana Bureau of Mines and Geology Special Publication 92, p. 147–209.

Todd, S. G., and 5 others, 1982, The J-M platinum-palladium reef of the Stillwater Complex, Montana; I. Stratigraphy and petrology: Economic Geology, v. 77, p. 1454–1480.

Wager, L. R., Brown, G. M., and Wadsworth, W. J., 1960, Types of igneous cumulates: Journal of Petrology, v. 1, p. 73–85.

Zientek, M. L., Czamanske, G. K., and Irvine, T. N., 1985, Stratigraphy and nomenclature of the Stillwater Complex, *in* Czamanske, G. K., and Zientek, M. L., The Stillwater Complex, Montana; Geology and guide: Montana Bureau of Mines and Geology Special Publication 92, p. 21–32.

A study in contrasts: Archean and Quaternary geology of the Beartooth Highway, Montana and Wyoming

Paul A. Mueller, *Department of Geology, University of Florida, Gainsville, Florida 32611*
William W. Locke, *Department of Earth Sciences, Montana State University, Bozeman, Montana 59717*
Joseph L. Wooden, *U.S. Geological Survey, 345 Middlefield Road, Menlo Park, California 94025*

LOCATION AND ACCESSIBILITY

The Beartooth Highway (U.S. 212) lies southwest of Billings, Montana (Fig. 1). The section of highway discussed here extends from Red Lodge, Montana, to Cooke City, Montana. The road is paved and well maintained, but it may be temporarily closed by snow as early as Labor Day. It closes for the season by middle to late September, and remains closed until mid-May. The stops discussed here (Fig. 1) are pullouts (4 to 6 cars) on the highway and are accessible by two-wheel drive cars. Other than pullouts the road is narrow, with little possibility of overflow parking. Land ownership is largely National Forest (Gallatin and Custer in Montana, Shoshone in Wyoming) with the exception of private holdings along Rock Creek below the switchbacks.

SIGNIFICANCE

Nowhere in the U.S. are the oldest *and* the most recent aspects of geology as spectacularly displayed as along the Beartooth Highway. The Beartooth Mountains are a block of largely Archean bedrock uplifted along high-angle reverse faults of Laramide age (Foose and others, 1961). The Archean rocks contain one of the best records of the early history of the igneous and metamorphic basement of the middle Rocky Mountains. The oldest rocks are the approximately 3,400-Ma granulite facies supracrustal rocks of the Quad Creek and Hellroaring Plateau areas that have been proposed as the products of continental collision (Henry and others, 1982). These were followed by an approximately 2,750-Ma group of calc-alkaline volcanic and plutonic rocks generated along an Archean continental margin (Mueller and others, 1985). The latest igneous rocks are a series of mafic dikes that range from approximately 2,550 Ma to 700 Ma (Mueller and Rodgers, 1973) and are exposed in many outcrops along the highway. This approximately 2,700-m.y. period of geologic time encompasses at least two extensive periods of crustal evolution and records more Archean geologic history at one location than any other place in the Wyoming Province.

The spectacular exposures of Archean rocks in the Beartooth Mountains are the result of substantial uplift of the range during Laramide time and subsequent Pleistocene glacial activity. The Beartooth Plateau, with an average elevation of about 10,000 ft (3,000 m), has been extensively sculpted by glacial processes during the Pleistocene. The highway crosses a classic locality of "biscuit-board topography"—plateau remnants partially dissected by cirques—as well as deposits left by the glaciers that etched the plateau. Features to be seen include glacio-fluvial

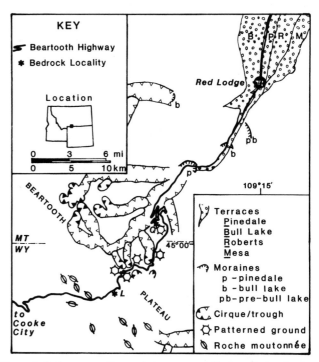

Figure 1. Index map to Precambrian outcrops (L, Long Lake; Q, Quad Creek) and geomorphic features along the Beartooth Highway southwest of Red Lodge, Montana.

terraces showing downstream effects of the glacial system and extensive areas of periglacial features that postdate glaciation.

PRECAMBRIAN GEOLOGY

Though Precambrian rocks are exposed almost continuously from Red Lodge to the Clarks Fork River east of Cooke City, the fundamental relations between major rock types can readily be seen at two outcrops (Fig. 1). The following sections briefly describe what can be seen at these two exposures. A more detailed description of the geology along the highway with an accompanying road log is available from the senior author (Mueller and Wooden, 1979).

Quad Creek area. Where the highway crosses Quad Creek at its highest point (immediately below the Forest Service Lookout) is an almost vertical mélange of sheared and boudinaged Archean metasupracrustal rocks (Fig. 2). Rock types present include mafic amphibolites, quartzite, tonalitic gneiss, granitic gneiss, ironstone, and a variety of metamorphosed, impure

QUAD CREEK SECTION

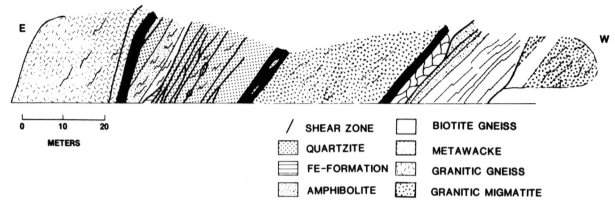

Figure 2. Schematic section of middle Archean rocks exposed along the switchbacks next to Quad Creek southwest of Red Lodge, Montana. Section depicted would be viewed from the canyon side of the highway with Quad Creek at the eastern (left) end of the outcrop.

quartz-rich and pelitic rocks. These are the oldest supracrustal rocks in the Wyoming Province. The presence of ironstone and quartzite in the section suggests deposition of the protoliths in a continental shelf environment. Such a shelf must have been part of a well developed continent similar to those of today.

This section is described in detail by Henry and others (1982) who demonstrate that these rocks have been subjected to granulite facies conditions of approximately 6 Kb and 800°C. These assemblages are best preserved in the ironstones (e.g., quartz, magnetite, ferrohypersthene, ferrosalite, almandine, and ferropargasite). Though not certain, it appears that this metamorphic event occurred at approximately 3,400 Ma. The difficulty of ascertaining the age of the granulite facies event is in large part a result of a later (approximately 2,800 Ma) upper amphibolite facies metamorphism that obliterated much of the petrologic and isotopic record of the earlier granulite facies event.

Long Lake area. Along U.S. 212, immediately southwest of the snow gate just above Long Lake, is a large outcrop of Long Lake Granite, which contains inclusions of a widespread amphibolite unit (andesitic amphibolite) and a granodiorite (Long Lake Granodiorite) that are found throughout this part of the range (Wooden and others, 1982; Warner and others, 1982). A late mafic dike (approximately 2.5 to 2.7 Ga) that has been quarried for road metal is also exposed here. The importance of these three main phases is that they represent the last major episode of crustal growth in this area and were produced over a time span of only about 50 m.y. This short time span, in conjunction with trace-element and isotopic data, suggest that these rocks were produced along an Archean subduction zone. If this interpretation is correct, these rocks represent the Archean equivalent of a modern magmatic arc, and the Beartooth Mountains must lie along a paleo-continental margin of Archean age.

The oldest unit is the andesitic amphibolite (AA). This am-

phibolite is composed of plagioclase, hornblende, biotite, and quartz, and has an overall andesitic composition. It varies from coarse to fine grained and occurs as inclusions in the Long Lake Granite and Long Lake Granodiorite. Though small here, larger (kilometer-sized) blocks are present in the region. This unit has a distinctive geochemistry highlighted by very high incompatible element contents (e.g., REE, Ba, Sr, etc.) as discussed in Mueller and others (1983). Field relations (intrusive contacts) here and elsewhere suggest it is the oldest of the three major units (approximately 2,790 Ma, U-Pb zircon).

The Long Lake Granodiorite (LLGd) is extremely difficult to recognize consistently in the field. The best exposures here are among the small outcrops on the north side of the highway directly across from the barrow pit. In general, this unit appears to have been a late synkinematic intrusion that varies from unfoliated to moderately foliated (as in this outcrop). Less foliated varieties are difficult to distinguish from the much more abundant Long Lake Granite series. The only consistent mineralogic difference is the higher abundance of sphene in the granodiorite. Geochemically, the unit is also enriched in incompatible elements, though the rock itself does not appear to be related directly to the andesitic amphibolite. Chronologically, U-Pb zircon determinations suggest a crystallization age of 2,780 Ma.

The Long Lake Granite (LLG) varies in composition from true granite to tonalite, with the major felsic minerals being quartz, plagioclase, and K-feldspar; biotite is the only mafic mineral. This unit is the most voluminous rock in the eastern Beartooth Mountains and is the principal unit exposed at this outcrop. Because bulk compositions of the LLG and LLGd overlap, the LLG is most easily distinguished from the Long Lake Granodiorite because it is generally less foliated and possesses a much lower incompatible element content despite its higher average SiO_2 content. Both units contain varying amounts of coarse-grained allanite, which is clearly visible in this area. U-Pb zircon analyses yield an age of 2,748 ± 41 Ma, which is indistinguishable from the U-Pb age of the granodiorite and andesite. The generally less foliated nature of the granite and its intrusive relationship to the granodiorite suggest it is a postkinematic intrusion.

In summary, the Long Lake outcrop contains evidence for an early volcanic phase (AA) that was subjected to amphibolite facies metamorphism at about 2,800 Ma and then injected by an incompatible element enriched granodiorite (LLGd). These two rocks were then intruded by the Long Lake Granite. This sequence of volcanic and metamorphic events represents the last major crust-forming event in this region and the second major one in the Beartooth Mountains. The presence of rocks representing both of these crust-forming cycles is unique to the Beartooth Mountains. Recent summaries of the Archean geology may be found in Mueller and Wooden (1982) and Mueller and others (1985). A detailed map of rocks from the general area of Long Lake is included in Warner and others (1982).

QUATERNARY GEOLOGY

The spectacular outcrops of Archean rocks along the highway are in large part outstanding examples of glacial features as well. The features of each aspect of glacial geology are best displayed along different segments of the highway (Fig. 1). They will be described from east (Red Lodge courthouse) to west (Cooke City).

Red Lodge. From northeast of Red Lodge to the mouth of Rock Creek Canyon, the most prominent features are the stream terraces (Ritter, 1967). U.S. 212 lies on the Pinedale terrace, which can be traced 9 mi (14.5 km) southwest to the Pinedale moraines. Above is the Bull Lake terrace, which can likewise be traced to a moraine. Well above the road are the Roberts and Mesa terraces, which may be glaciofluvial in origin but cannot be traced directly to a corresponding moraine.

Canyon. The Canyon section, from 5 to 10 mi (8 to 16 km) southwest of Red Lodge, includes the Bull Lake (5 mi; 8 km), Pinedale (8 mi; 13 km), and Pinedale recessional (10 mi; 16 km) moraines (Graf, 1971). They are not as well displayed here as at the mouths of some valleys where the glaciers reached the plains, but the pinchouts of the terraces into the moraines are striking. Beyond this section only a Holocene terrace remains.

Hairpins. The Hairpins section, from 15 to 25 mi (24 to 40 km) southwest of Red Lodge, offers some of the best displays of linear-erosional alpine glacial topography in the United States with hanging tributaries (Hellroaring Creek) feeding the U-shaped trough of Rock Creek. Best viewing is from the pullouts low in the switchbacks.

Plateau. In the Plateau section, from 25 to 32 mi (40 to 51 km) southwest of Red Lodge, views abound of the Rock Creek Valley, Beartooth Plateau, and Clarks Fork Valley. The Beartooth Plateau has been dissected into classic "biscuit board topography." In the foreground in all directions is the periglacial terrain of the plateau. Blockfield (felsenmeer) is common and gelifluction, sorted circles, sorted stripes, block streams, and nivation hollows are also present. These features are best viewed at the pullouts immediately west of the switchbacks and at Beartooth Summit. Permafrost is present in this region to elevations as low as 9,680 ft (2,950 m; Pierce, 1961).

Clarks Fork. The Clarks Fork section, from 32 to 57 mi (51 to 92 km) southwest of Red Lodge (Cooke City) lies within the upper portion of the basin of the Clarks Fork of the Yellowstone River. This valley has been glacially modified by areal scour rather than linear erosion. The scour probably largely represents the removal of Paleozoic sedimentary rocks, which outcrop along the south wall of the valley, from the Archean basement. An earlier U-shaped valley has been replaced by an asymmetrical broad valley transitional to the plateau surface on the northwest. Instead of cirques and hanging valleys, roches moutonnées (Fig. 1) and lake basins were scoured, apparently randomly, into the terrain. These features are best displayed on the Chain Lakes Plateau (Forest Service sign for Chain Lakes), shortly after leaving the upper plateau surface.

REFERENCES CITED

Foose, R. M., Wise, D. U., and Garbarini, G. S., 1961, Structural geology of the Beartooth Mountains, Montana and Wyoming: Geological Society of America Bulletin, v. 72, p. 1143–1172.

Graf, W. L., 1971, Quantitative analysis of Pinedale landforms, Beartooth Mountains, Montana and Wyoming: Arctic and Alpine Research, v. 3, p. 253–261.

Henry, D. J., Mueller, P. A., Wooden, J. L., Warner, J. L., and Lee-Berman, R., 1982, Granulite grade supracrustal assemblages of the Quad Creek area, eastern Beartooth Mountains, Montana: Montana Bureau of Mines and Geology Special Publication 84, p. 147–156.

Mueller, P. A., and Rogers, J.J.W., 1973, Secular chemical variations in a series of Precambrian mafic rocks, Beartooth Mountains, Montana and Wyoming: Geological Society of America Bulletin, v. 84, p. 3645–3652.

Mueller, P. A., and Wooden, J. L., eds., 1979, Guide to the Precambrian rocks of the Beartooth Mountains: 1979 Field Conference of the Archean, p. 86.

—— , eds., 1982, Precambrian geology of the Beartooth Mountains, Montana and Wyoming: Montana Bureau of Mines and Geology Special Publication 84, p. 167.

Mueller, P. A., Wooden, J. L., Schulz, K., and Bowes, D. R., 1983, Incompatible-element-rich andesitic amphibolites from the Archean of Montana and Wyoming; Evidence for mantle metasomatism: Geology, v. 11, p. 203–206.

Mueller, P. A., Wooden, J. L., Henry, D. J., and Bowes, D. R., 1985, Archean crustal evolution of the eastern Beartooth Mountains, Montana and Wyoming: Montana Bureau of Mines and Geology Special Publication 92, p. 9–20.

Pierce, W. G., 1961, Permafrost and thaw depressions in a peat deposit in the Beartooth Mountains, northwestern Wyoming: U.S. Geological Survey Professional Paper 424-B, p. 154–156.

Ritter, D. F., 1967, Terrace development along the front of the Beartooth Mountains, southern Montana: Geological Society of America Bulletin, v. 78, p. 467–484.

Warner, J. L., Lee-Berman, R., and Simonds, C. H., 1982, Field and petrologic relations of some Archean rocks near Long Lake, eastern Beartooth Mountains, Montana and Wyoming: Montana Bureau of Mines and Geology Special Publication 84, p. 57–68.

Wooden, J. L., Mueller, P. A., Hunt, D. K., and Bowes, D. R., 1982, Geochemistry and Rb-Sr geochronology of Archean rocks from the interior of the southeastern Beartooth Mountains, Montana and Wyoming: Montana Bureau of Mines and Geology Special Publication 84, p. 45–55.

18

Depositional surfaces in the Eagle Sandstone at Billings, Montana

William B. Hansen, Division of Mineral Resources, U.S. Bureau of Land Management, P.O. Box 36800, Billings, Montana 59107
Kevin R. Kendrick, Department of Geology, Rocky Mountain College, Billings, Montana 59102

LOCATION AND ACCESSIBILITY

The lower sandstone unit of the Upper Cretaceous Eagle Sandstone is extensively exposed in a 300-ft (100-m) high cliff which is dangerously sheer in places and rises above the north edge of Billings, Montana (Fig. 1). An excellent view of the cliffs from a distance can be seen from Rocky Mountain College. The safest and most accessible place to examine the outcrop is in Swords Park, a city park open to the public and accessible by two-wheel drive vehicles.

Billings is accessible from I-90, which skirts the south edge of the city. Take exit 450 (27th St. and Montana 3) northwest 3.5 mi (5.6 km) through the city. If entering Billings by air, exit the airport heading southeast toward the city on Montana 3 to 27th Street (Fig. 1). Directions from 27th Street to the stops are detailed in the site description.

SIGNIFICANCE

Marine shelf sandstones and barrier bars are commonly formed by lateral accretion of sand bodies with time lines inclined to the formation boundaries. Shelton (1965) described low-angle inclined bedding in the lowermost sandstone unit of the Eagle Sandstone at Billings, Montana. He recognized these beds as shoreface accretion surfaces of a barrier bar and likened them to those found on present-day Galveston Island, Texas. The Billings location provides an excellent opportunity to examine such surfaces in the field, as well as an opportunity to examine an important hydrocarbon reservoir rock of the Northern Rocky Mountain region.

SITE DESCRIPTION

General "Site" Characteristics. The Eagle Sandstone (Campanian) was deposited in the Western Interior Cretaceous epicontinental seaway (Shelton, 1965; Rice and Shurr, 1983). It is underlain by interstratified thinly bedded sandstone, siltstone, and shale of the Telegraph Creek Formation and overlain by the Claggett Shale. Underlying the Telegraph Creek Formation are Colorado Group shales. The underlying Colorado Group with the overlying Claggett shales provide good source beds and traps for the Eagle Sandstone reservoirs. The Eagle Sandstone is commonly divided into three informal units in this region: a lower sandstone, a middle shale, and an upper sandstone (Hancock, 1919). However, at Billings, the middle member contains mostly sandstone. This field guide concentrates on the lower sandstone unit and a portion of the middle unit.

The lower sandstone member is a coarsening-upward, fine-grained, chert-bearing lithic sandstone that contains minor glau-

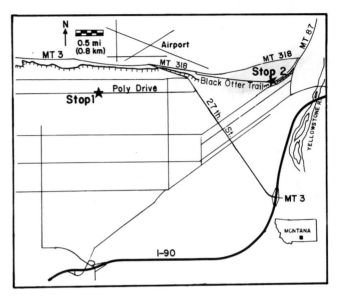

Figure 1. Location map, city of Billings, Montana.

conite, muscovite, and some carbonaceous material. No fossils are reported in the sandstone, but the shale below contains *Inoceramus* shells (Shelton, 1965).

Surface and subsurface information shows the lowermost member of the Eagle Sandstone to be a sand body that is approximately 40 mi (64 km) long and 20 to 30 mi (32–48 km) wide (Shelton, 1965). This elongate body trends N20W, is surrounded on all sides by shales, and is thickest at Billings (Shelton, 1965).

Shelton interpreted the lower sandstone unit near Billings as a barrier complex that grew to the west by beach and shoreface accretion and occurred some 50 mi (80 km) east of the mainland. This was thought to be in response to sand-laden currents and waves on the west (landward) side of the bar. This interpretation is based on fossils in the surrounding shales, sand-body trend, westward dip of the accretion surfaces, repetition of vertical sequence laterally, burrowing, and the large width-to-thickness ratio of the sand body. Recent interpretations of the Eagle Sandstone in central Montana by Rice and Shurr (1983) leave open the possibility that a sand body such as this was never above sea level. Although depth of water during deposition is debatable, interpretation of beach and shoreface accretion for the lowermost member at Billings remains unchallenged in the literature.

A detailed description of the middle member, which can also be seen at this site, has not been published. The description of this member presented here is based on the authors' observations.

One of the two stops provides a panoramic view of the

Figure 2. Photograph of accretion surfaces near Rocky Mountain College.

lower sandstone unit and a portion of the overlying middle
member. The other stop provides an opportunity to study the
outcrop.

Stop 1. To visit Stop 1 (the panorama), follow 27th Street
to Poly Drive. Turn west on Poly Drive and travel 1.7 mi
(2.7 km) to Rocky Mountain College. Park on the side of Poly
Drive at the entrance to the college between 13th and 17th
Streets West (SW¼NE¼NE¼,Sec.36,T.1N.,R.25/E., Billings
West U.S. Geological Survey 7½-minute Topographic Quadrangle). From this vantage point, careful examination of the cliffs
reveals large-scale depositional surfaces that dip at low angle to
the west (Fig. 2), are truncated at the top, and are tangential to
horizontal bedding at the bottom (Shelton, 1965). The darker
sandstone unit truncating the accretion surfaces is a remnant of
the middle member.

These structures led Shelton (1965) to conclude that the
lower sandstone unit was formed by westward (landward) accretion onto a barrier island analogous to the Galveston Island complex in Texas. Today there is some uncertainty regarding the
water depth at which this sandstone was formed (Shelton, 1965;
Rice and Shurr, 1983), but the process of westward accretion
remains unchallenged. Detailed examination of the accretion surfaces in the lower member, the middle member, and other sedimentary features of the Eagle Sandstone is possible at Swords
Park (Stop 2).

Stop 2. To get to Swords Park, return to 27th Street and
turn northwest toward the airport and the rimrocks formed by the
Eagle Sandstone. Turn east on Montana 318 at the airport intersection, drive 0.8 mi (1.3 km) to the entrance of Swords Park,
and turn south on Black Otter Trail. Proceed 1.1 mi (1.76 km)
and park at the gravel turnout on the south side of the road near
the electrical substation (NW¼NW¼NE¼,Sec.33, T.1N.,R.26E.,
Billings East U.S. Geological Survey 7½-minute Topographic

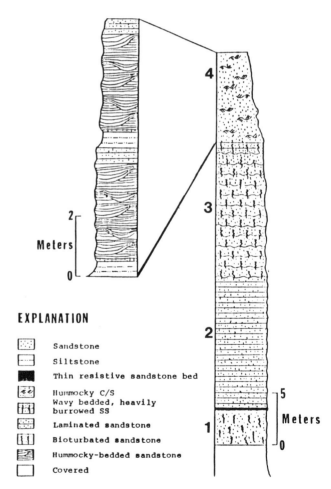

EXPLANATION

Sandstone	
Siltstone	
Thin resistive sandstone bed	
Hummocky C/S	
Wavy bedded, heavily burrowed SS	
Laminated sandstone	
Bioturbated sandstone	
Hummocky-bedded sandstone	
Covered	

Figure 3. Measured section for Black Otter Trail (Stop 2). Inset expands
hummocky cross-stratified unit for more detail. Numbers refer to units
discussed in text.

Figure 4. Photograph of measured section at Stop 2 showing burrowed (1) and laminated (2) units containing dipping accretion surfaces of lower member and burrowed (3) and cross-stratified (4) units of middle member of Eagle Sandstone. Units 1-4 shown in Fig. 3.

Quadrangle). Proceed on foot southeastward into the slight draw. Take care, since the dropoff is sheer.

The Eagle section can be divided into four parts at this location (Figs. 3, 4)—two units (1, 2) in the lower member and two (3, 4) in the middle member. The lower member contains a bioturbated fine-grained sandstone below and a well-sorted, fine-grained, light-tan sandstone unit above containing the laminated accretionary surfaces. Above this is the greenish-brown, medium-sorted, medium-grained bioturbated sandstone of the middle member. This unit grades upward into a cross-stratified unit of similar composition. The turnout for Stop 2 is at the top of the laminated unit, the contact between the lower and middle members of the Eagle Sandstone. The accretion surfaces dip 1 to 6 degrees S60W (Shelton, 1965). One of these surfaces can be traced for 1.75 mi (2.8 km) along the cliff to a point where it grades into a highly burrowed zone (Fig. 3).

Hike out of the draw and proceed east to the outcrop behind the high voltage power pole. Here one can see the sharp contact between the lower member—laminated sand with accretionary

surfaces—and the burrowed sand of the middle member above. Further research is needed to document the nature of this contact. *Ophiomorpha* burrows are common in the bioturbated zone but decrease upward into a cross-stratified unit. The cross bedding is interpreted locally to be hummocky cross-stratification, which indicates subaqueous storm-dominated deposition using criteria established by Dott and Bourgeois (1982).

The field relations in the rimrocks at Billings support the interpretation that the lower sandstone unit of the Eagle Sandstone was deposited by westward beach and shoreface accretion in response to sand-laden currents in the Campanian epicontinental seaway. The depth of water is less certain. These deposits may have been formed as sand accreted onto an exposed barrier complex whose subaerial deposits were removed by erosion (Shelton, 1965) or on unexposed sand ridges in deeper water on the floor of the seaway. It is possible that the two members at this site represent two separate environments of deposition. More field research, especially on the middle member, is needed to resolve this problem.

REFERENCES CITED

Dott, R. H., Jr., and Bourgeois, J., 1982, Hummocky stratification; Significance of its variable bedding sequences: Geological Society of America Bulletin, v. 93, p. 663–680.

Hancock, E. T., 1919, Geology and oil and gas prospects of the Lake Basin field, Montana: U.S. Geological Survey Bulletin 691-D, p. 101–147.

Rice, D. D., and Shurr, G. W., 1983, Patterns of sedimentation and paleogeography across the Western Interior seaway during the time of deposition of Upper Cretaceous Eagle Sandstone and equivalent rocks, Northern Great Plains, *in* Reynolds, M. W., and Dolly, E. D., Mesozoic paleogeography of west-central United States: Society of Economic Paleontologists and Mineralogists Rocky Mountain Paleogeography Symposium, v. 2, p. 337–358.

Shelton, J. W., 1965, Trend and genesis of the lowermost sandstone unit at Billings, Montana: American Association of Petroleum Geologists Bulletin, v. 49, p. 1385–1397.

Stratigraphy and sedimentology of the Paleocene Fort Union Formation along the Powder River, Montana and Wyoming

Romeo M. Flores, U.S. Geological Survey, Box 25046, MS 972, Denver Federal Center, Denver, Colorado 80225

LOCATION AND ACCESSIBILITY

Extensive areas of the Powder River Basin in southeast Montana and northeast Wyoming are underlain by the Paleocene Fort Union Formation (Fig. 1). The sites selected for the field guide (Fig. 1) are easily accessible via the Powder River road from U.S. 212 (Broadus, Montana) to U.S. 14-16 (northeast of Arvada, Wyoming). The Powder River road is a county road that has a hard gravel surface easily traversed by cars. The outcrops are located either immediately along the road or only 0.25 mi (0.4 km) from the road. The lands along the Powder River are either managed by the Bureau of Land Management or privately owned by ranchers, who reserve the right to grant access to these lands. However, numerous field trips sponsored by the American Association of Petroleum Geologists, U.S. Geological Survey, North Dakota Geological Survey, and Casper College have been permitted access to the study sites. Permission was provided with the agreement that field-trip participants respect the land by not littering, not leaving gates open to stock, and carefully observing fire prevention rules. The rancher to contact for sites 1 and 2 is William Gaye, and the rancher for site 3 is Richard Reese. Mr Gaye maintains a ranch house on the Powder River road; Mr. Reese has a residence in the vicinity of Leiter, Wyoming (Fig. 1).

SIGNIFICANCE

The Fort Union Formation in the Powder River Basin in Montana and Wyoming and adjoining areas (Fig. 1) contains a large number of very thick, closely spaced coal beds that make up some of the largest coal reserves in the United States. The study of the stratigraphy and sedimentology of an unnamed lower member and the overlying Tongue River Member of the Fort Union Formation in well-exposed outcrops along the Powder River and its tributaries provides an opportunity to understand the depositional environments of this coal resource. In addition, field localities along the Powder River permit a close examination of the fluvial facies of Tertiary rocks of Montana and Wyoming.

STRATIGRAPHIC-SEDIMENTOLOGIC SETTINGS OF SITES

The main objective of the field guide to the Powder River area is to provide a knowledge of the stratigraphy and depositional environments of part of the Fort Union Formation (Flores, 1981, 1983). In order to accomplish this goal, three sites (Fig. 1) that represent typical sections for analysis of the Fort Union Formation were selected for description.

The Fort Union Formation exposed along the Powder River

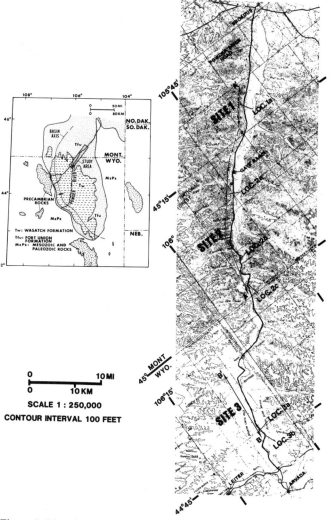

Figure 1. Map showing location of three sites (1, 2, and 3) along the Powder River. Locality numbers (loc. 1a to loc. 3b) refer to those in this paper that show facies successions in stratigraphic sections indicated in cross sections A–A' and B–B'. Inset map is a generalized geologic map of the Powder River Basin and location of study area in Montana and Wyoming.

is as much as 1,900 ft (580 m) thick. A composite stratigraphic section is shown in Figure 2. This formation includes a 400-ft-thick (120 m) unnamed lower member and the 1,500-ft-thick (450 m) Tongue River Member. The strata dip 1–3 degrees to the west with progressively younger rocks exposed along the Powder River from Broadus, Montana, to Arvada, Wyoming. The members are generally similar physically and consist of channel sandstones interbedded with floodplain sandstones, siltstones, and

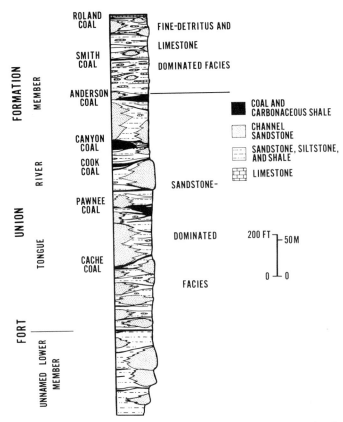

Figure 2. Composite stratigraphic section of Fort Union Formation in the Powder River area.

mudstones and backswamp coals and carbonaceous shales. Coal beds and roof rocks that were burned are present as red clinkers. Several miles southwest of Broadus, Montana, the lower unnamed member forms a generally treeless lowland of a partly forested plateau formed by the Tongue River Member. Here, the gray lower member grades upward into the yellowish Tongue River Member. The Fort Union Formation along the Powder River can be divided into two facies: a sandstone-dominated facies and a fine-detritus- and limestone-dominated facies. The sandstone-dominated facies, which includes the lower 1,100 ft (335 m) of the Tongue River Member (sites 1 and 2) and the upper 200 ft (60 m) of the lower unnamed member, consists of abundant channel sandstones and thick coal beds. The fine-detritus- and limestone-dominated facies, which characterizes the upper 400 ft (120 m) of the Tongue River Member (site 3), consists mainly of floodplain sediments and thin coal beds. This facies contains abundant freshwater mollusk fossils.

The channel sandstones are fine to medium grained, as much as 200 ft (60 m) thick, and have large-scale (as much as 8 ft—2.4 m—thick) festoon or trough crossbeds in the lower part and small-scale (as much as 1 ft—0.3 mi—thick) festoon and planar crossbeds and ripple cross laminations in the upper part. Channel sandstones consist of inclined sandstone bodies that are

separated by clay-silt drapes characteristic of point-bar accretion deposits of modern meandering rivers. The sandstones grade vertically upward and laterally into levee and channel clay-plug deposits consisting of interbedded mudstones, siltstones, and sandstones. The levee deposits show abundant root marks and minor ripple cross laminations that represent overbank deposition by river floods. Where the abandoned channel extends into the floodplain and received only clays during floods, it is identified as a clay-plug deposit. These deposits grade laterally into floodplain deposits consisting of crevasse-splay sandstones, siltstones, and mudstones. The floodplain deposits generally coarsen upward. This floodplain sequence is locally fossiliferous and interbedded with fossiliferous limestones; it was probably deposited by splays or lacustrine deltas debouching into lakes at the distal part of the floodplain. Lakes appear to have originated from local subsidence of the floodplain due to differential compaction of sediments.

The channel and floodplain lakes were bounded areally by backswamps within which coals and carbonaceous shales formed. The coals are subbituminous. The carbonaceous shales (>33 percent ash) are a mixture of mud and organic matter and are commonly interbedded with and grade laterally into coals. The coals (<33 percent ash) represent accumulations of organic matter that altered to vitrinite, inertinite, and exinite. Coal beds typically contain thin partings of mudstone and siltstone as well as siliceous, sulfide, carbonate, and other accessory minerals as dispersed grains, veinlets, and fracture infills. These minerals form either from detrital sediment deposited during floods or as diagenetic products. Ethridge and others (1981) subdivided the backswamp deposits into poorly drained and well-drained types. They interpreted the thick coal beds as deposits of poorly drained backswamps characterized by stagnant water and reducing conditions and relatively free from detrital influx. These settings developed on the floodplain away from inundation by overbank, crevasse-splay, and lake sediments. Where the backswamps were close to areas of detrital sedimentation derived from the drainage systems, highly oxidizing, well-drained conditions resulted.

SITE DESCRIPTION

Site 1. This site (Fig. 1) includes deposits of the lower part of the Tongue River Member of the Fort Union Formation. Variations in the stratigraphy and depositional environment of this member are shown in Figure 3. Multistory channel sandstone deposits are interbedded with floodplain and backswamp deposits. The channel deposits (Fig. 3) grade laterally into floodplain deposits, which in turn are interbedded with backswamp deposits. The vertical stacking of channel deposits at this site represents repeated occupation and/or superposition of channels of a meander belt, possibly due to subsidence resulting from differential compaction and/or tectonics. Reoccupation and superposition of stream channels promoted the accumulation of the thick floodplain deposits observed in the western part of the meander belt. This indicates that peak fluvial aggradation was accompanied by delivery of high suspended load into the interfluves. These sediment influxes across the floodplain commonly resulted in temporary well-drained conditions in backswamps as indicated by associated thin coal beds. The presence of freshwater mollusk fossils in the floodplain deposits suggests the presence of floodplain lakes, which likely served as catchments for crevasse sediments.

The facies succession of the site 1 stratigraphic interval is best observed at locality 1a. Location of the section is shown in Figure 1, and its stratigraphic position is indicated in Figure 3. The section consists mainly of three channel sandstone deposits that average 60 ft (18 m) thick and are interbedded with thin intervals of floodplain and backswamp deposits. The channel sandstones fine upward and contain festoon crossbeds in the lower part and ripple cross laminations in the upper part. Channel sandstones at the site are multistory bodies; the upper parts of these bodies are draped by fine-grained detritus indicating deposition in a meandering stream. The floodplain deposits include coarsening-upward crevasse mudstones, siltstones, and sandstones. The sandstones, which cap the sequence, are ripple cross laminated (asymmetrical and ripple drift laminations) and burrowed. The mudstones and siltstones in the lower part of the sequence are burrowed and ripple cross laminated and are interbedded with lenticular limestone and ironstone. These deposits are interbedded with coals and carbonaceous shales that are as much as 6 ft (2 m) thick.

Site 2. This site (Figs. 1 and 3) represents deposits of the middle part of the Tongue River Member of the Fort Union Formation. The vertical and lateral variations of this interval, shown in Figure 3, are similar to those at site 1. The stratigraphic interval includes multistory channel sandstones that are as much as 200 ft (60 m) thick and are laterally gradational into thick floodplain deposits. The channel sandstone bodies show multilateral characteristics indicating deposition in meandering streams. Each sandstone body contains festoon crossbeds as much as 8 ft (2.4 m) thick and basal lag conglomerates in the lower part and small-scale festoon and ripple lamination in the upper part. These are indicative of deposition in the channel floor and margin, respectively. The cluster of channel sandstones is as much as 8 mi (13 km) wide, suggesting a broad meander belt. The adjoining floodplain northeast of the meander belt, like that at site 1, received extensive detrital influx as indicated by thick floodplain deposits. However, unlike site 1, this floodplain deposit is associated with the thick Pawnee coal bed, which is as much as 40 ft (12 m) thick. The coal bed is in the upper part of the complex of floodplain deposits and is laterally equivalent to the upper part of the juxtaposed channel sandstones. The accumulation of thick floodplain deposits suggests prolonged infilling of the interfluve by sediments as a result of fluvial aggradation. The appearance of forest vegetation in the backswamp during the waning stage of fluvial aggradation retarded influx of detrital sediments into the interfluve. A slowdown of sedimentation, in turn, promoted extensive colonization of the backswamp by forest vegetation that generated raised or domed peat bog (Flores, 1981).

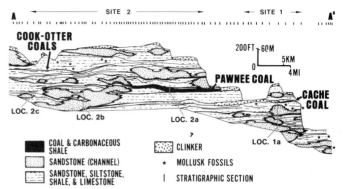

Figure 3. Stratigraphy and depositional environment of the Tongue River Member of the Fort Union Formation of sites and line of cross section A–A′ at sites 1 and 2. Representative facies sequences are best observed at localities 1a, 2a, 2b, and 2c.

The vertical variation at site 2 is represented by a series of stratigraphic sections best observed at localities 2a, 2b, and 2c (see Fig. 1 for location and Fig. 3 for stratigraphic position). Locality 2a represents floodplain and backswamp deposits at the western part of the Pawnee coal swamp-floodplain environment. The Pawnee coal bed is split in this section into a lower bed as much as 21 ft (6.4 m) thick and a upper bed as much as 7 ft (2.1 m) thick. These individual coal beds merge into a 40-ft-thick (12 m) coal bed to the north. Floodplain deposits consist of coarsening-upward and randomly arranged sequences of mudstones, siltstones, and sandstones. The fluvial channel sandstones that are a counterpart of these floodplain and backswamp deposits are shown at locality 2b. The channel sandstones consist of basal lag conglomerates (limestone, ironstone, sandstone, siltstone, and mudstone fragments), a festooned lower part, and a rippled upper part. It, in turn, is overlain by thin vertical accretion deposits and a channel sandstone with characteristics similar to those of the lower sandstone body. This stacking of the channel sandstones that are separated by thin vertical accretion deposits suggests abandonment and later reoccupancy of stream channel sites, a process that occurs by stream avulsion within a meander belt. Abandonment in this meander belt is illustrated at locality 2c by a scour base and a thick sequence of interbedded mudstones, siltstones, sandstones, and carbonaceous shales. The abandoned channel is incised into crevasse-splay and backswamp deposits. The scour is infilled by fine-grained detritus and muddy organic deposits typical of clay plugs. This clay-plug deposit marks the western margin of the meander belt.

A depositional model of the Pawnee coal and adjoining meander belt for site 2 is shown in Figure 4, which depicts the formation of a poorly drained backswamp in the eastern part of the alluvial plain that was stabilized by forest vegetation of broadleaved evergreens interspersed among grasslands consisting of such herbaceous plants as sedges, a few low shrubs, and rare small trees. The grassland may have resulted either from burning of swamp forest or as a natural climax of vegetation. The back-

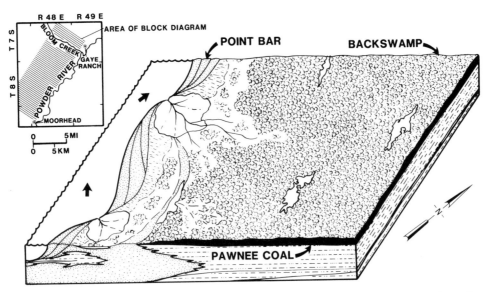

Figure 4. Depositional model of the Pawnee coal and associated deposits at site 2. Inset map shows the location of block diagram in study area.

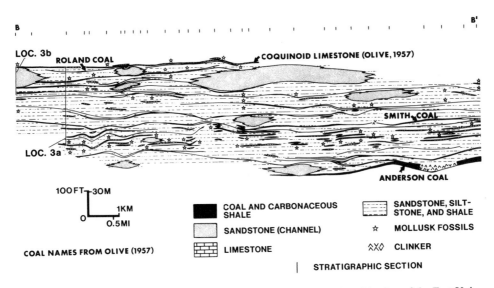

Figure 5. Stratigraphic and environmental framework of the Tongue River Member of the Fort Union Formation at site 3. See Figure 1 for location. Representative facies sequences are best observed at localities 3a and 3b.

swamp, which was formed away from areas of detrital influx, generated domed peat (Flores, 1981). The abandoned channel course at the eastern margin of the meander belt signifies stream avulsion accompanied by gradual incursion of swamp vegetation over parts of the meander belt. The meander belt consisted of a highly sinuous fluvial system that formed laterally shifting channels. Floodplain deposition is reflected by levee and crevasse sedimentation. These processes were most active in the western part of the alluvial plain where well-drained conditions resulted.

Site 3. This site (Fig. 1) contains deposits of the upper part

of the Tongue River Member. The vertical and lateral variations of the stratigraphic interval, which extends from the Anderson coal to the coquinoid limestone, are shown in Figure 5. Unlike the intervals of sites 1 and 2, the Anderson coal–coquinoid limestone interval is characterized by fine-detritus- and limestone-dominated facies. The interval may be divided into floodplain-lake-dominated deposits in the lower part (Anderson coal to Smith coal interval) and fluvial-channel and crevasse-splay-dominated deposits in the upper part (Smith coal to Roland coal interval). The backswamp deposits (excluding the

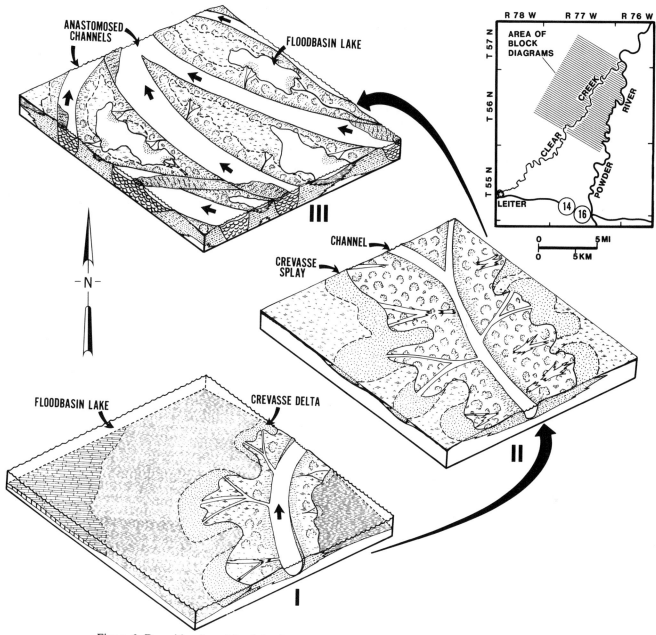

Figure 6. Depositional models of the Anderson coal to Roland coal interval of the Tongue River Member at site 3. Arrows indicate succession of development from I to III. Inset map shows the location of the block diagrams in study area. Modified from Flores and Hanley (1984).

Anderson coal) are generally thin but can be as much as 10 ft (3 m) thick. Lake and lacustrine delta deposits are characterized by common freshwater limestones and mudstones that contain abundant freshwater mollusk assemblages (Flores and Hanley, 1984). The mollusks include *Plesielliptio priscus* (Meek and Hayden), *Viviparus raynoldsanus* (Meek and Hayden), *Lioplacodes* sp., *Clenchiella* n. sp. A (Taylor), *Lioplacodes limneaformis* (Meek and Hayden), and Sphaeriidae (indeterminate). Lacustrine delta deposits consist of steel-gray, rippled, burrowed (e.g.,

pelecypod escape structures) sandstones that are interbedded with fossiliferous mudstones and siltstones. Locally, these deposits show inclined beds, slumps, and other types of syndepositional deformational structures. The crevasse channel and crevasse-splay and fluvial channel deposits contain abundant coarsening-upward mudstones, siltstones, sandstones, and fining-upward sandstones. The crevasse-splay sandstones are tabular rippled (asymmetrical and ripple drift) and burrowed (e.g., broad, tubular, pelecypod escape structure and long, curvilinear, narrow tubes probably formed by worms). A few of these crevasse-splay sequences contain silty limestones that grade laterally into ferruginous layers. These deposits contain mollusks such as those of the lacustrine deposits including *Hydrobia eulimoides* (Meek), *Lioplacodes* sp., and dwarfed species of *Plesielliptio*. The coarsening-upward sequences are locally scoured and infilled by crossbedded crevasse channel sandstones. The crevasse deposits laterally pass into fluvial channel sandstones that are lenticular and are as much as 80 ft (24 m) thick and 2.25 mi (3.36 km) wide. Channel sandstones are multistory, each body consisting of festoon and planar crossbeds with lag deposits of coal spar, sandstone, siltstone, mudstone, carbonaceous shale, ironstone, limestone, petrified tree logs, and mollusk shell fragments. Areal distribution of the fluvial channel sandstones in the upper part of the Smith-Roland coal interval (below the coquinoid limestone in Fig. 5) indicates the presence of contemporaneous bodies that converge and diverge and that were separated by small flood basins typical of anastomosed fluvial systems (Flores and Hanley, 1984). The anastomosed fluvial deposits are, in turn, overlain by a coquinoid lacustrine limestone that contains abundant bivalves, including *Plesielliptio priscus* (Meek and Hayden), *P. silberlingi* (Russell), Sphaeriidae (indeterminate), and the gastropods *Viviparus raynoldsanus* (Meek and Hayden), *Lioplacodes multistriata* (Meek and Hayden), *L. limneaformis* (Meek and Hayden), *L. tenuicarinata* (Meek and Hayden), *Hydrobia eulimoides* (Meek), *Clenchiella planospiralis* (Yen), and *Clenchiella* n. sp. A (Taylor) (J. H. Hanley, 1985, personal communication). Absence of economic coals in the Smith-Roland coal interval probably reflects burial of backswamps by detritus and perhaps drowning by lakes. Slow accumulation of suspended detrital influx promoted abundant high-ash organic deposits such as carbonaceous shales.

Vertical sequences of parts of the Anderson coal to coquinoid limestone interval are best observed at localities 3a and 3b (see Fig. 1 for location and Fig. 3 for stratigraphic position). The lower part of the interval includes the lake and lacustrine delta deposits of fossiliferous limestones, mudstones, siltstones, sandstones, and carbonaceous shales (locality 3a). This sequence is overlain by coarsening-upward crevasse-splay deposits locally dissected by crevasse-channel sandstones. More crevasse sequences follow and are interbedded with thin coal and carbonaceous shale beds. The crevasse channel-splay and backswamp sequence, in turn, is overlain by interchannel deposits of the anastomosed fluvial deposits. The interchannel deposits are well displayed in the upper part (locality 3a) where they include crevasse-splay sequences containing coarsening-upward mudstones, siltstones, and rippled sandstones. The lower part of the crevasse sequence contains abundant freshwater pelecypod and gastropod fossils and burrows at many levels and contains a few fossiliferous lacustrine limestones. When traced laterally, the crevasse deposits pass into a thin to thick, coeval, multistory channel sandstone, such as that observed at locality 3b, located 0.25 mi (0.4 km) to the southwest.

Reconstructions of depositional models for the lower, middle, and upper parts of the Anderson coal and Roland coal interval at site 3 are shown in Figure 6. The lower part, representing the Anderson coal and Smith coal interval, is dominated by a lacustrine environment that accumulated fossiliferous limestones, mudstones, siltstones, and sandstones, which can locally be traced into a fluvial channel sandstone to the east (model I). Sedimentation of flood-basin lakes by crevasse splays is illustrated by model II, representing deposition of the upper part of the Anderson coal and Smith coal interval and the lower part of the Smith coal and Roland coal interval. The crevasse-splay infilled lakes, in turn, were vegetated and formed well-drained backswamps. Crevasse channels were transformed into major conduits as they were stabilized by vegetated levees and reconnected with adjoining fluvial channels. This process may have promoted the anastomosed fluvial pattern (model III) that characterized the upper part of the Smith coal and Roland coal zone. Anastomosing was probably influenced by basin subsidence due to differential compaction and/or tectonism (Flores and Hanley, 1984). Subsidence caused lowering of stream gradient and promoted large floodplain lakes in which the coquinoid limestone accumulated.

REFERENCES

Ethridge, F. G., Jackson, T. J., and Youngberg, A. V., 1981, Floodbasin sequence of a fine-grained meanderbelt subsystem; The coal-bearing Lower Wasatch and Upper Fort Union Formations, southern Powder River Basin, Wyoming, *in* Ethridge, F. G., and Flores, R. M., eds., Recent and ancient nonmarine depositional environments; Models for exploration: Society of Economic Paleontologists and Mineralogists Special Publication 31, p. 191–209.

Flores, R. M., 1981, Coal deposition in fluvial paleoenvironments of the Paleocene Tongue River Member of the Fort Union Formation, Powder River area, Powder River Basin, Wyoming and Montana, *in* Ethridge, F. G., and Flores, R. M., eds., Recent and ancient nonmarine depositional environments; Models for exploration: Society of Economic Paleontologists and Mineralogists Special Publication 31, p. 169–190.

———, 1983, Basin facies analysis of coal-rich Tertiary fluvial deposits, northern Powder River Basin, Montana and Wyoming, *in* Collinson, J. D., and Lewin, J., eds., Modern and ancient fluvial systems; International Association of Sedimentologists Special Publication 6, p. 501–515.

Flores, R. M., and Hanley, J. H., 1984, Anastomosed and associated coal-bearing fluvial deposits; Upper Tongue River Member, Paleocene Fort Union Formation, northern Powder River Basin, Wyoming, U.S.A., *in* Rahmani, R. A., and Flores, R. M., eds., Sedimentology of coal and coal-bearing sequences: International Association of Sedimentologists Special Publication 7, p. 85–103.

Olive, W. W., 1957, The Spotted Horse coalfield, Sheridan and Campbell Counties, Wyoming: U.S. Geological Survey Bulletin 1050, 83 p.

The Hebgen Lake Earthquake Area, Montana and Wyoming

Irving J. Witkind, *U.S. Geological Survey, Box 25046, Denver Federal Center, Denver, Colorado 80225*
Michael C. Stickney, *Montana Bureau of Mines and Geology, Montana College of Mineral Science and Technology, Butte, Montana*
59701

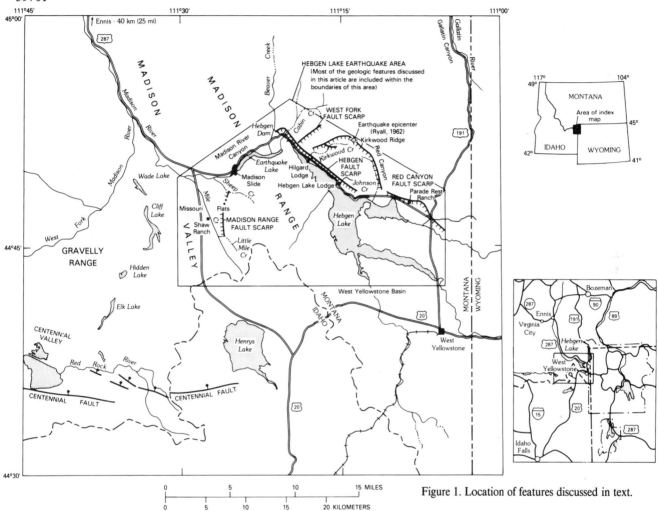

Figure 1. Location of features discussed in text.

LOCATION AND ACCESSIBILITY

The Hebgen Lake Earthquake Area, in southwestern Montana, extends westward from the east edge of Hebgen Lake through the Madison River Canyon, a narrow, deep gorge cut through the Madison Range, to Missouri Flats, west of the range. Hebgen Lake, about 8 mi (13 km) north of West Yellowstone, Montana, lies at the base of the east flank of the Madison Range directly west of the northwest corner of Yellowstone National Park, Wyoming. The lake is used as a storage reservoir impounded behind Hebgen Dam, an earth-filled structure built by Montana Power Company in 1914.

The earthquake area is easily reached from almost any direction by major all-weather surfaced highways (inset map, Fig. 1). The area is accessible from the east through Yellowstone Park's West Entrance, which leads directly into West Yellowstone; from

the south by multi-lane U.S. 20, which also passes through West Yellowstone; from the north via U.S. 191, which threads the narrow, deeply forested and scenic Gallatin Canyon; and from the west via U.S. 287, which passes through Ennis, a ranching center that still retains the flavor of the old west. The West Yellowstone airport also gives access to the area.

The Hebgen Lake area is a favorite tourist and resort area. Lodging is available in West Yellowstone and at several motels along the north shore of Hebgen Lake. Excellent campsites, well maintained by the U.S. Forest Service, are along U.S. 191 north of West Yellowstone, and in Madison River Canyon (Montana 287) west of Hebgen Dam. Most services are available throughout the year in West Yellowstone, although many motels and some restaurants close for the winter soon after Labor Day.

SIGNIFICANCE

Several high-angle normal faults north of Hebgen Lake, that have been recurrently active during much of Neogene time, reactivated in 1959. As a result, a series of impressive fault scarps were formed north of the lake when the structural basin that contained the lake subsided abruptly. These scarps, and the phenomena formed in response to this sudden displacement of the lake, lend a special significance to this sector of Montana.

The faults reactivated at 11:37 P.M. on August 17, 1959. Faulting was accompanied by a major earthquake centered in the Hebgen Lake area of southwestern Montana that shook the northwestern sector of the United States and adjacent parts of Canada. The earthquake was the strongest ever recorded in Montana, and was felt throughout a 600,000 mi^2 (1,555,000 km^2) area. Unusual geologic features were formed—the spectacular fault scarps, a large landslide, and a warped lake basin (and the resultant northeasterly displacement of a lake and damage to a dam)—each of which demonstrates the destructive power of a large earthquake. Each feature is an intriguing geologic site; as all, however, are interrelated, and were formed more or less synchronously during the earthquake, they are treated as a single geologic site—the Hebgen Lake Earthquake Area (Fig. 1).

A team composed of geologists, geophysicists, and hydrologists from various Federal agencies studied the effects of the earthquake. Their results have been published as separate chapters in U.S. Geological Survey Professional Paper 435.

Shortly after the earthquake, the U.S. Forest Service, recognizing the uniqueness of the earthquake area, established an educational and recreational site known as the "Madison River Canyon Earthquake Area." The area extends along the northeastern shore of Hebgen Lake and includes informative displays at many of the geologic features formed during the earthquake, as well as a Visitor Center at the Madison Slide.

Space limitations preclude any discussion in this article of the stratigraphy and structure of the Earthquake Area. In U.S.G.S. Professional Paper 435 the pre-Tertiary rocks have been described by Witkind, Hadley, and Nelson (Chapter R); the volcanic rocks by Hamilton (Chapter S); and the glacial deposits by Richmond (Chapter T).

SEISMOLOGICAL STUDIES

The Hebgen Lake earthquake (surface-wave magnitude 7.5) is the largest historic earthquake within the Intermountain Seismic Belt. Each year hundreds of small earthquakes occur in the Hebgen Lake area. Qamar and Stickney (1983), using cumulative plots to analyze historic earthquakes in the Hebgen Lake area, find that the recurrence time is less than 100 years for an earthquake of magnitude 7 or more. Indeed, outside of California, the Hebgen Lake–Yellowstone Park region is the most seismically active region in the conterminous United States.

Ryall (1962) made a detailed study of the epicenter location of the Hebgen Lake earthquake. He found that the preliminary location was inexact because seismic wave velocities in the region were not well known. After careful study of the seismological data Ryall determined an epicenter location that lies within the zone of surface faulting at lat. 44°52′N, long. 111°14′W (Fig. 1).

Recent work by Doser (1985) conclusively showed that the Hebgen Lake earthquake consisted of two large earthquakes, the larger occurring five seconds after the initial rupture. Fault rupture initiated 6 to 9 mi (10 to 15 km) below the surface. Comparisons between the seismological data and observed surface faulting indicate that fault rupture at depth occurred along one or more fault planes with strikes slightly discordant with the trace of surface faulting, suggesting reactivation of Laramide structures at the surface (Doser, 1985).

Using the method of joint epicenter determination, which locates earthquakes relative to a "master event," Dewey and others (1972) analyzed the epicenter locations of the larger Hebgen Lake aftershocks. Ryall's (1962) main shock epicenter location was used as the "master event." The calculated locations of 22 aftershocks, from which reliable data are available (including all four aftershocks greater than magnitude 6.0), form a west-trending zone 50 mi (80 km) long, extending significantly beyond the zone of surface faulting. The aftershock zone extends from north-central Yellowstone National Park westward through the Madison Valley, to the southern Gravelly Range (Fig. 1). The main shock epicenter lies midway along this aftershock zone, and the largest aftershocks are concentrated near the east and west ends of the zone.

The western end of the zone of aftershocks crosses the Madison Range and the Madison Valley and extends into the southern Gravelly Range. Its trend is difficult to reconcile with the surface geology. West-trending faults with Quaternary displacements are not known to cut the north-trending Madison Range and Madison Valley. In the Centennial Valley area the northern block of the Centennial fault is downdropped, whereas in the Hebgen Lake area the southern blocks of the reactivated faults were downdropped. However, slip at depth along the west-trending, north-dipping Centennial fault (Fig. 1) may explain the seismicity in the southern Gravelly Range—the west end of the aftershock zone.

Focal mechanisms for the Hebgen Lake main shock, determined both by Ryall (1962), and by Dewey and others (1972), indicate slip along a west-trending normal fault (southern block downdropped) as the source of the earthquake. Focal mechanisms of two large aftershocks with magnitudes greater than 6.0 at opposite ends of the aftershock zone, also suggest a similar sense of movement on west-trending normal faults. Other focal mechanisms showing strike-slip and thrust faulting within the aftershock zone indicate complex faulting and suggest a rapidly varying stress field in the region (Doser, 1985).

Recent studies using arrays of high-gain seismographs confirm that a very active 50 mi (80 km) long seismic zone trends N80°W in the Hebgen Lake area (Trimble and Smith, 1975). Accurately determined hypocenters along the zone have focal depths less than 12 mi (20 km). Fault-plane solutions for earth-

Figure 2. Colluvium offset by Red Canyon fault scarp. Scarp, about 8 ft (2.5 m) high, is in Red Canyon, north of Hebgen Lake. The trees on the downthrown block, and adjacent to the scarp, were slightly tilted locally but were not knocked down. Photograph taken several days after the earthquake.

quakes in the Hebgen Lake area indicate continued dip-slip movement along nearly west-trending normal faults.

REACTIVATED FAULTS

At least three major well-known high-angle normal faults were reactivated at the time of the earthquake, and spectacular fault scarps formed along their traces (U.S.G.S. Professional Paper 435, Witkind, Chapter G; Fig. 1). Two of the faults—the Red Canyon fault, and the Hebgen fault—are north of Hebgen Lake; the third fault—the Madison Range fault—delineates the west front of the Madison Range. Scarps also formed along other smaller faults north of Hebgen Lake. In all, about 18 mi (29 km) of new fault scarps were produced, mainly along the faults north of the lake. Ground displacement was large; the basin containing the lake tilted and dropped at least 22 ft (7 m) and was warped into a new shape (U.S. Geological Survey Professional Paper 435, Myers and Hamilton, Chapter I).

Red Canyon Fault Scarp. The Red Canyon fault scarp closely follows the approximate trace of the Red Canyon normal fault, and appears as a light-colored scar about midway up the slopes of the ridges north of Hebgen Lake. In plan view (Fig. 1), the scarp is shaped somewhat like a scythe. Along Kirkwood Ridge the scarp swings through an arc of about 115° to remain concordant with the bedrock units. The scarp extends northwestward for about 14 mi (23 km) from the edge of Yellowstone National Park to end near Kirkwood Creek (Fig. 1). The scarp dips steeply valleyward (southwestward), and the southwest fault block is downthrown. Small displacements at both ends increase to a maximum of about 22 ft (7 m) near the mid-point of the fault scarp. Downslope, minor scarps of limited extent parallel the major scarp. Commonly the main scarp appears as a steep scar that either cuts colluvium (Fig. 2), or locally conforms to the attitude of steeply dipping bedrock units. Seemingly everywhere along the scarp's trace, surficial deposits dropped, and in many places were juxtaposed against bedrock; nowhere along the surface is bedrock clearly offset by the scarp. When first exposed, the scarp was strongly marked by slickensides that tended to conform to the topography. For example, where the scarp crossed a ridge mantled with colluvium, the slickensides were vertical along the crest, but were aligned approximately with the fore and aft slopes. Apparently the colluvial material slumped away from the crest of the ridge and down the adjoining slopes as the scarp developed.

The scarp is best viewed in Red Canyon and also along the south flank of the curving Kirkwood Ridge, both areas reached via a secondary road that extends into Red Canyon from U.S. 287.

Hebgen Fault Scarp. The Hebgen fault scarp parallels the northeast shore of Hebgen Lake, and follows the approximate trace of the Hebgen normal fault (Fig. 1). The scarp extends northwestward for about 6 mi (10 km) from near Red Canyon to Beaver Creek, and parts of the scarp are visible upslope from U.S. 287. The scarp dips valleyward (southwestward), like the Red Canyon fault scarp, and the crustal block southwest of the scarp is downthrown. Where the scarp crosses the crest of Boat Mountain it appeared, shortly after the earthquake, as a near-vertical break marked by vertical slickensides. Slickensides east of the crest were inclined to the east, those west of the crest were inclined to the west. Here, too, as in the case of the Red Canyon fault scarp, the surficial material seemingly was detached and moved, in part, independently of the downthrown block. Small displacements at both ends of the fault scarp increase to a maximum near the middle. For most of its length the scarp is about 10 ft (3 m) high and is marked by downthrown colluvium juxtaposed against bedrock. In places, the main scarp appears as a gaping fissure.

The scarp is best viewed at the Forest Service's developed site near the mouth of Cabin Creek.

Madison Range Fault Scarp. The Madison Range fault, one of the major high-angle normal faults of western Montana, has determined the configuration of the west front of the Madison Range. The fault has probably been active since Miocene time and certainly during all of Quaternary time, since modern fault scarps that dip valleyward (westward) displace surficial deposits of Quaternary age. During the Hebgen Lake earthquake part of the fault reactivated to form a new scarp about 12.5 mi (4 km) long (Fig. 1), that is superimposed upon a previously formed older scarp extending along the base of the range (U.S.G.S. Professional Paper 435, Myers and Hamilton, Chapter I). The new scarp, which also dips westward, extends north from Little Mile Creek to a point about 1 mi (2 km) south of Sheep Creek. Surficial deposits west of the scarp are downthrown. The greatest displacement, about 3 ft (1 m), is near the mid-point of the scarp. The scarp breaks surficial deposits, and in several places is marked by a gap about 2 ft (0.6 m) wide.

HEBGEN LAKE

The Seiche. At the time of the earthquake the fault blocks that contained Hebgen Lake dropped abruptly and tilted northeastward in response to downward movement along the newly reactivated Hebgen and Red Canyon normal faults. As a result, the lake was thrown into a seiche (a periodic oscillation of a lake surface) that lasted for about 11½ hours and had a period of about 17 minutes (U.S.G.S. Professional Paper 435, Myers and Hamilton, Chapter I). The first three or four oscillations were so high that the lake crested Hebgen Dam each time. When the seiche ended, the lake reflected the uneven tilting of the downthrown crustal blocks—the northeast shore of the lake was submerged, even as the southwest shore was exposed. This northeasterly displacement of the lake was a direct result of the uneven subsidence and warpage of the Hebgen Lake and Red Canyon fault blocks.

The shape of Hebgen Lake influenced the effects of the seiche. Most of Hebgen Lake is within the broad northern part of the West Yellowstone Basin, a lowland filled with sand and gravel and encircled by imposing uplands and mountains (Fig. 1). The northwest arm of the lake, however, gradually narrows as it extends into the restricted and deep Madison River Canyon cut through the Madison Range. Hebgen Dam was constructed across the entryway to this canyon. When the seiche began, the waters of the lake were forced into this narrow arm; the lake shores were inundated, and the lake topped Hebgen Dam. As the waters withdrew (in response to the periodic oscillation of the lake) and moved away from the dam, the lake, no longer confined within the canyon, was able to spread outward, and boaters that night report that they felt the seiche only as a slight undulation of the lake. During this phase of the seiche the entire dam was exposed, and the caretaker of the dam, George Hungerford, recalls running out onto the dam, looking down on the upstream side of the dam and seeing no water. During the next oscillation of the lake the water returned and gradually rose onto the dam and then topped it again, rising to a height of about 3 to 4 ft (1 to 1.5 m) above the crest of the dam. The water kept flowing over the dam for 5 to 10 minutes according to Hungerford.

As a result of this repeated soaking, the shores of the lake slumped into the lake, disrupting highways and in one place destroying a motel known as Hilgard Lodge.

Figure 3. Aerial view southwestward (downstream) toward Madison Slide and the newly formed Earthquake Lake. Photograph was taken on August 21, 1959, four days after the earthquake and before work had begun on the construction of a spillway across the slide. At the time of the earthquake some 28 million yd^3 (21 million m^3) of rock broke free from the southeast canyon wall, slid into the Madison River Canyon at a speed of about 100 mi (160 km) per hour, and rode up the opposite canyon wall. Dolomite debris that formed the leading edge of slide appears as a white band along the northeast canyon wall. Main mass of the slide, composed chiefly of gneissic debris, is concealed beneath a layer of soil and fallen trees. Photograph by John R. Stacy, U.S. Geological Survey.

Submerged North Shore. By the next morning (August 18) the seiche had ended and the astounded residents and tourists discovered that the lake had been displaced northeasterly. The north shore of the lake was submerged. Boat docks were either below water or almost so. Boats that had been loosely anchored offshore were now snubbed tightly by near-vertical anchor chains. Motels and homes that had been as much as 200 ft (60 m) away from the lake now found lake waters lapping onto their front porches. Shoreline trees and shrubs were either wholly or partly submerged.

Stranded South Shore. Even as the people along the north shore were astounded by the position of the lake so close to their homes, those along the south shore were dismayed to discover that the lake had moved away from their homes. Homeowners who had once enjoyed a lake-front site, now discovered that long stretches of dry beach separated their property from the edge of the lake. Boat docks were high and dry; boats formerly anchored offshore were now stranded and lay on their sides.

Whereas the north shore of the lake contains both summer residences and motels, the south shore contains only summer

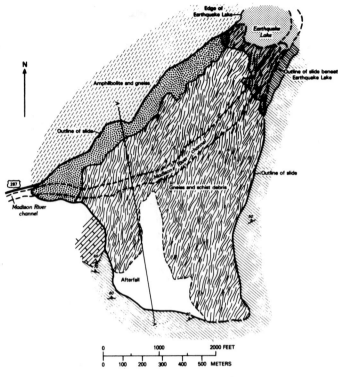

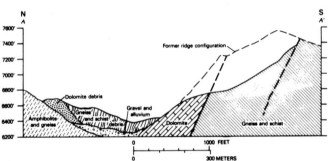

Figure 4. Geologic map and cross section of the Madison Slide showing conditions shortly after the slide was emplaced. After Hadley (U.S.G.S. Professional Paper 435, Chapter K, Figs. 59 and 60).

homes. The impact of the displacement of the lake was great and is perhaps best summed up by the question mostly commonly asked of the geologists who were examining the effects of the earthquake: "How soon will the lake come back to its former position?" After the earthquake, Montana Power Company raised the level of Hebgen Lake, restoring the lake to its pre-earthquake level along the south shore. Almost no hint of the lake's former displacement can be seen along the south shore. The north shore still displays some of the drowned features, chiefly boat docks that seem unusually deep in the water, and motel units that are rather close to the lake shore.

HEBGEN DAM

Hebgen Dam, an earth-filled structure with a concrete core that rests mostly on bedrock, was built across the entrance to the narrow and deep Madison River Canyon (Fig. 1). The dam is some 721 ft (220 m) long and rises 87 ft (27 m) above the river floor; it trends northeast. The southwest abutment of the dam is on bedrock; the northeast abutment, however, is on surficial material. The dam and the ground on which it rests were dropped about 10 ft (3 m) during the earthquake, and as a result, the surficial material uphill from and partly underlying the northeast abutment slumped valleyward during the earthquake, probably wedging the dam even tighter into its position. The drop heavily damaged the dam, and the subsequent repeated submergence of the dam during the seiche intensified the damage. The concrete core was cracked and locally tilted and twisted, the earthfill on both sides of the core slumped, and the earthfill on the downstream side of the dam was gullied. The concrete spillway was badly cracked and fragmented. Despite all this damage the dam held firm and is still in use. The Montana Power Company repaired the dam, and little evidence of the extensive damage done is visible today.

MADISON SLIDE

The Madison River Canyon area is a favorite site for camping and fishing, and at the time of the earthquake all camping sites were occupied. Within half a minute after the earthquake struck, a mass of rock some 5.5 mi (9 km) downstream from Hebgen Dam broke free from the southeast side of the canyon, raced across the canyon and up the opposite canyon wall (U.S. Geological Survey Professional Paper 435, Hadley, Chapter K, Fig. 3). This mass, now named the Madison Slide, killed 26 campers and buried U.S. 287 that traversed the canyon. When the slide came to rest, it acted as a natural dam across the Madison River (Fig. 4); as a result, a new lake, Earthquake Lake, was formed. The present course of U.S. 287 through Madison Canyon follows the northwest shore of Earthquake Lake. The Forest Service has constructed a Visitor Center near the toe of the slide from which one can gain excellent views of the mass of displaced material, Earthquake Lake, the spillways constructed across the landslide, and the steep southeast wall of the canyon, where the slide originated.

Geologic Setting. The core of the Madison Range, composed of crystalline metamorphic rocks, is exposed in the walls of the Madison River Canyon. These rocks are chiefly gneiss with small amounts of intercalated schist, amphibolite, and some dolomite. At the site of the Madison Slide the strata dip valleyward (northwestward) at very high angles (cross section, Fig. 4), and the near-surface rocks were thinly laminated and deeply weathered. Before the earthquake, the dolomite formed craggy bedrock spurs along the lower part of the southeast valley wall, and acted as a buttress holding the steeply inclined beds of gneiss in place much as a wedge of wood might hold tilted panes of glass. The shock of the earthquake broke the dolomite buttress obliquely, which gave way and then acted as the leading edge as the slide raced across the canyon floor and up the opposite valley wall. In general, the slide moved as an entity, for when it came to rest it was still covered with soil and trees, and the sequence of rocks in the slide reflected their former stratigraphic sequence on the southeast valley wall.

About 28 million yd^3 (21 million m^3) of rock slid into the canyon, and when the mass hit the Madison River the slide probably was traveling at a speed in excess of 100 m (160 km) per hour on a slope of about 30°. Volume of the slide when emplaced, as a result of added porosity, was about 37 million yd^3 (28 million m^3). The momentum of the slide carried it about 400 ft (122 m) up the opposite canyon wall. The slide is about 1 mi (1.6 km) long, about 0.75 mi (1.2 km) wide (Fig. 4), and is spread over an area of about 130 acres (53 hectares) in the bottom of the valley. When emplaced, it was, locally, as much as 225 ft (69 m) thick.

Current Aspects. The remnants of the dolomite buttress can still be seen near the base of the southeast valley wall as a light-brown, steeply inclined, jagged ridge that overlies the dark-gray thin-bedded gneiss (Fig. 3). The fragmented dolomite that formed the toe of the slide now appears as a continuous belt of white boulders along the distal edge of the slide. A plaque memorializing those killed during the earthquake is on one of the larger dolomite boulders. The dark-gray gneissic material that forms the bulk of the slide underlies the Visitor Center and the parking lot, and extends across the former course of the Madison River.

The Corps of Engineers cut a spillway for Earthquake Lake across the Madison Slide shortly after the slide was emplaced. Because Earthquake Lake was filling rapidly and time was short only a minimum operation was possible, and the initial gradient was too steep. When the downstream end of the spillway began to fray, the Corps cut a new spillway of lower gradient in the center of the older one. Both spillways are visible from the Visitor Center. Earthquake Lake is unusual in that its discharge is controlled by the spillway at Hebgen Dam.

MINOR PHENOMENA

At the time of the earthquake, various minor phenomena added to the confusion. Of these, the most interesting geologically was the development of a series of "sand boils" (sometimes called "sand spouts" or "sand blows") chiefly in sand and gravel along the north shore of Hebgen Lake. Water and sand began to spout from long gaping fissures that opened in these surficial deposits. Those boils that formed under the lake caused whirlpools on the surface of the lake, and the boaters that night were baffled by the sudden appearance of these whirlpools in the surrounding waters. One of the largest of these sand boils is at Hebgen Lake Lodge, near the mouth of Johnson Creek.

Other minor phenomena included many small earthflows and landslides, plus a multitude of changes in surface and groundwater. In U.S.G.S. Professional Paper 435 Stermitz (Chapter L) has discussed the effects of the earthquake on the surface water, Swenson (Chapter N) the groundwater phenomena, and Marler (Chapter Q) the effects on the hot springs of the Firehole Geyser Basins in Yellowstone Park. Other hydrologic aspects are described in Part 2 of U.S. Geological Survey Professional Paper 435.

INTERPRETING THE DEFORMATIONAL PATTERNS

The geologists who initially studied the effects of the earthquake recognized that large blocks of the Earth's crust had subsided but came to different views concerning the pattern of deformation impressed on those blocks. One group believed that the uneven subsidence of the crustal block resulted in the formation of a single basin (U.S.G.S. Professional Paper 435, Myers and Hamilton, Chapter I). The axis of this basin trends northeastward across the Madison Range and ends against the reactivated Red Canyon and Hebgen faults. This concept implies that the Madison Range subsided along with the flanking basins. Another group of geologists concluded that the pattern of deformation was best satisfied by the subsidence of two individual basins, one on each side of the tectonically stable Madison Range (U.S.G.S. Professional Paper 435, Fraser, Witkind, and Nelson, Chapter J). Each of these basins slopes toward the reactivated faults. The West Yellowstone Basin (east of the Madison Range) slopes northeastward toward the Hebgen and Red Canyon faults; the Missouri Flats (west of the range) slopes eastward toward the reactivated Madison Range fault. In this interpretation the Madison Range did not subside. Despite additional geologic work since 1959, the definitive evidence to prove or disprove the single basin or dual-basin concepts has not been found.

REFERENCES CITED

Dewey, J. W., Dillinger, W. H., Taggert, J., and Algermissen, S. T., 1972, A technique for seismic zoning; Analysis of earthquake locations and mechanisms in northern Utah, Wyoming, Idaho, and Montana, *in* Proceedings of the Internal Conference on Microzonation for Safer Construction in Research and Applications: Seattle, University of Washington, v. 2, p. 879–894.

Doser, D. I., 1985, Source parameters and faulting processes of the 1959 Hebgen Lake, Montana, earthquake sequence: Journal of Geophysical Research, v. 90, p. 4537–4555.

Qamar, A. I., and Stickney, M. C., 1983, Montana earthquakes 1869–1979; Historical seismicity and earthquake hazard: Montana Bureau of Mines and Geology Memoir 51, 79 p.

Ryall, A., 1962, The Hebgen Lake, Montana, earthquake of August 17, 1959, P-waves: Bulletin of the Seismological Society of America, v. 52, p. 235–271.

Trimble, A. B., and Smith, R. B., 1975, Seismicity and contemporary tectonics of the Hebgen Lake–Yellowstone Park region: Journal of Geophysical Research, v. 80, p. 733–741.

U.S. Geological Survey Professional Paper 435, 1964, The Hebgen Lake, Montana, earthquake of August 17, 1959: U.S. Geological Survey Professional Paper 435, 242 p.

Surface faulting associated with the 1983 Borah Peak earthquake at Doublespring Pass road, east-central Idaho

Anthony J. Crone, U.S. Geological Survey, Denver Federal Center, Denver, Colorado 80225

LOCATION

The Doublespring Pass road site is an excellent location at which to examine the surface faulting and ground breakage that accompanied the 1983 Borah Peak earthquake. The site can be reached by traveling 23 mi (37 km) northwest from Mackay, Idaho, or 30 mi (48 km) southeast from Challis, Idaho (Fig. 1), on U.S. 93 to the Doublespring Pass road turnoff. The Doublespring Pass road heads northeast from the highway. The intersection of the Doublespring Pass road with the highway is identified by a sign indicating the direction to the towns of May and Patterson. The turnoff is also identified at the intersection by a historical marker commemorating William E. Borah, after whom Borah Peak was named. The Doublespring Pass road crosses the fault scarps 2.5 mi (4 km) northeast of the intersection of U.S. 93 (Fig. 2).

Doublespring Pass road is a wide, well-maintained gravel road that can be safely traveled by passenger car and bus to the site (although the road is impassable at times in the winter and early spring when snow covers the area). The site lies within Challis National Forest and is open to the public.

SIGNIFICANCE

The M_s 7.3 Borah Peak earthquake on October 28, 1983, was the first earthquake in the intermountain west to produce surface faulting since the Hebgen Lake, Montana, earthquake on August 17, 1959. An impressive, Y-shaped zone of fault scarps and surface ruptures, approximately 22 mi (36 km) long, formed during the Borah Peak earthquake (Fig. 2). The scarps and ground ruptures occur primarily along the Lost River fault that separates the Thousand Springs and Warm Spring valleys from the Lost River Range to the northeast. The largest scarps and most complex patterns of ground rupture occur along the fault at the northeast margin of Thousand Springs Valley, between Elkhorn Creek on the southeast and Arentson Gulch on the northwest. At the Doublespring Pass road site, the ground breakage is typical of the surface faulting that accompanied the earthquake.

Trenching studies of the fault scarps conducted at this site are unique. In 1976 a trench, located about 200 ft (60 m) northwest of the road, was excavated across a Holocene fault scarp (Hait and Scott, 1978). During the 1983 earthquake, the backfilled trench was faulted, and a cross section of it was exposed in the newly formed fault scarps. Because of these rare circumstances, the 1976 trench was re-excavated and remapped in 1984 to study the structures associated with the new faulting and the stratigraphy of the colluvium associated with the new scarp (Schwartz and Crone, 1985).

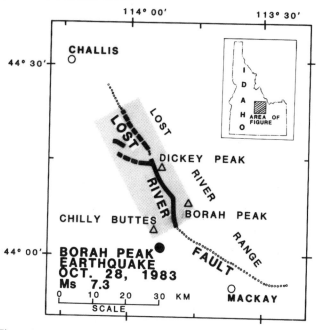

Figure 1. Location of the Borah Peak earthquake. Heavy line indicates surface faulting and ground ruptures associated with earthquake; dashed where discontinuous. Dotted line shows the unruptured part of Lost River fault. Solid dot is epicenter of M_s 7.3 earthquake. Area of Figure 2 is shaded.

GENERAL GEOLOGIC SETTING

The Borah Peak earthquake occurred in a region of typical basin and range topography in east-central Idaho. The Lost River Range and adjacent ranges to the northeast are composed of Paleozoic and Precambrian sedimentary rocks that were complexly folded and thrust faulted during the Mesozoic (Skipp and Hait, 1977). Cenozoic normal faults bound one or both flanks of the ranges (Skipp and Hait, 1977). Much of the present topography probably results from late Pliocene and Pleistocene displacements on the normal faults (M. H. Hait, Jr., 1984, written communication). The net Cenozoic vertical displacement on the Lost River fault in the vicinity of the 1983 earthquake is at least 1.6 mi (2.5 km) on the basis of 1.2 mi (1.9 km) of topographic relief between the summit of Borah Peak, at 12,662 ft (3,859 m) the highest point in Idaho, and the floor of Thousand Springs Valley, plus an estimated 0.4–0.6 mi (0.6–0.9 km) of Cenozoic fill in the valley.

The Doublespring Pass road site is located on the Willow Creek alluvial fan that slopes gently toward the valley. This broad, smooth fan, composed of upper Pleistocene (Pinedale) glacial outwash, was probably active until about 15,000 years ago (Pierce and Scott, 1982). At least several thousand years after

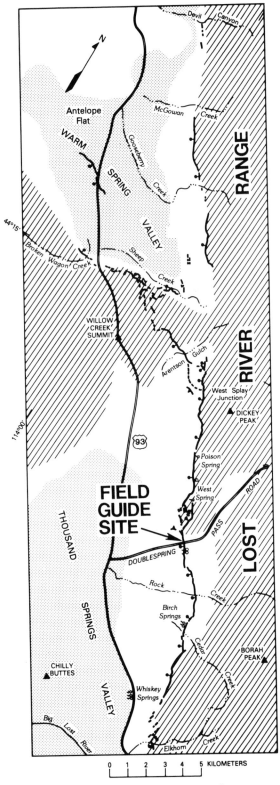

Figure 2. Generalized map of fault scarps and ground ruptures associated with the Borah Peak earthquake. Heavy solid lines indicate prominent scarps, bar and ball on downthrown side; dashed lines indicate poorly defined scarps or cracks. Stipple pattern shows valley bottoms; hachure pattern shows mountainous part of Lost River Range and hills near Willow Creek Summit.

outwash deposition ceased, a surface faulting event displaced the fan surface 4.9–6.6 ft (1.5–2 m) (Hait and Scott, 1978). The vegetated, rounded slope directly above the free-face of the main 1983 scarp is the remnant of the scarp from this older earthquake.

CHARACTERISTICS OF THE SURFACE FAULTING

The zone of surface faulting at the Doublespring Pass road site is 115 ft (35 m) wide and contains numerous en echelon scarps produced by displacements on both synthetic and antithetic faults. As a result, the ground in the fault zone is a broad depression internally broken into horsts and grabens (Fig. 3). The violent shaking during the earthquake locally shattered the ground surface in the fault zone into randomly tilted blocks up to several feet across. The height of individual scarps varies from a few inches to more than 6 ft (2 m) for the main scarp that bounds the upthrown block at the northeast edge of the fault zone. The height of the main scarp commonly exceeds the true tectonic displacement because, in most cases, a broad graben has formed in the adjacent downthrown block. Along strike, individual scarps typically diminish in height until the ground surface is warped; eventually the warping decreases to zero. As displacement on one scarp decreases, a corresponding increase usually occurs on an adjacent scarp.

The near-surface fault movement during the earthquake was dominantly normal slip with a subordinate amount of left-lateral slip (Crone and Machette, 1984). The best geologic measurements of these slip components are from the area between Elkhorn Creek and Arentson Gulch (Fig. 2). Here, at the base of large scarps, clasts dragged along the fault plane formed corrugations and grooves. The rakes of these grooves show about 7 in (17 cm) of left-lateral slip per 40 in (100 cm) of dip slip. At the Doublespring Pass road site, the left-lateral slip component can be observed by matching the graded drainage channel along the side of the old roadbed across the fault zone. Also at this site, as well as at many other locations, en echelon scarps and cracks typically form a pattern of right-stepping offsets in plan view that is characteristic of left-lateral strike-slip displacements (Fig. 3).

The focal mechanism, based on data from seismograph records, indicates that the displacement at depth had a larger component of lateral slip than that measured at the surface. This suggests that, as the rupture propagated upward from the focus, a smaller proportion of the lateral-slip component propagated to the surface than did the dip-slip component.

The throw (vertical component of slip across the entire fault zone) varies considerably along the fault zone and is difficult to measure accurately in places such as at Doublespring Pass road where there are multiple fault strands, local warping and backrotation, and extensive ground breakage. Measurements of the throw at this site, compensated for these complicating factors, are 5.2 ft (1.6 m) from geodetic data (Stein and Barrientos, 1985) and 7.9 ft (2.4 m) from displaced fluvial terraces (Vincent, 1985). The maximum throw along the entire fault zone, 8.2–8.9 ft (2.5–2.7 m) was measured near the base of Borah Peak about 2.2 mi (3.5 km) southeast of Doublespring Pass road.

The highest single scarp that formed in 1983, nearly 16 ft (5 m) high, is located about 1,180 ft (360 m) northwest of the Doublespring Pass road (Fig. 3). The height of the scarp exceeds the actual throw on the fault zone because of a small graben and backrotation of the ground surface on the downthrown block.

About 720 ft (220 m) northwest of Doublespring Pass road, the 1983 displacement is distributed on several closely spaced, synthetic strands of the fault, resulting in a stair-step series of scarps (Fig. 3). To the northwest, these strands merge into a single large, simple scarp. As the stair-step scarps erode, they will coalesce into a broad scarp with a more gentle slope than the adjacent simple scarp. In the future, the stair-step scarps will have a more subdued morphology and a geomorphically older appearance than the nearby simple scarp of the same age.

Most of the 1983 surface faulting followed pre-existing fault scarps, and in many locations the new ground breakage has mimicked the older scarps in amazing detail. This is especially apparent near the trench site just northwest of the road. Both the 1976 trench and pre-1983 aerial photographs show a well-developed horst within the graben adjacent to the main scarp. During the Borah Peak earthquake, the fault at the main scarp and the smaller faults bounding the horst were all reactivated, renewing the topographic expression of the horst (Fig. 3).

SIGNIFICANCE OF TRENCHING STUDIES

In recent years, trenching studies of prehistoric fault scarps have become increasingly valuable sources of paleoseismic data for earthquake-hazard assessments. However, often the stratigraphic relationships observed in trenches permit several plausible interpretations of the faulting history. Displacement of the 1976 trench in 1983 provided an unprecedented opportunity to re-excavate a previously mapped trench and to record, in detail, the effects of multiple surface-faulting events. The re-excavated trench clearly documents the stratigraphy of scarp-derived colluvial deposits that are used to interpret the history of faulting (Schwartz and Crone, 1985).

Detailed interpretations of the relationships in the re-excavated trench are still in progress, but two important conclusions about the behavior of the Lost River fault at this site are already clear. First, many small 1983 scarps, some only a few tens of inches high, overlie pre-1983 shear zones, indicating that even faults with small displacements were reactivated in 1983 (Fig. 4). This is additional evidence that the pattern of 1983 ground breakage was virtually identical to that associated with the prehistoric earthquake. Second, stratigraphic relationships in the trench show that the amounts of 1983 dip-slip displacement on most of these faults were similar to those associated with the earlier earthquake. The similarities in pattern of ground breakage and in amounts of displacement suggest that the prehistoric earthquake probably had a magnitude similar to the Borah Peak earthquake. These similarities support the concept of characteristic earthquakes in which a specific fault or fault segment tends to generate similar magnitude large earthquakes and the surface faulting dis-

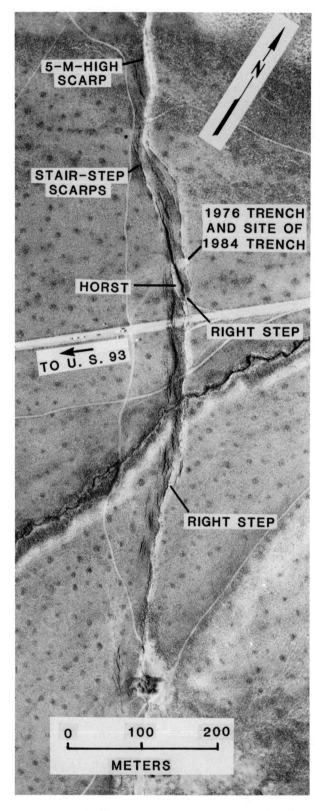

Figure 3. Vertical aerial photograph showing scarps and ground breakage near the Doublespring Pass road site. Labeled features are discussed in the text. Photograph taken October 29, 1983. (Print compliments of Robert Whitney, Mackay School of Mines, University of Nevada, Reno).

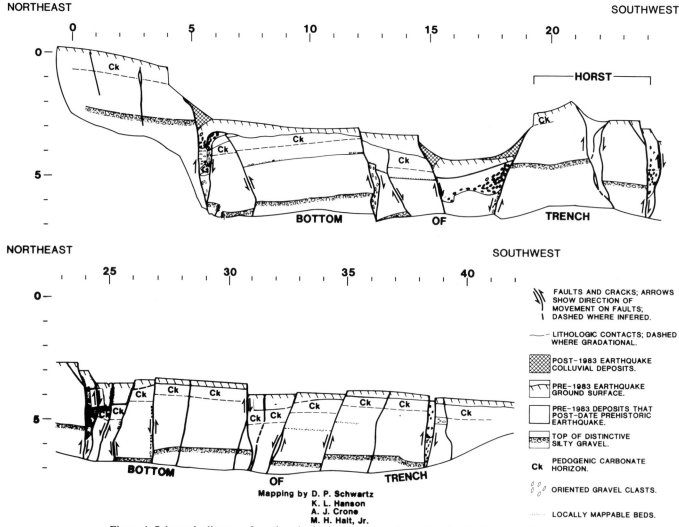

Figure 4. Schematic diagram of stratigraphy in the re-excavated trench at the Doublespring Pass road site. Map shows south wall of trench. Scale in meters; no vertical exaggeration. Offset of distinctive silty gravel shows net movement from 1983 and prehistoric earthquakes; offset of pre-1983 earthquake ground surface shows displacement from Borah Peak earthquake.

placements associated with these large earthquakes are similar. In addition, the trenching studies at the Doublespring Pass road site suggest that the pattern of ground breakage can be remarkably similar for successive large earthquakes. Thus, the trenching studies here have provided some new and valuable insight into the behavior of a seismogenic normal fault in an extensional tectonic environment.

REFERENCES

Crone, A. J., and Machette, M. N., 1984, Surface faulting accompanying the Borah Peak earthquake, central Idaho: Geology, v. 12, p. 664–667.

Hait, M. H., Jr., and Scott, W. E., 1978, Holocene faulting, Lost River Range, Idaho: Geological Society of America Abstracts with Programs, v. 10, p. 217.

Pierce, K. L., and Scott, W. E., 1982, Pleistocene episodes of alluvial-gravel deposition, southeastern Idaho, in Bonnichsen, B., and Breckenridge, R. M., eds., Cenozoic geology of Idaho: Idaho Bureau of Mines and Geology Bul-

letin 26, p. 695–702.

Schwartz, D. P., and Crone, A. J., 1985, The 1983 Borah Peak earthquake; A calibration event for quantifying earthquake recurrence and fault behavior on Great Basin normal faults, in Stein, R. S., and Bucknam, R. C., eds., Proceedings of Workshop XXVIII on the Borah Peak, Idaho earthquake: U.S. Geological Survey Open-File Report 85-290, p. 153–160.

Skipp, B., and Hait, M. H., Jr., 1977, Allochthons along the northeast margin of the Snake River Plain: Wyoming Geological Association, 29th Annual Field Conference Guidebook, p. 499–515.

Stein, R. S., and Barrientos, S. E., 1985, The 1983 Borah Peak, Idaho, earthquake; Geodetic evidence for deep rupture on a planar fault, in Stein, R. S., and Bucknam, R. C., eds., Proceedings of Workshop XXVIII on the Borah Peak, Idaho earthquake: U.S. Geological Survey Open-File Report 85-290, p. 459–484.

Vincent, K. R., 1985, Measurement of vertical tectonic offset using longitudinal profiles of faulted geomorphic surfaces near Borah Peak, Idaho; A preliminary report, in Stein, R. S., and Bucknam, R. C., eds., Proceedings of Workshop XXVIII on the Borah Peak, Idaho earthquake: U.S. Geological Survey Open-File Report 85-290, p. 76–96.

Southern Lemhi Range, east-central Idaho

M. H. Hait, Jr., 6421 Peppermill Drive, Las Vegas, Nevada 89102

LOCATION AND ACCESSIBILITY

The Lemhi Range, Idaho, is one of three northwest-trending basin-ranges along the northeastern margin of the eastern Snake River Plain (Fig. 1). The southern end of the range is about 55 mi (88 km) northwest of Idaho Falls. The small town of Howe is the closest location for food, water, gasoline, and a trailer park. The nearest motels and service stations are in Arco, about 22 mi (35 km) to the southwest. The area is covered by six 1:24,000 topographic quadrangles. They are listed in the caption of Figure 2A and will be essential for detailed planning of access.

Access to the outcrops is provided by graded dirt roads from Howe to the alluvial apron where rougher roads (traversable by pickups) lead to the seven major canyons in which the stratigraphic sections are exposed (Fig. 2A). From west to east, these canyons are South Creek Canyon, Black Canyon, Middle Canyon, East Canyon, Box Canyon (informal name), an unnamed canyon with a jeep road that crosses the range, and, on the east side of the range, Kyle Canyon.

SIGNIFICANCE

The range contains one of the few fairly well-exposed accessible continuous stratigraphic sections of Paleozoic rocks in Idaho north of the Snake River Plain. The rocks are displayed in a natural cross section that, in addition, shows examples of the entire array of structural forms characteristic of east-central Idaho and southwest Montana. The stratigraphic section extends from Precambrian rocks on the west to Permian rocks on the east, and contains Cambrian to Devonian shelf deposits, Lower Mississippian turbidites typical of basin-plain deposits, and middle Mississippian to Permian carbonate platform deposits. The Idaho thrust-belt structural features include the east-dipping Black Canyon thrust zone and associated complex folds in both quartzite and carbonate formations (Fig. 2B). Structures associated with Late Tertiary and Quaternary extension include normal faults along which late Tertiary conglomerate and rhyolytic-basaltic volcanics have been tilted eastward; the Lemhi range-front fault exposed in a trench through the scarp near the mouth of Black Canyon (Malde, 1985); and the South Creek block, a segment of a thrust sheet that has moved westward, apparently as a part of a gravity glide sheet from its original position atop the Lemhi Range (Fig. 2B). The major structures from Black Canyon to Box Canyon (Fig. 2B, Section A–A') can be easily seen through binoculars from viewpoints along Idaho Route 22/88 between Howe and the south tip of the Lemhi Range (Fig. 2A). Pick viewpoints that are about due south of the point you are interested in.

GEOLOGY

Regional geologic studies of the southern Lemhi Range are found in Ross (1961) and Beutner (1968). The most up-to-date geologic map is in Kuntz and others (1984). Skipp and Hait (1977) fit the southern Lemhi Range into the regional strati-

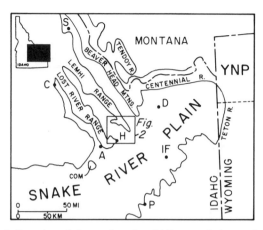

Figure 1. Location of the southern Lemhi Range relative to the major basin-ranges of east-central Idaho. A, Arco; C, Challis; COM, Craters of the Moon; D, Dubois; H, Howe; IF, Idaho Falls; P, Pocatello; S, Salmon; YNP, Yellowstone National Park.

graphic and tectonic framework of east-central Idaho and southwest Montana. Scott (1982) shows the distribution of surficial deposits that border the range.

Stratigraphic studies within the southern Lemhi Range have focused on particular parts of the stratigraphic section and are keyed to the major canyons.

South Creek Canyon. McCandless (1982) provides the type sections for the Cambrian Wilbert and Tyler Peak formations, and a revised section of the Lower Ordovician Summerhouse Formation. Beutner (1968) shows the structural complications that resulted from the westward gliding of the South Creek thrust plate. Hait and Hoggan (U.S. Geological Survey, 1977) suggested that the South Creek plate was a thrust distinct from the Black Canyon thrust, based on a thicker Devonian section in the South Creek plate than in the upper and lower plates of the Black Canyon thrust (Fig. 2).

Black Canyon. Ordovician rocks (Kinnikinic Quartzite) through Mississippian rocks (McGowan Creek Formation) crop out in this canyon. The McGowan Creek Formation was named by Sandberg (1975, reference in Skipp and Hall, 1980) to replace an earlier term, the Milligen Formation. Thin stratigraphic sections of Upper Ordovician Fish Haven Dolomite (Saturday Mountain Formation of many authors) and Devonian Jefferson Formation (Sloss, 1954; Churkin, 1962; Scholten and Hait, 1962; references in Mapel and Sandberg, 1968) were explained by pre-Devonian erosional thinning and Devonian depositional thinning over mid-Paleozoic tectonic features. In contrast, Mapel and Sandberg (1968) find that the thin Devonian rocks of earlier authors actually contain intervals with Middle or Late Ordovician conodonts and perhaps some Silurian Laketown Dolomite not recognized by earlier workers. Mapel and Sandberg (1968) explain this intermixing and thinning by post-Devonian thrusting.

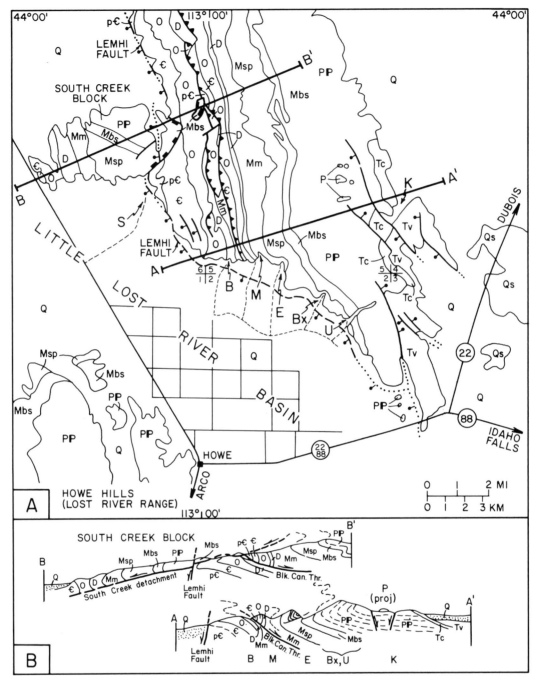

Figure 2. Geologic map and cross sections of the southern Lemhi Range. A = generalized geologic map. Access canyons: S, South Creek canyon; B, Black Canyon; M, Middle Canyon; E, East Canyon; Bx, Box Canyon; U, Unnamed Canyon; K, Kyle Canyon. Formation symbols: Q, surficial deposits; Qs, Snake River basalt; Tv, rhyolite and basalt; Tc, conglomerate; P, Phosphoria Formation; PP, Snaky Canyon Formation; Mbs, Bluebird Mountain, Arco Hills, Surrett Canyon, and South Creek formations; Msp, Scott Peak Formation; Mm, Middle Canyon and McGowan Creek formations; D, Three Forks and Jefferson formations; O, Fish Haven, Kinnikinnic, and Summerhouse formations; €, Tyler Peak and Wilbert formations; p€, sedimentary rock. Faults: bar and ball on down-thrown side of Lemhi range-front fault and other normal faults; square teeth on upper plate of South Creek detachment fault; sharp teeth on upper plate of Black Canyon thrust fault. Crosses with numbers are corners of 1:24,000 topographic quadrangles: 1, Howe; 2, Little Lost River Sinks; 3, Big Lost River Sinks; 4, Richard Butte; 5, Tyler Peak; 6, Howe NE. Modified from Kuntz and others (1984) and Hait, M. H., Jr., [unpublished mapping]. B = schematic cross sections showing general structure. Some vertical exaggeration. Modified from Beutner (1968).

The relative importance of depositional versus tectonic thinning or thickening is yet to be resolved. The resolution will take detailed mapping, section measuring, and fossil studies.

The east wall of Black Canyon is the site of the east-dipping Black Canyon thrust zone (Ross, 1961; Beutner, 1968), which brings Cambrian through Permian rocks over the McGowan Creek Formation. Repetitions of Ordovician and Devonian rocks cited by Mapel and Sandberg (1968) may be related to the emplacement of the overriding Black Canyon thrust plate.

Middle and East canyons. Huh (1967, reference in Skipp and Hall, 1980) established type sections for the Mississippian Middle Canyon, Scott Peak, South Creek, and Surrett Canyon formations in these canyons and postulated that this sequence represents a prograding depositional complex. Sando and Bamber (1985) summarize the major corals found in these formations, rename many of the corals, and set up a numbered zonation. Sando (1976), Rose (1976), and Gutschick and others (1980; all references in Skipp and Hall, 1980) discuss the regional Mississippian carbonate depositional patterns; Skipp, Sando, and Hall (1979, reference in Skipp and Hall, 1980) present the regional stratigraphic framework of the Mississippian and Pennsylvanian.

Box Canyon and unnamed canyon with jeep road. These canyons traverse the Scott Peak, South Creek, and Surrett Canyon formations, and give easy access to the Mississippian Arco Hills and Bluebird Mountain formations and the Pennsylvanian and Permian Snaky Canyon Formation. Skipp, Hoggan, and others (1979, reference in Skipp and Hall, 1980) give detailed stratigraphic and paleontologic descriptions of these limestone, quartzite, and sandy limestone units.

Kyle Canyon. A partial section of the Permian Phosphoria Formation is reached up this canyon in the hills west of Kyle Spring. Skipp, Hoggan, and others (1979, reference in Skipp and Hall, 1980) report the presence of the Rex Chert Member and phosphorite similar to the Retort Phosphatic Shale Member. The Phosphoria is conformable on the Snaky Canyon Formation.

Nearby, in Tertiary gravels, Skipp (Skipp, Hoggan, and others, 1979, reference in Skipp and Hall, 1980) has found fragments of Triassic Dinwoody Formation.

The Permian rocks are overlapped by Tertiary conglomerate, Miocene-Pliocene rhyolitic-basaltic rocks (McBroome, 1981), Snake River basaltic rocks (Kuntz and others, 1984), and surficial deposits (Scott, 1982).

STRATIGRAPHIC DESCRIPTIONS
(modified from Kuntz and others, 1984)

Surficial deposits (Holocene to Pleistocene) includes alluvial fans, lake deposits, eolian, playa, and colluvial deposits.

Snake River Basalt (Lower Pleistocene). Gray and reddish-oxidized weathered pahoehoe and aa flows, scoria, cinders, and ash.

Rhyolite and basalt volcanics (Pliocene and Miocene). Rhyolite welded ash-flow tuffs and ash interbedded with gray and black basalt flows. About 500 ft (150 m) thick.

Conglomerate (Oligocene?). Thick-bedded, limestone clasts, with pink silty matrix; local cobbles and boulders of quartzite;

relationships uncertain. About 100 ft (30 m) thick.

Phosphoria Formation (Permian). Limestone, chert, phosphatic siltstone, phosphorite, and dolomite. Limestone is dark brownish-gray, phosphatic, oolitic. Chert is grayish-black to medium dark gray, fine- to coarse-grained, thin- to medium-bedded. Phosphatic siltstone is brown, sandy or muddy, thin-bedded with phosphatic laminations and nodules. Phosphorite is brown-gray to dark gray, fine-grained, oolitic, pelletal and nodular, weathers light reddish-gray and is laminated to thin-bedded. Dolomite is brownish-gray to pale brown, and contains oolitic phosphatic laminations and reworked fragments and nodules of oolitic phosphorite, weathers pale brown. Thickness about 100 ft (30 m).

Snaky Canyon Formation (Lower Permian to Lower Pennsylvanian). Interbedded limestone, dolomite, and minor sandstone. Limestone and dolomite are medium gray to light gray, fine to coarse grained, sandy, fossiliferous (brachiopods, corals, mollusks, trilobites, calcareous microfossils); stromatolitic mounds are common in lower part; hydrozoan(?) and phylloid algae buildups are common in upper part; variably cherty; weather medium dark gray to light olive-gray. Sandstone is quartzose, calcareous, medium to light gray, very fine-grained, locally fossiliferous (calcareous microfossils and brachiopods), weathers pale brown to moderate yellowish-brown, medium-bedded, and forms low ledges. Thickness about 3,900 ft (1,200 m).

Bluebird Mountain Formation (Upper Mississippian). Sandstone with minor interbedded dolomite and limestone. Sandstone is quartzose, calcareous, medium light gray to light brown, very fine-grained, medium- to thick-bedded, weathers moderate brown. Dolomite and limestone are medium gray to medium light gray, fine- to coarse-grained, sandy, fossiliferous; weathers light brown to olive-gray. Unit forms cliffs and ledges. Thickness about 130 to 140 ft (40 to 50 m).

Arco Hills Formation (Upper Mississippian). Interbedded limestone, mudstone, siltstone, and minor sandstone. Limestone is medium gray to brownish-gray, fine- to coarse-grained, silty, sandy, variably cherty, fossiliferous, and weathers medium gray and yellowish-brown. Mudstone and siltstone are olive-gray to grayish-orange, calcareous, contain common limestone nodules, weather yellowish-gray. Sandstone is olive-gray and brownish-gray, very fine-grained, quartzose, calcareous, and weathers moderate brown. Formation forms slopes. Thickness 164 ft (50 m).

Surrett Canyon Formation (Mississippian). Ledge- and cliff-forming limestone, dark gray to medium gray, fine grained, thick bedded, with scattered white, sparry, bioclastic debris. Also contains minor waxy, light to medium gray limestone and brown, silty limestone, and is fossiliferous (corals, brachiopods, bryozoans, and abundant crinoidal debris). Thickness 245 ft (75 m).

South Creek Formation (Mississippian). Alternating thin-bedded, fine-grained, dark gray limestone that contains thin beds and nodules of light-tan–weathering chert with characteristic concentric banding; and dark gray, clayey, fissile limestone. Thickness 295 to 345 ft (90 to 105 m).

Scott Peak Formation (Mississipian). Upper 656 ft (200 m)

consists of medium- to thick-bedded, cyclic alternations of light gray calcarenite and fine-grained, dark gray, cherty limestone. Middle 900 ft (275 m) consists of very thick-bedded, dark gray, chert-free limestone. Lower 410 ft (125 m) consists of thick-bedded, coarse-grained calcarenite that is partially cross-laminated and interbedded with beds of dark, fine-grained, cherty limestone. Formation is fossiliferous (corals, brachiopods, bryozoans, crinoidal debris). Formation forms cliffs and ledges. Thickness about 2,000 to 2,130 ft (610 to 650 m).

Middle Canyon Formation (Mississippian). Interbedded limestone and calcareous siltstone. Limestone is dark gray to medium gray, thin-bedded, contains 10 to 50 percent black chert, weathers medium gray to medium light gray. Calcareous siltstone is pale yellowish-brown and dark yellowish-brown, thin-bedded, weathers pale yellowish-brown to moderate yellowish-brown. Contains sparse corals in upper part. Thickness about 1,083 ft (330 m).

McGowan Creek Formation (Lower Mississippian). Interbedded thin- to medium-bedded, medium to dark gray limestone, commonly sheared, light gray to grayish-black argillite, silty argillite and siltite. Thickness 200 ft (60 m).

Three Forks Formation (Upper Devonian). Interbedded, medium to dark gray limestone, silty limestone, and grayish black, silicified siltstone. Thickness about 100 ft (30 m).

Jefferson Formation (Devonian). Dolomite, dolomite-limestone breccia, and limestone with minor sandstone, siltstone, and mudstone. Dolomite and dolomite-limestone breccia are light gray, grayish-black, and yellowish-brown, finely to coarsely crystalline, sandy, silty, laminated in part, and locally petroliferous. Limestone is medium gray, fine-grained, and contains local black chert bands. Basal sandstone, siltstone, and mudstone are red to yellow-brown, laminated, and locally conglomeratic. Thickness 100 to 300 ft (30 to 100 m).

Fish Haven Dolomite (Lost River Member; Upper Ordovician). Chiefly medium to dark gray, massive, thick-bedded, crystalline dolomite; mottled, light, medium, and dark gray dolomite; and minor chert. Light gray dolomite on top. Lower 0 to 150 ft (45 m) is olive-gray to medium gray, argillaceous dolomite with fine, white, dolomite stringers and silicified brachiopods and corals. Thickness 65+ ft (10 m).

Kinnikinnic Quartzite (Middle Ordovician). Chiefly massive, medium- to thick-bedded, well-sorted, medium- to fine-grained, vitreous, white orthoquarzite; locally laminated and cross-bedded; rusty stains are common. Also contains some small lenses and ovoid patches of dolomitic sandstone, mostly in upper part. Thickness 325 to 1,080 ft (99 to 330 m).

Summerhouse Formation (Lower Ordovician). Quartzite and variably interbedded, calcareous sandstone and sandy dolomite. Dolomite at the top of the formation is grayish-red to reddish-brown, coarse-grained, and contains common siliceous laminae. Calcareous sandstones in middle of formation are yellowish-brown to reddish-brown and display common bioturbation and cross-laminations. Quartzite at the base of the formation is light gray to pale yellowish-brown, very fine- to medium-grained, and is thick-bedded to massive. Thickness 246 to 690 ft (75 to 210 m).

Tyler Peak Formation (Lower Cambrian). Interbedded sandstone, shale, and quartzite. Sandstone dominates upper part of formation and is yellowish-brown to dusky purple, fine- to coarse-grained, thin- to thick-bedded, and contains common bioturbations and local cross-laminations. Shale is greenish-gray, fissile, interbedded with sandy quartzite and silty sandstone, and occupies lower 100 ft (30 m) of formation. Quartzite is pure, vitreous, light gray to grayish-pink, fine- to coarse-grained, thin- to thick-bedded, and cross-laminated. Thickness 0 to 820 ft (1 to 250 m).

Wilbert Formation (Cambrian). Interbedded quartzite and conglomerate; white, light gray to reddish-purple in lower part, gray to yellow-brown in upper part, very fine- to coarse-grained, contains cross-laminations and bioturbations locally. Conglomerates are thick-bedded to massive, bioturbated beds are thin-bedded. Thickness 0 to 425 ft (0 to 130 m).

Precambrian Sedimentary rocks (Proterozoic). Red, pink, brownish-gray, impure sandstone; pink to purple and gray-green, phyllitic quartzite with heavy mineral laminations; gray-green slate and siltstone, reddish-gray to brown arkose; locally foliated. Thickness in excess of 655 ft (200 m).

REFERENCES CITED

Beutner, E. C., 1968, Structure and tectonics of the southern Lemhi Range, Idaho [Ph.D. thesis]: University Park, The Pennsylvania State University, 106 p.

Kuntz, M. A., Skipp, B., Scott, W. E., and Page, W. R., 1984, Preliminary geologic map of the Idaho National Engineering Laboratory and adjoining areas: U.S. Geological Survey Open-File Report 84-281, 25 p.

Malde, H. E., 1985, Quaternary faulting near Arco and Howe, Idaho, in Stein, R. S., and Bucknam, R. C., eds., Proceedings of Workshop XXIII on the Borah Peak, Idaho, earthquake: U.S. Geological Survey Open-File Report 85-290, p. 207–235.

Mapel, W. J., and Sandberg, C. A., 1968, Devonian paleotectonics in east-central Idaho and southwestern Montana: U.S. Geological Survey Professional Paper 600-D, p. D115–D125.

McBroome, L. A., 1981, Stratigraphy and origin of Neogene ash-flow tuffs on the north-central margin of the eastern Snake River Plain, Idaho [M.S. thesis]: Boulder, The University of Colorado, 74 p.

McCandless, D. A., 1982, A reevaluation of Cambrian through Middle Ordovician stratigraphy of the southern Lemhi Range [M.S. thesis]: University Park, The Pennsylvania State University, 157 p.

Ross, C. P., 1961, Geology of the southern part of the Lemhi Range, Idaho: U.S. Geological Survey Bulletin 1081-F, 260 p.

Sando, W. J., and Bamber, E. W., 1985, Coral zonation of the Mississippian system in the western interior province of North America: U.S. Geological Survey Professional Paper 1334, 61 p.

Scott, W. E., 1982, Surficial geologic map of the eastern Snake River Plain and adjacent areas, 111° to 115° W., Idaho and Wyoming: U.S. Geological Survey Miscellaneous Investigations Map I-1372, scale 1:250,000.

Skipp, Betty, and Hait, M. H., Jr., 1977, Allochthons along the northeast margin of the Snake River Plain, Idaho: 29th Annual Field Conference, Wyoming Geological Association Guidebook, p. 499–515.

Skipp, Betty, and Hall, W. E., 1980, Upper Paleozoic paleotectonics and paleogeography of Idaho, in Fouch, T. D., and Magathan, E. R., eds., Paleozoic paleogeography of the west-central United States: West-central United States. Paleogeography Symposium 1, Denver, Colorado, Society of Economic Paleontologists and Mineralogists, Rocky Mountain section, p. 387–422.

U.S. Geological Survey, 1977, Lemhi Range "detachment block" is a thrust fault: U.S. Geological Survey Professional Paper 1050, p. 65–66.

Island Park, Idaho; Transition from rhyolites of the Yellowstone Plateau to basalts of the Snake River Plain

Robert L. Christiansen, U.S. Geological Survey, 345 Middlefield Road, Menlo Park, California 94025
Glenn F. Embree, Ricks College, Rexburg, Idaho 83440

LOCATION AND ACCESSIBILITY

Island Park, situated between the northeastern end of the Snake River Plain and the western margin of the Yellowstone Plateau, is traversed by U.S. 20-191 between Ashton, Idaho, and West Yellowstone, Montana (Fig. 1) and, in part, by the Mesa Falls Road (Idaho 47). All of the physiographic and geologic features described in this guide can be reached during the summer months by passenger car on those two paved highways or on short gravel roads that extend from them. During the winter, U.S. 20-191 remains open, but other winter travel is possible only by skis or skimobile. One quarry locality described later is privately owned but has generally been freely accessible.

SIGNIFICANCE

For the past 10 to 15 million years a series of bimodal rhyolite-basalt systems characterized by voluminous caldera-forming rhyolitic ash-flow eruptions has migrated northeastward at 2 to 4 cm per year along the axis of the Snake River Plain (Christiansen and Lipman, 1972; Armstrong and others, 1975; Christiansen and McKee, 1978; Smith and Christiansen, 1980). The active focus at the northeastern end of this propagating volcanic zone probably has always been a magmatic system that was topographically high and characterized by thermal and seismic activity as well as by voluminous rhyolitic volcanism. The eastern Snake River Plain has formed by thermal contraction, subsidence, and episodic basaltic volcanism in the wake of the rhyolitic activity. The Yellowstone Plateau represents the current location of the major rhyolitic system, and the eastern Snake River Plain consists of older (Miocene and Pliocene) rhyolitic calderas buried beneath Pleistocene basalts (Embree and others, 1982; Morgan and others, 1984). Island Park is physiographically and geologically transitional between Yellowstone and the Snake River Plain and contains many accessible, well-exposed volcanic features of both regions, particularly the eruptive products and physiographic expression of three cycles of rhyolitic caldera-forming volcanism.

GEOLOGY

Introduction. The geology of the Island Park area is summarized here from Christiansen (1982). The topographic basin of Island Park (Fig. 1) is the product of events that occurred during each of three cycles of bimodal volcanism producing the Yellowstone Plateau volcanic field. Each of the cycles began with the eruption of relatively small volumes of both basalt and rhyolite,

then continued for as long as several hundred thousand years with eruptions of rhyolites from a growing system of arcuate fractures above a large and growing magma chamber. Each cycle then climaxed with the explosive eruption of a large volume (hundreds to thousands of cubic kilometers) of rhyolitic ejecta in an extremely short time, probably no more than a few days. Partial emptying of the magma chamber in each of these catastrophic eruptions caused the chamber roof to collapse, forming a large caldera. Further eruptions of rhyolites from the residual magma chambers over several hundred thousand years partly filled each caldera. After the rhyolitic magma had solidified in the older chambers and could be tectonically fractured, basaltic magmas erupted through fractures within the calderas. The Tertiary calderas of the eastern Snake River Plain southwest of Island Park have been buried by such basaltic activity.

The first of the volcanic cycles to affect the Island Park area climaxed about 2.0 Ma, forming the largest of three calderas of the Yellowstone Plateau region (56 by 19 mi; 90 by 30 km), extending from the west side of Island Park to the Central Plateau of Yellowstone National Park. Big Bend Ridge (Figs. 1 and 2A), the western and southwestern rim of Island Park, is a segment of the first-cycle caldera rim. The second volcanic cycle climaxed about 1.3 Ma to produce a somewhat smaller caldera (about 12 mi; 20 km across) nested against the northwest wall of the first-cycle caldera. The eastern margin of Island Park is formed by the flow-front scarps of younger rhyolitic lava flows of the third volcanic cycle that erupted from vents farther east, on the Yellowstone Plateau. These flows postdate the collapse that formed the third-cycle Yellowstone caldera. Thus, there is no "Island Park caldera" in the sense suggested by Hamilton (1965) but rather a complex basin formed by two separate caldera collapses 700,000 years apart and by later constructional volcanism; younger basalts floor the basin (Fig. 1).

Big Bend Ridge. Big Bend Ridge—the southern margin of the first-cycle caldera and of the Island Park basin—is well seen from the "Three Tetons" historical highway marker, 2.1 mi (3.4 km) north of Ashton on U.S. 20-191 (Fig. 1, Stop 1; Fig. 2A). At the east end of Big Bend Ridge lies Snake River Butte, a precollapse rhyolitic lava flow that may have vented along an incipient ring-fracture zone in the intrusively uplifted roof of the first-cycle magma chamber. The rhyolite of Snake River Butte (Figs. 1 and 2A), apparently a single lava flow, has a potassium-argon age of 2.0 Ma, analytically indistinguishable from that of the overlying Huckleberry Ridge Tuff.

Approximately 3.3 mi (5.3 km) north of Ashton (0.8 mi;

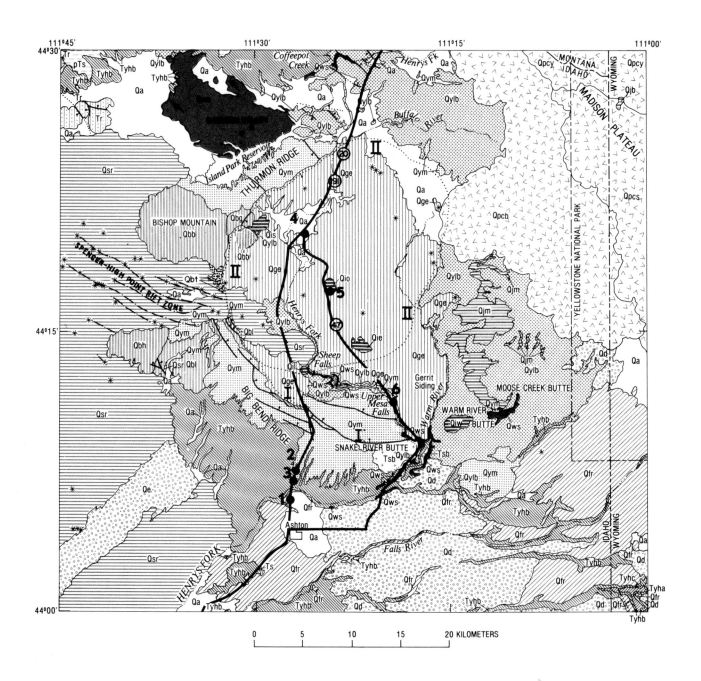

Figure 1. Geologic map of Island Park (modified from Christiansen, 1982), showing major roads and stops described in the text. I, Big Bend Ridge caldera segment; II, Henry's Fork caldera. Explanation on facing page.

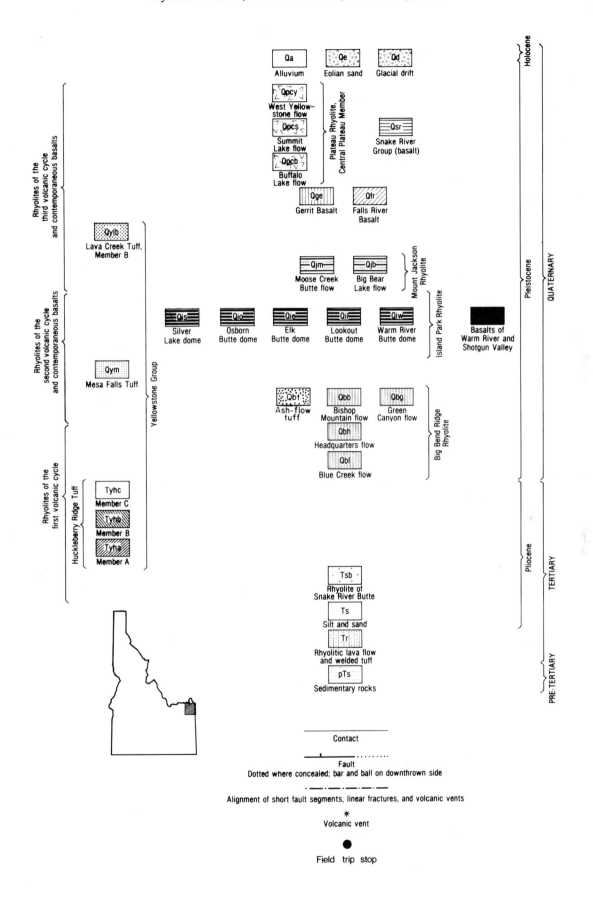

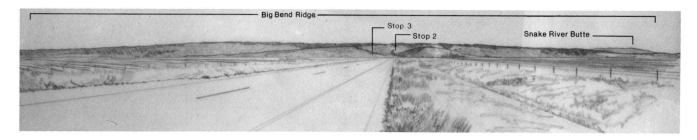

Figure 2. A. Panoramic view of Big Bend Ridge, as seen northward from the "Three Tetons" historical marker on U.S. 20-191, 2.1 mi (3.4 km) north of Ashton, Idaho (Fig. 1, Stop 1).

1.3 km north of the Henry's Fork Bridge) on U.S. 20-191 is the roadcut shown in Figure 3 of Hamilton (1965). This roadcut (Figs. 1 and 2A, Stop 2) exposes rhyolitic units laid down during the first two cycles of the volcanic field. At the base is the gray upper portion of the 2.0 Ma, densely welded Huckleberry Ridge Tuff. The Huckleberry Ridge Tuff covers an area of more than 5,800 mi^2 (15,000 km^2; Christiansen, 1982, Fig. 3) and has a total erupted volume of more than 588 mi^3 (2,450 km^3). Eruption of this unit was accompanied by collapse of the first-cycle Big Bend Ridge caldera segment (Fig. 1). Overlying the Huckleberry Ridge Tuff is a thin wedge of loess (now mostly covered by colluvium) deposited during the 700,000-year interval between the culminations of the first and second cycles.

Above are deposits of the Mesa Falls Tuff, related to the climactic eruptions of the second-cycle Henry's Fork caldera, nested against the northwestern edge of the Big Bend Ridge caldera segment (Fig. 1). At the base of the Mesa Falls is a thick deposit of bedded pumiceous ash laid down by fallout from a high eruption column during an early stage of the Mesa Falls eruption. The pink, partially welded, ash-flow tuff at the top of the roadcut represents the main eruptive unit of the 1.3-Ma Mesa Falls Tuff. This ash-flow sheet is inferred to have been deposited over more than 1,050 mi^2 (2,700 km^2) in and near Island Park (Christiansen, 1982, Fig. 3) and had a minimum initial volume of 67 mi^3 (280 km^3). The distinctive Mesa Falls Tuff contains abundant, very large, commonly euhedral phenocrysts of sanidine, quartz, and plagioclase and large pumice inclusions, commonly 4 to 12 in (10 to 30 cm) or more across.

A better exposure of the Mesa Falls fallout and ash-flow units can be observed in a pumice quarry, reached by backtracking 0.3 mi (0.5 km) down U.S. 20-191 toward Ashton and turning west on a short dirt road past a log store into the quarry (Figs. 1 and 2A, Stop 3). The basal bedded deposit in the quarry walls is white, very well sorted, generally has good planar bedding, and contains abundant crystals of the same types present in the overlying ash-flow tuff. The lower ash beds contain a marked concentration of crystals over glass. The upper, coarser beds consist predominantly of pumice. Bedding in much of the unit is parallel to the underlying slope and mantles local irregularities of the buried surface with a nearly uniform thickness of ash and

pumice, clearly indicating a fallout origin for most of the deposit. However, near the top are some thin, pink, nonwelded ash-flow tongues as well as lensoid and cross-stratified ash beds that may represent a phase of ground-surge deposition. Many of the nonwelded ash-flow tongues contain abundant large pumice blocks. In the upper part of the quarry walls, the multiple-ash-flow main unit of Mesa Falls Tuff can be seen. It is slightly welded near the base but becomes somewhat more welded upward.

Approximately 7.3 mi (12 km) north of Ashton, at the turnoff of Anderson Mill Canyon Road, U.S. 20-191 crosses the topographic rim of the Big Bend Ridge caldera segment. As the highway descends northward toward the floor of Island Park for a distance of 2 mi (3.2 km), a series of valleys and ridges parallel to the crest of Big Bend Ridge represents large slumped fault blocks formed during minor renewed collapse of Big Bend Ridge during formation of the second-cycle Henry's Fork caldera.

The highway reaches the floor of the caldera basin about 9 mi (14 km) north of Ashton. From here northward to Thurmon Ridge, most of Island Park has been flooded by upper Pleistocene Gerrit Basalt and Snake River Group basalts (Fig. 1). These basalts can be seen in numerous roadcuts and natural exposures between the Sheep Falls Road and Ponds Lodge at the Buffalo River.

Junction of U.S. 20-191 and Idaho 47. An excellent place to view the west rim of the Island Park basin is the junction of U.S. 20-191 and the Mesa Falls Road (Idaho 47), 1.3 mi (2.1 km) north of Osborne Bridge across Henry's Fork (Fig. 1, Stop 4; Fig. 2B). Looking southwest down the highway, the scarp of the Big Bend Ridge caldera segment can be seen 9 mi (15 km) in the distance. High Point, the vent for a potassic intermediate-composition eruption related to the Snake River Group basalts, occurs atop the ridge. This is one of many vents situated on the Spencer–High Point rift zone, which extends 30 mi (50 km) westward from Island Park onto the Snake River Plain (Fig. 1).

Westward from the viewpoint lies Bishop Mountain (Fig. 2B), one of four post–Huckleberry Ridge, pre–Mesa Falls rhyolitic lava flows and related tuff on the rim and outer flank of Island Park between Big Bend and Thurmon ridges. The topographic high points of most of these flows are near the rim of the Big Bend Ridge caldera segment, but each flow was truncated by

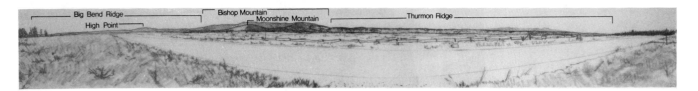

Figure 2. B. Panoramic view of the west rim of Island Park, as seen westward from the junction of U.S. 20-191 and Idaho 47 (Fig. 1, Stop 4).

second-cycle collapse. As a result, the actual source vents for the flows are unknown but must be buried somewhere within the Island Park basin, possibly along the ring-fracture zones of the first- and second-cycle calderas. Geochemical and isotopic data and remanent magnetic directions suggest that the two southern flows are postcaldera first-cycle rhyolites, but that Bishop Mountain and an adjacent flow erupted early in the second volcanic cycle (Hildreth and others, 1984). In front of Bishop Mountain is Moonshine Mountain (Fig. 2B), earlier thought to be one of several Island Park Rhyolite domes that crop out within or adjacent to the second-cycle Henry's Fork caldera (Christiansen, 1982), but now recognized as a downfaulted block of the Bishop Mountain flow.

Clearly seen to the northwest is Thurmon Ridge (Fig. 2B), formed of Mesa Falls Tuff. Thurmon Ridge is a segment of the topographic rim of the Henry's Fork caldera, produced by eruption of the second-cycle Mesa Falls Tuff.

Mesa Falls Road (Idaho 47). Basaltic lavas that form the floor of Island Park and the tops of a few older rhyolite domes that rise above them may be seen along Idaho 47 between the junction of U.S. 20-191 and Ashton. The Osborne Butte dome lies 4 mi (6 km) south of the U.S. 20-191 junction on the east side of Idaho 47. The dome is one of several that constitute the Island Park Rhyolite, a lithologically distinctive unit having very large (one- to several-centimeter) phenocrysts of quartz, sanidine, and plagioclase that constitute nearly half the rock. These steep-sided lava domes define a northwest-trending linear belt 19 mi long (30 km) and less than 4 mi wide (7 km) across Island Park (Fig. 1). The similar petrologic characteristics and paleomagnetic directions of the Mesa Falls Tuff and the Island Park domes, as well as the 1.3-Ma age of the Osborne Butte dome, indicate that

they are only slightly younger than the Mesa Falls Tuff. The crystal-rich rhyolite of Osborne Butte can best be seen by driving 0.4 mi (0.6 km) east of the paved road on the Hatchery Butte Road and walking a short distance to the southern edge of the dome (Fig. 1, Stop 5).

Upper Mesa Falls. A short drive 0.9 mi (1.4 km) westward from the pavement on an unmarked dirt road 8.7 mi (14 km) south of Hatchery Butte Road (or 0.7 mi; 1.1 km north of the Grand View overlook above Lower Mesa Falls) is rough but generally passable by passenger car. It leads to Upper Mesa Falls and a spectacular exposure of the Mesa Falls Tuff and younger basalts (Fig. 1, Stop 6). Park at the end of the dirt road and walk along any of the several gullied paths to the brink of the falls. (Special caution is needed on these unmaintained paths because of steep cliffs and swift water just below.)

At this locality, Henry's Fork of the Snake River has cut a gorge through the Gerrit Basalt that generally forms the flat floor of Island Park. The upper wall beneath the basalt on this side of the gorge consists of Lava Creek Tuff from the 600,000-year-old Yellowstone caldera and an upper nonwelded zone of the Mesa Falls Tuff, both of which are poorly exposed in the slumped roadcuts leading to the falls. The inner gorge walls expose welded ash flows lower in the Mesa Falls Tuff and a tongue of younger Gerrit Basalt that flowed down Henry's Fork when the gorge was about half its present depth; this basalt now maintains the river level upstream. Several flow units of the basalt are well exposed on this side of the gorge just downstream from the falls. A steep fisherman's trail can be followed below these basalt cliffs to an extraordinary riverside view of Upper Mesa Falls from below and a close look at welded Mesa Falls Tuff.

REFERENCES CITED

Armstrong, R. L., Leeman, W. P., and Malde, H. E., 1975, K-Ar dating, Quaternary and Neogene volcanic rocks of the Snake River Plain, Idaho: American Journal of Science, v. 275, p. 225–251.

Christiansen, R. L., 1982, Late Cenozoic volcanism of the Island Park area, eastern Idaho, *in* Bonnichsen, B., and Breckenridge, R. M., eds., Cenozoic geology of Idaho: Idaho Bureau of Mines and Geology Bulletin 26, p. 345–368.

Christiansen, R. L., and Lipman, P. W., 1972, Cenozoic volcanism and plate-tectonic evolution of the western United States; Part 2, Late Cenozoic: Royal Society of London Philosophical Transactions, Series A, v. 271, p. 249–284.

Christiansen, R. L., and McKee, E. H., 1978, Late Cenozoic volcanic and tectonic evolution of the Great Basin and Columbia intermontane regions, *in* Smith, R. B., and Eaton, G. P., eds., Cenozoic tectonics and regional geophysics of the western Cordillera: Geological Society of America Memoir 152, p. 283–311.

Embree, G. F., McBroome, L. A., and Doherty, D. J., 1982, Preliminary stratigraphic framework of the Pliocene and Miocene rhyolites, eastern Snake River Plain, Idaho, *in* Bonnichsen, B., and Breckenridge, R. M., eds., Cenozoic geology of Idaho: Idaho Bureau of Mines and Geology Bulletin 26, p. 333–343.

Hamilton, W., 1965, Geology and petrogenesis of the Island Park caldera of rhyolite and basalt, eastern Idaho: U.S. Geological Survey Professional Paper 504-C, 37 p.

Hildreth, W., Christiansen, R. L., and O'Neil, J. R., 1984, Catastrophic isotopic modification of rhyolitic magma at times of caldera subsidence, Yellowstone Plateau volcanic field: Journal of Geophysical Research, v. 89, p. 8339–8369.

Morgan, L. A., Doherty, D. J., and Leeman, W. P., 1984, Calderas on the eastern Snake River Plain; The Heise Volcanic Group and associated eruptive centers: Journal of Geophysical Research, v. 89, p. 8665–8678.

Smith, R. B., and Christiansen, R. L., 1980, Yellowstone Park as a window on the Earth's interior: Scientific American, v. 242, no. 2, p. 84–95.

Menan Buttes, southeastern Idaho

D. N. Creighton, Chevron U.S.A., Inc., P.O. Box 1635, Houston, Texas 77251

LOCATION AND ACCESSIBILITY

The Menan Buttes are located at the Madison County–Jefferson County border in southeast Idaho, 56 mi (90 km) southwest of Yellowstone National Park and 9.8 mi (15.7 km) west-southwest of Rexburg, Idaho (Fig. 1). From Rexburg, take Idaho 33 west and turn south at an improved light-duty road that is 2.6 mi (4.3 km) past the bridge that crosses the Henry's Fork (Fig. 1). North Menan Butte lies 1.9 mi (3 km) south of Idaho 33. From Idaho Falls, take U.S. 191/20 northeast to the town of Lorenzo; turn left at Idaho 80 and go past the town of Annis and a double bend in Idaho 80 where the road passes up and over two small volcanic cone remnants (Fig. 1). After passing the remnants, turn north at the second paved road that extends northward. North Menan Butte lies 3.9 mi (6.2 km) north of Idaho 80 on this road. As shown on Figure 1, paved light-duty and unimproved dirt roads encircle North Menan Butte and allow easy access for two-wheel drive vehicles. North Menan Butte is owned by the U.S. Bureau of Land Management and is open to the public, whereas South Menan Butte is privately owned, and permission is required from David South to gain entry. Since the two buttes are similar, it is advisable to study the one with public access rather than asking for access to the privately owned butte.

SIGNIFICANCE

The Menan Buttes are basaltic tuff cones that were formed by phreatomagmatic eruptions. They have been used in numerous textbooks as type examples of tuff cones and rank among the largest of this type of volcano. The buttes are middle to late Pleistocene in age, as established by overlap relations (LaPoint, 1977). Their dimensions and gross shapes are illustrated in Figure 1. In addition to their unusual size, the buttes display a number of geologic features resulting from soft-sediment deformation. Also exhibited are air-fall deposit variations, xenolithic fragments, and hydrothermally altered sideromelane ash. The most significant feature observable on the buttes is their morphology, which provides insight into their eruptional history.

GEOLOGY

Menan Buttes lie within the Eastern Snake River Plain (ESRP) volcanic subprovince, where tholeiitic basalt predominates in surface exposures. The primary expression of these basalt exposures is broad, flat shield volcanoes that were formed by low-viscosity flows. Frequent eruptions that occurred in the past resulted in a coalescence of individual flows, which slowly built the basaltic plain we see today. The flat expanse of the plain is broken in places by topographically distinct pyroclastic volcanoes, some of which are similar in composition to the tholeiitic basalt shield volcanoes. Phreatomagmatic processes produced many of these volcanoes, including the Menan Buttes; however, the buttes are not typical examples of phreatomagmatic volcanoes because of their unusually large sizes, the result of the amount of basalt erupted and the volume of water available to sustain the phreatomagmatic reactions. In some pyroclastic eruptions of this type, initial activity is dominated by phreatomagmatic processes, but after a significant amount of ash and scoria deposition, the water source becomes sealed off from the conduits and the eruptions evolve into quiescent outpourings of basaltic magma (Lorenz, 1973). Apparently the aquifer involved with eruption of Menan Buttes maintained its access to the vents throughout their eruptions, because no basalt flows are associated with their pyroclastic deposits.

An understanding of the aquifer involved with eruption of Menan Buttes is necessary to provide insight into the dynamics of their formation. A driller's log from an 850-ft (260-m) deep well at Menan indicates the well bottomed in gravel. The entire section penetrated by the well consists of fluvial clastics (mostly sand and gravel) and some volcanic ash. Apparently a considerable thickness of clastic material has been deposited within the meander belt of the Snake River, and this material is directly connected to the extensive *surface* drainage system of the Snake River. In addition, these porous sediments are charged by water from the unique Snake River Plain *subsurface* hydrologic system (Stearns and others, 1938). This aquifer is prolific and was able to continuously provide water to the conduits of the Menan Buttes throughout their eruptive periods.

With this understanding of the groundwater system at Menan, it is possible to evaluate the eruptive mechanisms of the buttes. Generally, the morphology of a phreatomagmatic volcano is controlled by the depth of the magma-water interaction (Lorenz, 1975). When the explosive contact occurs near the surface, particle trajectories tend to deviate from vertical, forming *tuff-rings* having a large diameter/height ratio. Conversely, the deeper magma-water contact creates a gunbarrel effect with particle trajectories that are closer to vertical. This causes the particles to be deposited closer to the vent, forming a *tuff-cone* that has a smaller diameter/height ratio. The important factors to consider are depth of magma-water contact, rate of magmatic effusion, and the ability of the water source to replenish itself at the zone of interaction. In order to simplify discussion of the eruptive conditions that formed Menan Buttes, the water influx rate will be considered to have been constant. The implication is that changes in the depth of magma-water interaction were caused by changes in the rate at which magma was extruded. Commonly, volcanic eruptions are strongest during initial activity and exhibit waning energy as the eruption progresses (Wood, 1980). This tendency appears to have been present during the formation of Menan Buttes.

D. N. Creighton

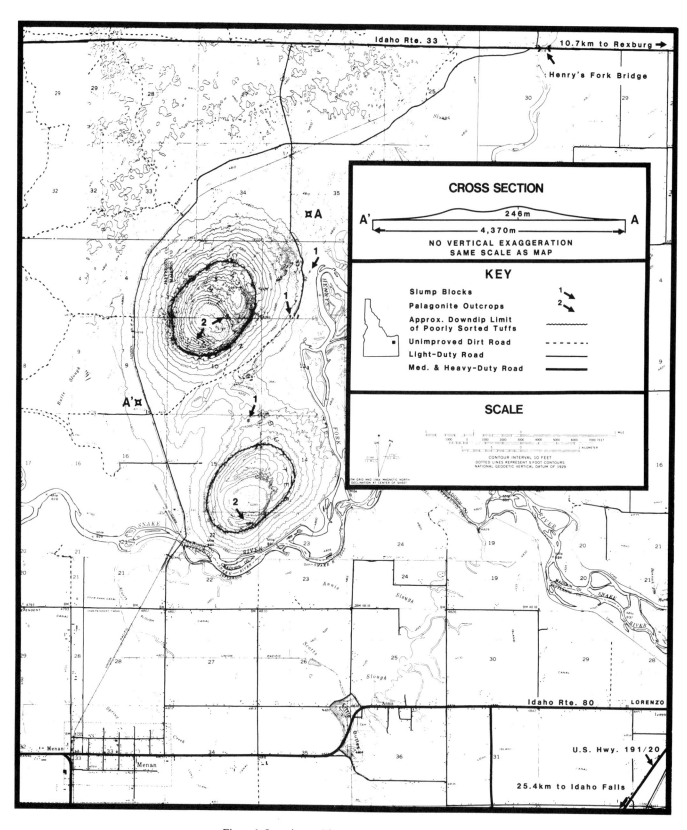

Figure 1. Location and features map of Menan Buttes.

Initial activity during the Menan Buttes eruptions appears to have been the most violent. Outcrops of tuff on the buttes indicate that the earliest deposits formed a low-relief, wide-diameter tuff ring. These tuffs are dark greenish-brown and have a resinous luster. They are composed principally of sideromelane ash with minor amounts of scoriaceous lapilli and are strongly consolidated. Individual beds are generally 0.4–4 in (1–10 cm) thick and exhibit both normal and reverse grading. The mode of deposition for these initial tuffs was probably air fall. Their well-stratified nature is most likely due to relatively long trajectories, which enabled efficient particle sorting. The presence of both normal and reverse grading can be attributed to eruptive pulsing.

As time progressed, eruptive energy decreased, allowing the magma-water contact to occur at greater depth. This change focused deposition within an area closer to the vents. The decrease in energy may have been abrupt at North Menan Butte because there is a sharp boundary between tuffs of this stage and the earlier deposited, well-stratified tuffs. The result was formation of a tuff-cone perched on top of the tuff-ring. These later stage tuffs are similar in color and luster to the earlier deposited tuffs but are poorly sorted and appear massive. They are strongly consolidated with individual beds ranging from 1.2 in to 3.3 ft (3 cm to 1 m) in thickness. Air fall was most likely their mode of deposition as well. Factors that may have affected sorting during this stage include shorter trajectories, collisions between falling and upward traveling particles, and more rapid pulsing. These poorly bedded tuffs were piled up until oversteepened bedding planes developed; downslope creep of the unconsolidated ash ensued. The result was the formation of large- and small-scale folds, which are visible within gullies that are incised into the flanks of the cones. The small-scale folds typically have amplitudes ranging from 2–4 in (5–10 cm) and wavelengths between 6 and 24 in (15 and 60 cm). The large-scale folds have amplitudes and wavelengths of 3–13 ft (1–4 m) and 66–98 ft (20–30 m) respectively. Folding is typically found only within tuffs that dip in excess of 35°. Maximum recorded dips on North Menan Butte are near 55°. Downslope creep often culminated in the detachment of large blocks of ash that slid downslope to positions on the lower flanks of the buttes (Fig. 1). These blocks are readily identified by their internal deformation, massive bedding, and the relief they display with respect to ash beds upon which they rest. One block was transported 0.9 mi (1.4 km) to its final resting place.

An interesting constituent of the deposits forming North and South Menan Buttes are xenoliths, which comprise a minor percentage of the overall ejecta. They range from sand-sized particles to large cobbles and are principally fluvial deposits that were carried up from the subsurface. Many of these xenoliths are basaltic or rhyolitic in composition, and sometimes they have an angular shape, indicating that these particles might represent pieces of country rock torn off the walls of the magmatic conduits.

One other distinctive feature found on the buttes is the occurrence of large masses of palagonite, which is an alteration product of sideromelane glass. At the Buttes, this palagonite is light greenish-brown with a vitreous luster. It exhibits a weak conchoidal fracture and is very hard. Significant outcrops of palagonite on North Menan Butte occur primarily as large, irregularly shaped masses within close proximity to the crater (Fig. 1). The discordant nature of these palagonite masses indicates they were formed by posteruptional hydrothermal alteration as evidenced by the lateral gradation of massive palagonite into unaltered, bedded-tuff.

REFERENCES

LaPoint, P.J.I., 1977, Preliminary photogeologic map of the eastern Snake River Plain, Idaho: U.S. Geological Survey Miscellaneous Field Studies Map MF-850.

Lorenz, V., 1973, On the formation of Maars: Bulletin of Volcanologique, v. 37, p. 183–204.

—— , 1975, Formation of phreatomagmatic Maar-diatreme volcanoes and its relevance to kimberlite diatremes: Physical Chemistry of the Earth, v. 9, p. 17–27.

Stearns, H. T., Crandall, L., and Steward, W., 1938, Geology and ground water resources of the Snake River Plain in eastern Idaho: U.S. Geological Survey Water Supply Paper 774, 268 p.

Wood, C. A., 1980, Morphometric evolution of cinder cones: Journal of Volcanology and Geothermal Research, v. 7, p. 387–413.

A Neogene(?) gravity-slide block and associated slide phenomena in Swan Valley graben, Wyoming and Idaho

David W. Moore and Steven S. Oriel, U.S. Geological Survey, Box 25046, Denver Federal Center, Denver, Colorado 80225
Don R. Mabey, Utah Geological and Mineral Survey, 606 Black Hawk Way, Salt Lake City, Utah 84108

LOCATION AND ACCESSIBILITY

The gravity-slide block lies below the steep west flank of the Snake River Range near Alpine, Wyoming. Alpine is 37 mi (60 km) southwest of Jackson, Wyoming, via U.S. 89-26 and 73 mi (117 km) east of Idaho Falls, Idaho, via U.S. 26. A Forest Service campground and several motels, restaurants, and stores are located at Alpine.

Rocks are well exposed at three stops on U.S. 26 north of Alpine along the east shore of Palisades Reservoir (Fig. 1). All stops are in the Targhee National Forest. No permission is required to visit them and all are accessible by passenger car when roads are dry. Stop 2 is reached by hiking 0.4 mi (0.7 km) overland and climbing 650 ft (200 m).

Stop 1 is in Long Springs Canyon (SW¼NW¼Sec.17,T.37N., R.118W. on the Alpine Quadrangle map; Albee and Cullins, 1975). Drive 1.9 mi (3 km) north on U.S. 26 from the junction of U.S. 26 and 89 to a gravel road at the U.S. Forest Service sign "4-H Camp." Turn east into Targhee National Forest and go 1.0 mi (1.6 km). Drive through the 4-H Camp and Idaho State University Geology Field Camp and across Long Springs Creek. An optional stop 1A is 0.2 mi (0.3 km) east and 0.2 mi (0.3 km) north.

Stop 2 is reached via an unimproved dirt road that is 1.3 mi (2.1 km) north of the 4-H Camp road intersection on U.S. 26. Drive east 0.5 mi (0.8 km) to the end of the dirt road. Walk upslope through the woods north-northeast to the base of the tan cliffs in the NE¼NE¼Sec.27,T.2S.,R.46E., about 650 ft (200 m) west of the state line.

Stop 3 is 2.2 mi (3.5 km) north of the dirt road entrance on U.S. 26, in the NW¼SE¼Sec.20,T.2S.,R.46E. where a white deposit is exposed in the highway cut.

SIGNIFICANCE

One of the largest ancient rockslides in the Idaho-Wyoming thrust belt is exposed near Alpine. The deposit, its structure, and stratigraphic relations to the surrounding sediments reveal much about Neogene tectonism of the region. During Neogene time, east-west extensional stresses broke Sevier thrust sheets composed of older layered rocks along major normal faults. A series of north-trending ranges and graben valleys, some tilted, formed in the Idaho-Wyoming thrust belt. Erosion of emerging ranges produced thick accumulations of volcaniclastic alluvium and lacustrine sediments that filled adjacent subsiding grabens.

Near Alpine, large-scale mass wasting accompanied the Neogene basin-range faulting. Parts of west-dipping thrust sheets

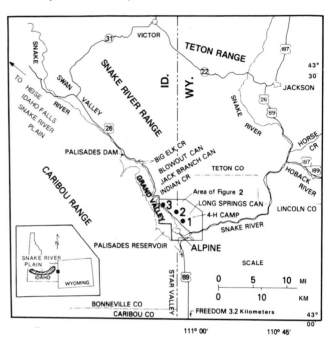

Figure 1. Index map. Numbers 1, 2, 3 are stops discussed in the text.

occasionally broke loose and slid into the valley, coming to rest as blocks or sheets of breccia that, after burial, were interbedded with graben-fill sediments. The strata were tilted eastward by incremental slip on the Grand Valley fault. In Quaternary time, the Snake River cut into the fill, exhuming some slide masses. Several sheetlike masses exposed near Palisades Dam (Fig. 1) slid during active regional volcanism about 8 to 4 m.y. ago. The large block near Alpine slid after 7 Ma. An exact upper time limit for the sliding is unknown, but is estimated to be late Pliocene or Pleistocene.

Relatively few Neogene slide masses are exposed in the thrust belt, but many were buried in subsiding grabens. Several ranges have thrusts that dip toward an adjacent valley-graben; extensional tectonism was long-lived, and exhumed slide masses are present in Swan Valley graben. Other recognized gravity-slide masses occur 1.9 mi (3 km) south of Freedom in Star Valley, Wyoming; at Horse Creek, Wyoming; and north of the Snake River Plain (Beutner, 1972).

Radiometric ages of volcanic deposits in Swan Valley graben provide a way to date gravity sliding and Neogene tectonism in the region. Sources of the volcanic deposits probably were calderas in the eastern Snake River Plain and at Yellowstone. These areas are nearby and radiometric ages of rocks there agree

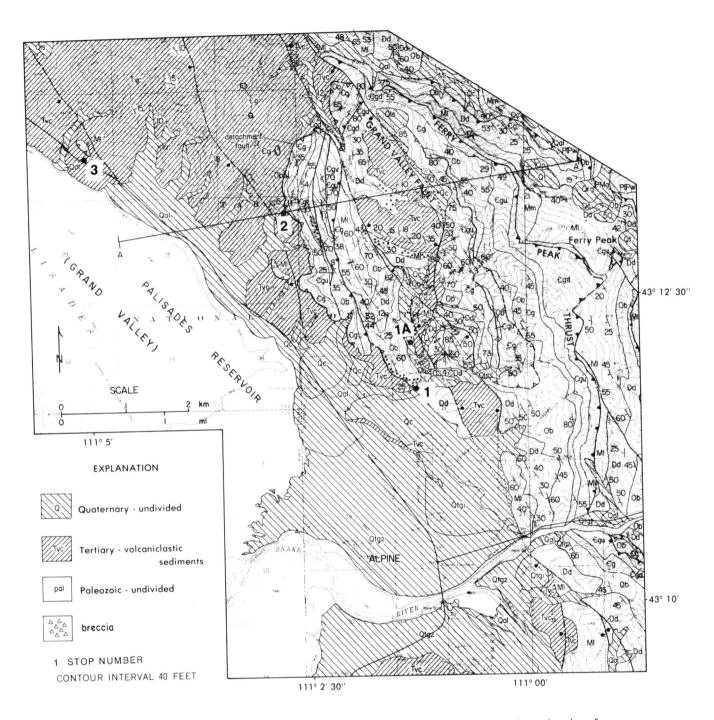

Figure 2. Geologic map simplified from Oriel and Moore, 1985 (see this reference for explanation of formations and expanded list of references). Contacts and faults dashed where inferred, dotted where concealed; bar and ball on downthrown block of normal fault; teeth on upper plate of thrust faults.

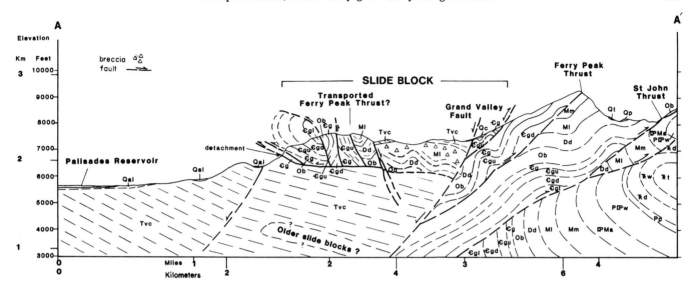

Figure 3. Geologic cross section across the slide block. Line of section is A–A′ on Figure 2. Symbols as in Figure 2.

with those in Swan Valley. One hypothesis to explain the volcanism is that a thermal plume in the mantle formed a hotspot that migrated northeastward under the eastern Snake River Plain as the continent drifted westward over it. Another explanation is that volcanism followed a northeastward propagating rift in the crust. Heating and thinning of the crust, whatever the cause, could have increased uplift and rotation of crustal blocks that began in early Neogene.

Besides being indicators of paleotectonism, slide deposits have economic and engineering significance. Buried breccia sheets could be host rocks for mineralizing hydrothermal solutions, especially near volcanic calderas as in the Snake River Plain. Sites for engineering projects such as reservoirs must be located to avoid geologic structures that might slide during earthquakes.

GEOLOGY

Stop 1. The best place to see the base of the block resting on fanglomerate valley fill is 164 ft (50 m) north of where the dirt access road crosses the stream in Long Springs Canyon. Here, 0.6 mi (1 km) west of the range-front fault, the base dips about 45° east, but along its west edge it averages 6° south-southeast. The local steeper dip is tectonic and records subsidence on a secondary high-angle fault in the Grand Valley fault system. This forms a small north-south graben under Long Spring Canyon (Figs. 2 and 3).

The planar basal contact dips about 45° east from the top of the ravine at the west end of the grassy hillslope that is the north wall of the canyon. Above the basal contact, cobbles and bouldery masses are monolithologic—brecciated light gray, nearly white, finely crystalline Bighorn Dolomite of Ordovician

age. A few masses of cemented breccia have rolled down to the canyon floor.

Below the contact is a polymictic fanglomerate/sedimentary breccia veneered by colluvium. This is a subtle clue proving that under the hillslope a tectonic breccia made of Paleozoic rock is above a fanglomerate and sedimentary breccia of Neogene age, an apparent exception to the law of superposition! Actually the law is followed because the Paleozoic rocks slid off the range sometime after about 7 Ma, and the younger Neogene slide-block breccia overlies the older (also Neogene) fanglomerate in proper stratigraphic order. Water-abraded pebbles from the lower unit are fossiliferous (horn corals, brachiopods, and bryozoans) gray limestone, white dolomite, and sparse white quartzite, that were transported by streams from Paleozoic formations in the range. Fossil snail shells from a fine-grained facies of this unit have been assigned a Pliocene age. Near the highway this unit contains layers of volcanic ash dated by fission-tracks in zircons at 6 to 7 Ma (late Miocene).

The view north from Stop 1A is an east-west cross section through the block (Fig. 3). East-dipping brown beds of conglomerate that cap the section were assigned a late Pliocene to Pleistocene age by Merritt (1956). From west to east where the beds abut the Grand Valley fault, dips steepen from 15° to almost 45°, indicating dip slip on the fault after block sliding and deposition of the conglomerate. The dipping beds formed by downward rotation around an imaginary hinge to the west characterize Neogene evolution of the larger Swan Valley graben. From relations seen here, latest movement on the fault can be bracketed only as late Neogene or Pleistocene.

If the capping conglomerate beds are as old as late Pliocene, the block must be exhumed. An exhumed block requires that Quaternary erosion must have speeded up or graben sinking must have slowed. Did glacial meltwater floods in the Snake River accelerate erosion in Swan Valley? Or did slowing of subsidence of the graben allow an increase in local erosion? In late Neogene time increased volcanism north of Swan Valley was attended by high rates of uplift. Rapid alluviation produced large alluvial fans and large-scale mass wasting into rapidly sinking grabens. As volcanic centers migrated northeastward, vertical uplift slowed, allowing erosion to "catch up." Could accelerated uplift of the range some 8 to 4 m.y. ago have triggered apparent maximum gravity sliding in the graben?

Stop 2. Three vertical zones of outcrop-sized structures are present in Cambrian limestone at the base of the block. These zones suggest that as the block slid west it overrode a basal layer of breccia that was mobile and flowed during or after the sliding. Shear stresses decreased upward in the block during the slide event. The largest shear was generated by frictional drag on the base. It formed a pervasively brecciated and sheared basal zone 10 to 26 ft (3 to 8 m) thick. Above the basal zone, limestone is broken by vertical fractures, some of which were injected by breccia dikes like those at Heart Mountain (Pierce, 1979). A recumbent isoclinal fold occurs in this zone, probably formed by horizontal shearing in a zone sandwiched between the opposing frictional drag on the base and the downhill momentum of the sliding block above. Above the breccia dikes, thick coherent limestone beds have been folded, overturned, and offset along nearly vertical faults. Geologic mapping of the west end of the block suggests that the basal breccia is part of the limb of an overturned fold that was overridden during transport (Figs. 2 and 3). The east-west order of formations in the block is the same as that of the range.

A detailed local gravity survey suggests a moderately thin, dense detached block overlying less dense conglomerate. One gravity model produced a block about 0.6 mi (1 km) thick using a measured bulk density of 2.46 g/cm^3 for the conglomerate and an assumed bulk density of 2.66 g/cm^3 for Paleozoic rocks.

Stop 3. At site 3 a bed of white volcanic ash 10- to 13-ft thick (3 to 4 m), is interbedded with fluvial pebbly sand and thin clay layers cut by channel gravel. A small slide block of limestone breccia juts from a hillslope above the highway. The ash is about 7 Ma and lies stratigraphically below the block. Sliding of this small block and the larger block at Stop 1 followed deposition of the ash.

Roadcuts north of Stop 3 along U.S. 26 expose Neogene volcaniclastic sediments, chiefly distal alluvial fan facies, some laminated midbasin lacustrine clays and silts, and layers of white rhyolitic volcanic ash.

A roadcut 656 ft (200 m) south of Indian Creek illustrates intragraben faulting. A bed of gray limestone breccia, at the top of the cut, overlies yellow and reddish brown soft sandstone beds that dip 35° north. The limestone breccia, offset by a normal fault, is intruded by one steeply dipping white ash bed and truncates another. The following sequence of Neogene events is indicated: volcanic ash fell on sandy alluvium on the distal part of fans; sand was deposited on the ash by streams; a block of limestone detached from the range and slid into the valley; layers were tilted northeastward by intermittent movement on the Grand Valley fault. Quaternary streams cut valleys into the Neogene sediments. Slumping on the valley wall of Indian Creek broke and rotated the gray limestone breccia to its north dip. During slumping, loading of water-saturated sand and ash layers by the overlying limestone caused them to fail by liquifaction, injecting the overlying rocks.

Northward 2.5 mi (4 km) from Indian Creek a white layer of volcanic ash 26 ft (8 m) thick is exposed in a cut of U.S. 26 on the north side of Jack Branch Canyon. Fission-tracks in zircons taken from this give a 6.68 ± 0.4 Ma age (R. G. Bohannon, cited in Moore and others, 1984).

Farther north and uphill from the highway sheetlike slide masses of limestone breccia are interbedded with conglomerate. One at Big Elk Creek (Fig. 1) is about 130 ft (40 m) thick and 2.8 mi (4.5 km) wide. They are a disaggregated type of mass-wasting deposit that spread more broadly on alluvial fan surfaces during emplacement than did the relatively coherent block at Alpine.

REFERENCES CITED

Albee, H. F., and Cullins, H. L., 1975, Geologic map of the Alpine Quadrangle, Bonneville County, Idaho, and Lincoln County, Wyoming: U.S. Geological Survey Map GQ-1259, scale 1:24,000.

Beutner, E. C., 1972, Reverse gravitative movement on earlier overthrusts, Lemhi Range, Idaho: Geological Society of America Bulletin, v. 83, p. 839–846.

Merritt, Z. S., 1956, Upper Tertiary sedimentary rocks of the Alpine, Idaho-Wyoming area: Wyoming Geological Association 11th Annual Field Conference Guidebook, p. 117–119.

Moore, D. W., Woodward, N. B., and Oriel, S. S., 1984, Preliminary geologic map of the Mount Baird Quadrangle, Bonneville County, Idaho, and Lincoln and Teton counties, Wyoming: U.S. Geological Survey Open-File Report 84-776, scale 1:24,000.

Oriel, S. S., and Moore, D. W., 1985, Geologic map of the West and East Palisades RARE II further planning areas, Idaho and Wyoming: U.S. Geological Survey Miscellaneous Field Studies Map MF-1619-B, scale 1:50,000.

Pierce, W. G., 1979, Clastic dikes of Heart Mountain fault breccia, northwestern Wyoming, and their significance: U.S. Geological Survey Professional Paper 1133, 25 p.

Geologic framework of the Boise Warm Springs geothermal area, Idaho

Spencer H. Wood, Department of Geology and Geophysics, Boise State University, Boise, Idaho 83725
Willis L. Burnham, 3220 Victory View, Boise, Idaho 83709

LOCATION

The Boise Warm Springs area is 2 mi (3.3 km) southeast of the Idaho State Capitol building (Fig. 1). Turn north from Warm Springs Avenue onto Old Penitentiary Road. Take the first left turn, about 200 ft (70 m), and proceed west about one block to Quarry View City Park. A major part of the area is a public park administered by the City of Boise and the Idaho Historical Society. A prominent outcrop of rhyolite, locally known as "Castle Rock" is on private land, but at the present time access is not restricted.

SIGNIFICANCE

The hydrogeology of an important geothermal-water resource is well displayed at the Boise Warm Springs area. Late Cenozoic volcanic and sedimentary strata exposed in the hill of an upthrown fault block are the same geologic layers as the subsurface aquifers from which 66°C (172°F) water has been produced from wells for nearly a century. Structure and volcanic stratigraphy at the site is typical of the western Snake River Plain, a 30-mi-wide (50 km) grabenlike basin filled with late Cenozoic silicic and basaltic volcanic rock and fluvial and lacustrine sediments. The fault system of the graben margin is the deep conduit for geothermal groundwater that supplies the oldest geothermal heating district in the United States. On the site are the original 1892 geothermal wells and pump house and the 1870 Idaho Territorial Penitentiary: both are listed on the National Register of Historic Places.

GEOLOGIC FRAMEWORK

The sequence of late Cenozoic rocks on the rocky hillside above Quarry View Park has been downfaulted to the south to a position 850 ft (260 m) deep beneath the park area. A structure section and map are shown in Figures 1 and 2. The stratigraphy of the rocks on the hillside is illustrated in Figure 3. These rocks are underlain by granite of the Idaho batholith, which is well exposed along the road in Cottonwood Creek Canyon, 2 mi (3 km) north of the park (Figs. 1 and 4). The oldest rock at the site is a body of glassy and stony rhyolite of the Idavada Group. The rhyolite is locally onlapped by a sequence of basalt flows, 50 ft (15 m) thick. Unconformably overlying the basalt and rhyolite is a poorly exposed 15 to 30 ft (5 to 10 m) sequence of red-and-green sandy clay of the Idaho Group. At the top of the section is a 20-ft-thick (6 m) bluff-forming sandstone that was extensively quarried in the nineteenth century for building stone.

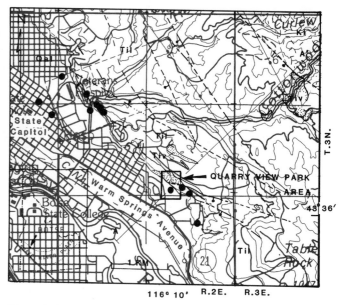

Figure 1. Geologic map of northeast Boise showing location of Quarry View City Park. Ki, granite of the Idaho batholith; Tiv, rhyolite of the Idavada Group; Til, mostly sand and siltstone of the lower Idaho Group, but contains a significant section of basaltic tuff and basalt in its lower part; Qal, alluvium and terrace gravels of the Boise River floodplain. Black dots show locations of geothermal water wells used in heating systems discussed by Wood and Burnham (1983).

Quarries and rock piles on the hill are a reminder of the prisoners' labor in procuring dimension stone to construct the penitentiary walls.

Main production of geothermal wells in the northeast Boise area is from fractures in subsurface rhyolite. The basalt and clay unit is comparatively thick in the subsurface and serves as a sealing aquitard on the rhyolite aquifer. Prior to uplift and exposure by erosion, the sandstone was also a geothermal aquifer fed by hot water ascending through high-angle faults. It became cemented as these waters travelled laterally through the sandstone and cooled, causing dissolved silica to be precipitated.

The main fault is obscured by talus and colluvium at the base of the hill, but its approximate location is shown in Figure 5. Several minor fault planes cut sandstone outcrops on the hillside. One fault, which dips about 60° south, drops the sandstone stratum down about 80 ft (25 m) to form a wedge-shaped plateau in the area east of the Warm Springs Water District pump house and north of the penitentiary. This fault is one of several within a

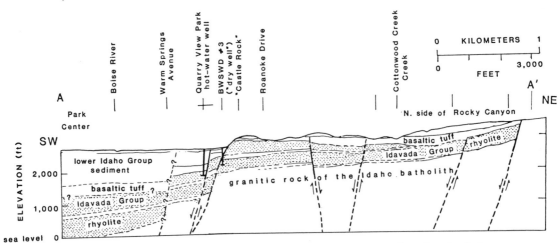

Figure 2. Geologic cross section through the well at Quarry View Park. Location of section is shown along line A–A' in Figure 1.

complex zone of normal faults that is characteristic of the northeast margin of the western Snake River Plain. The exposed fault zone in the Boise foothills is 2 mi (3 km) wide. Geophysical exploration has found other faults beneath the plain, just south of this area, and the total offset of the volcanic sequence on seismic sections is at least 0.9 mi (1.4 km; Wood, 1984).

LINDGREN'S LACCOLITH

Prior to our work, the only published work on the Boise Warm Springs area was an interpretation of this site by Lindgren (1898a) as a "laccolith in miniature." Lindgren's interpretation that the rhyolite is intrusive requires that it is younger than the sandstone, and a relatively young age for silicic igneous activity. The interpretation is important to understanding the geothermal resource, because the laccolith concept suggests a shallow magma body as the heat source, and our interpretation relies on the deep regional heat flow anomaly. However, his interpretation is interesting from a historical point of view. Laccolith intrusives of the Henry Mountains, Utah, described by G. K. Gilbert in the late 19th century were a relatively new concept that clearly influenced Lindgren.

In our interpretation, faulting and warping can account for the arched sandstone stratum (Fig. 5). The basalt strata clearly lapped upon the rhyolite as shown by a red-brick–colored soil at the bottom of the basalt sequence that was oxidized by the heat of the basalt flow.

Lindgren (1898a, 1898b) recognized faulting and orogenic movement as a process in the region, but he did not recognize faulting at this locality. He attributed the northwest-trending margins of the western plain to early Neogene faulting, which he must have believed to be buried by the deposits at this site. He assigned a late Neogene age to the north-south–trending set of faults that displace Miocene Columbia River basalt in the region north of the plain. In an early paper, H. E. Malde pointed out the

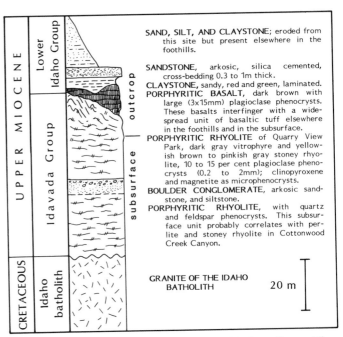

Figure 3. Stratigraphic section of outcrop on hill above Quarry View City Park and subsurface geologic units known from drilling beneath the park.

importance of large-displacement late Cenozoic northwest-trending faults that control the margins of the western plain (cited by Malde, 1987). Such faults apparently truncate the earlier Miocene north-south trend.

REGIONAL STRATIGRAPHY DISPLAYED AT SITE

Stratigraphic names of late Cenozoic rocks of the western Snake River Plain generally follow the work of Malde and his colleagues (cited by Malde, 1987). Interfingering volcanic, flu-

Figure 4. View toward the northeast of the Quarry View and Castle Rock area. It is this view of the sandstone-draped rhyolite mass that suggested the appearance of a laccolith structure to Lindgren in 1898a. Snow covered hills rise to 5,900 ft (1800 m) elevation above the plain, which is about 2,800 ft (850 m) in elevation.

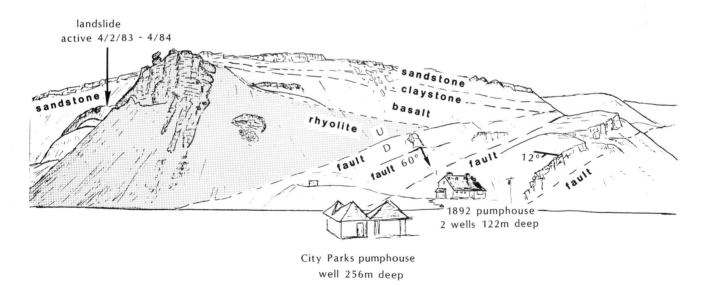

Figure 5. Field sketch of geologic features and the geothermal wells as seen from Quarry View City Park.

vial, and lacustrine facies complicate the assignment of formational names to these rocks. Armstrong and others (1975) described a systematic progression of volcanic and sedimentary facies that has migrated from west to east, showing that at any particular local section one sees (from oldest to youngest): (a) silicic volcanic rocks, mostly as ash-flow tuff deposits with minor basalt and sediment intercalations; (b) basalt flows with minor intercalated sediment and local silicic volcanic rocks; and (c) lacustrine and fluviatile sediments complexly interstratified with basalt flows. Associated with this facies progression is a systematic decrease in age of the rhyolite to the east, which has been interpreted by many as a record of the continental lithosphere sliding over a fixed "hot spot" in the mantle at an average rate of 3.5 cm/yr) (Leeman, 1982).

A similar facies sequence is exhibited on a much-compressed vertical scale in the Boise Warm Springs area (Fig. 3). The oldest rock here is of the upper Miocene Idavada Group of silicic volcanic rocks. The dark outcrops on the hill are a part of a tabular body of rhyolite which is known from recent drilling to be at least 300 ft (90 m) thick. It occurs in drill holes 850 ft (260 m) beneath the park area, and 2,000 ft (610 m) beneath the State Capitol Building, about 1.9 mi (3 km) west of here. The rhyolite has two lithologic variations here: a dark gray flow-banded vitrophyre, and a lighter-colored stoney rhyolite. We believe both lithologies are of about the same emplacement age; the differences are possibly produced by devitrification. Phenocryst mineralogy of the two lithologies is identical. In the stoney rhyolite, plagioclase phenocrysts have decomposed to a soft white chalky material that has washed out along joint surfaces, leaving a pitted surface. The rhyolite is flow banded on large and small scale, and has thin sheeting joints parallel to large-scale flow banding. Large-scale flow banding shows an asymmetric anticlinelike structure about 260 ft (80 m) wide. Drilling shows that the flow-banded rhyolite is part of a widespread flow or a tabular sheet of densely welded ash-flow tuff. However, the flow-banded rhyolite could also be part of a dome marking the source of the tabular flow. It is difficult in this region to distinguish the form of a rhyolite body at one locality, because densely welded tuffs are known to have completely remelted and remobilized as lavas (Ekren and others, 1984). A petrologically different rhyolite body underlies the ones exposed here. It was penetrated by the Boise Warm Springs Water District (BWSWD) no. 3 well (Fig. 2). This underlying rhyolite is distinguished by conspicuous quartz phenocrysts. This type of rhyolite is exposed as perlite and stoney rhyolite in Rocky Canyon about 1.5 mi (2.5 km) northwest of this area (Figs. 1 and 4). This type has a higher content of silica (76 percent), potassium, and sodium than the rhyolite mass exposed here at Quarry View City Park, which is 70 percent silica (S. H. Wood, unpublished analysis).

A tabular body of plagioclase-porphyritic basalt laps upon and, in places, overlies the rhyolite in outcrops on the hillside. Many vesicles and fractures are variously filled with zeolite minerals, calcite, clay, and chalcedony. Much of the basalt is discolored and altered, apparently by geothermal water. Chemical analysis of a little-altered sample is similar in composition to the group of "Snake River olivine tholeiites" described by Hart and others (1984).

The overlying sediments are fluvial and lacustrine rocks of the lower Idaho Group. Cross-beds in the sandstone stratum at the crest of the hill indicate deposition under river-bed conditions. A significant thickness of clayey siltstone originally covered this sand unit, but has since been stripped off of the uplifted fault block by erosion. It is preserved in the subsurface beneath the park area where the geothermal wells penetrate about 650 ft (200 m; Fig. 2).

The volcanic and sedimentary rocks on the northern margin of the western plain have not been dated. K-ar ages of the Idavada Rhyolite Group on the south side of the western plain span 9.8 to 13.5 Ma (Armstrong and others, 1975), and this span is certainly reasonable for the rhyolite exposed at this site. Affinity of the basalt to the "Snake River olivine tholeiites" is not helpful for age assignment because that geochemical group ranges from late Miocene to early Pleistocene (Hart and others, 1984). The overlying sediments are similar in lithology and setting to the Chalk Hills Formation of the lower Idaho Group on the south side of the Snake River Plain where intercalated volcanic ash layers have been dated by fission-track methods at 6.6 to 8.6 Ma or late Miocene age by Kimmel (1982).

THE GEOTHERMAL GROUNDWATER SYSTEM

This area is the site of the oldest space-heating development using geothermal water in the United States. Several wells were drilled here beginning in 1890. The first two, completed to 394 ft (120 m) and 404 ft (123 m), were strongly artesian and flowed at 170° F (77° C) water temperature at a combined rate of about 550 gallons per minute (35 l/s). In 1896 or early 1897, Lindgren (1898a) noted that the water was under moderate pressure and would rise not more than 50 ft (15 m) above the well mouth, or to about elevation 2,815 ft (858 m). This artesian flow of water was piped to a large natatorium, then to homes and businesses along Warm Springs Avenue; the unused flow was discharged to the Boise River. Credible records of the two producing wells of the Boise Warm Springs Water District are not available, but the present wells are believed to have been constructed in the early 1900s at or near the site of the original wells. The pumphouse (Fig. 5) incorporates the lower part of wooden derricks built to construct these wells. The house is on the National Register of Historic Places and was restored in 1982. The wells are 30 ft (9 m) apart and equipped with turbine pumps. According to owner records, the wells are lined with 16-in-diameter (41 cm) steel casing to 160 ft (49 m), and are an open 9-in (23 cm) hole to approximately 400 ft (122 m) depth. The wells are not capable of being shut in; however, based on tests in 1979, the calculated artesian head at the west well was at about elevation 2,783 ft (848 m), suggesting an artesian head loss of about 32 ft (10 m) since 1896. Both pump intakes are set at about elevation 2,606 ft (794 m), just above the 160-ft-deep (49 m) base of casing. This pump setting limits available drawdown to 160 ft (49 m) and consequently limits maximum available pump discharge. Artesian-head decline is expected to continue, and to be accelerated by post-1980 development of the Boise geothermal system. No change in temperature, or the physical and chemical quality of the water has been documented since the initial well construction of nearly 100 years ago. The water is slightly alkaline, with total dissolved solids of 320 mg/l, typical of the chemical quality throughout the system (Mayo and others, 1984).

The distribution system and pumps were upgraded in 1979–1980 to accommodate additional demand for hot-water service. During the 5-month heating season, the wells are pumped at 850–1,000 gallons per minute (54–63 l/s) creating drawdown to just above the pump intake. For about 5 months in the spring and fall, wells are pumped at about 200–500 gallons per minute (13–32 l/s). In the two late summer months, output is about 200 gallons per minute (13 l/s). The spent water is discharged through ditches and sewers to the Boise River. Precise values of discharge rates are not available because flow-measuring equipment and record keeping have been inadequate over past years. More than 250 homes and several buildings are currently served, those of long standing being billed a flat rate of about $400 per annum. Newer users in the district are on metered service at a present charge of $0.57½ per 100 cubic feet (2.8 cubic meters). This makes the cost of geothermal heat about $0.20 per therm if temperature is dropped 50°F through heat exchangers (1 therm = 100,000 BTU). This cost compares to a current cost of natural gas heat of about $0.36 per therm (based on $2.50 per 1000 cubic feet of natural gas, 70 percent efficiency.)

Much of own knowledge of the Boise geothermal system has been derived from drilling and testing post-1980 wells 1 to 2 mi (2 to 3 km) northwest of the Boise Warm Springs site (Fig. 1). The Boise Warm Springs wells are on the same major fault-fracture zone as are several of the newer wells. At this site the wells are clearly in or very near to fractures of a principal fault (Figs. 2 and 5). All producing wells have been completed in the rhyolite. Wells for the State of Idaho Capitol Mall system, 2 mi (3 km) northwest (Fig. 1), are much deeper and encounter thick rhyolite layers and intercalated sandstone and conglomerate (Wood and Burnham, 1983). The clayey basaltic tuffs, altered basalt, and siltstone of the Idaho Group compose the confining layers for the moderately to well-confined fault-fracture system. The geothermal "aquifer" system is not necessarily a layered unit, it may be more properly visualized as a linearly distributed system of numerous, variously interconnected zones of cracks, fractures, and open work. Hydraulic conductivity within this system of extreme range of permeability has not been determined with reasonable accuracy. Further, data thus far available are not adequate to reasonably characterize aquifer thickness, areal extent, or degree of interconnection between fractured blocks and the boundary fault fractures.

Data gathered from discharge-drawdown tests of several types, from artesian head response to barometric pressure change and from long-term response of artesian head to variable annual

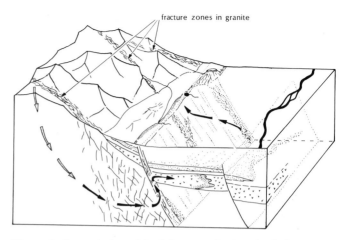

Figure 6. Conceptual model of the groundwater circulation system through fractured granite to the discharge area along the foothills fault zone of Boise and into the permeable rhyolite aquifers beneath the northeastern part of the city.

discharge rates, have been analyzed for engineering-design purposes. Such analyses suggest that within time limits of a few hours to a few days the system responds to a measured discharge-drawdown stress as though the average transmissivity is of the order of 240,000 gal/day/ft (32,000 ft^2/day or 3,000 m^2/day) and storativity is of the order of 5×10^{-4} (dimensionless). Large deviations from these gross engineering estimates occur in parts of the system. Here, near the Boise Warm Springs wells, analysis by C. J. Waag at Boise State University (personal communication, 1986) suggests transmissivities ranging from 3,500 gal/day/ft (43 m^2/day) to 25,000 gal/day/ft (435 m^2/day) and that the aquifer is confined or semiconfined. More exact values of aquifer parameters cannot be obtained until reliable observation wells are in place, discharge and injection activity by other well operators is controlled during testing, and better knowledge of fault locations and aquifer boundaries is in hand.

Several models may be developed to characterize both the heat supply and the areal geothermal groundwater circulation system. The one now thought most likely is a system of circulation to about 1 mi (2 km) depth over a path of about 6 mi (10 km) through north-to-northeast–trending deep fracture zones in the Idaho batholith (Wood and Burnham, 1983). Meteoric-water recharge at high elevations circulates with an indicated residence time (Carbon-14 activity) of 6,700 to 17,000 years (Mayo and others, 1984). Discharge at the western Snake River Plain margin is through seeps, springs, or by lateral migration through fractured silicic volcanic rock (Fig. 6). The regional high heat flow anomaly (>2.5 microcalories/cm^{2}s of the northern Basin and Range (Lachenbruch and Sass, 1978) apparently extends to the western Snake River Plain and the southern part of the Idaho batholith. Geothermal gradient in the plain and in the region of the geothermal wells is more than 40°C/km (Wood, 1984), which provides opportunity for the observed well-discharge temperatures with relatively shallow circulation. Further, heat flow along the margins of the western Snake River Plain is typically 3.0 microcalories/cm^{2}s, possibly localized by thermal refraction related to the faulted interface of relatively conductive granitic rocks of the margin with the more insulating basin fill deposits of the plain (Brott and others, 1978).

The Boise geothermal groundwater system has many geologic characteristics in common with other fault-related warm springs in the western United States. It is unique in that its potential for heating and culinary water was recognized and developed on a moderate scale nearly one century ago. Because of its prox-

imity to a center of population, demands on the system will presumably increase to its ultimate capacity. Since 1892, the Boise Warm Springs wells have averaged about 700,000 gal/day (2,650 m^3/day) for about 250 days of the year. Since 1981, withdrawals from the system have tripled to about 2,500,000 gal/day (9,500 m^3/day) for this heating season. Of this, about 700,000 gal/day is currently reinjected into the rhyolite aquifer. It is hoped that continued monitoring of the system as it is developed will avert expensive overdevelopment. Studies as the system is stressed will also greatly enhance our knowledge of the broader recharge mechanism of this type of groundwater system.

REFERENCES CITED

Armstrong, R. L., Leeman, W. P., and Malde, H. E., 1975, K-Ar dating of Quaternary and Neogene volcanic rocks of the Snake River Plain, Idaho: American Journal of Science, v. 275, p. 225–251.

Brott, C. A., Blackwell, D. D., and Mitchell, J. C., 1978, Tectonic implications of heat flow of the western Snake River Plain, Idaho: Geological Society of America Bulletin, v. 89, p. 1697–1707.

Ekren, E. B., McIntyre, D. H., and Bennett, E. H., 1984, High-temperature, large-volume, lavalike, ash-flow tuffs without calderas in southwestern Idaho: U.S. Geological Survey Professional Paper 1272, 76 p.

Hart, W. M., Aronson, J. L., and Mertzman, S. A., 1984, Areal distribution and age of low-K, high alumina olivine tholeite magmatism in the northwestern Great Basin: Geological Society of America Bulletin, v. 95, p. 186–195.

Kimmel, P. G., 1982, Stratigraphy, age, and tectonic setting of the Miocene-Pliocene lacustrine sediments of the western Snake River Plain, Oregon and Idaho, *in* Bonnichsen, B., and Breckenridge, R. M., eds., Cenozoic geology of Idaho: Idaho Geological Survey Bulletin 26, p. 559–578.

Lachenbruch, A. H., and Sass, J. H., 1978, Models of an extending lithosphere and heat flow in the Basin and Range Province, *in* Smith, R. B., and Eaton, G. P., eds., Cenozoic tectonics and regional geophysics of the western Cordillera: Geological Society of America Memoir 152, p. 209–250.

Leeman, W. P., 1982, Development of the Snake River Plain–Yellowstone

Plateau Province, Idaho and Wyoming; An overview and petrologic model, *in* Bonnichsen, B., and Breckenridge, R. M., eds., Cenozoic geology of Idaho: Idaho Geological Survey Bulletin 26, p. 155–177.

Lindgren, W., 1898a, Description of the Boise Quadrangle, Idaho: U.S. Geological Survey Geologic Atlas, Folio 103, 7 p.

—— , 1898b, The mining districts of the Idaho Basin and the Boise Ridge, Idaho: U.S. Geological Survey 18th Annual Report, pt. 3, p. 625–736.

Malde, H. E., 1987, Quaternary geology and structural history of Snake River Plain, Idaho and Oregon: *in* Morrison, R. B., ed., Quaternary Nonglacial Geology: Conterminous United States: Boulder, Colorado, Geological Society of America, The Geology of North America, v. K-2, in press.

Mayo, A. L., Muller, A. B., and Mitchell, J. C., 1984, Geochemical and isotopic investigations of thermal water occurrences of the Boise front area, Ada County, Idaho: Idaho Department of Water Resources Water Information Bulletin 30, pt. 14, 55 p.

Wood, S. H., 1984, Review of late Cenozoic tectonics, volcanism, and subsurface geology of the western Snake River Plain, *in* Beaver, P. C., ed., Geology, tectonics, mineral resources of western and southern Idaho: Dillon, Montana, Tobacco Root Geological Society, p. 48–60.

Wood, S. H., and Burnham, W. L., 1983, Boise, Idaho, geothermal system: Transactions of the Geothermal Resources Council, v. 7, p. 215–225.

Geology of the Craters of the Moon lava field, Idaho

Mel A. Kuntz, *U.S. Geological Survey, MS 913, Box 25046, Denver Federal Center, Denver, Colorado 80225*
Duane E. Champion, *U.S. Geological Survey, MS 937, 345 Middlefield Road, Menlo Park, California 94025*
Richard H. Lefebvre, *Department of Geology, Grand Valley State College, Allendale, Michigan 49401*

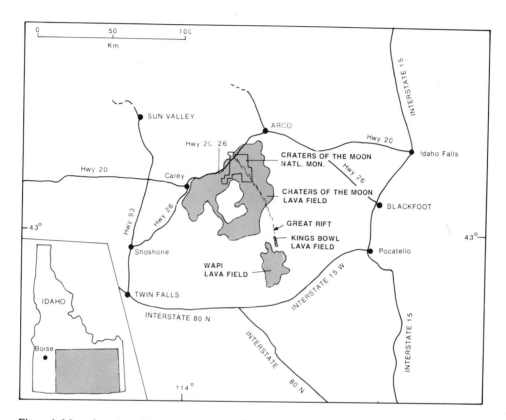

Figure 1. Map of southern Idaho showing location of Craters of the Moon, Kings Bowl, and Wapi lava fields and the Great Rift volcanic rift zone.

LOCATION AND ACCESSIBILITY

The most convenient place to see the Craters of the Moon (COM) lava field is at Craters of the Moon National Monument, reached by U.S. 20-26 from either Arco or Carey, Idaho (Fig. 1). A 7-mi (11-km) loop road and several hiking trails provide access to volcanic features within the Monument (Fig. 2). The 20-mi-long (32-km) Wilderness Trail is the longest of the trails and requires a permit for overnight trips; there is no potable water available along this trail. The east, south, and west edges of the COM lava field can be reached by four-wheel-drive vehicles on unimproved roads that are shown on U.S. Geological Survey 7½-minute quadrangle maps of the area. The geologic map of the Inferno Cone Quadrangle (Kuntz and others, 1986c) depicts the volcanic features that are most accessible to visitors to the monument. Please note that hammers may not be used and samples may not be taken in the monument.

SIGNIFICANCE

The COM lava field provides the most spectacular display of basaltic volcanic features in the conterminous U.S. This site provides geologists with an opportunity to examine the nature and style of basaltic volcanism that formed the flows on the surface of most of the Snake River Plain. The COM lava field is dominantly Holocene and consists of more than 60 basaltic lava flows that erupted from the northern part of the Great Rift, a 50-mi-long (80 km) by 1- to 5-mi-wide (1 to 8 km) volcanic rift zone (Fig. 1). The composite Craters of the Moon lava field consists of approximately 7 mi^3 (30 km^3) of basaltic flows that cover an area of about 600 mi^2 (1,600 km^2). Two other lava fields, Kings Bowl and Wapi, are located on the southern part of the Great Rift (Fig. 1). These fields are not discussed here but are shown on a map of the Great Rift by Kuntz and others (1986d).

The COM lava field formed during eight major eruptive

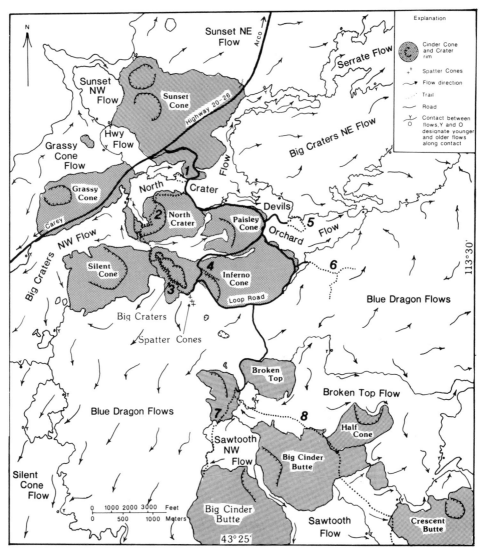

Figure 2. Simplified geologic map of the northern Great Rift area in Craters of the Moon National Monument. Numbers denote trails within the Monument: (1) North Crater flow, (2) North Crater, (3) Big Craters, (4) Inferno Cone, (5) Devils Orchard, (6) Caves, (7) Tree Molds, and (8) Wilderness.

periods (designated as A for the youngest, through H for the oldest) that occurred in the last 15,000 years (Kuntz and others, 1983, 1986a, 1986b). Most eruptive periods were probably only a few hundred years or less in length and were separated by repose intervals as short as 500 years and as long as 3,000 years. The eruptions of Eruptive Period A, which occurred about 2,100 years ago, formed most of the features present in the Monument area (Fig. 2).

VOLCANIC FEATURES OF THE GREAT RIFT

Cinder and spatter cones. Cones are the most striking features of vent systems along the Great Rift. About 25 cinder cones occur along the Great Rift in the COM lava field (Fig. 2);

most are composite or nested as a result of intermittent eruptions along several hundred yards (meters) of an eruptive fissure. Cinder cones formed early during eruption of gas-charged lava. At Big Craters, several vents are aligned along the rift system; at Grassy Cone, multiple vents show no strong alignment with the rift (Fig. 2). Some of the cones are breached with openings aligned with the Great Rift; examples are North Crater, Paisley Cone, Silent Cone, Inferno Cone, Half Cone, and Big Cinder Butte (Fig. 2).

Spatter cones produced during the late stages of Eruptive Period A are aligned along the Great Rift immediately south of the Big Craters cinder cone complex (Fig. 2). These cones formed from relatively degassed lava that produced globular, pasty masses of spatter.

Rafted blocks are parts of partially destroyed cinder cones that were transported by viscous a'a and block flows. Like cones, these blocks are composed of bedded tephra and spatter. The Devils Orchard block flow transported the rafted blocks that appear along the Devils Orchard Trail (Fig. 2, no. 5).

Lava Flows. Of the more than 60 lava flows within the Craters of the Moon lava field, about a dozen are within an easy walk of the Monument Loop Road. Craters of the Moon lava flows contain phenocrysts or microphenocrysts of olivine and/or plagioclase. The matrix consists of varying amounts of olivine, plagioclase, clinopyroxene, opaque minerals, and glass. In general, lava containing about 40 to 50 percent SiO_2 produces pahoehoe flows, flows with 50 to 60 percent SiO_2 produced a'a flows, and flows with more than 60 percent SiO_2 produced block flows.

Approximately 80 percent of the COM lava field consists of pahoehoe. Pahoehoe flows are about 30 ft (10 m) thick and are characterized by broad, flat pressure plateaus. They have ropy to filamented surfaces that locally have a blue or green glassy crust. The Blue Dragon and North Crater flows are good examples of flows with blue-crusted pahoehoe and the Big Craters flows consist of green-crusted pahoehoe. Weathered glassy crusts, more vegetation, and wind-blown sediment are more prevalent on older flows around the margins of the lava field.

Lava-tube systems fed most large-volume pahoehoe flows. As pahoehoe flows crust over, lava is delivered to the flow front by lava streams that flow in tunnels, forming tubes in the flow interior. As eruptions wane and the supply of lava diminishes, the level of lava in the tubes drops, creating a space between the roof of the tube and the surface of the flow. Lava tubes in the Blue Dragon flow in the "Caves" area (Fig. 2, no. 6) are popular attractions to visitors to the monument. Visitors to lava tubes should have flashlights, extra batteries, and hardhats.

A'a flows are rough-surfaced, have steep flow fronts, and are about 50 ft (15 m) thick. Surfaces are covered with blocks that range from a few centimeters to 1.5 ft (0.5 m) in diameter and have rough, spiny surfaces. Large a'a flows have surface waves with amplitudes of 10 ft (3 m) or more that are convex in the direction of movement. The waves appear much like the wrinkles (ogives) on the surface of some mountain glaciers. The edges of large a'a flows commonly have smoother, filamented, pahoehoe-like "squeeze-outs." Squeeze-outs formed along the margins of a'a flows as lobes of less viscous lava that were expelled from the flow interior. Unfortunately, most large a'a flows occur in remote sections of the COM lava field and are not easily accessible to monument visitors.

The Serrate, Devils Orchard, and Highway flows are block flows. These flows have smooth, rather than spiny blocks on their surfaces and, like a'a flows, have squeeze-outs at the base of steep fronts. The average thickness of block flows is about 60 ft (20 m). The Highway block flow is best seen north of U.S. 20-26 in the saddle between Sunset Cone and Grassy Cone (Fig. 2). The Devils Orchard block flow is easily reached by the Devils Orchard Trail (Fig. 2, no. 5).

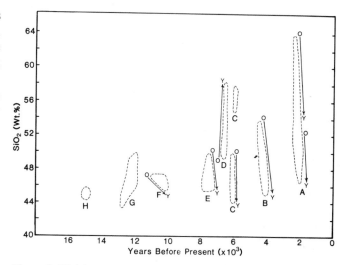

Figure 3. Weight percent SiO_2 versus age of lava flows for analyzed rocks of the Craters of the Moon lava field (after Kuntz and others, 1986a).

PETROLOGY OF COM LAVA FLOWS

Lava flows of the COM lava field have a broad range in chemical composition. The SiO_2 content ranges from about 44 percent to about 64 percent (Fig. 3). COM flows are higher in TiO_2, iron oxides, alkalis, P_2O_5, Zr_2O_3, and lower in MgO and CaO than most other basaltic suites. COM flows range from iron-, alkali-, and phosphorous-rich basalts to trachyandesite. The variation of SiO_2 content with time (Fig. 3) indicates that the compositional variability of COM magmas has increased with time, probably as a result of contamination of primitive basaltic magma by partial melts of lower crustal rocks (Leeman and others, 1976; Kuntz and others, 1986a). Most COM rocks containing more than about 52 percent SiO_2 show evidence of contamination in the form of xenoliths of granulite, gneiss, and disequilibrium minerals. The Devils Orchard flow contains abundant xenoliths of granulite and gneiss. COM rocks with SiO_2 contents of 44 to 52 percent generally lack petrographic evidence of contamination and the chemical and mineralogical variation can be accounted for mainly by crystal fractionation (Leeman and others, 1976; Kuntz and others, 1986a).

STEADY-STATE, VOLUME-PREDICTABLE VOLCANISM

The cumulative volume versus time plot for the COM lava field (Fig. 4), shows a linear, volume-predictable relationship, meaning that the volume of an eruption is proportional to the length of the prior repose interval and the long-term output rate. The long-term output rate was fairly uniform at about 0.35 $mi^3/1,000$ yr (1.5 $km^3/1,000$ yr) from Eruptive Period H through Eruptive Period E. About 6,000 years ago the output rate increased to about 0.67 $mi^3/1,000$ yr (2.8 $km^3/1,000$ yr), which

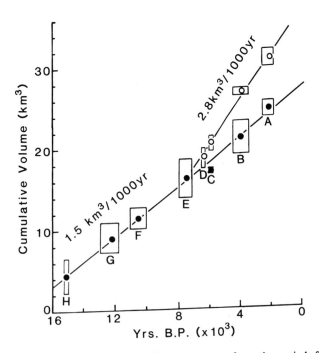

Figure 4. Cumulative volume of lava versus age of eruptive periods for the Craters of the Moon lava field. Solid dots show data for fractionated magma, open dots show data for fractionated plus contaminated magma, and letters refer to eruptive periods (after Kuntz and others, 1986a.

is largely a reflection of the addition of contaminated magma to the relatively constant output rate of fractionated magma.

On the basis of an average repose interval of about 2,000 yr and the fact that the last eruption was more than 2,000 yr ago, the Great Rift is "about due" for another eruption. However, the average repose interval has a large margin of error ($\pm$1,000 yr). If the next eruption should occur within the next few centuries, a volume of about 1.4 mi^3 (6 km^3) is predicted on the basis of cumulative volume versus age data (Fig. 4).

REFERENCES CITED

Kuntz, M. A., Lefebvre, R. H., Champion, D. E., King, J. S., and Covington, H. R., 1983, Holocene basaltic volcanism along the Great Rift, central and eastern Snake River Plain, Idaho: Utah Geological and Mineral Survey Special Studies 61, Guidebook, Part 3, p. 1–34.

Kuntz, M. A., Champion, D. E., Spiker, E. C., and Lefebvre, R. H., 1986a, Contrasting magma types and steady state, volume predictable, basaltic volcanism along the Great Rift, Idaho: Geological Society of America Bulletin, v. 97, p. 579–594.

Kuntz, M. A., Spiker, E. C., Rubin, M., Champion, D. E., and Lefebvre, R. H., 1986b, Radiocarbon studies of latest Pleistocene and Holocene lava flows of the Snake River Plain, Idaho: Data, lessons, interpretations: Journal of Quaternary Research, v. 25, p. 163–176.

Kuntz, M. A., Lefebvre, R. H., and Champion, D. E., 1986c, Geologic map of the Inferno Cone Quadrangle, Butte County, Idaho: U.S. Geological Survey Geologic Quadrangle Map GQ-1632, scale 1:24,000 (in press).

Kuntz, M. A., Champion, D. E., Lefebvre, R. H., 1986d, Geologic map of the Craters of the Moon, Kings Bowl, and Wapi lava fields, and the Great Rift volcanic rift zone, south-central Idaho: U.S. Geological Survey Miscellaneous Geologic Investigations Series Map I-1632, scale 1:100,000 (in press).

Leeman, W. P., Vitaliano, C. J., and Prinz, M. A., 1976, Evolved lavas from the Snake River Plain; Craters of the Moon National Monument, Idaho: Contributions to Mineralogy and Petrology, v. 56, p. 35–60.

The Montini Volcano; A lava dam on the ancestral Snake River, southwest Idaho

Harold E. Malde, U.S. Geological Survey, Box 25046, Denver Federal Center, Denver, Colorado 80225

LOCATION AND ACCESS

The name Montini Volcano is applied to a much-dissected basaltic vent where Sinker Creek joins the canyon of the Snake River southwest of Boise (Fig. 1). The name comes from Montini Ranch—now the Nahas Ranch—on the 1949 edition of the Sinker Butte Topographic Quadrangle (1:24,000). The volcano is also partly within the adjacent Wild Horse Butte Quadrangle. Downcutting by the Snake River and Sinker Creek has exposed the plug of the volcano and nearly 600 ft (180 m) of its surrounding tuff and lava.

The Montini Volcano can be reached from milepost 33 on Idaho 78 by following the road shown in Figure 1. Just before reaching Sinker Creek, a single-lane track negotiable by passenger cars leads down Sinker Creek to the central plug of the volcano (Fig. 2). Although a gate has been installed at the head of the track to control livestock, the track is open to the public. From the plug, the track continues to a pumping station one mi (1.6 km) down the Snake River (NW¼Sec.6,T.3S.,R.1E.). From here, looking upstream, one has a splendid view of the dissected volcano (Fig. 3). Those wanting to see other parts of the Montini Volcano must backtrack to the Nahas Ranch and ask permission to use the ranch road. From the ranch headquarters, this road leads southeast and then northeast to a point on the Snake River about 2 mi (3 km) above the mouth of Sinker Creek. A single-lane track on the north 1.7 mi (2.7 km) from the ranch climbs northwestward onto the volcano. From the end of the track, spectacular views of the Snake River Canyon and the canyon of Sinker Creek can be obtained by walking to the top of the volcano one mi (1.6 km) north (NE¼Sec.7,T.3S.,R.1E.).

The Montini Volcano can also be reached by boat from Swan Falls (Fig. 1) or by floating down the Snake River from Grand View, 25 mi (40 km) upstream.

The Snake River Canyon that cuts through the Montini Volcano is within the Snake River Birds of Prey Area, which was established in 1971 by the U.S. Department of the Interior to protect the nation's largest population of raptors. Countless ledges and crevices in the canyon walls provide protected nesting sites, and the nearby uplands are a productive hunting ground for the birds. Thus, visitors can expect to see many eagles, hawks, falcons, owls, and ospreys, particularly in the spring.

SIGNIFICANCE

The Montini Volcano is one of several closely related canyon-filling basalt eruptions that temporarily blocked the ancestral Snake River. It was erupted within a deep canyon of Pleistocene age, and some of its tuff intertongues upstream with

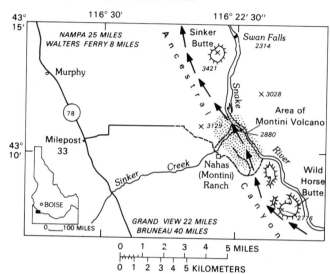

Figure 1. Map showing the location of the Montini Volcano with respect to the present Snake River and the course of the ancestral canyon. Spot elevations are given in feet.

several hundred feet of lake beds, which were impounded by the volcano. Collectively, the canyon-filling lavas and lake beds are known as the Bruneau Formation, a major Pleistocene stratigraphic unit in the western Snake River Plain (Malde and Powers, 1962). Near the Snake River, the Bruneau Formation is as much as 1,100 ft (335 m) thick and is composed of basalt and lake beds that fill four successive ancestral canyons (Malde, 1985). Each lava eruption dammed the Snake River, creating a reservoir for deposition of lake beds (or subaqueous basalt) in the canyon upstream. Eventually, another canyon would be cut around the lava dam and would, in turn, become dammed and filled with further Bruneau deposits.

By understanding the role of the Montini Volcano in damming and diverting the ancestral Snake River, one begins to comprehend the processes that account for the early Pleistocene physiographic evolution of the western Snake River Plain.

Basalt related to the first canyon-filling stage of the Bruneau Formation is represented near the Montini Volcano by lava flows on buttes in the southeast part of Figure 2 (Secs.16,21,28,T. 3S.,R.1E.). Deposits of the second canyon-filling stage have not been identified near the Montini Volcano but are widely found farther upstream. The process of damming and filling the third Bruneau canyon is vividly shown by the Montini Volcano and its associated basaltic deposits and lake beds. Basaltic lavas that fill the third canyon have K-Ar ages from 1.67 ± 0.09 to 1.37 ± 0.11 Ma (Hassan Amini *in* Malde, 1985). Younger lava flows of

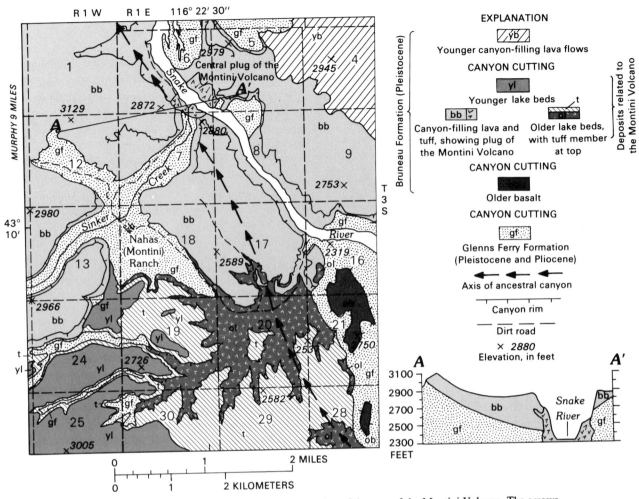

Figure 2. Generalized geologic map and cross-section of the area of the Montini Volcano. The arrows mark the course of the ancestral canyon of the Snake River.

the fourth canyon-filling stage of the Bruneau Formation overlie lava and tuff of the Montini Volcano in the northeast part of Figure 2 (Secs. 4,5,9,T.3S.,R.1E.).

Entrenchment of the four Bruneau canyons began soon after the Snake River was captured by Hells Canyon about 2 Ma, judging from physiographic relations (Wheeler and Cook, 1954), biogeographic evidence (Taylor, 1985), and the ages of stratigraphic units in the western Snake River Plain (Malde, 1985). A fifth canyon, which is occupied by the present Snake River, was cut when volcanic activity diverted the Snake River into its present course about 1 Ma.

OUTCROPS OF SPECIAL INTEREST

The form of the ancestral canyon occupied by the Montini Volcano is shown in Section A-A′, Figure 2. The canyon is cut in light-colored silt of the Plio-Pleistocene Glenns Ferry Formation,

a large body of lake and stream deposits several thousand feet thick. The southwest wall of the ancestral canyon is well exposed along a tributary canyon to Sinker Creek that heads about 1.5 mi (2.4 km) northwest of the Nahas Ranch (Sec.12,T.3S.,R.1W.). The lowest canyon-filling deposit is pillow lava, which forms a cascade that spills down the ancestral canyon wall. Tuff above the pillow lava can be traced down Sinker Creek to the center of the Montini Volcano.

The central plug of the Montini Volcano at the mouth of Sinker Creek is a striking feature. Most of the plug is dense lava, marked by joints that form more or less vertical, fan-shaped sets. Sills connected with the plug are exposed on both walls of the present Snake River Canyon. The craggy appearance of the plug is the result of erosion by the late Pleistocene Bonneville Flood, which was 435 ft (133 m) deep at this place (Malde, 1968, 1985).

About 100 ft (30 m) of lake beds capped by water-laid tuff (called "older lake beds with tuff member at top" on Fig. 2) can be seen one mi (1.6 km) southeast of the Nahas Ranch, just above

Figure 3. View of the dissected Montini Volcano from a point on the Snake River 1 mi (1.6 km) downstream. The banded layers are largely composed of palagonitized tuff. The crags at the center are part of the plug. Other remnants of the plug are preserved to the right and left of the river.

the road (near SE corner, Sec.18,T.3S.,R.1E.). The lowest bed is pebble gravel a few feet thick that represents a beach deposit, unconformable on the Glenns Ferry Formation. The gravel is overlain by several units of montmorillonitic clay, each 10 ft (3 m) or more thick, which show faint laminae when freshly exposed. The overlying tuff was produced by a subaqueous eruption from the Montini Volcano. It has been largely altered to palagonite—a yellow-brown clay derived from the basaltic glass that forms when lava explodes in water. Outcrops of the tuff in badlands 11 mi (18 km) south reach altitudes as high as 3,150 ft (960 m), or 270 ft (82 m) above the highest remnants of the Montini Volcano. In light of the lacustrine origin of the tuff, this distribution shows that the part of the Montini Volcano still

preserved must have been under water when the tuff was emplaced.

The tuff is also widespread in the south part of Figure 2. Here, younger lake beds on the tuff form rounded, light-colored hills that reach an altitude of 3,005 ft (916 m). The younger lake beds also extend as far as 80 mi (130 km) upstream, where they reach an altitude of 3,180 ft (970 m).

The northeast wall of the ancestral canyon, cut in the Glenns Ferry Formation, can be seen from the top of the Montini Volcano. The wall is exposed in "windows" eroded through canyon-filling tuffs and lavas of the volcano (SW¼Sec.5 and NW¼Sec.8,T.3S.,R.1W.).

By looking down the Snake River Canyon from the top of

the Montini Volcano, other components of the third canyon-filling stage of the Bruneau Formation can be seen. The components form a volcanic complex that extends at least 12 mi (19 km) downstream (Ekren and others, 1981; Malde, 1985). This large canyon-filling volcanic complex, including the Montini Volcano, formed a resistant obstacle in the path of the ancestral Snake River. The river was eventually diverted to the right (northeast), where it cut the fourth canyon of the Bruneau Formation. Pillow lava that fills the fourth canyon can be seen 8 mi (13 km) downstream.

REFERENCES

Ekren, E. B., McIntyre, D. H., Bennett, E. H., and Malde, H. E., 1981, Geologic map of Owyhee County, Idaho, west of longitude 116°W: U.S. Geological Survey Miscellaneous Investigations Map I-1256, scale 1:125,000.

Malde, H. E., 1968, The catastrophic late Pleistocene Bonneville Flood in the Snake River Plain, Idaho: U.S. Geological Survey Professional Paper 596, 52 p.

——— , 1985, Raft trip through the Snake River Birds of Prey Area to view canyon-filling volcanics and lake beds of the Pleistocene Snake River, *in* Composite Field Guide: Geological Society of America, Rocky Mountain Section, 38th Annual Meeting, April 22–24, 1985, Boise, Idaho, 48 p. (individually paginated).

Malde, H. E., and Powers, H. A., 1962, Upper Cenozoic stratigraphy of western Snake River Plain, Idaho: Geological Society of America Bulletin, v. 73, p. 1197–1220.

Taylor, D. W., 1985, Evolution of freshwater drainages and mollusks in western North America, *in* Smiley, C. J., and others, eds., Late Cenozoic history of the Pacific Northwest: San Francisco, American Association for the Advancement of Science, Pacific Division, p. 265–321.

Wheeler, H. E., and Cook, E. F., 1954, Structural and stratigraphic significance of the Snake River capture, Idaho–Oregon: Journal of Geology, v. 62, p. 525–536.

Thousand Springs area near Hagerman, Idaho

R. L. Whitehead, *Water Resources Division, U.S. Geological Survey, 230 Collins Road, Boise, Idaho 83702*
H. R. Covington, *Geologic Division, U.S. Geological Survey, Box 25046, Denver Federal Center, Denver, Colorado 80225*

LOCATION AND ACCESSIBILITY

Numerous spectacular springs discharge from the Snake Plain aquifer along the north side of the Snake River canyon near Hagerman, Idaho (Fig. 1). This stretch of canyon, known as the Thousand Springs area, contains 11 of the 65 springs in the United States that discharge more than 72,400 acre-ft annually (Meinzer, 1927, p. 27).

The springs can be reached by automobile or can be viewed from a distance. Most springs are on private land and permission is needed to enter the area, but some, such as Niagara Springs, are on public lands.

SIGNIFICANCE

Before development for irrigation and hydroelectric power generation near the turn of the century, majestic fountains of water gushed from vents in the canyon wall several hundred feet above the Snake River (Fig. 2). Some springs formed a nearly continuous wall of water that extended hundreds of feet along the canyon face (Russell, 1902, p. 163).

When the Snake River was dammed at Minidoka for irrigation on the Snake River Plain in about 1910, annual discharge from the springs averaged slightly less than 3,000,000 acre-ft. Owing to increased recharge from surface-water irrigation, annual discharge increased to about 4,920,000 acre-ft by the early 1950s. Since then, this trend has reversed as the result of increased use of groundwater for irrigation, decreased diversions of surface water, increased efficiency in water use, and changing precipitation patterns. In 1980, annual discharge from the springs was about 4,350,000 acre-ft.

From April to October each year, nearly all the water in the Snake River upstream from Twin Falls is diverted for irrigation, and the springs provide most of the water in the river for downstream agricultural and aquacultural use, domestic supply, and hydroelectric power generation. Although flumes and pipelines have diminished the previous grandeur of some springs, a few impressive ones remain.

GEOHYDROLOGIC SETTING

The Thousand Springs area is the result of unique geologic processes that were repeated several times during the late Quaternary Period when the Snake River Plain east of Hagerman, Idaho, was volcanically active. At times, lava flowed into the ancestral Snake River canyon, damming the river and forming a temporary lake. Lava flowing into the lake formed poorly consolidated pillow lava with relatively high permeability; lava in the obstructed channel downstream was dense and less permeable.

When the lava flows ceased, the river eroded a new canyon, generally south of its former course. Thousand Springs (Fig. 3) issue from pillow lava, which has been truncated by downcutting of the present canyon (Malde, 1971).

As shown in Figure 3, the emergence of springs is controlled by distribution of the following geologic units: Tertiary basalt, Quaternary and Tertiary sedimentary rocks, and talus and flood-plain deposits. The top of the Tertiary basalt defines the lowest elevation of springs in the present canyon. The overlying sediments are relatively impermeable and impede spring discharge near the canyon wall. In places, extensive talus deposits mask the true altitude of the water table by allowing water to migrate downward before it emerges as a spring. Finally, some water emerges from the flood-plain deposits underlying the canyon floor. The permeable flood-plain deposits, which were laid down during the late Pleistocene Bonneville flood, allow water to move laterally from the canyon wall and emerge as low-lying springs.

SPRING ACCESS AND DESCRIPTION

Niagara and Crystal springs. Take I-84 exit 157 at Wendell and proceed south on the county road 6.5 mi (10.5 km) to the canyon rim. Continue along the road about 2 mi (3.2 km) to the Idaho Power Company facilities on the canyon floor and park at the gate. Niagara Springs are on the left. Stairs east of the gate lead to the springs. Niagara Springs emerge from pillow lava high on the canyon wall, about 3,180 ft (970 m) above sea level, and discharge about 195,000 acre-ft/yr (Thomas, 1968, p. 35). Crystal Springs are located along the road 1.5 mi (2.4 km) east. Crystal Springs also emerge from pillow lava and discharge about 348,000 acre-ft/yr (Nace and others, 1958, p. 13). Early development of the springs for irrigation of pasture and orchards later changed to development for aquaculture and power generation. In 1965, Idaho Power Company developed the Niagara Springs site as part of their fish conservation program for replacing anadromous fish lost by construction of dams. Because the high-quality water has a constant temperature of 58°F, fish growth rate is about twice normal (Idaho Power Company, written communication, 1985).

Idaho Power Company Thousand Springs Powerplant. At Wendell, take I-84 exit 157 and proceed 0.5 mi (0.8 km) north to the center of town, turn left (west) toward Hagerman and proceed 4.5 mi (7.2 km), turn left (south) and proceed 2.5 mi (4.0 km), turn right (west) and proceed 2 mi (3.2 km), and finally, turn left and follow Idaho Power Company signs to the powerplant. During working hours, permission can be obtained

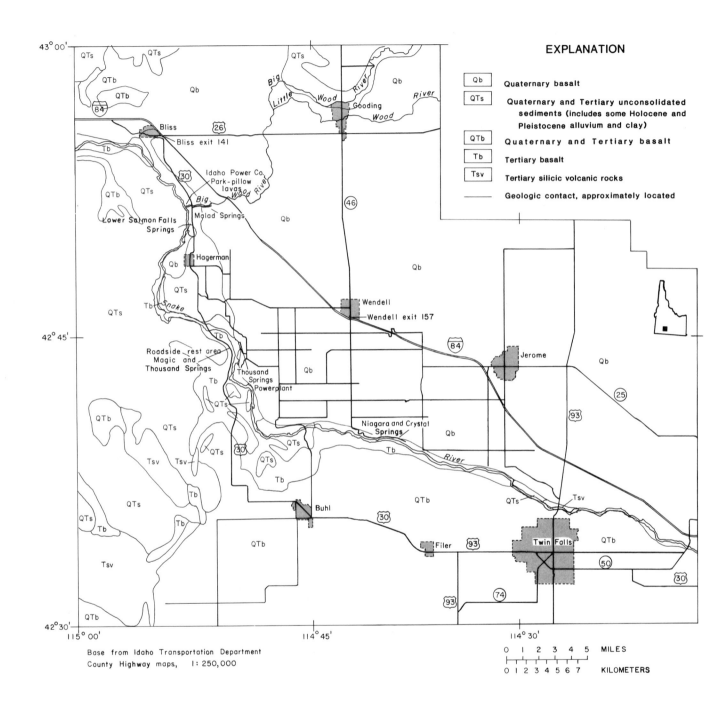

Figure 1. Selected spring locations and generalized geology in the Thousand Springs area near Hagerman, Idaho.

Figure 2. Thousand Springs, 7 mi (11 km) south of Hagerman, before being developed for hydroelectric power (from Russell, 1902, p. 26).

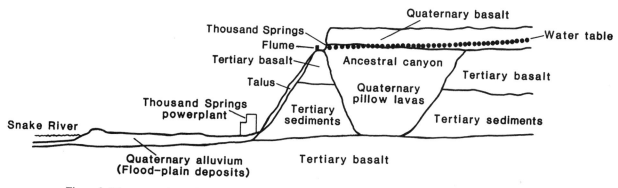

Figure 3. Diagrammatic section through the Thousand Springs powerplant area near Hagerman showing geohydrologic relations (not to scale).

to enter the flume area along the canyon wall. Thousand Springs discharge from pillow lava at the contact with Tertiary basalt at about 3,060 ft (930 m) above sea level. Annual discharge averages 870,000 acre-ft (Thomas, 1968, p. 47). In 1911, an Arizona-based company built a powerplant on this site. In 1917, Idaho Power Company acquired the site and expanded it by importing water from adjacent springs. The plant's unique feature is its nearly continuous output of hydroelectric power owing to the continuous water supply (Idaho Power Company, written communication, 1985).

Malad Springs. From Bliss at I-84 exit 141, take U.S. 30 south about 4 mi (6.4 km) to the south side of Malad Canyon (Big Wood River), turn left (east), and proceed about 1 mi (1.6 km) on the gravel road to a gate. Walk through the gate

opening and proceed 0.9 mi (1.4 km) upstream to the dam. Malad Springs, which emerge at river level, discharge about 869,000 acre-ft/yr (Thomas, 1968, p. 60) from pillow lava at and below the water surface. All the water from Malad Springs is used to generate electricity. Idaho Power Company acquired the site in 1916. At that time, only the lower facility was in operation. The upper facility was built in 1948 (Idaho Power Company, written communication, 1985). Very little water flows in the river channel upstream from the springs during irrigation season (April through October).

Lower Salmon Falls Springs. From U.S. 30 at Malad Canyon, proceed south about 2 mi (3.2 km). Turn right (west) at the Idaho Power Company sign and proceed toward the Lower Salmon Falls Dam. At the fork in the road upstream from the powerplant, bear right and park on the north side of the substation. Walk downstream along the riverbank road about 0.5 mi (0.8 km) to the U.S. Geological Survey gaging station. The Lower Salmon Falls Springs emerge from pillow lava a few feet above the river at the contact between pillow lava and sediments. The springs discharge about 14,500 acre-ft/yr (Thomas, 1968, p. 58).

Magic Springs and Thousand Springs. A roadside rest area on U.S. 30 about 6 mi (10 km) south of Hagerman provides a view of Magic Springs across the Snake River to the east and of Thousand Springs to the southeast. Magic Springs and Thousand Springs emerge from pillow lava high on the canyon wall and discharge about 123,000 and 847,000 acre-ft/yr, respectively (Nace and others, 1958, p. 12; Thomas, 1968, p. 47).

REFERENCES CITED

Covington, H. R., 1976, Geologic map of the Snake River canyon near Twin Falls, Idaho: U.S. Geological Survey Miscellaneous Field Studies Map MF-809, scale 1:24,000, 2 sheets.

Malde, H. E., 1963, Reconnaissance geologic map of west-central Snake River Plain, Idaho: U.S. Geological Survey Miscellaneous Geologic Investigations Map I-373, scale 1:125,000.

—— , 1971, History of the Snake River Canyon indicated by revised stratigraphy of Snake River Group near Hagerman and King Hill, Idaho, with a section on paleomagnetism by Allen Cox: U.S. Geological Survey Professional Paper 644-F, 21 p.

Meinzer, O. E., 1927, Large springs in the United States: U.S. Geological Survey Water-Supply Paper 557, 94 p.

Nace, R. L., McQueen, I. S., and Van t'Hul, A., 1958, Records of springs in the Snake River valley, Jerome and Gooding counties, Idaho: U.S. Geological Survey Water-Supply Paper 1463, 62 p.

Russell, I. C., 1902, Geology and water resources of the Snake River Plains of Idaho: U.S. Geological Survey Bulletin 199, 192 p.

Thomas, C. A., 1968, Records of north-side springs and other inflow to the Snake River between Milner and King Hill, Idaho, 1948–67: Idaho Department of Reclamation, Water Information Bulletin no. 6, 65 p.

Shoshone Falls, Idaho; A Pleistocene relic of the catastrophic Bonneville Flood

Harold E. Malde, U.S. Geological Survey, Box 25046, Denver Federal Center, Denver, Colorado 80225

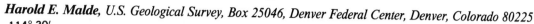

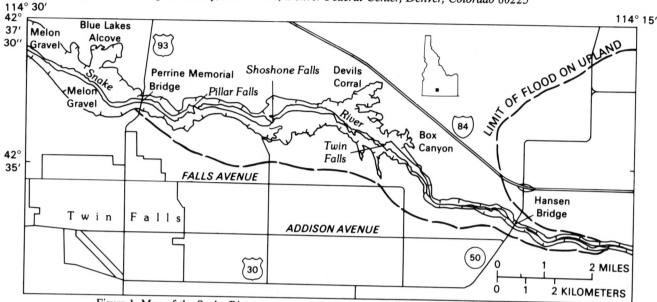

Figure 1. Map of the Snake River canyon near Twin Falls, Idaho, showing the enlarged canyon and other erosional features produced by the Bonneville Flood.

LOCATION AND ACCESS

Shoshone Falls is on the Snake River about 4 mi (6 km) northeast of the city of Twin Falls, near the midpoint of a chaotically eroded 14-mi (22-km) section of the Snake River canyon (Fig. 1). The south side of the falls, where the city maintains a picnic ground and park, offers the best views. Westbound travelers on I-84 should take the Kimberly–Twin Falls exit (no. 182), cross the Snake River on the Hansen Bridge, drive 5 mi (8 km) west on Addison Avenue, and then follow the signs northward to the falls. Those traveling east should take the Twin Falls exit at U.S. 93 (no. 173), cross the Snake River on the Perrine Memorial Bridge, and drive 3 mi (5 km) east on Falls Avenue to the Shoshone Falls road. Viewpoints on the south rim of the canyon at both bridges are convenient stopping places to see the narrow canyon above the falls and the wide canyon enlarged by flood erosion downstream. The north side of the falls can be reached by an improved road leading from U.S. 93.

SIGNIFICANCE

Shoshone Falls was an important landmark in the geological exploration of the West, but evidence that the falls are a relic of a great Pleistocene flood has been recognized only recently (Malde, 1960). As now understood, Shoshone Falls and its associated erosional features were formed at a time of catastrophic outflow from Lake Bonneville about 15,000 years ago (Scott and others,

1983), an event known as the Bonneville Flood (Malde, 1968). The significance of Shoshone Falls thus lies partly in its role in Western history and partly in modern knowledge about its geological origin.

PRESENT APPEARANCE

Those seeing Shoshone Falls for the first time may wonder why the falls are known as "the Niagara of the West." Since 1905, during all but the wettest years the entire flow of the Snake River has been diverted for irrigation by a dam a few miles upstream, and the falls are therefore dry. Furthermore, inflow below the diversion dam is diverted for power by a string of low dams linked by bedrock knobs at the top of the falls. The result, from April through October, and during some winters, is a spectacular, barren 212-ft (65-m) step in the canyon floor, formed by a massive layer of silicic volcanic rock.

When the full flow of the Snake River is unimpeded—in these days a rare event—Shoshone Falls is a thunderous display of plunging water, partly veiled in clouds of mist and surrounded by dark, hackly cliffs of basalt.

EARLY EXPLORATION

Among the earliest visitors to see the falls unfettered were Clarence King and his photographer, Timothy O'Sullivan, who paid a geological visit in 1868 (Fig. 2). O'Sullivan visited Sho-

Figure 2. View of Shoshone Falls from the south rim of the canyon, 0.3 mi (0.5 km) downstream, as photographed by Timothy O'Sullivan, October 1868. The falls plunge over a massive layer of silicic volcanic rock, which forms the lower story of the canyon. The upper part of the canyon is cut in basaltic lava flows. At the time of the photograph, discharge over the falls was between 7,000 and 8,000 cubic feet per second (cfs), according to an estimate made by Luther Kjelstrom, U.S. Geological Survey, Boise, Idaho. From National Archives.

shone Falls again in 1874, this time with the blessings of George M. Wheeler, making his last photographs of the West. King's descriptions of Shoshone Falls and O'Sullivan's photographs were significant elements in revealing the scientific importance and scenic splendor of the West.

King's words about the raw landscape of Shoshone Falls express a feeling of gloom: "Dead barrenness is the whole sentiment of the scene. . . . Black walls flank the abyss" (King, 1870). King nonetheless had considerable scientific interest in Shoshone Falls. He described its physical features, climbed down to the river below the falls to examine the "trachytes" and their relation to the overlying basalt, and followed the canyon 3 mi (5 km) upstream to a dual cascade now known as the Twin Falls, discovering on the way back "a labyrinth of side *crevasses* . . . usually in the form of an amphitheatre, with black walls a couple of hundred feet high, and a bottom filled with immense fragments of basalt rudely piled together" (King, 1870, p. 384). The crevasses, as noted below, are narrow alcoves eroded in the south rim of the canyon by the Bonneville Flood. King also tracked the continuity of the silicic volcanic rocks and the basalt several miles downstream (King, 1878, p. 592–593).

O'Sullivan's photographs of Shoshone Falls were greatly valued by King and Wheeler. He photographed Shoshone Falls from above, from the side, from the canyon wall downstream, and from the river below—making altogether in 1868 and 1874 about 20 full-plate views and more than 30 stereographs (Dingus, 1982, p. 89). In King's *Systematic Geology* (1878), three of O'Sullivan's 16 photographs (reproduced as lithographs) pertain to Shoshone Falls. Also, a set of 25 O'Sullivan photographs, four of Shoshone Falls, was published for the Wheeler survey to show the public "Landscapes, Geological and other Features . . . Obtained in Connection with . . . Surveys West of the 100th Meridian" (Snyder, 1981, p. 117). Wheeler's interest in the scientific use of photographs and the support of King and Wheeler for O'Sullivan's photography are now regarded as important highlights in the geological exploration of the West (Ostroff, 1981, p. 15).

FEATURES PRODUCED BY THE BONNEVILLE FLOOD

Cataracts, Alcoves, and Scabland. Shoshone Falls is at the junction of two channels of the Bonneville Flood, one along the narrow Snake River canyon and the other on the upland a few miles north (Fig. 3). Where the channels joined, flood erosion produced a chaotically eroded landscape of cataracts, spillway alcoves, and scabland (Malde, 1968, p. 25–30), and the original canyon was greatly enlarged (Fig. 1).

The main cataracts begin 2.5 mi (4 km) above Shoshone Falls, where the canyon is 230 ft (70 m) deep. These cataracts, called the Twin Falls, plunge 157 ft (48 m) from basalt to massive outcrops of silicic volcanic rocks in the lower story of the canyon. The Snake then drops another 212 ft (65 m) to the threshold of Pillar Falls, 1.5 mi (2.4 km) downstream. Pillar Falls, although only 20 ft (6 m) high, is surrounded by scabland crags that reach 175 ft (53 m) higher.

Alcoves and scabland along the north rim define a spillway 10 mi (16 km) wide. The largest alcoves are Blue Lakes alcove, Devils Corral, and Box Canyon (formerly called the Devils Washbowl). Further, as discovered by Clarence King, gashlike alcoves are found on the south side of the canyon just below the Twin Falls, and the south rim displays a ragged edge downstream to Perrine Memorial Bridge. These alcoves were overtopped by the flood.

Scabland is impressively developed in silicic volcanic rocks of the canyon floor between the Twin Falls and Blue Lakes alcove. Jagged and rounded knobs of the scabland rise 80 ft (24 m) above Shoshone Falls and 200 ft (60 m) above the river downstream.

This chaotic landscape has been interpreted as follows (Malde, 1968, p. 25, 27):

These erosional features indicate that this stretch of canyon was virtually brimful at peak discharge. . . . If this inference is correct, erosion along the north and south rims was not accomplished by water spilling as cascades in the ordinary sense, but was done by extraordinary turbulence in rapidly moving deep water. . . . These gigantic erosional features along the canyon floor [the falls] probably represent mainly the effect of turbulence in phenomenally deep water, although they must have made a series of rapids and cataracts as the flood waned in volume. Pillar Falls and Shoshone Falls coincide in a curious way with abrupt embayments on both sides of the canyon, which probably were eroded concurrently by turbulent floodwater when the canyon was full. . . . Because the present cataracts descend from bordering scabland, rather than from a lip upstream from scabland, they apparently mark the approximate limit of recessional erosion during the flood. That is, the topography suggests that recession of the falls since the flood has been comparatively negligible and that the present shape of the canyon floor is largely a relic of the Bonneville Flood.

Budget of Erosion and Deposition. In the 14 mi (22 km) from Hansen Bridge to the downstream end of the upland channel below Shoshone Falls, the volume of rock removed by the flood was between 36 and 50 billion ft^3 (1 and 1.4 billion m^3)—assuming dimensions for the original canyon based on the size of the canyon above and below (Malde, 1968, p. 44–45). Broken into gravel, the resulting volume of flood debris was therefore from 48 to 67 billion ft^3 (1.4 to 1.9 billion m^3), or from one-third to one-half a mi^3 (1.4 to 2.1 km^3).

The characteristics of the canyon downstream are such that

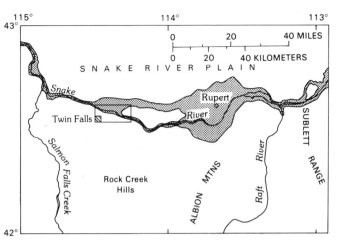

Figure 3. Path of the Bonneville Flood (pattern) in part of the eastern Snake River Plain, Idaho. The area outlined at Twin Falls is shown in greater detail in Figure 1. From Malde, 1968, Fig. 1.

all the flood debris, which is known as Melon Gravel, was trapped as boulder bars of enormous size in the next 125 mi (200 km) of canyon below Shoshone Falls. The volume of Melon Gravel in this downstream reach is about 84 billion ft^3 (2.4 billion m^3), some of which was derived from nearby sources along the canyon. About three-fourths of the gravel is found in the first 72 mi (116 km). Thus, "the enlarged canyon segment near Twin Falls, whatever is assumed about its original size, appears to have produced the major part of the Melon Gravel" (Malde, 1968, p. 45).

SUMMARY

In summary, Shoshone Falls is rarely seen today at full force, but it can be appreciated as splendid evidence of a geological catastrophe—the Bonneville Flood.

REFERENCES

Dingus, R., 1982, The photographic artifacts of Timothy O'Sullivan: Albuquerque, University of New Mexico Press, 158 p.

King, C., 1870, The falls of the Shoshone: The Overland Monthly, v. 5, p. 379–385.

—— , 1878, Systematic geology: U.S. Geological Exploration of the Fortieth Parallel, v. 1, 803 p.

Malde, H. E., 1960, Evidence in the Snake River Plain, Idaho, of a catastrophic flood from Pleistocene Lake Bonneville, *in* Short papers in the geological sciences: U.S. Geological Survey Professional Paper 400-B, p. B295–B297.

—— , 1968, The catastrophic late Pleistocene Bonneville Flood in the Snake River Plain: U.S. Geological Survey Professional Paper 596, 52 p.

Ostroff, E., 1981, Western views and eastern visions: Washington, D.C., Smithsonian Institution, 118 p.

Scott, W. E., and others, 1983, Reinterpretation of the exposed record of the last two cycles of Lake Bonneville, western United States: Quaternary Research, v. 20, no. 3, p. 261–285.

Snyder, J., 1981, American frontiers; The photographs of Timothy H. O'Sullivan, 1867–1874: Millerton, New York, Aperture, 119 p.

The Late Proterozoic Pocatello Formation; A record of continental rifting and glacial marine sedimentation, Portneuf Narrows, southeastern Idaho

Paul Karl Link, Department of Geology, Idaho State University, Pocatello, Idaho 83209

LOCATION AND ACCESSIBILITY

Portneuf Narrows is located on the Inkom, Idaho, 1:24,000 Topographic Quadrangle. The area is ideal for a one-day field trip because it is accessible from I-15 (Exit 63-Portneuf Area) and is only 3 mi (4.8 km) from Pocatello where tourist services and the Idaho State University Library are available. Stops 1 and 3 are on paved roads. Stop 2 is on a dirt road that can be driven by passenger cars in dry weather. The entire area is on public land.

Stop 1 is an overview of Portneuf Narrows (Fig. 1A); it is reached by heading southeast from Pocatello on I-15 and taking exit 63S. Turn west (right) onto the frontage road (South Fifth Avenue), proceed 0.7 mi (1.1 km), cross under the freeway, and stop on the east side of the frontage road. If approaching from the south (Salt Lake City), take exit 63N, turn right at the frontage road, and proceed 0.2 mi (0.3 km) to Stop 1 on the left side of the road.

The outcrops south of Portneuf Narrows (Stop 2) are reached by heading south on Fort Hall Mine Road from exit 63. Cross the Union Pacific Railroad and the Portneuf River and proceed 0.2 mi (0.3 km) to Portneuf Road. Cross it, and continue straight for 0.6 mi (1 km). There, bear left onto dirt road (avoiding road to sanitary landfill) and park at locked gate in Fort Hall Mine Canyon (Stop 2, Fig. 1A). The most complete stratigraphic sections can be seen in the base of the second canyon (Cyn B, Fig. 1A) and upper slopes of the first canyon (Cyn A) south of the gate.

The area north of the Narrows (Stop 3) is accessible by proceeding 1.3 mi (2.1 km) east on the frontage road from exit 63, to Blackrock Canyon Road. There, turn left (north), and proceed 0.3 mi (0.5 km). Take an immediate left (west) after going under the freeway and proceed 0.3 mi (0.5 km) on a dirt road to a parking place in the mouth of a small draw (Stop 3, Fig. 1A).

SIGNIFICANCE

Portneuf Narrows, 3 mi (4.8 km) southeast of Pocatello, Idaho, affords easy access to unique Late Proterozoic rocks of the Pocatello Formation. This formation provides evidence for Late Proterozoic rifting and glacial marine sedimentation. Exposures include greenschist facies metabasalt and volcanic breccia of the Bannock Volcanic Member and inferred glacial-marine diamictite of the Scout Mountain Member. These rocks are cut by low- and high-angle dip- and strike-slip faults, including the Portneuf Narrows fault, which separates overturned rocks to the north from their right-side-up equivalents to the south.

SITE DESCRIPTION

The Portneuf River cuts through the north-trending Bannock Range at Portneuf Narrows (Fig. 1A), in a superposed canyon thought to have been cut by the ancestral Bear River (Ore, 1982). This canyon was the route of the Lake Bonneville flood between Marsh Valley and the Snake River Plain (Bright and Ore, this volume). The route was also used by pioneers in the nineteenth century, and the narrow gauge Utah and Northern Railroad in 1878. The siting of Pocatello ("Gate City to the Northwest") was dictated by its location just west of this transportation funnel. Today the main line of the Union Pacific Railroad and highway I-15 pass through the narrow gap.

Geologic framework. Portneuf Narrows lies within the Pocatello 15-minute geologic map of Trimble (1976). The stratigraphic framework for the Late Proterozoic rocks of the area has been described most recently by Link (1983) and Crittenden and others (1983). A field trip roadlog (Link and LeFebre, 1983) contains a detailed description of the area. The area was part of the Fort Hall mining district in the early 1900s (Darling, 1985).

The rocks exposed at Portneuf Narrows are part of the Late Proterozoic (p∈Z) Pocatello Formation, which contains three members (Fig. 2): the basal Bannock Volcanic, medial Scout Mountain, and an informal upper member. The formation lies at the base of the strata of the Cordilleran miogeocline in southeastern Idaho. Among other lithologies, the Pocatello Formation contains mafic volcanic rock and glaciogenic diamictite, thought to have been deposited during a rifting event, possibly a continental separation (Harper and Link, 1986). The Pocatello Formation has not been dated directly, and its age is not well constrained. Radiometric dates on strata thought to be correlative range from 918 to 770 Ma. Extrapolation of tectonic subsidence curves for the Cordilleran miogeocline, however, suggests that continental separation may not have occurred until around 600 Ma. (Bond and others, 1985).

Portneuf Narrows is located at the north end of the Bannock Range, near the junction of the Idaho-Wyoming thrust belt, the Basin and Range, and the Snake River Plain (inset map, Fig. 1). The rocks were subjected to a complex deformational history involving Mesozoic folding, thrusting, and greenschist facies metamorphism, followed by Tertiary volcanism and normal faulting.

The Bannock Range lies in the hinterland of the Jurassic to Paleocene Idaho-Wyoming thrust belt. In the hinterland, shallowly dipping faults place younger strata on older. This anomalous relationship may have resulted in part from thrusting of previously folded rocks or by reactivation of thrust faults dur-

P. K. Link

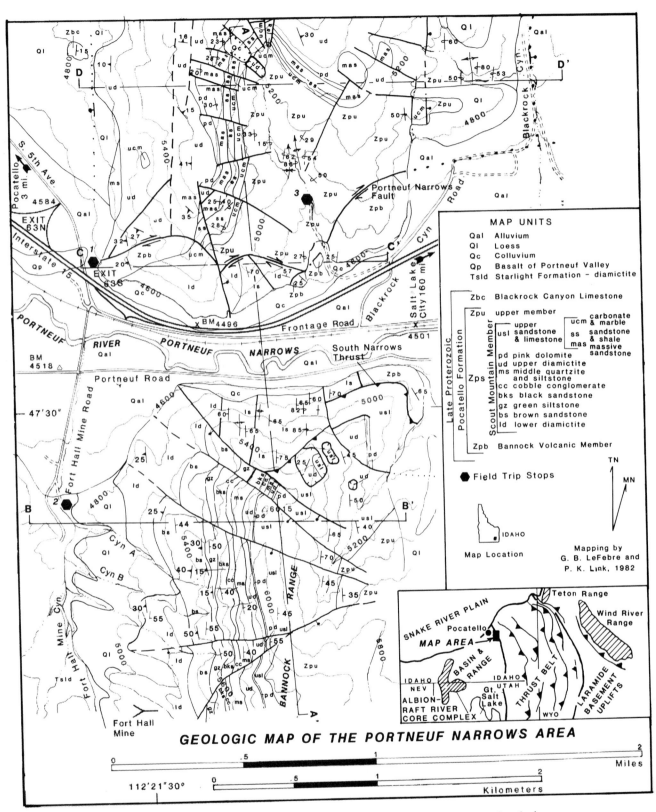

Figure 1A. Geologic Map of the Portneuf Narrows Area. Locations of field trip stops and geologic cross sections of Figure 3 are shown. Map after Link and LeFebre (1983).

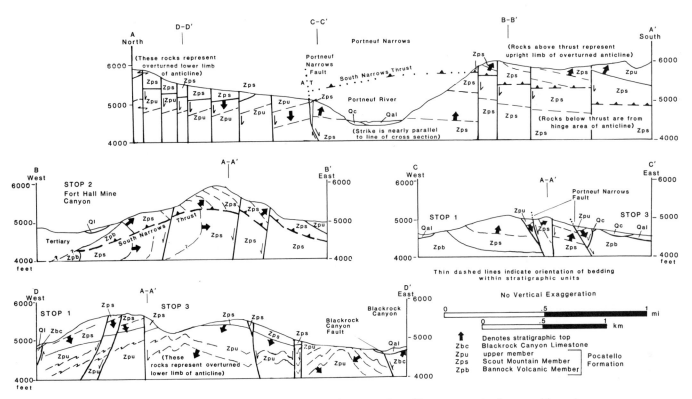

Figure 1B. Geologic cross sections of the Portneuf Narrows Area. Key to map units shown on Figure 1.

ing Tertiary extension (Platt, 1985). The early folds in the Portneuf Narrows area produced overturning of strata north of the Portneuf Narrows fault (see Fig. 1, geologic map, and Fig. 1B, cross sections).

The Pocatello Formation in the Portneuf Narrows area. The lowest exposed strata at Portneuf Narrows (seen just east of Stop 1) belong to the top of the Bannock Volcanic Member and consist of greenschist facies metabasalt and volcanic breccia (Fig. 2). Clasts in Bannock Volcanic breccia consist mainly of angular fragments of metabasalt, now mostly fine-grained chlorite and albite. A few clasts resemble fragments of basaltic pillows. South of the freeway (north of Canyon A, Stop 2), volcanic breccia of the Bannock Volcanic Member grades upward into diamictite of the Scout Mountain Member, the basal part of which contains quartzite clasts.

At Stop 2, a nearly complete section of the Scout Mountain Member can be observed (Fig. 2). The member contains a lower diamictite, a middle sequence of sandstones and siltstones with abundant graded beds, ripple marks, load casts, and edgewise conglomerate, a spectacular cobble-to-boulder conglomerate, an upper diamictite overlain by laminated pink dolomite, and a fining upward sequence of sandstone below a tan limestone bed at the top of the hill.

The upper part of the Scout Mountain Member is exposed in west-dipping overturned beds north of Portneuf Narrows at Stop 3. Walking from Stop 3 up the east side of the ridge north of Portneuf Narrows affords an opportunity to examine overturned

sedimentary structures within a sequence that fines stratigraphically upward, from the upper diamictite and overlying dolomite-chip breccia at the top of the hill, through a sequence of sandstones and siltstones, to a prominent marble bed. The upper diamictite forms a dip slope on the west side of the ridge.

The lower part of the Scout Mountain Member is also exposed in steeply dipping beds directly north and south of the Narrows. The northern beds are accessible by walking south and west from Stop 3, but the southern exposures are very steep and talus covered. They are best viewed from a distance, while standing north of the freeway.

The informal upper member of the Pocatello Formation is exposed as kink-folded phyllite near Stop 3, north of the Narrows. The rocks are structurally overturned, but stratigraphically overlie a prominent white marble bed at the top of the Scout Mountain Member with gradational contact. The upper member is also exposed in the canyon east of the ridge south of the narrows.

Interpretation of sedimentary facies in the Pocatello Formation. The Pocatello Formation was deposited during an episode of continental rifting characterized by block faulting and mafic volcanism (Harper and Link, 1986). The lower diamictite of the Scout Mountain Member has been interpreted to be a subaqueous mass flow deposit with input of both locally derived volcanic clasts and glacially derived extrabasinal clasts (Link, 1983). The upper diamictite is interpreted as a subaqueous lodgement or flow tillite, deposited close to the glacial margin.

Diamictites of the Scout Mountain Member have long been

thought to be of glacial origin. Although no dropstones have been located, the presence of isolated glacially striated quartzite cobbles, the lateral continuity for tens of miles (kilometers) of massive diamictite containing a wide variety of both intra- and extrabasinal clasts, and the stratigraphic and facies similarity with inferred glaciogenic strata in northern Utah all support the glacial hypothesis (Crittenden and others, 1983).

Other strata of the Scout Mountain Member are thought to represent a variety of sedimentary facies, including proximal turbidites (lower sandstone), subaqueous channel-fill cobble conglomerates, shallow marine and beach sandstones (upper sandstone), and intertidal carbonates (upper limestone). The pink dolomite, which rests, possibly disconformably, on the upper diamictite is a problematic unit, recording carbonate sedimentation immediately above an inferred glacial till. The shales of the upper member are interpreted to have been deposited in deep quiet water during the regional postglacial transgression.

Structural geology of the Portneuf Narrows area. Portneuf Narrows contains a recumbently folded rock sequence cut by numerous high- and low-angle faults. Structural synthesis by Link and LeFebre (1983) indicates a polyphase deformational history including: (1) formation of an east-vergent recumbent anticline; (2) movement along the South Narrows thrust, which cuts through the core of the recumbent anticline; (3) movement along the Portneuf Narrows fault, an east-trending, right lateral tear fault; and (4) movement along high-angle normal faults associated with Basin-Range development.

Stop 1 is located on the trace of the Portneuf Narrows tear fault. South of the tear fault, east-dipping right-side-up rocks (Stop 2; cross section B–B′; Fig. 1B) are cut by the South Narrows thrust. The steeply dipping rocks below the South Narrows thrust (cross section B–B′) are from the hinge of the overturned anticline. Rocks north of Portneuf Narrows at Stop 3 (cross section D–D′) represent the overturned west-dipping lower limb of the fold.

Cleavage is present in the lower diamictite in Canyon B near Stop 2 and in the inverted upper member and upper diamictite at Stop 3. This cleavage is believed to be axial planar to the recumbent anticline. Near Stop 3 the cleavage is folded into a north-plunging antiform (east end of cross section D–D′), which may be related to movement on the Portneuf Narrows tear fault or Basin-and-Range uplift of the Bannock Range.

REFERENCES CITED

Bond, G. C., Christie-Blick, N., Kominz, M. A., and Devlin, W. J., 1985, An early Cambrian rift to post-rift transition in the Cordillera of western North America: Nature, v. 316, p. 742–745.

Crittenden, M. D., Jr., Christie-Blick, N., and Link, P. K., 1983, Evidence for two pulses of glaciation during the Late Proterozoic in northern Utah and southeastern Idaho: Geological Society of America Bulletin, v. 94, p. 437–450.

Darling, R. S., 1985, Mineralization in the Fort Hall Mining District, Bannock County, Idaho, *in* Kerns, G. L., and Kerns, R. L., eds., Orogenic patterns and stratigraphy of north-central Utah and southeastern Idaho: Utah Geological Association Publication 14, p. 167–174.

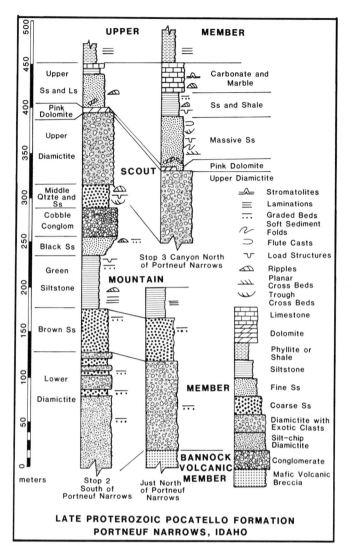

Figure 2. Stratigraphic sections of the Pocatello Formation in the Portneuf Narrows Area.

Harper, G. D., and Link, P. K., 1986, Geochemistry of upper Proterozoic rift-related volcanics in northern Utah and southeastern Idaho: Geology, v. 14, p. 864–867.

Link, P. K., 1983, Glacial and tectonically influenced sedimentation in the Upper Proterozoic Pocatello Formation, southeastern Idaho, *in* Miller, D. M., Todd, V. R., and Howard, K. A., eds., Stratigraphic and tectonic studies in the eastern Great Basin: Geological Society of America Memoir 153, p. 165–181.

Link, P. K., and LeFebre, G. B., 1983, Upper Proterozoic diamictites and volcanic rocks of the Pocatello Formation and correlative units, southeastern Idaho and northern Utah: Utah Geological and Mineral Survey Special Studies 60, p. 1–32.

Ore, H. T., 1982, Tertiary and Quaternary evolution of the landscape in the Pocatello, Idaho, area: Northwest Geology, v. 11, p. 31–36.

Platt, L. B., 1985, Geologic map of the Hawkins Quadrangle, Bannock County, Idaho: U.S. Geological Survey Miscellaneous Field Studies Map MF-1812, scale 1:24,000.

Trimble, D. E., 1976, Geology of the Michaud and Pocatello Quadrangles, Bannock and Power Counties, Idaho: U.S. Geological Survey Bulletin 1400, 88 p.

Evidence for the spillover of Lake Bonneville, southeastern Idaho

Robert C. Bright, *Bell Museum of Natural History, University of Minnesota, Minneapolis, Minnesota 55455*
H. Thomas Ore, *Department of Geology, Idaho State University, Pocatello, Idaho 83209*

LOCATION AND ACCESSIBILITY

The area described herein is in southeastern Idaho, in Franklin and Bannock counties. It is readily reached via I-15 and various paved state highways from any direction.

Each stop is easily accessible by two-wheel drive vehicles when roads are dry. However, the dirt roads leading to stops 1 to 6 can be extremely slick when soaked. Permission from property owners *is required* if you desire to wander beyond each stop.

The tour is keyed to the following 7½-minute quadrangles: Pocatello North, Inkom, Arimo, McCammon, Bonneville Peak, Downey West, Downey East, Oxford, and Swan Lake. The latter two are particularly important for locating features south of the overflow site. The Preston and Pocatello 2-degree quadrangles are also suggested.

The trip will take about six hours. It begins in northern Cache Valley at the hamlet of Oxford, proceeds northward through Red Rock Pass and Marsh Valley, and ends in the Pocatello area (Fig. 1). Food and lodging can be found in Preston (dry) and Pocatello.

SIGNIFICANCE

Lake Bonneville was a late Pleistocene lake about 20,000 mi^2 (52,000 km^2) in area in northwestern Utah, with major arms extending into southern Idaho. Study of its sediments and geomorphic features has contributed significantly to our knowledge of lacustrine processes. This tour allows first-hand examination of evidence for the existence of the lake and the notion that it overfilled its basin and spilled over a divide near Red Rock Pass. The site further provides a location where effects of the single catastrophic spillover of such a large lake can be observed in its outlet as well as far downstream. Finally the relationships that bracket the age of the spillover can be examined.

Much of the information herein is based on detailed geologic mapping of Bright and Meyer Rubin between Oxford and Inkom (in preparation).

SITE INFORMATION

Begin. North side of Oxford (Oxford Quadrangle) at intersection of oiled state highway and dirt road labeled "Oxford Loop Road" at the mouth of Oxford Creek. You are about 75 ft (23 m) above Provo Lake bottom. Turn west on the Loop Road and proceed up Oxford Creek 0.3 mi (0.5 km) to stop 1.

Stop 1. Gravel pit in Bonneville sediments on north side of road. Private property; view from road. As Lake Bonneville rose to its highest level (Bonneville), transgressive lacustrine sed-

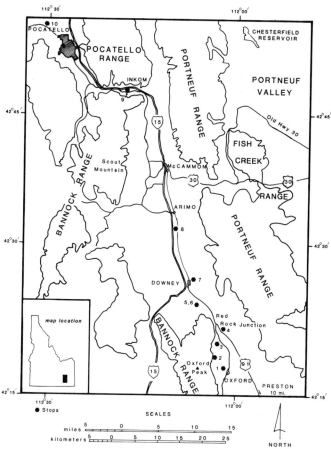

Figure 1. Location map, spillover of Lake Bonneville, Oxford to Pocatello, Idaho.

iments were deposited on the flanks of the mountains in Cache Valley. This pit exposes an example of mostly near-shore gravel and sand bars and beach deposits (the pebbles derived locally). In the upper left (west) end of the pit are some sandy and silty lagoonal sediments. This pit has produced a few specimens of the freshwater snail *Valvata utahensis.* Other deposits on this side of the valley include fine-grained, horizontal, deep-water units (some rhythmites), and locally small gravelly and sandy deltas at the heads of drowned canyons. A similar suite exists on the east side of northern Cache Valley but there the dominant surficial sediment is the fine-grained material supplied by the lake's major tributary, Bear River. These deltaic and lacustrine sediments are a distinctive brown to reddish brown, and thin to the north and south of the mouth of Bear River. In the foothills south of Swanlake they discontinuously cover bedrock and other transgressive lacustrine Bonneville deposits (best viewed in fall when fields are fallow).

Continue westward up Oxford Creek about 1.8 mi (3 km)

143

to the intersection with the north-south road (elevation 5,414 ft, Oxford Quadrangle), turn right (north) and drive 0.4 to 0.5 mi (0.6 to 0.8 km) to stop 2 (so you can see Oxford Reservoir).

Stop 2. General geology south of Red Rock Pass. The structure of Cache Valley here is essentially that of a complex faulted graben. To the west is a thick sequence of thrusted, primarily Precambrian rocks. The north-trending Oxford normal fault is somewhere beneath you, buried by a thick fan. East of U.S. 91 the rocks are mainly east-dipping Cambrian and Tertiary sediments that are displaced by northwest and east-trending normal faults (see Link, 1982). The relative ages of the two sets of faults are controversial, however, the major normal faulting in this area is certainly post Cache Valley Formation (Pliocene?). This faulting coupled with subsequent fan formation separated Cache Valley from Marsh Valley by producing the divide over which Lake Bonneville spilled near Red Rock Pass.

Fan formation created the bajada between Oxford Creek and a point about a mile south of the viaduct at Red Rock Junction. Another large fan almost covers Sec.33,T12S,R.38E (Oxford Quadrangle) on the east side of the valley. Ash beds in this fan indicate that it began to form before 738,000 B.P. (Izett, 1982) and continued growth after 610,000 B.P. A recent interpretation of the main basin Bonneville lake cycle (Scott and others, 1983, Fig. 5) indicates the lake began to rise about 26,000 B.P., entered Cache Valley through Cutler Gorge about 20,000 B.P., and "... probably reached the vicinity of the Bonneville shoreline ... about 16,000 years ago" (p. 277) where it stood for about 1,000 to 2,000 years. About 15,000 B.P. the lake spilled northward near Red Rock Pass and quickly dissected the divide 325 ft (100 m) to the Provo shoreline. Work in northern Cache Valley (Bright and M. Rubin, in preparation) substantially supports that chronology. The Bonneville shoreline, with voluminous shore deposits, is clearly visible on the east side of the valley south of Swanlake at about 5,090 ft (1,550 m) elevation. North of Swanlake the shoreline is somewhat obscure and is best determined by locating beach sediments. The prominent treeline on the hill (elevation 5,343 ft) just north of Swan Lake lies just above the lake level.

In Sec.33,T.12S,R.38E the Bonneville shoreline is inconspicuous, but beach deposits occur at about 5090 ft (1,550 m). To the north (west side of the valley), the obscure shoreline is cut into the bajada surface. It is not evident north of SE¼,SW¼, Sec.5,T.13S,R.38E (Oxford Quadrangle), above the terrace at about 5,000 ft (1,500 m).

Swan Lake occupies the valley bottom just northeast of Oxford Reservoir. Here basal sediments are 12,090 years old (Bright, 1966) providing one of the earlier limiting ages for stillstand at the Provo shoreline and indicating that the fan at the mouth of Stockton Creek, which dammed the valley and created the lake, began to form more than 12,090 B.P. Red Rock Pass may be seen in the distance to the north.

Proceed northward along the Loop Road, which in about 0.5 mi (0.8 km) veers to the northeast down Gooseberry Creek, to the first intersection (road elevation 4,998 ft, Oxford

Quadrangle) and turn north (left) on Chicken Creek Road. Proceed 1 mi (1.6 km) north to the next intersection (road elevation 4,936 ft). Turn northwest and proceed on Chicken Creek Road about 1 mi (1.6 km) to stop 3. An old homestead should be on your left, just northwest of road elevation 4,975 ft, Oxford Quadrangle).

Stop 3. Bedrock configuration of spillover area. The spillover scoured the steep bluffs to the west; narrow gulleys and fans postdate it. The reddish cliff about 0.25 mi (0.4 km) south of Red Rock Junction (just west of the state highway) is Nounan Dolomite (upper Є); Red Rock Butte (elevation 5,088 ft) due north of the viaduct, is composed of dolomite of the Upper Є. St. Charles Formation. The slumped and scoured limestone just east of the viaduct is Blacksmith Formation (middle Є). The butte in the NE¼,Sec.32,T.12S,R.38E (elevation 5,131 ft) is Nounan Dolomite and the nose of the juniper-covered hill just to the north is mostly Bloomington Formation (middle Є, much light brown shale).

The large fractured block of Nounan, about 100 yd (90 m) east of the state highway, in NW¼,Sec.32,T.12S,R.38E (Oxford Quadrangle), slid into the valley from the west, as did the scoured, rounded knobs of Tertiary conglomerate just to the north in SW¼,Sec.29,T.12S,R.38E. The knob at Red Rock Junction, atop which is a monument, is a highly fractured slide block of Worm Creek quartzite and dolomite derived from the west. Other such westerly derived slide blocks (Tertiary and Cambrian) lie on the east side of the valley along the railroad tracks. The time of their emplacement may be pre-spillover or contemporary with it.

North of Red Rock Butte the surface of the Marsh Creek fan is visible. An early Rancholabrean vertebrate fauna from a unit of this fan shows that it was active between 400,000 and 150,000 years ago (C. Repenning, written communication, 1984); the older date being most likely. The white beds that outcrop 0.5 mi (0.8 km) north of Red Rock Butte and on the east side of the Marsh Creek floodplain are tuffs and sandstones of the Cache Valley Formation.

Proceed northward to Red Rock Junction. Turn southeast, cross the viaduct, and then turn left at the second dirt road (Red Rock Road) about 50 yd (46 m) east of the viaduct (*danger* from speeding vehicles from ahead and behind!). Stop 4 is at the point where Red Rock Road turns east (road elevation 5,025 ft, Oxford Quadrangle), by the pinkish-red gate.

Stop 4. Spillover area. Across the valley to the west in the SW¼,Sec.30,T.12S.,R.38E., just below the skyline, is a dip slope of light gray conglomerates, sandstones, tuffs, and limestones of the Cache Valley Formation that overlie Cambrian rocks. In the N½,Sec.30 and adjacent Sec.19 is a large hummocky area of mass movement, the surficial debris consisting mostly of variously oriented blocks (50 to 300+ ft; 15 to 90+ m) of Cache Valley Formation. Between the mouth of Coalpit Creek and the SE¼,Sec.19 is a series of elongate slide blocks of Cache Valley Formation, the slide faces essentially paralleling the valley. A small landslide terrace is visible just southeast of Coalpit Creek.

Some of the sliding was contemporary with the spillover. Calvin Road (north of Red Rock Butte) traverses the west side of a flat stream terrace cut during the spillover; the terrace discontinuously extends down Marsh Creek Valley. Seismic evidence indicates that the bedrock divide is located beneath the west end of the viaduct beneath about 30 ft (9 m) of valley fill (Williams and Milligan, 1968). Note that the overflow area between Red Rock Butte and Downata Hot Springs was stripped of most debris by the high velocity flood.

Several controversial scenarios have been offered concerning the nature and precise location of the pre-spillover divide, the spillover, and rapid downcutting. As these details are somewhat peripheral here, they are not summarized, but for those interested we suggest reading the following: Gilbert (1890, p. 171–178), Bissell (1968), Williams and Milligan (1968, p. 67), and Malde (1968).

Return to U.S. 91 and proceed northward about 3.5 mi (5.6 km) to Downata Hot Springs Resort. Continue west past the resort to the fork and turn right (north) across the cattle guard onto Marsh Creek Road. Pass the post-spillover tufa deposits by the corral and continue northwesterly on Marsh Creek Road (Downey East Quadrangle) 1.4 mi (2.4 km). Stop 5 is the long roadcut beneath the two steel granaries.

Stop 5. The roadcut here contains spillover gravel with large, rounded boulders of various lithologies transported from the divide area; boulders in the road spoil were derived from the cut. Overflow flood deposits between McCammon and here are a complex sheet of clean pebble gravels and some with considerable fines; in some places the pebbles are imbricate. Many deposits contain rounded boulders (up to 6 ft; 1.8 m) supported in the gravel framework. These boulders tend to be larger in the center of the flood path and decrease in size from Downey to McCammon but there are exceptions. The dominant boulder lithologies are Cambrian dolomites, Cambrian and Precambrian quartzites, and Tertiary sandstones, conglomerates, tuffs, and sandy tuffs.

Continue northwest on Marsh Valley Road 0.9 mi (1.4 km) to the junction at the steal machine shed (road elevation 4,791 ft). Turn north (the boulders along the fence lines are lag) and stop 0.5 mi (0.8 km) north of the intersection at the gravel pit at the right at the brink of the terrace.

Stop 6. A complex bar sequence with abundant boulders. In the upper third of the east end of the pit are two distinct horizontal beds 3 to 6 in (8 to 15 cm) thick, which may indicate two episodes of decreased flood discharge. Boulders of conglomerate at the north edge of the pit are derived from the Cache Valley Formation upstream. Flood deposits on both sides of the valley overlie old fan and other fluviatile sediments.

Continue north to the intersection with U.S. 91, turn northwest and follow the highway to the yellow caution light in south Downey adjacent to the elevators. Find Center Street in Downey and drive to 8th East Street; there turn north (by the hangar) about 200 yd (180 m). A large gravel pit (stop 7) lies just west of the road.

Stop 7. Highest known flood deposit (approximately 4,900 ft; 1,500 m) in Marsh Valley. The northwest wall of the pit exposes the east flank of a large bar formed during spillover. To the right of the east-dipping beds is a small, backfilled channel. The bar gravels contain scattered rounded boulders up to 2 ft (0.6 m) in diameter. The plain to the west of Downey is covered with similar deposits. Return to the caution light on U.S. 91. Drive north 7.9 mi (13 km) (at the Virginia elevators continue *due north* on *old* U.S. 91, which parallels the railroad tracks) to stop 8, a large gravel pit just west of the highway in SW¼,SW¼, Sec.20,T10S,R37E, Arimo Quadrangle. Turn off highway into the pit.

Stop 8. Possibly thickest flood deposits in Marsh Valley, indicating pre-flood topography. Here rounded boulders are rare and not more than 1.5 ft (6.5 m) in size. One boulder of Tertiary tuffaceous sandstone was most likely derived from the Red Rock Pass area. A post overflow floodplain sequence near the top of the north wall of the quarry contains an organic horizon, snails, and bovid (*Bison*?) bones. The sequence indicates that the amount of post-spillover floodplain aggradation here was about 6 ft (2 m).

Return to old U.S. 91 and proceed north through Arimo, past Marsh Valley High School, and then ascend to the basalt surface. Scott and others (1982, p. 586) consider the flows in Marsh Valley to be about 583,000 B.P. From this point northward the upper vesicular part of the flow has been scoured away in many places by the Bonneville flood. Small potholes exist here and there and numerous scoured channels can be seen from the road. Thus the flood post dates the flows.

Drive through McCammon to the junction with U.S. 30N and turn west, proceeding across I-15, past a series of lake-filled, scoured depressions and channels, down the west edge of the lava flow, across Marsh Creek, and turn north on Marsh Creek Road.

Along the west side of Marsh Valley, particularly approaching Inkom, notice the numerous scoured alcoves in the basalt and the basaltic boulder bars in the valley bottom at the mouths of the alcoves. Stop 9 is located at roadside (north) on top of a scoured flow surface in NW¼,NE¼,Sec.28,T7S,R36E (Bonneville Peak Quadrangle), about 0.1 mi (0.16 km) southeast of the cement plant at Inkom.

Stop 9. Scoured basalt surface and effect of narrow canyon. Walk out onto the flow surface north of the road to examine the scoured and channeled surface. Exotic limestone and quartzite pebbles deposited by the flood can be found in depressions on the surface. Continue northwest to "T" junction with road to Portneuf; turn left. The road crosses the toe of Trough Canyon alluvial fan about 3 mi (5 km) west of junction. Across the valley to the northwest, the basalt outcrop terminates south of the mouth of Blackrock Canyon. Here Portneuf Canyon narrows sufficiently that resulting higher flood velocities tore the basalt from the canyon bottom. Continue west to Portneuf where you emerge from the narrows into Pocatello Valley and where basalt is again present. It did not wash away here because flood velocities were lower in the wider valley. Continue into Pocatello via old U.S. 91, I-15, or Portneuf Road. Transported boulders beneath downtown Pocatello are up to 8 ft (2 m) in diameter (Trimble, 1976, p. 60) and decrease to the northwest onto the Michaud Gravel, a fan-shaped deposit resulting partially from the flood path widening out onto the Snake River Plain. Also the flood debouched here into the Pleistocene American Falls Lake, a basalt-dammed lake extending up into the Pocatello area from 20 mi (32 km) downstream (see Carr and Trimble, 1963, p. 28).

To view a distal Michaud Gravel, drive northwest on Garrett Way past the J. R. Simplot plant on the left, (3 mi [4.8 km] west of Hawthorne Road), over the interstate, and take the first right (Frontage Road). Turn north around the sewage treatment plant, and take the first left into the FMC gravel pit (stop 10). Access has been approved but the caretaker might check you as a routine precaution.

Stop 10. FMC gravel pit, Michaud Gravel. The torrentially crossbedded gravels here contain boulders up to 4 ft (1.2 m) diameter (Trimble, 1976), and basalt boulders of 1 ft (0.3 m) diameter are commonly seen on the surface. Most of the gravel is derived from Pleistocene valley fill upstream.

REFERENCES CITED

Bissell, H. J., 1968, Bonneville; An ice age lake: Brigham Young University Geology Studies, v. 15, pt. 4, 66 p.

Bright, R. C., 1966, Pollen and seed stratigraphy of Swan Lake, southeastern Idaho; Its relation to regional vegetational history and to Lake Bonneville history: Tebiwa, Journal of the Idaho State University Museum, v. 9, no. 2, p. 1–47.

Carr, W. D., and Trimble, D. E., 1963, Geology of the American Falls Quadrangle, Idaho: U.S. Geological Survey Bulletin 1121-G, 44 p.

Gilbert, G. K., 1890, Lake Bonneville: U.S. Geological Survey Monograph 1, p. 171–176.

Izett, G. A., 1982, The Bishop Ash Bed and some older compositionally similar ash beds in California, Nevada, and Utah: U.S. Geological Survey Open-File Report, no. 82-582, 47 p.

Link, P. K., 1982, Structural Geology of the Oxford and Malad Summit Quadrangles, Bannock Range, Southeastern Idaho, *in* Powers, R. B., ed., Geologic studies of the Cordilleran Thrust Belt: Rocky Mountain Association of Geologists, v. II, p. 851–858.

Malde, H. H., 1968, The catastrophic late Pleistocene Bonneville Flood in the Snake River Plain, Idaho: U.S. Geological Survey Professional Paper 596, 52 p.

Scott, W. E., Pierce, F. L., Bradbury, J. P., and Forester, R. M., 1982, Revised Quaternary stratigraphy and chronology in the American Falls Area, southeastern Idaho, *in* Bonnichsen, B., and Breckenridge, R. M., eds., Cenozoic geology of Idaho: Idaho Bureau of Mines and Geology, Bulletin 26, p. 581–595.

Scott, W. E., McCoy, W. D., Shroba, R., and Rubin, M., 1983, Reinterpretation of the exposed record of the last two cycles of Lake Bonneville, western United States: Quaternary Research, v. 20, p. 261–285.

Trimble, D. E., 1976, Geology of the Michaud and Pocatello Quadrangles, Bannock and Power counties, Idaho: U.S. Geological Survey Bulletin 1400, 88 p.

Williams, J. S., and Milligan, J. H., 1968, Bedrock configuration and altitude, Red Rock Pass, Outlet of Lake Bonneville, southeastern Idaho: University of Wyoming Contributions to Geology, v. 7, no. 1, p. 67–72.

Heart Mountain detachment fault and clastic dikes of fault breccia, and Heart Mountain break-away fault, Wyoming and Montana

William G. Pierce, U.S. Geological Survey, 345 Middlefield Road, Menlo Park, California 94025

INTRODUCTION

This text includes two sites: the first is the Heart Mountain detachment fault and clastic dikes of fault breccia (Site 33 in Fig. 1); the second, located 1 mi (2 km) to the west of the first, is a related feature termed the Heart Mountain break-away fault (Site 33 in Fig. 1). If only one site can be examined, Site 33 should be selected: it is more accessible, and has better exposures and more features pertaining to the Heart Mountain fault.

Site 33 (Fig. 1), showing the Heart Mountain detachment fault and clastic dikes, is situated 0.5 mi (0.8 km) due south of Silver Gate, Montana. It can be reached on foot by climbing a steep mountain slope (600-vertical ft (180-m) gain in 1,800 ft (540 m), beginning at the end of the road 1,000 ft (300 m) south of the Post Office. The site is on public land but at the outset 300 ft (100 m) of private land must be crossed when taking the most direct route. The owner of the undeveloped private land does not object to scientists crossing her land to reach Site 33.

SITE 33. THE HEART MOUNTAIN DETACHMENT FAULT

Significance. The Heart Mountain fault is a detachment fault or décollement; it is a sheet of rocks that has broken loose along a basal shearing plane, has moved a long distance (in part by gravitational gliding without any push from the rear), and has been deformed independently from the rocks below the fault plane. At Site 33, several features pertaining to origin of the Heart Mountain detachment fault can be examined, including: (1) severely deformed upper plate rocks in contact with undeformed lower plate rocks; (2) the character and composition of the fault breccia; (3) contacts of volcanic rocks with upper plate blocks and the Heart Mountain fault; and (4) dikes of carbonate fault-breccia injected into both upper plate rocks and overlying volcanic rocks. (Two of these dikes contain either carbonized wood or angular xenoliths of Precambrian granitic rocks.) Significant questions to be asked about these features are: Are contacts between the upper plate and overlying volcanic rocks depositional or tectonic? and Do the dikes in the upper plate rocks indicate a fluid "flotation" mechanism in the mechanics of the Heart Mountain fault, or were they injected by lithostatic pressure imposed by a rapidly accumulating overburden of volcanic rock after the fault movement had ceased? These questions are relevant to the understanding of tectonic denudation and catastrophic movement on the Heart Mountain fault.

Site Information. Site 33 is in the bedding-plane part of the Heart Mountain detachment fault. Since it is close to the break-away fault, the upper plate blocks here have moved laterally only

about 1 mi (1.6 km) in contrast to those at McCulloch Peaks, which have moved laterally at least 30 mi (50 km) (Fig. 2). The bedding-plane part, covering an area of about 500 mi^2 (1,300 km^2), is bounded on the west by the break-away part of the fault, on the northeast by the Beartooth Mountains uplift, and on the southeast by the transgressive part of the Heart Mountain fault. The southwestern extent of the bedding plane part is concealed by volcanic rock cover. As the upper plate of the Heart Mountain detachment fault moved on the bedding-plane part, the upper plate broke up into numerous blocks that are now scattered over 1,300 mi^2 (3,400 km^2). The trace of the bedding plane part of the Heart Mountain fault is continuously exposed for 35 mi (56 km), except for intermittent slope-debris cover, from its junction with the break-away fault northwest of Cooke City, Montana, to the transgressive fault. At that point the fault passed upward to the Eocene land surface, and many large upper plate blocks, such as Heart Mountain, Logan Mountain, and Sheep Mountain, moved southeastward on that surface (Fig. 2).

Site 33 displays two upper plate fault blocks 200 to 500 ft (60 to 160 m) across, and four much smaller fault blocks. The largest block, on the right (west) side of Figure 3, is the most prominent; the left (east) side of that block is a good place to note the precise location of the fault. There, intensely sheared, brecciated, and faulted Paleozoic limestone (the high-angle faults generalized in Fig. 3) overlie the Heart Mountain fault. Immediately below the fault, the 8-ft (2.4-m)-thick basal bed of the Bighorn Dolomite occurs, undeformed except for vertical fractures; beneath the basal bed of the Bighorn Dolomite, limestone and shale of the Grove Creek Limestone Member of the Snowy Range Formation are interbedded. The unusual stratigraphic horizon at which the Heart Mountain fault occurs is a constraint on the origin of the fault. Melosh (1981) believed that the horizon at which it occurs is the result of an earthquake-generated acoustic lubrication mechanism. Pierce's view (1973) was that earthquake oscillations in conjunction with gravity constitute the only adequate explanation for movement on the Heart Mountain fault. Guth and others (1982) have concluded that simple gravity sliding under the influence of high pore pressure cannot explain the Heart Mountain fault.

The Eocene Wapiti Formation is mostly dark-brown, crudely bedded, andesitic volcanic breccia (dominant) and lenticular lava flows, with some volcanic conglomerate and sandstone. At Site 33 the contact of the Wapiti Formation with the Heart Mountain fault is poorly exposed, but 0.5 mi (0.8 km) to the northeast, on the east side of Falls Creek, the contact is very well exposed (see Pierce, 1979, p. 10–12). However, at Site 33

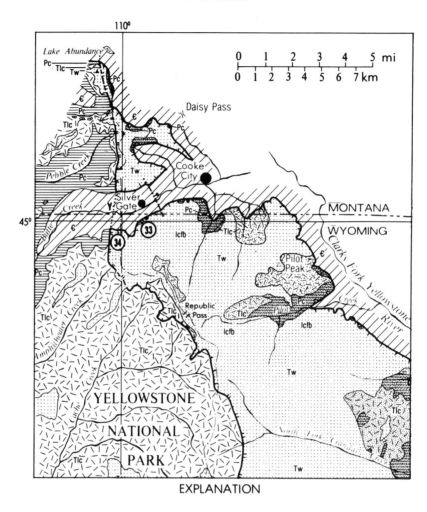

EXPLANATION

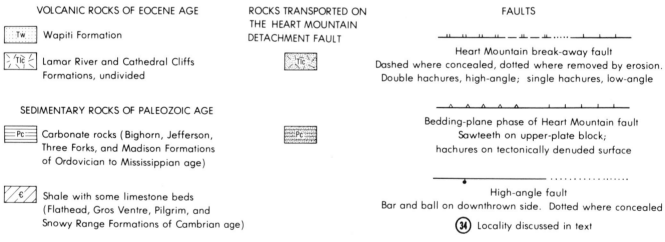

VOLCANIC ROCKS OF EOCENE AGE

| Tw | Wapiti Formation

| Tlc | Lamar River and Cathedral Cliffs
Formations, undivided

SEDIMENTARY ROCKS OF PALEOZOIC AGE

| Pc | Carbonate rocks (Bighorn, Jefferson,
Three Forks, and Madison Formations
of Ordovician to Mississippian age)

| € | Shale with some limestone beds
(Flathead, Gros Ventre, Pilgrim, and
Snowy Range Formations of Cambrian age)

ROCKS TRANSPORTED ON
THE HEART MOUNTAIN
DETACHMENT FAULT

| Tlc |

| Pc |

FAULTS

Heart Mountain break-away fault
Dashed where concealed, dotted where removed by erosion.
Double hachures, high-angle; single hachures, low-angle

Bedding-plane phase of Heart Mountain fault
Sawteeth on upper-plate block;
hachures on tectonically denuded surface

High-angle fault
Bar and ball on downthrown side. Dotted where concealed

(34) Locality discussed in text

lcfb, area within the bedding-plane phase of the Heart Mountain detachment fault
containing incompletely mapped, tectonically transported masses of undivided
Lamar River and Cathedral Cliffs Formations

Figure 1. Geologic map showing site localities and adjoining area. Surficial deposits and volcanic and
intrusive rocks younger than Wapiti Formation are not shown.

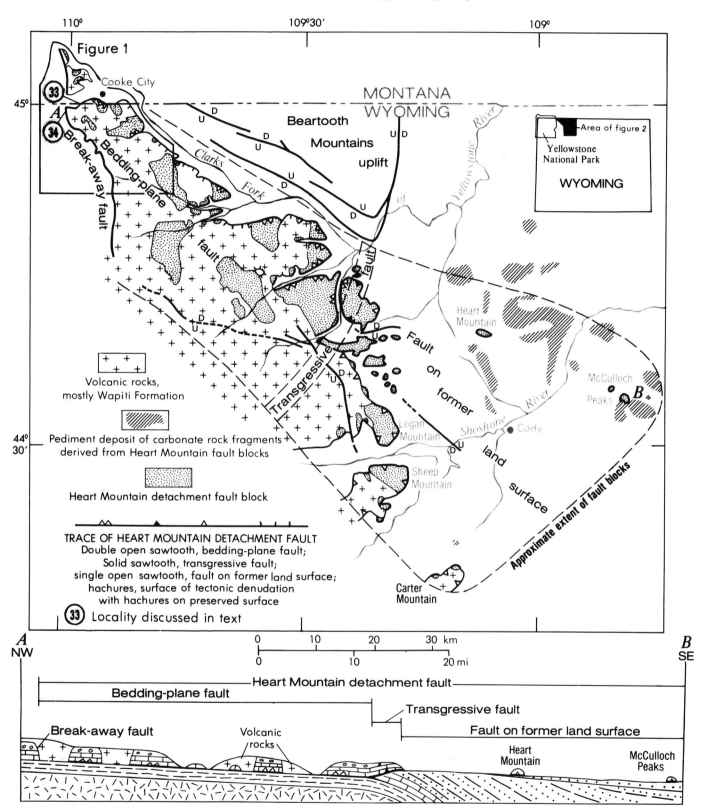

Figure 2. Sketch map showing distribution of the four phases of the Heart Mountain detachment fault and diagrammatic cross section from the break-away fault to McCulloch Peaks. Movement was to the southeast.

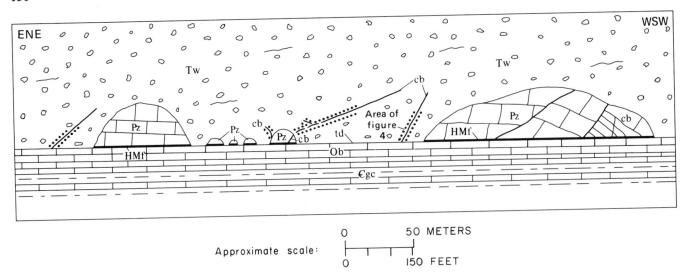

Figure 3. Diagrammatic cross section of relations at Site 33, south of Silver Gate, Montana, between upper plate Paleozoic rocks (Pz) and volcanic rocks of the Wapiti Formation (Tw), which may contain some undifferentiated Cathedral Cliffs and Lamar River rocks, and carbonate fault-breccia dikes (cb) in Paleozoic rocks of the upper plate (Pz) and in volcanic rocks of the Wapiti Formation. Dots indicate flow banding adjacent to dikes and intermixing of carbonate fault-breccia with volcanic rock; HMf, Heart Mountain fault; td, surface of tectonic denudation; Ob, basal bed of Bighorn Dolomite; €gc, Grove Creek Limestone Member of Snowy Range Formation. Faults in Paleozoic rocks do not extend into the Wapiti Formation. Diagram not to scale.

the contact of the Wapiti Formation with the upper plate Paleozoic carbonate rocks is well exposed. The relation of the Wapiti Formation to Heart Mountain fault has been interpreted both as depositional and tectonic. Hauge (1982, p. 178) has concluded that "field evidence indicates that all volcanic rock in contact with the Heart Mountain bedding-plane detachment fault was fault emplaced."

Pierce (1979) wrote that the contacts are depositional for the following reasons: (1) Wapiti Formation rests on large areas of the Heart Mountain fault, as shown in Figure 2. If the Wapiti had moved, along with the upper plate blocks, as part of the Heart Mountain fault movement, the fault breccia beneath some of the upper plate blocks presumably should contain some volcanic rock fragments. However, both here and beneath more than 30 other upper plate Paleozoic blocks, the fault breccia does not contain any volcanic fragments (Pierce and Nelson, 1970). (2) At Site 33 the Wapiti Formation overlying the upper plate block of Paleozoic rocks contains a few limestone fragments as much as 12 in (30 cm) across, which probably represent clasts incorporated into Wapiti Formation as it accumulated over the block. (3) The contact of the Wapiti volcanic rocks with the upper surface of this fault block is undulatory and irregular to jagged. It is firm and tight, with no brecciation or fault breccia along the contact. (4) The faults in the Paleozoic rocks do not extend into the overlying volcanic rocks. The westernmost of the two faults in the upper plate block (Fig. 3) has considerably displaced the Paleozoic rocks, but the fault does not offset the contact between the

Wapiti Formation and the Paleozoic rocks, nor does it extend into the Wapiti.

The clastic dike of fault breccia in the western part of the westernmost upper plate block in Figure 3 is in steeply inclined, thin-bedded shale of the Three Forks Formation. It is as much as 2 in (5 cm) thick and pinches and swells as it alternately follows and cuts across bedding. Its contact with the wallrock is sharp and distinct. No fragments of volcanic rock have been found in it. Voigt (1973) interpreted this kind of dike as "demanding a fluid 'flotation' mechanism in the mechanics of the Heart Mountain fault." Pierce (1979) believed they were injected by lithostatic pressure imposed by a rapidly accumulating overburden of volcanic rock after the fault movement had ceased.

Significance of fault-breccia dikes in volcanic rocks at Site 33. Immediately east of the westernmost upper plate block at Site 33, a lenticular dike of carbonate fault-breccia and volcanic breccia about 75 ft (25 m) long had intruded the Wapiti Formation. Rubble conceals a 20-ft (7-m) interval between the lowermost exposed part of the dike and the Heart Mountain fault. The dike strikes N30°E, has an arcuate shape, and is near vertical. It is made up of three components: carbonate fault-breccia, volcanic rocks of the Wapiti Formation, and a mixture of the two materials. In the central part of the dike is an irregular lenslike body of fine-grained carbonate fault-breccia that contains scattered angular clasts of volcanic rock ranging in diameter from less than 0.04 to 20 in (0.1 to 50 cm) or more. Most of the fragments, however, are 0.4 to 1 in (1 to 2 cm) across.

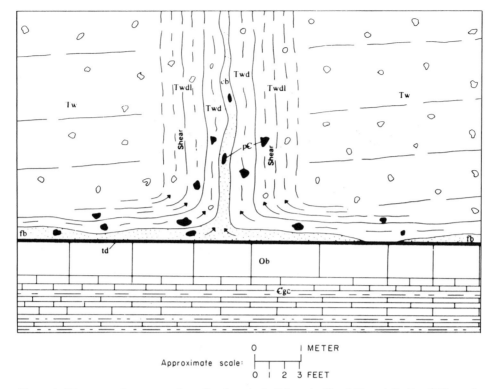

Figure 4, Diagrammatic cross section of carbonate fault-breccia dike (cb) on left side of Figure 3, showing its relation to the Heart Mountain fault carbonate breccia (fb), dike (Twd), dikelike (Twdl) bodies of Wapiti Formation, and bordering rocks of the Wapiti Formation (Tw). pϹ, angular xenoliths of Precambrian granitic rocks; Ob, basal bed of Bighorn Dolomite; Ϲgc, Grove Creek Limestone Member of Snowy Range Formation; td, Heart Mountain fault surface of tectonic denudation. Arrows indicate direction of flow.

Where the volcanic fragments are 0.4 to 1 in (1 to 2 cm) in size, they constitute no more than 10 percent of the carbonate fault-breccia; where they are smaller, they appear to form a lesser proportion of the carbonate fault-breccia. Within the carbonate fault-breccia, there is an irregular mass of breccia about 6 ft (2 m) long in the central part of the dike that is composed of about equal proportions of carbonate and volcanic rock fragments as much as 12 in (30 cm) across. This breccia body is elongated roughly parallel to the walls of the dike. At one place in the lower part of the carbonate fault-breccia is a limestone clast 6 in (15 cm) across (Fig. 4).

Volcanic breccia, much coarser than the carbonate fault-breccia, forms a zone 1 to 3 ft (0.3 to 1 m) wide on either side of the carbonate fault-breccia core and extends upward about 50 ft (15 m) above the core before pinching out. The volcanic breccia is intensely sheared and exhibits more or less vertically oriented, striated shear surfaces parallel to the dike walls. A specimen of the dike, taken at the boundary between the carbonate fault-breccia and volcanic breccia, shows a very narrow transitional contact 1 m or so thick between the two rock types; it indicates that the volcanic breccia was unconsolidated to some degree at the time the carbonate fault-breccia was injected.

Within the specimen, elongate grains and fragments of volcanic rock in both the carbonate fault-breccia and volcanic breccia show a preferential orientation parallel to the dike walls.

About 75 ft (25 m) above and to the southeast of this dike, a similar, less prominent and thinner body of carbonate fault-breccia occurs within a moderately distinct dike of volcanic breccia. This dike is about 350 ft (120 m) long, trends N55°E, and contains fragments of volcanic rock. Carbonate fault-breccia within the dike of volcanic breccia pinches and swells, is discontinuous, and is locally absent, ranging from less than 1 in (2 cm) to 1 ft (30 cm) in width. At one place in the dike are two thin 1- to 2-in (2- to 5-cm) tabular bodies of carbonate fault-breccia 1.5 ft (0.5 m) apart. Particles of carbonized wood, some as much as 1.5 in (4 cm) long, occur in the dike in a matrix of very fine-grained carbonate breccia. Although fossil wood has been found in many places in the Wapiti Formation, the wood in this dike evidently was not secondarily derived from the volcanic rocks because it is only associated with, and is entirely enclosed within, carbonate fault-breccia. Furthermore, the wood particles seem to have been incorporated into the carbonate fault-breccia after that breccia was formed because the wood is much too fragile to have withstood the deformation that produced the fault breccia.

The surface of tectonic denudation, created by movement along the Heart Mountain fault, was exposed for perhaps only a few hours or days after the fault movement ceased before being buried by volcanic rocks of the Wapiti Formation (Pierce, 1968). During this extremely brief interval, the carbonized wood became a constituent of the Heart Mountain fault breccia, presumably by aerial or aqueous transport from a short distance away. The conditions that permitted inclusion of this wood into the Heart Mountain fault breccia probably were extremely rare because carbonized wood has been found only in this one dike. However, similar fossil wood, albeit much silicified and presumably from the same source as that in the carbonate fault-breccia, has also been noted in volcanic breccia of the dike near its northeast end.

The carbonate fault-breccia dike with carbonized wood occurs within the seemingly flow-banded volcanic breccia of the Wapiti Formation that constitutes the main mass of the dike. The bodies of volcanic breccia in the dike that borders the carbonate fault-breccia are about 4 in (10 cm) wide, slightly darker than the adjoining wallrock, and contain a small amount of fine calcareous breccia. The core of carbonate fault-breccia at the southwest end of this dike lies about 75 ft (25 m) above the Heart Mountain fault. The enclosing breccia of the Wapiti Formation pinches out a few ft (m) higher up in the dike. At the northeast end of the dike, the discontinuous carbonate fault-breccia core can be traced down to within 10 ft (3 m) of the fault, below which it is concealed by talus. The contact between carbonate fault-breccia and volcanic breccia in the dike (Pierce, 1979, Fig. 1) is extremely irregular rather than planar, and the fragments in the carbonate fault-breccia do not show a preferred orientation parallel to the contact. Very thin stringers and protuberances of volcanic breccia into the carbonate fault-breccia indicate that the volcanic rock was in a mobile state: that is, it was incompletely consolidated at the time of dike injection, and the fault movement, which formed the carbonate fault-breccia, preceded deposition of the Wapiti Formation.

The carbonate fault-breccia dike shown farthest east in Figure 3 ranges to 18 in (45 cm) in width and is well exposed at the bottom and on the west side of a steep ravine. It pinches and swells and is irregularly distributed within another dike of volcanic breccia 3 to 6 ft (1 to 2 m) wide, which is Wapiti Formation: it is sheared and has fairly well defined shear borders with the enclosing volcanic rock (Pierce, 1979, Fig. 8). The contact between the carbonate fault-breccia and intrusive volcanic breccia is gradational. The carbonate fault-breccia is predominantly carbonate rock but contains abundant fragments of volcanic rock. At the outer border of the carbonate fault-breccia is a zone of fine-grained rock, 1 in (2 cm) wide, composed of volcanic breccia intermixed with a small amount of carbonate fault-breccia. The contact of the mixed breccia with the carbonate fault-breccia and the outer contact with the volcanic breccia are both gradational through a thin zone.

In addition to this irregular but more or less continuous body of carbonate fault-breccia, other discontinuous stringers less than 0.5 in (1 cm) to a few in (cm) wide occur within the dike.

The composite dike of volcanic breccia and carbonate fault-breccia is bordered by vertically sheared dikelike bodies of rock of the Wapiti Formation. The dikelike body on the south side is about 10 ft (3.5 m) wide, with the outer half less sheared and its outer border less well defined than its inner border. The dikelike body of the Wapiti on the north side of the carbonate fault-breccia dike is mostly covered. What can be observed is similar to its counterpart on the south side of the dike.

Fragments of Precambrian granitic rock occur in this dike and in the dikelike bodies of sheared Wapiti volcanic breccia that border it (see Pierce, 1979, Fig. 8). Angular xenoliths in the dike range from 0.5 to 1 in (1 to 2 cm) across in the carbonate fault-breccia to 3 ft (1 m) across in the adjacent dikelike bodies of Wapiti breccia. The xenoliths occur in the lowest exposed parts of the dike and adjacent breccia, about 10 ft (3 m) above the Heart Mountain fault, and are scattered upward for 50 ft (15 m) in the dike and in the Wapiti that borders it. The Precambrian rock fragments in the vertically sheared Wapiti breccia occur close to the dike.

Inasmuch as carbonate fault-breccia originally was irregularly distributed between the base of the Wapiti Formation and the surface of tectonic denudation, carbonate fault-breccia must have flowed laterally and upward to form the dikes, as illustrated in Figure 4. Vertical striations and shears in the rocks of the Wapiti Formation that enclose the carbonate fault-breccia, as well as the relation and mixing of the volcanic rock and carbonate fault-breccia, indicate that the Wapiti must have flowed with the carbonate fault breccia. The angularity of the granitic clasts, their wide range in size, and their lack of association with any sedimentary deposits seem to rule out any derivation by stream transport. A search was made for clasts of Precambrian rock in the basal part of the flat-lying Wapiti Formation; about 300 ft (100 m) to the southwest, granitic clasts were found less than 3 ft (1 m) above the base but not higher up in the sequence. The basal volcanic rocks, which overlie a 75-ft (25-m)-high mass of deformed Devonian rocks north of Republic Mountain (Pierce, 1979, p. 13), also contain clasts of Precambrian rock. So far as is known, the Wapiti Formation never rested directly on Precambrian rocks in this area, and Precambrian clasts have never been observed in the Heart Mountain fault breccia overlain by upper-plate rocks. Therefore, these clasts are interpreted to be xenoliths torn from the walls of a vent and incorporated into the basal part of the volcanic rocks immediately after the Heart Mountain fault movement.

SITE 34. THE HEART MOUNTAIN BREAK-AWAY FAULT

Access. The Heart Mountain break-away fault can be seen looking south from U.S. 12 from a point 300 ft (100 m) east of the northeast entrance to Yellowstone National Park; this is also the beginning of the best route to Site 34. The fault and access to it is on public land. It can be reached by wading across to the south side of Soda Butte Creek, which is easier to accomplish in mid- to late summer or fall when the water level is low. Access to

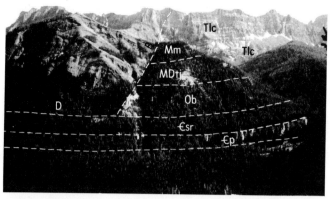

Figure 5. View looking south along the strike of the Heart Mountain break-away fault near northeast entrance to Yellowstone National Park. Break-away fault (B) terminates downward at bedding plane phase of Heart Mountain detachment fault (D). Both faults are surfaces of tectonic denudation on and against which volcanic rocks of the Eocene Wapiti Formation (Tw) were deposited. The rocks to the right of the break-away are the Lamar River and Cathedral Cliffs Formations (Tlc), Madison Limestone (Mm), Three Forks and Jefferson Formations (MDtj), and Bighorn Dolomite (Ob). Below the Heart Mountain detachment fault are the Snowy Range Formation (€sr) and Pilgrim Limestone (€p). Beyond the skyline, the break-away fault terminates upward against the Wapiti Formation, which overlies the Lamar River Formation (see Pierce, 1980, Fig. 6).

the base of the fault 700 ft (230 m) above the creek is restricted by a near-vertical cliff of Pilgrim Limestone directly below the fault, but a small ravine to the west does afford access. From the base of the break-away fault, which terminates downward at the top of the 8-ft (2.5-m)-thick basal bed of the Bighorn Dolomite, one can climb upward along the fault for 800 ft (270 m) to a saddle at the top of the Madison Limestone shown in Figure 5. Beyond (south of) the saddle, exposures are poor or lacking for the first 1,200 ft (400 m) along the fault but are excellent and accessible for the next 1,500 ft (500 m).

Significance. The Heart Mountain break-away fault is the only fault of its kind that thus far has been recognized and described (Pierce, 1960, 1980). The western face of the break-away fault is the footwall of a high-angle fault, but the eastern face of the break-away fault is a depositional contact of volcanic rocks of the Wapiti Formation against the footwall. Immediately after the fault was formed, near the end of early Eocene time, it was buried by volcanic rocks. The volcanic cover is now being removed by erosion, revealing the relation of the fault to the rocks adjoining it. This locality has the best exposures of the fault and is the most accessible part of its 23-mi (37-km) linear extent. The stratigraphic horizon at which the Heart Mountain detachment fault occurs also can be examined here for clues as to why the fault is always positioned above the base of the massive Bighorn Dolomite rather than in fissile Cambrian shale immediately below. The only suggestion that has been made for its position is one by Melosh, mentioned earlier.

Site information. The relation of the break-away fault (which dips eastward 55° to 60°) to the Heart Mountain bedding-plane fault (which is nearly horizontal) is shown in Figure 2 and 5. The stratigraphic and structural relations at the break-away fault are shown in the photograph of Figure 5, which was taken from a point higher than the highway viewpoint at the park entrance. The rocks on the right (west) side of the Figure are not faulted, and represent the normal sequence of Paleozoic and volcanic rocks in this area. On the left (east) side of the fault these rocks are missing because they have been displaced eastward along the Heart Mountain detachment plane. A large open space was created immediately to the east of the break-away fault, exposing the fault on a surface of tectonic denudation on which the Wapiti Formation then was deposited. Carbonate fault breccia was irregularly scattered on the fault surface, and before there was sufficient time for it to be removed by erosion, a thick sequence of volcanic rocks (Wapiti Formation) was deposited on it, against and over the top of the break-away fault, and on top of the Cathedral Cliffs–Lamar River volcanic rocks adjoining the west side of the break-away fault (Pierce, 1957, 1968; Pierce and others, 1973; Elliott, 1974; Prostka and others, 1975).

Adjoining the base of the break-away fault and extending eastward from it is the nearly horizontal Heart Mountain bedding-plane fault, which is concealed here by slope debris. The stratigraphic horizon at which detachment occurred is well-exposed in the unfaulted rocks just west of the base of the break-away fault. The detachment fault horizon is at the top of the basal bed of the Bighorn Dolomite, about 7 ft (2.25 m) above the base of the massive cliff-forming formation. For further description of this and other places along the break-away fault see Pierce, 1980.

REFERENCES CITED

Elliott, J. E., 1974, The geology of the Cooke City area, Montana and Wyoming; in Voight, B., and Voight, M. A., eds., Rock Mechanics: The American Northwest, 3rd Congress International Society of Rock Mechanics Expedition Guidebook: Special Publication Experiment Station, College of Earth and Mineral Sciences, Pennsylvania State University, p. 102–107.

—— , 1979, Geologic map of the southwest part of the Cooke City Quadrangle, Montana and Wyoming: U.S. Geological Survey Miscellaneous Geologic Investigations Map I-1084, scale 1:24,000.

Guth, P. L., Hodges, K. V., and Willemin, J. H., 1982, Limitation on the role of pore pressure in gravity gliding: Geological Society of America Bulletin,

v. 93, no. 7, p. 606–612.

Hauge, T. A., 1982, The Heart Mountain detachment fault, northwest Wyoming: Involvement of Absaroka volcanic rock: Wyoming Geological Association Guidebook, 1982, p. 175–179.

Melosh, H. J., 1981, Accoustically activated décollement: Mechanics of the Heart Mountain fault [abs.]: EOS, Transactions of the American Geophysical Union, v. 62, no. 45, p. 1046.

Pierce, W. G., 1957, Heart Mountain and South Fork detachment thrusts of Wyoming: American Association of Petroleum Geologists Bulletin, v. 41, no. 4, p. 591–626.

—— , 1960, The "break-away" point of the Heart Mountain detachment fault in northwestern Wyoming; in Short Papers in the Geological Sciences: U.S. Geological Survey Professional Paper 400-B, p. B236–B237.

—— , 1968, Tectonic denudation as exemplified by the Heart Mountain fault, Wyoming; in Orogenic Belts: 23rd International Geological Congress, Prague, Czechoslovakia, 1968, Report, Section 3, Proceedings, p. 191–197.

—— , 1973, Principal features of the Heart Mountain fault and the mechanism problem; in DeJong, K. A., and Scholten, R., eds., Gravity and Tectonics: New York, John Wiley & Sons, p. 457–471.

—— , 1979, Clastic dikes of Heart Mountain fault breccia, northwestern Wyoming, and their significance: U.S. Geological Survey Professional Paper 1133, p. 1–25.

—— , 1980, The Heart Mountain break-away fault, northwestern Wyoming: Geological Society of America Bulletin, Part 1, v. 91, p. 272–281.

Pierce, W. G., and Nelson, W. H., 1970, The Heart Mountain detachment fault— A volcanic phenomenon? A discussion: Journal of Geology, v. 78, p. 116–123.

Pierce, W. G., Nelson, W. H., and Prostka, H. J., 1973, Geologic map of the Pilot Peak Quadrangle, Park County, Wyoming: U.S. Geological Survey Miscellaneous Geologic Investigations Map I-816, scale 1:62,500.

Prostka, H. J., Ruppel, E. T., and Christiansen, R. L., 1975, Geologic map of the Abiathar Peak Quadrangle, Yellowstone National Park, Wyoming and Montana: U.S. Geological Survey Quadrangle Map GQ-1244, scale 1:62,500.

Voight, B., 1973, Role of fluid pressure in mechanics of South Fork, Reef Creek, and Heart Mountain rockslide: Geological Society of American Abstracts with Programs, v. 5, no. 2, p. 233–234.

Geology and evolution of the Grand Canyon of the Yellowstone, Yellowstone National Park, Wyoming

Gerald M. Richmond, U.S. Geological Survey, Box 25046, Denver Federal Center, Denver, Colorado 80225

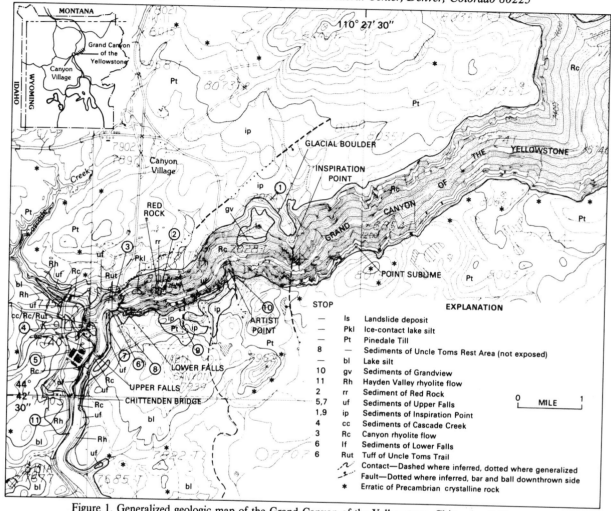

Figure 1. Generalized geologic map of the Grand Canyon of the Yellowstone, Chittenden Bridge to Sevenmile Hole.

LOCATION AND ACCESSIBILITY

The Grand Canyon of the Yellowstone is in east-central Yellowstone National Park. It is accessible from Canyon Village (Fig. 1, inset), where all facilities are available. Check in at the visitor center.

Observe park regulations: stay on established trails unless you have off-trail permission; do not break rocks or hot-spring deposits; take no samples, not even loose pieces (collecting is not permitted); leave hammers, picks, and shovels in your vehicle. Suggested time: one full day.

SIGNIFICANCE

Central Yellowstone National Park is in the Yellowstone Caldera, whose age, as determined by K-Ar dating of the contemporaneously extruded Lava Creek Tuff, is 610 ka. Since its formation, the caldera has been occupied alternately by lakes and ice caps into which and against which rhyolite flows have erupted. Stratigraphic studies of the lake and glacial deposits, and K-Ar dating of the flows and their associated volcanic ash beds, has permitted construction of a chronology of the successive glaciations and interglaciations (Richmond, 1986), and of the sequence of events in the development of the Grand Canyon (Fig. 2).

NOMENCLATURE

The nomenclature of glacial deposits and glaciations used on US Geological Survey surficial geologic maps and reports on the park is that of the Wind River Range, where the term "Pinedale"

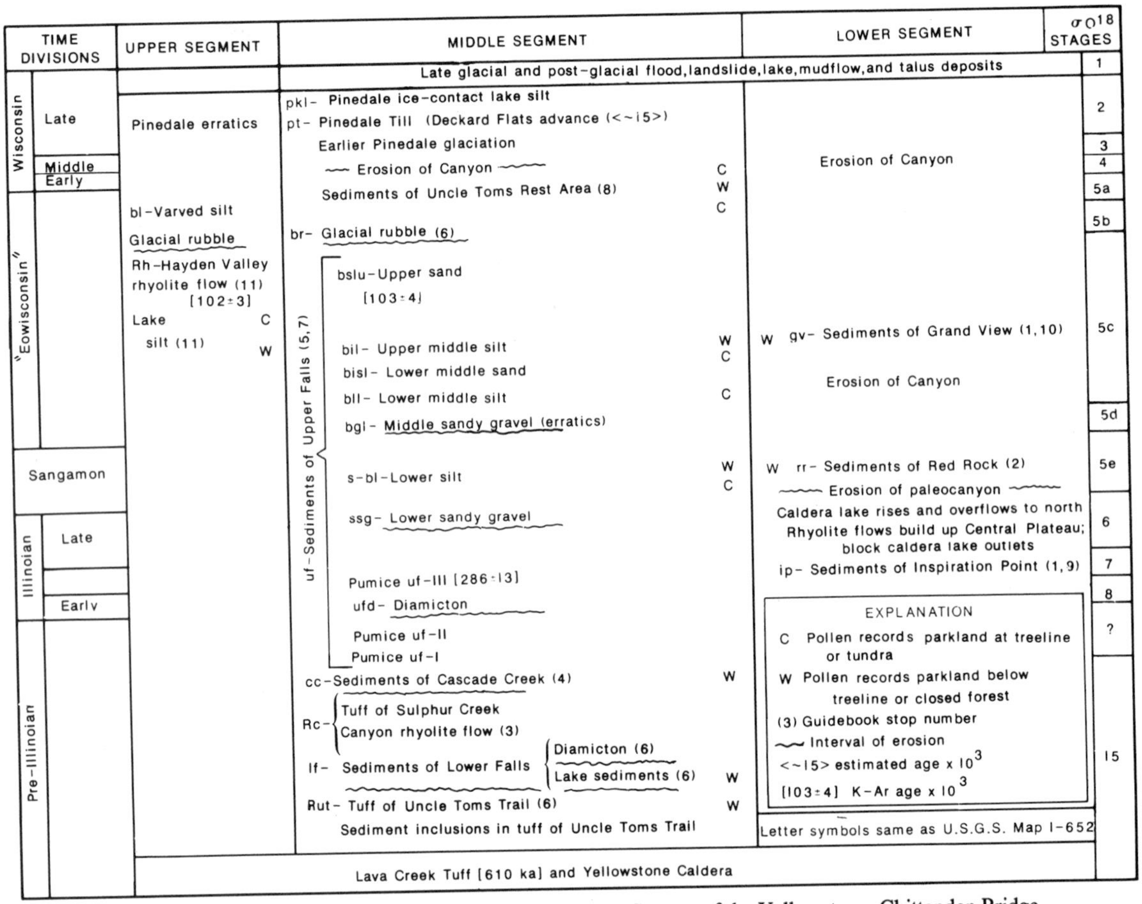

Figure 2. Stratigraphy and erosional history of the Grand Canyon of the Yellowstone, Chittenden Bridge to Sevenmile Hole.

has traditionally implied the last, or late Wisconsin, glaciation, and the terms "early" and "late Bull Lake" have implied the two next older glaciations. On that basis, deposits of two glaciations next older than the last (Pinedale) in the park were mapped as early and late Bull Lake (Richmond, 1976; 1977), and their maxima were K-Ar dated from ice-contact rhyolite flows as 117 ka and 90 ka, respectively. However, end moraines near West Yellowstone, dated by the obsidian-hydration technique as about 140 ka (Pierce and others, 1976), and till at Rock Point on Yellowstone Lake K-Ar dated as older than 153 ka and younger than 160 ka (Richmond, 1986); both are correlated with the Bull Lake glaciation and are Illinoian in age. The two younger glacial advances with maxima at about 117 ka and about 90 ka, therefore, intervened between the Bull Lake and the Pinedale glaciations (Richmond, 1986). Pollen records from sediments in the Grand Canyon and elsewhere show that the two younger glaciations were preceded, separated, and followed by "warm" intervals not unlike the present. The term "Eowisconsin" has been applied to the time, including the two glaciations and the two "warm" intervals that separated and followed them (Richmond and Fullerton, 1986). It extends from the end of the "warm" Sangamon interval (122 ka) to the beginning of Wisconsin time (79 ka).

SETTING

Yellowstone River flows northwest from Yellowstone Lake through Hayden Valley and turns abruptly northeast into the head of the Grand Canyon (Fig. 1). This guide concerns that part of the canyon seen by most visitors, between Chittenden Bridge and Sevenmile Hole, 0.5 mi (0.8 km) east of the map area (Fig. 1). This part of the canyon consists of three segments: A relatively shallow upper segment, cut in the Canyon rhyolite flow (Rc) (Christiansen, 1975) between Chittenden Bridge and Upper Falls; a middle segment, cut in relatively soft sediments and tuff between Upper Falls and Lower Falls; and a lower segment, cut entirely in the Canyon rhyolite flow, between Lower Falls and the edge of the caldera in Sevenmile Hole. Beyond, the canyon passes through Eocene volcanic rocks of the Washburn Range without significant change in gradient.

STRATIGRAPHY AND EROSIONAL HISTORY

It is impractical to visit the deposits and erosional features of the Grand Canyon in the order of their development because the localities where they are best observed are scattered in nonstratigraphic sequence on both sides of the canyon. The stratigraphy and erosional features, therefore, will be discussed from oldest to youngest, to provide a sequential geologic framework for reference in the course of the nonstratigraphically ordered tour stops. Figure 2 outlines the stratigraphy and erosional history. All of the deposits are within the caldera and thus are younger than 610 ka. Descriptions and map symbols are from Richmond (1977). Radiometric ages are from Richmond (1986); all K-Ar ages are by J. D. Obradovich.

Tuff of Uncle Toms Trail (Rut) (Stop 6). The oldest deposit in the Grand Canyon is the tuff of Uncle Toms Trail. It is a whitish, locally limonite-stained, partly welded pumiceous rhyolite tuff that was erupted into the caldera from the ring-fault system at the foot of the Washburn Range to the north. The lower part of the tuff contains irregular masses of sand and gravel several yards (meters) long and 3–6 ft (1–2 m) thick that are swirled irregularly upward in the tuff. Clasts are chiefly rhyolite, obsidian, basalt, and andesite; but some are quartzite, granite, and gneiss derived from a Pliocene conglomerate in the Washburn Range. At Lower Falls the tuff is in near-vertical contact with the outer crust of the Canyon rhyolite flow (Rc) and is baked and oxidized pinkish brown. At Stop 6 (Fig. 1) the tuff is about 148 ft (45 m) thick and is overlain disconformably by the lacustrine sequence of the sediments of Lower Falls (1f). Upstream, east of the river, the lacustrine sediments are missing and the tuff is overlain disconformably by the diamicton of the sediments of Lower Falls. West of the river, the diamicton is missing and the tuff is overlain disconformably by the canyon rhyolite flow (Rc). Pollen from a single sample of the tuff indicate a closed pine forest and a "warm" paleoclimate (E. B. Leopold, written communication, 1975).

Sediments of Lower Falls (1f) (Stop 6). The sediments of Lower Falls consist of a lower lacustrine sequence and a disconformably overlying diamicton. The lacustrine sediments fill steep gullies cut deeply into the tuff of Uncle Toms Trail. The lower 44 ft (13.4 m) is gray, medium to coarse, silty sand. The upper 47.9 ft (14.6 m) is compact, buff, thinly laminated, tuffaceous, fine sand and silt. Pollen from widely separated samples indicate a closed pine forest and "warm" paleoclimate (W. Mullenders, written communication, 1973) (E. B. Leopold, written communication, 1975).

The diamicton is a tough, unsorted, tuffaceous, clayey sand containing abundant subangular fragments of gray, fine-grained biotite diorite, as much as 2.0 ft (0.6 m) in diameter, derived from an Eocene stock 4.2 mi (6.4 km) to the northeast in the Washburn Range. It also contains a few pebbles of rhyolite, quartzite, granite, and gneiss. Thickness is 20 to 75 ft (6 to 23 m). Locally, the diamicton abuts a steep slope cut in the lacustrine sediments of Lower Falls and extends from the present canyon rim down to river level. The deposit was found in the subsurface nearly 8 mi (12 km) upstream from Upper Falls. The diamicton has been considered till, talus, and subaqueous tuffaceous mudflow or subaqueous lithic tuff.

Canyon Rhyolite Flow (Rc) (Seen throughout the Canyon; Detail at Stop 3). The Canyon rhyolite flow is a large rhyolite flow, overlain locally by the nearly indistinguishable tuff of Sulphur Creek. The Grand Canyon traverses the flow along a linear zone of hydrothermal alteration. Upper Falls, 109 ft (33 m) high, is formed where Yellowstone River flows over the outer margin of the flow into soft sediments downstream. Lower Falls, 319 ft (97 m) high, is formed where the river flows from the sediments across the outer crust of the flow and into its brilliantly oxidized and hydrothermally altered interior.

The Canyon rhyolite flow is in near vertical contact with the tuff of Uncle Toms Trail (Rut) at Lower Falls. It overlies the tuff along the lower sector of Cascade Creek (Stop 4) and at the same locality is overlain by the sediments of Cascade Creek (cc). The high bouldery front of the flow also is well exposed along lower Cascade Creek. Internal structures of th flow are well exposed along the trail to Lower Falls (Stop 3).

Sediments of Cascade Creek (cc) (Stop 4). The sediments of Cascade Creek include a bouldery alluvial fan facies and a tuffaceous, lacustrine facies. The alluvial fan facies consists of boulders to pebbles of rhyolite in a buff, crudely bedded, silica-cemented, rhyolitic sandy matrix. It is wholly derived from the surface of the canyon rhyolite flow. The alluvial fan facies interfingers abruptly eastward with the lacustrine facies, which consists of gray, moderately silica-cemented, thin-bedded, clayey silt, sand, and pebbly sand. Clasts are chiefly hydrothermally altered rhyolite and pumice. Numerous small voids represent casts of leached obsidian clasts. Pollen from the silt beds indicate a closed pine forest (E. B. Leopold, written communication, 1972; William Mullenders, written communication, 1973). The highest lake sediments are about 390 ft (120 m) below the pre-Grand Canyon drainage divide. The caldera lake, therefore, probably drained west or south.

Sediments of Inspiration Point (ip) (Stops 1, 9). The sediments of Inspiration Point consist mostly of pale buff, porcelaneous, intensely altered and silicified, alluvial and shallow lacustrine, rhyolitic silt, sand, and pebbly sand and, locally, layers of chalcedonic sinter. The deposits occur on both rims of the canyon, and commonly are so highly opalized that they are difficult to distinguish as sediment. However, they are less intensely altered at Stop 1 near Glacial Boulder (Fig. 1). Southeast of the canyon at Stop 9 layers of chalcedonic sinter are interbedded with silicified siltstone, sandstone, and conglomerate. The sinter forms near hot springs, initially as opaline sinter. It cannot be deposited under water (D. E. White, written communication, 1972). The caldera lake, therefore, must have been lower than the canyon rim when the sediments of Inspiration Point were deposited. Conversion of opaline sinter to chalcedonic sinter requires deep burial for a long time under high temperatures and near-surface ground water (White and others, 1956). The caldera lake, therefore, must have been higher at the time of conversion than during deposition of the sediments of Cascade Creek. Whatever deposits buried the sinter are now stripped away.

Cause of First Northward Overflow From the Caldera. The Central Plateau of Yellowstone National Park is underlain by a series of rhyolite flows, erupted between about 175 ka and 153 ka (Richmond, 1986). The flows eventually built up the plateau, impounding the caldera lake to the east and causing it to rise and overflow northward across the canyon rhyolite flow. The overflow probably occurred about 150 to 140 ka during latest Illinoian time. No late Illinoian glacial deposits have been found in the canyon between Lower Falls and Sevenmile Hole.

Following overflow, the outlet stream eroded a narrow canyon in the canyon rhyolite flow, within the limits of the present canyon. A remnant of this paleocanyon partly filled with sediments is preserved at Red Rock, a pinnacle jutting from the north wall of the present canyon (Stop 2). Its floor is about 318 ft (97 m) below the canyon rim and about 450 ft (137 m) above the present canyon floor.

Sediments of Red Rock (rr) (Stop 2). A series of sediments more than 197 ft (60 m) thick occurs in the paleocanyon at Red Rock. The series includes four units: a basal sequence of talus blocks mixed with gravel and beds of gravelly sand; a lower alluvial and lacustrine sequence of pebbly sand and silt; a middle deltaic sand dipping down canyon; and an upper lacustrine sand and silt sequence. The sediments are silica-cemented and all obsidian has been leached. Pollen from silt beds in the lower unit suggest a parkland below tree line; pollen from silt beds of the upper unit suggest a closed forest, thus a "warm" environment (William Mullenders, written communication, 1968). Impoundment of the lake by a glacier to the north as previously proposed (Howard, 1937; Richmond, 1976) seems unlikely. Possible damming by large block slides in the north end of Sevenmile Hole, by cross-canyon upthrown fault blocks, by volcanic doming, or by isostatic rebound have not been documented. The sediments of Red Rock are not dated but may be Sangamon in age (Fig. 2).

First Major Erosion of the Present Canyon. Displacement of Yellowstone River from the sediments of Red Rock to its present channel may have been the result of superposition or of diversion by a glacier. In either case, displacement occurred and a new channel was eroded to a depth of about 443 ft (135 m) below the canyon rim, or about 394 ft (120 m) above the present canyon floor. The channel is partly filled with the sediments of Grandview.

Sediments of Grand View (gv) (Seen From Stops 1, 10). This unit is chiefly gray altered rhyolite and obsidian sand, but includes some bluish gray silt in its upper and lower middle parts, and also local thin pebble layers and volcanic ash beds. Total thickness is about 148 ft (45 m). Pollen from a silt bed 5.9 ft (1.8 m) thick in the lower middle part suggests a parkland below treeline; pollen from a silt unit 1.6 ft (5.3 m) thick in the upper middle part suggests a dense pine forest, thus a "warm" paleoenvironment (William Mullenders, written communication, 1972). Impoundment of the sediments of Grand View by a glacier to the north (Howard, 1937; Richmond, 1976) seems unlikely. Damming by landslides or uplifted cross-canyon blocks, volcanic doming, or isostatic rebound, though possible, is not documented.

Sediments of Upper Falls (uf) (Stops 5, 7). The sediments of Upper Falls comprise deposits of three glaciations and intervening warm intervals whose age range probably includes the times of deposition of both the sediments of Red Rock and

the sediments of Grand View (Fig. 2). The deposits are exposed in the upper sector of the canyon between Upper Falls and Lower Falls, and locally downstream along the south rim to Artist Point. Most accumulated in a lake dammed by the Canyon rhyolite flow and by the diamicton of the sediments of Lower Falls. The deposits are grouped in one unit because they are not mappable separately and are distinguishable individually in only a few places. Their stratigraphy is most completely exposed at Stop 7, where they rest disconformably on the diamicton of the sediments of Lower Falls about 33 ft (10 m) above river level. The sediments of Upper Falls comprise the following units from oldest to youngest. For detailed descriptions see Richmond, 1976. Pollen interpretations are by William Mullenders (written communication, 1968).

Pumice Bed uf-I. Thickness 22.0 ft (6.7 m). Not dated. Overlain conformably by:

Pumice Bed uf-II. Thickness 7.9 ft (2.4 m). Overlain disconformably by:

Diamicton (ufd). Possible glaciolacustrine origin. Thickness 4.9 ft (1.5 m). Overlain conformably by:

Pumice Bed uf-III. Thickness 11.8 ft (3.6 m). K-Ar age of 286 ± 13 ka suggests that underlying diamicton is early Illinoian in age.

Disconformity. Hiatus inferred to represent time of late Illinoian glaciation (Fig. 2). Caldera lake is inferred to have risen and overflowed to the north near the end of this time.

Lower Sandy Gravel (ssg). Thickness 0.49–1.97 ft (0.15–0.6 m).

Lower Silt (s-bl). Thickness 3.12 ft (0.95 m). Pollen in lower part suggest "cold" environment. Pollen in upper part indicate environmental change upward from "cold" to "warm".

Disconformity. Hiatus inferred to represent time of early "Eowisconsin" glaciation (Fig. 2).

Middle Sandy Gravel (bgl). Thickness 6.9 ft (2.1 m).

Lower Middle Silt (bll). Thickness 12.8 ft (3.9 m). Pollen indicate deposition in a cold environment.

Lower Middle Sand (bisl). Thickness 7.2 ft (2.2 m).

Upper Middle Silt (bil). Thickness 11.2 ft (3.4 m). Pollen suggest environment became increasingly "cold," followed by a "warm" record in upper part where silt grades upward into sand.

Upper Sand (bslu). Thickness 31.8 ft (9.7 m). Along trail south of overlook, the sand contains a bed of volcanic ash K-Ar dated 103 ± 4 ka.

Disconformity. Hiatus inferred to represent time of late "Eowisconsin" glaciation.

Glacial Rubble and Outwash Gravel (br; Not Mapped) (Stop 6). Some erratics of basalt and andesite. Unsorted, silty, sand matrix. Thickness 4.9 ft (1.5 m).

Sediments of Uncle Toms Rest Area (Formerly Exposed in Temporary Excavation) (Stop 8). Sequence of lacustrine sediments overlying glacial rubble (br). The sediments consist, from bottom upward, of thin-bedded diatomite, pumiceous lake silt containing a bed of light-gray volcanic ash, silty sand, nd an upper lake silt. Thickness 8.2 ft (2.5 m). Pollen content indicates that the environment changed from "cold" during deposition of basal part to "warm" throughout middle part to "cold" during deposition of uppermost part (E. B. Leopold, written communication, 1975). Lake inferred to have been retained by margin of Canyon rhyolite flow at altitude of present canyon rim at Lower Falls.

Relation to Time Divisions. K-Ar ages from the sediments of Upper Falls in the section at Stop 7 indicate that deposition included the time between 286 ± 13 ka and 103 ± 4 ka. The date from pumice uf-III suggests that the underlying diamicton (ufd) is early Illinoian in age. As indicated above, late Illinoian and two "Eowisconsin" glacial maxima are K-Ar dated at 155–160 ka, 117 ka, and 90 ka, respectively. If the lower and middle sandy gravel units (ssg, bgl) and the glacial rubble (br) are inferred to be of late glacial origin, they are late Illinoian, early "Eowisconsin," and late "Eowisconsin," respectively, in age. Pollen from silt above each sandy gravel indicate that environment changed from "cold" to "warm." On this basis, I infer that the lower silt (s-bl) is early Sangamon in age and that the upper middle silt (bil) and sediments of Uncle Toms Rest Area, respectively, separate and follow the two "Eowisconsin" glaciations.

Lake Silt Invaded by the Hayden Valley Rhyolite Flow (Rh) (Stop 11). About 0.2 mi (0.3 km) south of Chittenden Bridge, the steep margin of the Hayden Valley rhyolite flow (Rh) is exposed against folded and disrupted, pale buff, lake silt containing abundant scattered perlite. A large swirl of the silt extends upward around and into the obsidian flow front, which is intensely perlitized as a result of instantaneous hydration when the flow entered the lake. The K-Ar age of the flow, 102 ± 19 ka, is sufficiently close to that of the pumice (103 ± 4 ka) in the upper sand (bslu) of the sediments of Upper Falls to indicate their eruptive association. The silt is, therefore, an offshore facies of the sand. Pollen from the silt beneath the flow at Chittenden Bridge indicate a gradual change in climate from "warm" to "cold" (E. B. Leopold, written communication, 1975) at the time of the eruption.

Lake Silt (bl) Over Glacial Rubble on Hayden Valley Rhyolite Flow (Rh). Varved silt overlies glacial rubble on the Hayden Valley rhyolite flow south of Cascade Creek and east of Yellowstone River (Fig. 1). Locally, the varves contain pumiceous volcanic ash and an undated pumiceous volcanic ash bed occurs in the silt east of Yellowstone River. The pumiceous ash probably was erupted in association with a nearby post-Hayden Valley rhyolite flow that was emplaced against an advancing late "Eowisconsin" glacier (Richmond, 1986). The varved silt, previously inferred to be a deposit of a proglacial Pinedale lake (Howard, 1937; Pierce, 1979), therefore, cannot be Pinedale (late Wisconsin) in age.

Pinedale Till and Erratics (pt). Pinedale Till and erratics are sparsely distributed throughout the area adjacent to the canyon (Fig. 1). Erratics of Precambrian crystalline rock show that the glacier flowed from north to south. The till is inferred to have been deposited during the Deckard Flats readjustment (Pierce, 1979) or readvance (Richmond, 1986) about 15,000 B.P. The

oldest of several ^{14}C ages from organic matter overlying deposits of that advance in the Yellowstone Lake area is 14,490 ± 350 (W-3183) B.P. At the Pinedale glacial maximum an ice cap covered most of the park. Although that ice filled and overrode the canyon, its erosional effects appear negligible.

Ice-contact Lake Deposit (pkl). A single deposit of blue-gray, steeply laminated, ice-contact silt is present on the north wall of the canyon within the spray of Lower Falls (Fig. 1).

Late Glacial and Post-Glacial Deposits (ls). Flood, landslide, lake, mudflow, and talus deposits are present in places along the lower walls or floor of the canyon, notably in Sevenmile Hole. Discussion and some ^{14}C ages of these deposits are given by Richmond (1976).

SUGGESTED ROUTE AND STOPS

Canyon Village Visitor Center. Be sure to check in and discuss your plans. Obtain parking and other instructions (Stops 9, 11). Request permission for any desired off-trail observations (Stops 2, 4, 6, 7). The USGS topographic map of the Canyon Village quadrangle is available at the Visitor Center. The bedrock, and surficial geologic quadrangle maps are not available.

Drive to Inspiration Point (Fig. 1) for orientation. Upstream: Lower Falls, Artist Point (across canyon) and sediments of Grand View. Note the youthfulness of the canyon, absence of glacial scour, and colors of the hydrothermally altered Canyon rhyolite flow. Downstream: Sevenmile Hole is beyond the distant bend of the river to the north. Walk 150 ft (50 m) upstream along the North Rim Trail to a look-out over sediments of Grand View and, upslope, to sediments of Inspiration Point.

Stop 1—Sediments of Inspiration Point (ip). Near Glacial Boulder (Fig. 1), the road is cut in little-altered sediments. Note low outcrops of the white, opalized phase of the sediments along the road as far west as Lookout Point, from which the trail descends to Red Rock.

Stop 2—Sediments of Red Rock (rr). Descend the trail. Observe beds of upper sequence along the lower part of the trail and in the gully to the west. From the lookout, view Lower Falls; observe deposits of lower sequence and basal talus breccias in foreground; note pinnacle of Red Rock and sediments between it and the canyon wall. Delta sand unit is exposed off trail beneath the platform.

Stop 3—Lower Falls, and Canyon Rhyolite Flow (Rh). Descend the trail. Note exposures of texture and structure of the Canyon rhyolite flow (Rh). From the platform observe the reddish tuff of Uncle Toms Trail (Rut) across the river in near-vertical contact with the flow. View the river flowing from the tuff over the hard margin of the flow and falling into its soft interior to form Lower Falls.

Stop 4—Sediments of Cascade Creek (cc). Park at the upper end of the Upper Falls parking area. Follow the North Rim Trail about 150 ft (50 m) from the trailhead and take the fork to the right. Note the bouldery front of the Canyon rhyolite flow and its white matrix of re-worked tuff of Uncle Toms Trail in the steep slope to the north. Bouldery alluvial facies of sediments of

Cascade Creek underlie the south slope of the ridge. Descend about 75 ft (25 m), observe and return. With permission, continue down the trail beyond the blocked point to vertical exposures of tuffaceous lacustrine facies to the left. Observe, and return to the North Rim Trail.

To see the Canyon rhyolite flow and underlying tuff of Uncle Toms Trail beneath the sediments of Cascade Creek, follow the North Rim Trail north across the bridge over Cascade Creek above Crystal Falls, with permission, descend to Cascade Creek to about 300 ft (100 m) below the falls. In the steep slope across the creek, about 20 ft (6 m) of tuff of Uncle Toms Trail are overlain unconformably by about 9 m of steeply dipping Canyon rhyolite flow, which is overlain in turn by the flat-bedded lacustrine sediments of Cascade Creek. Return to the parking area and walk to:

Stop 5—Upper Falls Lookout. View falls retained by the hard margin of the Canyon rhyolite flow (Rh). Observe the sequence of sediments of Upper Falls (uf) in the gully across the river. Drive south, cross Yellowstone River at Chittenden Bridge and continue to Uncle Toms parking area.

Stop 6—Tuff of Uncle Toms Trail (Rut); sediments of Lower Falls (lf), and glacial rubble and gravel (br). walk straight across to the canyon rim and follow the South Rim Trail downstream 300 ft (100 m) to a steep reentrant. Observe white tuff of Uncle Toms Trail (Rut) at bottom. It is overlain by the silty lake sediments of Lower Falls (lf) which, in turn, are overlain disconformably by the greenish diamicton of the sediments of Lower Falls. Glacial rubble and gravel (br) form a cemented ledge at the rim.

With permission, descend the glacial rubble (br) ledge to the left of reentrant and walk left 150 ft (50 m) along the slope to a good diamicton outcrop. Is it a till, subaqueous mudflow, or subaqueous lithic tuff? Upstream, note that it extends down to the river. Return to the trail and walk to the Upper Falls Overlook.

Stop 7—Upper Falls Overlook and Sediments of Upper Falls (uf). View the Upper Falls. Observe the sediments of Upper Falls exposed in the gully beneath the overlook. The upper sand (bslu) at the top is underlain by middle silt and sand units (bil, bisl, bll). The middle sandy gravel (bgl) forms a ledge protruding from the slope. Lower units probably cannot be seen. Do not descend unless accompanied by a ranger! With a ranger, descend the left side of the outcrop. Note pumice uf-III, dated 286 ± 13 ka, on diamicton (ufd), and the underlying pumice beds (uf—II and uf-I) that rest on the diamicton of the sediments of Lower Falls about 30 ft (10 m) above the river. About 300 ft (100 m) along the slope upstream, observe the near-vertical contact of the diamicton with the margin of Canyon rhyolite flow.

Return to Lookout Trail and proceed south about 300 ft (100 m). Look upslope for the whitish pumice bed, dated 103 ± 4 ka, in upper sand (bslu). Return to the overlook noting thin-bedded brittle diatomite debris in trail cuts.

Stop 8—Sediments of Uncle Toms Rest Area (not exposed). Review discussion under stratigraphy, and Figure 2.

Look for Pinedale erratics of Precambrian crystalline rocks around parking areas or across the road. Drive about 0.5 mi (0.8 km) toward Artist Point to road cuts 30 to 50 ft (9 to 15 m) high in highly altered sediments on the right.

Stop 9—Sediments of Inspiration Point (ip). Park out of traffic. Note opalized silt, sand, and conglomerate. Examine white, thinly layered masses of chalcedonic sinter. Do not collect!

Continue toward Artist Point. At curve to the left, note (on right) weakly silicified, varved silt of Pinedale age.

Stop 10—Artist Point. From the lookout, view the Lower Falls and Red Rock upstream. Downstream, observe sediments of Grand View (gv) beneath reddish landslide deposit (ls) in the basin across the canyon. Drive back to Chittenden Bridge, and south (left) along river about 0.2 mi (0.3 km).

Stop 11—Lake Sediments Invaded by Hayden Valley Rhyolite Flow (rh). A large cut on the right exposes dark obsidian at the upper left over buff silt at the lower right. Park ahead on the left and walk back. Observe perlitized obsidian and baked lake silt containing abundant pumice. See discussion under stratigraphy. Silt is offshore facies of near-shore upper sand (bslu) of sediments of the Upper Falls.

See stratigraphy for discussion of glacial rubble on Hayden Valley rhyolite flow, overlying varved silt (bl), and Pinedale Till and erratics.

REFERENCES

Christiansen, R. L., 1975, Geologic map of the Canyon Village quadrangle, Yellowstone National Park, Wyoming: U.S. Geological Survey Geologic Quadrangle Map GQ-1192, scale 1:62,500.

Howard, A. D., 1937, History of the Grand Canyon of the Yellowstone: Geological Society of America Special Paper 6, 159 p.

Pierce, K. L., 1979, History and dynamics of glaciation in the northern Yellowstone National Park area: U.S. Geological Survey Professional Paper 729-F, 89 p.

Pierce, K. L., Obradovich, J. D., and Friedman, Irving, 1976, Obsidian hydration dating and correlation of Bull Lake and Pinedale Glaciations near West Yellowstone, Montana: Geological Society of America Bulletin, v. 87, p. 703–710.

Richmond, G. M., 1976, Surficial geologic history of the Canyon Village quadrangle, Yellowstone National Park, Wyoming, for use with Map I-652: U.S. Geological Survey Bulletin 1427, 35 p.

——, 1977, Surficial geologic map of the Canyon Village quadrangle, Yellowstone National Park, Wyoming: U.S. Geological Survey Miscellaneous Geologic Investigations Map I—652, scale 1:62,500.

——, 1986, Stratigraphy and chronology of glaciations in Yellowstone National Park, in Richmond, G. M., and Fullerton, D. S., eds., Quaternary glaciations in the United States of America, Part II, Quaternary glaciations in the Northern Hemisphere: London, Pergamon Press (in press).

White, D. E., Brannock, W. W., and Murata, K. J., 1956, Silica in hot-spring waters: Geochimica et Cosmochimica Acta, v. 10, nos. 1-2, p. 27–59.

The Devils Tower, Bear Lodge Mountains, Cenozoic igneous complex, northeastern Wyoming

Frank R. Karner and Don L. Halvorson, Department of Geology and Geological Engineering, University of North Dakota, Grand Forks, North Dakota 58202

LOCATION AND ACCESS

The Devils Tower sites (Lisenbee and others, 1981) are located in northeastern Wyoming, adjacent to and within the Devils Tower National Monument. Sundance, Wyoming is the nearest town of any size; access to the sites is via U.S. 14 from Sundance. The Bear Lodge Mountain sites (Lisenbee and others, 1981), with the exception of Sundance Mountain, are all within the Black Hills National Forest. Access to these sites is by state and county highways from Sundance, Wyoming. The Sundance Mountain site is accessible by walking from the ranch just south of the intersection of U.S. 14 and Wyoming 116. Permission must be obtained before entering this site. These sites are all shown on topographic maps of the U.S. Geological Survey 15-minute series. Sites in the Devils Tower region may be located on the Devils Tower and Nefsy Divide 15-minute Quadrangles; points of interest in the Bear Lodge Mountains are found in the Sundance 15-minute Quadrangle.

INTRODUCTION

Devils Tower is the most widely known landmark of the northern Great Plains. It has been a sacred place for Native American peoples, a communal recreation site for early white settlers, and a must for traveling geologists since the Custer expedition of 1875 (Newton and Jenny, 1880). Devils Tower is recognized as the world's finest example of columnar jointing in a phonolite monolith. The Bear Lodge Mountains center forms a classic example of a carbonatite-pyroxenite-alkalic rock intrusive and extrusive continental igneous association. The center also shows thorium and rare-earth mineralization, and remarkable potassium fenitization. This scenic uplift forms the northwestern extension of the Black Hills; the geology is prominently displayed throughout the area.

SIGNIFICANCE

The Black Hills and Bear Lodge Mountains regions are similar to the Central Rocky Mountains and related smaller ranges that contain large exposures of uplifted Precambrian rocks. This uplift may be the easternmost evidence of major Laramide foreland deformation involving the movement of large Precambrian blocks. The uplift could also be considered as a northern extension of the Southern Rocky Mountains and the intimately related Rio Grande Rift, which is more fully developed to the south (Karner, 1981). Viewed in this way, the Black Hills–Bear Lodge region may be the locus of incipient or aborted rifting.

Shallow intrusive and extrusive rocks of Cenozoic age form a continental phonolite-trachyte-quartz latite association; these rocks are distributed in an east-west zone consisting of approxi-

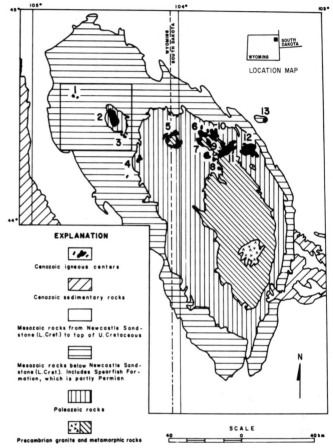

Figure 1. Generalized geology of the northern Black Hills Cenozoic igneous province (Karner, 1981). The 13 principal igneous centers are 1, Devils Tower–Missouri Buttes; 2, Bear Lodge Mountains; 3, Sundance Mountain–Sugarloaf; 4, Inyan Kara–Black Buttes; 5, Mineral Hill–Tinton; 6, Spearfish–Carbonate; 7, Terry Peak; 8, Strawberry Hill–Custer; 9, Cutting; 10, Mount Theodore Roosevelt; 11, Gilt Edge–Galena; 12, Vanocker; 13, Bear Butte.

mately 13 centers (Fig. 1). Sources of the igneous variation (Fig. 2) include contamination of magma by crustal Precambrian basement rocks, as suggested by petrographic evidence including partly melted granitic xenoliths in breccia. These suites are well displayed in the Devils Tower region. In the central Bear Lodge Mountains, carbonatite intrusions, altered pyroxenites, and igneous and metasomatic alkalic rocks form an intriguing complex. Potassium fenites, together with thorium and rare-earth mineralization, complete this classic alkalic suite.

DEVILS TOWER

Devils Tower and the Missouri Buttes are major features in

the westernmost igneous center of the northern Black Hills Province (Fig. 1). Devils Tower is a steep-sided igneous body that shows spectacular columnar jointing. It consists of phonolite porphyry surrounded by talus and the remnants of a large volume of alloclastic breccia. The obelisk rises 1,255 ft (382 m) above the Belle Fourche River, to a maximum elevation of 5,070 ft (1,545 m). Devils Tower is surrounded by a talus sheet, which radiates outward some 1,100 ft (335 m). Talus is about 150 ft (46 m) thick near the base of the tower, but thins rapidly. Three possible outcrops of phonolite have been noted within the talus sheet; these could be radiating dikes, or part of the original uneroded igneous body.

An elliptical knoll of alloclastic breccia is located northwest of the tower. This breccia may at one time have been plastered against the phonolite of the tower, in the same fashion as the volcanic necks in the Mount Taylor region of New Mexico, which have phonolite interiors sheathed in agglomerate.

The Devils Tower phonolite is coarsely porphyritic with a gray to olive gray aphanitic groundmass. Phenocrysts consist of anorthoclase, aegerine-augite, and sphene in a trachytic groundmass of albite, microcline, analcime, aegerine, nepheline, and nosean. Common alteration or replacement products are calcite, zeolites, hematite, clay, and analcime. The principal mafic mineral is a green aegerine-augite, often found as phenocrysts with a dark rim of aegerine, and as groundmass needles. Analcime is one of the major constituents of the groundmass. Both nosean and nepheline occur as microphenocrysts; nosean appears as very small euhedral to subhedral grains that range in color from pale bluish gray to brown. Nosean may contain minute white barite inclusions.

Both Devils Tower and the Missouri Buttes are the erosional remnants of volcanic necks (Halvorsen, 1980). They originally vented to the surface through several hundred feet (more than a hundred meters) of sediments. Most of the pyroclastic material was blown out through these vents; some remained as a lining for the vent walls. After the more fluid phases of the eruption had ceased, cooling began from the surface downward, and from the periphery inward; this cooling history is responsible for the series of joints that parallels the circumference of the igneous mass, and for the set of vertical radial cracks, which are normal to the peripheral jointing. Continued marginal cooling formed the horizontal columnar joints; this set was almost immediately turned upward as temperature gradients from the surface became greater than horizontal temperature gradients.

The Missouri Buttes are smaller bodies of columnar-jointed trachyte and phonolite, and are exposed northwest of Devils Tower. Eocene ages of 40 Ma (for Devils Tower) and 50 Ma (for the Missouri Buttes) have been determined (Halvorsen, 1980).

The sedimentary rocks exposed in the Devils Tower–Missouri Buttes area range in age from Triassic to Quaternary; total thickness of the sediments is about 950 ft (290 m). Ten formations are represented, exclusive of surficial deposits. Two cycles of marine transgression are apparent in this sequence; hence, the proportion of marine and nonmarine sediments is

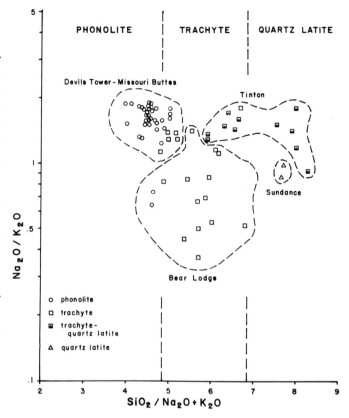

Figure 2. Silica-alkali variation in preliminary chemical analyses of rocks of the phonolite trachyte and quartz latite in the Devils Tower–Missouri Buttes, Bear Lodge, Sundance, and Tinton areas.

about equal. The oldest rocks are the red sandstones of the Permo-Triassic Spearfish Formation. The best exposures are east of Devil's Tower, where 150 ft (46 m) of reddish brown interbedded sandstone, siltstone, and claystone are present. Gypsum occurs throughout this sequence as thin stringers along bedding planes and in fractures. The late Jurassic Sundance Formation is 200 to 420 ft (60 to 130 m) thick and consists of five marine sandstone and shale members. The nonmarine Jurassic Morrison Formation is up to 210 ft (64 m) thick and is overlain by the Cretaceous Lakota and Fall River sandstones and siltstones (240 to 380 ft; 75 to 120 m thick), the Skull Creek Shale (40 ft; 12 m thick), the Newcastle Sandstone (20 ft; 6 m thick), and the Mowry Shale (30 ft; 9 m thick). The youngest unit is the Oligocene White River Formation, a coarse feldspathic sandstone and gray calcareous claystone unit that caps large areas west and north of the Missouri Buttes.

The sedimentary beds all dip westward at a shallow angle (less than ½ degree).

To reach the first site, proceed west from Sundance, Wyoming, approximately 14 mi (22.7 km) on U.S. 14, to Wonderview. Devils Tower can be seen 11 mi (18 km) to the north-northwest. The Missouri Buttes are visible some 3 mi (5 km) northwest of Devils Tower, where five peaks of phonolite and trachyte rise above the plateau formed by the Fall River

Formation. These are arranged in a roughly rectangular pattern that tapers southward; three buttes are along the western edge, and two are on the eastern edge. Alloclastic breccia occurs south of the northwest butte, between the two northernmost buttes, and on a plateau in the triangular area between the three southern buttes.

Continue west on U.S. 14 to the Devils Tower National Monument. The phonolite porphyry of the tower is intruded through and appears to rest upon a "platform" of the Sundance Formation, which dips gently inward toward the tower. This platform stands as a resistant hill, surrounded on all sides by streams, and capped by an armor of igneous talus, which has impeded further erosion. The tower itself rises almost vertically 740 ft (225 m) above the Jurassic platform. The upper 585 ft (180 m) is characterized by distinctive columnar jointing. The lower section is up to 120 ft (37 m) in height and forms the massive base of the tower. The slightly elliptical perimeter measures 740 ft by 990 ft (225 by 300 m); the top is 180 by 290 ft (55 by 90 m). From a distance, the top appears to be flat, though it is actually rounded, particularly toward the southeast. The top of the basal section (the "shoulder") is defined by outward-flaring columns, which dip toward the perimeter about 15° and merge with the blocky jointing of the base. Above the shoulder stand huge columns which taper from 6 to 10 ft diameter (2 to 3 m) at the base to a 5-ft (1.5-m) diameter at the top. The effect of the upward tapering is to yield an overall slope angle of from 75° to 85°. While most of the columns are five-sided, some have four or six sides. In many places, columns merge or split. The top 150 ft (45 m) of the tower is extensively horizontally jointed. Small joint blocks that have fallen from this area show the beginnings of spheroidal weathering. The lower limit of the horizontal jointing approximates the former upper level of the Fall River Formation. These resistant beds impeded erosion so that only the top section of the tower was exposed to weathering. When the resistant beds were finally removed, the rest of the tower was rapidly denuded.

Devils Tower is surrounded by a talus sheet, which thins rapidly outward from the tower base. Although most of the present talus is from the upper columns, it does not necessarily follow that the original size of the tower is constrained by the dimensions of the present base. At Hulett, 10 mi (16 km) to the northeast, old stream terraces of the Belle Fourche River are littered with rock from Devils Tower; this indicates the removal of a considerable amount of talus material through the Quaternary. Evidence of four stages of talus alteration is present in the talus sheet; these, together with relict soil horizons, probably coincide with climatic changes that occurred during the Pleistocene Epoch. Such periods of talus development and removal, together with the possible outliers of phonolite and the presence of related alloclastic breccia, imply that the original igneous body may have been somewhat larger and of a different shape than the presently exposed tower.

BEAR LODGE MOUNTAINS

The basement uplift of the Bear Lodge Mountains exposes a continental phonolite-trachyte-quartz latite association of shallow intrusive and volcanic igneous rocks, which is cored by a carbonatite-pyroxenite-alkalic rock intrusive association showing prominent brecciation and mineralization (Karner, 1981; Jenner, 1984). A central dome of alkalic igneous rocks, 5.5 mi (9 km) long and 2.5 mi (4 km) wide is surrounded by smaller igneous plugs, dikes, and sills, which were intruded into upturned Paleozoic and Mesozoic strata around the flanks of the uplift. The oldest exposed rock is a Precambrian granite, dated at 2.6 Ga (Staatz, 1983). The granite occurs as xenoliths up to 4,200 ft (1,300 m) across in the central dome. Paleozoic and Mesozoic strata having a combined thickness of 3,500 ft (1,070 m) dip radially away from the uplift. The surface igneous rocks include subvolcanic latite, trachyte, phonolite, syenite, carbonatite, lamprophyre, and intrusive breccia. Shallow emplacement occurred during Eocene time; potassium-argon techniques give ages of 38 Ma to 50 Ma.

Alkalic igneous activity represented by the formation of latite, trachyte, and phonolite porphyries, natrolite-garnet syenites and malignites, was followed by extensive potassium fenitization prior to the emplacement of carbonatites as dikes and veins. Fenitization involved the introduction of potassium, iron, sulfur, carbon dioxide, and fluorine. Rocks associated with the fenitization contain significant deposits of copper, lead, zinc, horium, cerium, lanthanum, and gold. Chemistry and isotopic data for the carbonatites suggest origins both by mantle derivation and by fusion of Paleozoic limestone (Jenner, 1984).

Begin examination of the Bear Lodge Mountains on the north side of Sundance Mountain, directly south of the town of Sundance, Wyoming. One reaches the northwest base of the mountain by walking from a ranch, just south of the intersection of U.S. 14 and Wyoming 116. Sundance Mountain is at the southern end of the Bear Lodge complex.

Sundance Mountain is a dissected volcanic cone of quartz latite. It occupies an area of approximately 0.75 mi^2 (2 km^2), and shows local relief of 550 ft (170 m). The most notable exposures are on the north and northwest sides of the mountain, where sheer cliffs rise above a prominent talus slope. Numerous grass- and pine-covered lobes of colluvium extend out from the base. Prominent joint sets can be identified on the northwest cliff face. Columnar jointing is most strongly developed in the lower two-thirds of the exposure. Subhorizontal joint sets, dominant near the top, are subparallel to a faint layering within the rock. The rock is a light to dark gray, slightly porphyritic quartz latite, with a predominantly cryptocrystalline groundmass consisting of alkali feldspar and quartz. Oligoclase occurs as zoned phenocrysts and microlites. The phenocrysts have distinct oscillatory zones, and show a narrow compositional range.

The quartz latite appears in several different varieties; layered sequences, monolithic breccias, and massive types all occur. Brecciated units are most common near the base of Sundance Mountain. Individual breccia units tend to be thin (5 to 16 ft; 1.5 to 5 m in thickness). The breccias consist of more than 60 percent angular to subangular clasts of quartz latite, set in a weakly isotropic groundmass. Breccias may be interbedded with massive and layered quartz latites; the contact between a given breccia unit and overlying layered rock is always a sharp disconformity. The layered sequences consist of alternating light and dark gray laminae a few hundredths of an inch (1 mm) thick. The layers tend to dip outward from the interior of Sundance Mountain, and also tend to fill in spaces between the angular breccia clasts. Structures in the layering often resemble soft-sediment deformation structures; small-scale folding may also be present. Folds are asymmetrical, and are overturned in the direction of dip. Layered sequences often grade upward into massive units. Massive quartz latite is most common in the middle part of the exposed section of Sundance Mountain. Such massively bedded units are generally 33 to 120 ft (10 to 36 m) in thickness.

Return to U.S. 14 and continue west from Sundance 1.5 mi (2.4 km) to the intersection with Wyoming 838 (Taylor Divide Road). Proceed northwest on Wyoming 838 into the south flank of the Bear Lodge Mountains. Outcrops of sedimentary units dip away from the core of the uplift to the north; hence, in driving the 7 mi (11 km) to Warren Peaks, one traverses a section of the Bear Lodge uplift. The units from youngest to oldest include: red beds of the Permo-Triassic Spearfish Formation, limestone of the Permian Minnekahta Formation, red beds of the Permian Opeche Formation, sandstone of the Pennsylvanian Minnelusa Formation, the Pahaspa Limestone (Mississippian), the Devonian Englewood Limestone, the Ordovician Whitewood(?) Limestone, and sandstones of the Cambrian Deadwood Formation.

The older sedimentary rocks are cut by dikes, sills, and plugs of altered latite and trachyte porphyries (good exposure 1.7 mi; 2.7 km north of Reuter Campground). These rocks are gray to tan to reddish brown in color. They contain abundant large phenocrysts of alkali feldspar, and smaller phenocrysts of amphibole and pyroxene, in an aphanitic groundmass.

A good exposure of feldspathoidal rocks may be seen to the west of Wyoming 838 on a logging road 0.75 mi (1.2 km) north of the altered latite dikes. The dark green phonolite is porphyritic with an aphanitic groundmass. Phenocrysts of sodalite stand out as violet to pale blue aggregates on fresh surfaces; they weather to a pale brown color. Melanite garnet is commonly associated with sodalite. The groundmass is composed of alkali feldspar laths and acircular aegerine crystals.

The most abundant rock type in the central core of the Bear Lodge Mountains is trachyte porphyry; this appears to underlie much of the Warren Peaks area. The rock may be white to pinkish, but most often is iron stained. It consists of euhedral sanidine phenocrysts in a trachytic aphanitic groundmass. Clusters of altered amphiboles and pyroxenes are present, as well as pyrite in accessory amounts. Good exposures may be seen in a cut at the base of the road leading to the fire tower, on the peaks across the road to the northwest, and at the fire tower.

Granitic rocks, which are interpreted as xenolithic inclusions, occur in many areas throughout the intrusion. One such block forms four of the Warren Peaks; it is 3,900 by 780 ft (1,200 by 240 m) in outcrop area. Much granitic rock also occurs as clasts in breccias, along with other xenoliths in a fine-grained igneous or fragmental groundmass. The granites are gray to pink and weather to a yellowish color. Some have been altered by alkali metasomatism (fenitization) during which silica was selectively removed, leaving a porous alkali feldspar-rich syenitic rock.

A phonolite dike exposed in the roadcut crosscuts both trachyte porphyry and granite. The phonolite consists of phenocrysts of sanidine, aegerine-augite, and altered nepheline in a trachytic groundmass.

From the Warren Peaks area, continue northwest 2.5 mi (4 km) on Wyoming 838 to the "Four Corners" intersection with Wyoming 851 and 847. Proceed north on Wyoming 838 0.1 mi (0.2 km) to a small prospect pit on the southeast side of the road. Fenitized trachyte porphyry, which shows copper mineralization (turquoise?) is exposed in the pit. A carbonatite dike has been exposed in trenches at the east end of a ridge that runs just northwest of the prospect pit. The carbonatite is a brown, fine to medium-grained equigranular rock, which contains calcite, minor fluorite, alkali feldspar, and altered mineral grains containing rare-earth elements. Breccias and other calcite-bearing rocks are associated with the carbonatite.

Continue north on Wyoming 838 for 1.5 mi (2.4 km) to the junction with Wyoming 897. Follow Wyoming 897 east for 0.1 mi (0.2 km) to an exposure of pseudoleucite trachyte porphyry. At this location, dikes of the porphyry (3 to 7 ft; 1 to 2 m thick) cut trachyte. Contacts with the trachyte are well defined and are marked by Fe-oxide staining. Trapezohedral phenocrysts of pseudoleucite, 0.02 to 0.5 in (0.5 to 12 mm) in diameter are glomerophyrically arranged in a cryptocrystalline groundmass of alkali feldspar, natrolite, iron oxides, and carbonates. The eight-sided, euhedral to anhedral crystals are composed of orthoclase, which is rimmed by radial-fibrous natrolite.

REFERENCES CITED

Halvorsen, D. L., 1980, Geology and petrology of the Devil's Tower, Missouri Buttes, and Barlow Canyon area, Crook County, Wyoming [Ph.D. thesis]: Grand Forks, University of North Dakota, 218 p.

Jenner, G. A., 1984, Tertiary alkalic igneous activity, potassic fenitization, carbonatitic magmatism, and hydrothermal activity in the central and southeastern Bear Lodge Mountains, Crook County, Wyoming [M.S. thesis]: Grand Forks, University of North Dakota, 232 p.

Karner, F., 1981, Geologic relationships in the western centers of the northern Black Hills Cenozoic igneous province, in Rich, F. J., ed., Geology of the Black Hills, South Dakota and Wyoming: Field Trip Guidebook for G.S.A. Rocky Mountain Section Meeting, American Geological Institute, p. 126–133.

Lisenbee, A., Karner, F., Fashbaugh, E., Halvorsen, D., O'Toole, F., White, S., Wilkinson, W., and Kirchner, J., 1981, Geology of the Tertiary intrusive province of the northern Black Hills, South Dakota, and Wyoming, Field Trip 2, in Rich, F. J., ed., Geology of the Black Hills, South Dakota and Wyoming: Field Trip Guidebook for G.S.A. Rocky Mountain Section Meeting, American Geological Institute, 221 p.

Newton, H., and Jenny, N. P., 1880, Report on the geology and resources of the Black Hills of Dakota, with atlas: U.S. Geographical and Geological Survey, Rocky Mountain Region, 556 p.

Staatz, M. H., 1983, Geology and description of thorium and rare-earth deposits in the southern Bear Lodge Mountains, northeasten Wyoming: U.S. Geological Survey Professional Paper 1049-D, 52 p.

Rhyolite-basalt volcanism of the Yellowstone Plateau and hydrothermal activity of Yellowstone National Park, Wyoming

Robert L. Christiansen, *U.S. Geological Survey, 345 Middlefield Road, Menlo Park, California 94025*
Roderick A. Hutchinson, *National Park Service, Yellowstone National Park, Wyoming 82190*

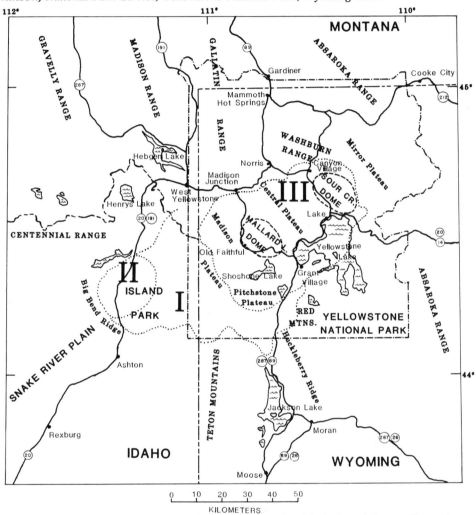

Figure 1. Location and major features of Yellowstone National Park. Dotted lines outline calderas; dashed lines, resurgent domes; I, first-cycle caldera; II, second-cycle caldera; III, third-cycle (Yellowstone) caldera.

LOCATION AND ACCESSIBILITY

A volcanic plateau constructed by late Pliocene and Quaternary eruptions spans the Continental Divide and occupies the central part of Yellowstone National Park, mainly in northwestern Wyoming but overlapping into eastern Idaho and southwestern Montana. The National Park is accessible by road through five entrances, reached by U.S. 20 from the west and east, U.S. 89 from the north and south, and U.S. 212 from the northeast (Fig. 1). The localities described in this guide are all on or adjacent to the park road system and can be reached, during the summer months (usually early May through late October), by

passenger car or, during the winter, by skis or skimobile. A nominal entry fee for vehicles is valid in both Yellowstone and Grand Teton national parks.

Special caution should be exercised in visiting the thermal areas; they are both delicate and dangerous. Designated walkways are provided in the thermal areas described here, and posted park regulations should be followed. Specimen collecting in the national park is allowed only by permit, which can be issued by the National Park Service only in advance of a collecting trip.

SIGNIFICANCE

In 1872 the Yellowstone region became the first national park in the world, largely through the recognition that it displays some of the largest and most spectacular hot-spring and geyser activity on Earth. Early scientists recognized this hydrothermal activity as a manifestation of recent volcanism, but only in the last few decades has it been realized that the compositionally bimodal rhyolite-basalt volcanism of the Yellowstone Plateau volcanic field (Boyd, 1961; Christiansen and Blank, 1972; Christiansen, 1984) is younger than about 2.2 Ma and completely distinct from the predominantly andesitic volcanism of the Eocene Absaroka volcanic field (Smedes and Prostka, 1972). The major hydrothermal activity of Yellowstone National Park (White and others, 1975) is related in time and origin to postcaldera magmatism in the youngest of three calderas that resulted from voluminous rhyolitic ash-flow eruptions at 2.0, 1.3, and 0.6 Ma. More than 1,560 mi^3 (6,500 km^3) of erupted magma make this one of the major Quaternary volcanic fields on Earth. Current activity includes not only the high-temperature hydrothermal systems but distinctive seismicity and rapid uplift, probably indicating that magma still remains in the Yellowstone Plateau system (Smith and Braile, 1984).

GEOLOGY

Introduction. This guide stresses only the young volcanic and hydrothermal features of Yellowstone National Park, sampling geologic features of the Yellowstone caldera and its margin and related hydrothermal activity in a northwestern sector from Old Faithful to Mammoth Hot Springs (Fig. 1). Other aspects of the complex geologic history of the Park, from Archaean to the present, are depicted on a geologic map (U.S. Geological Survey, 1972) and are reviewed briefly by Keefer (1971). A more general geologic road guide is by Fritz (1985).

The Yellowstone Plateau includes the Madison, Pitchstone, and Central plateaus and a series of broad uplands that extend irregularly beyond them to the surrounding mountains on the northwest, north, east, and south and to Island Park and the Snake River Plain on the west and southwest (Fig. 1). This tectonically and erosionally dissected plateau is the product of three cycles of mainly rhyolitic volcanism. Each cycle began with relatively small basalt and rhyolite eruptions, then continued for as long as several hundred thousand years with rhyolitic lava eruptions from a system of arcuate fractures above a large and growing magma chamber. Each cycle climaxed with the brief explosive eruption of tens to hundreds of cubic miles (hundreds of thousands of cubic kilometers) of rhyolitic ejecta, probably in no more than a few days. Collapse of the roof with partial emptying of the magma chamber in each of these catastrophic eruptions formed a large caldera, and further rhyolitic eruptions from residual magma chambers over several hundred thousand years partly filled each of them. Basalts continued to erupt episodically around the margins of each rhyolitic system and, ultimately,

through tectonic fractures within the two older calderas after solidification of their rhyolitic magmas.

The first volcanic cycle climaxed about 2.0 Ma with eruption of the 600-mi^3 (2,500-km^3) Huckleberry Ridge Tuff, forming a caldera that extended 56 mi (90 km) from Island Park, at the east margin of the Snake River Plain, to the Central Plateau (Fig. 1). The Mesa Falls Tuff, about 67 mi^3 (280 km^3), erupted at 1.3 Ma to climax the second volcanic cycle and produce a 12-mi (20-km)-diameter caldera, nested within the first in the northern part of Island Park. No deposits of the second cycle are exposed in Yellowstone National Park. The third cycle began about 1.2 Ma and continued with the eruption of rhyolitic lava flows from all sectors of a growing ring-fracture system extending from the Madison Plateau to the Mirror Plateau, and from the Washburn Range to the Red Mountains (Fig. 1). The climactic eruption of 240 mi^3 (1,000 km^3) of Lava Creek Tuff, at 0.6 Ma, resulted in collapse to form the Yellowstone caldera (28 by 47 mi; 45 by 75 km). The Lava Creek Tuff erupted through two distinct segments of the ring-fracture system to produce two stratigraphic members that form a single compound cooling unit. Early postcollapse magmatic resurgence uplifted the floors of the two caldera segments to form the Mallard Lake and Sour Creek structural domes (Fig. 1). Intracaldera lava flows that postdate this resurgence demonstrate that it occurred so soon after formation of the caldera that K-Ar dating cannot resolve the time—much less than 100,000 years. Renewed rhyolitic volcanism in the Yellowstone caldera during the past 150,000 years has produced rhyolitic lavas that aggregate more than 216 mi^2 (900 km^3). These youngest lavas erupted in three episodes, about 150,000, 110,000, and 70,000 yr B.P.; individual events within each episode are not clearly resolved by K-Ar dating. Rejuvenated uplift of the Mallard Lake dome in the western part of the caldera occurred early in this period. The young rhyolite flows are exceptionally large, many of them more than 12 mi^3 (50 km^3). They extruded through two linear fracture systems across the caldera beneath the Pitchstone-Madison and Central plateaus (Fig. 1) and filled much of the caldera basin, overflowing it on the southwest.

Hydrothermal activity occurs widely in the park, much of it linked to structure and topography of the Yellowstone caldera. Most hot-water discharge is in topographically low areas above the buried ring-fracture zone of the caldera, but one major hydrothermal zone extends northward, radial to the caldera, in a narrow corridor from Norris Geyser Basin to Mammoth Hot Springs.

The Firehole River and Upper Geyser Basin. The Firehole River flows northward through a series of large open valleys, including Upper and Lower Geyser basins, and narrower constrictions, reflecting the constructional topography of the Madison and Central plateaus. The road along the river between Old Faithful and Madison Junction provides access to the most outstanding display of hydrothermal features in the world, including more than two-thirds of all known naturally erupting geysers.

Stop 1 (Fig. 2) is in the Old Faithful area of Upper Geyser

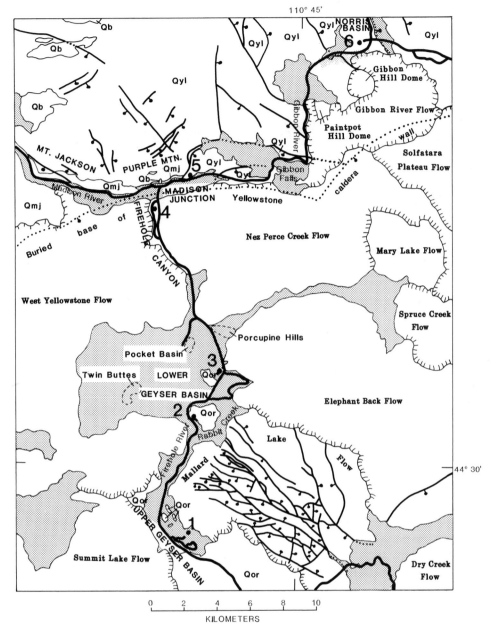

Figure 2. Geologic map showing route and stops between Upper Geyser Basin (Old Faithful) and Norris Geyser Basin. Qmj, Mount Jackson Rhyolite (pre–Lava Creek flows); Qyl, Lava Creek Tuff; Qor, older intracaldera rhyolitic lavas; younger rhyolitic lava flows and domes are named, and hachures mark the younger sides of contacts; Qb, basalts; stipple, surficial deposits.

Basin. Views from Old Faithful Inn and from the walkway between it and the Park Service Visitor Center show the setting of the basin between rhyolitic lava flows of the Madison Plateau to the west and the Mallard Lake structural dome to the east. A graben that branches from the axis of the Mallard Lake dome is apparent on the skyline east of Old Faithful. This dome, rejuvenating the postcollapse western resurgent dome, rose early in the renewed voluminous rhyolitic volcanism of the past 150,000 years. Plagioclase-rich older postcollapse rhyolites crop out around the west and south base of the structural dome and form the poorly exposed bedrock of Upper Geyser Basin, but the uplifted 150,000-year sanidine-rich Mallard Lake flow tops the dome (Fig. 2). The Elephant Back flow, also about 150,000 years old, partly buries the east flank of the dome, demonstrating its rapid rise early in the period of renewed rhyolitic activity.

Old Faithful, Yellowstone's best known geyser, is but one of dozens in the Upper Geyser Basin. Geysers are boiling springs that intermittently erupt, fed by high-temperature deeper reservoirs in which substantial amounts of silica are dissolved from the surrounding rocks. Some understanding of subsurface hydrothermal processes, including the mechanism of geyser eruptions, is gained through the resolutions of two seeming anomalies. First, neighboring springs may have nearly identical temperatures and water compositions but greatly different water levels; for example, Solitary Geyser discharges about 130 ft (40 m) above most springs in Upper Geyser Basin. Second, some hot springs erupt as geysers, but most do not. Research drilling in Yellowstone (White and others, 1975) showed that most reservoirs are "overpressured" below about 100 ft (30 m) through partial "self sealing" of the water-bearing fractures by silica minerals and zeolites. Thus, water levels are controlled by permeabilities and flow rates in the upflow channels, regulated through the competing effects of seismic fracturing and self sealing, and by subsurface outflow near ground level. A typical spring, discharging at 90°C, rises from a subsurface reservoir near 200°C; the rising water cools by more-or-less steady-state conduction to the channel walls and convection, evaporation, and discharge from the hot-spring pool. A geyser may form where waters rising through heated rocks in the self-sealed zone lose little heat conductively, maintaining slightly hotter water near the surface. With higher overpressures, recharge rate through a self-sealed zone can be nearly constant regardless of the eruptive state of any geyser above. Deep recharge fills a shallow water-holding reservoir immediately below the geyser and heats it above its surface boiling temperature. If boiling and rapid convection can eject some water through the spring orifice to reduce the head in the holding system, flash boiling may be induced, expelling large quantities of water and reducing pressures deeper in the system to initiate a chain reaction of rapid boiling. During an eruption, water and heat are discharged much faster from the shallow holding reservoir than they are recharged, requiring a repose time between eruptions to "reload" the system.

A variety of behavior in major geysers can be observed in Upper Geyser Basin, including eruptions from the large sinter cone at Castle Geyser (average 90 ft; 27 m height), from Grand Geyser, which fountains through a broad pool to average heights of 200 ft (60 m), and from Riverside Geyser (average 70 ft; 22 m), which jets a water column at an angle over the Firehole River. Old Faithful has particularly interesting eruptive intervals; unlike those of most other geysers, they are a function of the length of play of the preceding eruption. Durations are bimodally distributed between 1.5 and 5 minutes; long eruptions are followed by long intervals (70 to 120 minutes), short eruptions by short intervals (32 to 70 minutes). The 130-ft (40-m) average height of Old Faithful eruptions is independent of eruptive interval or duration.

Midway and Lower Geyser Basins. The road northward down the Firehole River from Old Faithful toward Madison Junction passes through additional areas of Upper Geyser Basin. The Firehole valley narrows around the northwestern margin of the Mallard Lake dome and again opens out into the Rabbit

Creek–Midway Geyser Basin area, which is topographically part of Lower Geyser Basin.

A viewpoint for several features of the Yellowstone caldera (Stop 2, Fig. 2) can be reached by a short walk from the Midway parking area, 4.5 mi (7.2 km) from the main-road entrance to Old Faithful. Across the road from the parking area, a conspicuous bluff exposes perlitic glassy rhyolite with conspicuous flow layering. This plagioclase-rich early postcollapse rhyolite probably was emplaced into a caldera lake. Most of the glassy intracaldera rhyolites of Yellowstone are virtually nonhydrated and obsidianlike, though commonly fractured. The hydrated perlitic glass of this outcrop is suggestive of high-temperature hydration, probably during emplacement of the flow, suggesting the lake interpretation. (Similar perlites can be seen farther east in the park, where the younger Hayden Valley flow plowed into and was diked by the plastic sediments of a glacial lake.)

From the north end of this bluff, follow an anastamosing group of game trails along the rim for a few hundred yards (meters) to a point that extends out from the trees. From this vantage, the Mallard Lake dome and its axial graben are visible to the southeast. Rhyolitic flow-front scarps of Madison Plateau lavas form the valley walls to the west. Just across the park road is the Midway Geyser Basin, discharging boiling waters from Excelsior Geyser Crater to the Firehole River. The crater was enlarged to its present size by eruptions of water and debris, some out to as far as 500 ft (150 m) from the crater, during 1878 and the 1880s. The first eruptions of Excelsior since then were in the summer of 1985. Grand Prismatic Spring, just beyond, is nearly 300 ft (90 m) across and reveals graded colors from the clear blue, boiling waters of its center to lower-temperature bands around its margins that are brightly colored by microorganisms. To the northwest is Lower Geyser Basin, in the distant southwestern corner of which is Twin Buttes, formed of hydrothermally cemented Pleistocene englacial sediments deposited by localized hydrothermal melting of the ice. The buttes enclose a complex hydrothermal explosion crater. The crater eruptions were not hydromagmatic but, rather, strictly hydrothermal, probably occurring as the hydrostatic head in an englacial thermal lake was catastrophically lowered during a glacial-outburst flood (Muffler and others, 1971). Beyond Lower Geyser Basin, the skyline ridge along Mount Jackson and Purple Mountain is the northwest rim of the Yellowstone caldera (Fig. 2).

About 1 mi (1.6 km) past the Midway parking area, the Old Faithful–Madison Junction Road passes the entrance to a one-way, 3.3 mi (5.3-km) loop road past Firehole Lake, not described here but leading past additional hydrothermal features of interest. The main road, 1.1 mi (1.8 km) farther along, rejoins the Firehole Lake Loop Road at the Fountain Paint Pot area of Lower Geyser Basin (Stop 3, Fig. 2).

The Fountain area is the most convenient place in the park where one short walk displays good examples of the four principal types of hydrothermal discharge features—hot springs, geysers, fumaroles, and mud pots. Distributions of these types allow relationships between the ground surface and local static water levels in the boiling system to be readily inferred. The large, clear hot springs are chloride-rich and nearly neutral, representing unrestricted discharge of hot waters from a reservoir. Geysers represent the effects of variations in plumbing geometry, discharge rate, and heat loss in the hot-spring system. Fumaroles (steam vents) discharge through fractures where only vapor finds access to the surface. New fumaroles were opened here during the magnitude-7.5 Hebgen Lake earthquake of 1959. During dry seasons, the fumaroles vent only vapors. In wet seasons, shallow condensation of steam and the oxidation of H_2S can drown the fumaroles and form small acid-sulfate hot springs. The acid waters attack their enclosing rocks and, with enough decomposition of the rocks, can form mudpots or (with ferric-oxide coloration) paintpots.

The boardwalk trail around the Fountain Paint Pot area leads to a group of geysers, which share episodic exchanges of discharge function. These include Fountain, Morning, Clepsydra, Jet, Spasm, and Jelly geysers. For most of the time since the 1959 earthquake, Clepsydra has been a perpetual spouter rather than a true (episodically erupting) geyser. Also visible from this walk are other features of Lower Geyser Basin and its surroundings (Fig. 2). The wide meadows of the basin are bounded by flow-front scarps of young rhyolitic lava flows of the Madison Plateau on the west and the Central Plateau on the east. In the west part of the basin is Twin Buttes. The Pocket Basin hydrothermal-explosion crater (Muffler and others, 1971) is visible to the north as a low tree-covered rim near the center of the basin, rising as much as 70 ft (21 m) above the basin floor. To the northeast are the ragged Porcupine Hills, another group of hydrothermally cemented englacial hot-spring deposits. Visible 8.1 mi (13 km) to the north are the northwestern wall and rim of the Yellowstone caldera.

Firehole Canyon. About 7.5 mi (12.1 km) past Fountain Paint Pot (0.5 mi; 0.8 km south of Madison Junction) the 2-mi (3.2-km) one-way Firehole Canyon scenic loop road branches westward from the main road. The course of the river through the canyon is topographically controlled along the contact between two rhyolitic lava flows—the 150,000-yr-old Nez Perce Creek flow that vented on the Central Plateau and flowed westward across the area, and the 110,000-yr-old west Yellowstone flow that forms the marginal scarp of the Madison Plateau, across the river to the west.

Outcrops along the road demonstrate especially clearly some internal features typical of viscous rhyolitic lavas. Just past the main road, the one-way loop approaches the little-modified flow-front scarp of the West Yellowstone flow across the Firehole River. The gray glassy carapace and steeply dipping contorted flow layering are conspicuous, even from across the river. A short distance beyond, at Stop 4 (Fig. 2) where the road turns southward and climbs above the river, is a noteworthy outcrop of the upper flow breccia of the Nez Perce Creek flow. Viscous lavas generally flow by marginal shear and pluglike movement of a more or less fluidal core, their chilled surfaces breaking up into rafts of blocky rubble, some of which is dumped at the front and

partly overridden by the still moving flow in tractor-tread fashion. The result is an enveloping flow breccia. The outcrop along the south side of the roadcut at Stop 4 demonstrates fluidal layering, locally arrested in the process of brecciation. The fluidal layers themselves are seen to intrude the glassy upper flow breccia formed in earlier stages of brecciation. The blocks have various textures, some hydrated, others more obsidianlike, and some partly devitrified. These porphyritic rhyolite glasses are generally fractured on a centimeter scale, resulting in their rapid reduction to obsidian-rich sand and gravel. Following the loop road southward from the stop, further examples of fluidal layering and varied degrees of brecciation in the upper part of the Nez Perce Creek flow can be followed through a series of roadcuts. Firehole Falls, below a large parking area, is held up by nearly vertical, erosionally resistant flow layers in the Nez Perce Creek flow. The marginal scarp of the younger West Yellowstone flow continues to form the west rim of the canyon across the river.

Madison Junction and the Gibbon River. Madison Junction marks the confluence of the Firehole and Gibbon rivers to form the Madison River (Fig. 2). For 7.5 mi (12.1 km) eastward, from Madison Junction to Gibbon Canyon, the park road follows the wall of the Yellowstone caldera. The cliffs of Purple Mountain rise nearly 1,600 ft (500 m) above the river near Madison Junction, exposing pre–Lava Creek rhyolitic lava flows and a thick basal section of the Lava Creek Tuff. Although no specific stop is designated here, the trail to Purple Mountain, shown on topographic maps of Yellowstone National Park or the Madison Junction Quadrangle, passes upward through this section. Better exposurs can be found in parts of the gully that crosses the trail just above the old gravel pit near its base.

Eastward the road passes Terrace Spring, one of the few significant travertine-depositing hot springs within the caldera, and onward along the base of the caldera wall. From a short distance past Terrace Spring nearly to Gibbon Falls, the eastward-trending road lies at the base of a slumped block of the caldera rim (Fig. 2) so that the topographic wall near the road is generally only a few tens of yards (meters) high. Near the west end of this slumped block, 1.6 mi (2.6 km) from Madison Junction, is a small parking and picnic area, not generally well marked by road signs but designated as Tuff Cliff on an explanatory sign behind the picnic area, away from the road (Stop 5, Fig. 2). A steep scramble over sandy scree uphill from this sign takes one to the base of the cliff, which exposes nearly nonwelded Lava Creek Tuff with large relict pumice inclusions modified by crystallization from a high-temperature vapor phase. Some of the individual pyroclastic-flow units that comprise the lower part of the Lava Creek Tuff can be seen in the cliff, demonstrating the double grading characteristic of such flows (i.e., normally graded lithic inclusions and reversely graded large pumice inclusions.) Some of the flow units are separated by bedded tuffs. The top of the cliff consists of phenocryst-rich, more densely welded tuff, most readily examined in large talus boulders at the edge of the trees below the cliff. The abundance of both lithic inclusions and phenocrysts in the nonwelded and welded tuffs of this locality is characteristic of near-source ash flows; more distal exposures of the Lava Creek Tuff have fewer and smaller lithics and crystals.

Eastward beyond Tuff Cliff, the road continues for 3 mi (4.8 km) along the front of the slumped caldera-rim block, nearly to Gibbon Falls, then turns abruptly northward to cross the block. The falls, held up by densely welded, vertically jointed Lava Creek Tuff, have retreated from the scarp at the front of the slumped block. A short way above the falls the road again changes back to an eastward course at the base of the main, outer caldera wall (Fig. 2), whose high cliffs rise to the north. About a mile farther east, the road crosses the river and once again changes course northward, out of the caldera, through Gibbon Canyon. Here the course of the river was controlled by a buried fault that dropped the rim eastward, only to be buried by subsequently erupted rhyolitic lavas. Thus, the west wall of the canyon consists mainly of Lava Creek Tuff while the east wall is the 80,000-yr-old rhyolitic Gibbon River flow.

North of the canyon, the road passes along the Gibbon River for about 5 mi (8 km) to Norris Geyser Basin through a series of broad alluviated valleys and forested bedrock ridges, reflecting a sequence of normal-fault blocks that displace the Lava Creek Tuff and tilt it gently westward. Two prominences on the skyline south of the open valleys are Gibbon Hill and Paintpot Hill (Fig. 2), post–Lava Creek rhyolitic lava domes that are flanked by the younger Gibbon River flow.

Norris Geyser Basin. Norris Geyser Basin (Stop 6, Figs. 2 and 3) is a large area of varied hydrothermal activity that is easily examined on two loop trails from the visitor center. Norris discharges hot water from the highest-temperature reservoir in Yellowstone—270°C, compared to about 200°C for other geyser basins.

Walking through the Visitor Center leads to Porcelain Basin, the northern area, marked by numerous fumaroles, hot springs, and geysers, one or another of which can be seen in eruption much of the time. In the eastern part of Porcelain Basin are large siliceous sinter terraces, some of which are the most rapidly growing sinter deposits in the park today.

The springs of Norris are unusual for discharging waters of two main types. One is from a deep reservoir, nearly neutral in pH and high in dissolved chloride and silica. The other is slightly acidic, probably reflecting the mixing of deep chloride-rich waters with acid-sulfate waters formed by shallow condensation of steam and oxidation of H_2S. As a result, some of Norris' features are different from those of the Upper and Lower Geyser basins. For example, much of what at first appear to be white hot-spring deposits around the basin are actually acid-leached volcanic rocks. Bedrock of the Norris area is Lava Creek Tuff, and its quartz phenocrysts survive even intense alteration, clearly marking the origin of the bleached rocks as altered rhyolites rather than hydrothermal sinter deposits. Also, note the bright green color of a few of the hot-water discharge channels. This color is produced by microorganisms that require low pH environments, in contrast to the yellow, red, and brown organisms and precipitates of the near-neutral discharge channels.

Returning through the museum and visitor center leads to the southern or "Back Basin" area of Norris. Of particular interest on this loop trail are Steamboat Geyser—the world's highest-erupting active geyser, eruptions of which are infrequent, irregular, and unpredictable—and Echinus Geyser. Echinus is named for its spiney sinter that results from the deposition of SiO_2 along with Fe, As, and Mn that are transported in the acidic splash of its eruptions. The geyser erupts fairly predictably about every 50 to 60 minutes, but also has some long periods of inactivity. Fragments of banded pyrite and marcasite as large as several centimeters are sometimes ejected in these eruptions.

The Norris-Mammoth Corridor. North of Norris Geyser Basin the park road traverses 21 mi (34 km) to Mammoth Hot Springs; along much of the way, the bedrock—though poorly exposed—is Lava Creek Tuff. Along this corridor are vents for both rhyolitic and basaltic eruptions as well as Yellowstone's major zone of extracaldera hydrothermal activity, consisting largely of acid-sulfate altered areas that reflect the oxidation of acidic gases. About 4 mi (6.4 km) past Norris Geyser Basin the road passes Roaring Mountain, a steep bluff of hydrothermally altered Lava Creek Tuff with numerous fumaroles and small acid-sulfate seeps. The fumaroles are generally inconspicuous on warm sunny days but can be impressive in cool cloudy weather.

East of the road, 3.5 mi (5.6 km) past Roaring Mountain (8 mi; 13 km south of Mammoth Hot Springs), is Obsidian Cliff (Stop 7, Fig. 3); a parking area is about 330 ft (100 m) north of the cliff. From the parking area the cliff can be seen to lie along the margin of a rhyolitic lava flow that abuts the wall of a paleovalley near here. Large columnar joints mark the base of the flow near the road, and steep contorted flow layering is visible near the top. Obsidian Cliff is an excellent locality to examine rhyolitic obsidian and its crystallization textures. (Specimen collecting is not allowed.) Some visitors to this locality are somewhat disappointed to find that clear, undevitrified blocks of obsidian are uncommon; sites once used by native Americans to quarry crystal-free obsidian for implements are located away from the road along other margins of the same flow.

Characteristic textural features are readily seen in talus boulders at the base of the cliff near the road. The rhyolite here is entirely free of phenocrysts, indicating that the magma erupted above its liquidus temperature. The obsidian occurs in basal and marginal chilled zones of the flow, which has a pumiceous top and a crystallized interior that is seen in a single gully exposure. Fluidal laminations are distinct in the talus boulders, generally marked by bands of small gray spherulites that formed by partial high-temperature devitrification of the rhyolitic glass. Further devitrification is represented by larger pinkish spherulites and lithophysae, generally 0.4 to 2 in (1 to 5 cm) across. Lithophysae are hollow spherulites lined by concentric shells of small crystals—generally trydimite, alkali feldspar, and minor accessories such as fayalite—separated by open interspaces. The lithophysae represent high-temperature vapor exsolution and crystallization from the vapor phase.

North of Obsidian Cliff the Norris-Mammoth Road passes

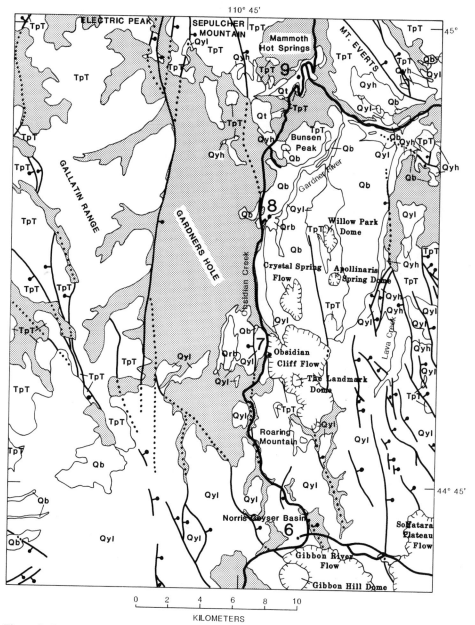

Figure 3. Geologic map showing route and stops between Norris Geyser Basin and Mammoth Hot Springs. TpT, pre-Tertiary and Tertiary rocks (older than Yellowstone Plateau volcanic field); Qyh, Huckleberry Ridge Tuff; Qyl, Lava Creek Tuff; rhyolitic lava flows and domes are named, and hachures mark the younger sides of contacts; Qb, basalts; Qt, travertine; stipple, surficial deposits.

along Obsidian Creek for 4.6 mi (7.4 km) to its junction with the Gardner River. About 0.1 mi (0.16 km) north of the Gardner River is a short spur road to the east, ending at a parking and picnic area with an explanatory sign for "Sheepeater Cliffs" (Stop 8, Fig. 3). The bluff behind the sign (the main Sheepeater Cliffs are about 3 mi; 5 km downstream) is a post–Lava Creek, columnar-jointed tholeiitic basalt flow. Note that the talus here consists largely of broken columns that fell intact and can still be

recognized in the arrangement of talus blocks. A fishermen's trail downstream from here leads past more exposures of this basalt flow to a cataract about 1,600 ft (500 m) from the parking area where a well-known mixed-lava complex of rhyolite and basalt is exposed. The first exposure of the complex, where the trail climbs a few yards (meters) above river level, is glassy porphyritic rhyolite. Beyond are basalts, rhyolites, and variously mixed rocks with disequilibrium assemblages of basaltic and rhyolitic phenocrysts, numerous included fragments, and streaky textures. Wilcox (1944) described the locality in detail and demonstrated that the basalt and rhyolite existed simultaneously as liquids, the higher-temperature basaltic liquid tending to chill against lower-temperature rhyolitic liquid. Subsequent studies favor partial mixing of the contrasted magmas during preeruptive ascent rather than as separate lava flows as Wilcox envisioned (Christiansen, 1984).

North of the Gardner River, the Norris-Mammoth Road traverses Swan Lake Flat, underlain mainly by post–Lava Creek basalts, nearly buried in glacial deposits. To the west is the Gallatin Range, exposing Precambrian rocks, Paleozoic and Mesozoic strata, and Eocene intrusive bodies; the section is generally younger northward. Electric Peak near the north border of the park comprises Cretaceous marine strata and an Eocene sill complex; Sepulcher Mountain, east of Electric Peak (Fig. 3), is mainly Eocene volcanic rocks. East of Swan Lake is Bunsen Peak, an Eocene dacitic stock.

As the road approaches Bunsen Peak to leave Swan Lake Flat, 3.3 mi (5.3 km) north of the Gardner River (4.9 mi; 7.9 km) south of Mammoth Hot Springs), it drops into the narrow valley of Glen Creek at Golden Gate. Just past Rustic Falls, where Glen Creek drops 46 ft (14 m) across a cliff of welded tuff, is a wide parking area at the roadside, with a view of Rustic Falls, the north side of Bunsen Peak, and a striking cliff exposure of the 2.0-Ma Huckleberry Ridge Tuff at the roadside. This is the first exposure of Huckleberry Ridge Tuff, the climactic deposit of the first volcanic cycle, seen on the route of this guide. (If you examine this exposure, use special care; rockfalls from the cliff are frequent, and traffic along the road is heavy.) Looking down the canyon of Glen Creek and across the valley of the Gardner River, the mesalike Mount Everts is prominent. Its main cliffs comprise a thick marine Cretaceous section, and the prominent caprock is the same Huckleberry Ridge Tuff seen at Golden Gate but also

exposing a black basal vitrophyre that is buried in talus here. The Mount Everts section is an excellent and instructive exposure, best seen on a half-day cross-country hike from the road crossing at Lava Creek, 4.5 mi (7.2 km) east of Mammoth Hot Springs.

Just a mile past Golden Gate is Silver Gate, a blocky landslide deposit of bedded Pleistocene travertine from Terrace Mountain, which rises above the road to the west. The road continues through landslides and glacial deposits to Mammoth Hot Springs.

Mammoth Hot Springs. Mammoth Hot Springs is situated on the steep western side of the Gardner River valley. The bedrock on most of the valley walls is somber-colored landslide-prone Cretaceous marine sedimentary rock. To the south is the Bunsen Peak dacitic stock. Pleistocene travertine caps Terrace Mountain to the west. Mount Everts, capped by the Huckleberry Ridge Tuff lies to the east. Northward the valley opens toward the Yellowstone River, beyond which are other Pleistocene travertine terraces above the town of Gardiner, Montana, and glaciated Precambrian to Eocene rocks of the Snowy Range beyond.

Mammoth Hot Springs (Barger, 1978) is different from any of the other thermal areas of this guide. In contrast to the silica-depositing hot waters described previously, Mammoth is presently the world's largest carbonate-depositing spring system. The characteristic form of the deposits is a large travertine terrace. Older travertine underlies the village and much of the slope all the way down to the Gardner River. The waters of Mammoth, with a subsurface reservoir temperature of 73°C, are cooler than those of the geyser basins. They rise through Paleozoic and Mesozoic limestones, are high in dissolved carbonate, and release excess CO_2 as they discharge at lower pressure. The deposited travertine initially forms ridgelike mounds above the water-bearing fractures. The water pools and evaporates behind these mounds, losing CO_2 to the atmosphere and depositing carbonate as it flows across the rims of the pools to enlarge the terraces.

REFERENCES CITED

Barger, K. E., 1978, Geology and thermal history of Mammoth Hot Springs, Yellowstone National Park Wyoming: U.S. Geological Survey Bulletin 1444, 55 p.

Boyd, F. R., 1961, Welded tuffs and flows in the rhyolite plateau of Yellowstone National Park, Wyoming: Geological Society of America Bulletin, v. 72, p. 387–426.

Christiansen, R. L., 1984, Yellowstone magmatic evolution: Its bearing on understanding large-volume explosive volcanism, *in* Explosive volcanism; Inception, evolution, and hazards: Washington, D.C., National Academy of Sciences, p. 84–95.

Christiansen, R. L., and Blank, H. R., Jr., 1972, Volcanic stratigraphy of the Quaternary rhyolite plateau in Yellowstone National Park: U.S. Geological Survey Professional Paper 729-B, 18 p.

Fritz, W. J., 1985, Roadside geology of the Yellowstone country: Missoula, Montana, Mountain Press Publishing Company, 144 p.

Keefer, W. R., 1971, The geologic story of Yellowstone National Park: U.S. Geological Survey Bulletin 1347, 92 p. (Reprinted, 1976, by Yellowstone Library and Museum Association.)

Muffler, L.J.P., White, D. E., and Truesdell, A. H., 1971, Hydrothermal explosion craters in Yellowstone National Park: Geological Society of America Bulletin, v. 82, p. 723–740.

Smedes, H. W., and Prostka, H. J., 1972, Stratigraphic framework of the Absaroka Volcanica Supergroup in the Yellowstone National Park region: U.S. Geological Survey Professional Paper 729-C, 33 p.

Smith, R. B., and Braile, L. W., 1984, Crustal structure and evolution of an explosive silicic volcanic system at Yellowstone National Park, *in* Explosive volcanism; Inception, evolution, and hazards: Washington, D.C., National Academy of Sciences, p. 111–144.

U.S. Geological Survey, 1972, Geologic map of Yellowstone National Park: U.S. Geological Survey Miscellaneous Geologic Investigations Map I-711, scale 1:125,000.

White, D. E., Fournier, R. O., Muffler, L.J.P., and Truesdell, A. H., 1975, Physical results of research drilling in thermal areas of Yellowstone National Park, Wyoming: U.S. Geological Survey Professional Paper 892, 70 p.

Wilcox, R. E., 1944, Rhyolite-basalt complex on Gardiner River, Yellowstone National Park, Wyoming: Geological Society of America Bulletin, v. 55, p. 1047–1080.

Teton mountain front, Wyoming

J. D. Love, P.O. Box 3007, University Station, Laramie, Wyoming 82071

LOCATION

Three vantage points from which to view the Teton mountain front are discussed (Figs. 1, 2) all in Grand Teton National Park, Teton County, Wyoming. Site 1 is in the NW¼NW¼ Sec.26,T.44N.,R.115W. (U.S.G.S. 7½-minute topographic map of Moran Quadrangle, 1968). Site 2 is in the NW¼SW¼SW¼ Sec.1,T.43N.,R.116W. (U.S.G.S. 7½-minute topographic map of Moose Quadrangle, 1968). Site 3 is in the center SE¼ Sec.12,T.44N.,R.116W. (U.S.G.S. 7½-minute topographic map of the Jenny Lake Quadrangle, 1968).

ACCESSIBILITY

All three sites are accessible by paved highways. Site 1 is at the Snake River Overlook on the west side of U.S. 26-89-187; Site 2 is 200 ft (60 m) east of the Park Service paved road between Moose and South Jenny Lake junction, on the west rim of Timbered Island; and Site 3 is at the Cathedral Group Scenic Turnout on the north side of the paved road (one-way traffic, west) 0.8 mi (1.3 km) northeast of the String Lake turnoff.

SIGNIFICANCE

The Teton mountain front is the youngest (less than 9 Ma) and steepest of the high mountain fronts in the Rocky Mountain chain. The unique scenery of the Tetons and adjacent areas is the product of an unusual geologic history. The record of sedimentation and tectonism, especially that beginning with Late Cretaceous time, is one of the thickest (more than 50,000 ft; 15,000 m), most complete, and most complex in North America. Tectonism did not end with the Laramide Revolution in Eocene time but has continued intermittently to the present. This, in part, explains the fresh jagged appearance of the Teton Range, the anomalous drainages, the abundant landslides, the narrow canyons, and other distinctive features.

These and additional phenomena in the Teton area have been discussed in detail and presented graphically elsewhere (Love and others, 1978; Love and Love, 1983). Site 1 was selected because it shows the broad panorama of the Teton mountain front as contrasted with the flat valley floor of Jackson Hole. Sites 2 and 3 are closer views of the high peaks, the mountain front, and the bounding fault along the base of the mountains.

SITE DESCRIPTIONS

Site 1. This viewpoint is at the Snake River Overlook, on the south margin of the Burned Ridge moraine. The moraine is marked by trees and extends across the floor of Jackson Hole except where it has been breached by the Snake River in the foreground. Note that trees are common on morainal debris but sparse or absent from the outwash debris. They are very selective as to where they will or will not grow. The mountain front rises

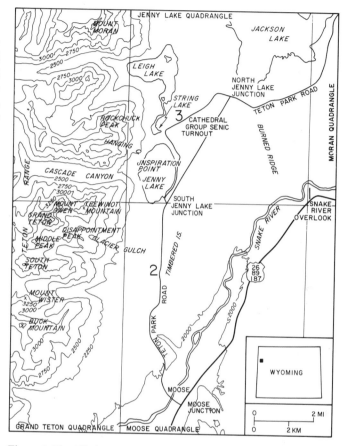

Figure 1. Simplified topographic map showing major part of the Teton mountain front and Sites 1, 2 and 3. Map from Jackson Lake 1:100,000 scale series, U.S.G.S., 1981.

5,000 to 7,000 ft (1,500 to 2,100 m) directly from the outwash plain; there are no intermediate foothills. The break in slope is the approximate trace of the Teton normal fault system, which has a possible combined displacement of about 35,000 ft (11,000 m) down to the east (Fig. 3). Most or all of this displacement has occurred within the last 9 m.y. All the mountain front in this panorama is composed of Precambrian crystalline and metamorphic rocks. These are discussed at Sites 2 and 3.

The outwash plain that extends for 5 mi (8 km) to the south of Site 1 was deposited by meltwater from the Burned Ridge ice, which built up the moraine and occupied the cavity to the north and northwest. The meltwater channelways, which can best be seen from the air, trend southward; yet today there is a westward component to the tilt of this surface. For example, this surface rises to 7,040 ft (2,146 m) directly east of site 1, the site is at 6,905 ft (2,105 m), and the surface lowers to 6,800 ft (2,093 m)

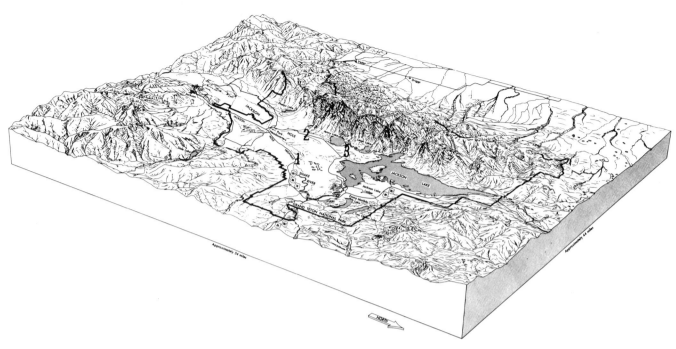

Figure 2. Block diagram view to the southwest showing the Teton Range, Jackson Hole, and Sites 1, 2 and 3 (after Love and Reed, 1971).

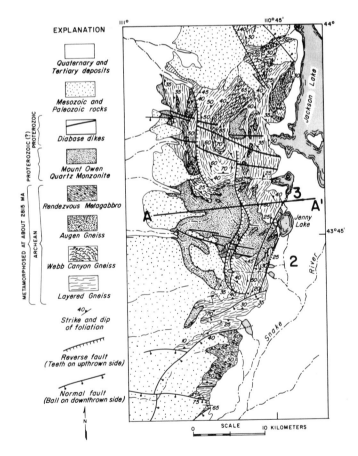

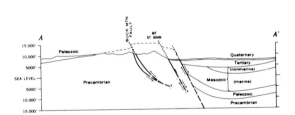

Figure 3. Generalized geologic map and cross section of Precambrian and younger rocks of the Teton Range (after Reed and Zartman, 1973; Love and Reed, 1971). Positions of Sites 2 and 3 and general position of line of cross section A–A′ shown.

Figure 4. Cathedral Group of Teton peaks. View west from Site 2 on Timbered Island. The most important geologic and topographic features are numbered as follows: 1, Middle Teton, elevation 12,804 ft (3,903 m); 2, Garnet Canyon; 3, Disappointment Peak; 4, Grand Teton, elevation 13,770 ft (4,197 m); 5, Teton Glacier; 6, Glacier Gulch; 7, Mount Owen, elevation 12,928 ft (3,940 m); 8, Mount Teewinot, elevation 12,325 ft (3,757 m); 9, Pinedale glacial moraine; 10, outwash plain along highway. Photo by J. D. Love, June, 1955.

4 mi (6.4 km) directly to the west. Similarly, a traverse 4 mi (6.4 km) south, again at right angles to the meltwater channelways, shows that the west margin of the surface is 50 ft (15 m) lower than the east margin. The age of the Burned Ridge moraine and related outwash is not definitely known, but the southern part of the outwash appears to bury loess that is thought to be 15,000 to 20,000 years old. Thus, if it is assumed that the age of the surface at Site 1 and 4 mi (6.4 km) to the south is no more than 20,000 years, the westward rate of tilt would range from 1 ft (0.3 m) in 80 years to 1 ft (0.3 m) in 400 years. (For additional profiles showing westward tilting, see Love and Montagne, 1956, p. 169–178.)

After this outwash surface was developed, and probably within the last 15,000 years, the Snake River cut the channel seen in the foreground, 285 ft (89 m) down, through the Burned Ridge

moraine and the adjacent outwash gravel. The river is no longer down cutting, so it must have been much larger for some interval of time during the last 15,000 years.

From a position 25 ft (8 m) south of the Park Service panorama sign, it is possible to see ragged brown ledges projecting out into the river. These ledges are the southernmost remnant of the Huckleberry Ridge Tuff, 2.05 Ma, a pyroclastic flow that originated in Yellowstone Park. These rhyolitic welded tuffs were deposited with a horizontal east-west strike, but are now tilted west 13° as a result of downhinging of the floor of Jackson Hole after the tuffs were emplaced. Average westward tilt is about 1 ft (0.3 m) every 1,600 years at this site.

The inner terraces along the Snake River represent stages of downcutting through Quaternary gold-bearing quartzite gravel. Looking northwest from the north parapet, one can see a broad

Figure 5. Air-oblique view of east face of Teton Range opposite Jenny Lake: S, South Teton; M, Middle Teton; G, Grand Teton; T, Mount Teewinot; R and F, Stair-stepped fault scarps indicating postglacial movement along the Teton normal fault zone; star symbol, general area of upright trees on the bottom of Jenny Lake; C, Hanging Canyon. Jenny Lake, 247 ft (75 m) deep, encircled by a tree-covered terminal moraine, is in the middle ground. Ice that gouged out the lake basin flowed eastward toward the camera from the U-shaped Cascade Canyon (not labelled) at the right center. This is the deepest canyon in the range and heads behind the Teton peaks. Channelways of treeless quartzite gravel outwash from the Jackson Lake glaciation are visible in the foreground. National Park Service photograph by Bryan Harry, 1965.

treeless plain with tufts of conifers that mark glacial kettles (locally known as "potholes"). The treeless land surface indented by the kettles is an outwash plain deposited by meltwater from the Jackson Lake moraine, which is marked by the elevated line of trees at the far edge of the plain. The Jackson Lake moraine is the youngest on the floor of Jackson Hole, and is perhaps 15,000 years old. Here again, conifers grow abundantly on moraines from the granitic rocks of the Tetons, but do not grow on the quartzite outwash debris.

To the north, 6 mi (10 km) away, the prominent wedge-shaped uplift is Signal Mountain, an ice-carved butte of resistant rocks. The prominent ledge on the bare east-facing slope is the Huckleberry Ridge Tuff, which dips westward about 7° as a result of westward tilting of the floor of Jackson Hole following emplacement at 2.05 Ma.

Looking east from site 1, the white outcrops on the densely wooded hillside are west-dipping tuffs in the Teewinot Formation of late Miocene age. In the foreground to the east, the lumpy hills are remnants of the Burned Ridge moraine partly buried by quartzite outwash gravel.

Site 2. Site 2 is at the top of the west rim of Timbered Island, about 200 ft (60 m) east of the paved highway. Timbered Island is a lateral moraine and is somewhat older than the Burned Ridge moraine discussed at Site 1. The morainal debris was derived largely from the Precambrian rocks of the Teton Range; these provide a favorable environment for the growth of

Figure 6. Postglacial fault scarp (arrows) about 115 ft (35 m) high offsetting alluvial fan near base of Rockchuck Peak. View west from Site 3. National Park Service photograph.

conifers—hence the name, Timbered Island. This feature is surrounded by quartzite gravel outwash, which is unfavorable for the growth of conifers.

The ground view of the Cathedral Group of peaks is shown in Figure 4; their geology is shown in Figure 3. The higher parts of all the peaks except Mount Teewinot are composed of light gray medium- to fine-grained biotitic Mount Owen Quartz Monzonite (late Archean—about 2.5 Ga; Reed and Zartman, 1973). The lower slopes of Glacier Gulch and Mount Teewinot are dark to light gray layered gneiss and migmatite, and are some of the oldest rocks in the Teton Range, perhaps 2.8 Ga or more. They have been down-dropped an unknown amount along several north-northeast–trending normal faults. The visible face of Mount Teewinot is largely layered gneiss and migmatite; the back side is quartz monzonite. The conspicuous black dike cutting vertically up through the Middle Teton is diabase. It is very even sided, 40–60 ft (12–18 m) wide, and extends west of the peak for several miles. It is the youngest Precambrian rock in the Teton

Range. Conflicting data from a siilar diabase dike on Mount Moran discussed by Reed and Zartman (1973) suggest a possible age range from 775 Ma to 1.35 Ga. Another black dike is visible on the south side of the Grand Teton.

Figure 4 shows the Teton glacier and the gray terminal moraine. This is the largest glacier in the Teton Range. The rate of ice movement in the center averages about 28 ft (9 m) per year (Reed, 1967).

About 2 mi (3.2 km) north of Site 2, along the lower slopes of the mountain front, just south of Jenny Lake (locality R, Fig. 5), are several stair-stepped faults that cut Quaternary deposits, both talus and glacial debris. These and others to the north (west of Site 3, Fig. 6), were first described by Fryxell (1938). Profiles of those southwest of Jenny Lake (loc. R, Fig. 5) were constructed by Love and Montagne (1956) and studied in more detail by Gilbert and others (1983). Gilbert did not determine the displacement on the stair-stepped faults, but estimated 63 ft (19 m) on the southernmost individual fault.

At locality F (Fig. 5), on the partly wooded slope above the northwest margin of Jenny Lake are three fault scarps cutting the glacial moraine. From the uppermost down, they have heights of 180, 85, and 140 ft (55, 26, and 43 m)—total displacements are difficult to determine because of the steep slumped slopes.

Figure 5 shows that the top of the encircling moraine around Jenny Lake is nearly the same distance above the lake on the north, east, and south sides. This contrasts with the lumpy moraines dipping steeply eastward where they emerge from the mountain front to the north and south of Jenny Lake. Could the Jackson Lake moraine be flat because of westward downtilting along the postglacial mountain-front faults described above? Possibly related to this question is the occurrence at the star symbol on Jenny Lake (Fig. 5) of about a dozen trees in upright positions on the bottom of the lake, in 35–80 ft (11–24 m) of water. One Engelmann spruce tree is 70 ft (21 m) tall, broken off at the top, and is in 80 ft (24 m) of water. ^{14}C dates on the wood of this and one other tree are 600 ± 100 years (Meyer Rubin, written communication, 1983, samples W53-16 and W53-18). There is a controversy as to whether some, or none, of the trees are rooted, or if they all slid into the lake and maintained or later assumed upright positions. If rooted, then the lake bottom has dropped at least 80 ft (24 m) in the last 600 years. On the other hand, if none is rooted, where did they slide from? Cascade Canyon has a gradient of less than 100 ft per mi (20 m per km) for 3 mi (5 km) above Inspiration Point (Figs. 1, 5), too gentle for an avalanche of the required size. This leaves Hanging Canyon as the only slide source. The bedrock here is not susceptible to sliding, so the trees would have to be deposited by snow avalanches. This is possible, but no large Engelmann spruce trees have been observed along the undisturbed margins of the canyon.

Site 3. Site 3 is at the Cathedral Group Scenic Turnout. This site is on a relatively treeless quartzite gravel outwash deposited by meltwater from the Jackson Lake moraine. The forested margin of the moraine is directly to the north. To the west, on the green slope near the base of Rockchuck Peak is a postglacial fault scarp approximately 115 ft (35 m) high (Fig. 6). A small parallel subsidiary fault is downhill from the main fault. Rocks on the upper slopes are migmatitic biotite gneiss. Note that there are no foothills along the mountain front.

From the site point, the outwash plain drops about 50 ft (15 m) westward to String Lake (Fig. 1). This westward tilting is, at least in part, later than the development of the outwash plain, for the meltwater channelways trend southwestward. Thus, much of the 50 ft (15 m) of westward tilting must have occurred in the last 15,000 years. The break in slope between the mountain front and the flat plain to the east marks the approximate trace of the Teton fault system (Fig. 3).

A good view of Mount Moran (elevation 12,605 ft; 3,842 m), 4.5 mi (7 km) to the northwest, is visible from this site. Most of the darker rock is migmatitic biotite gneiss (>2,875 Ma). It is cut by many light-colored dikes of Mount Owen Quartz Monzonite (about 2,500 Ma). The vertical black diabase dike, about 150 ft (46 m) thick, is 775–1350 Ma. This is the youngest Precambrian rock in the Teton Range. The dike has been traced westward across the range for more than 7 mi (11 km). Partly capping the black dike on the summit of Mount Moran is a gray mound consisting of about 50 ft (15 m) of nearly horizontal Flathead Sandstone of Middle Cambrian age (about 550 Ma). This occurrence is especially significant from a structural standpoint because it provides the highest control point on the Cambrian-Precambrian surface in the northern part of the Teton Range and helps to determine the general amount of displacement on the Teton fault system. East of a line between Mount Moran and the Grand Teton, the total displacement is thought to be between 30,000 and 35,000 ft (9,100–10,700 m). Most or all of this movement probably occurred during the last 9 m.y.

REFERENCES CITED

Fryxell, F. M., 1938, Post-glacial faulting in the Teton Range, Wyoming [abs.]: Geological Society of America Bulletin, v. 30, p. 85.

Gilbert, J. D., Ostenaa, D., and Wood, C., 1983, Seismotectonic study, Jackson Lake dam and reservoir, Minidoka project, Idaho-Wyoming: U.S. Bureau of Reclamation Seismotectonic Report 83-8, 123 p.

Love, J. D., and de la Montagne, J. M., 1956, Pleistocene and recent tilting of Jackson Hole, Teton County, Wyoming: Wyoming Geological Association Guidebook, 11th Annual Field Conference, p. 169–178.

Love, J. D., and Love, J. M., 1983, Road log, Jackson to Dinwoody and return: Geological Survey of Wyoming Public Information Circular no. 20, 34 p.

Love, J. D., and Reed, J. C., Jr., 1971, Creation of the Teton landscape; The geologic story of Grand Teton National Park: Grand Teton Natural History Association, 120 p.

Love, J. D., Leopold, E. B., and Love, D. W., 1978, Eocene rocks, fossils, and geologic history, Teton Range, northwestern Wyoming: U.S. Geological Survey Professional Paper 932-B, 40 p.

Reed, J. C., Jr., 1967, Observations on the Teton Glacier, Grand Teton National Park, Wyoming, 1965 and 1966: U.S. Geological Survey Professional Paper 575-C, p. C154–C159.

Reed, J. C., Jr., and Zartman, R. E., 1973, Geochronology of Precambrian rocks of the Teton Range, Wyoming: Geological Society of America Bulletin, v. 84, p. 561–582.

The Rhodes allochthon of the Enos Creek–Owl Creek Debris-Avalanche, northwestern Wyoming

Thomas M. Bown, U.S. Geological Survey, Box 25046, Denver Federal Center, Denver, Colorado 80225
J. David Love, Box 3007, University Station, Laramie, Wyoming 82071

LOCATION

SE¼SE¼NE¼SW¼ sec. 17, NE¼NE¼SE¼SW¼ Sec. 17, W¾N½SW¼SE¼ sec. 17, and W¾S½NW¼SE¼ Sec. 17, T.44N., R.99W., Hot Springs County, Wyoming (Fig. 1; Bown, 1982a, Pl. 1). Twentyone Creek, Wyoming, Quadrangle (U.S. Geological Survey 7½ minute topographic series, 1956).

ACCESSIBILITY

A vantage point on Putney Flat terrace of Cottonwood Creek overlooking the Rhodes allochthon (center, SW¼ sec. 20, T.44N., R.99W.) is easily reached by passenger car from Thermopolis, Wyoming, on paved, gravel, and improved dirt roads (see road log). This vantage point is on private land and permission to stop there as well as to proceed to the allochthon through the cow camp in the N½SW¼ and NW¼ sec. 20, T.44N., R.99W. is obtainable by contacting the Rhodes Ranch Company, Hamilton Dome Route, Wyoming. The Rhodes allochthon in sec. 17 is on land administered by the Bureau of Land Management. A pickup truck or similar vehicle is advised for crossings of Cottonwood and Twentyone creeks.

SIGNIFICANCE

The Rhodes allochthon is one of more than 160 erosional remnants of debris of the Enos Creek–Owl Creek Debris-avalanche, emplaced about the time of the Tertiary-Quaternary boundary. The triggering mechanism for the debris-avalanching is believed to have been violent earthquake activity, perhaps that accompanying deposition of the approximately 2.05 Ma Huckleberry Ridge Tuff in the Yellowstone National Park–Teton region of Wyoming (Love, 1977; Love and others, 1978). Remnants of the Enos Creek–Owl Creek Debris-avalanche vary from 15–1,300 ft (5–400 m) in thickness and are distributed across an area of 350 mi² (more than 900 km²). The original lateral displacement of rocks was at least 28 mi (45 km); the pre-erosion width of the deposit was more than 25 mi (40 km), and the original area covered by debris and the original volume of displaced rock are conservatively estimated at 470 mi² (1,220 km²) and 45 mi³ (~187 km³), respectively (Bown, 1982a, b, c). These figures indicate a displaced volume exceeding by a factor of seven that of the debris-avalanche deposit adjacent to ancestral Mount Shasta volcano in California (Crandell and others, 1984). Because there is no field evidence supporting more than a single episode of displacement, the Enos Creek–Owl Creek Debris-avalanche deposit appears to be the largest mass of debris of this

origin yet recognized in the solar system, even exceeding in size the immense Martian landslides described by Lucchitta (1978, 1979). In addition, the considerable lateral reach and the vast distribution of great thicknesses of debris are more consistent mechanically with a single displacement event. The compound nature of the displacement mechanisms of the debris-avalanching is reflected in the varying character of the internal structure of the deposit. From areas most proximal to debris source to those more distal, these mechanisms appear to have included block slumping, retrogressive block-gliding, gravity sliding, and landsliding.

SITE INFORMATION

The Rhodes allochthon is situated at the extreme southeast margin of the Enos Creek–Owl Creek Debris-avalanche deposit (Fig. 1). The debris field extends some 24 mi (38 km) northwest of the Rhodes allochthon (to north of the valley of the Wood River), and about 7 mi (11 km) north to Adam Weiss Peak. About 50–100 ft (15–30 m) of debris capping Squaw Teats, nearly 25 mi (40 km) north-northwest of the allochthon (Fig. 1) marks all that is preserved of the eastern limits of the debris field. The southwest margin of the source area of the displaced rocks is at least 9 mi (14 km) to the west-southwest, above the valley of the North Fork of Owl Creek.

From the vantage point on Putney Flat terrace overlooking Cottonwood Creek, the Rhodes allochthon is clearly displayed as a mass of dark reddish-brown volcanic sandstone breccia and contorted volcanic sandstone occupying the upper two-thirds of a butte 360 ft (110 m) high and situated about one mi (1.6 km) to the north-northeast (Fig. 2). The source of these displaced sandstones is the upper part of the Tepee Trail Formation and the lower part of the Wiggins Formation, at least 9 mi (14 km) to the west-southwest. At the Rhodes allochthon, the displaced rocks lie atop about 165 ft (50 m) of the middle Eocene alluvial Aycross Formation that comprises the sequence of light gray, green, and brown volcanic sandstones and mudstones forming the top of the bedrock terrace across Cottonwood Creek, up to the base of the allochthon. Beneath the displaced rocks, the Aycross Formation locally strikes to the northeast, dips steeply to the northwest, and makes up the upper plate of a minor low-angle overthrust fault. Cross sections of other minor thrusts in Aycross strata are visible to the northeast in the ridge east of the allochthon (NE¼-NE¼SE¼ sec. 17), and occur as well farther north (in Dugout Draw) and west (in the drainage of Twentyone Creek). The brightly variegated mudstones beneath the top of the bedrock terrace along Cottonwood Creek belong to the non-volcanic lower Eocene Willwood Formation, also an alluvial unit and

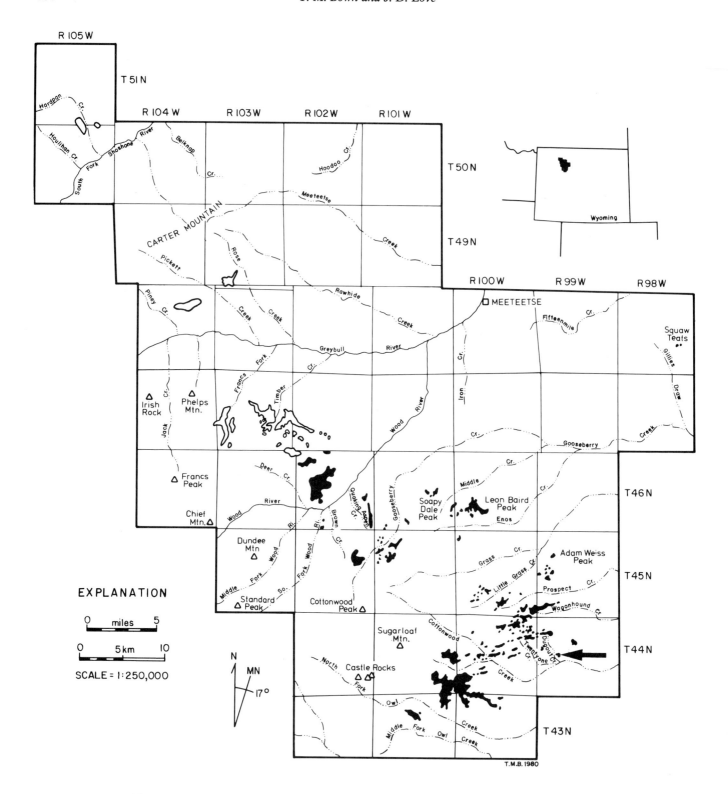

Figure 1. Distribution of allochthonous remnants of Late Cenozoic Enos Creek–Owl Creek Debris-avalanche in the southeast Absaroka Range and southwest Bighorn Basin (*black*). Rhodes allochthon designated by *arrow*. *Unshaded outline*: remnants are of middle Eocene debris-avalanche deposits.

Figure 2. Rhodes Butte above Dugout Draw. Allochthonous mass is darker sandstones forming upper two-thirds of left (north) part of butte. View to northeast from point in NE¼NW¼ sec. 20, T.44N., R.99W. (Photo by T. M. Bown, July 7, 1977.)

Figure 3. Schematic graphic representation of breccia and sandstone deformation on west face of Rhodes Butte. View to northeast from point slightly north of that in Figures 2 and 3.

here separated from Aycross rocks by a minor erosional unconformity.

The internal structure of different parts of the Enos Creek–Owl Creek Debris-avalanche deposit indicates that mechanically distinct types of movement occurred during the displacement; these are spatially related to the distance from the source area. Displaced rocks most proximal to source consist of immense, basinward rotated slump blocks, whereas deposits adjacent to these slump blocks but situated farther basinward show features of retrogressive block-gliding (Voight, 1973). Loss of internal cohesion of the mobile mass was produced even farther from the source by gravity gliding for several mi (km) across a high-level basin margin erosion surface (Bown, 1982a, 1982b, 1982c), and by landsliding of debris into paleovalleys cut into this surface. The most thoroughly comminuted debris (reduced completely to breccia) is landslide in origin and fills paleovalleys developed on the Aycross Formation throughout the Cottonwood Creek, Twentyone Creek, Dugout Creek, and Prospect Creek areas (Fig. 1).

The structure of the Rhodes allochthon exemplifies the initial internal disordering of the displaced stratified rock during the gravity-glide phase of movement. This disordering resulted in the peeling off of beds from one another, shingling of more resistant units, plastic folding due to intense confining pressure in the interior of the mobile mass, and, with cessation of movement, the stranding of the shingled beds and rotated S-shaped segments of folds in a matrix of breccia (Fig. 3). A walk around the Rhodes allochthon occupies about an hour and provides excellent close-up views of breccia and plastically folded and shingled sandstones.

Some minor gouging of the underlying Aycross Formation is seen at the Rhodes allochthon, but there gouging is not nearly as intense as it is under those parts of the deposit emplaced by retrogressive block-gliding. In place rocks that bound landslide materials of the debris-avalanche are rarely gouged, even where made up of incompetent volcanic mudstones that bound debris

filled paleovalleys. These relationships suggest that the moving landslide debris trapped air in the overridden valleys and raced down the valleys on cushions of compressed air; the air then was gradually dissipated from beneath the mass, as described by Kent (1966) and Shreve (1968). Examples of landslide debris filling paleovalleys with little or no gouging of surrounding rocks occur north and northwest of the Rhodes allochthon, in secs. 31 and 32, T.45N., R.99W. and in the SE¼ sec. 3, T.44N., R.100W. (Bown, 1982c, Figs. 5 and 8, respectively).

Several other gravity-glide-emplaced allochthonous remnants of the Enos Creek–Owl Creek Debris-avalanche are visible from the Putney Flat vantage point. The most obvious of these are: (1) about 2.5 mi (4 km) west, in the SW¼SE¼ sec. 23, T.44N., R.100W.; (2) about 1.25 mi (2 km) northwest, in the NE¼SE¼NW¼ sec. 18, T.44N., R.99W; (3) several remnants about 2–3 mi (3–5 km) north-northwest, in the north halves of secs. 7 and 8, T.44N., R.99W.; and (4) about 1.9 mi (3 km) northeast, in the SE¼ sec. 9, T.44N., R.99W. (Cottonwood Creek allochthon).

ROAD LOG—THERMOPOLIS, WYOMING, TO RHODES ALLOCHTHON

0.00 mi (0.0 km). Thermopolis, Wyoming. Begin log at junction of U.S. 20 and Wyoming 120. Proceed west on Wyoming 120 toward Meeteetse.

9.2 (14.8). Junction with Wyoming Highway 170. Turn left (west) up Owl Creek valley. Owl Creek Mountains form skyline to south and southwest; southeast Absaroka Range forms distant skyline straight ahead.

19.2 (30.9). Junction of Wyoming Highways 174 and 170. Continue west on Highway 174.

30.3 (48.7). Junction with Blondie Pass road (on left), providing access to Wind River Basin over Owl Creek Mountains. Continue west on Wyoming Highway 174.

31.2 (50.2). Absaroka Range on skyline straight ahead. High tree-covered foothills just beneath skyline comprise Enos Creek–Owl Creek Debris-avalanche deposits emplaced by retrogressive block-gliding.

33.4 (53.7). Good view to west and northwest of steeply dipping blocks of retrogressive block-gliding phase of Enos Creek–Owl Creek Debris-Avalanche.

36.5 (58.7). Pavement ends, junction with South Fork Owl Creek and Cottonwood Creek roads; turn right. Good view straight ahead of southern limit of retrogressive block-gliding phase of Enos Creek–Owl Creek Debris-Avalanche and tree covered blocks of displaced middle Eocene Tepee Trail Formation.

37.6 (60.5). Junction of Cottonwood Creek road and North Fork Owl Creek road; bear to right. Persons interested in examining debris of the Enos Creek–Owl Creek Debris-Avalanche emplaced by retrogressive block-gliding and gravity-gliding (as well as extensively gouged rocks of the underlying Aycross Formation) should return to this junction and proceed west on the North Fork Owl Creek road.

39.2 (63.1). Junction. To left (west) is east face of detached Tepee Trail rocks; straight ahead are Chimney Rocks, gravity-glide outliers of the Enos Creek–Owl Creek Debris-Avalanche. Turn right and continue on Cottonwood Creek road.

40.7 (65.5). Badlands to left (north) are in lower part of volcaniclastic middle Eocene Aycross Formation.

41.7 (67.1). Junction with Cottonwood Creek road on left and Putney Flat road on right. Drainage to left is Cottonwood Creek, and several minor normal faults are visible in the cliff of Aycross rocks to the left. The Rhodes allochthon is the prominent butte straight ahead across Cottonwood Creek, and the Cottonwood Creek allochthon (a much larger debris-avalanche remnant) forms the hummocky skyline behind and to the right of the Rhodes allochthon. Turn right and cross Putney Flat, a Pleistocene terrace of Cottonwood Creek.

42.2 (67.9). Large volcanic boulder 60 ft (18 m) to left (north) of road and other similar boulders on Putney Flat are unusually large clasts of probable Pleistocene glacial outwash.

43.8 (70.5). Good views of Rhodes (left) and Cottonwood Creek (right) allochthons across Putney Flat and Cottonwood Creek to left (north).

44.8 (72.1). Fence straight ahead; before reaching fence, turn left (north) on road crossing Putney Flat.

46.4 (74.7) Stop 1. Edge of Putney Flat on Cottonwood Creek terrace. Vantage point for viewing Rhodes allochthon (see text). Proceed on road over Cottonwood Creek terrace.

46.8 (75.3). Olds Rhodes Ranch, now a cow camp. Be certain to close all gates.

47.0 (75.6). Cross Twentyone Creek; turn right after 300 ft (91 m) and pass through gate in fence.

47.4 (76.3) Stop 2. Edge of Dugout Draw; Rhodes allochthon across the draw straight ahead. Proceed on foot across Dugout Draw, circumnavigating the allochthon. This hike takes about one hour or less and provides excellent views of deformation of gravity-glide materials of Enos Creek–Owl Creek Debris-Avalanche. Note several minor thrust faults in underlying Aycross Formation.

End of Road Log—Please respect private property.

REFERENCES CITED

Bown, T. M., 1982a, Geology, paleontology, and correlation of Eocene volcaniclastic rocks, southeast Absaroka Range, Hot Springs County, Wyoming: U.S. Geological Survey Professional Paper 1201-A, p. 1–75.

Bown, T. M., 1982b, Catastrophic large-scale Late Cenozoic detachment faulting, Absaroka Range, northwest Wyoming: Geological Society of America Abstracts with Programs, v. 14, no. 7, p. 449.

Bown, T. M., 1982c, Catastrophic large-scale Late Cenozoic detachment faulting of Eocene volcanic rocks, southeast Absaroka Range, northwest Wyoming: Wyoming Geological Association 33rd Annual Field Conference Guidebook, p. 185–201.

Crandell, D. R., Miller, C. D., Glicken, H. X., Christiansen, R. L., and Newhall, C. G., 1984, Catastrophic debris-avalanche from ancestral Mount Shasta volcano, California: Geology, v. 12, p. 143–146.

Kent, P. E., 1966, The transport mechanism in catastrophic rock falls: Journal of Geology, v. 74, p. 79–83.

Love, J. D., 1977, Summary of Upper Cretaceous and Cenozoic stratigraphy, and of tectonic and glacial events in Jackson Hole, northwestern Wyoming: Wyoming Geological Association 29th Annual Field Conference Guidebook, p. 585–593.

Love, J. D., Leopold, E. B., and Love, D. W., 1978, Eocene rocks, fossils, and geologic history, Teton Range, northwestern Wyoming: U.S Geological Survey Professional Paper 932-B, p. B1–B40.

Lucchitta, B. K., 1978, A large landslide on Mars: Geological Society of America Bulletin, v. 89, p. 1601–1609.

Lucchitta, B. K., 1979, Landslides in Valles Marineris, Mars: Journal of Geophysical Research, v. 84, p. 8097–8113.

Shreve, R. L., 1968, The Blackhawk landslide: Geological Society of America Special Paper 108, 47 p.

Voight, B., 1973, The mechanics of retrogressive block-gliding, with emphasis on the evolution of the Turnagain Heights landslide, Anchorage, Alaska, *in* DeJong, K. A., and Scholten, R., eds., Gravity and Tectonics: New York, John Wiley and Sons, Interscience, p. 97–121.

Idaho-Wyoming thrust belt: Teton Pass, Hoback Canyon, Snake River Canyon

H. Thomas Ore, Geology Department, Idaho State University, Pocatello, Idaho 83209
A. A. Kopania, Champlin Petroleum Company, P.O. Box 1257, Englewood, Colorado 80150

LOCATION AND ACCESSIBILITY

This guide describes two areas of observation in the Idaho-Wyoming thrust belt. The first is a 3-mi (4.9-km) long section on Teton Pass, west of Jackson, Wyoming, along Wyoming 22 (Fig. 1). The geologic map (Fig. 2) and discussion below refer to a section that stretches from 2.5 mi (4.0 km) west of the summit of Teton Pass to 0.5 mi (8.0 km) to the east. The area is depicted on the Teton Pass and Rendezvous Peak 7½-minute Geologic Quadrangle Maps (Schroeder, 1969, 1972). The second area is along the Hoback Valley and Canyon, about 10 mi (16 km) southeast of Jackson. Observations there are along U.S. 187-189, from Hoback Junction to Granite Creek, about 11.5 mi (18.5 km) east. The area is depicted on the Camp Davis and Bull Creek 7½-minute Geologic Quadrangles (Schroeder, 1974, 1976). Two other locations in the area are briefly described. They are both on U.S. 26-89 in the Grand Canyon on the Snake River. The first is about 3 mi (4.8 km) southwest of Hoback Junction, where the Darby thrust is exposed, the other is at Keyser Creek, about 8 mi (12.9 km) east of Alpine Junction, where the Little Greys Anticline is exposed. All locations described herein are on public land.

SIGNIFICANCE

The Teton Pass area, Hoback Canyon, and several other specific sites listed below illustrate the structural style of the eastern Idaho–western Wyoming thrust belt. The west-dipping, Sevier-type Absaroka, Darby, and Jackson thrusts (in order from west to east) are exposed in the Snake River Range and are dissected by the Snake River between Alpine and Jackson. "Sevier" refers to thrust faults with displacements of tens of miles (km), rising from major décollements, and which locally ramp upward through the section in the direction of transport. The eastward-dipping Cache Creek thrust, marking the southwest margin of the Teton–Gros Ventre foreland uplift of western Wyoming, meets the Jackson thrust in the Teton Pass area where they have a common footwall. (We use foreland in the same sense as discussed by Wiltschko and Dorr [1983, p. 1308]). Minor high-angle normal faults cut the thrusts at several locations on Teton Pass. Major listric normal faults define the west end of the Snake River Range (east edge of the Snake River Graben), and the west edge of the coupled Hoback–Gros Ventre blocks (the major block southeast of Jackson Hole, see Fig. 1). In the Teton Pass area, the north-south trend of the southerly portion of the thrust belt curves toward the northwest. The Teton block just to the north may have acted as a buttress to eastward movement of the thrusts.

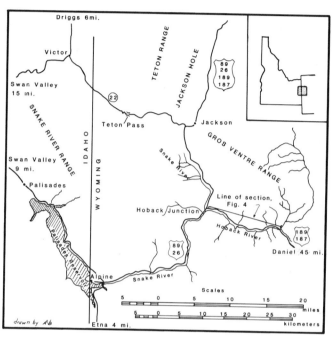

Figure 1. Index map of thrust belt area.

TETON PASS AREA

A generalized geologic map of the Teton Pass area is provided (Fig. 2); comparison of the shape of the road on the map and ground make location identification easy. Note the proximity of the west-dipping Jackson thrust trace to the east-dipping Cache Creek trace. There is some uncertainty about the relative ages of the two thrusts. The Jackson thrust in this area dips 45 to 70° SW in contrast to the 20 to 30° dips on the fault elsewhere. This fact has been used to suggest that the Cache Creek fault occurred later, and that wherever the two thrusts impinge closely on one another, the Cache Creek thrust has steepened the Jackson thrust. Other evidence, some using subsurface data from Teton Valley and Hoback Basin, suggests that at least the earliest movement on the Cache Creek thrust predates the Jackson thrust. Moreover, a large, Laramide-type, basement-cored uplift along the Cache Creek thrust (ancestral Teton–Gros Ventre uplift) is indicated by restoration of the Plio-Pleistocene Teton fault. This additionally suggests early major movement on the Cache Creek.

On the pass, the Cambrian Gros Ventre and Gallatin Formations are thrust onto a Cretaceous (Frontier?) formation, indicating a stratigraphic displacement of at least 10,800 ft (3,290 m). At least seven splays, each less than 300 ft (100 m) wide, mark the leading edge of the Jackson thrust in the Teton Pass area.

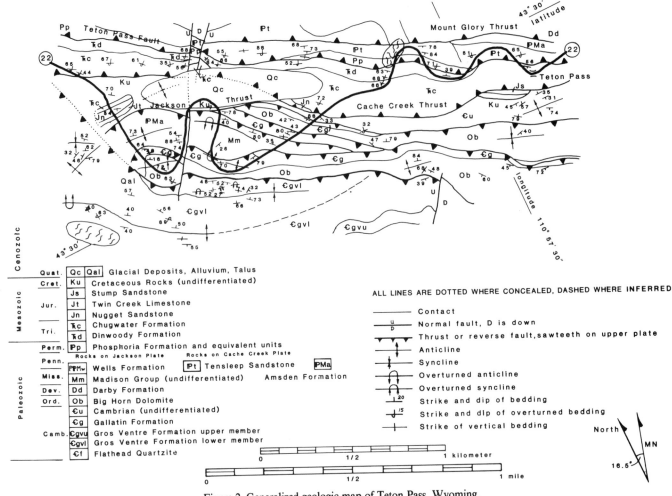

Figure 2. Generalized geologic map of Teton Pass, Wyoming.

They are exposed in roadcuts near the radio towers along the ridge south of the summit.

The Cache Creek thrust dips about 45 to 55°N in the Teton Pass area. The thrust is marked here by the southern edge of reddish Triassic Chugwater Formation on the hanging wall. The trace of the thrust is covered by colluvium just east of the summit, but is present along the southern end of the west and east Gros Ventre buttes in the valley between the summit and the Jackson thrust (Fig. 3).

Thrusting of the Chugwater onto the common Frontier footwall on the Cache Creek fault involves at least 4,800 ft (1,460 m) of stratigraphic displacement. Displacement decreases to the west. Precambrian granite gneiss is exposed only 200 ft (61 m) from the thrust in a valley east of the summit, so that the basement rocks are apparently thrust as well.

HOBACK CANYON AREA

Hoback Canyon is here divided into two segments: that west of the Hoback Range front (7 mi [11.3 km] west of Granite Creek, 4.5 mi [7.2 km] east of Hoback Junction), and that east of

the range front. Figure 4 is a cross section that parallels the highway (the location of section is indicated in Fig. 1).

Segment west of Hoback Range Front. As the valley east of the Hoback Range was downdropped along the Hoback listric normal fault, coarse clastic detritus was deposited along its eastern margin and gradually tilted back toward the range as evolution of the Hoback fault proceeded. This sequence is the upper Pliocene Camp Davis Formation, which is composed of three members. The lowest is about 295 ft (90 m) of braided stream fanglomerate that contains clasts from the Hoback Range only, with an upper Paleozoic-Mesozoic provenance. The middle member is a lacustrine sequence, about 197 ft (60 m) thick, presumably resulting from damming of the ancestral Snake River drainage and flooding of the basin. The upper member, about 4,750 ft (1,450 m) thick, contains clasts from the Gros Ventre Range to the east, representing a new eastern provenance of lower Paleozoic and Precambrian rocks, exposed as headward erosion ate into the Gros Ventre Range. The Camp Davis is exposed on the northeast limb of the east-southeast–trending Willow Creek anticline, probably formed during evolution of the Hoback listric normal fault. The cliff-forming lower member is

Figure 3. View looking east from Teton Pass turnout toward Jackson Hole and Gros Ventre Mountains. The Jackson (= Prospect) and Cache Creek thrusts parallel each other in Cache Creek, east of the town of Jackson, and across Jackson Hole. On Teton Pass (out of the photograph to the right; see Fig. 2) these two thrusts come into contact. Initial movement of the Cache Creek thrust predates the Jackson thrust; however, Tertiary reactivation of the former has caused an imbricate to possibly overlie the latter. The Teton Range and Gros Ventre Mountains were once a large continuous uplift of Laramide style. Subsequent movement on the Teton fault (at the base of the Tetons, out of the photograph) down-dropped Jackson Hole, leaving the Teton Range high and hinging the Gros Ventre Mountains so that they plunge west beneath Jackson Hole.

visible along U.S. 187-189 between 2.5 and 3.5 mi (4.0 and 5.6 km) west of the Hoback Mountain front (2 and 3 mi ([3.2 and 4.8 km] east of Hoback Junction), and for several mi (km) along U.S. 89 from Hoback Junction toward Jackson. Tuffaceous white beds of the middle member are visible at the top of the cliff 2 mi (3.2 km) east of Hoback Junction (2.5 mi [4.0 km] west of the Hoback Range front). The upper member underlies the reddish slopes from the cliffs to the Hoback Range.

Cretaceous rocks of the Bear River and Aspen Formations are exposed north and south of the road just east of Hoback Junction. They dip west on the west limb of the Willow Creek anticline. The University of Michigan geology summer field station is the large group of buildings located south of the road 3.5 mi (5.6 km) east of Hoback Junction (1 mi [1.6 km] west of the Hoback front).

Segment east of the Hoback Range front. This section of the canyon extends from the Hoback normal fault, 4.5 mi (7.2 km) east of Hoback Junction to Granite Creek Road, 7 mi (11.3 km) east. At the Stinking Springs turnout, on the north side of U.S. 187-189, a cross section of the core of an anticline in the footwall of the Hoback fault is exposed. Here, Devonian Darby Formation is overlain by Mississippian Madison Group limestone. These limestones dip easterly for several mi (km) east of the pullout. At 2.5 mi (4.0 km) east of the Hoback front, the Pennsylvanian Wells Formation is exposed at the top of the hill to the east and south, along the eastern leading edge of the anticline. Here the Wells comprises the hanging wall of the west-dipping

Bear thrust, which crosses the highway 0.1 mi (0.2 km) east of the Hoback River highway crossing. The footwall is the Triassic Chugwater, whose redbed lithology is visible to the northeast 0.3 mi (0.5 km) east of the fault (0.4 mi [0.6 km] east of the highway bridge). Another 0.3 mi (0.5 km) east, the overlying Jurassic Nugget Formation is exposed dipping easterly, and another 0.4 mi (0.6 km) east, the overlying Jurassic Twin Creek is present. The rocks here are in the west limb of a syncline in the footwall of the Bear thrust. At 1.5 mi (2.4 km) east of the river crossing, one crosses the axis of this fold, as indicated by west-dipping Cretaceous Gannett Group rocks. At 5 mi (8 km) east of the Stinking Springs turnout (0.8 mi [1.3 km] west of the Granite Creek Road), Bull Creek enters the Hoback River from the north.

East of Bull Creek is an anticline-syncline couplet with Gannett Group (Ephraim Formation) in the core of the syncline in the first major reentrant east of the Creek. East of that, the following occur on the east limb of the syncline: thin, cliff-forming, glauconitic Jurassic Stump Formation; thin, poorly exposed Jurassic bPreuss; thick Twin Creek; and finally, steep dip slope on the Nugget. The Nugget comprises the higher topography east of the small stream entering from the north (northeast of the guard rail on the south side of the road). The anticlinal axis of this anticline-syncline couplet is present just west of the reentrant referred to above. Battle Mountain is visible to the east, where the Prospect thrust places Mesozoic rocks over the Lower Tertiary Hoback Formation.

OTHER SITES

Two other easily accessible sites (Fig. 1) in the Snake River Canyon are near Astoria Hot Springs, where the Darby thrust is exposed, and Keyser Creek, where Mesozoic rocks are folded into the tight, asymmetric Little Greys anticline just east of the Absaroka thrust.

The first location is on the north side of U.S. 26-89, about 3 mi (4.8 km) southwest of Hoback Junction, and is illustrated on the Munger Mountain Quadrangle. The Darby thrust is in the bottom of a valley on the north side of the road with a large alluvial fan at its mouth, 0.5 mi (0.8 km) east of the Astoria Hot Springs bridge across the Snake River. The hanging wall on the west side of the valley contains Wells, Phosphoria, Dinwoody, Woodside, and Thaynes (Pennsylvanian through Triassic) rocks folded into an anticline, whose east limb forms the west side of the valley. The footwall has been steeply tilted to the east, and contains rocks of the Jurassic Twin Creek, Preuss, Stump, and Cretaceous Gannett Group. The thrust trace cuts up section and generally follows the deepest portion of the valley. The white, almost vertical, cliff former in the footwall is the Peterson Limestone of the Gannett Group.

The second location is at Keyser Creek, 8.3 mi (13.4 km) east of Alpine Junction, shown on the Pine Creek Quadrangle. The Little Greys anticline exposed here is highly asymmetric, with its eastern limb vertical to overturned, and the western limb dipping 25 to 30°. Jurassic Twin Creek limestone is in the core of

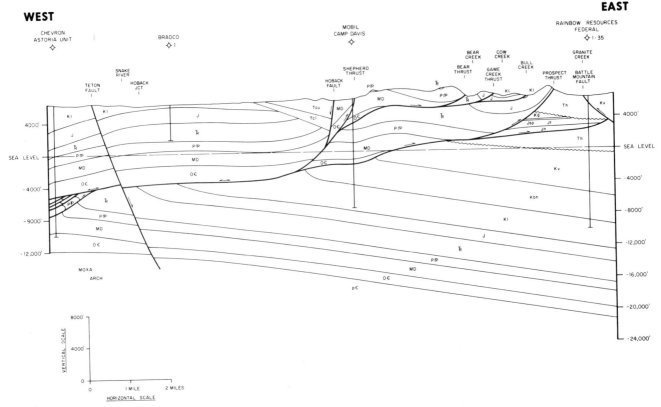

Figure 4. East-west cross section drawn parallel to, and just north of, U.S. 187-189, from the Hoback Junction area to Granite Creek. Important points to note are: (1) the position of the Moxa Arch; (2) the basement-cutting Teton fault, still present this far south, although its displacement here is very minor; (3) increasing dips toward the east in the Camp Davis Formation (Tcv, Tcl), implying the listric nature of the Hoback fault; (4) the position of the Hoback fault above a ramp in a Prospect sheet imbricate stack; (5) the deeper trend of the Prospect; and (6) the deeper thrust beneath the Prospect (observed in the Rainbow Reservoir Federal I-35 well), which must predate the Prospect (= Cliff Creek) thrust and Battle Mountain fault.

the structure. The dark, nearly vertical bed on the east limb is the Jurassic Stump Formation and the white unit just to the east is the Peterson Limestone of the Gannett Group. On the west limb, the prominent sandstone is the Tygee Sandstone member of the lower Bear River Formation (Lower Cretaceous). Stratigraphic units referred to throughout are summarized in Armstrong and Oriel (1965).

REFERENCES CITED

Armstrong, F. C., and Oriel, S. S., 1965, Tectonic development of the Idaho-Wyoming thrust belt: American Association of Petroleum Geologists Bulletin, v. 49, p. 1847–1866.

Dorr, J. A., Jr., Spearing, D. R., and Steidtmann, J. R., 1977, Deformation and deposition between a foreland uplift and an impinging thrust belt, Hoback Basin, Wyoming: Geological Society of America Special Paper 177, 82 p.

Schroeder, J. L., 1969, Geologic map of the Teton Pass Quadrangle, Teton County, Wyoming: U.S. Geological Survey Geological Quadrangle Map GQ-793.

——, 1972, Geologic map of the Rendezvous Peak Quadrangle, Teton County, Wyoming: U.S. Geological Survey Geologic Quadrangle Map GQ-980.

——, 1974, Geologic map of the Camp Davis Quadrangle, Teton County, Wyoming: U.S. Geological Survey Geological Quadrangle Map GQ-1160.

——, 1976, Geologic map of the Bull Creek Quadrangle, Teton and Sublette Counties, Wyoming: U.S. Geological Survey Quadrangle Map GQ-1300.

Wiltschko, D. V., and Dorr, J. A., Jr., 1983, Timing of deformation in overthrust belt and foreland of Idaho, Wyoming, and Utah: American Association of Petroleum Geologists Bulletin, v. 67, p. 1304–1322.

Phanerozoic stratigraphy and structures in the northwestern Wind River Basin at Red Grade, Wyoming

J. D. Love, U.S. Geological Survey, University of Wyoming, Laramie, Wyoming 82071

LOCATION

Red Grade viewpoint is in the SW¼ NW¼ SE¼ Sec. 8, T.5N., R.5W., Wind River Meridian, 17.3 mi (28 km) southeast of Dubois, Fremont County, Wyoming, Blue Holes and Wilderness 7½-minute Quadrangles (Fig. 1).

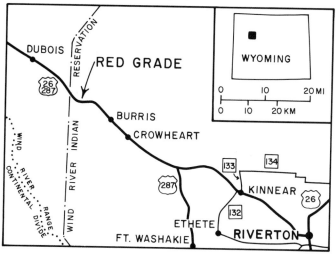

Figure 1. Index map showing location of Red Grade site. Geology of area between Dubois and the Red Grade site is described on the road log.

ACCESSIBILITY

The best vantage point is on a knoll 30 ft (9 m) north of paved highway U.S. 26-287, at a broad pullout on the north side of the highway. The knoll is on the Wind River Indian Reservation, and if there are problems of access, the geology can be viewed from the highway right-of-way.

SIGNIFICANCE

This is the best of the easily accessible places to see Paleozoic, Mesozoic, and Cenozoic stratigraphy and structure in the northwestern part of the Wind River Basin, northwest-central Wyoming. The Triassic, Jurassic, Lower Cretaceous, and lower Eocene strata are spectacularly colorful and well exposed. Other significant features are an angular unconformity between lower Eocene and Lower Cretaceous strata; several northeast-dipping thrust faults that put lower Eocene strata on both Upper and Lower Cretaceous rocks; Paleozoic klippen(?) within lower Eocene strata; and 600,000-year-old ash and travertine, the surface of which has been incised as much as 700 ft (215 m) by later stream erosion.

SITE INFORMATION

Parts of the geologic panorama visible to the west, north, and northeast of this site were mapped by Love (1939). More detailed unpublished mapping was done by W. R. Keefer (see also regional map by Keefer, 1970) and J. F. Murphy of the U.S. Geological Survey and by G. F. Winterfeld as part of a Ph.D. study, University of Wyoming. Much of the site description and the photographs are published in Love and Love (1983).

The viewpoint is on part of a moraine of Pinedale age that emerged from Dinwoody Canyon to the southwest and spread out on the floor of the Wind River Basin. An extensive panorama of complex Paleozoic, Mesozoic, Tertiary, and Quaternary stratigraphy, structure, and geologic history can be seen. The annotated photographs (Figs. 2–10) serve as guides for the discussion at the Red Grade Viewpoint in the accompanying road log.

ROAD LOG

The following road log covers the area from Red Rock Lodge (15.6 mi—25 km—southeast of Dubois, Wyoming) to Red Grade Site.

Red Rock Lodge (0.0 mi; 0.0 km). Across the river at 9 o'clock is a completely exposed pink to tan cliff of Nugget Sandstone, 135 ft (40 m) thick (Fig. 2). At the base of the cliff, the softer green, red, and yellow strata are part of the Popo Agie Member of the Chugwater Formation. The red beds overlying the Nugget are part of the Gypsum Spring Formation of Middle

Figure 2. View northeast of entire 135 ft (40 m) of Nugget Sandstone (between arrows) in Red Grade measured section. Evenly bedded red, green, and orange strata below Nugget are in the upper part of the Popo Agie Member of the Chugwater Formation. Wind River flows along base of cliff. Photograph by J. D. Love, November 1944.

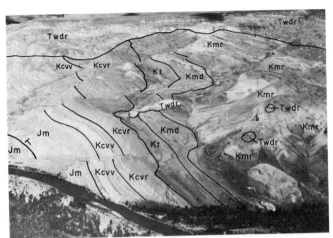

Figure 3. Air oblique view north across Wind River, showing Jurassic and Lower Cretaceous part of the Red Grade measured section. The nearly horizontal Wind River Formation unconformably overlaps the Cretaceous rocks at the top of the photograph. Formations indicated are: Jm = Morrison Formation (Upper Jurassic); the Lower Cretaceous units Kcvv = variegated part of the Cloverly Formation, Kcvr = rusty beds member of the Cloverly Formation, Kt = Thermopolis Shale; Kmd = Muddy Sandstone Member of the Thermopolis Shale, Kmr = Mowry Shale; and Twdr = the Lower Eocene Wind River Formation. A thrust fault (T on overriding block) is at lower left. Photograph by W. B. Hall, T. H. Walsh, and J. D. Love, July 7, 1974.

Figure 4. View west from Red Grade, toward Wind River Mountains on skyline. Indicated are: Mm = Madison Limestone (Mississippian); ℙt = Tensleep Sandstone (Pennsylvanian); Pp = Phosphoria Formation (Permian); ₸c = Chugwater Formation; (J₸n) = Nugget Sandstone (Jurassic [?] and Triassic [?]); and localities of Pleistocene ash. Photograph by J. D. Love, October 25, 1979.

Jurassic age, from which the lower thick gypsum bed has been leached at and near the surface. The gray ledges above the red are the Sundance Formation of Middle and Late Jurassic age. At 2 o'clock, Nugget Sandstone extends up toward the skyline. The flat surface between 2:30 and 3 o'clock is capped by horizontal travertine interlayered with ash at the highest skyline exposures, 700 ft (200 m) above Wind River, which is less than 2 mi (3.2

km) to the north. This ash, "Pearlette type O," has a K-Ar age of about 600,000 years at this site (G. A. Izett, written communication, 1977). Thus, if we assume that the ash and travertine were deposited nearly at river level, then the rate of downcutting in the last 600,000 years would be about 1 ft (0.3 m) in 850 years. However, a similar ash less than 1 mi (1.6 km) to the east is only 460 ft (140 m) above the river. If this ash has the same age (which has not yet been determined), the rate of downcutting would be 1 ft (0.3 m) in about 1,300 years.

(0.8 mi; 1.8 km). At 9 o'clock is the colorful panorama shown in Figure 3. This is the upper part of the Red Grade measured section (Love and others, 1947; also other references cited below at the Red Grade viewpoint). Greenish-gray, thin-bedded strata of the Sundance Formation are overlain by dully variegated claystones and sandstones in the Morrison Formation. Overlying the Morrison is brightly variegated red, lilac, and white claystone capped by the rusty beds member of the Cloverly Formation. Overlying the rusty beds member is the black Thermopolis Shale, and at 10 o'clock the gray-banded Mowry Shale is visible. On the skyline at 9 o'clock, the candy-striped red and white badlands are in the horizontal Wind River Formation.

Red Grade Viewpoint (1.7 mi; 2.5 km). Go through the fence to the top of the knoll 30 ft (10 m) north of fence. You are standing on part of the moraine of Pinedale age that emerged from Dinwoody Canyon to the southwest and spread out on the floor of the Wind River Basin.

An extensive panorama of complex Paleozoic, Mesozoic, Tertiary, and Quaternary stratigraphy, structure, and geologic history can be seen. Figures 4 through 10 serve as guides for the following discussion. Beginning with the Wind River Mountains to the west, the bare part of the skyline is composed of northeast-dipping Madison Limestone. The black zone of trees marks the outcrops of Amsden and Tensleep Formations. The bare flatiron-shaped lower dip slopes are Phosphoria Formation. The red beds we have just passed are Chugwater Formation. The two ash localities are shown on Figure 4, as well as the overlying travertine and gravel surface that has been incised during the last 600,000 years.

Coming clockwise to the northwest is the panorama shown in Figure 5. The rocks on the far skyline are Precambrian granite and gneiss in the Union Peak area, part of the core of the Wind River Mountains. The cliff in the left center is the Nugget Sandstone shown in Figure 2. The prominent terrace about 200 ft (60 m) above Wind River is the Circle terrace of Blackwelder (1915). The Jurassic and Lower Cretaceous rocks just across Wind River to the northwest (Fig. 6) comprise one of the most colorful sections in this region that can be seen from the highway. In the distance, the candy-striped horizontal badlands are part of the Wind River Formation unconformably overlapping the Cretaceous and older rocks (Fig. 7). To the north and north-northeast on the skyline is our only glimpse of the Aycross Formation, of middle Eocene age, on the South Table (Fig. 8).

Just across Wind River is a series of northwest-trending thrust faults, the nearest of which is clearly visible. It is the Wind

Figure 5. View west-northwest from Red Grade, toward Wind River Mountains on skyline. Indicated are: pЄ = Precambrian rocks on Union Peak in the core of the Wind River Mountains; Pt = Tensleep Sandstone (Pennsylvanian); Pp = Phosphoria Formation (Permian); Tc = Chugwater Formation (Triassic); JTkn = Nugget Sandstone (Jurassic [?] and Triassic [?]); Jgs = Gypsum Spring Formation, Js = Sundance Formation, and Jm = Morrison Formation, all of Jurassic age. Photograph by J. D. Love, October 5, 1979.

Figure 7. View north-northwest from Red Grade. Indicated are Kmr = Mowry Shale (Lower Cretaceous), unconformably overlain by Twdr = Wind River Formation (Eocene). Thrust onto the Wind River is Kf = Frontier Formation (Upper Cretaceous). The horizontal T is on the upper thrust plate. Photograph by J. D. Love, October 25, 1979.

Figure 6. View northwest from Red Grade. Wind River Mountains are on left skyline. Indicated are Jm = Morrison Formation (Jurassic); Kcvr = lilac claystone zone in Cloverly formation, Kcvr = rusty beds member of Cloverly Formation, Kt = Thermopolis Shale, Kmd = Muddy Sandstone Member of Thermopolis Shale, and Kmr = Mowry Shale, all of Cretaceous age; and Twdr = Wind River Formation of early Eocene age. Photograph by J. D. Love, October 25, 1979.

Figure 8. Telephoto view north from Red Grade. Indicated are: Twdr = Wind River Formation (lower Eocene), overridden in the middle distance by Kf = Frontier Formation (Cretaceous). The Frontier is overlain by Kc = Cody Shale (Cretaceous). Tim = Indian Meadows Formation (lowest Eocene); Tac = Aycross Formation (middle Eocene); Twi = Wiggins Formation (middle and/or upper Eocene). South Table is also shown on the 1906 U.S.G.S. topographic map as Coulee Mesa and on the 1967 topographic map as Lower Table; it is known by some residents as South Mesa. The horizontal T is on the upper thrust plate. Photograph by J. D. Love, October 25, 1979.

Ridge thrust (Love, 1939), which placed the Frontier Formation and overlying basin facies of the Indian Meadows Formation (the dark ragged cliffs) on top of the horizontal Wind River. Thus, it is a post-early Eocene thrust, as are all the others in this general area. Black dumps of two coal mines are plastered across the Wind River Formation, but the coal is actually along the fault plane in the Frontier Formation. On the lumpy skyline to the right of the power pole in the foreground (Fig. 9) are several of 10

or more masses (klippen?) of Paleozoic rocks, chiefly Bighorn Dolomite and Madison Limestone resting on bright salmon-red strata of the Indian Meadows Formation and dull-gray Cody Shale, and overlain by more variegated strata of the Indian Meadows Formation. For a detailed map, see Love, 1939, Plate 17.

The far skyline peaks to the northeast are the Wiggins Formation in the Absaroka Range. This sequence overlaps the southeast end of the Washakie Range (Fig. 10), which is marked by

190   *J. D. Love*

Figure 9. Normal lens view north from Red Grade. Indicated on left skyline are masses (klippen?) of Paleozoic rocks, chiefly Bighorn Dolomite (Ordovician) and Madison Limestone (Mississippian) that were emplaced by thrusting or gravity sliding onto the Indian Meadows Formation (lowest Eocene) and then buried by more Indian Meadows strata. Intricate structure in the middle distance is shown on telephoto, Figure 8. Photograph by J. D. Love, October 25, 1979.

the high bare summit of Black Mountain. To the left of Black Mountain is Crow Creek Canyon, about 1000 ft (300 m) deep, cut in Precambrian granite. In the middle distance, the Coulee Mesa thrust puts gray Cody Shale on variegated Wind River and bright red Indian Meadows Formations. To the right of Black Mountain, on the lower skyline, all rocks are part of the Wind River Formation in the main Wind River Basin. To the south, the lumpy topography is on Pinedale moraines left by ice that emerged from Dinwoody Canyon in the Wind River Mountains.

REFERENCES

Blackwelder, E., 1915, Post-Cretaceous history of the mountains of central western Wyoming: Journal of Geology, v. 23, p. 97–117, 193–217, 307–340.

Keefer, W. R., 1970, Structural geology of the Wind River Basin, Wyoming: U.S. Geological Survey Professional Paper 495-D, 35 p.

Love, J. D., 1939, Geology along the southern margin of the Absaroka Range, Wyoming: Geological Society of America Special Paper 20, 137 p.

Love, J. D. and Love, J. M., 1983, Road log, Jackson to Dinwoody and return: Geological Survey of Wyoming, Public Information Circular 20, 34 p.

Love, J. D., Tourtelot, H. A., and others, 1947, Stratigraphic sections of Mesozoic rocks in central Wyoming: Wyoming Geological Survey Bulletin, v. 38, 59 p.

Figure 10. Panoramic view northeast from Red Grade, showing an intricate series of lower Eocene thrust faults and their relation to the Washakie Range and the Wind River Basin. Also shown is the overlapping of the upper and middle Eocene volcaniclastic rocks of the Absaroka Range onto Precambrian and Paleozoic rocks near the southeast end of the Washakie Range. Indicated are: Kf = Frontier Formation and Kc = Cody Shale, both of Cretaceous age; Tim = Indian Meadows Formation, Twdr = Wind River Formation, Tac = Aycross Formation, and Twi = Wiggins Formation, all of Tertiary age; Ob = Ordovician Bighorn Dolomite; and pC = Precambrian granite. The horizontal T is on upper thrust plate. Photograph by J. D. Love, October 25, 1979.

Wind River Canyon, Wyoming

Edwin K. Maughan, U.S. Geological Survey, MS 939, Box 25046, Denver Federal Center, Denver, Colorado 80225

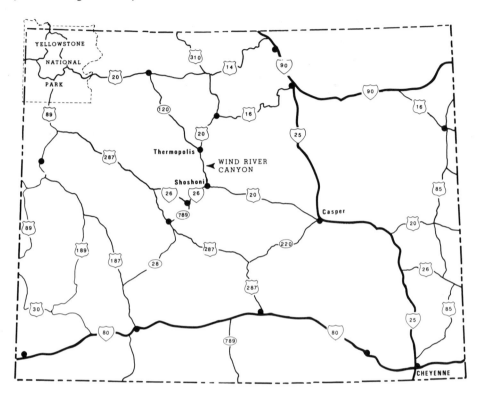

Figure 1. Map of Wyoming showing location of Wind River Canyon in relation to principal highways and towns.

LOCATION AND ACCESS

The Wind River Canyon transects the Owl Creek Mountains near the center of Wyoming. The canyon is in the northwest quadrant of the state (Fig. 1) and within the northeast corner of the Wind River Indian Reservation. U.S. 20 traverses the canyon from a point about 15 mi (24 km) north of the town of Shoshoni in the Wind River Basin to the mouth of the canyon about 5 mi (8 km) south of Thermopolis in the Bighorn Basin. The Wyoming Geological Association and the State Highway Department have erected signs that label the geological formations within the canyon; many of the features that will be noted are referenced to mileposts along the highway.

SIGNIFICANCE OF THE SITE

The Wind River has eroded a spectacular gorge that provides excellent exposures of Paleozoic rocks, Laramide-style tectonic structures, and other geologic features. Wind River Canyon, although much smaller, is one of the few places outside of the Grand Canyon in Arizona where a significant part of the North American geological record is well exposed, and it is very readily

accessible. The principal geologic features displayed here are: uplifting and faulting related to the Laramide orogeny, Archean metavolcanic rocks and related intrusions, the Cambrian-Precambrian unconformity, Paleozoic strata showing a variety of carbonate and siliciclastic facies deposited in a central area of the Wyoming shelf, and the feather-edge of an Upper Devonian unit between Ordovician and Mississippian rocks. Exposures of Mesozoic through Tertiary rocks adjacent to the canyon augment the stratigraphic record and provide a broad spectrum of geologic history within central Wyoming. The major geologic features are here briefly described in the context of a highway journey northward from Shoshoni to Thermopolis.

GENERAL SITE DESCRIPTION

The approach to Wind River Canyon from the south is via U.S. 20. Between Shoshoni and the head of the canyon, near Boysen, the highway crosses terraces and gravel-veneered pediment surfaces established on the lower Eocene Wind River Formation. A down-faulted remnant of the middle and upper Eocene

S

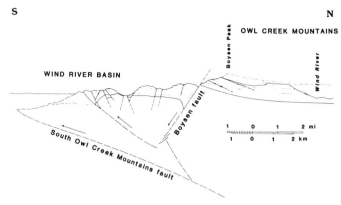

N

OWL CREEK MOUNTAINS

WIND RIVER BASIN

South Owl Creek Mountains fault

Boysen fault

1   0   1   2 mi
1   0   1   2 km

Figure 2. Cross section of the structural collapse complex at the south end of the Wind River Canyon adapted from Fanshawe (1939).

Wagon Bed Formation occurs north of the highway between mile 110.5 and milepost 111. These Eocene rocks represent braided stream and lacustrine volcaniclastic and terrigenous sediments derived from Laramide highlands to the west and north (the Absaroka Range and the Owl Creek Mountains) and deposited in syntectonically generated intermontane basins. Eventually, the superimposed Wind River not only eroded much of the soft Tertiary sediment that had filled the Wind River and Bighorn basins, but also cut the gorge in the well-indurated Paleozoic and Archean rocks of the exhumed Owl Creek uplift.

Archean rocks are exposed north of the Boysen fault and define the south edge of the northward-dipping Owl Creek Mountains block. North of the Boysen fault, the Paleozoic strata above the Precambrian rocks dip 4° to 10° into the Bighorn Basin (Fig. 2).

Boysen Area Structural Complex

The highway crosses the trace of the northward-dipping South Owl Creek Mountains fault (Keefer, 1970) between milepost 110 and approximately the boundary of the Wind River Indian Reservation (milepost 112). The uplifted Owl Creek Mountains block has been thrusted southward over the deepest part of the Wind River Basin along this fault (Fig. 2). There is about 33,000 ft (10,000 m) of structural relief between basin and uplift. The valley of the Wind River narrows progressively to the north from the reservation boundary as the river, in its approach to the Boysen fault, takes its course between a complex jumble of faulted and rotated structural blocks composed of Mesozoic and Paleozoic strata (Fig. 3).

Strata within individual fault blocks are increasingly older from south to north in the complex, so that younger Mesozoic rocks are adjacent to the South Owl Creek Mountains fault, and lower Paleozoic rocks are adjacent to the Boysen fault. The Boysen Dam, situated near the middle of the structural complex,

is tied at its north abutment into a small, down-faulted block of the Cambrian Gros Ventre Formation. Numerous, mostly normal, high-angle faults, which evidently are the result of basinward collapse of the thrusted lip of the Owl Creek Mountains block (Sales, 1983), separate the blocks within the complex. The collapse of a thrusted hanging-wall lip into an adjacent basin is a common feature of many Laramide uplifts in the Rocky Mountain region. Fanshawe (1939) has ably illustrated this classic example of structural collapse (reproduced, with slight modifications, as Fig. 2). Gravity sliding is also a component of the offset seen in the structures in the Boysen area (Wise, 1963).

The Boysen fault, which crosses the highway at mile 116.1 and marks the entrance to the narrow part of the canyon, dips 50° to 70° to the south. Stratigraphic displacement on this east-west–striking normal fault is some 2,000 ft (610 m) at the canyon, increasing to about 2,500 ft (760 m) east of the canyon.

Precambrian Rocks

Precambrian rocks occur north of the Boysen fault as part of the Owl Creek Mountains uplift. Mafic Archean amphibolites and schists are cut by dikes of a quartz monzonite intrusion. Exposures occur at road level northward from the Boysen fault at about milepost 116 to the unconformable contact with overlying Cambrian rocks at about mile 117.6. Similar rocks occur in several blocks of the overthrust collapse zone south of the fault (Fanshawe, 1939), but those exposures are now covered by waters of the Boysen Reservoir.

The Archean rocks are dominantly dark gray to greenish black hornblende amphibolite and biotite-quartz schists (Condie, 1967). The protoliths may have been fine- to medium-grained clastic sedimentary rocks (Condie, 1967), or tholeiitic basalts and interlayered high silica basalt and basaltic andesite (Mueller and others, 1985). This sequence contains the oldest rocks exposed in Wyoming; they were formed about 2,900 Ma and were metamorphosed at about 2,750 Ma. The light-colored, pinkish gray to white intrusive dikes and pegmatites, which may be part of a larger intrusive body (Condie, 1967), are primarily quartz monzonite (Gwynne, 1938) that contrasts markedly with the darker rocks.

The Great Unconformity

Archean and Proterozoic rocks have been erosionally beveled and are unconformably overlain by the Cambrian Flathead Sandstone. The unconformable contact is at road level in the vicinity of mile 117.6 and dips gently northward. Some 1.5 to 3 ft (0.5 to 1 m) of weathered granite is found along this segment of the contact. Locally, there is little evidence of topographic relief on the Precambrian surface except for slight thinning of the basal strata of the Flathead. However, regional thicknesses of the basal Cambrian beds vary considerably, owing to deposition on the cratonic shelf where the Precambrian surface had been reduced to a gently undulating plain and scattered monadnocks.

Figure 3. Aerial view looking north into Wind River Canyon reproduced from Keefer and Van Lieu (1966). Boysen Dam and north end of Boysen Reservoir are at bottom of photograph. PPu, Park City, Tensleep, and Amsden formations; MOu, Madison Limestone and Bighorn Dolomite; Єu, Gallatin Limestone, Gros Ventre Formation, and Flathead Sandstone; pЄr, Precambrian igneous and metamorphic rocks; f–f, Boysen normal fault. Faulted complex of Paleozoic and Mesozoic rocks occupies lower half of view. Photograph courtesy of P. T. Jenkins and L. P. House.

Paleozoic Rocks

A view of a nearly complete sequence of the regional Paleozoic stratigraphy occurs opposite mile 117.6 in the northwest wall of the canyon. The upper part of the Paleozoic sequence (Pennsylvanian and Permian strata) is not readily seen in this vista; these strata are better viewed farther north in the canyon. The Paleozoic rocks (Fig. 4), which dip northward into the canyon, are briefly described where they are exposed at road level.

Flathead Sandstone. The Middle Cambrian Flathead Sandstone (mile 117.6–118.3) is at the base of the Paleozoic sequence in this region. The Flathead is an arkosic and micaceous quartz arenite, conglomeratic near the base with angular to subangular clasts of quartz, feldspar, and lithic fragments. Siltstone beds are common, especially near the top of the sequence. Glauconite grains occur, and they are common in the upper part of the formation where some beds are colored gray or gray-green by the mineral. Overall, the sandstone is yellowish gray to pale orange in color, but some hematitic beds are colored pinkish gray to dark reddish brown. The thin-bedded formation forms ragged ledges or cliffs. Thickness (about 260 ft; 80 m) varies somewhat and the

Flathead pinches out against topographic relief on the Precambrian surface nearby (Thaden, 1976). The Flathead is a classic example of sandstone deposited in lower foreshore and nearshore subtidal environments during eastward transgression of the Cambrian sea (Keefer and Van Lieu, 1966).

Gros Ventre Formation. The Gros Ventre Formation, of Middle to Late Cambrian age, conformably overlies the Flathead. The contact between the two formations is at mile 118.3. The Gros Ventre is dominantly medium to dark greenish gray, moderately fissile, micaceous and glauconitic mudstone. Argillaceous and silty sandstone occurs in the lower part of the formation, whereas calcareous mudstone and argillaceous limestone are common in the upper part. Flaser bedding, low-amplitude ripples on some bedding surfaces, and horizontal burrows and tracks on bedding planes indicate that deposition was in a shallow-water, offshore marine environment. The Gros Ventre generally forms slopes, but at many places along the steep-walled Wind River Canyon, the more resistant sequences form ragged cliffs.

Gallatin Limestone. The contact of the Gros Ventre Formation with the overlying Gallatin Limestone is gradational; this contact is obscured at road level by talus and colluvium. An exposure of the cliff-forming Gallatin Limestone occurs at mile 120.5. This Late Cambrian formation comprises medium gray to greenish gray, thin-bedded to laminated limestone, argillaceous limestone, and silty calcareous claystone. Glauconite is common in the lower part, whereas oolitic limestone beds and finely crystalline intraclastic flat-pebble conglomerate are common in the upper part. Many of the silty beds are micaceous. The lower part of the Gallatin stands as a banded cliff between the Gros Ventre and argillaceous and silty beds of the upper Gallatin, which closely resembles the upper part of the Gros Ventre. Thickness of the Gallatin is about 450 ft (140 m), but is variable due to erosion along the unconformable contact with the overlying Bighorn Dolomite. Deposition was generally in quiet offshore marine water near wave base; but rip-up clasts, present as intraformational flat-pebble conglomerate, testify to occasional periods of greater bottom agitation (storms or low sea-level stands).

Bighorn Dolomite. The Middle and Late Ordovician Bighorn Dolomite forms the reticulated weathered lower part of the conspicuous carbonate rock cliffs in the canyon (mile 123.6–124.4). This formation is massive, mottled light to medium gray, medium to coarsely crystalline dolomite. The contact with the underlying Gallatin Limestone is sharp, and the basal beds are generally sandy and conglomeratic, containing clasts derived from the underlying beds. Thickness of the Bighorn generally increases from south to north and ranges from 85 to 140 ft (26 to 43 m). Deposition occurred in a regionally extensive shelf environment.

Darby Formation. The contact between the Bighorn Dolomite and the overlying Madison Limestone is unconformable; it occurs through most of the canyon within a slight, tree-covered shelf along the carbonate rock cliff. However, for about 0.5 mi (0.8 km) north from milepost 124, the Upper Devonian Darby Formation intervenes between the Madison and the Bighorn. This marks the depositional/preservational limit of Devonian rocks in central Wyoming. At mile 124.4, the unit consists of 3.3 ft (1 m) of greenish gray clayey to sandy dolomitic siltstone. It is lithologically similar to the Darby Formation (southwest in the Wind River Basin) and to the Jefferson Formation (to the northwest in the Wind River Basin).

Madison Limestone. The upper part of the conspicuous carbonate cliffs, which dominate the Wind River Canyon, is composed of the Mississippian Madison Limestone. The Madison consists of thin to massive beds of limestone and dolomitic limestone, interbedded with dolomite. Thickness ranges from about 440 to 550 ft (135 to 170 m). Initially, deposition occurred in deepening water during the Mississippian marine transgression; but shallow-water deposits in the upper part of the formation resulted as carbonate sediments prograded across the shelf. The uppermost beds (the Bull Ridge Member) include oolitic and stromatolitic dolomitic limestone and evidence of dissolved, evaporite beds that indicate intertidal to supratidal deposits within a shallowing shelf environment. The beds in the uppermost 3 ft (1 m) of the Madison Limestone are well exposed in a tributary canyon about 1 mi (1.6 km) southwest of milepost 122;

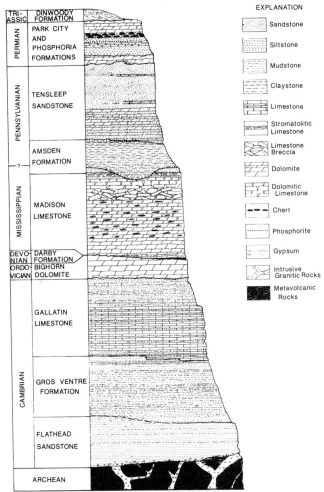

Figure 4. Stratigraphic column of Paleozoic rocks in Wind River Canyon, Wyoming.

at this point interbedded sandstone and limestone indicate that the contact is conformable and transitional into the overlying Darwin Sandstone Member of the Amsden Formation (Maughan, 1972). At road level, the top of the Madison may be seen in a railroad cut opposite mile 125. 9.

Amsden Formation. The Amsden Formation forms slopes between the cliffs of the Madison below, and of the Tensleep Sandstone above. The Amsden contains three lithologically distinct members: the Darwin Sandstone Member at the base, and the Horseshoe Shale and Ranchester Limestone members successively above it.

The Upper Mississippian Darwin Member is fine- to medium-grained, mostly cross-bedded sandstone. This member is commonly missing, however, and the Madison is generally overlain by the Horseshoe Member. The Darwin Sandstone was deposited chiefly as eolian dunes in part of a sabkha complex that succeeded the carbonate shelf deposits of the Madison.

The Lower and Middle Pennsylvanian Horseshoe Shale Member consists of 100 to 200 ft (30 to 60 m) of moderate red to reddish orange mudstone that is conglomeratic at its base along the erosional unconformity that separates it from Mississippian rocks. The conglomerate incorporates rounded chert, limestone, and sandstone pebbles derived from the underlying Madison and Darwin. The Horseshoe represents terra rosa resedimented in peritidal environments during transgression of the Pennsylvanian sea. The upper part of the Horseshoe, where it is thick in the canyon, includes thin limestone beds that are transitional into the overlying Ranchester Limestone Member; however, the Ranchester is not otherwise present in Wind River Canyon, and the Horseshoe, also, is locally missing along the erosional unconformity beneath the Tensleep.

Tensleep Sandstone. The Middle Pennsylvanian Tensleep Sandstone comprises an upper sandy member that forms a prominent, near-vertical sandstone cliff and a lower dolomitic member that forms an underlying slope of sandy dolomite. The dolomitic member is generally poorly exposed, though it may be seen between mile 126.2 and 126.6. The top of the sandy member occurs at milepost 127. The Tensleep unconformably overlies the Amsden Formation and locally overlies the Madison Limestone.

The dolomitic member consists of thin to medium beds of yellowish gray silty and sandy dolomite, dolomitic sandstone, and greenish gray mudstone. Sand content increases upward, until there is an abrupt, but conformable, transition into the overlying sandy member. Thickness of the lower member ranges from 100 to 275 ft (30 to 85 m). The dolomitic member, in its thin-bedded aspect, resembles the Ranchester and commonly has been incorrectly identified as such (Maughan, *in* Lageson and others, 1979). The upper member consists of a yellowish gray, medium to massive and commonly cross-bedded sandstone in low-angle, wedge-planar sets. Some cross-beds are high angle and may be contorted. A few thin and medium beds of sandy dolomite also occur. Thickness of the unit ranges from 195 to 290 ft (60 to 90 m). Deposition occurred in a shallow-water marine environment in the lower member and ranged into eolian dune and interdune pond environments as sand, transported from north to south, prograded across the shallowly submerged shelf.

Park City and Phosphoria Formations. The uppermost strata at the rim are the intertongued beds of the Permian Park City and Phosphoria Formations. These compose a cliff- and slope-forming sequence. The lower part is cliff-forming sandy dolomite of the Lower Permian (Leonardian) Grandeur Member of the Park City Formation, which is only subtly distinct from the underlying Tensleep. The Grandeur Member, about 40 ft (13 m) thick, lies above an erosional unconformity on the Tensleep with as much as 10 ft (3 m) of relief locally. The rocks are mostly yellowish gray, sandy and cherty dolomite, and contain scattered, broken shells, angularly fragmented chert, and locally abundant sand derived from erosion of the Tensleep. Deposition probably occurred during storm surge deposits in intertidal to supratidal environments. Above an erosional surface, the Grandeur is overlain in ascending sequence by the Meade Peak Phosphatic Shale and Rex Chert Members of the Phosphoria Formation, the Franson Limestone Member of the Park City Formation, the Retort Phosphatic Shale and Tosi Chert Members of the Phosphoria Formation, and the Ervay Limestone Member of the Park City Formation. The lower members of this sequence are not exposed at road level. The Retort Member (1.5 ft; 0.5 m thick) and the Tosi Member (12 ft; 4 m thick) are slope-forming units, which are poorly exposed at mile 127.6; they are overlain by the Guadalupian Ervay Limestone Member, which forms a near-vertical, 55-ft (17-m) cliff of light gray dolomitic limestone and dolomite. Oolitic and algal stromatolitic beds in the upper part of the Ervay indicate a shoaling-upward carbonate shelf sequence and deposition in probable intertidal and supratidal environments. Silcrete at the top surface of this limestone member evidences weathering during the Late Permian. The Ervay forms a spectacular dip slope on the north flank of the Owl Creek Mountains.

Mesozoic Rocks

At the point where the Wind River flows from the canyon into the Bighorn Basin, it becomes the Bighorn River. The change in name, at mile 127.7, is poetically known as the Wedding of the Waters. The highway crosses the Bighorn River at mile 128.3 in a valley at the southern edge of the Bighorn Basin. Mesozoic rocks are moderately well exposed in the bluffs along the valley walls; they include the Triassic Dinwoody Formation, Red Peak Formation, Crow Mountain Sandstone, and Popo Agie Formation (the Red Peak to Popo Agie sequence is also known as the Chugwater Formation or Group). Jurassic rocks are present in the Gypsum Springs, Sundance, and Morrison Formations; Cretaceous rocks occur beyond the Jurassic exposures and are not readily seen from the highway. The strata dip northward to an east-west normal fault at the north edge of the town of Thermopolis. This fault defines the northern limit of the Owl Creek Mountains structural block and it localizes groundwater that issues at the Bighorn Hot Springs.

REFERENCES CITED

Condie, K. C., 1967, Petrologic reconnaissance of the Precambrian rocks in Wind River Canyon, central Owl Creek Mountains, Wyoming: Wyoming University Contributions to Geology, v. 6, no. 2, p. 123–129.

Fanshawe, J. R., 1939, Structural geology of Wind River Canyon area, Wyoming: American Association of Petroleum Geologists Bulletin, v. 23, no. 10, p. 1439–1492.

Gwynne, C. S., 1938, Granite in the Wind River Canyon: Geological Society of America Bulletin, v. 49, no. 9, p. 1417–1424.

Keefer, W. R., 1970, Structural geology of the Wind River Basin, Wyoming: U.S. Geological Survey Professional Paper 495-D, 35 p.

Keefer, W. R., and Van Lieu, J. A., 1966, Paleozoic formations in the Wind River Basin, Wyoming: U.S. Geological Survey Professional Paper 495-B, 60 p.

Lageson, D. R., Maughan, E. K., and Sando, W. J., 1979, The Mississippian and Pennsylvanian (Carboniferous Systems) in the United States—Wyoming: U.S. Geological Survey Professional Paper 1110-U, 38 p.

Maughan, E. K., 1972, Geologic map of the Wedding of the Waters Quadrangle, Hot Springs County, Wyoming: U.S. Geological Survey Geologic Quadrangle Map GQ-1042, scale 1:24,000.

Mueller, P. A., Peterman, Z. E., and Granath, J. W., 1985, A bimodal Archean volcanic series, Owl Creek Mountains, Wyoming: Journal of Geology, v. 93, p. 701–712.

Sales, J. K., 1983, Collapse of Rocky Mountain basement uplifts, *in* Lowell, J. D., ed., Rocky Mountain foreland basins and uplifts: Denver, Colorado, Rocky Mountain Association of Geologists, p. 79–97.

Thaden, R. E., 1976, Geologic map of the Birdseye Pass Quadrangle, Fremont and Hot Springs counties, Wyoming: U.S. Geological Survey Geologic Quadrangle Map GQ-1537, scale 1:24,000.

Wise, D. U., 1963, Keystone faulting and gravity sliding driven by basement uplift of Owl Creek Mountains, Wyoming: American Association of Petroleum Geologists Bulletin, v. 47, no. 4, p. 586–598.

Hoback River Canyon, central western Wyoming

John A. Dorr, Jr., *Department of Geological Sciences, University of Michigan, Ann Arbor, Michigan, 48109*
Darwin R. Spearing, *Marathon Oil Company, Cody, Wyoming 82414*
James R. Steidtmann, *Department of Geology and Geophysics, University of Wyoming, Laramie, Wyoming 82070*
David V. Wiltschko and John P. Craddock, *Center for Tectonophysics, Texas A & M University, College Station, Texas 77843*

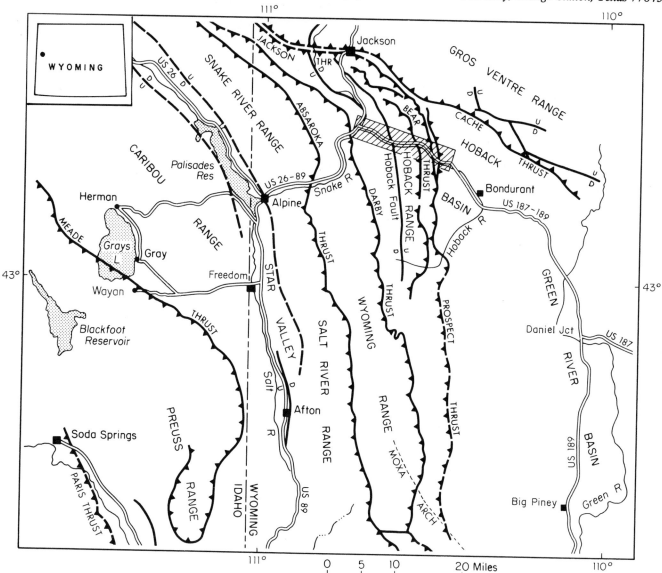

Figure 1. Map of major regional features. Site cross-hatched.

INTRODUCTION

Geologic features along the Hoback River Canyon in central western Wyoming illustrate tectonic effects and stratigraphic and structural evidence for dating thrusts and later listric normal faults in part of the Idaho-Wyoming Thrust Belt and adjacent Foreland. Figure 1 and Table 1 combine to show that major thrusts in the Thrust Belt young in the direction of tectonic transport, in this case from west to east, the oldest being the Paris Thrust and the youngest the Prospect. The preeminent features, logged below

and visible from the highway (Fig. 2) are, west to east, the Hoback Fault (listric normal), Bear Thrust, Prospect (=Cliff Creek= Jackson) Thrust and Game Hill Thrust. These are capitalized on Table 1.

The site description begins on the west at Hoback Junction, 13 mi (21 Km) south of Jackson, Wyoming, on U.S. 187-189. It follows U.S. 187-189 east from there for 16 mi (26 km) with four stops enroute (Fig. 2). It crosses from south-central Teton

TABLE 1.   DIRECTION OF MOVEMENT OF MAJOR THRUSTS
                    IN THE THRUST BELT

```
===============================================================
WEST (Overthrust Belt)                    EAST (Foreland)
                                                          ↑
┬─HOBACK FAULT.
↓      Listric normal.   Late Miocene-Pliocene
       (Episodic).

--►BEAR THRUST.
       Imbricate splay from Prospect Thrust.    Late
       Early Eocene.
     Cache Thrust.
       Earliest Eocene.     Gros Ventre Range.          ◄---
--►PROSPECT THRUST (and possibly late movement on
       Darby Thrust).
       Paleocene-Eocene boundary.
     GAME HILL THRUST. Initial movement.   Middle        ◄---
       Paleocene.
     Gros Ventre Range.                                   ↑
       Initial independent uplift, early part of Late
       Paleocene.
--►Darby Thrust.
       Initial movement.   Middle Paleocene.
--►Absaroka Thrust.   Late, minor movement. Cretaceous-
       Paleocene boundary.
--►Absaroka Thrust.
       Major movement.   Late Campanian (Late
       Cretaceous).
     Ancestral Teton-Gros Ventre-Wind River Uplift.   ↑
       Mid-Campanian (Late Cretaceous).
↑  Moxa Arch formed.
       Late Santonian (Late Cretaceous).
--►Absaroka Thrust.
       Early, minor movement. Mid-Santonian (Late
       Cretaceous).
     Targhee-Blacktail-Snowcrest Uplift in northwest  ↑
     Wyoming and southwest Montana.
       Early Santonian (Late Cretaceous).
--►Meade and Crawford thrusts.
       Coniacian (early in the Late Cretaceous).
     Paris Thrust.
       Several movements in Idaho:
--►      Final movement. Mid-Turonian (Late Cretaceous).
--►      Renewed movement. Late Albian (Early
           Cretaceous).
--►      Renewed movement. Early Albian (Early
           Cretaceous).
--►      Renewed movement. Aptian (Early Cretaceous).
--►      Initial movement. Jurassic-Cretaceous boundary.
───────────────────────────────────────────────────
```

County into northwestern Sublette County within U.S. Geological Survey Camp Davis and Bull Creek 7½-minute topographic and geologic map quadrangles. Small-scale colored geologic map and cross-section sheets, which include the site, are in the pocket at the back of Dorr and others, 1977. A detailed road log of this area is given by Spearing and Steidtmann (1977).

Stop 1. Hoback Fault (5.7 mi [9 km] From Hoback Junction). Turn left into parking area. This stop is nearly on the trace of the north-south trending Hoback Normal Fault. The west side was downdropped about 10,000 ft (3050 m) in Late Miocene and Pliocene, and movement was recurrent. High angle at surface, the fault flattens at depth. Seismic evidence (Royse and others, 1975) indicates the fault is listric normal with rotational movement, in part, along the preexisting Bear and Prospect thrusts (Fig. 3). More than 4,000 ft (1200 m) of conglomerates of the Camp Davis Formation, eroded from upthrown block, filled the trough created by faulting. A single fossil tooth of the late Miocene and early Pliocene horse *Pliohippus* sp., from about 200 ft (61 m) above the base of the formation is the only indicator of

age found thus far. However, this is sufficient to show that the Hoback Fault was first activated in late Miocene or early Pliocene. During normal faulting, the Hoback and Gros Ventre ranges were coupled together, in spite of their previously distinct Laramide structural history, and behaved as an integrated block along the fault, standing high relative to the Camp Davis trough to west. Erosion of the higher block began on terrain of Tertiary, Mesozoic, and Late Paleozoic strata, which provided coarse clasts to the lower member of Camp Davis Formation. The Precambrian was breached in the Gros Ventre Range by Late Camp Davis time, providing some clasts to the upper part of the formation. Continued downward rotation of the graben, and damming by volcanism to north, ponded water against the fault to produce the middle member of the formation. Continued faulting and rotation reversed the original westward depositional dip of lower beds, tilting them downward toward fault. Relative upward displacement of the Hoback Range along the fault produced steep downdrag toward the west in strata along the face of the range. Farther north along the fault (not in view from this site) in the Horse Creek area, large gravity slide blocks were detached from these downdragged beds and slid into the trough where, as deposition continued, they were buried and overlapped by conglomerate. Slide blocks piled up in reverse stratigraphic order as erosional unroofing of the fault scarp and adjacent range progressed. Continued deposition of conglomerate, after cessation of faulting, overlapped the Camp Davis Formation across the fault trace, concealing the latter in most places. In the high cliffs across the river just east of this site, the Darby (Devonian) and Madison (Mississippian) limestones and dolomites are complexly fractured, and faulted on a small scale, probably because they went over a ramp (Fig. 3) in the Prospect Thrust during much earlier Laramide thrusting.

Stop 2. Bear Thrust (9.2 mi [14.8 km] from Hoback Junction). Turn left off highway into parking area. Across the river, the Tensleep Sandstone high on the hill is thrust over highly deformed Triassic redbeds along the Bear Thrust. A leading-edge anticline in the Tensleep can be seen on the south side of the canyon approximately 0.5 mi (0.8 km) west. The Bear Thrust is a subsidiary thrust slice within Prospect (=Cliff Creek= Jackson) overthrust plate. No Tertiary deposits have been found involved with or derived from the Bear Thrust; therefore its age is uncertain. However, its geometric similarity to, and possible southward continuation into the Lookout Mountain Thrust, which has been dated about 15 mi (24 km) farther south, suggest it is younger

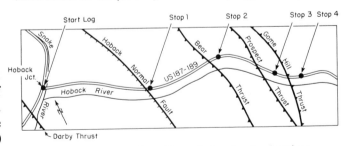

Figure 2. Site enlargement from Figure 1, showing stop locations.

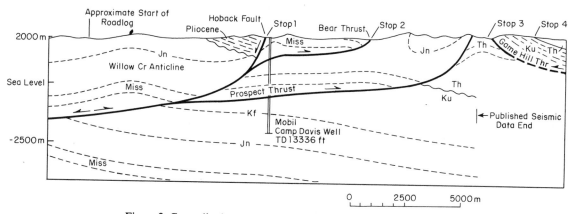

Figure 3. Generalized cross section, in part from seismic and well data, with roadlog stops shown. (In part from Royse and others, 1975).

than the Prospect Thrust (i.e., latest Paleocene or Early Eocene; see Dorr and Steidtmann, 1977; Wiltschko and Dorr, 1983, p. 1306, 1313).

Stop 3. Prospect and Game Hill Thrusts (14.4 mi [23.2 km] from Hoback Junction). Turn left into the flat meadow and look back north along the highway to Battle Mountain, which is the low, half-dome hill with a steep cliff of sandstone and truncated right (east) face. Beneath the grassy hills to the right (east) of Battle Mountain and beneath Battle Mountain itself is the Late Paleocene Hoback Formation, dated there by fossil mammals (Dorr and others, 1977; Wiltschko and Dorr, 1983). The Hoback Formation was overridden by the Prospect (=Cliff Creek=Jackson) Thrust here. About 15 mi (24 km) to the south, the same thrust is overlapped by earliest Eocene synorogenic deposits also dated by fossil mammals (Dorr and Steidtmann, 1977). This pair of relationships establishes the age of the Prospect thrust here as falling on the Paleocene-Eocene boundary.

Across the river on hills east of this site are Upper Cretaceous sandstones and shales. Stop 3 is approximately on the trace of the Game Hill Thrust (Figs. 2, 4 and 5). This fault dips east and brings Cretaceous rocks up on the east against the younger, Late Paleocene Hoback Formation on the West. The Prospect Thrust here collided with the upthrown block of the Game Hill Thrust, which had risen slightly earlier (Table 1). The Prospect Thrust is the frontal thrust of the thrust belt in this area. The Game Hill Thrust cuts beds within the Hoback Basin just east of the Prospect Thrust. Several lines of evidence suggest that the Game Hill Thrust predates the Prospect Thrust. First, between the junction of Little Cliff and Cliff creeks on the north, and Sandy Marshall Creek on the south, the Prospect Thrust trace is offset by several tear faults. If the Game Hill Thrust had formed after the Prospect Thrust and had cut trace of the latter, those tears would now be truncated; but they are not. Second, Cretaceous beds on the upthrown side of the Game Hill Thrust dip away from, rather than into the fault. This anomalous east dip on the upthrown side is not due to drag along the fault, but can be accounted for by assuming that uplift on the Game Hill Thrust produced a buttress against which the Prospect Thrust rode, truncating beds in Game

Hill Thrust block and reversing their dip. Third, between Little Cliff Creek and Sandy Marshall Creek, beds in the Prospect Thrust plate turn steeply upward and are highly brecciated and sheared where they struck the uplifted Game Hill block. The Game Hill Thrust is the earliest major fault that involves Tertiary and older rocks within the Hoback Basin. The east dipping, high angle reverse attitude of the fault plane at the surface indicates a response to forces similar to those that acted along the Cache Thrust in the Gros Ventre Range farther east. Because the Game Hill Thrust preceded the Prospect Thrust, but involved the Paleocene Hoback formation, it appears to be closely related in age to the Cache Thrust. The Game Hill Thrust created a buttress against movement on the Prospect Thrust that accounts for a great amount of folding and tearing of beds in the front of the Prospect Thrust plate, just as the Gros Ventre uplift produced a buttress which deflected thrusts farther north. Rotations in the Prospect Thrust plate were caused by these impeding buttresses as shown by paleomagnetic rotation data (Grubbs and Van der Voo, 1976; Wiltschko and Dorr, 1983, p. 1319).

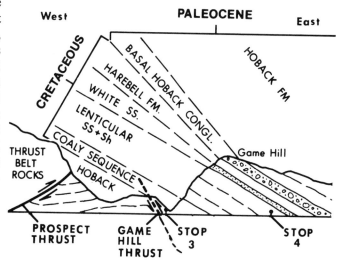

Figure 4. Generalized cross section for *Stops 3* and *4*, looking north.

Figure 5. Aerial view of Stop 3 area (upper photo) and closer view of the Battle Mountain area showing sites that yielded fossil mammals (lower photo). *BM,* Battle Mountain; *CC,* Cliff Creek; *PT,* Prospect Thrust; *CF,* Cache Thrust; *CV,* Little Granite–Cache Creek Valley; *GH,* Game Hill; *GHT,* Game Hill Thrust; *GVR,* Gros Ventre Range; *HR,* Hoback River; Hoback Range (skyline); *Jn,* Jurassic Nugget Sandstone; *Ku,* Upper Cretaceous strata; *Th,* Hoback Formation (Late Paleocene); *F,* Late Paleocene fossil mammal localities (21 and 22) in Hoback Formation at Battle Mountain.

Stop 4. Upper Cretaceous–Paleocene Boundary (16.6 mi [26.7 km] from Hoback Junction). Pull off into parking area on right. This site is at the western edge of the physiographic Hoback Basin that forms the northern tip of the Green River structural basin (Fig. 1). To the southwest, shales of the Upper Cretaceous Harebell Formation are exposed in an erosional scar near the skyline. White sandstone beneath this shale forms the tree-covered dip slope to the right of the scar. The basal conglomerate of the Paleocene Hoback Formation is on the left side of the scar. Just to the east, this same basal Hoback conglomerate outcrops at road level and continues to the top of the hill. Sandstones and shales of the Hoback Formation form hills to the east.

For the next 17 mi (27 km) southward to the "Rim", which is the topographic divide between the Hoback and Green River physiographic basins, the highway passes upward through exposures of the Paleocene Hoback Formation, Early Eocene Chappo Formation, and Early Eocene Pass Peak formation. Details of the geology are in Dorr and others, (1977; especially see map in pocket). Numerous well-developed Pleistocene terraces also are visible.

REFERENCES CITED

Dorr, J. A., Jr., and Steidtmann, J. R., 1977, Stratigraphic-Tectonic implications of a new, earliest Eocene, mammalian faunule from central western Wyoming: Wyoming Geological Association, Guidebook, 29th Annual Field Conference, p. 327–337.

Dorr, J. A., Jr., Spearing, D. R., and Steidtmann, J. R., 1977, Deformation and deposition between a foreland uplift and an impinging thrust belt, Hoback Basin, Wyoming: Geological Society of America Special Paper 177, 82 p. (Geologic map and cross sections in pocket.)

Grubbs, K. L., and Van der Voo, R., 1976, Structural deformation of the Idaho-Wyoming overthrust belt (U.S.A.) as determined by Triassic paleomagnetism: Tectonophysics, v. 33, p. 321–336.

Royse, F., Jr., Warner, J. A., and Reese, D. L., 1975, Thrust belt structural geometry and related stratigraphic problems, Wyoming-Idaho–northern Utah *in* Bolyard, D. W., ed., Deep drilling frontiers of the central Rocky Mountains, Rocky Mountain Association of Geologists, Symposium Volume: p. 41–54.

Spearing, D. R., and Steidtmann, J. R., 1977, Front of thrust belt to foreland: Wyoming Geological Association, 29th Annual Field Conference, Guidebook. Roadlogs for Trip Number One, p. 763–767.

Wiltschko, D. V., and Dorr, J. A., Jr., 1983, Timing of deformation in Overthrust Belt and Foreland of Idaho, Wyoming and Utah: American Association of Petroleum Geologists Bulletin, v. 67, no. 8, p. 1304–1322.

Type Pinedale Till in the Fremont Lake area, Wind River Range, Wyoming

Gerald M. Richmond, U.S. Geological Survey, Box 25046, Denver Federal Center, Denver, Colorado 80225

LOCATION AND ACCESSIBILITY

The Fremont Lake area is at the southwest base of the Wind River Range, northeast of the town of Pinedale in Sublette County, Wyoming, (Fig. 1). It is included in the Fremont Lake South Quadrangle (scale 1:24,000) for which a topographic map, a geologic map (Richmond, 1973), and a soils map (Sorenson, 1986) are available.

The area is accessible from Pinedale on U.S. 191, about 100 mi (160 km) north of Rock Springs and 70 mi (112 km) southeast of Jackson. County Road 23-111 (paved) leads northeast from Pinedale into the area. Most of the terrain is in Bridger National Forest, but some is private land. All maintained roads are accessible by passenger car. Many unimproved roads require four-wheel drive. Obtain permission to enter gated roads. Respect signs; close all gates unless found open; and do not drive into areas under irrigation.

For a short tour, take County Road 23-111 from the east end of Main Street in Pinedale toward Fremont Lake. The road leads up a terrace front, passes a cemetery, and extends northeast across gently rolling Bull Lake end moraines to an S-curve 2.2 mi (3.5 km) from U.S. 191. Here, the road ascends the steep, very bouldery front of the outermost Pinedale end moraine. Park at the bottom and walk 50 yd (46 m) up the road to a cut where Pinedale Till is well exposed (locality A, Fig. 1). Drive on to the moraine crest, from which an outstanding overview of Fremont Lake, the enclosing Pinedale moraines, and the mountains is available. Beyond the crest, the road descends into the lake basin where two well-marked, paved side roads lead to the foot of the lake and to a campground along its eastern side. Paved road 23-111 continues northward between lateral moraines to a side road east to Half Moon Lake. From that junction, it ascends to a winter sports area at Surveyor Park and, 4.3 mi (6.9 km) farther on, to Elkhart Park and campground (altitude 9,500 ft; 2,900 m), from which trails lead into the Bridger Wilderness.

To compare the surface features and soils of Pinedale Till with those of Bull Lake Till (see text), return to the foot of the S-curve and take the gated road east 0.3 mi (0.5 km) to the Pinedale town dump (locality C, Fig. 1; open 10 A.M. to 6 P.M., closed Tuesdays and Wednesdays). Bull Lake Till is also well exposed in a washout along Highland Ditch (map locality B). Contact U.S. Soil Conservation Service in Pinedale to determine accessibility.

SIGNIFICANCE

The Fremont Lake area is one of several areas of glacial deposits described as typical Pinedale drift by Blackwelder (1915), now Pinedale Till (Richmond, 1964). It is the most accessible of those areas, and nearest to the town of Pinedale, from which the name of the till is derived.

GEOLOGIC SETTING

Fremont Lake is enclosed by both Bull Lake (outer) and Pinedale (inner) nested marines (Fig. 1). It is 9 mi (14.6 km) long and lies in a glacial basin, the southern two-thirds of which is excavated in the Wasatch and Green River formations of Eocene age. The Wasatch Formation is overlapped by a boulder conglomerate of younger Tertiary age that is well exposed on the west slope of Half Moon Mountain. The northern third of the lake basin is in a deep canyon cut in Precambrian rocks, against which the Tertiary formations are downfaulted. The canyon heads in a broad glaciated upland on which an ice cap existed during Pinedale and earlier glaciations. A tongue of ice extended from the cap down the canyon into the basin of Fremont Lake. Here it deposited end moraines that outline a maximum advance and a succession of six rejuvenations during recession of the ice from the piedmont. Deposits of younger Wisconsin glacial readvances on the uplands and in the cirques are not discussed here.

NOMENCLATURE

The terms Pinedale drift and Pinedale stage were assigned by Blackwelder (1915, p. 324) to "the youngest moraines . . . around each of the large lakes near Pinedale on the southwest side of the Wind River Range" and to deposits elsewhere in northwest Wyoming. He compared these deposits (Blackwelder, 1931) to deposits of the former Wisconsin glacial stage in Iowa and Illinois. The Wisconsin has been redefined to include older deposits. However, the Pinedale has not been redefined, nor has the existence of end moraines of middle or early Wisconsin age been proven in areas of Pinedale Till. The lithostratigraphic term Pinedale Till and the climate-stratigraphic term Pinedale Glaciation were introduced by Richmond (1964). Climate-stratigraphic terminology is now abandoned, but the term Pinedale glaciation is still useful as an informal nonstratigraphic event. Pinedale Till is a valid lithostratigraphic unit.

Three informal subdivisions of Pinedale Till have been proposed (Richmond, 1964): a "lower till" (early stage), represented by the outermost end moraine; a "middle till" (middle stage), represented by deposits of a group of succeeding end moraines; and an "upper till" (late stage), represented by the innermost end moraine. This informal nomenclature is here abandoned in favor of local individual end moraine designations because of problems in identifying and correlating moraines from range to range, because of confusion between early, middle, and late Pinedale and early, middle, and late Wisconsin, and because no criteria for lithostratigraphic subdivision of Pinedale Till are presently available in the Fremont Lake area.

STRATIGRAPHIC POSITION

Pinedale Till in the Fremont Lake area most commonly overlies Precambrian or Tertiary rocks and local, pre-Pinedale

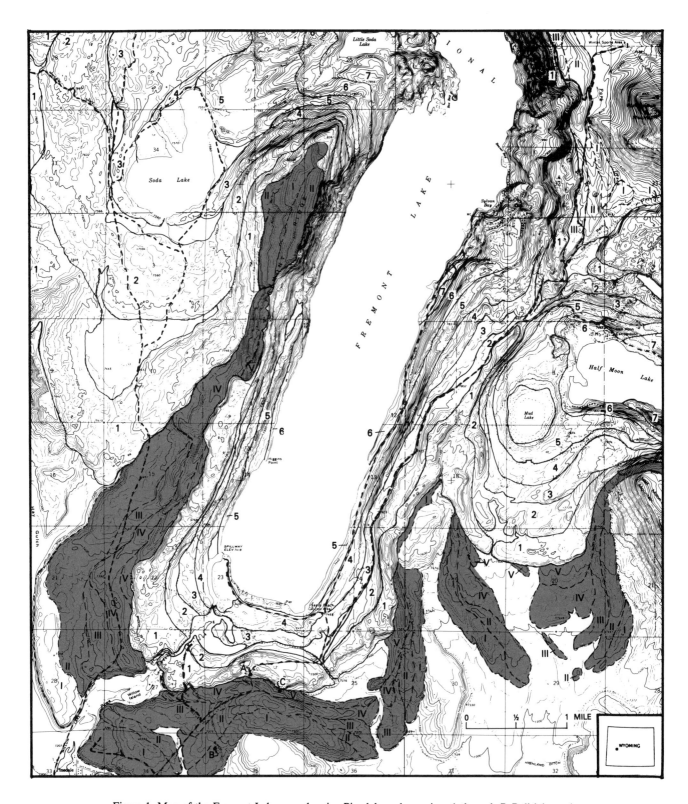

Figure 1. Map of the Fremont Lake area showing Pinedale and moraines 1 through 7, Bull lake end moraines I to V, localities A to C, and access road (heavy short-dashed lines). Modified after Richmond (1973).

lacustrine sand. In places, Pinedale end moraines override and crosscut Bull Lake end moraines (Fig. 1). In the Cora Quadrangle (adjacent to the west), a soil developed on Bull Lake Till was temporarily exposed beneath till of the outermost Pinedale end moraine.

Pinedale Till is overlain by younger colluvium and loess. It is also trenched by stream channels containing late glacial and postglacial alluvium.

TILL CHARACTERISTICS

Surface Form. The Pinedale end moraines nested around Fremont Lake are characterized by steep, irregular, and extremely bouldery slopes, which give them a fresh and youthful appearance. The slopes range from 30 to 60 percent. Outer slopes commonly exhibit secondary lobate forms, some of which appear to be slumped masses. A small apron of outwash sand is present locally on the lower slope of the outermost moraine.

The end moraines range in height from about 23 ft (7 m) to nearly 164 ft (50 m); in general, the outermost is highest. Commonly, moraine height reflects till thickness, but locally, Pinedale moraines overlie Bull Lake moraines. Crests are mostly narrow, but, where broad, are pitted by numerous kettles and other irregularities. Kettles in the northern part of the area contain water, but elsewhere most are dry. Some contain 3 to 5 ft (1.0 to 1.5 m) of loess. The largest at Mud Lake (Fig. 1; Halls Lake on some maps), is underlain by varved silt (Sorenson, 1986). Commonly, a belt of kettles surrounded by low ridges immediately back of the outermost Pinedale end moraine represents a zone of stagnation. Lateral moraines are banked in steplike fashion against both the east and west sides of the Fremont Lake basin. The gradient of the crest of the highest is approximately 1°.

Abundant boulders occur on the moraines. Most are 1.5 to 3 ft (0.5 to 1 m) in diameter, but many are 6.5 ft (2 m), and some are as large as 32 ft (10 m). Striated stones are rare, and ventifacts are seldom found, though both are present on Pinedale moraines on the east side of the Wind River Range. Troughs and swales between moraines generally are underlain by till, locally pocked by kettles, but some are underlain by outwash sand. The outermost Pinedale end moraine and one or more of the recessional moraines are narrowly breached by outwash channels. However, Pinedale moraines generally are little dissected except by principal drainageways.

Lithology. Almost all clasts in Pinedale Till in the Fremont Lake area are crystalline rock. Most are granite, granodiorite, and gneiss, but some are biotite gneiss, amphibolite, granulite, and vein quartz. A few in the outermost moraine are quartzite. Most clasts probably were derived from the interior of the range, but some may be reworked from older till, and some large weathered boulders appear to be joint-core boulders derived from a Tertiary regolith, locally preserved in the northeast part of the quadrangle. Other large boulders appear to be derived from the Tertiary conglomerate, for they are deeply stained yellowish to reddish brown, as are boulders in the conglomerate. The till matrix is derived in part from the same sources and in part from the Wasatch and Green River formations.

Texture. Till texture ranges from stony sandy loam to stony loamy sand (Shroba, 1977; Mahaney, 1978). Clay content is 2 to 3 percent (Shroba, 1977); clasts are mostly subangular to subround, but some are angular and some round. In some places, the till contains so little matrix that digging or augering is difficult. "Bullet-shaped" stones, common in till derived from Paleozoic limestones or sandstones, are lacking; glacially soled or faceted stones are rare.

Structure. The till is remarkably homogeneous. Stratification, fissility, and clast fabric are rare.

WEATHERING CHARACTERISTICS

Surface Boulders. Most cobbles and boulders at the surface of Pinedale moraines are fresh or only slightly weathered. Some are cracked or broken. Locally, the shape of a fragment permits recognition of its source in a broken boulder upslope. In places, boulders display the spalling effects of past forest fires. Mahaney (1978) indicates that 91 to 98 percent of the surface clasts are unweathered and 2 to 9 percent are weathered. His data show that weathering rinds range in average maximum thickness from .4 to .6 in (10.3 to 14.7 mm) to an average minimum thickness of .03 to .09 in (0.8 to 2.4 mm); and that weathering pits range from .46 to .67 in (11.8 to 16.9 mm) in depth and 1.6 to 2.2 in (40.0 to 55.5 mm) in width.

Soils. The soils on the till have been studied by Shroba (1977) and Mahaney (1978). Their analyses are detailed, but based on relatively few test sites.

With respect to thickness and color, they indicate that: A horizons are 3.9 to 9.1 in (10 to 23 cm) thick and 10*YR* 4/2 to 5/3 in color. B horizons are 5.9 to 10.2 in (15 to 26 cm) thick and 10*YR* 4/3 to 6/3 in color. Cox horizons are 15 to 20 in (38 to 50 cm) thick and 2.5*Y* to 10*YR* 6/3 in color. Unweathered till is 2.5*Y* 7/2 to 10*YR* 8/1 in color.

According to Shroba (1977), silt and clay are about 2 to 4 percent more abundant in the upper 40 in (100 cm) of the profile than in the parent till. Silt is most abundant in the A and B horizons, clay most abundant in the B horizon. From this distribution, he inferred a mixture of loess. Cca horizons contain less than 2 percent carbonate, whereas the underlying till is noncalcareous. Stones in the soil commonly are not coated with carbonate, but thin films of carbonate are present on them in places. Clay mineral analyses are given by Shroba (1977), and by Mahaney (1978).

Sorenson (1986) has mapped the soils in the Fremont Lake South Quadrangle. Most of his map units are complexes of two or more soil groups whose individual areal distribution is too small to show at the scale of the map. The soils map shows Burnt Lake stony sandy loam, a Typic Cyroboroll, extensively developed on Pinedale Till under sagebrush and grass in the zone of end moraines enclosing Fremont Lake, Soda Lake, and Half Moon Lake (Fig. 1). Its distribution includes the soil test sites of Shroba (1977) and Mahaney (1978). The Sublette sandy loam, an Argic Pachic Cryoboroll, is extensively developed on colluvium and eolian silt on the lower slopes of Pinedale end moraines and in intermorainal swales.

COMPARISON WITH BULL LAKE TILL

Nomenclature and age range of Bull Lake Till. Bull Lake drift was defined by Blackwelder (1915) from Bull Lake at the northeast base of the Wind River Range in Fremont County, Wyoming. Subsequently the deposits were designated Bull Lake Till (Richmond, 1964). Study of the type deposits showed that Bull Lake Till includes two locally superposed tills separated by a soil, and on that basis, early and late Bull Lake glaciations were defined. The upper till underlies two inner Bull Lake moraines; the lower till underlies the outer Bull Lake end moraine. Recently, I have distinguished five Bull Lake end moraines at Bull Lake, and five are recognized at Fremont Lake (Fig. 1). At Fremont Lake, till in the outermost Bull Lake end moraines (moraine I, Fig. 1), sampled at locality B (Fig. 1), has been dated by the uranium-trend method at 160,000 ± 50,000 years (Rosholt and others, 1985), and is therefore late Illinoian in age. Based on similarity of surface form, end moraine II (Fig. 1) is also probably late Illinoian in age. End moraines III, IV, and V (Fig. 1) may be Illinoian, but their locally steep and irregular form suggests that they may represent three post-Illinoian, pre–late Wisconsin (pre-Pinedale) glacial advances.

Contrast between Bull Lake Till and Pinedale Till. Bull Lake and moraines in the Fremont Lake area are broader, and have more gentle and more regular slopes than Pinedale moraines. Closed depressions are lacking. The end moraines are more widely breached than Pinedale end moraines and lateral moraines are more dissected. Surface boulders are less abundant than on Pinedale moraines. Weathering rinds are thicker (average maximum 0.5 to 0.6 in; 12.6 to 14.7 mm; Shroba, 1977) and weathering pits slightly larger (average depth 0.7 to 0.8 in; 18.6 to 21.0 mm; Mahaney, 1978). More boulders are split or fractured than on Pinedale moraines.

Loess, commonly 8.2 ft (2.5 m) thick and locally 16.4 ft (5 m) thick (Shroba, 1977), is widespread on lee slopes of Bull Lake moraines. Patches, less than 0.9 ft (0.3 m) thick, occur leeward of low knolls.

Bull Lake Till contains the same rock types as Pinedale Till, but the matrix is more compact, includes 2 to 6 percent more silt and up to 20 percent more clay (Shroba, 1977). Completely rotted clasts, generally micaceous, are more abundant in Bull Lake Till.

Post–Bull Lake soils, where preserved, are more maturely developed than post-Pinedale soils. B horizons are thicker and their clay content (18 to 22 percent) is greater than the clay content (8.5 to 11 percent) of B horizons of soils on Pinedale Till (Shroba, 1977). Cca horizons are thicker and much more intensely and more uniformly impregnated with carbonate. Carbonate coatings on clasts in post–Bull Lake Cca horizons are as much as 0.08 in (2 mm) thick. Carbonate content is up to 27 percent as compared to less than 2 percent in soils on Pinedale Till (Shroba, 1977). Further data are given by Shroba (1977) and Mahaney (1978).

Sorenson (1986) mapped three soil series under sagebrush and grass on Bull Lake Till in the Fremont Lake area: Gelkie stony sandy loam, present only on Bull Lake deposits; Burnt Lake stony sandy loam, developed on both Bull Lake Till and Pinedale Till; and Subletter sandy loam, developed on colluvium on both Bull Lake and Pinedale moraines. The Gelkie stony sandy loam is post–Bull Lake soil. The Burnt Lake stony sandy loam on Bull Lake Till is a post-Pinedale soil formed where Gelkie soil has been removed by mass wasting or sheetflood erosion.

Shroba (1977) considered Bull Lake end moraine V (Fig. 1), immediately south of locality A, to be "early Pinedale" in age because the soils on the crest and upper slope of the moraine are similar to soils on Pinedale moraines. I consider these soils to be the Burnt Lake Soils of Sorenson (1986) formed where an original Gelkie soil has been removed. An old road cut (locality C, Fig. 1) along the lower north slope of moraine V, immediately south of the Pinedale dump, exposes a carbonate horizon, about 4 ft (1.2 m) thick, developed on till and disconformably overlain by colluvium in which a Sublette soil is developed. The colluvium contains fragments of both carbonate coatings and carbonate-enriched soil matrix, which demonstrates that transport of debris from a former mature soil higher on the slope has taken place. Westward along the old road, the buried soil is thinned and convoluted indicating that the erosion was accompanied by mass movement. I conclude that the moraine is pre-Pinedale in age and properly belongs in the group of moraines mapped as Bull Lake (Richmond, 1973). Shroba suggested that its age is greater than 45,000 years. If Bull Lake end moraines III, IV, and V are younger than 120,000 years, moraine V may be 60 to 70,000 years old.

REFERENCES CITED

Blackwelder, E., 1915, Post-Cretaceous history of the mountains of central western Wyoming: Journal of Geology, v. 23, p. 307–340.

——— , 1931, Pleistocene glaciation in the Sierra Nevada and Basin ranges: Geological Society of America Bulletin, v. 42, no. 4, p. 865–922.

Mahaney, W. C., 1978, Late Quaternary stratigraphy and soils in the Wind River Mountains, western Wyoming, *in* Mahaney, W. C., ed., Quaternary soils: Norwich, England, Geo Abstracts, p. 223–264.

Richmond, G. M., 1964, Three pre–Bull Lake Tills in the Wind River Mountains, Wyoming; A reinterpretation: U.S. Geological Survey Professional Paper 501-D, p. D104–D109.

——— , 1973, Geologic map of the Fremont Lake South Quadrangle, Sublette County, Wyoming: U.S. Geological Survey Geologic Quadrangle Map GQ—1138, scale 1:24,000.

Rosholt, J. N., Bush, C. A., Shroba, R. R., Pierce, K. L., and Richmond, G. M., 1985, Uranium-trend dating and calibrations for Quaternary sediments: U.S. Geological Survey Open-File Report 75-299, 48 p.

Shroba, R. R., 1977, Soil development of Quaternary tills, rock glacier deposits, and taluses, southern and central Rocky Mountains, [Ph.D. thesis]: Boulder, University of Colorado, 424 p.

Sorenson, C. J., 1986, Soils map of the Fremont Lake South Quadrangle, Sublette County, Wyoming: U.S. Geological Survey Miscellaneous Investigations Map I-1800, scale 1:24,000 (in press).

Tertiary subsurface solution versus Paleozoic karst solution, Guernsey, Wyoming

L. W. (Dan) Bridges, *1925 South Vaughn Way, #207, Aurora, Colorado 80014*

LOCATION AND ACCESS

Guernsey is located in southeastern Wyoming on the Hartville Uplift between Torrington and Douglas on U.S. 26. Guernsey State Park, which is 1 mi (1.6 km) north of U.S. 26, has an average elevation of 4500 ft (1400 m). It is readily accessible by car or bus any time of the year because this is one of the warmest parts of Wyoming. Most of the outcrops are located in Guernsey State Park, which is in the Guernsey Reservoir quadrangle, Platte County, Sec.4,T.27N.,R.66W., approximately 42°17′N,104°46′W.

Key locations to examine the problem of the nature of the solution surface on the top of the Guernsey Limestone are shown on Figure 1. An excellent view of the principal locality (B) (Fig. 2) on the east side of the North Platte River can be obtained from area A on the west side. The paleo-sinkhole at the south end of the exposure, and the solution surface to the north can be reached for closer examination by means of arduous hikes from parking lot C.

To reach the sinkhole area, walk southeasterly from the parking lot on the maintenance road, through the maintenance yards, down into a tributary valley and south along the river.

To see several 3 to 5 ft (1 to 1.5 m) solution arches, and a chert breccia (Fig. 3), on the solution surface, walk from the parking lot down to the top of the Guernsey Limestone and southward along this unconformity. Hiking here should be done with care because you are near the upper edge of a cliff.

A more accessible exposure of the solution surface, with different details, can be seen at locality D by driving south on the rough dirt road shown on Figure 1.

SIGNIFICANCE OF LOCALITY

This Paleozoic sedimentary outcrop locality is chosen because it contains highly controversial, complex geology that is superbly exposed, readily accessible, and ideal for a lively field trip discussion stop. Did the solution take place in Mississippian or Early Pennsylvanian time on an exposed karst surface? If so, we are dealing with a single solution surface. Or, was the solution caused by heated subsurface waters in Tertiary time? If so, we are dealing with two solution surfaces, major amounts of late solution, redeposition, recementation, and Tertiary chert.

GEOLOGIC DESCRIPTION

Figures 2, 4, and 5 provide the essential ingredients for

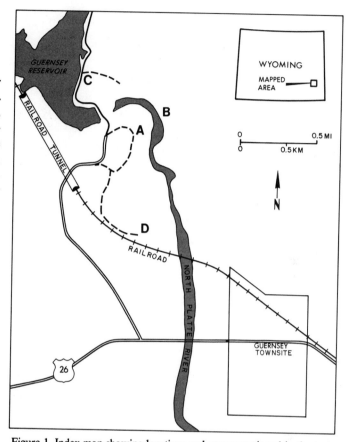

Figure 1. Index map showing locations and areas mentioned in the text.

discussion and interpretation at this locality. Figure 2 is a photo of the Mississippian (Osagean or Meramecian) Guernsey Limestone and overlying Pennsylvanian (Morrowan) Hartville Sandstone at locality B on Figure 1. Figures 4 and 5, at the same general scale as Figure 2, show two possible geologic interpretations of Figure 2.

The Paleozoic karst solution interpretation in Figure 5 has been the accepted interpretation and is figured prominently by Craig (1972). According to this interpretation, there was a considerable period of solution on the top of the exposed Guernsey Limestone in Mississippian time, and possibly earliest Pennsylvanian time, prior to the deposition of the overlying Hartville Sandstone. If this interpretation is correct, it seems most logical that the rubble on top of the Guernsey Limestone unconformity should grade into the overlying Hartville sandstone (This is not the case). Furthermore, if the vertical contact between the Hartville Sandstone and the Guernsey Limestone on the south side of the sinkhole (see Figs. 2, 4, and 5) is an unconformity, there should be limestone clasts in the sandstone. (There are none.)

Figure 2. View at the sinkhole outcrop at locality B from area A.

Figure 3. Slumped chert breccia at solution surface north of locality B.

Tertiary subsurface solution is my interpretation of the Guernsey/Hartville relationships (Bridges, 1982). As shown schematically in Figure 4, this interpretation requires 2 unconformities (solution surfaces): one lies above the redeposited sandstone, limestone, and chert breccia; and one unconformity (solution surface) lies below it.

The most compelling evidence favoring late solution is the vertical contact between the Hartville Sandstone and the Guernsey Limestone on the south (downstream) side of the sinkhole. The Hartville Sandstone gradually sags southward toward the sinkhole; this evidence suggests that the Hartville is faulted down against the Guernsey. Thus the solution must be post-Hartville.

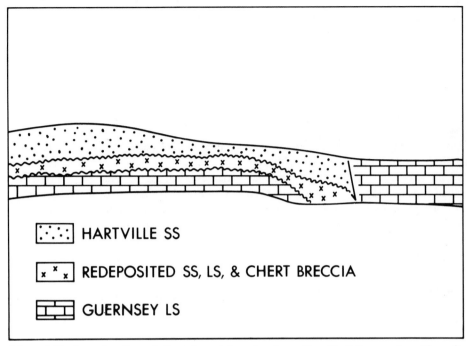

Figure 4. Tertiary subsurface solution interpretation of geology at locality B. Same scale as Figure 2.

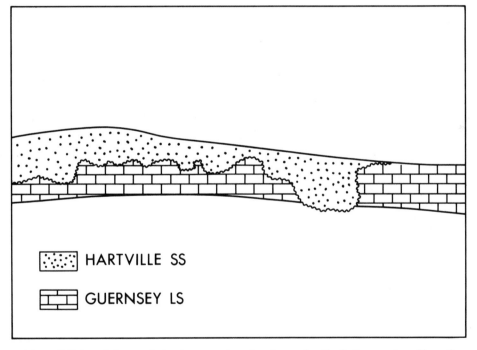

Figure 5. Paleozoic karst solution interpretation of geology at locality B. Same scale as Figure 2.

Generally speaking, both the Guernsey Limestone and Hartville Sandstone have the same low dip directions, except where the Hartville sags into the sinkhole. Thus the upper solution surface, which is cut into the Hartville Sandstone, is so irregular that it cannot be a depositional unconformity.

The three types of solution deposits, which range in thick-ness from 20 to 70 ft (6 to 21 m) are: (1) Non-bedded carbonate that has been dissolved from Guernsey Limestone and redeposited; (2) Red, non-bedded sandstone that has been dissolved from the Hartville Sandstone and redeposited; and (3) Chert breccia composed of 5% to 10% angular chert fragments in slumped sandstone (see Fig. 3). In Guernsey State Park, upstream from the sinkhole, the oldest solution deposits are mostly carbonate; and the chert breccia is less abundant and younger. At locality D, redeposited sandstone is most abundant. (For a more detailed cross section of the spatial relationship of these solution deposits see Bridges, 1982, Fig. 6.)

The most complicated geology occurs in the lower part of the sinkhole (locality B). In most areas, the active solution appears to have progressed upward; but below the sinkhole the solution appears to have progressed downward. The northern corner of the sinkhole displays some nearly vertical, highly contorted sandstone beds with some chert. Apparently solution beneath these partly lithified beds caused them to slump into their present attitude. Bretz (1940) described some similar penecontemporaneous slumping in northeastern Illinois.

In the lower part of the sinkhole is a prominent black-stained bedding-plane that dips 20° toward the river. This bedding-plane surface has grooves resembling slickensides that Bretz (1940) called slickolites, noting they must have formed under the pressure of considerable rock overburden. Continued solution below this bed apparently caused its present trough-shaped attitude.

Above the prominent bedding plane are micro-laminated, micro-fractured, thin-bedded sandstones. The fracturing is interpreted to indicate penecontemporaneous deformation from continued settling into the solution cavity below. The upper solution surface above the redeposited sandstone but beneath the Hartville

Sandstone is very obscure, but generally marked by bleaching of the red sandstone by heated subsurface waters.

The angular chert fragments, which show no signs of abrasion, pose an additional problem. If they had been eroded from the Guernsey Limestone, why are there no limestone clasts with them? Contrary to popular expectations, the chert appears to be part of the solution-redeposition process here. In fact, at one spot in area D, a chert bed can be seen extending from undisturbed limestone into redeposited sandstone.

Solution temperatures of at least 150°C seem necessary to dissolve such a large volume of the Hartville Sandstone. This area may even have been the site of hot spring activity. There is no direct evidence for the time of solution. However, it seems most likely that during regional Laramide-Tertiary tectonism, volcanism, and mineralization, geothermal gradients were abnormally high. Therefore, Tertiary subsurface solution is proposed.

REFERENCES CITED

Bretz, J. H., 1940, Solution cavities in the Joliet Limestone of northeastern Illinois: Journal of Geology, v. 48, p. 337–384.

Bridges, L.W.D., 1982, Rocky Mountain Laramide-Tertiary subsurface solution vs. Paleozoic karst in Mississippian carbonates, *in* Geology of Yellowstone Park Area: Wyoming Geological Association Guidebook, p. 251–264.

Craig, L. C., 1972, Mississippian system, *in* Mallory, W. W., ed., Geologic Atlas of the Rocky Mountain Region: Denver, Rocky Mountain Association of Geologists, p. 100–110.

Dunes of the Wyoming Wind Corridor, southern Wyoming

Ronald W. Marrs, *Department of Geology, University of Wyoming, Laramie, Wyoming 82071*
Kenneth E. Kolm, *Department of Geology, Colorado School of Mines, Golden, Colorado 80401*
David R. Gaylord, *Department of Geology, Washington State University, Pullman Washington 99164*

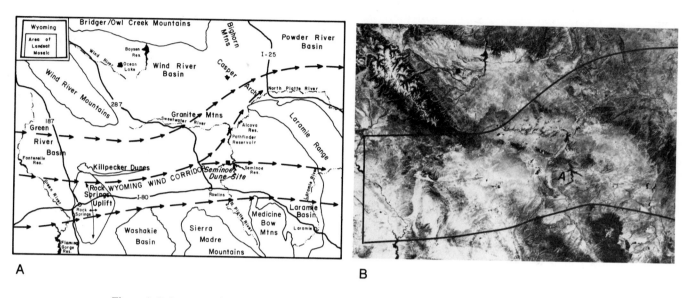

A

B

Figure 1. Index map of Wyoming Wind Corridor area. A. Features shown on LANDSAT mosaic (part B) and location of Seminoe Dunes site.

LOCATION

This geological field site, notorious for strong, persistent, unidirectional winds that have created classic eolian landforms, is located in the SW¼Sec.30,T.25N.,R.84W., Carbon County, Wyoming (Figs. 1, 2). The area is readily accessible via paved road by traveling north from the town of Sinclair, Wyoming, a distance of 30 mi (48 km) on the Seminoe Road. This highway is the main access route to Seminoe State Park. The locale is on public land, administered by the Bureau of Land Management, but it borders on private ranch land to the north and state-owned land to the west.

The area is within the broad, windswept zone of central Wyoming (Fig. 1) known as the Wyoming Wind Corridor (Marrs and Kolm, 1982). The large active dunes at this site and older stabilized dunes in the area are part of the Seminoe Dune Field (Fig. 2). Most dunes in the area are marginally stabilized parabolic dunes. The few active dunes are relatively large and slow moving with active fronts and partially stabilized tails.

SITE DESCRIPTION

The dune at the Seminoe Road location is approximately 130 ft (40 m) high and covers an area approximately 0.25 mi² (0.5 km²). It is a large, complex, parabolic dune of deflation (following the terminology of Hack, 1941) that is presently encroaching upon the road from the west (Fig. 3). This dune, and a smaller dune located upwind (to the west), displays the classic features of dune tails that are partially stabilized by vegetation, an upwind blowout or erosional face, and a downwind slope of deposition. A closer inspection of the dune reveals eolian crossbeds. There is also an interdunal depression aligned with the two dunes (Fig. 2).

DISCUSSION

The Wyoming Wind Corridor extends from the Green River Basin of western Wyoming to the Nebraska Sand Hills nearly 375 mi (600 km) to the east, and covers a swath 90 to 125 mi (150 to 200 km) wide in its north–south dimension (Fig. 1). Large ares of stable dunes are found within the corridor at Fontinelle Reservoir, Table Rock, Doty Mountain, Killpecker Creek, Seminoe Reservoir, Pathfinder Reservoir, Ferris, Casper, Chugwater, and Torrington. Small isolated patches of dune sand and eolian erosional features (blowout playas and scour streaks) are also common throughout the region. Active dunes are largely confined to the Killpecker, Ferris/Seminoe, and Casper Dune Fields at the present time. Geologic evidence suggests that active sand has periodically covered a much broader region within the Wyoming Wind Corridor in the recent geologic past.

Geologic work on the northern segment of the Ferris/Seminoe Dune Field (Gaylord, 1982) shows that the dunes have been

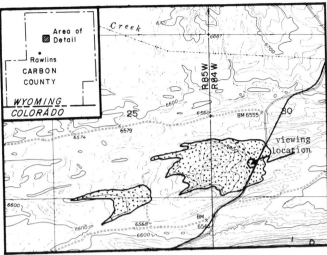

Figure 2. A. Part of Seminoe Dune Southwest 7½-minute Quadrangle, prepared from 1949 aerial photographs, showing the Seminoe Dune site. Active sand areas are shown in stippled pattern and on accompanying aerial photograph (part B; from NASA Mission 213, September, 1972).

quite active with periods of intense activity separated by periods of stabilization and soil development (much like we see today). Stratigraphic and sedimentologic work and ^{14}C age dates reveal a rather precise climatic history of central Wyoming for the last 10,000 years (Gaylord, 1982). Figure 4 is a composite stratigraphic column of dune sand and interdune sediments constructed from data gathered in the Clear Creek area to the northwest of Seminoe Reservoir. These data reveal that the accumulation of dune sands was most rapid prior to 7666 B.P. and during the approximate period of 7660 to 6460 B.P. These periods are interpreted as more arid times when the dunes were most active. Sedimentologic analysis of sand samples suggests that the winds have not changed significantly during the 10,000-year history of these dunes. Changes in dune activity are attributed to changes in temperature and/or annual precipitation patterns.

Winds in this area are strong, persistent, and nearly unidirectional from the west (Martner and Marwitz, 1982). The strength and persistence of these winds, particularly during the winter months, is such that wind energy may prove a useful supplement to hydroelectric power in this area. The winds are strongest in winter, when water must be retained in the reservoirs for later use in irrigation or urban water supply. In summer, when the water can be released from the reservoirs and can be used to generate hydroelectric power, the winds are much weaker and do not provide nearly as much power-generating potential. The economic potential of the wind in this area has been tested by the U.S. Department of Interior, Bureau of Reclamation. Two of the world's largest wind turbines were installed near Medicine Bow, Wyoming, about 35 mi (60 km) to the southeast of this site in the central Wyoming Wind Corridor (Martner, 1983).

The active dunes present today represent a hazard to agriculture, transportation, and urban development in some areas. Active dunes in tis area migrate from 10 to 100 ft (3 to 30 m) each

year, averaged over a 30-year period (Gaylord, 1982). Rate of migration largely depends upon the size of the dune (larger dunes generally move more slowly), vegetation, moisture, and exposure to the wind. The dune which is crossing the Seminoe Road is migrating at an average rate of 16 ft (5 m) each year (Fig. 3). At this rate (assuming it is not artificially stabilized), it will cover the road in two years and will reach the ID Ranch buildings (0.6 mi; 1 km to the east, downwind) in roughly 200 years. In other areas of the Wyoming Wind Corridor, sand dune migration presents a problem at building sites (north side of Casper, Wyoming) and at reservoirs (increased sedimentation of Pathfinder and Seminoe Reservoirs).

Figure 3. Seminoe dune encroaching eastward on New Seminoe Road (photo by R. Marrs).

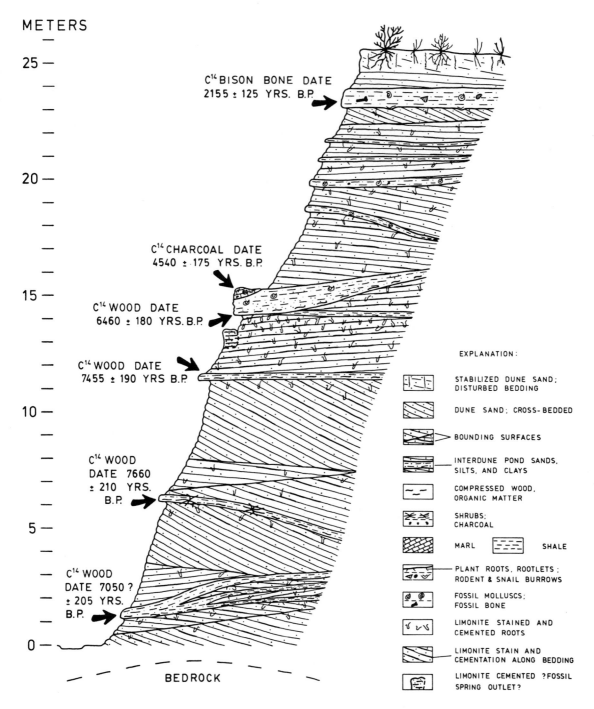

Figure 4. Composite stratigraphic column of eolian deposits in the Clear Creek area, Ferris/Seminoe Dune Field, Wyoming (after D. R. Gaylord, 1982).

REFERENCES

Gaylord, D. R., 1982, Geologic history of the Ferris Dune Field, south-central Wyoming, *in* Marrs, R. W., and Kolm, K. E., eds., Interpretation of windflow characteristics from eolian landforms: Geological Society of America Special Paper 192, p. 65–82.

Hack, J. T., 1941, Dunes of the western Navajo country: Geographical Review, v. 31, no. 3, p. 240–263.

Marrs, R. W., and Kolm, K. E., eds., 1982, Interpretation of windflow characteristics from eolian landforms: Geological Society of America Special Paper 192, 109 p.

Martner, B. E., 1983, Giant wind turbines dedicated at Medicine Bow: Bulletin of the American Metrological Society, v. 64, no. 1, p. 29–30.

Martner, B. E., and Marwitz, J. D., 1982, Airflow through the Wind Corridor in southern Wyoming, *in* Proceedings of the Second Conference on Mountain Metrology, Steamboat Springs, Colorado: Boston, American Meteorological Society, p. 309–315.

Little Muddy Creek area, Lincoln County, Wyoming

Frank Royse and M. A. Warner *Chevron U.S.A., Inc., 700 South Colorado Blvd., Denver, Colorado 80222*

INTRODUCTION

Location. The Little Muddy Creek area is located in a highly deformed frontal (eastern) zone of the Absaroka thrust fault system in the thrust belt of southwestern Wyoming. The locality is in the SW¼, T.19N., R117.W., approximately 28 mi (45 km) north-northeast of Evanston, Wyoming (Fig. 1).

Access. The area can be easily reached by passenger car by driving about 6 mi (9.7 km) to the west of U.S. 189 on a dirt road. The turn-off from U.S. 189 is about 300 ft (100 m) north of the sign marking the boundary between Uinta and Lincoln Counties. The road approximately follows the old Emigrant Trail, a wagon trail used by settlers, which was a branch of the famous Oregon Trail. Monuments mark the position of the Emigrant Trail. A recommended foot traverse is shown on Figure 1.

SIGNIFICANCE

The most significant feature of the Little Muddy Creek area is a remarkable outcrop of a thick, coarse, synorogenic conglomerate (Figs. 1 and 2). Study of this conglomerate and adjacent rocks indicates there were two distinct periods of major fault motion on the Absaroka thrust fault system here, one in pre-early Campanian time and a lesser one between early Campanian and early middle Paleocene time. The study also shows that there was an eastward progression in age of thrust faults, and that fault motion was episodic in nature. In addition, the sedimentology of the conglomerate should be of special interest to stratigraphers because it offers an excellent example of inverted stratigraphy and pronounced lateral and vertical changes in clast size and sorting.

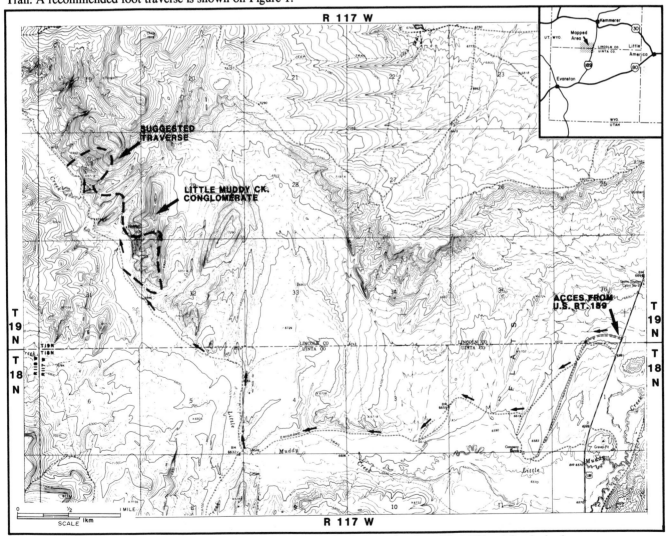

Figure 1. Topographic map of the Little Muddy Creek Area taken from Elkol SW and Cumberland Gap 7½-minute quadrangles. Contour interval = 20 ft. (6 m.). Refer to Figure 2 (geologic map).

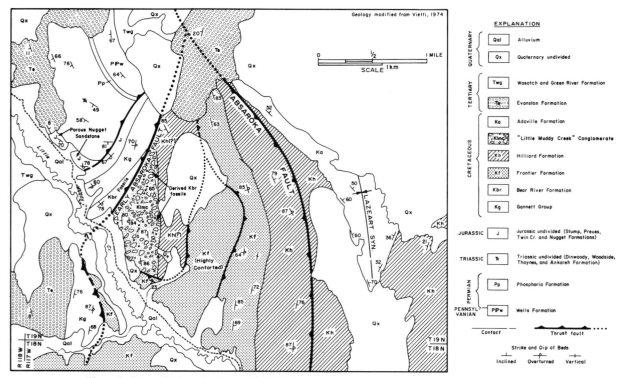

Figure 2. Generalized geologic map of the Little Muddy Creek area, T.19N.,R.117W., Lincoln County, Wyoming. Structural details in highly deformed areas have been omitted. Refer to Figure 1 (topographic map).

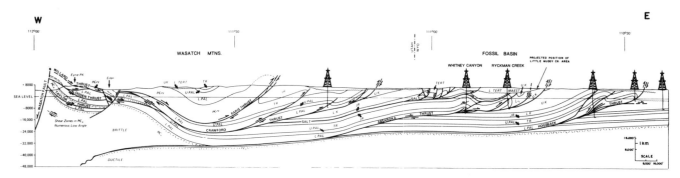

Figure 3. Generalized structural cross section of the Wyoming–N. Utah thrust belt. The Little Muddy Creek area lies about 10 mi (16 km) north of the indicated projected position. Major oil and gas fields are at Whitney Canyon and Ryckman Creek.

DISCUSSION

The general structural position of the Little Muddy Creek area within the thrust belt is shown on the regional cross section (Fig. 3). The sedimentary record shows that thrusting in western Wyoming and northern Utah occurred episodically over 95 m.y., from about 145 Ma in late Jurassic to 50 Ma in early Tertiary. Generally, age of thrusting is oldest in the west and progressively younger toward the east. Four major thrust systems are recognized on the basis of geometry and age, and are noted on the cross section. They are, from west to east, the Willard-Paris, the Crawford, the Absaroka, and the Hogsback. Each of these thrust systems includes more than one significant thrust fault.

Motion on these thrust systems resulted in uplift and consequent deposition of major synorogenic conglomerate units. Paleontologic dating, together with a correlation of times of deposition of these conglomerates to times of motion on specific thrust systems, allow a reconstruction of the geological evolution of the thrust belt (Royse, Warner, and Reese, 1975). The conglomerate at Little Muddy Creek is one such synorogenic unit which was deposited and deformed as a result of Absaroka thrust fault motion.

The exposure of conglomerate at Little Muddy Creek is unique in the Wyoming thrust belt. Vietti (1974) measured and described the upper 1,300 ft (395 m) of the unit in near vertical beds exposed on a ridge in the southwest part of Section 29

a

b

Figure 4. Photographs of the Little Muddy Creek Conglomerate outcrop. (a) Look east from road at near vertical strata in NW¼ of NW¼ of Section 32, T19N-R117W. Marine sandstones in foreground conformably overlie the conglomerate. (b) Look south at same outcrop as in 4a. Larger clasts are about 5 ft (1.5 m) across.

(Fig. 1). Here the top of the unit is to the west and its total thickness appears to be about 2,000 ft (610 m) (Fig. 4a, b). Rounded boulders with long dimensions of 7 ft (2.1 m) or more are present in a massive conglomerate near the middle third of the unit. The coarser clastics grade into sandstones and mudstones both above and below. Vertical changes in composition and abundance of clasts in the conglomeratic unit demonstrate an excellent example of inverted stratigraphy (Fig. 5). Clasts derived from distinctive Mesozoic formations such as the Aspen, Bear River, Ephraim, Twin Creek, and Thaynes, can be identified. Many of the larger boulders are sandstone and appear to have been derived from the Triassic Nugget Formation. Distinctive silicious shale clasts derived from the lower Cretaceous Aspen Formation are present near the base of the conglomeratic unit. The first appearance of clasts from formations older than the Aspen which include derived Lower Cretaceous Bear River Formation gastropods, occurs at successively higher stratigraphic positions in an order opposite that in which the parent rocks were deposited. The sequence clearly indicates deposition concurrent with uplift of the source area on which progressively older formations were being bared to erosion. This source area is exposed immediately northwest of the conglomerate outcrop (Fig. 2). Palynomorphs recovered from samples collected below, within and just above the conglomerate unit provide a basis for dating this unit and indicate that it is late Santonian-early Campanian in age (Jacobson and Nichols, 1983) and probably equivalent to part of the upper Hilliard shale or lower Adaville sandstone sequence exposed below the Absaroka thrust sheet on the Hogsback thrust plate 2 mi (3.2 km) to the east (Fig. 2).

The conglomeratic unit was water laid in an environment near sea level. Shales within and immediately above the upper part of the unit contain marine microfossils. Oyster shells and shark teeth are found in sandstones in the same stratigraphic interval. The large size of the sandstone boulders and their lack of resistance to abrasion attest to the relative short distance of sedimentary transport from, and probable abrupt topographic relief of, the nearby uplift.

Deposition and subsequent deformation of the conglomerate resulted from motion on the Absaroka thrust fault system. Two faults with significant stratigraphic displacement of 5,000 ft (1,500 m) or more have been mapped as parts of the Absaroka fault system in the Little Muddy Creek area. Seismic and drill hole data indicate that these faults merge into a single detachment plane in the subsurface to the west. To avoid a lengthy discussion over which of these two faults should be called Absaroka, we shall use the terms early Absaroka fault and Absaroka fault in an informal designation on the geologic map (Fig. 2). The early Absaroka fault places strata of the Lower Cretaceous Bear River Formation and Gannet Group in contact with the Upper Cretaceous Little Muddy Creek Conglomerate. The stratigraphic displacement is on the order of 13,000 to 15,000 ft (4,000-4,600 m). The eastern fault, labeled Absaroka fault on Figure 2, has a stratigraphic displacement of 5,000 to 6,000 ft (1,500-1,800 m) where it brings the Frontier Formation to within a few hundred

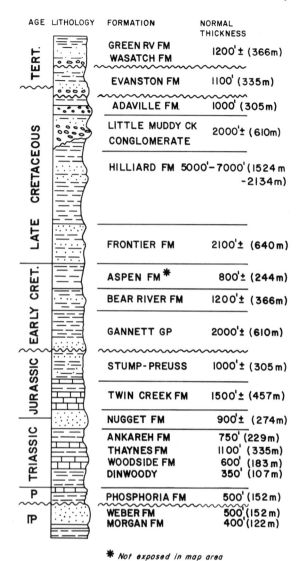

Figure 5. Stratigraphic column of rocks exposed in the Little Muddy Creek Area.

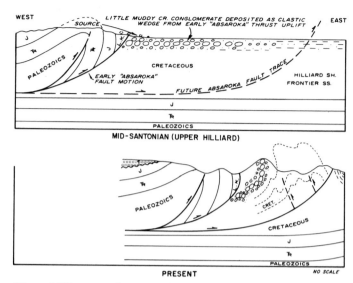

Figure 6. Diagrammatic cross sections in Little Muddy Creek area showing sequence of thrusting in Absaroka fault system. Upper diagram depicts a time shortly after movement on early Absaroka fault and deposition of the conglomeratic unit. Lower diagram illustrates the present structure after movement on Absaroka fault in latest Cretaceous.

(Fig. 2). Therefore, final motion on this fault must have occurred during latest Cretaceous time.

In the area west and north of Little Muddy Creek, where surface, drill hole, and seismic data show only one fault in the Absaroka system, motion must have occurred in at least two phases on the same fault plane. At Little Muddy Creek, motion was expressed on two separate faults in the frontal zone. The combined horizontal motion is about 27 mi (44 km). Thus, exposures of synorogenic deposits in this unique area not only can be used to demonstrate the episodic nature of thrusting and the general eastward progression of age of thrusting in southwest Wyoming, but also are of special sedimentologic interest.

An additional feature of interest in this area is the excellent exposure of Nugget sandstone which is uncharacteristically very porous. This is notable to petroleum geologists because the Nugget sandstone is one of the principal oil and gas reservoirs found in thrust-faulted structures buried beneath Tertiary deposits immediately south of the Little Muddy Creek area (Fig. 3).

REFERENCES

Jacobson, S. R., and Nichols, D. J., 1983, Palynological dating of syntectonic units in the Utah-Wyoming thrust belt: The Evanston Formation, Echo Canyon Conglomerate, and Little Muddy Creek conglomerate in Powers, R. L., ed., Geologic Studies of the Cordilleran Thrust Belt: Rocky Mountain Association of Geologists, v. II, p. 735–750.

Royse, F., Warner, M. A., and Reese, D. L., 1975, Thrust belt structural geometry and related stratigraphic problems, Wyoming-Idaho-Northern Utah: Rocky Mountain Association of Geologists, 1975 Symposium, p. 41–54.

Vietti, J. S., 1974, Structural geology of the Ryckman Creek Anticline area, Lincoln and Uinta Counties, Wyoming [M.S. thesis]: Laramie, University of Wyoming.

feet of the base of the Adaville Formation. Neither of the two faults can be precisely dated, but the field evidence shows that motion occurred in two separate stages. The diagrammatic cross sections in Figure 6 show an interpretation of these events. Thrust motion on the early Absaroka fault in late Hilliard (late Santonian) time created a fold uplift of Mesozoic and Paleozoic rocks which supplied the coarse clastics in the conglomeratic unit. Horizontal movement at this time was probably on the order of 21 mi (34 km). Later movement on the Absaroka thrust below and in front of the early Absaroka fault caused the intense folding and minor thrusting now evident in the conglomeratic unit and older strata on the Absaroka hanging wall. The Absaroka fault cuts the west flank of the Lazeart syncline. There it involves strata as young as the Adaville Formation of Campanian age, and is overlain by the lower Paleocene part of the Evanston Formation

Sandstones of the Casper Formation of the southern Laramie Basin, Wyoming: Type locality for festoon cross-lamination

James R. Steidtmann, Department of Geology and Geophysics, University of Wyoming, Laramie, Wyoming 82071

LOCATION

Very large scale, trough cross-stratification in sandstone units of the Casper Formation is well exposed along Sand Creek in the southernmost Laramie Basin, Wyoming (Fig. 1). The exposures of features described herein are located in the SW¼,Sec.31,T.13N.,R.74W. and the E½,Sec.1,T.12N.,R.75W. (Fig. 1). This locality is included in a field trip guide by Steidtmann and Weimer (1976); photographs of the features mentioned herein are in Steidtmann (1974, 1976).

ACCESSIBILITY

This locality is accessible for much of the year by passenger car or bus by driving approximately 21 mi (34 km) southwest of Laramie on Albany County Road 34. The only time four-wheel drive becomes necessary is in the spring after heavy snow melt, or immediately after heavy rain or snow. Once in the area, vehicles can be parked along the main road and specific outcrops can be reached by foot (usually within 1,600 ft [500 m]). There are several places where buses can be turned around. Most of the exposures are on land belong to the Chimney Rock Grazing Association; permission to visit the area can be obtained in advance from Mr. Frank Lilley, Foreman, 1910 Sand Creek Road, Laramie, Wyoming 82070, (307) 745-9575.

SIGNIFICANCE

Knight (1929) was the first to describe the spectacular exposures of trough cross-stratification in the Casper Formation of the southern Laramie Basin; he proposed the term "festoon cross-lamination" to designate this structure. Since then these structures have been frequently cited as the classical occurrence of large-scale trough cross-stratification. Knight postulated a submarine origin for the cross-stratification because it seemed to him that associated deformational structures in the sandstone suggested saturated conditions and because laterally associated limestones only 5.6 mi (9 km) to the northeast contain marine fossils. Since this work by Knight, hundreds of geologists have examined these outcrops. Opinions on the origin of the cross-stratification have varied widely, the most common alternative explanation being a subaerial dune environment. Hanley and others (1971), Hanley and Steidtmann (1973), Steidtmann (1974, 1976), and Steidtmann and Haywood (1982) are among those who have presented evidence for the eolian interpretation.

GEOLOGICAL INFORMATION

Regional Setting. In the southern Laramie Basin, the Casper Formation is underlain by, and intercalated with, the Fountain Formation (Pennsylvanian), a partially contemporaneous coarse, arkosic sandstone and conglomerate; it is overlain by red siltstone and shale of the Satanka Formation (Permian). In

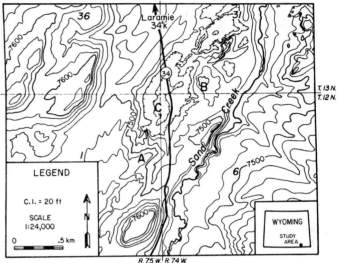

Figure 1. Topographic map showing the location of the type locality for festoon cross-stratification in the Casper sandstone of the southernmost Laramie Basin, Albany County, Wyoming. Location A displays the cross-stratification and ripples; location B, exposures of lenticular, nonmarine limestone; location C, an example of a lag-grain surface. Map base is the Johnson Ranch and Downey Lakes 7½-minute Quadrangles.

general the Casper Formation consists of interbedded sandstone and limestone. The sandstone is a fine-grained, well-sorted, calcareous subarkose or orthoquartzite exhibiting large-scale cross-stratification. Permian fusulinids and other reworked and unreworked marine fossils and fossil fragments are present in the sandstone at several localities. The interbedded limestone units are between 8 and 29 ft (2.5 and 9.0 m) thick, are only locally continuous, and are best exposed along the west flank of the Laramie Range to the east. The limestone is dense, sandy micrite containing fusulinids, nautiloid cephalopods, trilobites, brachiopods, crinoids, and bryozoa.

At this locality in the southernmost part of the basin, the Casper Formation differs somewhat from its character to the north and east. There are no interbedded marine limestone beds; instead there are thin, lenticular, nonmarine limestone lithosomes at several stratigraphic levels within the cross-stratified sandstone. Large-scale contortion and brecciation of sandstone laminae occur throughout much of the area.

Cross-stratification. The cross-stratification (Fig. 1, location A) is trough-shaped with the troughs ranging in size from 5 ft (1.5 m) wide, 24 ft (7.5 m) long, and 1 ft (0.3 m) deep to 1,000 ft (300 m) wide, several times as long, and at least 50 ft (15 m) deep. They are roughly symmetrical, both symmetrically and asymmetrically filled, and plunge southwest. Approximately 75 percent of the cross-strata dips in the Casper are between 10 and 25°, with the average being around 18° (Knight, 1929). Few if any cross-strata dip at the maximum angle of repose for sand of 34°. The presence of ripples parallel to the dip of cross-strata, subcritically climbing translent strata, and the lack of avalanche features in sandstone of the Casper Formation indicate lateral sand transport across the lee side of a dune.

Fossils. Hanley and others (1971) described trace fossils from cross-stratified sandstone in the Casper Formation. Prior to this discovery all interpretations concerning the origin of the cross-strata were based on physical evidence because no biogenic remains had been found (a fact cited by some observers as proof of eolian origin). The trace fossils are rare and occur on the upper bounding surfaces of sets of small-scale cross-stratification within the large-scale trough structures. A small fossil aucarian or lycopod cone was found wedged in the bottom of a filled mudcrack. To date, this is the only preserved organic remain found in Casper sandstone at this locality. Although lack of preserved detail prohibits specific identification, its terrestrial origin and mode of entrapment and preservation suggest eolian conditions.

Limestone Lenses. Hanley and Steidtmann (1973) described limestone lenses in the Casper Formation at this locality (Fig. 1, location B). The limestone occurs along truncation surfaces of low relief. The maximum extent of continuous outcrop observed is approximately 1,500 by 500 ft (460 by 150 m). The maximum thickness of limestone is 2 ft (60 cm). The lithosomes are distinctly lenticular, thinning from as much as 24 to 0.5 in (60 to 1 cm) over a distance of 10 ft (3 m). The upper and lower contacts are generally regular and gradational through a 2-in (5-cm) zone of red and almost white calcareous siltstone. Some limestone lenses merge laterally with zones of red silty sandstone up to 1 ft (30 cm) thick, which thin irregularly away from the limestone lenses and contain small, angular limestone fragments and mudcrack fillings. The limestone is a peloidal microsparite. It is wavy bedded, reflecting extensive stylolitization, and contains disrupted laminae, which are probably a result of desiccation. Fresh-water ostracodes are the only fossils present. Their frequency ranges from scattered individuals to nearly coquinal layers.

Ripples. Low-amplitude ripples occur on cross-stratification surfaces in the vicinity of location A (Fig. 1). Almost everywhere they are oriented with their crests parallel to the dip of the cross-strata. Their crests are relatively straight, parallel, slightly asymmetric, and spaced at 2 to 3.5 in (5.5 to 9.0 cm). Ripple height is from 0.06 to 0.12 in (1.5 to 3 mm), giving indices of 36 to 30. Ripples of this type are similar in regularity and ripple index to those developed on lee slopes of modern dunes where sand is transported across the face and subcritically climbing translent strata (Hunter, 1977) are developed. Many of the ripples in the Casper display truncated crests that appear as lineations where the foreset bedding of the ripple intersects the bounding surface. It is doubtful that this truncation is caused by recent erosion of complete ripples at the outcrop, since it can be traced under overlying strata. Truncation of modern ripples takes

place when sand, blown into ripples while dry, subsequently becomes moist. Greater exposure of certain parts of the crests to wind and sun causes drying and remobilization. Portions of the crests are blown away while other parts of the crests and almost all of the troughs remain wet and intact.

Lag Grains. Uniformly distributed coarse grains occur along surfaces at the base of troughs that are otherwise filled with well-sorted, fine-grained sand (Fig. 1, location C). These grains are composed primarily of quartz and feldspar, are up to 0.2 in (5 mm) in diameter and are spread 0.2 to 0.4 in (5 to 10 mm) apart. Uniform spacing of the lag grains is apparently uniquely related to an eolian process described by Bagnold (1954). Grains with diameters more than six times those of the grains in saltation are nearly immobile. Where these large grains are scattered among fine sand, as is the case in the Casper, they become concentrated as lag grains on deflation surfaces as they are progressively exposed. A stable surface is thus formed of uniformly spaced lag grains.

"Soft-Sediment" Deformation. Deformation of cross-strata in sandstone of the Casper Formation (Fig. 1, location A) ranges in scale from complex folds with amplitudes of as much as 24 ft (7.5 m) to minute crinkles and brecciations. The deformation occurs at several stratigraphic levels and in discontinuous zones at the same level. In general, it is more common in the upper part of sets of cross-strata just beneath the superjacent truncation surface, but this is not true in every case. Relations between deformed strata and overlying undeformed strata or overlying truncation surfaces show that the deformation was contemporaneous or penecontemporaneous with sand deposition. In no instances are the major truncations surfaces deformed.

REFERENCES CITED

Bagnold, R. A., 1954, The Physics of Blown Sand and Desert Dunes: London, Methuen and Co, 265 p.

Hanley, J. H., and Steidtmann, J. R., 1973, Petrology of limestone lenses in the Casper Formation, southernmost Laramie Basin, Wyoming and Colorado: Journal of Sedimentary Petrology, v. 43, p. 428–434.

Hanley, J. H., Steidtmann, J. R., and Toots, H., 1971, Trace fossils from the Casper sandstone (Permian), southern Laramie Basin, Wyoming and Colorado: Journal of Sedimentary Petrology, v. 41, p. 1065–1068.

Hunter, R. E., 1977, Basic types of stratifications in small eolian dunes: Sedimentology, v. 24, p. 362–387.

Knight, S. H., 1929, The Fountain and the Casper Formations of the Laramie Basin: A study of the genesis of sediments: Wyoming University Publishers in Science, v. 1, p. 1–82.

Steidtmann, J. R., 1974, Evidence for eolian origin of cross-stratification in sandstone of the Casper Formation, southernmost Laramie Basin, Wyoming: Geological Society of America Bulletin, v. 85, p. 1835–1842.

——— , 1976, Eolian origin of sandstone in the Casper Formation, southernmost Laramie Basin, Wyoming, in Epis, R. C., and Weimer, R. J., eds., Studies in Colorado field geology: Professional Contributions of Colorado School of Mines, no. 8, p. 86–95.

Steidtmann, J. R., and Haywood, H. C., 1982, Settling velocities of quartz and tourmaline in eolian sandstone strata: Journal of Sedimentary Petrology, v. 52, p. 395–399.

Stiedtmann, J. R., and Weimer, R. J., 1976, Paleozoic depositional environments of the northern Front Range, Colorado and Wyoming, in Epis, R. C., and Weimer, R. J., eds., Studies in Colorado field geology: Professional Contributions of Colorado School of Mines, no. 8, p. 78–85.

Early Proterozoic and Precambrian-Cambrian unconformities of the Nemo area, Black Hills, South Dakota

Jack A. Redden, South Dakota School of Mines and Technology, Rapid City, South Dakota 57701

INTRODUCTION

Excellent exposures of both an early Proterozoic unconformity and the Precambrian-Cambrian unconformity occur in the Nemo area along the eastern side of the Precambrian core of the Black Hills approximately 18 mi (30 km) northwest of Rapid City (Fig. 1). Although the two unconformities are exposed relatively near one another, there is approximately 1.5 Ga difference in their ages. The lithologies and structure of these localities permit significant inferences about tectonic events and geologic history of the Black Hills.

SITE 49, EARLY PROTEROZOIC UNCONFORMITY

Location

One of the best exposures of the early Proterozoic unconformity is approximately 2 mi (3 km) south-southwest of the village of Nemo (Fig. 2). The site can be reached in high clearance vehicles (vans, pickups) by following an access road (shown on part of the Nemo 7½-minute Quadrangle in Fig. 1) or by leaving conventional vehicles along Estes Creek and walking approximately 1,600 ft (500 m) north-northwesterly. The area is public land and is part of the Black Hills National Forest. Unpatented mining claims exist on the area shown in Figure 2, but no mining development is anticipated, and access is permissible under usual regulations of the U.S. Forest Service.

The unconformity in the Precambrian rocks was first described by Runner (1934). A more complete description and interpretation of the geologic relationships in the Nemo area is in Redden (1980; 1981).

Significance

The exposures at this site document not only the existence of a well-exposed overturned angular unconformity in the low grade early Proterozoic metamorphic rocks, but also the local source of very coarse clastic fanglomerate occurring above the unconformity. In addition, careful examination of the rock distribution indicates the following: (1) The older rocks were strongly folded before uplift and the erosional surface locally had a relief of approximately 400 ft (125 m) and a nearly vertical slope. Most of the relief developed on the steep limb of a pre-unconformity fold in resistant banded iron formation. (2) Folding and the development of a strong northwest foliation occurred following the deposition of younger rocks and overturning of both younger and older rocks.

The observations possible at this site help confirm that the site is adjacent to one of the major growth faults [displacement

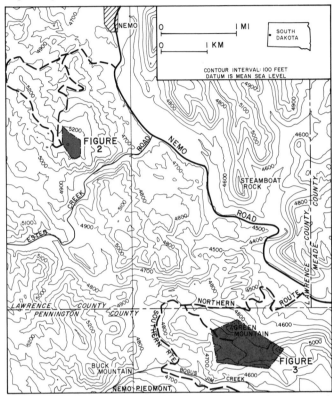

Figure 1. Parts of the Nemo and Piedmont 7½-minute Quadrangles, showing the locations of the areas covered in Figures 2 and 3.

approximately 6 mi (10 km)] of the Nemo region and that the region represents a tensional tectonic regime that developed in an intracontinental rifting environment.

Site Information

The accompanying geologic map (Fig. 2) shows most of the critical relationships of the various rock units. The Precambrian rocks are exposed on a small flat hill whose upper surface virtually corresponds to the Cambrian-Precambrian unconformity, but conglomerate from the Cambrian Deadwood Formation is present only in a small prospect pit shown in the northwest part of Figure 2. The oldest rocks are poorly exposed quartzites of the Boxelder Creek Formation. However, if the unconformity is followed approximmtely 2,000 ft (600 m) to the north-northeast of the area shown on the map, there are excellent exposures of easterly trending, steep northerly dipping quartzite. These exposures contain well-developed small-scale cross beds that indicate the beds are overturned. The beds also strike into and are truncated by the unconformity. These exposures document that the quartzites are lithologically identical to large clasts found above the unconformity.

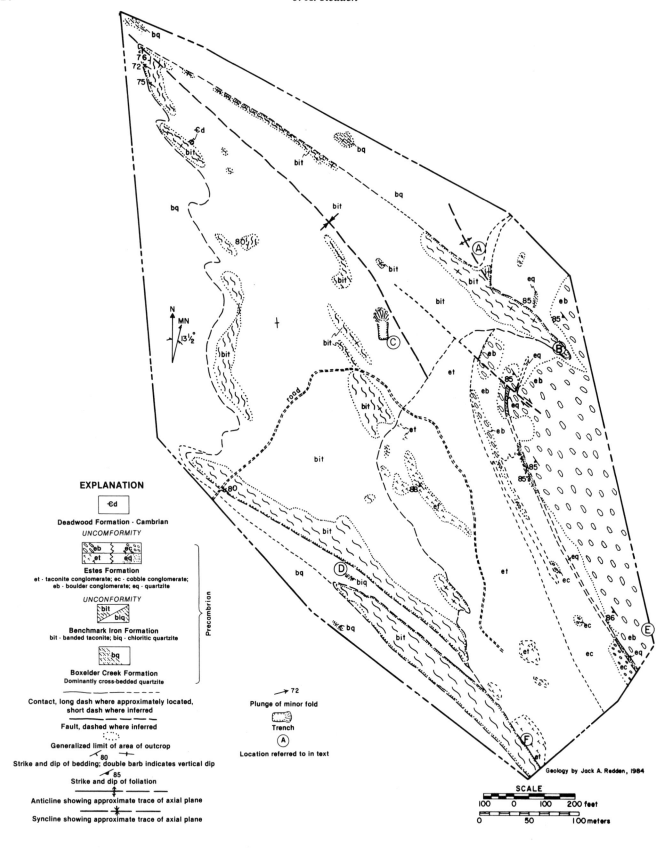

Figure 2. Geologic map of the Early Proterozoic unconformity near Estes Creek, Nemo quadrangle, Black Hills, South Dakota.

Although the contact of the Boxelder Creek Formation and the overlying Benchmark Iron Formation is not exposed, a few exposures shown in Figure 2 indicate the location to within about 80 ft (25 m). Scattered float suggests the contact is a transitional zone of quartzose chloritic schist and cherty beds.

The Benchmark Iron Formation is virtually all uniformly banded taconite consisting of hematite and recrystallized chert bands typical of Precambrian banded iron formations. Included near the middle of the unit are a few thin beds of chloritic quartzite. A few thin beds of pebbly taconite conglomerate occur along the west side near locality D (Fig. 2). Immediately south of locality D, taconite is separated from the main body by poorly exposed chloritic quartzite. Because there is no evidence of folding, this taconite is interpreted to be a lens of the main taconite mass and is included therein. However, there may be some other structural explanation for this isolated occurrence. The general distribution of the Benchmark indicates that it is exposed in a syncline that has been truncated by the unconformity to the southeast. The core of this structure (locality C, Fig. 2) should be the upper part of the Benchmark or possibly a younger unit. Unfortunately, there are no exposures in this area and the available float could equally well be derived from taconite, a younger taconite conglomerate, or from mixtures of the two. The only in situ rock is a very small outcrop of chloritic schist. Lacking any subsurface information, I interpret this area of no outcrop to be the upper part of the Benchmark; it probably consists of mixed taconite, chloritic schist, and chert.

The unconformable contact as shown in Figure 2 follows a relatively sinuous course, highlighted by the "peak" at locality B. Rock units above the unconformity are part of the Estes Formation, which forms an extensive fan along a major north-trending growth fault located approximately 0.6 mi (1 km) east of the area shown in Figure 2. The different lithologies in the Estes occurring along the unconformable contact include boulder conglomerate, taconite conglomerate, cobble conglomerate, and quartzite.

The oldest unit is taconite conglomerate, which consists almost entirely of pebble- to cobble-sized, subangular to subrounded, generally flat, irregularly shaped clasts of taconite. These are oriented with their smallest dimension subperpendicular to a north-northwesterly trending well-developed foliation. The longer dimensions of the clasts are subparallel to bedding in the few exposures where bedding is recognizable. The conglomerate matrix is chloritic quartzite or smaller fragments of taconite and chert. Magnetite is relatively abundant and where present tends to rim the individual taconite clasts. Bedding is virtually unrecognizable in the northernmost exposures of the unit, but noticeable interbeds of coarse-grained chloritic quartzite are located near the south end of the area, and the general clast size diminishes as does the quantity of taconite clasts. The unit thickness is generally about 200 ft (60 m), but pinches to zero against the unconformity midway between localities B and C (Fig. 2).

The cobble conglomerate consists dominantly of ellipsoidal quartzite clasts and fewer taconite clasts. The latter tend to be more abundant to the south where the general clast size dimin-

ishes. To the north the cobble conglomerate grades laterally into boulder conglomerates. Clasts are ellipsoidal and generally oriented with their long dimensions plunging steeply northwestward and short dimensions perpendicular to the northwest foliation. Bedding is not recognizable in the relatively small outcrops characteristic of the unit. The thickness of the cobble conglomerate is approximately 100 ft (30 m).

The boulder conglomerate lithology is exceptionally well exposed in both irregular craggy outcrops and flat ground-level slopes marked by depressions between boulders. The large clasts are dominantly light tan to dark gray quartzite that are lithologically identical to the Boxelder Creek quartzite, complete with crossbedding. The largest clasts in the area of Figure 2 are more than 3 ft (1 m) in longest dimension, although clasts more than 6 ft (2 m) across occur very close to the unconformity about 1,600 ft (500 m) northeast of the mapped area. A few boulders consist of taconite, but most of the taconite is pebble-sized material that, along with quartzite, forms the matrix of the boulders. The quartzite matrix formed from sand derived from the Boxelder Creek Quartzite, and thus it is difficult locally to distinguish the clasts from the matrix. The conglomerate is almost universally tightly packed, and individual ellipsoidal boulders tend to wrap around and mold between adjacent boulders. Although the initial shapes of the boulders are not known, the pronounced molding of the elliptical shapes clearly indicates considerable deformation and flattening in the plane of foliation. The long axes of the boulders also tend to plunge steeply northwest in agreement with the other conglomerate clasts in the Estes Formation.

Stratigraphically, the boulder conglomerate is separated into two units by a thin, 10 to 20 ft (3 to 6 m) thick unit of grayish white to tan quartzite that also contains isolated clasts and lenses of conglomerate. This quartzite unit forms the lower contact of an upper thick boulder conglomerate that is more than 330 ft (100 m) thick southeast of locality B (Fig. 2). This unit changes laterally to the southeast in that clast size and packing decrease somewhat; at the extreme southeast part of the area shown on Figure 2, bedding is recognizable as thin lenses of quartzite interspersed with the conglomerate.

The quartzite unit shown in Figure 2 can be traced from the south to within approximately 50 ft (15 m) laterally from the unconformity northwest of locality B (Fig. 2). Here the outcrop is concealed, but a small outcrop of quartzite lying on the northeast side of the prong of Benchmark taconite is interpreted to be the same stratigraphic unit.

Structural Interpretation

The dominant pre-unconformity structure, as shown by the lower contact of the Benchmark, is the syncline previously mentioned whose axial trace is shown on Figure 2. The plunge of the syncline is approximately 70° to the northwest, but because the plunge has been tilted past 90° during the overturning of the unconformity and adjacent rocks, the syncline is now antiformal. Because of the steep plunge, inverted stratigraphy, and near vertical dips it is possible to see a reasonable approximation of the fold's cross section if the map (Fig. 2) is viewed upside down. The northwest limb of the fold has several minor folds as indicated by the general outcrop distribution and trends of bedding. These also are interpreted to be pre-unconformity in age because of the lack of digitations in rocks above the unconformity. However, some geologists might favor the interpretation that the folds formed at the time of formation of the foliation and that the failure of their expresson in overlying younger rocks is the result of the different physical properties of the taconite and those of the younger conglomerates. This alternate interpretation cannot be evaluated because of the lack of exposures. An anticlinal axis is believed to occur at locality A (Fig. 2), where taconite trends northerly from the main mass and where one very small isolated outcrop believed to be Benchmark taconite occurs about 160 ft (50 m) from the main mass. The Benchmark has been eroded to the east except for a small north-closing outcrop of taconite about 650 ft (200 m) northeast of the area shown in Figure 2. This is truncated by the unconformity and is interpreted to be another pre-unconformity syncline whose keel was preserved below the unconformity.

Folding and the development of the strong foliation also occurred following the unconformity and deposition of younger Estes rocks. Nearly all the ellipsoidal clasts are oriented with longest and intermediate dimensions in the plane of the foliation, and although bedding is generally not recognizable, the plunge of the longest dimension is approximately that expected from the intersection of the foliation and the steeply dipping bedding. Younger folds are not readily visible on the map but are visible along Estes Creek to the south of Figure 2. Their axial planes are parallel to the northwest foliation. Throughout the Nemo area there are right-handed, steep northwesterly plunging minor folds. These minor folds and the elongate clasts also plunge about 70° northwest. Plots of the various fold plunges and bedding attitudes of rocks below the unconformity versus those above the unconformity indicate no significant differences in the plunges or of bedding attitudes. It is thus inferred from the data that the pre- and post-unconformity foldings (excluding the overturning of the rocks) are coaxial. Because the fold plunges are about the same for both episodes of folding, it can be inferred that the plunge of the pre-unconformity folding was most likely subhorizontal before overturning.

There is almost no evidence of the event that led to the overturning of the rocks prior to the final folding. Elsewhere in the Black Hills, however, apparently early northeasterly trending folds occur, which predate the metamorphism and northwest foliation.

The steep-plunging, post-unconformity folding is related to the development of the northwest-trending foliation and was probably caused by either NE-SW compression or a large-scale, approximately north-south trending, right lateral shear couple.

Summary of Events

Following the deposition of quartzite in the Boxelder

Creek Formation and banded iron-formation (taconite) in the Benchmark Iron Formation, folding occurred, probably along subhorizontal fold axes. The trend of these early folds is not known for certain, although, if it is assumed that the subsequent overturning of these and younger rocks was along northeasterly trending subhorizontal axes, then the original trend was northwesterly and the folds were asymmetric and possibly slightly overturned to the southwest. Following this early folding, the rocks were uplifted and eroded to a surface of locally steep relief whose topography was related to previous structure. A major growth fault developed just east of the site area, and a fan of very coarse clastic material formed adjacent to this fault. Local relief on the unconformity controlled the deposition of the clastic deposits and influenced the rapid lateral facies changes. The original direction of transport is unknown but in the area of Figure 2 was probably either up or down the present dip rather than parallel to the strike. Somewhat younger parts of the fan are probably marine in origin, because they contain subunits of dolomite matrix conglomerate, but the lowermost clastic units at the site may have been valley fill deposits.

The next known structural event involved the overturning of both pre- and post-unconformity rocks along possible northeast-trending horizontal folds. Very little is known about the events leading to the overturning, but it was followed by metamorphism and deformation that produced a northwest-trending, near vertical foliation and steep northwesterly trending folds that are now essentially coaxial with the early pre-unconformity folding. Considerable deformation, including both flattening and elongation of clasts as well as rotation of clasts, accompanied the development of the foliation.

The age of the uplift and development of the unconformity is considered to be younger than 2.1 Ga, which is the preliminary age obtained by R. E. Zartman from zircon that occurs in the upper part of a 3,250 ft (1,000-m) thick gravity-differentiated sill. The sill intrudes the Boxelder Creek quartzites approximately 0.6 mi (1 km) north of the area shown in Figure 2 (Redden, 1981).

SITE 50, PRECAMBRIAN-CAMBRIAN UNCONFORMITY

Location

Excellent exposures of this unconformity occur on a small, relatively flat table near Green Mountain in the southwestern corner of the Piedmont 7½-minute Quadrangle (Fig. 1). The site can be reached by following either of the two routes indicated in Figure 1. The area is part of the Black Hills National Forest and is public land.

Significance

The angular unconformity between the Paleozoic and Precambrian rocks in the Black Hills has long been recognized, but except in a few valleys, the contact is commonly concealed or poorly exposed. The local relief on the unconformity has gener-

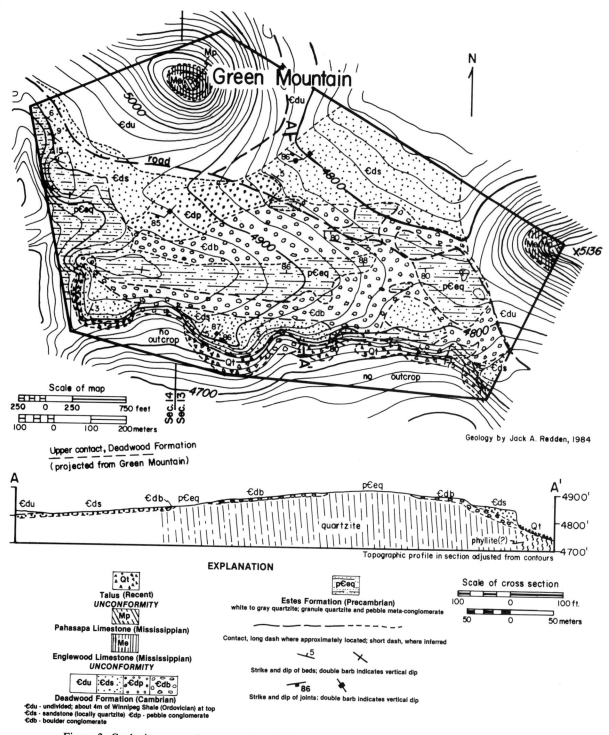

Figure 3. Geologic map and cross section of Green Mountain area, Piedmont quadrangle, South Dakota.

ally been believed to be no more than about 100 ft (32 m). At Green Mountain the unconformable contact is exceptionally well exposed as are the lower facies of the Deadwood Formation. The data on the accompanying geologic map (Fig. 3) and nearby geologic relationships demonstrate that local relief on the unconformity exceeds 200 ft (65 m) above a buried Precambrian quartzite ridge. The exposures also demonstrate the steep and very sharp contact between wave-deposited beach conglomerates and low-dipping beach sandstones. These are known to occur only on the south side of the Precambrian ridge and are interpreted to have formed on the lee side of an island created by advancing Precambrian seas.

Textures of the beach sandstone of the lower Deadwood also confirm that the Cambrian seas advanced over a semitropical weathering profile that had resulted in the disaggregation of the Precambrian quartzites and thus permitted the local deposition of supermature sandstones just above the unconformity. Because of the relief and lateral facies changes, it is also likely that much of the beach sandstone facies at Green Mountain is laterally equivalent to green glauconitic shales, carbonate-rich clastic rocks, and siltstones characteristic of the middle part of the Deadwood Formation.

Site Information

The accompanying geologic map (Fig. 3) shows the general distribution of the different facies at the lower contact of the Deadwood Formation and the unconformable contact with the Precambrian rocks. The map also shows younger Paleozoic formations, but these are not particularly well exposed and are not discussed further.

The exposed Precambrian rocks are dominantly thick bedded, white to nearly black, coarse-grained quartzites that dip nearly vertically and whose stratigraphic tops are to the south. They are part of the Estes Formation present at the Estes Creek Precambrian unconformity described in the previous section, but these outcrops occur on the upthrown side of the large growth fault described at the Estes Creek locality. The large growth fault is approximately 0.6 mi (1 km) west of the area shown in Figure 3; the Proterozoic unconformity previously described is partly concealed by the Cambrian rocks but follows an irregular but approximate east-west trend near the Pennington County line north of Figure 3. Another north-trending Precambrian growth fault occurs approximately 0.3 mi (0.5 km) east of the common corner of Lawrence, Meade, and Pennington Counties (Fig. 1). Fanglomerate is abundant in the Estes Formation near this fault, but there are only a few beds of slightly deformed pebble conglomerate and granule quartzite in the Estes at Green Mountain. Hence the Estes Formation rocks at Green Mountain are also part of a marine fan that thickens to the east near the second fault.

Although not exposed in the area shown on Figure 3, the Precambrian rocks along the extreme south edge of the map and in the valley of the adjacent Bogus Jim Creek are nonresistant phyllite and slightly recrystallized dolomite of the Roberts Draw Formation. The intensity of deformation is considerably less than at Estes Creek, although the northwest-trending, near vertical foliation is evident in the less resistant rocks.

The lower part of the Deadwood at Green Mountain consists of three main facies including boulder conglomerate, pebble conglomerate, and sandstone/quartzite. The boulder conglomerate occurs adjacent to the Precambrian quartzite exposures and is thickest and best developed along steep slopes on the original erosion surface. It consists of rounded to subangular boulders, as much as 3 ft (1 m) across, of various types of quartzite derived from the adjacent or underlying Precambrian rocks. A slight downslope imbrication of clasts is noticeable in one exposure.

The matrix is coarse sand-sized material and debris typical of the same quartzite forming the boulders. Packing is commonly tight, although locally the clasts may be matrix supported. The thickness ranges from as little as one layer of boulders to possibly as much as 45 ft (13 m).

Pebble conglomerate is widespread on the north side of the quartzite ridge as a thin dip slope veneer that tends to overlie sandstone but locally may overlie boulder conglomerate. The clasts are similar in lithology to those in the boulder conglomerate and range from angular to well rounded. The exact thickness of the pebble conglomerate generally cannot be determined, but it is nowhere greater than 6 ft (2 m) and commonly is less than 3 ft (1 m).

The sandstone/quartzite lithology occurs as thick, wedge-shaped deposits overlying the boulder conglomerate along the south side of the Precambrian ridge and as dip slope exposures north of the ridge. The southern exposures form south-facing ledges as much as 50 ft (15 m) high, which are fringed by talus blocks of the quartzite (Fig. 4). The rock is a coarse-grained, well-sorted, tan to white, well-cemented sandstone that locally grades into orthoquartzite. The lower contact with boulder conglomerate is sharp and dips as steeply as 40° to the south, but beds in the sandstone typically dip less than 6°, generally to the southeast. Locally, the uppermost part of the sandstone is relatively coarse grained and grades into a small pebble conglomerate. The thickness of the unit ranges from zero to at least 50 ft (15 m) in the areas shown in Figure 3 but may be somewhat thicker in areas to the southeast.

The second area of outcrop of sandstone/quartzite occurs as a large dip-slope area east of Green Mountain. The sandstone underlies approximately 2 ft (0.6 m) of pebble conglomerate, which crops out closer to the Precambrian contact. The sandstone lithologically resembles the ledge-forming sandstone, but it is inferred to be only about 10 ft (3 m) thick. The thickest section known is about 4,200 ft (1,300 m) N80°W from Green Mountain, where the sandstone fills a 65-ft (20-m) deep, 500-ft (150-m) wide depression along the unconformity.

Depositional Environment

The restricted distribution of the boulder conglomerate adjacent to the Precambrian quartzite ridge and the sharp, steep contact with overlying units indicate that it formed as a wave-reworked beach deposit. Furthermore, the failure of wave action to move boulders away from the shore suggests that this was the lee side of the quartzite island rather than the ocean side. On the north side of the ridge, the boulders are loosely packed, and the thin dip slope of pebble conglomerate suggests a lower slope, open ocean environment where the full force of the waves was active. The sandstone overlying the boulder conglomerate on the south side of the ridge formed contemporaneously with the conglomerate but probably at approximately low tide elevation. The extremely well-rounded sand grains are characteristic of the quartzites that make up the Boxelder Creek Formation and that

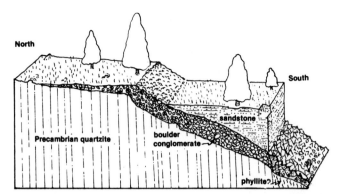

Figure 4. Block diagram showing Cambrian/Precambrian unconformity and distribution of facies in the lower part of the Deadwood Formation.

formed largely when sea level had risen to such elevations that storm waves were breaking across the nearly buried island.

The height of the original hill cannot be measured exactly but the total thickness of the Deadwood is less than 200 ft (62 m), whereas the thickness of the Deadwood about 1.2 mi (2 km) N40°E from the Green Mountain is approximately 500 ft (150 m). The local relief was therefore approximately 300 ft (90 m). Probably a valley existed above nonresistant phyllites that underlie Bogus Jim Creek as inferred by Daly (1981), but there are no exposures of the unconformity immediately south of Figure 3 to permit either a determination of the valley depth or an estimation of how much of that valley was filled with the lower beach facies of the Deadwood. It is likely that the beach-forming sandstone does not extend far from the buried ridge and that it undergoes a lateral facies change to glauconitic shales, siltstones, and flat limestone pebble conglomerate characteristic of the middle part of the Deadwood Formation.

were redeposited in the Estes fans. These grains are extremely well cemented in the Precambrian rocks, and the single prograding cycle of encroachment by the Cambrian sea would be inadequate to round the quartzite fragments if the rock had only been mechanically disrupted. However, elsewhere in the Black Hills a semitropical weathered zone at least 49 ft (15 m) deep is known to exist locally in the Precambrian rocks at the unconformity (Redden, 1963, p. 246). Hence it seems most likely that the original quartzite ridge at Green Mountain was covered by a regolith of disaggregated quartz sand and partially decomposed blocks of quartzite when the sea gradually advanced over the surface. The sand was readily freed by wave and current action and transported into depressions such as the one described west of Green Mountain.

The distribution of the pebble conglomerate on top of most of the sandstone unit on the north side of the area and the gradual change to pebbly sandstone in the uppermost part of the sandstone on the south side of the Precambrian ridge, suggest that it

REFERENCES CITED

Daly, W. E., 1981, Basement control of the deposition of the Cambrian Deadwood Formation in the eastern Black Hills, South Dakota [M.S. thesis] Rapid City, South Dakota School of Mines and Technology, 89 p.

Redden, J. A., 1963, Geology and pegmatites of the Fourmile Quadrangle, Black Hills, South Dakota: U.S. Geological Survey Professional Paper 297-D, p. 199–291.

——— , 1980, Geology and uranium resources in Precambrian conglomerates of the Nemo area, Black Hills, South Dakota: National Uranium Resource Evaluation Report (GJBX 127(80), 147 p.

——— , 1981, Summary of the geology of the Nemo area, *in* Rich, F. J., ed., Geology of the Black Hills, South Dakota and Wyoming: Washington, D.C., American Geological Institute, p. 193–290.

Runner, J. J., 1934, Pre-Cambrian geology of the Nemo district, Black Hills, South Dakota: American Journal of Science, 5th series, v. 28, p. 353–372.

Harney Peak Granite and associated pegmatites, Black Hills, South Dakota

C. K. Shearer and J. J. Papike, Institute for the Study of Mineral Deposits, South Dakota School of Mines and Technology, Rapid City, South Dakota 57701

LOCATION AND ACCESS

The Harney Peak Granite is located in the south-central portion of the Precambrian core of the Black Hills, South Dakota. This field study includes five sites near Keystone, South Dakota (Figs. 1 and 2). All sites are accessible by passenger car. At Site 1, the Harney Peak Granite is exposed at the Mount Rushmore National Memorial. (Large groups visiting at Mount Rushmore are expected to check in at the memorial headquarters.) *Hammers or sample collecting are not allowed at the Rushmore site.* A suite of Harney Peak Granite may be collected to the west of the memorial (Site 1a). Sites 2 through 5 are pegmatites (Diamond Mica, Etta, Peerless, and Dan Patch) associated with the Harney Peak Granite. Permission for access to the Etta pegmatite may be obtained from the Pacer Corporation, Box 912, Custer, South Dakota 57730 (605/673-4458). Permission for access to the Peerless pegmatite may be obtained from the Rushmore Borglum Museum in Keystone, South Dakota (605/666-4449). The other sites are on public land.

SIGNIFICANCE

The close spatial relationship between granites and pegmatites has been taken to indicate their common petrogenetic origins, with the pegmatite being derived from the granite. The Harney Peak Granite and associated pegmatites are a classic example of this type of spatial-petrogenetic relationship. This geological site provides clear illustration of the critical spatial and genetic relationships between a granite intrusive and its associated barren and mineralized pegmatites.

SITE INFORMATION

The Harney Peak Granite was intruded into country rock consisting chiefly of metamorphosed graywackes, shales, and quartzites; it has been dated at 1.7 Ga (Riley, 1970). Metamorphic grade in the vicinity of the intrusion reaches second sillimanite conditions. The main body of the Harney Peak Granite is in the form of a structural dome, and consists of several large sills and a multitude of smaller sills and dikes. The major mineral phases present are quartz, microcline, plagioclase, and muscovite. Tourmaline is nearly ubiquitous, while other common accessory minerals include garnet, apatite, sillimanite, and biotite. The Harney Peak Granite is considered to be an "S-type" granite, based on its peraluminous chemistry (Shearer and others, 1985a).

Approximately 20,000 pegmatites are distributed through the country rock surrounding the Harney Peak Granite. A majority of these are unzoned pegmatites, and are mineralogically and geochemically similar to the Harney Peak Granite. Zoned, or complex pegmatites make up approximately 1 percent of the total pegmatite population. This pegmatite field has been classified as a rare-element type, with mineralogical characteristics ranging from barren to Li-, Rb-, Cs-, Be-, Ta-, and Nb-enriched types (Cerny', 1982). These complex pegmatites exhibit mineralogical and geochemical zonal distribution, with barren pegmatites commonly occurring near the Harney Peak Granite, while Li-enriched pegmatites occur on the periphery of the field. This zonation has been observed in many other pegmatite districts, and has been attributed to the ability of highly fractionated, volatile-enriched granitic melt to migrate some distance from the parent intrusion (Cerny' and others, 1981; Goad and Cerny', 1981). This zonal distribution suggests a genetic link between pegmatites in the district and the central granitic intrusion. In the Black Hills, such regional zoning is particularly evident to the south of the Harney Peak Granite in the vicinity of Custer, South Dakota; zonation is more obscure in the Keystone area to the northeast of the Harney Peak Granite. (The district is shown in Fig. 1; the Keystone area, to be discussed, is enlarged in Fig. 2.)

Stop 1. Harney Peak Granite. The intricate intrusive relationships between the Harney Peak Granite and Precambrian country rock can be observed at the western boundary of the Mount Rushmore Memorial, the first stop. In the Mount Rushmore Memorial, the Harney Peak Granite consists of numerous, irregularly shaped sills and dikes, which intrude country rock of quartz-mica-feldspar schist and sulfide-bearing garnet schist. Schist blocks are occasionally incorporated as xenoliths within the intrusive. Metamorphic grade at this stop is above the low sillimanite isograd (Fig. 2).

The Harney Peak Granite is compositionally and texturally varied. This site illustrates many of the variations. The large north-south–trending sill in the memorial consists of layered aplitic, granitic, pegmatitic, and monomineralic units. Two common assemblages are present. The first is a medium- to fine-grained hypidiomorphic-granular rock, which exhibits an aplitic, banded texture. Layering in the sill is defined by modal differences in albite, quartz, potassium feldspar, and muscovite. Elsewhere in the Harney Peak Granite, tourmaline or garnet may define a microlayering superimposed on the larger scale layering. Layering is generally parallel to the contacts of the sill, and is discontinuous along strike. Other common forms are lenticular pegmatitic granite segregations, which consist of coarse graphic microcline in a fine-grained quartz-albite-muscovite groundmass. Massive quartz may sometimes be found within these segregations (Fig. 3).

Numerous small intrusions of varying composition may be

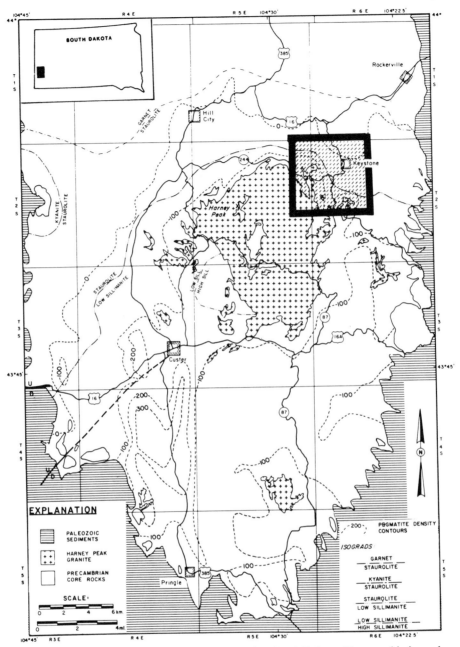

Figure 1. Generalized geology of the southern Black Hills, South Dakota. Metamorphic isograds and pegmatite density contours (pegmatite bodies per square mile) are shown (after Norton, 1975). Shaded area outlined is enlarged in Figure 2.

seen within the large granitic body in the Mount Rushmore Memorial. Four mineral assemblages, which show compositional gradation, are observed: (1) plagioclase + perthite + quartz; (2) perthite + quartz; (3) perthite; and (4) quartz. These smaller intrusions may cut the primary layering in the granite, and so disrupt it; or they may intrude along the layering, and so disguise their intrusive origin.

Stop 2. Diamond Mica pegmatite. The Diamond Mica pegmatite is a large simple pegmatite; it is ellipsoidal in shape and consists primarily of an albite-quartz-perthite assemblage. A discontinuous, fine-grained albite-quartz-muscovite assemblage occurs along the perimeter of the pegmatite. This unit varies from 1.2 in to 5 ft (3 cm to 1.5 m) in thickness. Tourmaline, garnet, and biotite occur as accessory minerals throughout the pegmatite. Coarse-grained quartz and microcline are also present in a small fracture-filling unit that cuts the main pegmatite body.

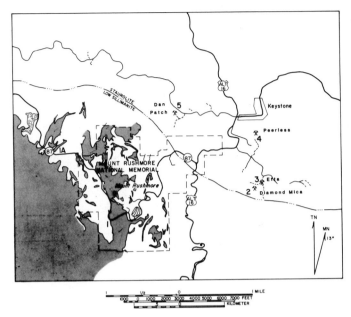

Figure 2. Location of 5 stops at this site. Also shown are the distribution of the Harney Peak Granite and metamorphic isograds. The outline of the Harney Peak Granite is from Norton (1976).

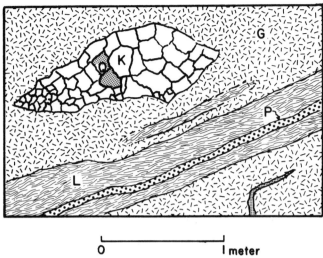

Figure 3. Outcrop scale compositional variation in the Harney Peak Granite. G, medium-grained granite; K, perthite segregation; Q, massive quartz occurring as veins and within perthite segregations, L, fine- to medium-grained line rock showing well-developed compositional layering; P, perthite + quartz assemblage with an ambiguous relationship with line rock.

In the coarse-grained quartz-albite-microcline assemblage, megacrysts of perthitic microcline are present as megacrysts in a finer-grained groundmass of quartz + albite + muscovite. Microcline commonly is graphically intergrown with quartz.

This pegmatite shows a change in the modal mineralogy from the contacts inward: enrichment in potassium is indicated by increasing abundance of microcline and a simultaneous decrease in the amount of albite. The Rb and Cs contents of the muscovite and microcline also increase from the footwall to the hanging wall. Alkali enrichment of this type is also commonly observed in more complex, zoned pegmatites. Here, however, such enrichment is gradational and is obscure in outcrop.

Stop 3. Etta pegmatite. The Diamond Mica pegmatite is simple in character and is similar in composition and mineralogy to the Harney Peak Granite; the Etta pegmatite, on the other hand, is a large, complexly zoned pegmatite with an exotic mineral assemblage. Structural complexity and trace element ratios suggest that the Diamond Mica pegmatite and the Etta pegmatite represent the two ends of the petrogenetic spectrum in this pegmatite district (Shearer and others, 1985b).

The Etta pegmatite has been the largest producer of spodumene in the Black Hills. Additionally, potassium feldspar, beryl, columbite-tantalite, cassiterite, and mica have been mined. Much of the original pegmatite is inaccessible due to flooding in the abandoned mine works. However, many of the zones, showing important textural and mineralogical features, can still be seen. Obvious contrasts to the Diamond Mica pegmatite can be made.

The Etta pegmatite is shaped like an inverted teardrop, plunging steeply to the north. Li, Rb, and Cs dispersion halos extend from the pegmatite into the country rock, which is predominantly a quartz-mica schist. The pegmatite consists of six different zones: (1) microcline-biotite zones; (2) quartz-muscovite-albite zone; (3) perthite-quartz-spodumene zone; (4) quartz-cleavelandite-spodumene zone; (5) perthite-quartz zone; (6) quartz-spodumene zone; and (7) quartz core. This zonal distribution is shown in Fig. 4.

The most accessible zones in the Etta pegmatite are the quartz-muscovite-albite zone and the quartz-cleavelandite-spodumene zone. When one enters the pegmatite through the main adit, one sees the quartz-muscovite-albite zone in direct contact with the country rock. This zone is a fine- to medium-grained pegmatite, and ranges in thickness from 4 in to 7.5 ft (0.1 to 2.3 m). The mineral assemblages consists primarily of quartz (50 to 60%), muscovite (15 to 25%), and albite (10 to 20%). Accessory minerals include microcline, tourmaline, beryl, spodumene, and triphylite-lithiophilite. The quartz-muscovite-albite zone grades inward to the coarse- to very-coarse-grained quartz-cleavelandite-spodumene zone. Modal percentages of minerals in this zone are approximately: quartz (30 to 40%), cleavelandite (30 to 40%), and spodumene (20 to 30%). Spodumene crystals are commonly 3.5 to 10.5 ft (1 to 3 m) long, although crystals up to 40 ft (12 m) in length have been observed in the Etta pegmatite (Page and others, 1953). Accessory minerals within this zone include muscovite, beryl, cassiterite, columbite-tantalite, fluorapatite, tourmaline, and uraninite. Other zones in this pegmatite are less accessible.

Stop 4. Peerless pegmatite. The Peerless pegmatite, a large, complex pegmatite, has an anticlinal shape in cross section;

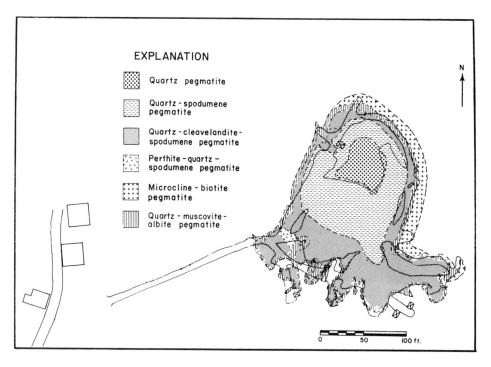

Figure 4. Geologic map of the Etta pegmatite (Page and others, 1953).

it is partly discordant with the quartz-mica schist country rock. The pegmatite consists of seven coherent zones, two replacement units, and two types of fracture filling. The distribution of these units, and the dominant mineralogy of each, are shown in Figure 5.

The border and wall zones of the Peerless pegmatite are best observed just outside the main adit. The border zone is very fine grained, and has an average thickness of 0.4 in (10 mm). The mineral assemblage in the border zone is typical of a complex pegmatite, consisting of albite, quartz, and muscovite, with accessory tourmaline, apatite, garnet, beryl, and columbite-tantalite. The next innermost zone, the wall zone, has an average thickness of 5 ft (1.5 m), but may range in thickness from 0 to 20 ft (0 to 6 m). The wall zone exhibits layering, which is generally parallel to the contact with the country rock. The layers are texturally diverse, ranging from aplitic to pegmatitic. The general mineral assemblage is quartz-albite, with varying proportions of muscovite and perthite. Beryl and tourmaline may be associated with pegmatitic layers, while apatite, garnet, amblygonite, tantalite-columbite, and cassiterite are also common accessories.

The cleavelandite-quartz-muscovite zone and the cleavelandite-quartz zone are accessible via the main adit. These zones are mineralogically similar, except that as one progresses inward, muscovite becomes minor or absent. In both zones, cleavelandite is the dominant mineral phase, although there is great modal variability in the quartz to cleavelandite ratio. Within the cleavelandite-quartz zone are large ovoidal masses of radiating cleavelandite.

The perthite-cleavelandite-quartz zone is most easily acces-

sible at the southeastern side of the Peerless pegmatite. At this locality it is adjacent to the border zone. This zone is approximately 8.5 ft (2.5 m) here, but thickens to 33 ft (10 m) where it forms a hood in the upper part of the northeastern limbs of the pegmatite. This zone consists of large perthite crystals (to 8.5 ft; 2.5 m in length) in a quartz-cleavelandite-muscovite groundmass. Beryl, tourmaline, and apatite are common accessory minerals.

The quartz-microcline zone and the lithia mica-cleavelandite core are not accessible at the present time. They have been extensively mined, and are precariously located deep within the pegmatite. The mineral assemblage is primarily quartz and microcline at the extremities of these zones; the mode of lithia mica increases inward to a high of 82 percent at the core (Sheridan and others, 1957).

Although the Peerless pegmatite was mined primarily for scrap mica, it was known as the largest producer of beryl in the Black Hills.

Stop 5. Dan Patch pegmatite. This pegmatite consists of two interconnected bodies, each nearly spherical in shape, which are separated by a roll in the quartz-mica schist country rock. The pegmatite has been divided into seven units: (1) quartz-albite-muscovite wall zone, (2) quartz-albite-perthite zone, (3) perthite-quartz-albite zone, (4) quartz-albite zone, (5) albite-quartz-phosphate zone, (6) perthite-quartz zone, and (7) quartz core (Norton and others, 1964). Only zones (1), (2), and (3), on the north-northwest side of the pegmatite, are readily accessible for detailed observation. The zonal distribution of minerals within the pegmatite is shown in Figure 6.

One enters the pegmatite on the northwest side. At this point

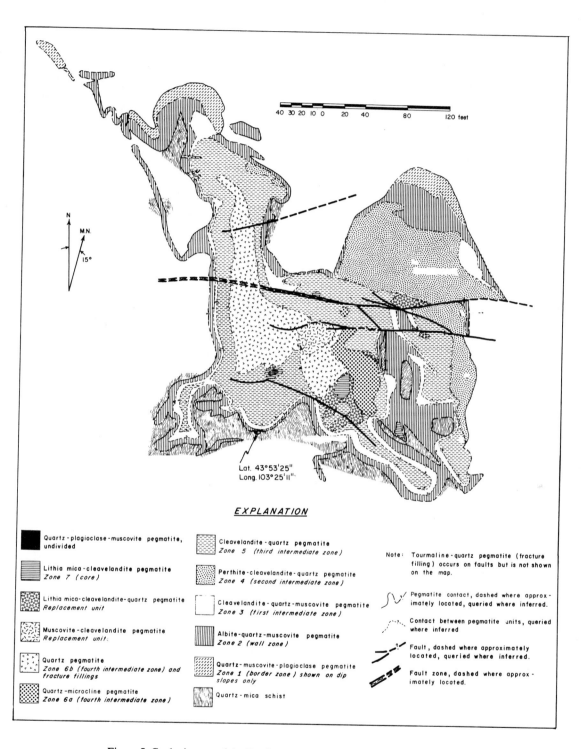

EXPLANATION

Quartz-plagioclase-muscovite pegmatite, undivided

Lithia mica-cleavelandite pegmatite
Zone 7 (core)

Lithia mica-cleavelandite-quartz pegmatite
Replacement unit

Muscovite-cleavelandite pegmatite
Replacement unit

Quartz pegmatite
Zone 6b (fourth intermediate zone) and
fracture fillings

Quartz-microcline pegmatite
Zone 6a (fourth intermediate zone)

Cleavelandite-quartz pegmatite
Zone 5 (third intermediate zone)

Perthite-cleavelandite-quartz pegmatite
Zone 4 (second intermediate zone)

Cleavelandite-quartz-muscovite pegmatite
Zone 3 (first intermediate zone)

Albite-quartz-muscovite pegmatite
Zone 2 (wall zone)

Quartz-muscovite-plagioclase pegmatite
Zone 1 (border zone) shown on dip
slopes only

Quartz-mica schist

Note: Tourmaline-quartz pegmatite (fracture filling) occurs on faults but is not shown on the map.

Pegmatite contact, dashed where approximately located, queried where inferred.

Contact between pegmatite units, queried where inferred

Fault, dashed where approximately located, queried where inferred.

Fault zone, dashed where approximately located.

Lat. 43°53'25"
Long. 103°25'11"

Figure 5. Geologic map of the Peerless pegmatite (Sheridan and others, 1957).

the mineralogical and textural transition between the wall zone and the perthite-quartz-albite zone is continuously exposed. The wall zone completely surrounds the pegmatite; at the northwest entrance it is approximately 1 to 2 ft (30 to 60 cm) wide, and dips 50° to 65° to the east. The unit consists primarily of quartz, albite, and muscovite, with accessory tourmaline, beryl, apatite, perthite, and columbite. Average grain size is 2 in (5 cm).

On the northwest side of the pegmatite, the quartz-albite-perthite zone is transitional between the wall zone and the perthite-quartz-albite zone. The modal abundance and average grain size of the perthite increase inward from the wall zone. The next inner unit, the perthite-quartz-albite zone, is volumetrically the dominant unit in the pegmatite; it forms a hood over most of the other units. The texture of the rock in this zone consists of subhedral to euhedral megacrysts of perthite immersed in a finer-grained groundmass of quartz and albite. Common accessory minerals are beryl, tourmaline, apatite, and muscovite.

The septum of country rock, which structurally divides the pegmatite into two separate spherical bodies, is present in the north wall of the pegmatite. Large perthite crystals extend into the quartz-albite groundmass at the base of the schist.

The Dan Patch pegmatite was mined for potassium feldspar in the 1940s. Mica and beryl were recovered as economic by-products.

REFERENCES CITED

Cerny', P., 1982, The pegmatite field of the southern Black Hills; Geology of the pegmatite field: Geological Association of Canada; Mineralogical Association of Canada Field Trip Guidebook 12, p. 3–8.

Cerny', P., Trueman, D. L., Ziehlke, D. V., Goad, B. E., and Paul, B. J., 1981, The Cat Lake–Winnipeg River and the Wekusko Lake pegmatite fields, Manitoba: Manitoba Department of Energy and Mines, Mineral Resources Division, Economic Geology Report ER80-1, 234 p.

Goad, B. E., and Cerny', P., 1981, Peraluminous pegmatitic granites and their pegmatitic aureoles in the Winnipeg River district, southeastern Manitoba: Canadian Mineralogist, v. 19, p. 177–194.

Norton, J. J., 1975, Pegmatite minerals; in Mineral and water resources of South Dakota, 2nd edition: South Dakota Geological Survey Bulletin 16, p. 132–149.

——, 1976, Field compilation map of the geology of the Keystone pegmatite area, Black Hills, South Dakota: U.S. Geological Survey Open File Map 76-297, scale 1:24,000.

Norton, J. J., and others, 1964, Geology and mineral deposits of some pegmatites in the southern Black Hills, South Dakota: U.S. Geological Survey Professional Paper 297-E, p. 293–341.

Page, L. R., and others, 1953, Pegmatite investigations, 1942–1945, Black Hills, South Dakota: U.S. Geological Survey Professional Paper 247, 228 p.

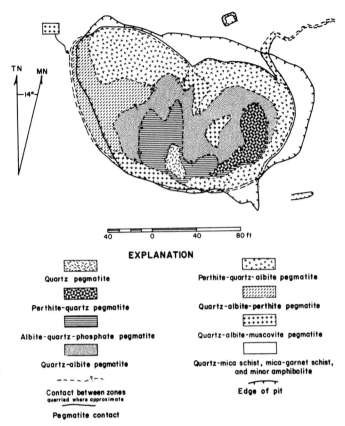

EXPLANATION

Quartz pegmatite

Perthite-quartz pegmatite

Albite-quartz-phosphate pegmatite

Quartz-albite pegmatite

Contact between zones
queried where approximate

Pegmatite contact

Perthite-quartz-albite pegmatite

Quartz-albite-perthite pegmatite

Quartz-albite-muscovite pegmatite

Quartz-mica schist, mica-garnet schist, and minor amphibolite

Edge of pit

Figure 6. Geologic map of the Dan Patch pegmatite (Norton and others, 1964).

Riley, G. H., 1970, Isotopic discrepancies in zoned pegmatites, Black Hills, South Dakota: Geochimica et Cosmochimica Acta, v. 34, p. 713–725.

Shearer, C. K., Papike, J. J., and Walker, R. J., 1985a, Mineral chemistry and geochemistry of the Harney Peak Granite and associated pegmatites, in Rich, F. J., ed., Geology of the Black Hills, South Dakota and Wyoming: Field Trip Guidebook for G.S.A. Rocky Mountain Section Meeting, American Geological Institute, p. 241–260.

Shearer, C. K., Papike, J. J., and Laul, J. C., 1985b, Chemistry of potassium feldspars from three zoned pegmatites, Black Hills, South Dakota; Implications concerning pegmatite evolution: Geochimica et Cosmochimica Acta, p. 663–673.

Sheridan, D. M., Stephens, H. G., Staatz, M. H., and Norton, J. J., 1957, Geology and beryl deposits of the Peerless pegmatite, Pennington County, South Dakota: U.S. Geological Survey Professional Paper 297-A, p. 1–47.

The White River Badlands of South Dakota

James E. Martin, South Dakota Geological Survey and South Dakota School of Mines and Technology, Rapid City, South Dakota
57701

INTRODUCTION

One of the most famous and spectacular Tertiary deposits in North America is the White River Badlands. The colorfully banded, fossiliferous, highly dissected landscapes are exposed across much of southwestern South Dakota; a specific landmark is difficult to designate. Most visitors to the Badlands traverse the area in the Badlands National Park between the towns of Interior and Wall in Pennington and Jackson Counties (Fig. 1). Most geological features may be viewed along South Dakota 240, the Badlands Loop; therefore this area has been designated a landmark. Geology is well-exposed along the route at such localities (Fig. 1) as the Pinnacles, Yellow Mounds, Rainbow Overlook, Banded Buttes Overlook, and Cedar Pass. Due to space limitations herein, frequent references are made to the road log, photographs, and bibliographic references of the area by Martin (1985).

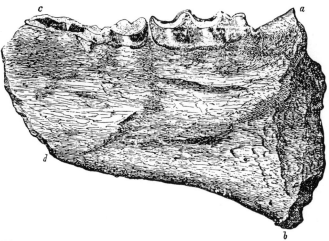

Figure 2. The first vertebrate fossil illustrated and described from the White River Badlands, a fragment of a lower jaw of a brontothere by Prout (1846).

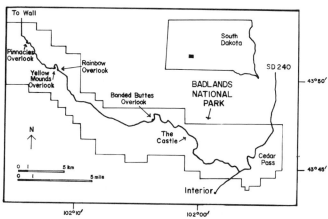

Figure 1. Eastern portion of Badlands National Park, South Dakota. The locations of prominent landmarks discussed in the text are illustrated.

HISTORY OF INVESTIGATION

The Badlands typify the Oligocene sedimentary rocks of North America. (The erosional landscape term, badlands, is based on these exposures.) In 1823, the Jedediah Smith party became the first white men to enter the Badlands (*Mako sica* of the Indians; *Le Mauvaises terres á traverser* of French trappers). The first reported vertebrate fossil was that of a brontothere (Fig. 2) described by Prout (1846), a St. Louis physician. The following year, a skull of a fossil camel was reported by Leidy (1847), who described many vertebrates initially collected from the Badlands, and summarized them in Leidy (1869). Government surveys (including those of Owen, 1852 [Fig. 3]; Meek and Hayden, 1858, who first named the White River Badlands; and Darton, 1899), together with investigations by universities and museums,

amassed large vertebrate collections and geological information during the following century. Much of this data is summarized by O'Harra (1920) and by Scott and Jepsen in the White River monographs (1936–1941). The work of these many groups remains the standard for Oligocene faunas, whereas ongoing research enhances our understanding of taxa present, their biostratigraphic distribution, and concurrent depositional environments.

GEOLOGY OF THE WHITE RIVER BADLANDS

The Badlands succession in this area (Fig. 4) lies disconformably over the black, marine Pierre Shale (Late Cretaceous). Where covered, the succession is commonly overlain disconformably by Quaternary aeolian deposits. Between the Pierre Shale and the Chadron Formation is the thick, yellow and red Interior paleosol. In the Badlands region, the paleosol represents the missing Paleocene and Eocene interval, part of which is preserved in northwestern South Dakota. (See Martin, 1983, for a composite of Tertiary formations in western South Dakota.)

The flat-lying succession of the White River Badlands is best exposed in this area of South Dakota, but smaller, equivalent sequences exist elsewhere in the state and in surrounding western states. Overall, the Badlands are composed of volcaniclastic and fluvial sediments eroded from the Black Hills and Rocky Mountains to the west and deposited by easterly flowing rivers. Between episodes of sedimentation, soil horizons were formed and impure ash beds were deposited. This resulted in the banded appearance of the Badlands succession. Each formation has definite characteristics. The Chadron Formation is usually a gray

Figure 3. The first illustration of the Badlands, published in the Owen report (1852).

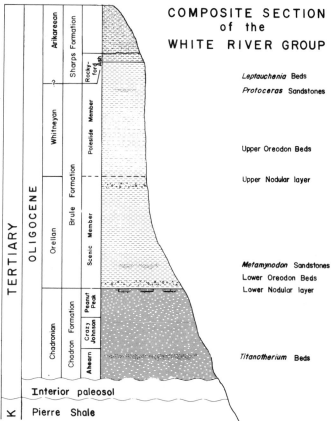

Figure 4. Diagrammatic stratigraphic section of the White River Group (derived from Bump, 1956; Clark, 1954; and Darton, 1899). Listed on the right are the faunal horizons and lithological markers utilized by Wortman and Wanless.

claystone which may have discrete limestone lenses, a rounded weathering profile, channel sandstones, and is between 30 and 50 ft (9–15 m) thick in this area due to deposition over a paleosurface. [Elsewhere, thickness ranges to 175 ft (55 m) (Harksen and Macdonald, 1969)]. The overlying Brule Formation may be 460 ft (140 m) thick, but in the landmark area it is composed of approximately 330 ft (100 m) of pink and tan siltstone and claystone containing volcaniclastics and channel sandstones. In contrast to the Chadron Formation, the Brule Formation exhibits a steeper weathering profile, coarser texture, and distinct color bands. The succeeding Sharps Formation is about 100 ft (30 m) thick in this area and is composed of a pink, sandy siltstone with concretions, channel sandstones, and impure ash layers. Overall, these fine-grained deposits are very susceptible to erosion; the result is a highly dissected landform.

PALEONTOLOGY OF THE WHITE RIVER BADLANDS

These deposits have produced exceedingly large collections of fossil vertebrates, so well preserved and prolific that the faunas have been considered a standard for world comparison. Additionally, there are successive faunal assemblages that may be utilized for biostratigraphic purposes. For instance, the large, archaic perissodactyls (brontotheres) are characteristic of the Chadron Formation and, with few exceptions, do not survive into the Brule Formation. Above the Chadron Formation, the primitive artiodactyls, merycoidodonts (oreodonts), predominate, particularly in the Scenic Member. The large, hippopotamus-like, aquatic rhinoceros, *Metamynodon,* also occurs in this member and is associated with channel sandstones which bear its name. Another series of channels in the overlying Poleslide Member have been termed the *Protoceras* beds, named for an unusual artiodactyl

with cranial ornamentation. The upper portion of the section is characterized by the small merycoidodont, *Leptauchenia.* Many other taxa may be utilized to characterize portions of the section, but those listed above have traditionally been considered diagnostic.

In 1941, Wood and others formulated successive land mammal ages for North America and included faunas from the White River Group in the definitions (Fig. 4). For the lower *Titanotherium* beds, the Chadronian Age was defined, for the Lower *Oreodon* (-*Merycoidodon*) beds which now include the vertebrates from the Scenic Member, the Orellan Age was designated, and for the Upper *Oreodon* and *Protoceras-Leptauchenia* beds of the Poleslide Member, the Whitneyan Age was defined. Faunas collected from the Sharps Formation were assigned to the overlying Arikareean Age by Macdonald (1963), but some investigators believe that at least the lower portion of the formation, including those rocks exposed in this landmark area, should be assigned to the older Whitneyan Age (Tedford and others, 1985). Overall, faunas from each age within the Oligocene are represented in the White River Badlands and comprise one of the best known terrestrial faunal successions in the world.

AGE OF THE WHITE RIVER BADLANDS

Correlation of the White River Badlands with the European Oligocene section was based primarily on stratigraphic position and faunal similarity. Age refinements have been relatively recent. Only one radiometric date of 36.3 ± 0.7 Ma from the top of the Ahearn Member of the Chadron Formation has been reported (McDowell and others, 1973). However, in the past three years the results of extensive magnetostratigraphic investigations have been published. These were summarized by Prothero (1985), who suggested the following dates for land mammal age boundaries: Duchesnean/Chadronian—36.5 Ma, Chadronian/Orellan—32.4 Ma, Orellan/Whitneyan—30.7 Ma, and Whitneyan/Arikareean—28.5 Ma.

GEOLOGICAL LOCALITIES ALONG THE BADLANDS LOOP

The Pinnacles. From the Pinnacles Overlook (Fig. 1) south of Wall, almost the entire succession of rock units present in the Badlands may be observed to the south (Fig. 5). Erosion has not exhumed the Interior paleosol, and only the top of the gray Chadron Formation is exposed in the lowest valleys. The red- and tan-banded Scenic Member of the Brule Formation lies upon the Chadron Formation, and where the regular banding grades into pink and gray thick-bedded vertical outcrops, the Poleslide Member is present. At the level of the overlook trail, the white, resistant Rockyford Ash Member of the Sharps Formation occurs; the overlying siltstones of this unit are exposed along the highway to the east before it descends into the Yellow Mounds area. Retallack (1983) derived much of his data concerning paleosols from the Badlands in this area.

Vertical clastic dikes transect the Poleslide and Rockyford Members at the end of the overlook (see Martin, 1985, Figs. 5-7). These dikes formed as clastic filling of vertical cracks and are well exposed at the Clastic Dikes Overlook 5.5 mi (9 km) to the east (Martin, 1985, Fig. 12). These dikes are filled with either chalcedony or detritus; the latter commonly exhibits compressional evidence of slickensides and vertical folds.

Yellow Mounds Overlook. Two major geological features may be observed in this area (Fig. 1). The first, the Interior paleosol, is the brilliant yellow and red unit comprising the lower portion of the basin. The base of the paleosol is exposed just south of the highway, 0.2 mi (0.3 km) to the west of the overlook turnoff. Here, the black Pierre Shale is exposed at the base of the section (Martin, 1985, Fig. 10). The yellow rocks are weathered Pierre Shale, and typical concretions and marine fossils including *Baculites* and mosasaurs have been found in the unit. Retallack (1983) suggested that weathered Cretaceous Fox Hills Sandstone and Eocene Slim Buttes Formation may also comprise the Interior paleosol (indicated by differential texture and vertical color change to red). Undoubtedly, the Interior paleosol is the result of multiple erosional cycles. Overlying the Interior paleosol is the gray Chadron Formation, characterized by a rounded "haystack"

Figure 5. View to the south from the Pinnacles Overlook. The exposures in the distance are the Scenic Member of the Brule Formation, those in the central portion of the figure above the trees are the Poleslide Member, and the white rocks in the foreground are the Rockyford Ash Member of the Sharps Formation.

weathering profile. At the top of the formation 0.05 mi (0.08 km) further along the road are white, calcareous lenses which have been used to characterize the top of the Chadron Formation. These carbonate lenses, however, may be found at other horizons.

The second geological feature in this area is the Dillon Pass Fault (Martin, 1985, Fig. 10) exposed just to the west of the overlook. The southern block of this gravity fault is down; this places the Pierre Shale against the Interior paleosol, the Interior paleosol against the Chadron Formation, and the Chadron Formation against the Scenic Member of the Brule Formation. This fault is exposed for 3 mi (5 km), exhibits about 50 ft (15 m) of displacement, and like other faults in the area, trends northwesterly.

Rainbow Overlook. Much of the area described above may be seen in panorama from this viewpoint (Fig. 6). All of the units exposed in the Badlands are present, as well as the Dillon Pass Fault. The southern fault block may be observed dipping gently to the south in contrast to the horizontal northern block. At the base of the section is the Interior paleosol disconformably overlain by the gray, low outcrops of the Chadron Formation. The overlying banded section across the valley is the Scenic Member overlain by vertical exposures of the Poleslide Member. The uppermost white layer just below the peaks on the horizon is the Rockyford Ash (Fig. 6).

Banded Buttes Overlook. The red banding of the Scenic Member of the Brule Formation is well developed in this area (Martin, 1985, Fig. 14). The color is due to oxidized iron derived during paleosol formation (Retallack, 1983). Along the road 3.5 mi (5.6 km) to the southeast is the erosional remnant known as the Castle. The plain from here to the Castle is developed approximately at the Scenic-Poleslide contact, and the Castle is composed of the Poleslide Member and the lower Sharps Formation. Clastic dikes and extensive landslides composed of huge blocks of

Figure 6. View to the west from Rainbow Overlook. Abbreviations: C = Chadron Formation, S = Scenic Member, P = Poleslide Member, and Sh = Sharps Formation.

Rockyford Ash and green channel sandstones may be seen at the Castle (Martin, 1985, Fig. 15).

From the Castle the road passes by Fossil Exhibit Trail, crosses through Norbeck Pass, and descends the Scenic Member onto a plain developed on the Chadron Formation. As the road winds to the east, good outcrops of the Scenic-Sharps section may be observed as well as the Pass Creek Fault system (See Martin, 1985, pp. 32-35, Fig. 16).

Cedar Pass. Just to the north of the Visitor's Center is the Cedar Pass-Millard Ridge area where much of the upper portion of the stratigraphic section is well exposed. Prothero (1985) obtained one suite of samples for his magnetostratigraphic analysis of the Oligocene from this area. The Scenic Member crops out at the base of the cliffs below the vertical exposures of the Poleslide Member. The prominent white band one-fourth of the distance below the top of the cliffs is the Rockyford Ash, and the siltstones and channel sandstones of the Sharps Formation lie above. Many of these channels have been observed incised into the Brule Formation (Harksen, 1974) and may have been confused with the *Protoceras* channels of the Poleslide Member.

REFERENCES

Bump, J. D., 1956, Geographic names for the members of the Brule Formation of the Big Badlands of South Dakota: American Journal of Science, v. 254, p. 429-432.

Clark, J., 1954, Geographic designation of the members of the Chadron Formation in South Dakota: Annals of the Carnegie Museum, v. 33, p. 197-198.

Darton, N. H., 1899, Preliminary report on the geology and water resources of Nebraska west of the one hundred and third meridian: U.S. Geological Survey, 19th Annual Report, p. 719-785.

Harksen, J. C., 1974, Miocene channels in the Cedar Pass area, Jackson County, South Dakota: South Dakota Geological Survey, Report of Investigations, v. 111, p. 1-10.

Harksen, J. C. and Macdonald, J. R., 1969, Guidebook to the major Cenozoic deposits of southwestern South Dakota: South Dakota, Geological Survey, Guidebook, n. 2, p. 1-103.

Leidy, J., 1847, On a new genus and species of fossil Ruminantia; *Poebrotherium wilsoni*: Proceedings of the Philadelphia Academy of Sciences, v. 3, p. 322-326.

Leidy, J., 1869, The extinct mammalian fauna of Dakota and Nebraska, including an account of some allied forms from other localities, together with a synopsis of mammalian remains of North America: Journal of the Philadelphia Academy of Sciences, v. 7, p. 23-472.

Macdonald, J. R., 1963, The Miocene faunas from the Wounded Knee area of South Dakota: Bulletin of the American Museum of Natural History, v. 125, p. 139-238.

Martin, J. E., 1983, Composite stratigraphic section of the Tertiary deposits in western South Dakota: Museum of Geology, South Dakota School of Mines, Dakoterra, v. 2, no. 1, p. 1-8.

Martin, J. E., 1985, Geological and paleontological road log from Rapid City, through the Oligocene White River Badlands and Miocene deposits, to Pine Ridge, South Dakota: *In* Martin, J. E. (ed.), *Fossiliferous Cenozoic Deposits of Western South Dakota and Northwestern Nebraska,* Museum of Geology, South Dakota, School of Mines, Dakoterra, v. 2, no. 2, p. 13-59.

McDowell, F. W., Wilson, J. A., and Clark, J., 1973, K-Ar dates from biotite from two paleontologically significant localities: Duchesne River Formation and Chadron Formation: Isochron/West, v. 7, p. 11-12.

Meek, F. B. and Hayden, F. V., 1858, Descriptions of new species and genera of fossils collected by Dr. F. V. Hayden in Nebraska Territory under the direction of Lieut. G. K. Warren, U.S. Topographical Engineer; with some remarks on the Tertiary and Cretaceous formations of the northwest and the parallelism of the latter with those of other portions of the United States and Territories: Proceedings of the Philadelphia Academy of Sciences, 1857, v. 9, p. 117-148.

O'Harra, C. C., 1920, The White River Badlands: Bulletin of the Department of Geology, South Dakota School of Mines, v. 13, p. 1-181.

Owen, D. D., 1852, Report of a geological survey of Wisconsin, Iowa, and Minnesota; and incidentally of a portion of Nebraska Territory: Lippincott, Grambo & Co., Philadelphia, Pennsylvania, 638 p.

Prothero, D. R., 1985, Correlation of the White River Group by magnetostratigraphy: *In* Martin, J. E. (ed.), *Fossiliferous Cenozoic Deposits of Western South Dakota and Northwestern Nebraska,* Museum of Geology, South Dakota School of Mines, Dakoterra, v. 2, no. 2, p. 265-276.

Prout, H. A., 1846, Gigantic *Palaeotherium*: American Journal of Sciences, series 2, v. 2, p. 288-289.

Retallack, G. J., 1983, Late Eocene and Oligocene paleosols from Badlands National Park, South Dakota: Geological Society of America, Special Paper, 193, p. 1-82.

Scott, W. B. and Jepsen, G. L., 1936, 1937, 1940, 1941, "The White River Monographs": Proceedings of the American Philosophical Society, v. 28, no's. 1-5.

Tedford, R. H., Swinehart, J. B., Hunt, R. M., Jr., and Voorhies, M. R., 1985, Uppermost White River and lowermost Arikaree rocks and faunas, White River Valley, northwestern Nebraska, and their correlation with South Dakota: *In* Martin, J. E. (ed.), *Fossiliferous Cenozoic Deposits of Western South Dakota and Northwestern Nebraska,* Museum of Geology, South Dakota School of Mines, Dakoterra, v. 2, no. 2, p. 335-352.

Wanless, H. R., 1923, The stratigraphy of the White River beds of South Dakota: Proceedings of the American Philosophical Society, v. 62, p. 190-269.

Wood, H. E., Chaney, R. W., Clark, J., Colbert, E. H., Jepsen, G. L., Reeside, J. B., Jr., and Stock, C., 1941, Nomenclature and correlation of the North American continental Tertiary: Bulletin of the Geological Society of America, v. 52, p. 1-48.

Wortman, J. L., 1893, On the divisions of the White River or lower Miocene of Dakota: Bulletin of the American Museum of Natural History, v. 5, p. 95-106.

Metamorphic, igneous, and sedimentary relationships on the Sioux Quartzite Ridge, South Dakota

Richard F. Bretz, HCR 2, Box 8A, Wallace, Kansas 67761

LOCATION AND ACCESS

The Sioux Quartzite Ridge geological site consists of three stops east and northeast of Sioux Falls, Minnehaha County, South Dakota, on the Sioux Falls 1° × 2° topographic map (Fig. 1). Sioux Falls is the starting point for a visit to these stops, which are east of the city on I-90. All of the stops are on all-weather roads, and can be reached by large groups traveling by bus. These stops are included in a longer field trip guide (Brenner and others, 1981), which contains more detail and further references.

Stop 1, the I-90 section, is about 0.5 mi (0.8 km) east of the Corson-Brandon interchange, where the Interstate crosses Split Rock Creek (NE¼NE¼SW¼Sec.26,T.102N.,R.48W.; Brandon 7½-minute Quadrangle). Since parking is not allowed on the interstate highway, park at the interchange and walk to the exposures, which occur both north and south of the highway. The outcrops are on private land owned by Howard R. Johnson, Corson, South Dakota 57019, telephone (605) 582-3898.

To reach the second stop, proceed 3.5 mi (5.7 km) to Exit 410; leave I-90, turn left (north) onto the county road. Travel 5.8 mi (9.5 km) north and turn left (west) onto a gravel road. Proceed for 1 mi (1.6 km) and turn left (south), following signs to Palisades State Park. Drive 0.1 mi (0.2 km) farther and turn right (west) onto the road that leads into the state park.

Stop 2. Palisades State Park is publicly-owned land, so access is unlimited. A series of roads and foot trails provide numerous vantage points from which the gorge cut into Sioux Quartzite can be viewed. State park land is located in the S½Sec.30 and N½Sec.31,T.103N.,R.47W., Garretson West 7½-minute Quadrangle.

After leaving the park, turn left (north) onto a gravel road and proceed 1.2 mi (2 km). Turn left (west) and follow the curving road 0.2 mi (0.3 km) to a stop sign; turn left (west) onto blacktop South Dakota 11. Proceed 2.7 mi (4.4 km) to a junction, and bear left (south), remaining on South Dakota 11. Continue south 5.2 mi (8.5 km), then turn right (west) onto a gravel road. Go 0.4 mi (0.7 km) west to the bridge across Split Rock Creek.

Stop 3 consists of the Risty and Rovang farm sections (Fig. 2). The Risty section occurs on the east bank of the creek on either side of the bridge (SW¼SW¼SE¼Sec.15, and NW¼NW¼ NE¼Sec.22,T.102N.,R.48W., Garretson West 7½-minute Quadrangle), and can be reached on foot from the roadway. It is on private land owned by Albert Risty, Rural Route, Brandon, South Dakota 57005, telephone (605) 582-3891. The Risty residence is just northeast of the bridge at the Risty section.

The Rovang section occurs on the west bank of the creek

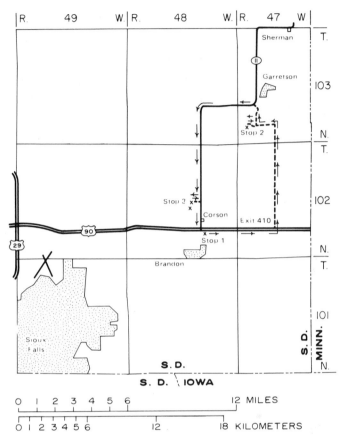

Figure 1. Map showing route and locations of stops.

about 0.1 mi (0.2 km) south of the roadway (SE¼NE¼NW¼ and NE¼SE¼NW¼Sec.22,T.102N.,R.48W., Garretson West 7½-minute Quadrangle), and can be reached on foot. This section is on private land owned by Ordell Rovang, Corson, South Dakota 57019, telephone (605) 582-6139. The Rovang residence is on the hill above the creek, south of the section.

After leaving Stop 3, return to South Dakota 11. Turn right (south) and proceed 1.5 mi (2.5 km) to the interchange (I-90 and South Dakota 11).

SIGNIFICANCE

These three localities show relationships of the Proterozoic Sioux Quartzite, the southwesternmost of the known Baraboo Interval quartzites; the Corson Diabase, which is intruded into the Sioux Quartzite; and the Upper Cretaceous Split Rock Creek Formation, a sedimentary unit of unusual lithology, which was deposited in a protected embayment in the Sioux Quartzite ridge.

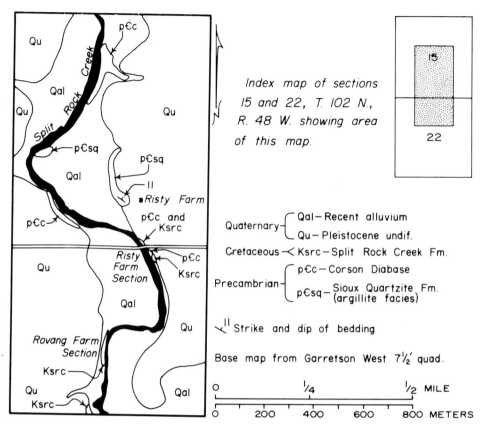

Figure 2. Geologic map of the Risty and Rovang farm area showing locations at Stop 3. Modified from Stach, 1970; from Brenner and others, 1981.

The Sioux Quartzite, one of three known major Proterozoic red quartzites of the Baraboo Interval (Dott, 1983), is currently of interest because of a controversy regarding the depositional and tectonic setting of the quartzites of the north-central United States (Dott, 1983; Greenberg and Brown, 1984).

All three units are economically important. The Sioux Quartzite has been suggested as a possible host for unconformity vein-type uranium deposits (Cheney, 1981) because of its similarities to the Athabasca basins of northern Saskatchewan. The intrusion of the Corson Diabase into the Sioux Quartzite may have promoted magmatic and/or hydrothermal enrichment near the contact zone.

Recent authors have suggested the Cretaceous rocks of the Sioux Quartzite ridge area as a potential exploration target for high-grade shallow-marine manganese deposits (Cannon and Force, 1983). The Sioux Ridge is similar to the Groote Eylandt, Australia, manganese deposits (Frakes and Bolton, 1984), in that wave-cut benches occur on Proterozoic quartzite (Shurr, in Brenner and others, 1981) and to the Nikopol manganese deposits in the USSR (Varentsov, 1964), which contain rocks that are similar in lithology to the Split Rock Creek Formation.

SITE DESCRIPTION

Stop 1. The I-90 section. At this stop one views exposures of the Upper Cretaceous Split Rock Creek Formation, a small basin-filling deposit of basal sandy diamictites and quartz sandstones, which grade upward into laminated organic claystones, laminated organic to nonorganic opaline spiculites, bioturbated organic to nonorganic opaline spiculites, and finally into an upper clay unit. This sequence rests unconformably on Sioux Quartzite and Corson Diabase; the complete sequence is known only from subsurface information. Only the bioturbated to laminated chalcedonic to opaline spiculites, with the upper clay member, can be seen at this stop. None of the exposures contain the laminated claystone sequence.

This section (Fig. 3) has two exposures. The northern exposure is a 2.6-ft-thick (0.8 m) section, composed of a variably cherty opaline spiculite, overlain by a noncalcareous light gray to yellow-orange clay (upper clay member), which is possibly a weathered phase of the spiculite. The contact between the spiculite and the clay is sharp. The 9-ft-thick (2.7 m) southern exposure is light gray to white, thinly to massively bedded, laminated

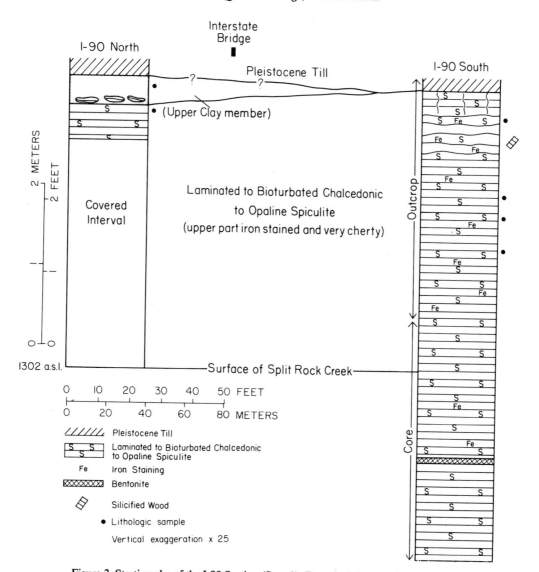

Figure 3. Stratigraphy of the I-90 Section (Stop 1). From Ludvigson and others, 1981.

to bioturbated chalcedonic opaline spiculite, with iron liesegang staining at irregular intervals. The bioturbated portion of the outcrop is composed of chalcedonic chert tubes, while the laminated portion is opaline spiculite. The cherty portion is distinctly vuggy, due to irregularly-distributed chalcedonization controlled by burrowing.

The spiculites are composed of siliceous sponge spicules, which are probably derived from demospongean sponges (indicated by the monaxon and tetraxon forms). These sponges were apparently very successful in Late Cretaceous time, and preferred hard substrate and low wave activity. The protected embayment in which the Split Rock Creek Formation was deposited probably provided this type of environment.

Recent drilling has revealed the subsurface presence of siliceous rocks similar to those of the Split Rock Creek Formation in

southeastern Lake County and in southern Moody County, South Dakota. These subsurface occurrences are, at least in part, stratigraphically equivalent to the Upper Cretaceous Niobrara Formation.

Stop 2. Palisades State Park. The gorge cut by Split Rock Creek provides one of the more spectacular and extensive exposures of the Precambrian (Middle Proterozoic) Sioux Quartzite. This red orthoquartzite, together with the Baraboo and Barron quartzites, is one of the major Baraboo interval quartzites, which date within the time span between 1,450 and 1,750 Ma (and possibly in the narrower range between 1,615 and 1,750 Ma). Deposition of the Sioux Quartzite occurred sometime between 1,200 and 1,850 Ma (probably between 1,530 and 1,750 Ma; Dott, 1983).

Various interpretations of the sedimentary history of the Bara-

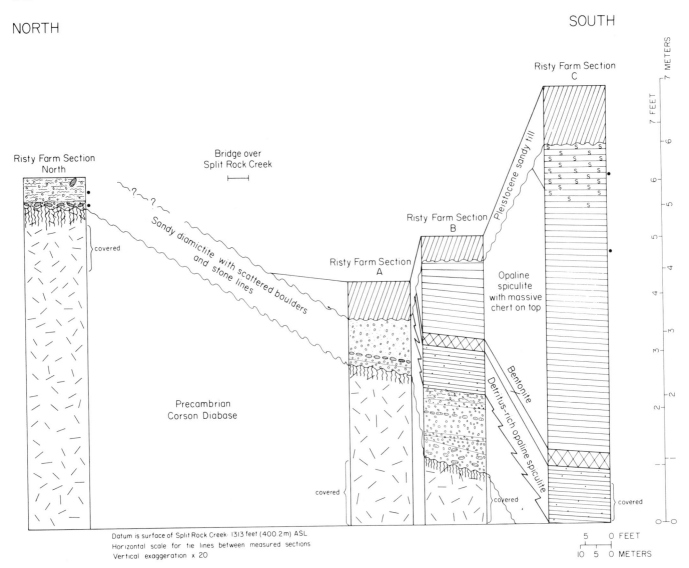

Figure 4. Stratigraphy of the Risty farm section (Stop 3). Symbols not explained on section shown in Figure 3. From Ludvigson and others, 1981. All units between Corson Diabase and Pleistocene till are Cretaceous Split Rock Creek Formation.

boo Interval quartzites have been presented. Until recently, a shallow nearshore marine interpretation was favored (Dott and Dalziel, 1972). Modern sedimentological analyses favor fluvial deposition with possible shallow marine incursions recorded near the top of some of the quartzite sequences (Dott, 1983; Greenberg and Brown, 1984).

The sedimentary/tectonic environment of the Baraboo Interval quartzites has been the subject of recent controversy. Dott (1983) has proposed a model that begins with marine deposition on a stable, passive continent margin (miogeoclinal sedimentation), followed by deformation due to plate collision. In contrast, Greenberg and Brown (1984) favor a cratonic depositional sys-

tem (complex basin model), later deformed by magmatism and epeirogenic uplift.

The Sioux Quartzite has no known stratigraphic top; its thickness has been estimated at 3,000 to 5,000 ft (900 to 1,500 m; Baldwin, cited in Gries, 1983). These great thicknesses have been cited as evidence for some interpretations of the Baraboo Interval quartzites. However, Gries (1983) proposed that the thickness of the Sioux Quartzite has been overestimated by misinterpretation of exposures and by use of questionable drillhole data; he stated that the thickness does not exceed 1,000 ft (300 m), and is probably less than 500 ft (150 m).

The Sioux Quartzite is very resistant to erosion and forms a highland at the Palisades. This highland existed during deposition of the Split Rock Creek Formation; together with similar highland areas this was probably a partial source terrane for the sandy diamictites and quartz sandstones deposited at the base of the Split Rock Creek Formation.

Stop 3. Risty and Rovang farm sections. (See Figs. 2, 4, and 5.) This stop shows the Split Rock Creek Formation resting on weathered Corson Diabase. Fresher diabase and Sioux Quartzite can also be seen upstream, north and northwest of these exposures (Fig. 2).

The Corson Diabase, and other diabases in the region, are titaniferous olivine diabases, which were intruded as sills into the Sioux Quartzite. These probably represent distal occurrences of the late Precambrian (middle to upper Keweenawan) tecto-volcanic event, which was responsible for the emplacement of the Duluth Gabbro and other mafic rocks associated with midcontinent rifting 1.1 to 1.2 Ga. Differential weathering of the Corson Diabase is probably a principal cause of the formation of the basin in which most of the Split Rock Creek Formation was deposited.

At the Risty farm section (Fig. 4), the Split Rock Creek Formation rests on an erosional unconformity, developed on weathered Corson Diabase, which dips as much as 15° to the south. The bottom portion of the outcrop is not well exposed, and will require some excavation. The basal unit of this section is a sandy diamictite, which is up to 4 ft (1.2 m) thick; it consists of very fine to coarse quartz and weathered feldspar grains in a yellow-brown clay matrix. Flat pebble clasts of weathered argillite up to 2 in (5 cm) long occur in a thin bed at the base. Weathered quartzite clasts up to 7 in (18 cm) long are present throughout the entire basal unit. The unit probably represents either southward-flowing subaqueous slides or subaerial mudflows.

Twenty-one feet (6.4 m) of white opaline spiculite occur above the basal diamictite unit. The spiculite contains abundant detrital quartz sand grains in the lowermost 3.3 ft (1 m), a calcium bentonite 7 to 9.5 in (17.8 to 24 cm) thick 2.2 ft (0.7 m) above the base, and a massive chert unit in the uppermost 3.3 ft (1 m). These spiculites have a high degree of moldic porosity, which evidently makes the rocks good insulators; water contained in them can be found as ice into late spring.

The Rovang farm section (Fig. 5) occurs about 700 to 1,000 ft (200 to 300 m) southwest of the Risty section; it has a similar stratigraphy. The contact with the deeply weathered Corson Diabase, which is often altered to a saprolite, undulates but maintains a generally southward dip. The basal diamictite is 2.4 ft (0.7 m) thick and consists of very fine to very coarse quartz and weathered feldspar sand grains in a tan olive green clay matrix. The overlying white opaline spiculite is 7 ft (2.1 m) thick and contains detrital quartz grains in the lowermost 3.3 ft (1 m). Two calcium bentonite layers are present. The uppermost, thicker bentonite contains burrows filled with opaline spiculite. This section is fossiliferous, and contains mollusk molds, fish debris, and plant

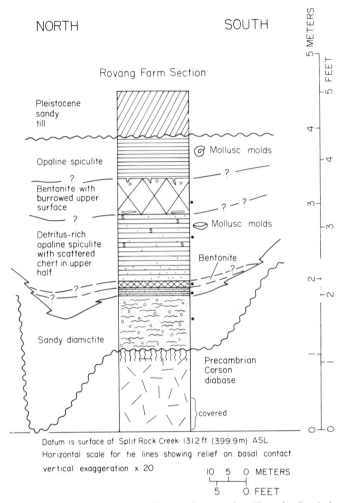

Figure 5. Stratigraphy of the Rovang farm section (Stop 3). Symbols explained above or shown on Figure 3. From Ludvigson and others, 1981.

leaf and stem molds, some of which have been identified as sequoia twigs.

The Cretaceous age assigned to these deposits is based on (1) foraminifera, from the organic-rich laminated claystone in the middle of the formation, which are comparable to those of the Niobrara Formation and lower Pierre Shale (Witzke and others, 1983); (2) the facies relationships with the Niobrara Formation in southern Lake County; and (3) the possible correlation of the Risty and Rovang section bentonites with the Ardmore Bentonite Suite of the Sharon Springs Member of the lower Pierre Shale (Witzke and others, 1983).

In summary, the fluvially-deposited, 1,530 to 1,750 Ma (Precambrian, Middle Proterozoic) Sioux Quartzite was intruded by the 1,100 to 1,200 Ma (late Precambrian, middle to late Keweenawan) Corson Diabase. Rapid differential weathering of the diabase formed a small protected basin, surrounded by quartzite highlands, in which the major part of the Upper Cretaceous Split Rock Creek Formation was deposited.

REFERENCES CITED

Brenner, R. L., and 7 others, 1981, Cretaceous stratigraphy and sedimentation in northwest Iowa, northeast Nebraska, and southeast South Dakota: Iowa Geological Survey Guidebook Series Number 4, 172 p.

Cannon, W. F., and Force, E. R., 1983, Potential for high-grade shallow-marine manganese deposits in North America, *in* Shanks, W. C., III, ed., Cameron volume on unconventional mineral deposits: New York, Society of Mining Engineers of the American Institute of Mining, Metallurgical, and Petroleum Engineers, p. 175–189.

Cheney, E. S., 1981, The hunt for giant uranium deposits: American Scientist, v. 69, p. 37–48.

Dott, R. H., 1983, The Proterozoic red quartzite enigma in the north-central United States; Resolved by plate collision?, *in* Medaris, L. G., Jr., ed., Early Proterozoic geology of the Great Lakes region: Geological Society of America Memoir 160, p. 129–141.

Dott, R. H., Jr., and Dalziel, I.W.D., 1972, Age and correlation of the Precambrian Baraboo Quartzite of Wisconsin: Journal of Geology, v. 80, p. 552–568.

Frakes, L. A., and Bolton, B. R., 1984, Origin of manganese giants; Sea level change and anoxic-oxic history: Geology, v. 12, p. 83–86.

Greenberg, J. K., and Brown, B. A., 1984, Cratonic sedimentation during the Proterozoic; An anorogenic connection in Wisconsin and the upper Midwest: Journal of Geology, v. 92, p. 159–171.

Gries, J. P., 1983, Geometry and stratigraphic relations of the Sioux Quartzite: Proceedings of the South Dakota Academy of Science, v. 62, p. 64–74.

Ludvigson, G. A., McKay, R. M., Iles, D., and Bretz, R. F., 1981, Lithostratigraphy and sedimentary petrology of the Split Rock Creek Formation, Late Cretaceous, of southeastern South Dakota, *in* Brenner, R. L. and 7 others, Cretaceous stratigraphy and sedimentation in northwest Iowa, northeast Nebraska, and southeast South Dakota: Iowa Geological Survey Guidebook Series Number 4, p. 77–104.

Stach, R. L., 1970, A weathering surface and associated sedimentary rocks near Sioux Falls, South Dakota: Vermillion, unpublished report in files of South Dakota Geological Survey, 15 p.

Varentsov, I. M., 1964, Sedimentary manganese ores: Amsterdam, Elsevier, 119 p.

Witzke, B. J., Ludvigson, G. A., Poppe, J. R., and Ravn, R. L., 1983, Cretaceous paleogeography along the eastern margin of the Western Interior Seaway, Iowa, southern Minnesota, and eastern Nebraska and South Dakota, *in* Reynolds, M. W., and Dolly, E. D., eds., Mesozoic paleogeography of west-central United States: Denver, Rocky Mountain Section, Society of Economic Paleontologists and Mineralogists, p. 225–252.

Precambrian metamorphic rocks of the Devil's Gate–Weber Canyon area, northern Utah

Frank L. DeCourten, Utah Museum of Natural History, University of Utah, Salt Lake City, Utah 84112

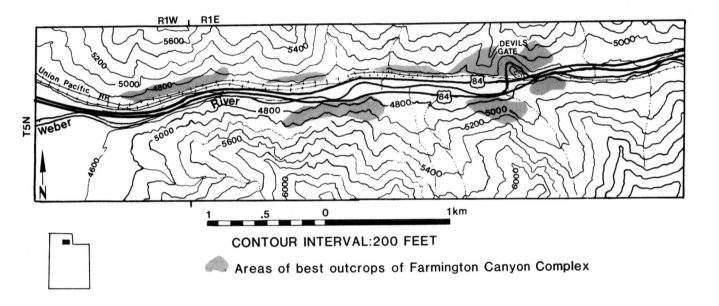

CONTOUR INTERVAL: 200 FEET

Areas of best outcrops of Farmington Canyon Complex

Figure 1. Locality Map of the Devil's Gate–Weber Canyon site, 1:24000.

LOCATION

The site discussed herein is located in the central Wasatch Mountains near the mouth of Weber Canyon, 9 mi (15 km) southeast of Ogden, Utah. Numerous excellent exposures of Precambrian metamorphic and igneous rocks occur in Weber Canyon from approximately Mountain Green, Utah, downstream to the mouth, in an area of roughly 4 mi² (10.5 km²) (Fig. 1). The area of principal interest is in sections 26 and 36, T.5N., R.1W., and in Sec. 28–32, T.5N., R.1E. (Salt Lake Meridian). The Devil's Gate–Weber Canyon locality occupies the southwest corner of the Snow Basin 7½-minute Quadrangle and the southeast corner of the Ogden, Utah 7½-minute Quadrangle. I-84 bisects the area and offers easy access to many excellent outcrops.

SIGNIFICANCE OF SITE

Metamorphic and igneous basement rocks of Archean and Proterozoic age are exposed along the western foothills of the north-trending Wasatch Range from approximately Brigham City southward to Bountiful, a distance of over 46 mi (75 km). This essentially continuous exposure constitutes the largest and most accessible outcrop of ancient basement rocks between localities in the Wyoming province, to the north and northwest, and the small, isolated outcrops of correlative rocks in the deeper canyons of the Colorado River system to the southeast. The

outcrop belt of crystalline basement rocks in the Wasatch Range is located immediately east of the zone of late Proterozoic continental rifting and represents one of the westernmost exposures of rocks along the rifted edge of the Precambrian North American craton. Exposures at Devil's Gate in Weber Canyon and at other nearby localities have attracted the attention of numerous investigators over the past forty years, resulting in a significant body of data concerning the geochronology, geochemistry, petrology, and metamorphic history of this complex and heterogeneous sequence. The site discussed herein is thus extremely important in elucidating regional patterns of metamorphism and tectonism in the crystalline basement rocks of western North America.

GEOLOGIC INFORMATION

Precambrian crystalline basement rocks exposed in the northern Wasatch Range are referred to as the Farmington Canyon Complex (Eardley and Hatch, 1940). These rocks comprise an extremely heterogeneous assemblage including several types of gneiss, varied schists, amphibolite, migmatite, and quartzite cut by later bodies of pegmatite. Near the mouth of Weber Canyon, the Farmington Canyon Complex is overlain unconformably by the basal Cambrian Tintic Quartzite that, in turn, is succeeded by a thick sequence of younger Paleozoic strata principally consisting of shallow marine carbonates.

The crystalline basement rocks were locally affected by deformation during the late Jurassic–Cretaceous Sevier Orogeny, when thin slices of the Farmington Canyon Formation were transported several tens of miles to the east along thrust faults. Exposures in the immediate vicinity of Devil's Gate probably lie beneath the basal thrust of the Sevier Orogenic Belt and are at least para-autochthonous. However, numerous shear zones are observed at this locality, and these may represent Sevier-age deformation; the whole complex is thought by some investigators to be allochthonous and underlain by a concealed regional decollement that seems to be required by the geometry of Sevier thrusts farther east (Hedge and others, 1983). The present exposures of the crystalline basement complex in Weber Canyon are the result of late Tertiary to Quaternary uplift and east-tilting of the Wasatch Range, achieved mainly through displacement on the Wasatch Fault, which is for the most part concealed along the western foothills of the range.

Petrology of the Farmington Canyon Complex at Weber Canyon

The Farmington Canyon Complex is an extremely heterogeneous assemblage of igneous and metamorphic basement rocks. Of the four units recognized by Bryant (1979), the migmatite-gneiss-schist assemblage is the most varied and is particularly well exposed at the mouth of Weber Canyon. At this locality, the Farmington Canyon complex consists of banded migmatite containing amphibolite lenses and pegmatite pods, biotite-hornblende-quartz-feldspar gneiss, garnet-quartz-feldspar gneiss, and biotite-garnet-quartz-feldspar gneiss. The gneissic materials are interlayered with well-foliated schists of two principal mineral assemblages: biotite-garnet schist and sillimanite-biotite-garnet schist. Both types of schist commonly contain minor amounts of microcline (Hedge and others, 1983). Pegmatite pods and lenses are common throughout the metamorphic sequence and are composed mainly of varying ratios of quartz and alkali feldspar. Hornblende phenocrysts and inclusions up to 6 in. (15 cm) in maximum dimension are present in the pegmatite at some outcrops.

In the Devil's Gate area of Weber Canyon, the Farmington Canyon Complex is cut by numerous shear zones that range in width from 3 ft (1 m) to several hundred feet. Rocks within the pervasive shear zones consist of dark greenish-black mylonite and phyllonite composed of greenschist facies minerals. Textures of the shear-zone rock indicate eastward transport of the upper sheet relative to the lower sheet (Bruhn and others, 1983). See Figure 2 for examples of lithologies common in the Farmington Canyon Complex at the Weber Canyon site.

Metamorphic and Tectonic History of the Farmington Canyon Complex

The textural and compositional heterogeneity of the Farmington Canyon Complex implies a long and complex history of deformation and metamorphism. The migmatites, schists, and amphibolites probably represent an originally sedimentary sequence intruded by basalt and gabbro, which is possibly as old as 3000 m.y. Initial metamorphism took place about 2600 m.y. under granulite facies conditions (Hedge and others, 1983). A later period of metamorphism associated with quartz monzonite plutonism in areas north of Weber Canyon occurred at about 1,800 Ma (Hedge and others, 1983). This second metamorphic event took place under amphibolite-facies conditions at a burial depth approaching 6 mi (10 km) and resulted in the present mineral assemblages observed in the non-sheared metamorphic rocks of the Devil's Gate area.

The final period of deformation probably took place during the Sevier Orogeny (late Jurassic–Cretaceous), when compressive forces resulted in the formation of pervasive shear zones in which retrograde metamorphism produced the mylonites and phyllonites under greenschist-facies conditions. Bryant (1979) has observed shear zones in the Farmington Canyon Complex west of Weber Canyon that do not cut overlying late Proterozoic strata, indicating that some of the shearing in that area may reflect Precambrian deformation. The pegmatites in the Farmington Canyon Complex yield radiometric ages ranging from 535 Ma to 1,580 Ma (Whelan, 1969) and cannot be precisely linked to any of the three broad phases of deformation described above.

SITE INFORMATION

The Weber River flows from a headwater region in the western Uinta Mountains and follows a northwesterly course for approximately 59 mi (95 km) to a point near Mountain Green, Utah. As it crosses the axis of the Wasach Range, the Weber River follows a generally due-west course along which it has carved a deep rugged canyon through the Precambrian crystalline rocks of the Farmington Canyon Complex. At Devil's Gate, the Weber River makes a sharp loop to the north, the result of a landslide that moved down the prominent canyon to the south, temporarily blocking the Weber River and forcing a migration of the former channel. A new channel was then incised around the toe of the slide, and the deep meander loop is known as Devil's Gate. This was a formidable barrier to travel through the lower Weber Canyon until the old highway grade was cut along the course of the river. More recently, the interstate highway and rail lines have been constructed along the former course of the Weber River and now pass south of Devil's Gate. Interestingly, the new highway and railroad grades were constructed by cutting away a portion of the toe of the slide, providing a mechanism for possible future slope stability problems. Ancient landslide debris can be seen on the floor of the canyon immediately south of Devil's Gate.

Many excellent exposures of the Farmington Canyon Complex are available for study from just upstream of Devil's Gate to the mouth of Weber Canyon. Outcrops along the old highway loop through Devil's Gate are especially interesting because a great variety of rock types and structures may be observed. Amphibolite, biotite-garnet schist, varied gneiss, and pegmatite bear-

Figure 2. Exposures of metamorphic and igneous rocks of the Farmington Canyon Complex at Devil's Gate, Weber Canyon. (a) Biotite-hornblende-quartz-feldspar gneiss overlain by biotite schist. Note quartz-rich pegmatite cutting gneiss at bottom of photograph. (b) Shear zone in metamorphic rocks exposed in northwest wall of Devil's Gate. (c) Hornblende phenocrysts and inclusions in pegmatite dike. (d) Migmatitic banding in injection gneiss at road cut along east-bound lanes of I-84, south of Devil's Gate. (e) Outcrop of hornblende-biotite-quartz-feldspar gneiss at Devil's Gate showing deformed bands. Length of scale in all photographs is 6 in (15 cm)).

F. L. DeCourten

ing hornblende inclusions up to 4 in (10 cm) in length are all exposed in Devil's Gate. Several shear zones within this assemblage, along which mylonites have formed, may be observed in the steep walls north of the old highway.

The best access to Devil's Gate is from the west-bound lanes of I-84, via a dirt parking area immediately west of the bridge crossing the Weber River. From this area, the old highway may be walked through Devil's Gate, along which an essentially continuous exposure of the Farmington Canyon Complex is observable.

Additional outcrops are present to the west in road cuts along the west-bound lanes of I-84 and just above the double railroad grades immediately to the north. These outcrops are directly accessible from the interstate highway and provide especially good exposures of contact relationships between the biotite schists, pegmatites, and biotite-hornblende-feldspar gneiss.

From the east-bound lanes of I-84, the best exposures are located directly south from Devil's Gate and downstream for a distance of 1.2 mi (2 km). Along this segment of the highway, numerous prominent road cuts provide excellent exposures of pegmatite dikes and lenses, migmatitic injection gneiss, and a varied set of intercalated schists. The shaded areas on the locality map (Fig. 1) indicate areas of best outcrops in the Devil's Gate–Weber Canyon area.

The Devil's Gate–Weber Canyon locality is accessible year round, but winter snows, which are commonly very heavy, not only obscure outcrops but also eliminate parking areas on the highway shoulder. Early spring (late April–early June) run-off in the Weber River is often extremely high and may result in bank erosion and local flooding, which could restrict access during this period. The best times to visit the Weber Canyon area are from late June through early October.

REFERENCES

Bruhn, R., Picard, M. D., and Beck, S., 1983, Mesozoic and early Tertiary structure and sedimentology of the central Wasatch Mountains, Uinta Mountains, and Uinta Basin, *in* Gurgal, K. D., ed., Geologic excursions in the Overthrust Belt and metamorphic core complexes of the Intermountain Region, Utah Geological and Mineral Survey Special Studies 59, p. 63–107.

Bryant, B., 1979, Reconnaissance map of the Precambrian Farmington Canyon Complex and surrounding rocks in the Wasatch Mountains between Ogden and Bountiful, Utah: U.S. Geological Survey Open-File Report, OF 79-0709, p. 79–709.

Eardley, A. J., and Hatch, R. A., 1940, Pre-cambrian crystalline rocks of north-central Utah: Journal of Geology, v. 48, p. 58–72.

Hedge, C. E., Stacey, J. S., and Bryant, B., 1983, Geochronology of the Farmington Canyon Complex, Wasatch Mountains, Utah, *in* Miller, D. M., Todd, V. R., and Howard, K. A., eds., Tectonic and stratigraphic studies in the Eastern Great Basin, Geological Society of America Memoir 157, p. 37–44.

Whelan, J. A., 1969, Geochronology of some Utah rocks, *in* Jensen, M. L., (ed.), Guidebook of Northern Utah: Utah Geological and Mineralogical Survey Bulletin 82, p. 97–104.

Green River Formation (Eocene), central Utah: A classic example of fluvial-lacustrine environments

James L. Baer, *Geology Department, Brigham Young University, Provo, Utah 84602*

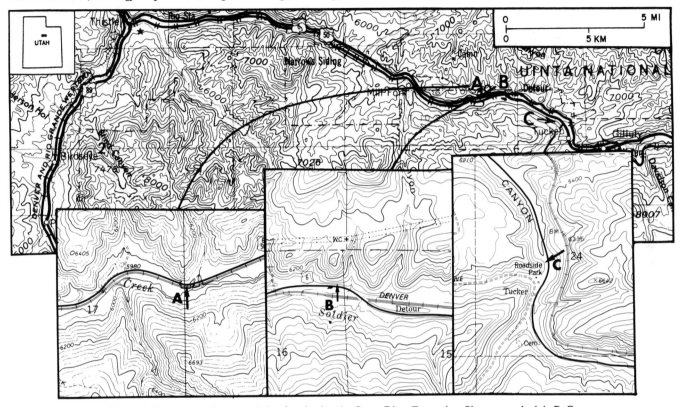

Figure 1. Index map of suggested sites for viewing the Green River Formation. Sites are marked A, B, C on the large-scale map with the corresponding inset map showing the site in detail (1:24,000 scale).

LOCATION AND ACCESSIBILITY

The Green River Formation is well exposed in roadcuts along U.S. 50-6 in central Utah. This highway is an all-weather road, and provides easy access to these excellent exposures during the normal field season (Fig. 1).

SIGNIFICANCE

Central Utah was the site of the fluvial and lacustrine deposition for nearly the entire Tertiary, and the record of these depositional environments is no better preserved than in the Green River Formation. The history of the Eocene lake and associated environments is recorded in interfingering freshwater lake, saline lake, mudflat, shoreline, and deltaic sedimentary units. Regional correlations are made possible by a number of tuff-bentonite beds and confirm that the Green River Formation accumulated over approximately 14,000 mi^2 (36,000 km^2). This lake, named Lake Uinta by Bradley (1931), was the eventual site for deposition of more than 3,300 ft (1,000 m) of lacustrine rocks.

These rocks have received a great deal of recent attention because of synfuels research as the Green River Formation contains abundant "oil shales." "Oil shales" is a misnomer, for they are not shales and contain no oil, but are kerogen-rich marlstones.

These "oil shales" are organic rich mainly because of their slow deposition in a stratified lake in which the anoxic conditions persisted over very long periods of time and allowed laminated organic-rich rocks to accumulate. These oil shales, following the popular terminology, formed in the deep parts of Lake Uinta, where it was stratified, in local areas along the shallow shorelines where protected lagoons developed and a flotant provided abundant organic material, and in locally restricted evaporative sites. Most oil shales seen in the selected sites described here are of the second type.

The Green River Formation contains a rich fossil flora and fauna. The famous Green River fish are found in a variety of locations but most notably in Fossil Basin near Kemmerer, Wyoming. In addition, excellent examples of fossil crocodiles, birds, turtles, palms, ferns, charophytes, algae, insects, ostracodes, and mollusks have been recovered from these lacustrine beds.

SITE INFORMATION

All three sites are along U.S. 50-6. Figure 1 indicates distances to each site and location.

Site A. Approximately 82 ft (25 m) of lacustrine deposits are exposed in this roadcut (Fig. 2). Basal outcrops near the

Figure 2. Site A - This site has approximately 82 ft (25 m) of open lacustrine rocks. Note cyclic sedimentary units. Greenish units near the base have occasional vertebrate fossil remains.

highway and upwards for 20 ft (6 m) are green-gray thin-bedded shales and claystones, with occasional ironstone concretions. Thin, organic-rich beds occur near the top of this unit. Well-preserved fish scales, reed fragments, and casts of gastropods are found throughout. The overlying 6.5 ft (2 m) are light tan to gray-white beds that appear at first glance to be limestones, but are mainly siltstones with a few calcareous stringers. These beds are dark brown on the fresh surface and weather gray-white. Some units are organic-rich and contain plant fragments and coprolites. The next 26 ft (8 m) of green-gray siltstone and claystone are much like the lower unit. The prominent gray-white

medium-bedded, blocky unit near the top (arrow, Fig. 2) is an organic-rich unit that contains abundant fossil remains. At least two thin units are rich enough in organic matter to be classified as low-grade oil shale.

Site B. This near-vertical outcrop is approximately 260 ft (80 m) thick and can be divided into four general units (Fig. 3). The lowermost unit is 26 ft (8 m) of green-gray to brown-green mudstones and shales with some thin dolomite and gypsum interbeds, which form minor ledges. Some fragments of fish, plants, and crocodile bones occur in this unit. The second unit is a prominent, light gray mudstone that is nearly 10 ft (3 m) thick.

Figure 3. Site B - This spectacular display of open lacustrine and lagoonal rocks is indicative of the majority of the Green River Formation all along the highway. For orientation, an arrow marks the marshy, peat-rich unit.

The third unit is nearly 100 ft (30 m) thick and is composed of many thin units that range from siltstones, coquinas, to a meter-thick tan sandstone at the top. This sandstone is thought to be part of a distributary accumulation, and beds below it are probably of lagoonal or shallow open-lake origin. The fourth unit, above the sandstone, is a series of evenly bedded mudstones that contain abundant fragments of fish and plants. Lower mudstones of the unit grade upward into prominent dark brown beds that

are paper shales and peat of a suspected marsh. The brown unit is capped by a yellowish-gray mudstone that contains abundant fish fragments. The top of the exposed section is a series of greenish-gray mudstones. A well-exposed fault on the west side of this outcrop may be of interest.

Site C. Rocks at this site are accessible for close examination (Fig. 4) and include mainly three lithologic types. The first is a gray-green mudstone that contains thin, organic-rich brown mud-

Figure 4. Site C - Ease of access to this outcrop will allow for closer examination of the rocks. Characteristic fossils can be found in the talus of the base and on the east slope.

stones. These mudstones contain a variety of fish and crocodile fragments. The second is thin-bedded tan to light gray, blocky, calcareous mudstones. These mudstones are dark gray to dark brown on the fresh surface and often commonly organic. The third lithology is also tan to light gray, medium-bedded and is a low-grade oil shale. Occasional bird tracks and plant remains are found associated with the thin-bedded mudstones.

REFERENCES CITED

Baer, J. L., 1969, Paleoecology of cyclic sediments of the lower Green River Formation, central Utah: Brigham Young University Geology Studies, v. 16, p. 3–95.

Bradley, W. H., 1931, Origin and microfossils of the oil shale of the Green River Formation of Colorado and Utah: U.S. Geological Survey Professional Paper 168, 58 p.

——— 1964, Geology of Green River Formation and associated Eocene rocks in southwestern Wyoming and adjacent parts of Colorado and Utah: U.S. Geological Survey Professional Paper 496–A, 86 p.

Picard, M. Dane, 1985, Hypothesis of oil shale genesis, Green River Formation, Utah, Colorado, and Wyoming, *in* Picard, M. D., ed., Geology and Energy Resource Unit Basin of Utah: Utah Geological Association Publication 12, p. 193–210.

Rigby, J. K., 1968, Guide to the geology and scenery of Spanish Fork Canyon along U.S. Highways 50 and 6 through the southern Wasatch Mountains, Utah: Brigham Young University Geology Studies, v. 15, pt. 3, 31 p.

The Book Cliffs Cretaceous section: Western edge of the Interior Seaway

J. Keith Rigby, Michael P. Russon, and Richard E. Carroll, *Department of Geology, Brigham Young University, Provo, Utah 84602*

INTRODUCTION

Location and Access. A classic cross section through Cretaceous and early Tertiary rocks is well exposed along U.S. 50-6 in Price Canyon, a gorge cut through the Book Cliffs of eastern Utah (Fig. 1). Price Canyon occurs, in part, on the Helper, Standardville, and Kyune 7½-minute Quadrangles, in northern Carbon County and southeastern Utah County.

Most significant outcrops are readily available along the highway on land controlled by the Utah State Highway Commission. Land beyond the right-of-way is privately owned in the lower part of the canyon, from Castlegate southward, but above Castlegate it is controlled by the Bureau of Land Management.

Significance. Roadside exposures up Price Canyon provide an almost totally exposed section from the middle of the Late Cretaceous Mancos Shale up to the Paleocene-Eocene Flagstaff and Colton Formations (Fig. 2). These outcrops show three-

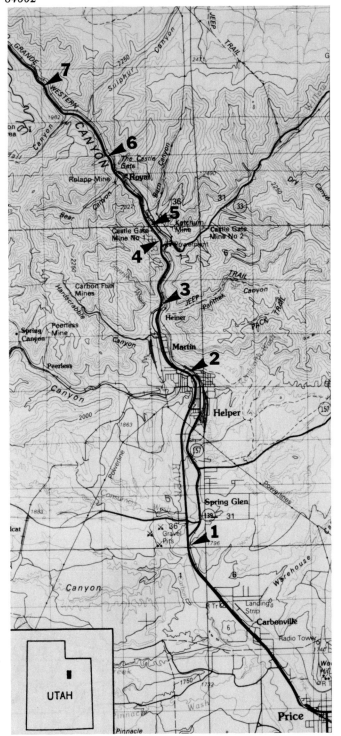

Figure 1. Index map to the Price Canyon site. Stops are indicated by numbered arrows. Base from Price 1:100,000 Quadrangle.

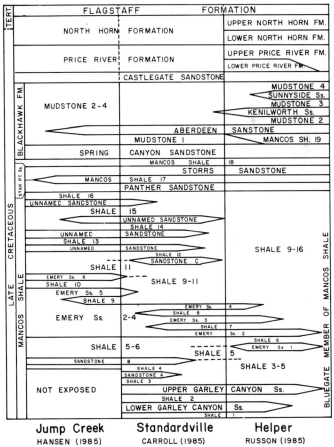

Figure 2. Stratigraphic nomenclature of units exposed in the Price Canyon area. Not to scale. (Modified from unpublished theses, R. E. Carroll, M. P. Russon, and C. D. Hansen, 1985, Brigham Young University.)

dimensional relationships of open marine, deltaic and piedmont systems of the western margin of the Cretaceous Interior Seaway. The section provides excellent models for interpreting deposition in such environments and records a generally regressive clastic sequence. Relationships of major coal accumulation to these depositional environments are also clearly exposed in and near Price Canyon.

Previous Work. Geologic investigation in the area began more than 80 years ago. Numerous reports that concern the area are listed in Cross et al. (1975); Balsley (1982); Fouch and others (1983); and Russon (1986). Fouch and others (1983) summarized regional patterns of deposition of Late Cretaceous rocks of the area. Howard (1972) analyzed facies of the Panther Sandstone Member of the Star Point Formation and provided an example of trace fossil facies in sandstone tongues. More recent analyses by Balsley (1982) added to understanding of the intertonguing marine-nonmarine Cretaceous rocks, with emphasis on the coal-producing Black Hawk Formation. Russon (1985) has mapped the lower part of Price Canyon in the Helper and Castlegate areas.

STRUCTURE

Rocks within the site dip homoclinally northward off the large San Rafael Swell dome into the Uinta Basin. Dips of only a few degrees allow analysis of three-dimensional stratigraphic relationships, for the section in Price Canyon is essentially unbroken by faults and uncomplicated by folds.

STRATIGRAPHY

Cretaceous and Tertiary rocks are magnificently exposed in the Rock Cliffs near Price, Utah, and range from the upper part of the Mancos Shale, in pediment-marked lowlands near Helper, up to the Tertiary Flagstaff and Colton Formations in the steep headwaters of Price Canyon (Fig. 2). In general, this sequence documents the final regression of the Western Interior Sea from the area.

Coarse clastic marine tongues within the Mancos Shale mark pulses of sediment swept from uplifts to the west. Such tongues are thicker and more common in the upper Mancos Shale and the overlying Star Point Formation than in the lower Mancos Shale. The still younger Black Hawk Formation, however, is dominantly nonmarine, but still contains interdigitations of marine rocks in wave-dominated deltaic and lagoonal sequences in exposures along Price Canyon. The Black Hawk Formation is the principal coal-bearing Cretaceous unit at this longitude.

The Castlegate Sandstone forms a prominent cliff and functions as a key unit for regional stratigraphic analysis. It accumulated in braided stream complexes; roadside exposures now show characteristic lenticular bedding. The overlying Price River and North Horn Formations are piedmont alluvial plain deposits and include sediments that accumulated in braided and meandering streams, flood plains, and lacustrine environments. The North

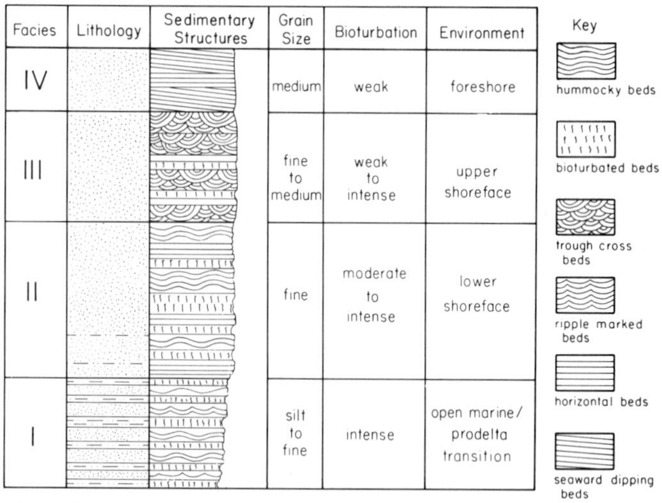

Figure 3. Stratigraphic and lithologic relationships of Facies 1 to 4, as developed in Cretaceous rocks in the Book Cliffs area. (From unpublished thesis, R. E. Carroll, 1985, Brigham Young University.)

Horn Formation bridges the Cretaceous-Tertiary boundary, although the exact position of that boundary is uncertain in Price Canyon.

The Paleocene Flagstaff Formation caps dip slopes in upper parts of the canyon and is a lacustrine and lacustrine-deltaic unit. Reddish Colton Formation erodes to form the broad strike valley of Emma Park that separates the Book Cliffs region from the high, gray rim of the Green River Formation on the south margin of the Uinta Basin.

FACIES

General Statement. Rhythmic repetition of facies associated with transgressive and regressive shorelines and clastic wedges in these rocks have been studied by many workers (see Fouch and others, 1983; Russon, 1986). Balsley (1982) differentiated four major facies within the sandstones (Fig. 3). Facies 1 marks the outer edge of coarse clastic deposition and is transitional from Mancos Shale. At the other extreme, Facies 4 marks the beach and shore zone.

Sandstone Tongues. Rocks of Facies 1 consist of interbedded siltstone and sandstone in layers from only a few centimeters up to a meter thick. Lowermost beds grade from open-marine Mancos Shale upwards to interbedded siltstone and sandstone. Parallel lamination is typical, although ripple marks and a variety of trace fossils may also occur. Shale and siltstone are usually dark gray because of abundant organic detritus and woody debris. Sole marks are common where floods of clastic debris have washed out onto semi-consolidated soft clays. Facies 1 accumulated near the outer edge of sand invasion into the muddy environment.

Facies 2 is generally composed of thin to medium beds of fine-grained quartz sandstone with silty partings. Each sandstone shows an upward increase in grain size and degree of bioturbation. Most beds are horizontally laminated but hummocky stratification is also common. Fine-grained partings may be internally bioturbated. Trace fossils include *Ophiomorpha, Thalassinoides, Asterosoma, Teichichnus, Chondrites,* and other trails and burrows. Facies 2 is interpreted to represent lower shore face environments, with deposition dominated by pulses of accumulation during storms. Fine-grained layers mark periods of relative quiet, during which bottom-dwelling organisms reworked the sediments.

Facies 3 is characterized by cliff-forming, orange-gray to light gray, fine- to medium-grained sandstone that is generally well cemented. High-angle trough cross stratification is characteristic of the facies, with individual sets up to 1 m thick, although some planar cross-beds also occur. Bioturbation ranges from limited to moderate, with *Ophiomorpha* and cylindrical burrows common in zones that have been reworked. The lower contact with Facies 2 is gradational to sharp and contact with overlying Facies 4 rocks is generally sharp. Facies 3 beds accumulated in high-energy environments of the upper shore face and were influenced by breaking waves and longshore currents.

Facies 4 rocks in the Price Canyon area are generally fine- to medium-grained sandstone, peppered with minor grains of black chert and locally may contain feldspar fragments. Facies 4 beds range up to 2 m thick and generally consist of tabular-bedded sandstone sets, 10-30 cm thick that dip seaward 2°-6°. Trace fossils are rare. These rocks generally form a striking white cap over the yellowish sandstone of Facies 3. Origin of the white cap is somewhat conjectural. It may have resulted from leaching beneath coal beds where acid swamp waters circulated through the sand, or it may have resulted from sedimentary differentiation within the shore zone during accumulation. Facies 4 rocks commonly cap major sandstone tongues. The top of the facies corresponds to the landward extent of clastic shoreline deposition. The gently inclined laminated sets record essentially wave swash and back wash zones within the foreshore. Facies 4 beds are accretionary sheets of the beach. The upper surface may have been ridged, with swells and shales in "rolls" so common in coal mines of the district.

Coal-bearing swamp and lagoonal sediments culminate progradation of the shoreline and commonly overlie Facies 4 rocks. Upper surfaces of Facies 4 beds may be rooted where plants have penetrated into the sandstone. Coal commonly rests directly on Facies 4 rocks of the Blackhawk Formation in the Price Canyon area.

Complete sequences of Facies 1 through 4 may not be present in every sandstone tongue nor at every locality. Facies 4 is commonly not present in seaward parts of tongues where repeated cycles of Facies 1 and 2 or of 2 and 3 may occur. For example, only Facies 1 and 2 are well developed in the Panther Sandstone at Stop 3 but all four facies are clearly defined in the Aberdeen Sandstone in cuts of Stops 4 and 5 near the old townsite of Castlegate. Facies 1 and 2 occur in the two Garley Canyon Sandstone tongues at Stop 1.

Mancos Shale Tongues. Interbedded tongues of the Mancos Shale are calcareous, dark bluish-gray or dark brownish-gray, silty shale. Each tongue becomes increasingly silty vertically and horizontally toward interdigitating sandstones. Occasional thin interbedded sandstones are probable deposits of shallow turbidites. Thin lenses of dark limestone and beds of septarian nodules occur but are rare. The Mancos Shale, Star Point and Black Hawk Formations have been interpreted as a record of interaction of barrier island complexes or of wave-dominated deltas and shallow open seas.

Mudstone Members. Piedmont alluvial plain deposits are well shown in road cuts near Castlegate where interbedded carbonaceous mudstone, fine- to medium-grained sandstone and coal crop out. Mudstone members in the Black Hawk Formation are generally slope-forming units that have sharp basal contacts on underlying thick sandstones. Mudstones are generally rich in plant debris and locally contain well-preserved fossil floras (Parker, 1976). Lenticular sandstones of cross-bedded point bars document complex meandering channel accumulations in many exposures.

Interbedded coals are generally thickest over beach ridges of wave-dominated deltas and thin lagoonward or northwestward. In the Price Canyon area, seven coal seams in the Black Hawk Formation are potentially minable. Individual seams may range up to approximately 5 m thick and are generally high volatile B bituminous coals. Relationships of sandstone and mudstone tongues and the economically important coals are well shown in exposures of the Black Hawk Formation at Stops 4 and 5.

SITE DESCRIPTIONS

Mileages to aid in locating exact outcrops are in reference to the official Utah State Highway markers.

Stop 1. Railroad cuts and highway exposures, east of the highway at approximately Mile 236.2, are in the Garley Canyon Sandstone, one of the major sandstone tongues in the middle part of the Bluegate Member of the Mancos Shale (Fig. 2). The member includes two prominent cliffs of sandstone, each approximately 49 ft (15 m) thick, separated by a tongue of Mancos Shale. Each sandstone shows a classic coarsening-upward sequence. Lower Bluegate Shale is exposed along the base of both cliffs, toward the southeast, and grades upward into Facies 1 in each sandstone. Facies 2 rocks, above, are cross-bedded, bioturbated, thin-bedded sandstones with minor interbedded shale. Such beds of the lower sandstone are well exposed in the southern end of the railroad cut.

The top of the Garley Canyon Sandstone is inaccessible here, but comes down to highway level 0.3 mile to the north, where the highway crosses the Price River at approximately Mile 235.9. Fossils are relatively rare throughout the Mancos Shale, particularly near detrital wedges like the Garley Canyon Sandstone.

Stop 2. Panoramic view of the Helper face, to the north, where upper Mancos Shale, Star Point Formation, and the lower Black Hawk Formation are well exposed (Figs. 2, 4). Mancos

Figure 4. View north to the Helper face from Stop 2. M, Mancos Shale; P, Panther Sandstone; S, Storrs Sandstone; SC, Spring Canyon Sandstone.

Shale erodes to low badland topography near the base. It is overlain by the Panther Sandstone Member of the Star Point Formation, which holds up the first prominent cliff and is composed of Facies 1 and 2. Upper beds of Facies 2 are the blocky, well-cemented, angular-weathering sandstones at the top of the Panther Cliff. The member is composed of Facies 1 beds toward the east, near where Panther beds disappear on the skyline. It grades into Mancos Shale a few miles farther to the east. The Panther Sandstone becomes increasingly prominent toward the west and forms the first major cliff above the lower slope and pediment cut across the Mancos Shale.

The overlying Storrs Member is a less resistant sandstone and is made of Facies 1 here. Eastward, it holds up only a silty shoulder.

The Storrs Member is overlain by a thick tongue of Mancos Shale that forms the prominent slope up to the base of the Black Hawk Formation, here marked by the sheer wall of the thick Spring Canyon Sandstone.

The Spring Canyon Sandstone, in the western face, appears moderately massive but it breaks up into several stratified units when traced eastward. These are separated by thin siltstone bands that become increasingly prominent toward the east. The cliff-forming part of the sandstone is cross-bedded Facies 3 beds. The massive Spring Canyon Sandstone is overlain by a semi-slope zone that includes a thin eastward extension of the Spring Canyon coal zone.

The massive Aberdeen Sandstone caps exposures on the skyline.

Stop 3. Panther Canyon (east) and Gentile Wash (west) at Mile 231.2. The type section of the Panther Sandstone is in the bold bluff to the east. The member is characterized here by well-bedded sandstone layers that dip slightly to the south and southeast. Beds may be traced diagonally downward and laterally from the top of the bluff to near its base, where stratification becomes horizontal and obscure. Most of the type Panther Sandstone here is composed of Facies 2 over thin lower silty units of Facies 1. The Sandstone is medium grained, cross laminated, and separated by thin, dark gray, silty partings. Ripple marks and a variety of sole marks and trace fossils are common in virtually every exposure and are particularly well shown in exposures west of Stop 2.

A thin sandstone rests with angularity across truncated upper edges of the foreshore or delta front beds of Facies 2. The sandstone accumulated as upper units were reworked by a subsequent marine transgression.

Each bed within the Panther Sandstone is a small coarsening-up sequence that grades from heavily bioturbated shale and siltstone upward through a thin unit of siltstone and fine-grained sandstone to coarser sandstone near the top. Uppermost sandstone has stratification obscured by intense bioturbation, which also affects the overlying siltstones. Numerous well-preserved trace fossils include a wide variety of burrows, tracks, trails, feeding marks, and resting marks (Howard, 1972).

Westward up Gentile Wash, the Storrs Sandstone forms the moderately well-defined bluff above angular Panther Sandstone exposures. Ledge-and-slope outcrops in the headwaters of the canyon are the lower part of the Black Hawk Formation. The uppermost cliff with a distinctive white cap is the Aberdeen Sandstone Member of the Black Hawk Formation (Fig. 2). The somewhat more ragged exposures below are of the Spring Canyon Member. Two of these sandstones also have well-developed white caps.

Stop 4. Deep, double road cuts through the lower part of the Black Hawk Formation at Mile 229.8. Uppermost Spring Canyon Sandstone is exposed in the lower cuts. Older beds of the member, in Facies 1 and 2, are exposed south of the stop. Upper beds in Facies 3 and 4 are well exposed on the west side of the

Figure 5. Road cut of part of the Black Hawk Formation at Stop 5. Aberdeen Sandstone in Facies 4 underlies the coal bed over the truck. Most of road cut is in mudstone member 2, in piedmont facies. A, Aberdeen Sandstone; K, Kenilworth Sandstone.

highway, south of the cut. These Facies 3 sandstones are massive to thick bedded, moderately well cemented and form the steep cliffs. *Ophiomorpha* and cylindrical vertical burrows are common. Approximately the upper 1 m of the Spring Canyon Sandstone is transitional into Facies 4 where a white cap is moderately well developed. Bioturbation is less extensive in Facies 4 and stratification is more planar and dips gently toward the east and southeast.

Some of the logs on top of the Facies 4 sandstone in the cut show *Teredo* borings (Cross and others, 1975, p. 20). The carbonaceous shale, here, is equivalent to the A sub 3 seam of the Spring Canyon coal group. The B sub 2 and C sub 1 seams occur higher in the road cuts. These coals are mined in the Wasatch Plateau west of Price Canyon. The lower half of the road cut is in a mudstone member between the prominent Spring Canyon and overlying Aberdeen sandstones.

The Aberdeen Sandstone is an upward-coarsening sequence and grades upward from a thin marine shale through Facies 1 to 4. The marine shale is the uppermost tongue of Mancos Shale in Price Canyon and marks the last time marine waters covered this part of central Utah.

Thin Facies 1 beds grade upward to somewhat thicker Facies 2 and Facies 3 units in road cut exposures. Facies 3 is particularly well exposed in the upper part of highway cuts on the west side of the road. Trough cross-beds, hummocky stratification, and distinctive trace fossils appear almost identical to Facies 3 beds of the Spring Canyon Member exposed below the lower

road cut. Facies 1 through 4 are repetitive in each sandstone tongue where the deltaic wedge prograded into the Mancos Sea far enough that beaches, Facies 4, developed.

Stop 5. Roadcut at Mile 229.7, west of the tipple complex of the Price River Coal Company at the old townsite of Castlegate. Lower road cut exposures are of the upper meter or so of the massive Aberdeen Sandstone, in Facies 4 (Fig. 5). Aberdeen beds are overlain by mudstone member 2, the irregularly interbedded lenticular sandstone, coal, carbonaceous shale that make most of the cut. The Kenilworth Sandstone is the pink clinkered sandstone at the apex of the cut. The moderately thick coal above the Aberdeen Sandstone is the Aberdeen or Castlegate A-seam. It is as much as 6 m thick beyond the canyon to the east.

Topographic expressions of Facies 1 to 4 are well shown in outcrops of the Aberdeen Sandstone along the east side of the canyon (Fig. 6). Recessive Facies 1 forms a dark gray slope zone and a thin, well-bedded Facies 2 occurs above that. Most of the cliff is made of moderately massive, cross-bedded Facies 3. Light-colored Facies 4 caps the cliff below the slope-forming mudstone member.

Lenticular lagoonal or lower delta plain sandstones show scour-and-fill features in mudstone member 2 beds in the big cut west of the highway. Many show prominent point bars even though the channels are only 1 or 2 m deep. In general, these streams flowed towards the east and southeast, although with sinuous courses. Interbedded gray mudstones commonly show root traces, particularly beneath the thin, discontinuous fluvial coals.

Red beds in the immediate vicinity all have been produced by baking when underlying coals burned.

Toward the north, up canyon, the prominent cliff-former is the Castlegate Sandstone that overlies the Black Hawk Formation

Figure 6. View southeastward from Stop 5 showing Facies 1 to 4 in the Aberdeen Sandstone at Castlegate. Numbers indicate facies.

with local unconformity. Some angularity is visible at the base of the cliff immediately east of The Castle Gate, the type locality.

Stop 6. Castlegate view area, at approximately Mile 228.0. The Castle Gate is the turreted cliff down canyon to the southeast. The mileage marker on the west side of the highway is at the base of the Castlegate Sandstone. Exposures across Price River show the complex lenticularity distinctive of the sandstone throughout much of central Utah.

The sandstone contains irregular shaly partings or lenses of coal. Bases of many lenses are defined by clay-pebble conglomerates made of rip-up clasts. These sandstones are not bioturbated although root casts occur in some carbonaceous units. The formation accumulated as deposits in a coarsely braided fluvial system. It marks an abrupt flush of sediment off the rising Sevier highlands to the west and more or less marks the end of coastal plain deposition and the beginning of piedmont, fluvially dominated deposition.

Interbedded slopes and ledges in the upper half of the canyon wall are lower Price River Formation. These sandstones are also complexly lenticular and show a continuation of braided stream deposition in individual units. Shaly interbeds are lacustrine or back swamp accumulations.

Stop 7. Price River Formation, at Utah State Highway Mile 225.7. Perennial damage along the highway is produced by slumping of interbedded, massive, fluvial sandstones, like those exposed in road cuts at the highway bend, and a gray, shaly lacustrine or back swamp section well exposed in cuts along the old highway, on the canyon wall to the west. The massive sandstones are braided stream deposits, like the Castlegate Sandstone below.

Gray-green, lacustrine and gray-brown back swamp deposits show root impressions and bleached zones. Thin, interbedded calcareous sandstones and very sandy limestones are deposits of ponds or flat-bottomed splays that spread out into interfluve basins.

REFERENCES CITED

Balsley, J. K., 1982, Cretaceous wave-dominated delta systems: Book Cliffs, East Central Utah: Field Guide published for the American Association of Petroleum Geologists, 219 p.

Cross, A. T., Maxfield, E. B., Cotter, E., and Cross, C. C., 1975, Field guide and road log to the Western Book Cliffs, Castle Valley, and parts of the Wasatch Plateau: Brigham Young University Geology Studies, v. 22, pt. 2, 132 p.

Fouch, T. D., Lawton, T. F., Nichols, D. J., Cashion, W. B., and Cobban, W. A., 1983, Patterns and timing of synorogenic sedimentation in Upper Cretaceous rocks of central and northeast Utah, *in* Mesozoic Paleogeography of the West-central United States, Rocky Mountain Paleogeography Symposium 2: Rocky Mountain section, Society of Economic Paleontologists and Mineralogists, Denver, Colorado, p. 305–336.

Howard, J. D., 1972, Trace fossils as criteria for recognizing shorelines in the stratigraphic record, *in* Recognition of ancient sedimentary environments: Society of Economic Paleontologists and Mineralogists, Special Publication 16, p. 215–226.

Parker, L. R., 1976, The paleoecology of the fluvial coal-forming swamps and associated floodplain environments in the Blackhawk Formation (Upper Cretaceous) of central Utah: Brigham Young University Geology Studies, v. 22, pt. 3, p. 99–116.

Russon, M. P., 1985, Geology, depositional environments, and coal resources of the Helper 7½′ quadrangle, Carbon County, Utah: Brigham Young University Geology Studies, in press.

57

The House Range, western Utah: Cambrian mecca

Lehi F. Hintze, *Department of Geology, Brigham Young University, Provo, Utah 84602*
Richard A. Robison, *Department of Geology, University of Kansas, Lawrence, Kansas 66045*

LOCATION AND ACCESSIBILITY

The House Range is one of the principal ranges in western Utah (Fig. 1). It trends north-south for 60 mi (100 km) along the west edge of the Sevier Desert and lies 40 mi (65 km) west of Delta, Utah. The range has an asymmetric profile: Its eastern slopes, formed mostly on dipslopes of Cambrian strata, rise gradually up from the Sevier Desert floor; its western face, formed by block faulting, rises abruptly from the floor of Tule Valley (also known as White Valley) at an elevation of 4,400 ft (1,300 m) to its highest point, Notch Peak, elevation 9,655 ft (2,943 m). The northwest face of Notch Peak forms one of the tallest sheer cliffs in the west.

Delta, an irrigation-farming community situated in the center of the Sevier Desert, is the nearest source of gas, food, and supplies. U.S. 6-50 cuts southwestward from Delta to cross the southern part of the House Range at Skull Rock Pass. Graded gravel roads extend the entire length of the range on either side; roads cross the range only at Skull Rock Pass, at Marjum Pass in the middle, and at Death (Dome) Canyon a few miles north of Marjum Pass.

People planning to visit the House Range should inquire in Delta about road conditions. Although most of the time the main gravel roads are suitable even for passenger cars, after storms these roads may be washed out and passable only with four-wheel-drive vehicles.

SIGNIFICANCE

A magnificent continuous sequence of shallow-water marine Cambrian strata more than 10,000 ft (3,000 m) thick is exposed in the House Range (Fig. 2) (Hintze, 1973; Hintze and Robison, 1975). From a scientific standpoint, the most important feature is that these sedimentary layers contain one of the most diverse sequences of Middle and Upper Cambrian faunas known from anywhere in the world (Robison, 1976, 1984). The House Range is also famous among fossil collectors because one of its fossil localities in the Wheeler Shale near Antelope Springs has yielded more whole trilobite specimens than any other locality known. *Elrathia kingii* from this site can be found in most fossil collections that include trilobites.

Charles D. Walcott, then director of the U.S. Geological Survey, was among the first to study Cambrian rocks in the House Range during the early part of this century, and he described many trilobites, brachiopods, and sponges from here. Several decades later, Robison (1984, and earlier papers) restudied Walcott's primary Middle Cambrian localities from a modern biostratigraphical standpoint, describing many new species and finding particular value in agnostoid trilobites for worldwide

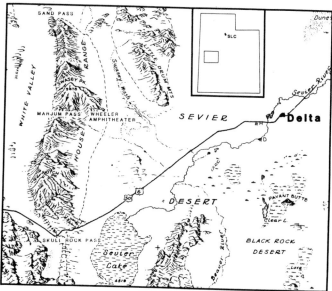

Figure 1. Landforms of west-central Utah showing features in the vicinity of the House Range (modified from part of the Landforms Map of Utah by M. K. Ridd).

correlation. Paleoenvironmental studies of Cambrian strata in the House Range helped fix the position of House Range strata relative to the trailing continental margin of western North America during late Middle and early Late Cambrian time.

One of the most recent scientific activities in the House Range has been a detailed collecting program aimed at obtaining fossils of sponges (Rigby, 1984 and earlier) and soft-bodied organisms similar to those from the famous Burgess Shale in the Canadian Rockies. Much field collecting was done by the Lloyd Gunther family, whose thoroughness and patience in working bed-by-bed through the Wheeler Shale and lower Marjum Formation resulted in finding a number of new taxa, some of which have been illustrated by Gunther and Gunther (1981).

WHAT TO SEE IN THE HOUSE RANGE

A guidebook (Hintze, 1973) with geologic road logs of the principal routes from Provo, Utah, to the House Range and beyond may be purchased from the Geology Department at Brigham Young Univeristy. Geologic quadrangle maps of the central House Range (Hintze 1981a, 1981b) are available from U.S. Geological Survey sales offices (e.g., Federal Building, Salt Lake City, Utah).

Wheeler Shale at Antelope Springs

Surely this is one of the best-known fossil localities any-

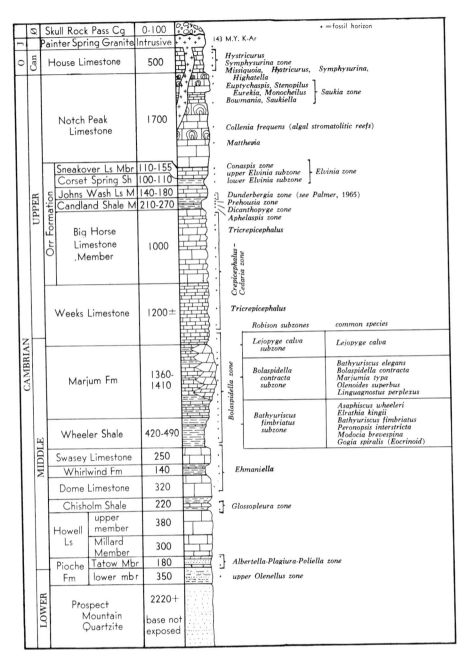

Figure 2. Stratigraphic column for the House Range. Thickness figures are in feet. Fossil names are mostly for trilobites.

where. Indians made amulets from trilobites found here during prehistoric times, and the locality's popularity among modern rockhounds and commercial collectors is attested to by the activity of collecting parties at the site almost every day. Antelope Springs lies at the south base of Swasey Peak, the second highest prominence in the House Range. The Wheeler Shale, being comparatively less resistant than adjacent formations, has been eroded here to form a 1-mi (1.6-km) wide cove called Wheeler Amphitheater.

Trilobites can be obtained from many horizons within the Wheeler Shale, which is about 500 ft (150 m) thick here (Hintze, 1981a). The lowest third is especially rich in the small agnostoid trilobites *Peronopsis* and *Ptychagnostus.* Most collectors work the upper part of the formation to obtain *Elrathia kingii* and *Asaphiscus wheeleri,* specimens of which range up to 2 in (5 cm) long. Trilobites from Antelope Springs are prized because they are preserved with unusual incrustations of secondary cone-in-cone calcite, 0.04 to 0.12 in (1 to 3 mm) thick, on the under

Figure 3. West face of the House Range viewed northeastward from the Marjum Canyon–Tule Valley road junction at mile 6.7 in the road log. Darker beds low on the slope are Pioche Formation; lowest massive cliff is Howell Limestone; highest point is Swasey Limestone; bench-forming units near the top of the mountain are Chisholm Shale and Whirlwind Formation.

surface. This calcite strengthens the fossil, rendering it more durable, so that specimens may weather free or be readily broken out of the matrix of calcareous shale.

Marjum Canyon

The easiest way to view much of the Cambrian stratigraphic sequence of the House Range is to travel down Marjum Canyon from Marjum Pass to the west range front. In the following short road log covering this route, mileage is cumulative from Marjum Pass.

0.0. STOP 1. Marjum Pass Summit. Wheeler Shale is exposed south of the road; Swasey Limestone forms the dip slope north of the road. The Wheeler Shale here contains sparse trilobites, but they do not separate from the rock as freely as those at the Wheeler Amphitheater. The Wheeler Shale occupies only the lower slopes here; higher alternating limestone and shale ledges belong to the Marjum Formation.

1.5. Rainbow Valley. A Jurassic granitic intrusion, which is out of view to the south, has caused chemical alteration of the Wheeler and Marjum formations, resulting in vivid red and yellow colors.

1.9. STOP 2. Head of Marjum Canyon. The Swasey Limestone forms a medium-gray cliff underlain by slope-forming yellow shaly thin-bedded limestones of the Whirlwind Formation. Certain olive-weathering limestone layers in the Whirlwind are a coquina of small heads of the trilobite *Ehmaniella.* The darker limestone beneath the Whirlwind is the top of the Dome Limestone.

2.1. A good view of the Dome Limestone cliff can be seen ahead at the base of the hill; Whirlwind Formation makes the

bench covered with juniper trees, and the cliff-forming Swasey Limestone is at the top.

2.7. Large wash enters Marjum Canyon from Rainbow Valley to the south. Hillside to the south shows Swasey limestone (top cliff), Whirlwind Formation (slope), and Dome Limestone (at road level). Minor faults offset the beds.

3.7. STOP 3. Most of the Middle Cambrian sequence can be seen on the hillside to the west and south:

Swasey Limestone	top cliff
Whirlwind Formation	upper bench
Dome Limestone	cliffs with reddish alterations
Chisholm Shale	bench
Howell Limestone	ledges near road level

Northward, the two members of the Howell can be seen. The Millard Member is dark gray and forms the base of the Howell cliffs; the overlying Howell weathers very light gray and forms the upper part of the Howell cliffs.

From a fossil-collecting standpoint, the massive cliff-forming units are mostly barren; they commonly consist of lime-mudstone but contain some oolitic and pisolitic horizons. Many massive beds are mottled in shades of gray; this may represent bioturbation of the lime mud at the time of deposition. The shaly bench-forming formations contain some fossil-bearing layers. The fossils commonly are in thin limestone beds that weather olive-gray. The Chisholm Shale here yields coquinas of the trilobite *Glossopleura* in a few horizons.

3.8. STOP 4. On the north side of the canyon, the dark-brownish-orange Tatow Member of the Pioche Formation is underlain by green and brown quartzite and phyllite of the lower member of the Pioche Formation. Thin limestone beds in the Tatow Member bear an *Albertella* Zone trilobite fauna.

4.2. STOP 5. Cross wash at the mouth of Marjum Canyon at the head of its alluvial fan. The lower member of the Pioche Formation just north of the road has yielded the oldest trilobites yet collected in Utah; large *Olenellus* impressions in the quartzites were described by Robison and Hintze (1972).

6.7. ROAD JUNCTION. Turn south toward Painter Spring. Looking back from here one has a fine view of the Cambrian strata in the House Range (Fig. 3).

7.9. Descend off Lake Bonneville shore terrace. The low hills in the bottom of Tule Valley are fault blocks of Upper Cambrian Notch Peak Formation, relatively displaced about 1 mi (1.6 km) downward from their occurrence in the House Range. Tule Valley is so named because of the rushes that grow around small saline springs in the valley bottom.

11.6. STOP 6. Painter Spring water tank. The road eastward leads to Painter Spring in Jurassic granitic rocks in the canyon. The lowest bedded unit to the east is contact-metamorphosed Marjum Formation, here invaded by pink granite sills. The cliff-forming Notch Peak Formation forms the upper part of Notch Peak. The ledgy slopes beneath are the shaly members of the upper third of the Orr Formation, below which are the more resistant lower Orr and Weeks formations. All strata surrounding the Jurassic granite have been contact-metamorphosed; the metamorphic aureole extends away from the contact for about 1 mi (1.6 km). The prominent white bed on the west face of Notch Peak is marble.

The shortest way from the Painter Spring water tank to the paved U.S. 6-50 is to proceed southward for about 10 mi (16 km) to join the highway at the valley bottom. From this point it is 33 mi (53 km) westward to the Utah-Nevada state line where there is a gas station, or 55 mi (88 km) eastward to Delta, Utah.

SELECTED REFERENCES

Only most recent references are cited. Earlier works, commonly major contributions, may be identified by citations in these listed papers.

Gunther, L. F., and Gunther, V. G., 1981, Some Middle Cambrian fossils of Utah: Brigham Young University Geology Studies, v. 28, pt. 1, 81 p.

Hintze, L. F., 1973, Geologic road logs of western Utah and eastern Nevada: Brigham Young University Geology Studies, v. 20, pt. 2, 66 p.

——1981a, Preliminary geologic map of the Marjum Pass and Swasey Peak SW quadrangles, Millard County, Utah: U.S. Geological Survey Miscellaneous Field Studies Map MF-1332.

——1981b, Preliminary geologic map of the Swasey Peak and Swasey Peak NW quadrangles, Millard County, Utah: U.S. Geological Survey Miscellaneous Field Studies Map MF-1333.

Hintze, L. F., and Robison, R. A., 1975, Middle Cambrian stratigraphy of the House, Wah Wah, and adjacent ranges in western Utah: Geological Society of America Bulletin, v. 86, p. 881–891.

Rigby, J. K., 1984, Sponges of the Middle Cambrian Marjum Limestone from the House Range and Drum Mountains of western Millard County, Utah: Journal of Paleontology, v. 57, p. 240–270.

Robison, R. A., 1976, Middle Cambrian trilobite biostratigraphy of the Great Basin: Brigham Young University Geology Studies, v. 23, pt. 2, p. 93–109.

——1984, Cambrian Agnostida of North America and Greenland, Part I, Ptychagnostidae: University of Kansas Paleontological Contributions Paper 109, 59 p.

Robison, R. A., and Hintze, L. F., 1972, An Early Cambrian trilobite faunule from Utah: Brigham Young University Geology Studies, v. 19, p. 3–13.

Exceptionally fossiliferous Lower Ordovician strata in the Ibex area, western Millard County, Utah

Lehi F. Hintze, Brigham Young University, Provo, Utah 84602

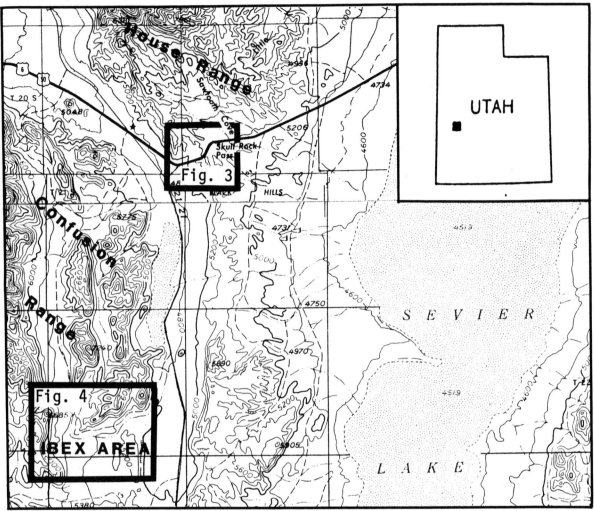

Figure 1. Ibex area and vicinity. Skull Rock Pass is about 50 miles (80 km) west of the nearest source of gasoline or supplies at Delta, Utah. This map is taken from parts of the Delta and Richfield 1:250,000 topographic quadrangles. The squares are townships, six miles on each side.

LOCATION AND ACCESSIBILITY

Exposures of the most richly fossiliferous Lower Ordovician strata in the United States occur about 50 mi (80 km) west of Delta, Utah, and extend over an area 6 by 8 mi (15 by 20 km). These rocks form low-lying hills in the southern part of the House and Confusion ranges (Fig. 1) and are reached by paved U.S. 6-50, which passes along the northern edge of the Ibex area, named for a small ranch abandoned in the 1930s. The area is neither populated nor fenced and is used primarily for winter sheep grazing under administration of the U.S. Bureau of Land Management, which maintains graded dirt roads that lead to within easy walking distance of most outcrops. There is no water.

SIGNIFICANCE

Interest in the Ibex area stems primarily from the well-exposed stratigraphic section of shallow-water marine deposits more than 4,700 ft (1,430 m) thick (Fig. 2) that contains fossils in various degrees of abundance and diversity throughout. Trilobites, conodonts, brachiopods, graptolites, cephalopods, bryozoans, bivalves, gastropods, ostracods, echinoderms, sponges, and corals have all been recovered from these strata and their study has permitted the establishment of a sequence of fossil zones useful to Ordovician students throughout the world.

Except for graptolites that are found in thin shale interbeds most of the fossils occur as fragments in thin-bedded muddy and

261

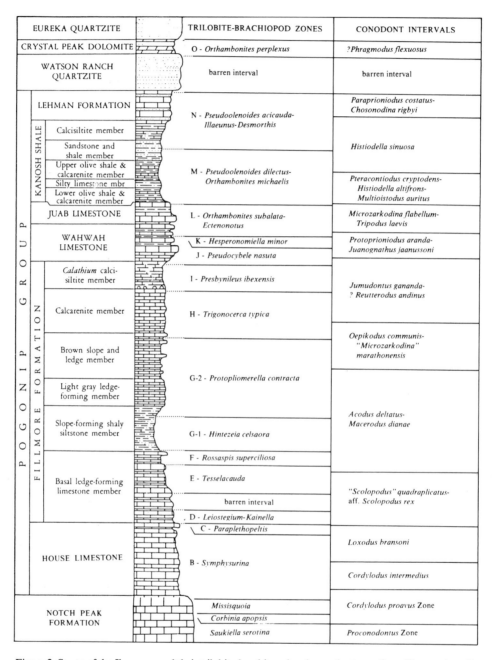

EUREKA QUARTZITE			TRILOBITE-BRACHIOPOD ZONES	CONODONT INTERVALS
CRYSTAL PEAK DOLOMITE			O - Orthambonites perplexus	?Phragmodus flexuosus
WATSON RANCH QUARTZITE			barren interval	barren interval
	LEHMAN FORMATION		N - Pseudoolenoides acicauda-Illaeunus-Desmorthis	Paraprioniodus costatus-Chosonodina rigbyi
	KANOSH SHALE	Calcisiltite member		Histiodella sinuosa
		Sandstone and shale member		
		Upper olive shale & calcarenite member	M - Pseudoolenoides dilectus-Orthambonites michaelis	Pteracontiodus cryptodens-Histiodella altifrons-Multioistodus auritus
		Silty limestone mbr		
		Lower olive shale & calcarenite member		
P O G O N I P G R O U P	JUAB LIMESTONE		L - Orthambonites subalata-Ectenonotus	Microzarkodina flabellum-Tripodus laevis
	WAHWAH LIMESTONE		K - Hesperonomiella minor	Protoprioniodus aranda-Juanognathus jaanussoni
			J - Pseudocybele nasuta	
	FILLMORE FORMATION	Calathium calcisiltite member	I - Presbynileus ibexensis	Jumudontus gananda-? Reutterodus andinus
		Calcarenite member	H - Trigonocerca typica	
		Brown slope and ledge member		Oepikodus communis-"Microzarkodina" marathonensis
			G-2 - Protopliomerella contracta	
		Light gray ledge-forming member		Acodus deltatus-Macerodus dianae
		Slope-forming shaly siltstone member	G-1 - Hintezeia celsaora	
			F - Rossaspis superciliosa	
		Basal ledge-forming limestone member	E - Tesselacauda	"Scolopodus" quadraplicatus-aff. Scolopodus rex
			barren interval	
			D - Leiostegium-Kainella	
	HOUSE LIMESTONE		C - Paraplethopeltis	
			B - Symphysurina	Loxodus bransoni
				Cordylodus intermedius
	NOTCH PEAK FORMATION		Missisquoia	Cordylodus proavus Zone
			Corbinia apopsis	
			Saukiella serotina	Proconodontus Zone

Figure 2. Strata of the Ibex area and their trilobite-brachiopod and conodont zonations. The stratigraphic column shown totals about 4,700 ft (1,430 m) thick. For scale, the House Limestone is 500 ft (152 m) thick.

silty bioclastic limestone and intraformational flat-pebble conglomerate beds. Trilobites and echinoderms are almost never found articulated or whole; brachiopods and ostracods often form coquinas; almost all fossils reflect, in their fragmented occurrence, the energy of a wave-washed, near-shore, shallow-water environment. Persons seeking trophy-type fossils here must be prepared for a long search amidst the debris to find the perfect specimen.

The abundance of Ordovician fossils in the Ibex area became common knowledge during the early part of this century shortly after Jack Watson and his wife established a clapboard ranchhouse at Ibex, 40 mi (65 km) west of their nearest neighbors at Black Rock. The Watsons developed their water supply from rainwater caught in enlarged potholes on a dipslope of Eureka Quartzite; they sold their water, at 25 cents a gallon, to thirsty passersby. F. F. Hintze and Frederick J. Pack, professors of geology at the University of Utah during the twenties and early thirties, brought paleontology classes to Ibex. Fossil Mountain, shown on The Barn 15-minute Quadrangle, (see Fig. 4) obtained its name during this period.

The first description of sections near Ibex rocks and their trilobite faunas was done by Hintze (1951, 1952). A decade later no other Lower Ordovician area in the United States had been discovered that could compare in faunal richness with the Ibex area. We undertook a three-year, more detailed study that permitted additional collecting. This study emphasized the non-trilobite elements, particularly, and aimed at documenting all of the faunal elements completely. More than 20 scientific papers have been published by various authors describing the various taxa collected in this coordinated study. From these papers I summarized (Hintze, 1979) the zonal distributions established for 10 phyla, the most significant of which are trilobites, conodonts, graptolites and brachiopods.

The trilobite-brachiopod zones A through O, established by Ross (1951) and Hintze (1952) continue to be widely used in Ordovician correlations. They form a principal basis for subdividing the Lower Ordovician in North America and have been so utilized on the latest correlation tables (Ross and Bergstrom, 1982). These tables, in fact, use the name Ibexian to replace Canadian as a series designation for the Lower Ordovician in thhis country. In the Ibex area the trilobites are silicified throughout most of the stratigraphic sequence. Their fragments can be recovered in quantity by dissolving the host rocks in acid. Thus the stratigraphic range of each trilobite species is well documented from Ibex. Brachiopod occurrences were carefully recorded by Jensen (1967) as a result of several months of detailed collecting at Ibex. He identified 25 species which he grouped into nine brachiopod zones. Braithwaite (1976) spend most of two summers collecting graptolites in the Ibex area. He obtained beautifully preserved material by excavating into shale horizons beyond the deleterious effects of surface weathering. He recognized 45 species from 13 genera and established seven graptolite zones in the Ibex area. Conodonts were studied by Ethington and Clark (1981) who sampled the 3,300 ft (1,000 m) of Ordovician strata at 5-10 ft (2-3 m) intervals; they processed more than 600 samples and obtained more than 25,000 conodont elements which they grouped into 13 zones or intervals. Comparison of the trilobite-brachiopod and conodont zonation intervals is shown on Figure 2.

WHAT TO SEE AT IBEX

As mentioned above, the Ibex area is not a place to readily collect unbroken fossils; but if your interest lies with paleoenvironments, as evidenced by unusual rock types, and with the occurrence of a wide variety of early life forms, then the area has much for you.

Along the Highway. If you are traveling in a low-clearance passenger car, you should not venture from the paved highway. Figure 3 shows that portion of U.S. 6-50 that transects the

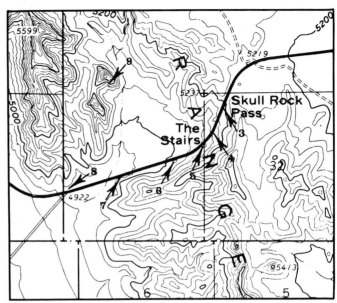

Figure 3. Paved U.S. Highway 6-50 from Skull Rock Pass to the Ibex turnoff (4922). Numbered arrows identify Braithwaite's graptolite localities in roadcuts into the Fillmore Formation along the highway.

northern part of the area. Braithwaite's (1976) graptolite collecting localities in highway roadcuts are indicated by numbered arrows. All of these cuts are in the Fillmore Formation whose most conspicuous rock type is intraformational flat-pebble conglomerate made up of silty or fine quartzose sandy limestone containing fragments of trilobites, brachiopods, echinoderms, and occasionally other organisms. Interbedded with these ledge-forming beds are less resistant, light-olive-gray shales that contain graptolites.

At the east end of the roadcut at Skull Rock Pass (marked 3 on Fig. 3), the light-gray lenses 3-6 ft (1-2 m) in diameter are sponge-algal reefs interbedded in the Fillmore Formation. This horizon is about 400 ft (120 m) above the base of the formation and it contains the *Tesselacauda* trilobite fauna of Ross-Hintze zone E. Church (1974) studied these patch reefs and determined that stromatolitic algae stabilized the substrate providing a base for a sponge reef framework which harbored trilobites, brachiopods, gastropods, and burrowing organisms. Braithwaite (1976, p. 57) identified *Mastigograptus* sp., *Dictyonema* sp., and *Dendrograptus* sp. from locality 3. Localities 4-8 yielded *Dictyonema, Adelograptus, Clonograptus,* and *Phyllograptus.* In terms of Ross-Hintze trilobite zones, localities 4-6 contain zone F trilobites, locality 7 contains zone G-1 trilobites, and locality 8 includes zone G-2 trilobites.

Off the Highway. With a high-clearance vehicle you can drive south, leaving U.S. 6-50 at the Ibex Well turnoff (4922 on Fig. 3), and follow the main graded road, passing by the Ibex Well pumphouse at the south end of the playa about 7 mi (11 km) from the turnoff. Continue south past the well for 1.2 mi (2 km) and then leave the main graded road to take a less prominent road that angles to the west and southwest. Traveling 2.7 mi (4.3 km) on this road brings you around the south face of

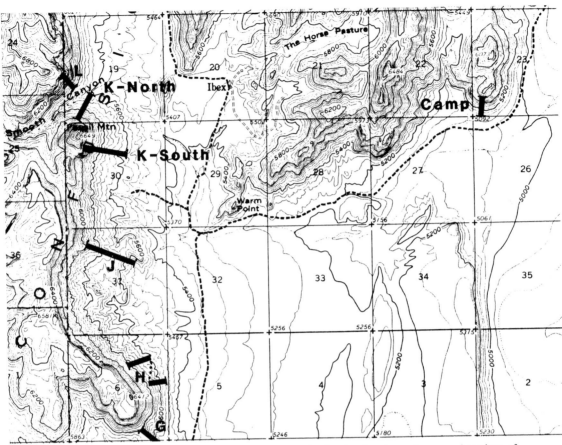

Figure 4. A portion of The Barn 15-minute Quadrangle showing locations of measured sections of Ordovician strata described by Hintze (1973). Each square represents a square mile.

an orange-weathering hill capped with Eureka Quartzite. This face is the Camp section (Hintze, 1973) shown on Figure 4. The base of this hill is in the Lehman Formation. The hillslope is fossiliferous.

Continuing westward from the Camp section for another 3.5 mi (5.5 km) brings you to Warm Point where the road divides as shown on Figure 4. From there you can go south to sections G and H (Hintze, 1951), which traverse the Fillmore

Formation, or west to section J (Hintze, 1951), which crosses the Wah Wah, Juab and lower Kanosh formations, or northwestward to the K and L (Hintze, 1951) sections, which cover the Kanosh, Lehman, Watson Ranch, Crystal Peak, and Eureka formations.

The serious student would be helped by access to measured sections (Hintze, 1973) and the geologic map (Hintze, 1974) that covers the area.

REFERENCES CITED

Braithwaite, L. F., 1976, Graptolites from the Lower Ordovician Pogonip Group of western Utah: Geological Society of America Special Paper 166, 106 p.

Church, S. B., 1974, Lower Ordovician patch reefs in western Utah: Brigham Young University Geology Studies, v. 21, part 3, p. 41–62.

Ethington, R. L., and Clark, D. L., 1981, Lower and Middle Ordovician conodonts from the Ibex area, western Millard County, Utah: Brigham Young University Geology Studies, v. 28, part 2, 155 p.

Hintze, L. F., 1951, Lower Ordovician detailed stratigraphic sections for western Utah and eastern Nevada: Utah Geological and Mineralogical Survey Bulletin 39, 99 p.

——1952, Lower Ordovician trilobites from western Utah and eastern Nevada: Utah Geological and Mineralogical Survey Bulletin 48, 249 p.

——1973, Lower and Middle Ordovician stratigraphic sections in the Ibex area, Millard County, Utah: Brigham Young University Geology Studies, v. 20, part 4, p. 3–36.

——1974, Preliminary geologic map of The Barn quadrangle, Millard County, Utah: U.S. Geological Survey Miscellaneous Field Studies Map MF–633.

——1979, Preliminary zonations of Lower Ordovician of western Utah by various taxa: Brigham Young University Geology Studies, v. 26, part 2, p. 13–19.

Jensen, R. G., 1967, Ordovician brachiopods from the Pogonip Group of Millard County, western Utah: Brigham Young University Geology Studies, v. 14, p. 67–100.

Ross, R. J., Jr., 1951, Stratigraphy of the Garden City Formation in northeastern Utah, and its trilobite faunas: Yale University, Peabody Museum of Natural History Bulletin 6, 161 p.

——and Bergstrom, S. M., eds., The Ordovician System in the United States—Correlation Chart and Explanatory Notes: International Union of Geological Sciences Publication No. 12, 73 p., 3 charts.

The geology of Salina Canyon, Utah

Timothy F. Lawton, Department of Earth Sciences, New Mexico State University, Las Cruces, New Mexico 88003
Grant C. Willis, Utah Geological and Mineral Survey, 606 Blackhawk Way, Salt Lake City, Utah 84108

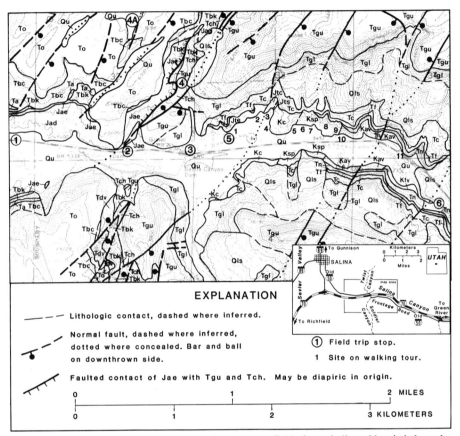

Figure 1. Geologic map of Salina Canyon locality. Stops on field trip are indicated by circled numbers. Sites on walking traverse (Stop 5) are indicated by numerals. Unit symbols are explained in Figure 3. Inset is a location map for the Salina Canyon site.

INTRODUCTION

The spectacular exposures in the lower part of Salina Canyon (Figs. 1, 2) lie in the Basin and Range to Colorado Plateau transition zone to the east of thrust-faulted Paleozoic rocks in the Pavant and Canyon ranges. The stratigraphic and structural relations exposed along a 4-mi (6.5-km) east-west corridor here provide unsurpassed insight into the late Mesozoic to middle Tertiary geologic history of central Utah. A well-exposed Middle Jurassic to Upper Cretaceous sedimentary section is unconformably overlain by one of the most complete lower Tertiary sequences found in this part of the state.

The site can be reached by following old U.S. 50 (called the "frontage road" in this text) from Salina or from the Gooseberry Valley exit off I-70 east of the site (Fig. 1, inset). Examination of the area requires two short hikes, one of about 1 mi (1.6 km) round trip (depending upon your vehicle) and one of 2.5 mi (4.0 km).

STRATIGRAPHY OF ROCKS EXPOSED IN SALINA CANYON

Three distinct stratigraphic sequences, separated by major unconformities, occur in Salina Canyon. Individual formations of the three sequences are depicted in Figure 3. The lower sequence, which includes units from the Arapien Shale through the Indianola Group, ranges in age from Middle Jurassic (Bajocian) through Late Cretaceous (Turonian). The Jurassic part of the lower sequence (Arapien Shale and Twist Gulch Formation) contains red to gray marine and marginal-marine clastic rocks and fluvial deposits, probably deposited in a back-arc basin. The Cretaceous section (Cedar Mountain Formation and Indianola Group) represents conglomeratic to shaly molasse-type clastic material shed eastward from the Sevier orogenic belt in western Utah and Nevada into a foreland basin.

The Jurassic to Cretaceous sequence was folded late in the Cretaceous to form a subaerial topographic feature that existed

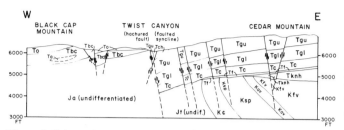

Figure 2. Schematic east-west cross section of Salina Canyon locality, about 0.2 mi (0.3 km) south of the north boundary of Figure 1.

into Eocene time. It was onlapped from the east by red- to gray-weathering fluvial and lacustrine clastics and carbonates of Paleocene to Eocene age (North Horn, Flagstaff, Colton, Green River formations). These units form the lower part of the middle sequence. The Late Eocene Crazy Hollow Formation, a fluvial deposit derived from the west, disconformably overlies the Green River Formation and is gradationally overlain by the lacustrine Bald Knoll Formation, which includes ash-flow tuffs with a northern source.

Following a subsequent hiatus, the youngest sequence, consisting of Oligocene volcaniclastics and ash-flow tuffs derived from the Marysvale volcanic field to the south, was deposited on a pre-Oligocene topographic surface that had significant relief.

GEOLOGIC FIELD GUIDE

Outcrops described here occur on the north and south sides of I-70 between 2.4 and 4.8 mi (4–8 km) east of downtown Salina. All road distances and directions indicated are from the traffic light at U.S. 89 and Center Street in Salina.

Begin at the intersection of U.S. 89 and Center Street. Proceed east for 0.3 mi (0.5 km), then turn south (right) on 300 East and proceed south on old U.S. 50.

Stop 1 (mile 2.4): Axis of Sanpete–Sevier Valley anticline. Outcrops adjacent to the road are gray calcareous shale and siltstone of Unit D of the Arapien Shale. The thin, discontinuous, brownish-red outcrop on the south side of the canyon, which parallels the volcanics, is Unit E, the salt-bearing member of the Arapien Shale. Because the Arapien is complexly folded and faulted in most exposures, its original stratigraphic thickness is uncertain. Recent subsurface data suggest that it was probably about 5,690 ft (1,735 m) thick (Standlee, 1982). The Arapien Shale here occupies the axis of the Sanpete–Sevier Valley anticline, an extensive north-northeast trending feature with at least 20,000 ft (7,100 m) of structural relief. The origin of the anticline has been attributed both to salt and shale diapirism (Witkind, 1982) and to thrust faulting (Standlee, 1982). Thin beds of the Arapien Shale typically show tight small-scale folds in the vicinity of such large-scale structures. An eastward-verging recumbent fold is exposed in the low hillside directly to the south.

Pale gray slopes underlain by the Bald Knoll Formation occur above the Arapien Shale in the vicinity of Stop 1. The Bald

Knoll Formation thins across the crest of the anticline. Oligocene volcanic units overlying the Bald Knoll Formation were derived from the south and deposited in a paleovalley eroded into the soft Arapien Shale. The lowest unit is the "formation of Black Cap Mountain" (Willis, 1986), a slope-forming bluish-gray volcaniclastic sandstone. The small resistant knoll to the north is composed of undifferentiated tuff of Albinus Canyon and the Antimony Tuff Member of the Mount Dutton Formation. The tuffs locally interfinger with the "formation of Black Cap Mountain" a few feet above the lower contact. The Osiris Tuff caps the volcanic hills.

Stop 2 (mile 3.4): Faulted Wasatch Monocline. On the north side of the road, the lower slopes west of the small drainage are underlain by Arapien Shale. Salt mines in Unit E of the Arapien Shale are visible near the head of the gulch. The Arapien Shale is unconformably overlain by Oligocene volcanic units.

A small part of the Crazy Hollow Formation (orangish and brownish-red) is visible in a small gulch south of the freeway. The light gray exposures above the Crazy Hollow Formation just west of the gulch are the Bald Knoll and Dipping Vat formations. Tuffs in the Bald Knoll Formation probably had a northern source, whereas the Dipping Vat Formation, which is 5 m.y. younger, had a southern source.

Gentle west dips within the Green River section mark the expression of the north-trending Wasatch monocline. Northward along the Wasatch Plateau, the structural dip increases and the monocline attains structural relief of 5,000 ft (1,525 m). Willis (1986) suggests that monocline development is bracketed by deposition of the Bald Knoll and Dipping Vat formations.

Stop 3 (mile 3.6): Mouth of Twist Canyon. To the north, west-dipping beds of the Green River Formation define the Wasatch monocline. The Green River Formation, particularly well exposed here, consists of a greenish-gray weathering lower unit composed of shale and shaly limestone and an upper tan weathering cherty limestone unit. The change in dip at the monocline is also visible on the south side of Salina Canyon from this point.

Stops 4 and 4A. (mile 3.6): A side trip up Twist Canyon. A round trip of 0.8 mi (1.3 km) up Twist Canyon to the north may be made by high-clearance vehicle or on foot (a car will make it part of the way). Directly south from Stop 4 (Fig. 1), the major fault described at Stop 2 can be seen. The fault appears to show downward drag on both the east and west sides. Mapping by Willis (1985) has shown that the structural relations here can best be explained by faulting of a pre-existing south-trending syncline (Fig. 1). Typical drag does occur here, but fails to compensate for the original synclinal dip. The syncline was folded by renewed upward movement on the Sanpete–Sevier Valley anticline, perhaps a result of flowage in evaporites of the Arapien Shale.

Steep east-dipping Green River and Crazy Hollow formations of the west limb of the syncline are visible directly to the west and northwest of Stop 4. These units are intensely brecciated and are cut by a second high-angle fault that cuts bedding at a

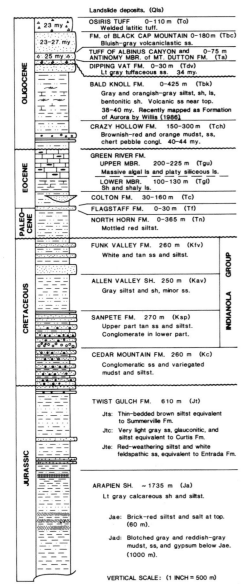

Figure 3. Stratigraphic column showing units exposed in Salina Canyon with map figures used in Figure 1. Numbers in upper units are radiometric ages from Willis (1986).

shallow oblique angle (shown by hachures in Fig. 1). This fault is up to the west and places salt-bearing Unit E of the Arapien Shale against Green River and Crazy Hollow formations. It is probably the bounding fault of salt diapirism in the Arapien Shale.

Salt dissolution and subsequent collapse of overlying units is an important agent of deformation in this area. The rigid volcanic units to the west and north of Stop 4 have been tilted to various angles by this mechanism. A graben begins near stop 4A on the map and runs north for about 3 mi (5 km) directly over the salt-bearing portion of the Arapien. A modern collapse feature with an internal drainage pattern developed in alluvium overlying the salt can be seen at Stop 4A. An optional visit to the site requires a strenuous hike of about 1 mi (1.6 km) round trip.

Stop 5 (mile 4.0): Traverse of Jurassic-Cretaceous sec-

tion. Turn left onto a diverging dirt road (3.7 miles—6.0 km— from start) before the tunnel under the freeway. The route for this walking traverse follows the base of the slope for 1.2 mi (2.0 km). An alternate way to see the outcrops from a greater distance is to proceed through the tunnel under I-70, turn left immediately beyond the tunnel, and view the outcrops from points along the frontage road. For convenience, the following distances are reported along the base of the slope from the road leading to the prospects.

North of Stop Five, vertical red-weathering beds of the Twist Gulch Formation occur beneath the nearly flat Tertiary section at the mine entrances. The Arapien–Twist Gulch contact, covered by alluvium a short distance to the west, is conformable throughout the region. The Twist Gulch section has been correlated with the San Rafael Group on the Colorado Plateau. Contacts between the lithostratigraphic equivalents of the Entrada, Curtis, and Summerville formations are noted on this traverse (see also Fig. 1).

Discontinuous red channel sandstone above the contact here is assigned to the Flagstaff Formation. Siltstone and shale immediately overlying the unconformity elsewhere in this vicinity are included in the Colton Formation. The tunnels at the beginning of this traverse were used to prospect and exploit lead-zinc ores with minor copper and little or no silver. The metals were probably emplaced by ore-bearing fluids that ascended along vertical bedding planes of the Twist Gulch Formation.

The following are the site distances from Stop 5 on the walking traverse.

1. 0–1,475 ft (0–450 m): Twist Gulch Formation (Entrada Sandstone lithostratigraphic equivalent). The Twist Gulch Formation consists of thin-bedded red-brown shale and siltstone and thin beds of white quartzose sandstone. Uncommon lenticular beds of coarse sandstone about 6.5 ft (2 m) thick occur in the lower part. The sandstone contains abundant granule-size red chert and some potassium feldspar grains. A 20-ft (6-m) upward-fining tidal-flat sequence beginning in trough-cross-bedded and wave-rippled red sandstone occurs at 655 ft (200 m).

2. 1,320 ft (400 m): Base of Curtis-equivalent beds within Twist Gulch Formation. A chert-pebble conglomerate marks the contact above a section of red-brown siltstone on the crest of a small south-projecting ridge. Chert pebbles reach 0.5 in (1.5 cm) in diameter; the sand contains some potassium feldspar grains. The conglomeratic sandstone grades within 6.5 ft (2 m) into fine-grained glauconitic sandstone that contains small bivalves. Associated shale beds are rich in plant fragments.

3. 1,560 ft (475 m): Base of Summerville Formation equivalent lithology within Twist Gulch Formation. Very light gray sandstone of the Curtis is overlain by 10 ft (3 m) of thin-bedded brown siltstone and very light gray rippled quartzose sandstone exposed in small drainages on the slope to the east. The Colton Formation overlies the unconformity at this point.

4. 1,640 ft (500 m): Base of Cedar Mountain Formation. Conglomeratic float lying upslope from the brown siltstone marks the base of the Cedar Mountain Formation, a sequence of interbedded conglomeratic coarse-grained sandstone and pink and gray siltstone exposed from 1,640 to 1,970 ft (500 to 600 m) east of Stop Five. The unit crops out as conglomeratic ribs along the flank of the hill that projects south to the freeway. Interbedded siltstone and shale are exposed in the freeway cut at the south end of the ridge. The discordant bedding attitudes within the road cut are a result of slumping of the hillside. The unit is approximately 850 ft (260 m) thick at this locality. This unit was called the Morrison (?) Formation by Spieker (1949) based on its stratigraphic position and lithologic similarity to Morrison beds exposed 33 mi (55 km) to the east. However, zircon grains from mudstone intervals in the lower 60 ft (18 m) of the unit have recently yielded Albian to Cenomanian fission-track ages (96.2 ± 5.0 Ma; 90.3 ± 4.8 Ma; Willis, 1986). These dates corroborate palynomorph data from the Gunnison Plateau to the northwest that show similar beds to be Early Cretaceous (Standlee, 1982). The new data indicate that the contact between Jurassic and Cretaceous beds in Salina Canyon marks a hiatus of approximately 60 m.y.

5. 2,295 ft (700 m): Base of Sanpete Formation. Although not specified by previous workers, we place the base of the Sanpete Formation at the base of a prominent white-weathering bed of quartzite-cobble conglomerate. This unit, exposed on both the north and south sides of Salina Creek, rests on a scoured base and contains clasts of white quartzite to 9 in (20 cm) in diameter.

6. 2,380 ft (725 m): Fluvial sandstone beds of lower Sanpete Formation (Indianola Group). Tan-weathering pebbly sandstone beds of fluvial origin occur up section to the east. Quartzite, carbonate, and chert clasts occur in conglomeratic lags at the bases of the sandstone beds, which grade upward into poorly exposed intervals of siltstone and shale.

7. 2,790 ft (850 m): Abrupt dip change in Sanpete beds. Sanpete beds change abruptly from steep (80–90°) to moderate (30–35°) east dips. A change in curvature near the base of the easternmost steep sandstone bed suggests that the change in dip is a result of folding rather than faulting.

8. 3,050 ft (1,000 m): Delta plain and distributary or tidal channel deposits of Sanpete Formation. Sandstones of the Sanpete Formation fine upward to coal-bearing siltstone and shale of this interval. A 6.5-ft (2-m) upward-coarsening sandstone bed on the west side of the small side canyon yielded a freshwater gastropod. Immediately to the east, channelform sandstone beds are overlain by a bed of oyster shells (*Crassostrea soleniscus*) in life position. Burrows are common in the sandstone. Up section from the channelform sandstone, shale content increases, and thin sandstone beds are fine-grained with symmetric ripples.

REFERENCES

Lawton, T. F., 1982, Lithofacies correlations within the Upper Cretaceous Indianola Group, central Utah, in Nielson, D. L., ed., Overthrust Belt of Utah: Utah Geological Association Publication 10, p. 199–213.

Spieker, E. M., 1949, The transition between the Colorado Plateaus and the Great Basin in central Utah: Utah Geological Society Guidebook to the Geology of Utah, no. 4, 106 p.

9. 3,445 ft (1,050 m): Marine sandstone and siltstone of upper part of Sanpete Formation. A 1.5 to 3.3 ft (0.5 to 1 m) thick, poorly sorted sandstone bed containing shell and bone fragments at this point is overlain by gray siltstone. The siltstone interval grades upsection into a prominent ledge of very fine grained sandstone, hummocky stratified near its base and extensively bioturbated above. A similar, thinner upward-coarsening cycle of siltstone and sandstone occurs immediately upsection. This section has produced an extensive marine fauna of early Turonian age.

10. 3,925 ft (1,200 m): Base of Allen Valley Shale. The Sanpete Formation here grades up into the Allen Valley Shale, which consists of gray shale and siltstone with thin beds of sandstone. The unit contains the middle Turonian ammonite *Collignoniceras woollgari* and is best exposed in a highway cut at 5,085 ft (1,550 m). Above the unconformity, strata assigned to the North Horn Formation pinch out from the east at the Sanpete–Allen Valley contact.

11. 5,905 ft (1,800 m): Funk Valley Formation exposed at base of road cut below North Horn Formation. White-weathering sandstone of the Funk Valley Formation (Indianola Group) is exposed in the road cut east of a small side canyon. A complete upward-coarsening sandstone sequence occurs 820 ft (250 m) to the southwest on the south side of Salina Creek. The sequence is plate-laminated to bioturbated, contains bivalves, and represents deposition in shoreface and foreshore environments (Lawton, 1982). The reddish-brown weathering siltstones that compose the lower part of the road cut here are lacustrine deposits of the North Horn Formation. Sandstone and shale above the siltstones are included in the Flagstaff Formation (Willis, 1986).

At this point, return to the turnoff for Stop 5, continue under the tunnel, turn left at the junction immediately after exiting the south end of the tunnel, and proceed to the narrow gorge at Stop 6.

Stop 6 (mile 5.4): Landslide and lake deposits. A large landslide composed primarily of material from the Colton and Green River Formations dammed the narrow gorge at this point. A lake at least 150 ft (50 m) deep formed behind the dam and backed up about 2 mi (3.2 km) in Salina Canyon, depositing up to 12 ft (4 m) of medium-gray, well-bedded organic-rich clay. Plant fragments near the base of the clay yielded a radiocarbon age of 6,460 ± 80 years B.P. The plant fragments show essentially the same flora that is present in the area today. The clays can be seen at water level along Salina Creek in several places for the next 1.5 mi (2.4 km) eastward.

Standlee, L. A., 1982, Structure and stratigraphy of Jurassic rocks in central Utah: Their influence on tectonic development of the Cordilleran fold and thrust belt, in Powers, R. B., editor-in-chief, Geologic Studies of the Cordilleran Thrust Belt: Rocky Mountain Association of Geologists, v. 1, p. 357–382.

Willis, G. C., 1986, Geologic map of the Salina Quadrangle, Sevier County, Utah: Utah Geological and Mineral Survey, Map Series no. 83, 2 sheets, 16 p., scale 1:24,000.

Witkind, I. J., 1982, Salt diapirism in central Utah, in Nielson, D. L., ed., Overthrust Belt of Utah: Utah Geological Association Publication 10, p. 13–30.

Stratigraphy and structure of the San Rafael Reef, Utah; A major monocline of the Colorado Plateau

J. Keith Rigby, Department of Geology, Brigham Young University, Provo, Utah 84602

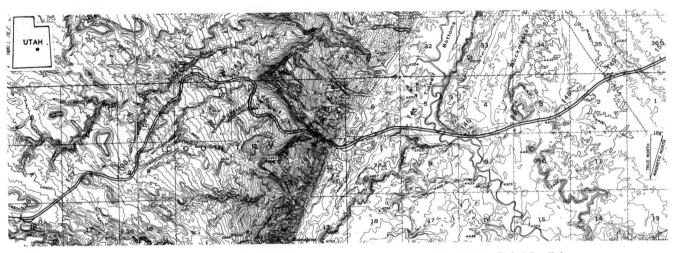

Figure 1. Index map to the San Rafael Reef area along the eastern margin of the San Rafael Swell, in eastern Emery County, Utah, on the Tidwell Bottoms and the Wickiup 15-minute Quadrangles. Alignment of I-70 is approximate. Mileposts are indicated by numbers along the route.

INTRODUCTION

Location. The San Rafael Swell is a major domal uplift in the northwestern Colorado Plateau and dominates the structure of east-central Utah. The San Rafael Reef site is on its eastern edge, approximately 13 mi (21 km) west of Green River, in Emery County, Utah (Fig. 1). I-70 cuts more or less directly across the structure (Rigby and others, 1974).

Accessibility. Roadcuts are on right-of-way controlled by the Utah State Highway Commission. Land along side the route is generally controlled by the U.S. Bureau of Land Management. Alluvial flats along the San Rafael River are on private land.

Significance of Site. The San Rafael Swell is characteristic of many asymmetric large uplifts of the Colorado Plateau; and its eastern margin, the San Rafael Reef, is representative of monoclines of the province. A complete stratigraphic sequence from the Permian White Rim-Cedar Mesa Sandstone up through Triassic and Jurassic rocks to the lower part of the Cretaceous Mancos Shale (Fig. 2) is exposed along the highway. These rocks accumulated in a wide variety of depositional environments, ranging from marine carbonate and clastic environments to deserts (Rigby, 1978).

GENERAL STRUCTURE

The San Rafael Swell is a large dome, affecting an area approximately 70 mi (115 km) wide, east-west, and 130 mi (210 km) long, north-south. Flank dips on the fold are only a few degrees, except in the sharp monocline along the eastern margin where beds dip up to 55° (Fig. 3). Dips flatten abruptly both to the east and west. Rocks in the eastern part of the site dip only 3° to 4° to the east. Western exposures are of nearly flat beds near the anticlinal crest of the San Rafael Swell. The monocline is interpreted as a drape fold over a basement fault, in which upper beds have folded rather than faulted. Although the structure is mainly a fault-related feature, some folding also must have taken place because width of the moderately steep part of the structure is 3 to 4 mi (4.8-6.4 km).

STRATIGRAPHY

Rocks from the Permian White Rim-Cedar Mesa Sandstone up to the Cretaceous Mancos Shale are exposed on the east flank of the San Rafael Swell in that area included in the site (Fig. 2). Measured sections and a summary of stratigraphy of the San Rafael area were given by Trimble and Doelling (1978, p. 5-77), but they treated formations only from the Moenkopi Formation up the Mancos Shale.

At total thickness of approximately 4,600 feet (1,400 m) of beds is exposed in the site. An additional 4,000 feet (1,200 m) of Upper Cretaceous and Tertiary rocks must have extended over the breached anticline but have been removed and are now preserved only around the periphery of the uplift (Hintze, 1973, p. 155–156). These Mesozoic and Cenozoic rocks accumulated in the Rocky Mountain geosyncline, a major trough on the craton border in front of the Sevier orogenic belt.

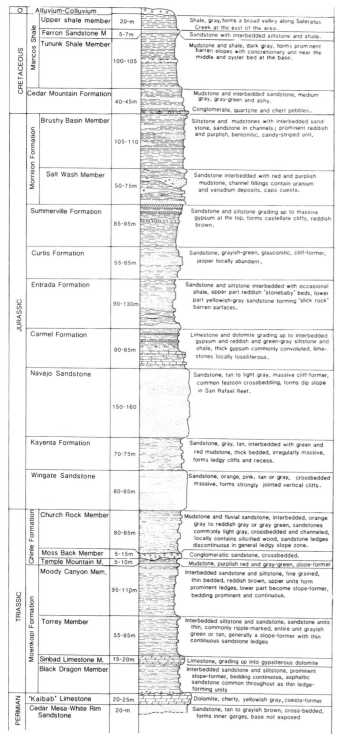

Figure 2. Generalized stratigraphic section of rocks exposed in the San Rafael Reef area, eastern Emery County, Utah.

Figure 3. Low-angle oblique photograph of the San Rafael Reef area, a monocline on the eastern flank of the San Rafael Swell, as seen looking southward. The rest areas at approximately Mile 144 along I-70 are in the left foreground. M, Morrison Formation; S, Summerville Formation; Cu, Curtis Sandstone; E, Entrada Sandstone; Ca, Carmel Formation; N, Navajo Sandstone; K, Kayenta Formation; W, Wingate Sandstone; Ch, Chinle Formation; Mo, Moenkopi Formation. Mount Ellen in the Henry Mountains is visible on the left skyline (Photograph by W. K. Hamblin).

SITE GUIDE

Stops are keyed to official highway mileposts and to geomorphic features along I-70.

Stop 1 (Mile 136). West end of site at the west end of the brake test area. Prominent ridges on the skyline to the east are the LaSal Mountains, beyond Canyonlands and Arches National Parks, along the Utah-Colorado border. These are cored by stock-like intrusions and are part of a major Tertiary intrusive complex that includes, among others, the Henry Mountains visible along the distant horizon to the south, and the Abajo Mountains to the southeast near the Four Corners region. Far to the northeast, Eocene Green River Formation caps the Roan Cliffs along the south flank of the Uinta Basin.

The brake test area is a short distance above the contact of the Triassic Moenkopi Formation on the Permian Kaibab Limestone. Fractured Kaibab Limestone is "forested" with juniper trees, in contrast to the moderately barren Moenkopi beds above. Rattlesnake Bench, the mesa to the north, is capped by the marine Sinbad Limestone Member of the Moenkopi Formation. Rocks here are essentially flat lying, almost at the crest of the San Rafael Dome. At the east end of the test area, rocks dip abruptly to the east into the San Rafael Reef Monocline.

Stop 2 (Mile 138.5-138.6). Classic roadcut of the disconformity between the Leonardian Kaibab Limestone, in double roadcuts, and the middle Lower Triassic Moenkopi Formation, in tan exposures above. The erosional surface, 6.5 to 10 ft (2 to 3 m) above the edge of the road, shows shallow channels filled with angular fragments of white chert above the tan dolomitic Kaibab Limestone and below the evenly bedded, slope-forming, siltstone and sandstone of the basal Moenkopi Formation.

Stop 3 (Mile 140.4). Exit to view area in the eastbound

Figure 4. View eastward along Spotted Wolf Canyon from the east-bound rest area at approximately Mile 141. Mile 142 is in double cuts along the highway in the Kaibab Limestone (K) in the middle distance. S, Sinbad Limestone Member of the Moenkopi Formation; M, Moss Back Sandstone Member of the Chinle Formation; W, Wingate Limestone.

Figure 5. View northeastward from the west-bound lane rest area at approximately Mile 141. B, Black Dragon Member; S, Sinbad Limestone Member; T, Torrey Member; M (black), Moody Canyon Member, all of the Moenkopi Formation, M (white), Moss Back Member; C, Church Rock Member, both of the Chinle Formation. W, Wingate Sandstone. The West Tavaputs Plateau is visible in gaps in the distance. Kaibab Limestone (K) is exposed in juniper-covered ledges in the foreground and middle distance.

lane, opposite junction of the access road from the viewpoint of the westbound lane. Pyrite-stained, weathered surfaces on the tan sandstone and limestone show well on roadcuts opposite the junction and in the access road leading to the eastbound view area. Thin-bedded, tidal flat-dominated sediments are exposed in roadcuts. Elsewhere in the Colorado Plateau, the Moenkopi Formation is principally a reddish brown but here is thought to have been bleached by hydrocarbons that may have produced a strongly reducing environment. A cross-bedded asphaltic sandstone is well exposed in the double roadcuts immediately beyond where the rest area routes divide.

View Area of Eastbound Lane. Along the viewpoint access route, 0.4 mi (0.64 km) beyond the freeway, thin-bedded, ripple-marked sandstone and siltstone of the Black Dragon Member of the Moenkopi Formation grade up into basal limestones of the Sinbad Member. Upper roadcuts along the access road show nearly the full thickness of the Sinbad Limestone, which grades upward from shallow-water, burrowed limestone at its base through oolitic and bioclastic beds, where bivalve and gastropod fragments are abundant, into irregularly undulating beds of gypsiferous dolomite in the upper part of the member. Marine-dominated Sinbad rocks grade up into the tidal-dominated overlying Torrey Member of the Moenkopi Formation.

The eastbound rest area is essentially on the top of the Sinbad Limestone and provides a view to the east of the steep monocline in uppermost Permian and overlying Triassic and Jurassic rocks of the east margin of the San Rafael Swell. Kaibab Limestone is exposed in the gorge along Spotted Wolf Canyon, immediately below the rest area where junipers are common (Fig. 4). The reddish Moss Back Sandstone is the lowest Chinle cliff-former above the well-bedded, reddish upper Moenkopi Formation and is one of the principal uranium-bearing units in

the San Rafael Swell. Eastward, beyond the monocline, the LaSal Mountains rise in the distance beyond the north end of the Monument Valley Upwarp. Toward the north and northeast, through gaps, Tertiary Green River beds form the skyline along the south margin of the Uinta Basin. Segments of the Book Cliffs are visible through narrow notches in the Wingate-Navajo cliffs, as are parts of the low "race track" on the Mancos Shale.

View Area of the Westbound Lane. The rest area of the westbound lane is on the cherty upper Kaibab Limestone. Tan White Rim or Cedar Mesa Sandstone forms the inner gorge in Black Dragon Canyon to the west. Numerous potholes are carved in the cross-bedded, eolian to marginal marine sandstone.

The rest area is at the stripped disconformable surface where the lower part of the Triassic Moenkopi Formation rests on the Permian Kaibab Limestone. Kaibab beds are dolomitic and very cherty. They locally contain moderately abundant brachiopod fragments and sponges, although most readily accessible fossils have been collected from the rest area. Bluffs beyond the freeway, south of the rest area, are capped by the Sinbad Limestone. The minor ledge approximately halfway between the crest and the freeway is an asphaltic sandstone.

Views through the gorge of Black Dragon Canyon, east of the rest area (Fig. 5), show the subsequent Saleratus Valley on the Mancos Shale, in front of the Book Cliffs and Roan Cliffs of the West Tavaputs Plateau.

Black Dragon Wash, to the northeast, is bordered by stripped juniper-covered surfaces on the Kaibab Limestone overlain by the slope-forming Moenkopi Formation. The ledgy slopes of lower Moenkopi Formation are well exposed above that, but below reddish slopes and cliffs of the Chinle Formation. Cliffs of Wingate Sandstone cap the ledges and are overlain by moderately well-bedded Kayenta rocks and the more rounded, overlying, light-colored Navajo Sandstone. Rocks dip eastward along the monocline but flex even more sharply in the San Rafael Reef to the east.

Stop 4 (Mile 141.9). Bridge across narrow Spotted Wolf Canyon. Upper White Rim-Cedar Mesa Sandstone is well exposed in the gully, above and below the freeway. These are some of the oldest rocks exposed in the San Rafael Swell. Sheer-walled double roadcuts are in upper dolomitic, tan Kaibab Limestone. Chert masses contain brachiopods, corals, bryozoans, and crinoid debris, as well as occasional sponge fragments. The channeled surface of the Moenkopi rocks on the Kaibab Limestone is littered with white chert debris. The highway crosses the disconformable surface a short distance east of Milepost 142.

The Sinbad Limestone Member of the Moenkopi Formation caps the moderately steeply dipping cuestas to the northeast and holds up a stripped slope to the southeast. Bleached Torrey Member is exposed in semiledge slopes south of the freeway, east of Stop 4, and consists dominantly of interbedded siltstone and sandstone. Prominent ripple marks often show currents in two directions, suggesting a tidal-flat origin for the member.

Stop 5 (Mile 143). Westernmost exposures in the cut are orange-brown, well-bedded, uppermost Moody Canyon Member of the Moenkopi Formation (Fig. 3). It is thin-bedded, silty, and probably of tidal-flat origin. The steep road on the north approximates the Moenkopi contact with overlying purplish Temple Mountain Member of the Chinle Formation. The plant *Stanleya* is common here and is a uranium-vanadium indicator plant. The massive cuesta-forming sandstone is the Moss Back Member of the Chinle Formation. Local concentrations of uranium salts occur widely in lower parts of such sandstones, elsewhere. The Moss Back Sandstone accumulated in braided stream systems and locally contains abundant fossil wood, which was long ago collected from the immediate vicinity of the freeway.

The Church Rock Member of the Chinle Formation is exposed near the milepost and consists of interbedded brick-red, reddish-purple to green-gray siltstone and mudstone, with prominent, thin, cross-bedded fluvial sandstones.

Stop 6 (Mile 143.1). Wide spot in the canyon is in shale at the top of the Chinle Formation, beneath cliff-forming, cross-bedded Wingate Sandstone. Cross bedding shows well in outcrops south of the road. The Wingate Sandstone is a moderately and uniformly cemented sandstone that forms angular jointed outcrops, in contrast to rounded ledges of the overlying Navajo Sandstone. Wingate Sandstone appears better bedded than Navajo Sandstone largely because cross-bed sets are smaller, but both units are interpreted as eolian accumulations.

A short distance farther east, narrow Spotted Wolf Canyon widens out again, particularly along the south side, where fluvial Kayenta Formation occurs between reddish Wingate Sandstone, below, and light-colored massive Navajo Sandstone, above.

Stop 7—Rest Areas of Both Eastbound and Westbound Lanes. Steeply dipping beds in the monocline contrast with more gently dipping Upper Jurassic and Cretaceous rocks

exposed to the east. Massive, light-colored Navajo Sandstone forms the prominent dip slope on the San Rafael Reef. It is overlain by dark flatirons carved from well-bedded, lower Carmel Limestone and dolomite well-exposed in a major roadcut to the west.

Upper Carmel rocks show considerable syndepositional folding, related to gypsum flowage. Northwest of the westbound rest area a rim of yellowish basal Entrada Sandstone above reddish Carmel beds is folded into small anticlines and synclines. Convolution of underlying gypsum beds shows well in shallow canyons southwest and northwest of both rest areas. Overlying Entrada, Curtis, and Summerville rocks show related deformation, but in decreasing intensity upward.

Cliffs of well-bedded reddish Summerville Formation to the east (Fig. 3) are capped by basal light-colored sandstone of the Salt Wash Member of the Morrison Formation. Gently folded grayish-green marine Curtis Formation is exposed below the castellate Summerville beds and appears as an undulating, semiresistant, unit above the reddish Entrada Formation.

Mile 145. Curtis Formation exposed in cliffs to the south is greenish glauconitic marine sandstone that grades upward from the "stone baby" pink rocks of the Entrada Formation. This gentle transition indicates a moderately slow transgression of the Curtis seaway into the tidal flat-dominated environment represented by the Entrada Formation.

The surface between the Entrada and Curtis Formations is locally unconformable, probably related to slight contemporaneous salt movement. These must have occurred during Curtis accumulation because beds low in the section are interrupted whereas those high in the section are not. Similar, although larger, structures form breccia units in isolated butte-like exposures north of the freeway.

Stop 8 (Mile 145.3). Upper greenish Curtis beds exposed just beyond the fence, east of the prominent cliff area, show ripple marks with two directions of tidal current motion. Reddish castellate cliffs to the south are the gypsiferous upper part of the Summerville Formation and record gradual withdrawal of the Curtis seaway, with re-establishment of subaerial tidal flats and later accumulation of evaporites. Veinlets of gypsum are abundant in upper Summerville beds.

Stop 9 (Mile 146). Debris from early uranium operations dribble down over thick lenticular fluvial sandstones of the Jurassic Salt Wash Member of the Morrison Formation, a short distance east of the bridge over the San Rafael River. These are the principal host rocks for uranium produced to the north and south. The sandstones accumulated as fills of braided and meandering stream channels that generally flowed eastward and northeastward. Uranium salts replaced plant debris trapped in the fluvial system.

Salt Wash fluvial sandstones are interbedded with slope-forming reddish to purplish-gray lacustrine or marshy mudstones, both well exposed in deep double roadcuts to the east. North and south of the highway, resistant ridges are exhumed paleochannels in upper Salt Wash.

Stop 10 (Mile 147). Bridge of Hanksville Interchange. The mileage marker is immediately west of the bridge. Cross-bedded fluvial sandstones are exposed in the small rounded hill between the two lanes and are interbedded with red, popcorn-like, bentonitic ashy mudstone. Contact of the younger lacustrine-dominated Brushy Basin Member on the underlying Salt Wash Member of the Morrison Formation is just west of the interchange.

The unimproved road to the north leads to uranium mines of the San Rafael District, where yellow carnotite ores are still exposed in old workings in upper sandstones of the Salt Wash Member. The access highway to the south leads to Hanksville and the upper end of Lake Powell. Both roads follow the subsequent valley carved from the Brushy Basin Member of the Morrison Formation. Candy-striped Brushy Basin beds immediately northeast and southeast of the interchange include light-colored fluvial and lacustrine-deltaic sandstones interbedded with bentonitic reddish lacustrine and marshy beds.

Stop 11 (Mile 147.4). A well exposed "inverted valley" channel-fill sandstone occurs in flats a short distance south of the highway fence in the middle part of the Brush Basin Member. Two prominent sandstones cap cliffs beyond. The lower of these, Buckhorn Conglomerate, is the basal member of the Lower Cretaceous Cedar Mountain Formation and separates the grayish and gray-green Cedar Mountain Formation from the underlying more brightly colored, reddish Brushy Basin Member of the Morrison Formation. The Cedar Mountain was considered as part of the Morrison Formation until Cretaceous dinosaurs were recovered out of the unit.

Stop 12 (Mile 148.6). Tan *Gryphaea newberryi* beds cap the small ridge between the two lanes and a low cuesta to the north and south. The oyster-bearing shale is the basal unit of the Mancos Shale, which here rests directly on the nonmarine Cedar Mountain Formation without intervening Dakota Sandstone. A tan, selenite-bearing, volcanic ash forms a popcorn-weathering zone at the contact.

A panorama of the San Rafael Swell and the east-dipping monocline forms the skyline on the west, capped by Triassic Moenkopi Formation. Prominent flatirons of Carmel Limestone and Navajo Sandstone form the serrated topography along the monocline.

Stop 13 (Mile 149.8). East end of the site. Double roadcuts through uppermost Tununk Shale and Ferron Sandstone Members. Three major stratigraphic units are visible in the cuts. The lower one is massive, bioturbated, silvery gray Tununk Shale. Interbedded dark gray shale and medium to dark gray limestone and sandstone form intermediate slopes and contain ammonoids and oysters. Limestones consist of bioclastic debris, principally prisms of the oyster *Inoceramus* and show a westward onshore drift in ripple marks.

Top of the gray Tununk Shale and base of the overlying tan Ferron Sandstone Member are marked by a moderately thick volcanic ash that forms a light recess. The tan, fine-grained Ferron Sandstone is thin bedded and shows an array of trace fossils. The Ferron Sandstone caps the gentle cuesta at the divide.

Ferron Sandstone beds are gently folded, particularly in the eastern part of the roadcuts, and show some minor faults. A low-angle reverse fault is well shown in the eastbound lane, where the volcanic ash and lower Ferron Sandstone have been duplicated.

Eastward, toward Green River, rocks dip gently into a broad north-plunging syncline between the San Rafael Swell and Nequoia Arch, a northwestern arm of the Monument Uplift. Prominent dip slopes on middle and upper units of the Mancos Shale are well exposed in barren badlands along Saleratus Creek and between the lower cuestas and steep walls of the Book Cliffs that rise beyond Saleratus Creek, along the margin of the Tavaputs and Red Plateaus.

REFERENCES CITED

Only a few recent papers are listed. Earlier, often locality–specific, papers are cited in these references.

Hintze, L. F., 1973, Geologic history of Utah: Brigham Young University Geology Studies, v. 20, Pt. 3, 181 p.

Rigby, J. K., 1978, Mesozoic and Cenozoic sedimentary environments of the northern Colorado Plateau: Brigham Young University Geology Studies, v. 25, Pt. 1, p.47–65.

Rigby, J. K., Hintze, L. F., and Welsh, S. L., 1974, Studies for students no. 9; Geologic guide to the northwestern Colorado Plateau: Brigham Young University Geology Studies, v. 21, Pt. 2, 117 p.

Trimble, L. M., and Doelling, H. M., 1978, Geology and uranium-vanadium deposits of the San Rafael River area, Emery County, Utah: Utah Geological and Mineralogical Survey, Bulletin 113, 122 p.

Moab salt-intruded anticline, east-central Utah

D. L. Baars, *29056 Histead Drive, Evergreen, Colorado 80439*
H. H. Doelling, *Utah Geological and Mineralogical Survey, 606 Black Hawk Way, Salt Lake City, Utah 84108*

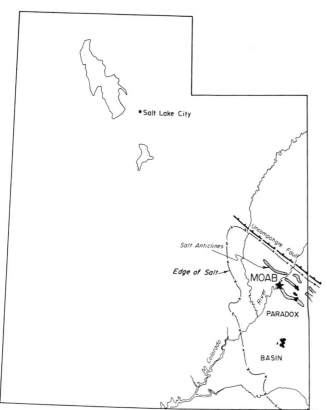

Figure 1. Index map of Moab and vicinity, Utah.

LOCATION AND ACCESS

Moab, Utah, lies on the northwestern end of the Moab Valley salt-intruded anticline on the banks of the Colorado River in east-central Utah, in T.25 and 26S., R.21E. (38°36′N., 109°32′W.), on the Moab 15′ Quadrangle, in southern Grand County, Utah (Fig. 1). U.S. 163 passes through Moab and along the length of Moab Valley. The elevation at Moab is 4,000 ft (1,467 m). The area is readily accessible with any vehicle, but off-pavement lands are almost entirely privately owned and permission should be secured for specific locality access.

SIGNIFICANCE

Moab Valley (Fig. 2) lies in the eastern and deepest part of the Paradox evaporite basin of Pennsylvanian age. That basin area is a unique part of the Colorado Plateau's physiographic province in which structures and stratigraphy have been influenced by salt tectonics. The Moab salt anticline and fault are good examples and provide some of the better, accessible exposures that demonstrate intimate relationships between salt flow—

age and related sedimentation. Salt diapirism occurred from Middle Pennsylvanian through Jurassic time, and sedimentary accumulations detail growth history of the folds, local angular unconformities and thickness variations within the stratigraphic column of those geologic periods.

GEOLOGY

The Paradox Basin was formed as the pre-Pennsylvanian "basement" collapsed, mainly along pre-existing northwest-trending faults. The most active period of subsidence extended from mid-Pennsylvanian to Late Triassic time. Thick cyclic accumulations of salt were deposited early and were influenced by irregular and episodic movements of the faults. Zones of least confining pressure developed along the active faults, and the salt migrated toward them, assisted in part by differential thickness variations in the overburden, thus forming northwest-trending bulges—"salt anticlines"—overlying the faulted basement fabric. Succeeding formations were deposited more thinly over the growing structures, some were never deposited, and others were deposited only to be later removed by erosion. The salt-bearing Pennsylvanian Paradox Formation was locally thickened to more than 10,000 ft (3,667 m) along these zones. The salt adjacent to the "salt anticlines" migrated to supply the growing structures, and troughs in adjacent areas were filled with "thicker than normal" amounts of sediment.

Along the Moab Valley salt anticline much of the Pennsylvanian Honaker Trail Formation, all of the Permian Cutler and Lower Triassic Moenkopi Formations, and the lower part of the Upper Triassic Chinle Formation are missing in outcrop (Fig. 3). North of the Colorado River, along the southwest flank of the anticline, these units begin to reappear and demonstrate the irregular deposition and erosion that occurred along the "salt anticlines."

Migration of the salt continued after Chinle time, but gradually diminished. Sedimentary units of Jurassic age thin over the core of the neighboring Salt Valley anticline, but it is difficult to prove continued growth into Cretaceous time. There are few indications that angular unconformities developed over the anticlines after Chinle time. Eventually the Upper Cretaceous Mancos sea covered the entire region, and a thick section of marine dark-gray shale was deposited. Collapsed sections of the lower Mancos Shale and Ferron Member, in the southern extension of the Moab Valley anticline (Pack Creek syncline), are of normal thickness.

At some later time, probably during the Laramide orogeny of Late Cretaceous or Early Tertiary age, the area was submitted to a west-to-east compressional tectonic event. The northwest-trending zones of thick salt were reshaped into true anticlines,

Figure 2. Aerial view toward the southeast of the Moab fault in the foreground, Moab Valley salt-intruded anticline in the middle distance, and the Tertiary igneous-intrusive La Sal Mountains in the distance. The broad valley resulted from salt intrusion in Paleozoic-Mesozoic time and subsequent collapse caused by groundwater solution of the salt cap (photo by D. L. Baars).

while adjacent areas became synclines. Salt movement into the anticlines may have been rejuvenated for a short time. The Moab salt-intruded anticline was complete by Early Tertiary time, and the Moab fault, an apparently normal fault, formed parallel to the present southwest escarpment of the valley. If the Moab fault is normal, then a relaxation of compression had to have occurred some time later. The relative displacement on the fault north of Moab, near the Arches National Park visitors center, is about 2600 ft (953 m), down to the northeast (Fig. 3).

STRATIGRAPHY

The sedimentary column exposed in and near Moab Valley ranges from the leached evaporites of the Middle Pennsylvanian Paradox Formation upward to the Jurassic Entrada Sandstone. All of the formations are in some way distorted by paleo-growth of the Moab salt-intruded anticline and/or later dissolution-induced collapse of the structure. Pre-Pennsylvanian sedimentary rocks are known only in the subsurface of the Moab region and are variously thinned and otherwise affected by early penecontemporaneous movement of the basement structure. The older rocks are diagrammatically illustrated on Figure 4.

Jumbled masses of gypsum, black shale, and some dolomite of the leached cap of the Paradox evaporites, contorted by diapiric intrusion and later dissolution and collapse, are exposed along the lower margins of Moab Valley. The true nature of such caprock is more evident along Onion Creek, a few miles east of Moab. Some 800 ft (293 m) of this caprock underlie Moab Valley (Fig. 4). Drilling has revealed that over 4198 ft (1540 m) of allochthonous evaporites, including salt, occur beneath the valley, and about 15,000 ft (5,500 m) of Paradox evaporites were penetrated in another salt structure nearby (Paradox Valley). The probable thickness in Moab Valley is estimated in Figure 5.

The Honaker Trail Formation of Middle to Late Pennsyl-

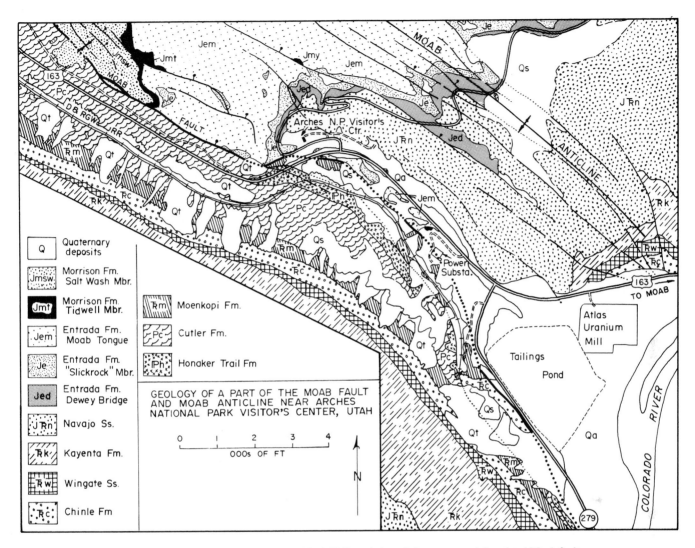

Figure 3. Geologic map of northwestern Moab Valley salt-intruded structure and the related Moab fault in the vicinity of Arches National Park Visitor's Center.

vanian age stratigraphically overlies the Paradox Formation. Only the upper few feet of the formation are exposed in the flanks of Moab Valley. The best exposures occur along U.S. Highway 163 near the Arches National Park visitors center, where the formation consists of interbedded fossiliferous, marine, gray limestone and brown Cutler-like arkoses. Here the limestones are highly fossiliferous and contain abundant brachiopods, bryozoa, rugose corals and crinoids. Rare fusulinids, trilobites, and crinoid calyces date the limestones as Virgilian (latest Pennsylvanian) age. Similar strata of Lower Permian age, the Elephant Canyon Formation, may be seen on the Cane Creek anticline, west of Moab, but were either not deposited or were removed by erosion from the Moab anticline.

The overlying Permian Cutler Formation is a coarse-grained arkosic sandstone in the Moab area. The clasts were derived from the "Ancestral Rockies" Uncompahgre uplift, to the east, and were distributed by fluvial processes. The formation thickens

northward from a wedge-edge north of the Colorado River at the "West Portal" to more than 1,100 ft (357 m) in about 4 mi (6.5 km). It also thickens markedly in the subsurface on both flanks of the Moab structure.

Rocks of Triassic age are mainly mudstones, siltstones, and sandstones of the Moenkopi and Chinle Formations in the Moab area. The section is best displayed along the high cliffs bordering the Moab fault north of the Colorado River and in synclines, both upstream and downstream, along the river. The Moenkopi Formation is a dark brown, slope-forming unit that contains abundant ripples, mudcracks, and compactional structures. The Chinle Formation is a varicolored slope-forming shaly and ledgy unit directly beneath the massive Wingate Sandstone. A basal member, the Moss Back, is not well developed, but its stratigraphic position at the base of the Chinle slopes is marked by a thin white layer and numerous bulldozer scars, made in the search for uranium in the 1950s.

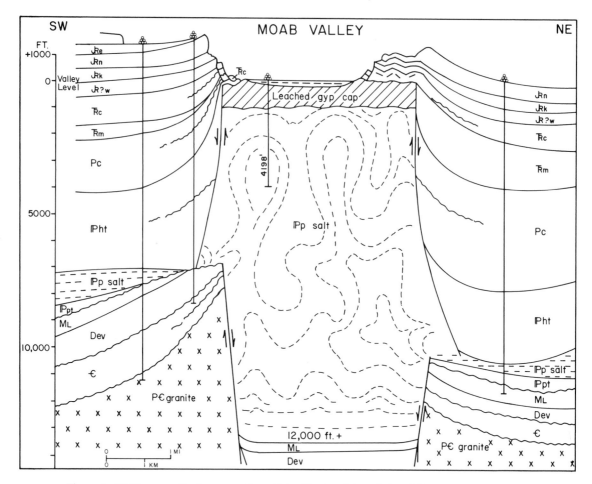

Figure 4. Highly generalized cross section of the Moab salt-intruded anticline. Deep-seated faults originated in Late Precambrian time and were episodically rejuvenated throughout the Phanerozoic. Cambrian through Mississippian strata thin locally and display fault-related facies changes near the faults. Middle Pennsylvanian evaporites, including salt, were deposited within the paleograben and buried the structure as salt flowage was initiated. Upward growth of the salt bulge continued through late Paleozoic and Mesozoic time, causing excess thicknesses of clastic sediments to accumulate in synclines and thinning by deposition and local unconformities to occur along the rising salt core. Tertiary to Recent near-surface groundwater dissolution of salt created a residual "leached gypsum cap" and subsequent collapse of overlying strata. The valley surface is now largely covered by fluvial and eolian Recent deposits. Flowed salt thickness may exceed 15,000 ft (5,500 m).

The Triassic-Jurassic boundary is not well defined on the eastern Colorado Plateaus province, as fossil data are meager at best. Some workers now place it at the base of the Wingate Sandstone rather than in the middle of the Navajo Sandstone on the basis of palynomorphs. Wherever it may be, the rocks overlying the Triassic Chinle Formation may be summarized as follows:

Wingate Sandstone. Vertical cliff-forming brown sandstone, highly cross-bedded, forms the lower massive cliffs high on the southwestern walls of Moab Valley, along the Moab fault north of Moab and in the Colorado River canyons above and below Moab. Most workers consider it to be eolian in origin.

Kayenta Formation. Ledgy sandstone cliffs at the top of the Wingate Sandstone. It displays small- to medium-scale cross-stratification in channels and irregular lenticular beds; fluvial in origin.

Navajo Sandstone. Light brown to white, massive cliff-forming unit, with well-developed large-scale cross-stratification of an eolian origin; forms rounded cliffs and knobs above the Wingate-Kayenta cliffs. Best seen in Arches National Park, but widely exposed along the margins of Moab Valley, high above the Moab fault north of the Colorado River, and in the Colorado River canyons.

Entrada Sandstone. Best known as the massive sandstone cliffs containing the arches in Arches National Park and elsewhere; now subdivided into the reddish basal, crinkly-bedded mudstones, siltstones, and silty sandstones of the Dewey Bridge Member that forms the notch at the top of the Navajo Sandstone; the Slick Rock Member of massive cliffs containing many of the arches; and the highest massive sandstone cliffs of the Moab Tongue.

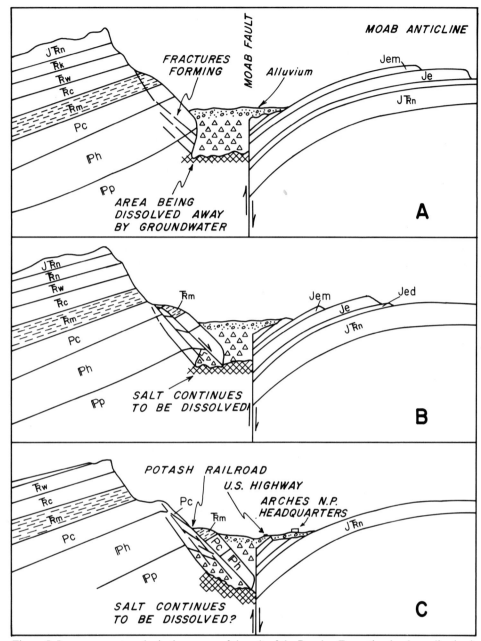

Figure 5. In more recent geologic time some of the salt of the Paradox Formation has been dissolved away by ground water, causing collapse of the top of Moab Valley salt-intruded anticline. The above series of diagrams illustrates how this may have occurred in the vicinity of the D&RGW railroad tunnel and Arches National Park visitors center.

Morrison Formation. Consists of the basal thin reddish siltstone and sandstone of the Tidwell Member that contains large, white, siliceous concretions; the medium-bedded, fluvial sandstones of the Salt Wash Member, forming ledgy cliffs above the reddish-orange Entrada cliffs and containing vanadium-uranium deposits in the general Moab region; and an upper mudstone-siltstone Brushy Basin Shale Member, widely exposed along U.S. Highway 163 north of Moab, where it is mainly a bright green.

Cretaceous Formations. Seen only in and along the Book Cliffs near Crescent Junction north of Moab and in the southern extension of Moab Valley. Best recognized by the dark gray shaly badlands exposures of the Mancos Shale. Consist of the Lower Cretaceous Burro Canyon–Cedar Mountain Formation and the Upper Cretaceous Dakota Formation and Mancos Shale. Cretaceous strata occur in well-developed dissolution collapse structures in the southern part of Moab Valley (Pack Creek syncline) and in Salt Valley in Arches National Park.

DISSOLUTION COLLAPSE STRUCTURES

In more recent geologic time, since about the mid-Tertiary, the Colorado Plateaus were elevated and are currently being eroded by the Colorado River and its tributaries. The erosional regime includes groundwater activity, and wherever natural conduits to the salt are available, dissolution of salt can and does occur. The removal of the salt has resulted in the massive collapse of overlying strata, some of which have been removed at the surface by normal erosion and some of which are now buried under a veneer of alluvium. The overall result has been the formation of elongate, northwest-trending valleys above the salt-intruded anticlines. Superimposed stream courses, such as that of the Colorado River, cross the salt anticlines and their collapsed valleys nearly at right angles, rather than running along the valleys. These valleys, essentially unrelated to the river courses, were considered to be "paradoxical" by early pioneers of the region, and therefore the names of "Paradox Valley," "Paradox Formation," and finally "Paradox Basin" were originated.

Gypsum plug outcrops of Paradox Formation are exposed along the flanks of Moab Valley south of the Colorado River, the associated salt having long been dissolved away. North of the river, near the tunnel portal of the Potash spur of the D&RGW Railway, a relatively thin sandstone slab of Cutler Formation, dipping 35° to the NE, is faulted against oppositely dipping edges of Honaker Trail Formation (Fig. 5C). This peculiar structure was probably produced by dissolution collapse. In this case, the Moab fault provided the conduit for the surface water to move underground to the salt. North of the tunnel, opposite the Arches National Park visitors center, an unrotated downdropped block of Moenkopi Formation is faulted against the Entrada Sandstone by the Moab fault. In this case, collapse may have been essentially vertical, at least at depth.

SELECTED REFERENCES

Baars, D. L., 1966, The Pre-Pennsylvanian paleotectonics—key to basin evolution and petroleum occurrences in Paradox basin, Utah and Colorado: American Association of Petroleum Geologists Bulletin, v. 50, p. 2082–2111.

—— 1983, The geology in and near Canyonlands and Arches National Parks, Utah: Utah Geological and Mineralogical Survey Special Studies 60, p. 75–92.

Doelling, H. H., 1984, Geologic map of Arches National Park and vicinity, Grand County, Utah: Utah Geological and Mineralogical Survey Map 74.

Lohman, S. W., 1975, The geologic story of Arches National Park: U.S. Geological Survey Bulletin 1393, 113 p.

Late Paleozoic-Mesozoic stratigraphy, Hite region, Utah

D. L. Baars, 29056 Histead Drive, Evergreen, Colorado 80439

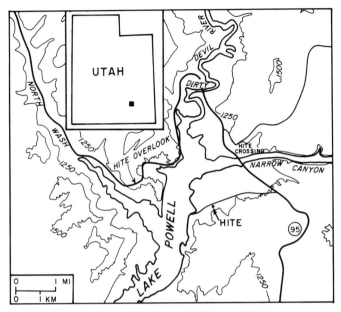

Figure 1. Index map to the Hite area, Utah.

LOCATION

Present-day Hite, Utah, is on the east bank of Lake Powell in the Glen Canyon National Recreation Area of southeastern Utah (Fig. 1F). It is in T.33 S., R. 14 E., on the Browns Rim 15-minute Quadrangle, San Juan County, Utah. Hite Crossing bridge over the Colorado River extension of Lake Powell (Fig. 2) is about 75 mi (120 km) west of Blanding, Utah, and 48 mi (77 km) southeast of Hanksville, Utah, along Utah 95. The present site of Hite, relocated when Lake Powell flooded the Colorado River canyons, is at about 3,960 ft (1,200 m) near the mouths of White Canyon, entering the Colorado River from the southeast, and the Dirty Devil River and North Wash, entering from the northwest. Access is best shown on the Hite Crossing 1:100,000 map. The area is readily accessible via Utah 95 and is entirely on public lands, either Glen Canyon National Recreation Area or Bureau of Land Management administered lands.

SIGNIFICANCE

Hite lies on the west flank of the Monument Upwarp, the crest of which has been largely stripped of latest Paleozoic and Mesozoic cover by late Tertiary to Recent erosion. The mid-Permian through Cretaceous age sedimentary column (Fig. 3) is well exposed west of the Colorado River (Lake Powell) where the west flank of the major Upwarp dips gently into the Henry Basin. The view west from Hite is of this sedimentary column in magnificently colored cliffs, where rocks from the Permian Cedar Mesa Sandstone, at lake level, to the Jurassic Navajo Sandstone

(Fig. 4), on the skyline, stand stark and awesome in the high desert climate. Utah 95 climbs gradually upward to the northwest through the stratigraphic section, revealing excellent exposures of all formations of this part of the eastern Colorado Plateau of Permian through Jurassic ages. Nowhere else in this region is there a better exposure, more ready access, or a more complete section of these strata.

STRATIGRAPHIC SECTION

The Hite Marina and Hite Crossing lie near the top of the Cedar Mesa Sandstone of Wolfcampian (Lower Permian) age. Utah 95, westward from Comb Wash near Blanding, follows the contour of the upper Cedar Mesa contact, so older formations are only seen in the deep canyons of the San Juan River to the south and along the Colorado River (Cataract Canyon) to the north, both inaccessible by road. In those canyons, rocks as old as Middle Pennsylvanian Paradox Formation crop out and are known to underlie the Hite area in the subsurface. A section of Cambrian through Mississippian strata, much like the Grand Canyon section, is known only in the subsurface of the Hite area. Younger formations are well exposed in the cliffs west of Hite and northwestward along Utah 95 through North Wash toward Hanksville, Utah.

Cedar Mesa Sandstone (Permian)

The light colored cliffs and rolling hummocky slickrock country along the lake near Hite (Fig. 5) are in the upper Cedar Mesa Sandstone (Cutler Group). The white, light gray, and tan sandstones occur in thin to massive ledges of highly cross-stratified depositional sets, separated by originally nearly horizontal bedding planes. Baker (1946) described this 1,000-ft (370-m) thick sandstone as of eolian origin because of its massive, cross-bedded appearance. The nature of the cross-bedding, ripple marks, and contained marine fossils, both rare megafossils and abundant sandsize skeletal constituents, prompted Baars (1962) to conclude that much of the formation was of nearshore, marine origin. Recent observations by Loope (1984) have rekindled the controversy by his interpretation that all light-colored sandstones of the Permian System of the region are of eolian origin. Regional relationships suggest, however, that a combination of nearshore marine and coastal dune accumulations constitute the formation. Studies of cross stratification in the Cedar Mesa reveal that depositional currents were strongly oriented from the northwest, and prevailing winds and/or longshore currents were directed toward the southeast during Cedar Mesa time. The Cedar Mesa correlates with the Esplanade Sandstone (Supai Group) in Grand Canyon.

Figure 2. Hite Crossing (bridge over Colorado River crossing of Lake Powell). Light-colored sandstone in foreground is the Permian Cedar Mesa Sandstone (Pcm); reddish-brown beds in upper slopes are in the Permian Organ Rock Shale (Por) (note the stray sandstone in the middle of the formation); upper white cliff-forming unit is the White Rim Sandstone of Leonardian age (Pwr); butte is capped by a thin remnant of the Triassic Moenkopi Formation (Trm).

Organ Rock Shale (Permian)

Reddish-brown slopes of siltstone, mudstone, and sandstone above the Cedar Mesa cliffs and bench (Figs. 2, 5) are in the Organ Rock Shale (Cutler Group). The tidal flat to coastal lowland deposits gradationally overlie the Cedar Mesa and interfinger upward with the overlying light-colored White Rim Sandstone. The Organ Rock red beds are less resistant to erosion than adjacent sandstones and consequently form slopes and benches. Fossil plant debris and vertebrate remains found in the Organ Rock of the Monument Valley region indicate that the formation is of Leonardian (Lower Permian) age. The Organ Rock is the stratigraphic equivalent of the Hermit Shale of Grand

Canyon. Red clastic material of the Organ Rock was derived from the Uncompahgre Uplift (Ancestral Rockies) to the east and northeast and delivered to the Hite area by fluvial processes. The formation averages about 300 ft (110 m) thick in this area.

White Rim Sandstone (Permian).

The prominent light-colored sandstone cliff at the top of the Organ Rock red beds at Hite Crossing and forming the caps of buttes and ledges to the north (Fig. 2) is the White Rim Sandstone (Cutler Group). The massive single bed is composed of medium- to large-scale cross strata, much like the Cedar Mesa Sandstone, and is believed to be largely water deposited in this region (Baars and Seager, 1970). A large tar-impregnated sand bar is exposed about 40 mi (65 km) north of Hite in Teapot Rock and Elaterite Basin, forming an exhumed stratigraphic oil trap. The White Rim appears to interfinger eastward, in White Canyon, with red siltstones of the Hoskinnini Member; it definitely grades to red beds eastward (shoreward) near Elaterite Basin. The White Rim is readily accessible in road cuts along Utah Highway 95 about 5.6 mi (9 km) west of Hite Crossing, where it is about 75 ft (30 m) thick. The White Rim Sandstone represents the eastern coastal deposits of the Toroweap Formation of the Grand Canyon.

Hoskinnini Member (Triassic? or Permian?), Moenkopi Formation

The Hoskinnini Member is a reddish-brown sandy siltstone that overlies the Organ Rock Shale to the east in White Canyon (Thaden et al., 1964). Exact relationships with the White Rim Sandstone, DeChelly Sandstone of Monument Valley, and the Triassic Moenkopi Formation remain enigmatic. Although the relationships are critical to a solution of Permian and Triassic correlations, the controversy remains unsolved. No Hoskinnini beds have been mapped in the immediate Hite area.

Moenkopi Formation (Lower and Middle? Triassic)

Chocolate-brown slopes and cliffs of siltstone and mudstone overlying the White Rim cliffs constitute the Moenkopi Formation. As elsewhere on the Colorado Plateau, Moenkopi beds here are replete with myriad sedimentary structures of a tidal flat nature; mud cracks, small-scale ripples, raindrop impressions, burrows, and compactional features abound. A basal conglomerate of white chert pebbles, probably derived from the erosion of the Permian Kaibab Limestone to the west, is locally present in the Hite area. It occurs in erosional channels cut into the underlying White Rim Sandstone, marking the Permo-Triassic boundary in road cuts west of Hite Crossing.

Chinle Formation (Upper Triassic).

Varicolored mudstones, siltstones, and sandstones of the Chinle Formation constitute the slope- and ledge-forming sequence above the Moenkopi and below the towering massive cliffs of the Wingate Sandstone (Fig. 4). All colors of the rainbow—red, purple, gray, brown, green, white—occur randomly throughout the 900-ft (330-m) thick formation and its seven members. Because of its high bentonite content, the Chinle Formation makes unpaved roads hazardous to impassable when wet. A basal disconformity marks the contact with the underlying Moenkopi Formation on a regional basis, and the scoured channels and infilled fluvial sandstones are usually obvious.

West of Hite and in White Canyon, the Shinarump (pronounced properly shĭn-ār/-rŭmp) Member is composed of local channels of conglomeratic sandstone and is usually the basal unit

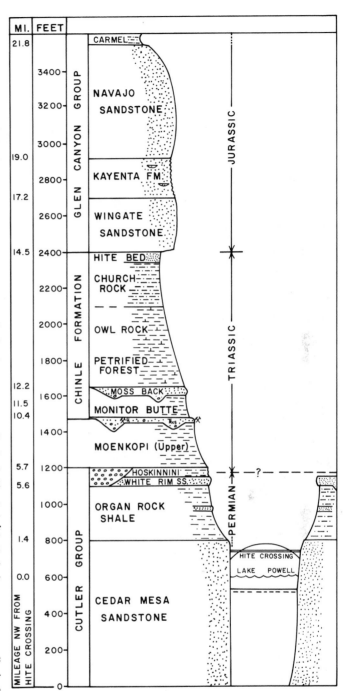

Figure 3. Permian-Jurassic stratigraphic section of the Hite–North Wash area, Utah.

of the Chinle Formation. This fluvial member is widespread south of the latitude of Hite and contains most of the uranium deposits of the White Canyon mining district (Thaden et al., 1964). As such, it is readily distinguished by its "index" trace fossils—bulldozer scars. The locally prominent Shinarump cliffs are overlain by varicolored slopes of the mudstone-siltstone Monitor Butte Member. Another channel-filling conglomeratic sandstone, the Moss Back Member, occurs locally at the top of the Monitor Butte slopes, forming a double topographic bench. The

Figure 4. View west from Hite Overlook. Upper siltstones of the Triassic Moenkopi Formation (Trm) in the foreground; varicolored slopes and ledges are in the Triassic Chinle Formation (Trc). The upper cliffs are the Glen Canyon Group, consisting of the lower vertical cliffs of the Wingate Sandstone (Jw), the middle ledgy cliffs of the Kayenta Formation (Jk), and the light-colored rounded cliffs of the Navajo Sandstone at the skyline (Jn). Note the prominent northwest-trending joints in the Navajo escarpment.

Moss Back is widespread north of Hite, where it is usually the basal member of the Chinle Formation and also often contains uraniferous mineralization. The three-member sequence of the lower Chinle—the Shinarump–Monitor Butte–Moss Back—constitutes about the lower one-third of the formation. The complete threesome of members occurs together only in the White Canyon–North Wash area.

The middle one-third of the Chinle Formation is made up of pale yellow, brown, blue, and purple mudstone of the Petrified Forest Member and the overlying light reddish brown claystone-limestone unit of the Owl Rock Member. The Petrified Forest Member, as the name implies, contains abundant large silicified logs and common fossil leaf and limb fragments. It is usually quite bentonitic. The Owl Rock Member, above the Petrified Forest Member, contains numerous light red beds of limestone up to 20 ft (7.3 m) thick within the claystone slopes. Together, the two members were considered as the "limy unit" of Thadin et al. (1964).

The upper "sandstone-siltstone unit" of Thadin et al. (1964) is a light reddish brown sequence of predominantly siltstone, the Church Rock Member, and an uppermost grayish- and purplish-red arkosic and conglomeratic sandstone known as the Hite bed. Together these two units form ledgy slopes in the upper Chinle Formation. The Hite bed is only locally present, but is underlain and overlain by erosional surfaces and forms an obvious upper cap for the formation south of Hite.

Figure 5. View eastward from Hite Overlook, about 6.5 mi (10.5 km) west of Hite Crossing near mouth of North Wash. Lake Powell is in the foreground, with Narrow Canyon in the middle distance to the left. The broad expanse of light-colored slickrock is the upper Cedar Mesa Sandstone of Permian age (Pcm), capped by thin remnants of Organ Rock Shale, also Permian (Por). Small buttes in the distance are capped by the White Rim Sandstone (Pwr) and Moenkopi Formation.

Glen Canyon Group (Lower Jurassic)

Massive cliff-forming sandstones that hold up the high escarpment west of Hite (Fig. 4) and the extensive facade guarding the inner "canyonlands" to the north and along Utah Highway 95 through North Wash are composed of the Glen Canyon Group. The lower massive unit, forming the prominent 300-ft (110-m) high vertical escarpment, is the Wingate Sandstone. Although massive in appearance from a distance, the Wingate Sandstone is a highly cross-stratified unit that is probably of eolian origin.

The extensive, usually impassable, Wingate cliffs are invariably capped by lower portions of ledgy, fluvial sandstones of the intermediate Kayenta Formation. Thin to massive, highly channeled and cross stratified, fine- to coarse-grained and locally conglomeratic sandstones that are commonly separated by flat-bedded mudstones and siltstones typify Kayenta beds.

The Kayenta Formation is, in turn, overlain by the light-colored, massive and rounded cliffs of the Navajo Sandstone. Although less resistant to weathering processes than the Wingate and Kayenta formations, the Navajo Sandstone completes the triplet "Great Wall" of canyonlands country. It is a striking formation because of its large-scale, highly cross-stratified nature, making the Navajo Sandstone the "type" eolian sand deposit of the Colorado Plateau. Thus, the Glen Canyon Group constitutes a vast desert sand accumulation, seen at Hite as the capping cliffs of the multicolored escarpment to the west. The cliffs mark the division between the Monument Upwarp and Henry Basin structural elements that dominate the region.

SELECTED REFERENCES

Baars, D. L., 1962, Permian System of Colorado Plateau: American Association Petroleum Geologists Bulletin, v. 46, p. 149–218.

Baars, D. L., and Seager, W. R., 1970, Stratigraphic control of petroleum in White Rim Sandstone (Permian) in and near Canyonlands National Park, Utah: American Association Petroleum Geologists Bulletin, v. 54, p. 709–718.

Baker, A. A., 1946, Geology of the Green River Desert–Cataract Canyon region, Emery, Wayne, and Garfield Counties, Utah: U.S. Geological Survey Bulletin 951, 122 p.

Loope, D. B., 1984, Eolian origin of upper Paleozoic sandstones, southeastern Utah: Journal of Sedimentary Petrology, v. 54, p. 563–580.

Thaden, R. E., Trites, A. F., and Finnell, T. L., 1964, Geology and ore deposits of the White Canyon area, San Juan and Garfield Counties, Utah: U.S. Geological Survey Bulletin 1125, 166 p.

Kolob Canyons, Utah: Structure and stratigraphy

Sheldon Kerry Grant, Department of Geology, University of Missouri–Rolla, Rolla, Missouri 65401

INTRODUCTION

The site encompasses that portion of the Kolob Canyons section of Zion National Park, in southwestern Utah, that is visible from the 5-mi-long (8-km) park highway (Fig. 1). The park road is suitable for passenger cars and other vehicles.

Kolob Canyons offer spectacular views of slopes that show the effects of erosion on the thick massive Navajo Sandstone. In addition, exposures in Taylor Creek, LaVerkin Creek, and their tributaries reveal details of stratigraphy and structure. Much of the Mesozoic section is exposed, from the basal Timpoweap Limestone of Early Triassic age to the Jurassic Carmel Formation, inaccessible at the tops of the Navajo cliffs. Structurally, two episodes of tectonic activity are evident. The Late Cretaceous Sevier orogeny produced the Kanarra fold, with effects throughout the site. Most interesting perhaps is the Taylor Creek thrust zone, on the eastern limb of this fold. The Late Cenozoic Hurricane fault scarp forms the western boundary of the area, and uplift along this fault set up conditions that led to the erosional phenomena.

STRATIGRAPHY

Stratigraphic units of the site are shown in the lithologic legend for Figure 1. In addition to the dominant sedimentary rocks, the descriptions include basalt and mappable zones of jumbled fault slices, too chaotic to subdivide. Some items of lithology are presented in discussions of panoramic views from selected stopping points.

The distribution of stratigraphic units is shown on Figure 1. The Kaibab Formation occurs in the high cliffs on the west margin of the site, and the Timpoweap Member forms a rounded cap to these cliffs. Eastward, successively younger Mesozoic rocks are exposed on the limb of the Kanarra fold. Dips are gentle in eastern and western parts of the area but are steep in the middle. Ridge-forming units, from west to east, are the Virgin Member, Shinarump Member, and Springdale Member. The Navajo Sandstone forms gigantic cliffs on the east margin, and Late Cenozoic basalt, landslides, and alluvium are scattered in and near the site. Basalt is exposed on Horse Ranch Mountain, to the northeast, and Pace Knoll, to the southwest, both just off the map. Landslides occur along the Taylor Creek fault zone and at the base of the Navajo cliffs. Alluvium is present in the creek valleys, but it is not mapped, except where thick, in order to improve clarity of the figure.

VIEWPOINTS

Four viewpoints along the park highway are selected for discussion.

Stop 1

Stop 1 is 0.2 mi (0.3 km) west of the park entrance on the underpass road of I-15, where it intersects old U.S. 91. At this and other viewpoints, directions are indicated using twelve o'clock (12:00) as north, supplemented by the azimuth directions in parentheses. Beginning at 12:40 (20), lower slopes at the base of the Hurricane Cliffs expose chaotic, fault-sliced fragments of the Moenkopi Formation as horses within the Hurricane fault zone, which trends north-south along the mountain front. Clockwise and upward are exposed Moenkopi beds as a cliff of yellowish Timpoweap Member, a red slope of lower red member, and the prominent point of Virgin Limestone supporting the local TV booster at 1:00(30). These strata dip southeastward on the east limb of the Wayne Canyon culmination of the Kanarra fold. At 1:20(40) the gorge at the mouth of Taylor Creek is cut in Timpoweap beds. Beyond are the flat-lying lower red and Virgin members, on the crest of the Kanarra fold, in the Taylor Creek saddle.

Continuing clockwise, the Kolob Canyons road is constructed within the Hurricane fault zone and, higher up, within relatively unbroken Timpoweap beds. The low, craggy cliff at 3:00(90) is held up by the Fossil Mountain Member of the Kaibab Formation, the host for the inactive Epsolon Mine, from 3:40–4:40 (110–140). The next higher cliff is held up by Timpoweap beds, which are dropped down on the north side of a northeast-trending fault—3:00(90). On the distant skyline, beyond the mine workings, a section of the Springdale Member is repeated by the middle fault of the Taylor Creek thrust zone—3:40(110).

From 3:40 to 5:40 (110–170), the Kaibab Limestone forms a continuous cliff immediately east of the Hurricane fault zone, where red horses of Moenkopi beds form slopes between the valleys and on the south end of Kaibab exposures—5:50(175). The upper part of the Toroweap Member is exposed at the base of the mountain in each valley. In this same general direction, several knobs on the skyline are the Virgin Limestone.

At 5:20(160) a flat part of the skyline is Pace Knoll, capped by a basalt flow remnant. The same basalts are seen farther south, on Black Ridge at 5:50(175) and near Pintura at 6:20(190). These nearly horizontal basalts flowed across a flat erosional surface cut on progressively older Mesozoic and Permian strata toward the south. That relationship is a reflection of northward regional dip and plunge of the Kanarra fold. The basalt flows are downthrown west of the Hurricane fault and are visible from 6:10 to 8:10 (185–245) as a juniper-covered plain. The skyline from southwest to northwest is made of intrusive and extrusive rocks of Middle Tertiary ages.

Stop 2

Stop 2 is the trailhead of the Middle Fork or Taylor Creek, immediately north of the designated parking area, 2.1 mi (3.4

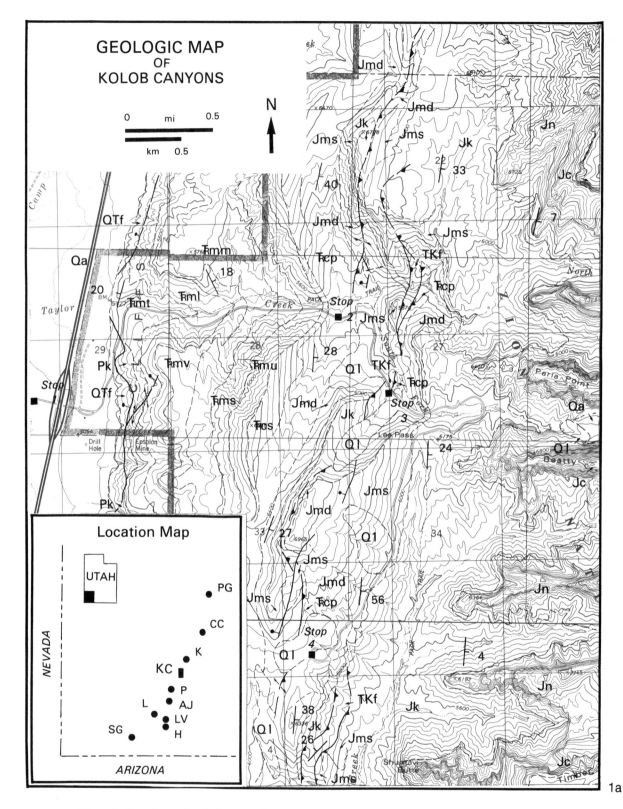

1a

Figure 1a. Location and geologic maps of the Kolob Canyons field site. Figure 1b explains the lithologic units and their map symbols. Contacts shown by short dashes, faults and fault zones by heavy solid lines (dashed where inferred). Fault teeth on upper plate of thrusts; lollipops on downthrown side of high-angle faults. Parowan Gap, PG: Cedar City, CC; Kanarraville, K; Kolob Canyons, KC; Pintura, P: Andersons Junction, AJ; LaVerkin, LV; Hurricane, H; Leeds, L; St. George, SG.

km) from the park entrance. At 12:10(5), in the middle ground, the reddish-brown basal ledge of the Dinosaur Canyon Member of the Moenave Formation begins a northerly trace to the skyline saddle. At 12:40(20) a grove of tall pines lies below and to the right of the western of three fault-repeated ledges of Springdale Sandstone. This particular ledge is terminated by an east-dipping thrust fault that follows the gully of the pine grove. From the grove, this western flank thrust passes behind and then to the right of the mound of purplish mudstone of the Petrified Forest Member of the Chinle Formation at 12:40(20). The fault trace then swings toward the viewpoint, following the gully between the two hills that exposes a displaced white sandstone bed, at 1:00(30). The western sandstone is downthrown.

East of the fault, the upthrown basal Dinosaur Canyon ledge appears at 1:40(50), and above it, at 1:30(45), the middle of three fault-repeated Springdale Sandstone ledges forms the skyline. This midle Springdale unit can be traced downhill to a point where it is folded into a small open anticline, just before terminating at 2:00(60) at the middle flank thrust. This particular Springdale Sandstone appears again on the south side of the Middle Fork of Taylor Creek at 2:30(75) and can be followed to 3:10(95), where it is faulted out again against the middle thrust. The three terminations, at 60, 75, and 95, form a line of intersection (fault and bed) that plunges north at 10°, an expression of post-fault tilting of Kanarra fold elements into the structurally low Taylor Creek saddle.

At 2:30(75) the middle Springfield Sandstone seems to line up with the distant eastern Springdale bed but this is an accident of perspective. The distant bed passes southward behind the middle Springdale ledge and appears again at 3:10(95) on the skyline between the Middle and South Forks of Taylor Creek. This eastern Springdale ledge is traceable north of Taylor Creek to a point on the skyline at 1:00(30), where it ends against the middle flank thrust. The pattern of terminations is illustrated in Figure

2a. The faults are steeper than the beds, and hanging walls moved westward over footwalls.

Immediately south of this viewpoint, smectite-bearing mudstones of the Petrified Forest Member are visible in the roadcut. The high ridge at 6:10 to 6:20(185–190) is the upper (eastern) Springdale Sandstone, thrust over the middle Springdale unit that forms a continuous ledge below, from 5:20 to 6:20(160–190). The ledge at 5:10(155) on the skyline is the eastern Springdale unit. In this same direction, the nearer middle Springdale Sandstone ledge ends down dip against the middle thrust.

In saddles at 6:30(195) and 12:10(5), the sandstone of the middle Petrified Forest Member of the Chinle Formation passes from view. Shinarump Member forms much of the western skyline, north and south of Taylor Creek gap. To the northeast, basalt-flow mounds rest on eroded Carmel Limestone, at 2:00 and 2:10(60 and 65), above the Navajo Sandstone cliff. The base of the Carmel Formation is sharply defined where the brushy slope meets the top of the cliff. The Navajo Sandstone can be more closely observed by hiking to beautiful Leda's Cave, 2 mi (3.2 km) up Taylor Creek. The cave is the lower part of a stacked double alcove formed at the tip of an entrenched meander.

Stop 3

The third stop is at a road cut northwest of a parking area, 2.8 mi (4.5 km) from the park entrance. This cut through a small hill exposes red and white beds of the Dinosaur Canyon Member of the Chinle Formation on both sides of the road. A minor thrust fault of the Taylor Creek zone in the northeast wall has deformed and cut the laminated, white upper Dinosaur Canyon sandstone ("albino Dino"). The fault dips eastward, nearly parallel to beds in the hanging wall and nearly perpendicular to deformed footwall beds that dip westward (Fig. 2b). The footwall drag fold plunges northerly, consistent with other linear features on the Kanarra fold.

Stop 4

The fourth viewpoint is at the end of the park highway at the turnaround loop, 5.4 mi (8.7 km) from the park entrance. To the northwest, Springdale Sandstone is repeated by the middle flank thrust. The middle Springdale outcrop caps the low ridge from 10:00 to 11:20(300–340), but the high point at 11:30(345) is within the upper (eastern) Springdale unit. Eastward views are of cliffs of massive Navajo Sandstone as closely jointed rock at 12:40(20) and extending to 5:30(165) in the far distant skyline, southwest of Zion National park. Throughout the Navajo Sandstone, south-facing cross-strata in 50-ft (15-m) sets indicate a northerly source. At 3:30(105), midway up the cliff, is one of two hanging valleys related to stream capture of west-flowing streams by south-flowing Timber Creek, which has a lower base level. At 4:20(130), a July 1983 rockfall lies beneath the scar it caused. At the base of the Navajo cliff, a white bed in the uppermost Kayenta Formation serves as a convenient marker of the gradational contact. Several sandstones occur in the Kayenta Formation below this bed. The vegetated slopes above the Navajo cliffs are on the Carmel Formation.

Southward, Smith Mesa, from 5:20 to 6:00(160–180), has a

Qa	Alluvium	
Ql	Landslide Breccia	QTf — Hurricane Fault Zone
QTb	Basalt Flows	
		TKf — Taylor Creek Thrust Fault Zone
Jc	Carmel Fm	
Jn	Navajo Ss	
Jk	Kayenta Fm	
Jms	Springdale Mbr	Moenave Fm
Jmd	Dinosaur Canyon Mbr	
Ŧcp	Petrified Forest Mbr	Chinle Fm
Ŧcs	Shinarump Mbr	
Ŧmu	Upper Red Mbr	
Ŧms	Shnabkaib Mbr	
Ŧmm	Middle Red Mbr	Moenkopi Fm
Ŧmv	Virgin Mbr	
Ŧml	Lower Red Mbr	
Ŧmt	Timpoweap Mbr	
Pk	Kaibab Fm	

1b

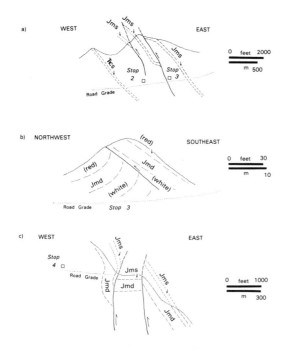

Figure 2. Diagrammatic structure sections along the Taylor Creek thrust zone (not to exact scale). (a) View northward at the north side of Ŧcs, Shinarump Member of the Chinle Formation, and Jms, Springdale Sandstone of the Moenave Formation, Taylor Creek. (b) View northeastward from the third viewpoint at the east wall of the road cut. All outcrops are in the Dinosaur Canyon Member of the Moenave Formation. (c) View northward along the eastern thrust branch, just east of the fourth viewpoint. Jms, Springdale Sandstone Member; Jmd, Dinosaur Canyon Member; both in the Moenave Formation.

cliff of undisturbed Springdale Sandstone, just below its top. In the tree-covered area north of the mesa, a large sliding klippe of Springdale Sandstone overlies undisturbed Springdale rocks. The skyline at 6:40(200) is the eastern Springdale Sandstone of other viewpoints. At 7:30(225) is Pace Knoll, capped by flat basalt over east-dipping Moenkopi and Chinle formations.

TECTONIC ANALYSIS

Regional information indicates that two tectonic episodes affected this area. The first was folding and low-angle faulting of the Sevier Orogeny during the Late Cretaceous or Early Tertiary. That folding and related faulting can be subdivided into four stages: (1) anticlinal folding to a moderate angle of closure; (2) flank thrusting on fold limbs; (3) warping of the fold crest into

culminations and saddles, with thrusting in the culminations; and (4) overturning in some of the culminations, with formation of zig-zag faults and rollover break thrusts. The first three stages are reached in the Virgin anticline south of the site. The last stage is illustrated on the Kanarra fold between Kanarra and Cedar City. At Kolob Canyons the flank thrusts are en echelon, and each branch is often a complex zone with multiple fault surfaces separating small blocks. Each branch developed first as an asymmetrical kink-fold set, which later faulted in the two hinge areas, including the shorter limb within the fault-bounded zone (Fig. 2c).

Two types of fractures in the Kolob Canyons attest to their nearness to intense folding that occurs near Kanarraville. Northerly trending joints on the western edge of the Navajo Sandstone cliff are a dwindling expression of a more intense zone that extends to the break thrust to the north. Minor strike faults in Kayenta and Dinosaur Canyon units are up on the west and eliminate beds, both characteristics of the strike leg of zig-zag faults at Kanarraville.

The second tectonic episode to affect the Kolob Canyons is the Late Cenozoic movement along the Hurricane fault. Although opinions differ, the simplest interpretation of the structure is that of a normal fault along which plateaus on the east were uplifted several thousand feet with respect to the Basin and Range province on the west. The faulting is later than the Middle Tertiary Leach Canyon Tuff, which is uplifted east of Cedar City by an amount equal to the known throw on the fault zone. An average uplift rate of 1,280 ft (390 m)/m.y. has been determined by dating basalts offset by the fault, leading to an age of 5–6 m.y. for the calculated throw.

The throw can be determined with confidence at LaVerkin and at Andersons Junction. At LaVerkin, the throw is 6,100 ft (1,860 m), determined by offset on the Navajo Sandstone. At Andersons Junction the throw is 7,000 ft (2,100 m), determined by reconstructing the fold on top of the Moenkopi Formation, over a distance of 12 mi (19 km) on either side of the fault. Normal drag on the upthrown block and "reverse drag" on the downthrown block may affect some definitions of throw.

The Hurricane fault is a zone with complex internal structure. Rocks within the zone are of intermediate age, relative to those outside the zone, and dip westerly. This pattern is that of a pre-fault monocline, such as exists between Cedar City and Parowan. The fault zone follows the axial region of the Kanarra fold from Andersons Junction to Cedar City, complicating the interpretation of both features.

REFERENCES

Anderson, R. E., and Mehnert, H. H., 1979, Reinterpretation of the history of the Hurricane fault in Utah, *in* Newman, G. W., and Goode, H. D., eds., Basin and Range Symposium: Rocky Mountain Association of Geologists and Utah Geological Association, p. 145–165.

Gregory, H. E., 1950, Geology and geography of the Zion Park region, Utah and Arizona: U.S. Geological Survey Professional Paper 220, 200 p.

Gregory, H. E., and Williams, N. C., 1947, Zion National Monument, Utah: Geological Society of America Bulletin, v. 58, p. 211–244.

Hamblin, W. K., Damon, P. E., and Bull, W. B., 1981, Estimates of vertical crustal strain rates along the western margins of the Colorado Plateau: Geology,

v. 9, p. 293–298.

Hamilton, W. L., 1978, Geological map of Zion National Park, Utah: Springville, Zion Natural History Association, scale 1:31,680.

Kurie, A. E., 1966, Recurrent structural disturbance of Colorado Plateau margin near Zion National Park, Utah: Geological Society of America Bulletin, v. 77, p. 867–872.

Lovejoy, E.M.P., 1973, Major Early Cenozoic deformation along Hurricane fault zone, Utah and Arizona: American Association of Petroleum Geologists Bulletin, v. 57, p. 510–519.

Late Cenozoic volcanism in the St. George basin, Utah

W. Kenneth Hamblin, Department of Geology, Brigham Young University, Provo, Utah 84602

INTRODUCTION

One of the most spectacular features in the St. George basin is the series of late Cenozoic basaltic flows that were extruded near the base of the Pine Valley Mountains and flowed southward into the ancestral drainage of the Virgin River. These flows are now preserved as south-trending inverted valleys up to 400 ft (120 m) high which stand in bold contrast to the brilliant red and white east-west trending cliffs and slopes formed on the older Mesozoic rocks. The arid climate and sparse vegetation result in excellent exposures of the striking red, white, and black rocks of Utah's "color country" and provide scenic emphasis for what is one of the most impressive examples of topographic inversion in North America.

LOCATION AND ACCESSIBILITY

The volcanic features of the St. George basin are readily accessible from I-15 and U.S. 91 (Fig. 1). In addition, Utah 18 follows the top of a major inverted valley from St. George to Diamond Valley and maintained gravel roads are present near most other volcanic features (Fig. 1).

SIGNIFICANCE

Basaltic eruptions have occurred in the St. George basin intermittently throughout the last 2 m.y. leaving excellent examples of flows in various stages of dissection. The area thus provides an exceptional opportunity to study geologic processes and sequence of events involved in volcanic, tectonic, and erosional development of inverted topography. In addition, the basalt flows of the inverted valleys provide a unique method of determining amounts and rates of tectonic uplift (Hamblin and others, 1981).

CLASSIFICATION OF RELATIVE AGE OF BASALTIC FLOWS

Geomorphic relationships provide the basic criteria by which the relative age of the lava flows is determined. Erosion has been the dominant process throughout the region during late Cenozoic time; consequently, the present surface is the result of a relatively long sequence of events in the process of denudation. The extrusion of lava during any period of time in the erosional process would preserve the surface beneath the basalt from further destruction. The relation of this buried surface to the present drainage therefore provides a reasonable criterion by which the relative age of the flows can be determined. According to this classification, four major periods, or stages of extrusion, can be recognized in the St. George Basin.

Stage I Flows. Stage I flows are those deposited on an erosional surface which exhibits no apparent relation to the pres-

ent drainage system. All original margins of Stage I flows have been removed by erosion. In the St. George region, Stage I flows cap the high mesas on the skyline to the south which rise more than 1,000 ft (300 m) above the surrounding area. They range in age from 3 Ma to 6 Ma.

Stage II Flows. Stage II flows were deposited on an erosional surface which now stands 200-500 ft (60-150 m) above the present drainage. These flows were extruded during an early or ancestral stage in the development of the present rivers. Hence the erosional surface beneath the Stage II flows slopes in the same direction and has a similar gradient and pattern to the present adjacent drainage. Like the older Stage I flows, the original margins and surface features of Stage II flows have been destroyed by erosion or covered with sediment or soil. In the St. George area, two major periods of Stage II extrusion are recorded. The oldest flows (Stage IIa) are approximately 400 ft (120 m) above the surrounding surface and have been eroded into small elongate remnants of high inverted valleys. The younger (Stage IIb) flows are more extensive and form three long north-south trending ridges. These flows have been named, from east to west, the Washington flow, the Middleton flow, and the Airport flow. Ages for Stage II flows range from 1 Ma to 2 Ma.

Stage III Flows. Stage III flows are those which were deposited on a surface formed by the present drainage. The present margins of the flows come close to representing the configuration of the original flow unit. The flows can be traced to their source, and cones associated with the extrusion are only slightly eroded. Most of the original surface features of Stage III flows have been modified somewhat by weathering and erosion or have been partly covered with alluvium. Ages of Stage III flows are less than 0.5 Ma.

Stage IV Flow. Stage IV flows represent the most recent volcanic activity within the region and are readily distinguished by their fresh and relatively unweathered appearance. Original surface features such as pressure ridges, aa crust, and flow structures are preserved. Stage IV flows partly fill present stream valleys and have locally disrupted the drainage system. Judging from the degree of weathering and erosion the Stage IV flow in the St. George basin is only about 1 ka.

SITE DESCRIPTIONS

The Santa Clara Flow, Snow Canyon (road to Snow Canyon from Santa Clara and Ivins). The Santa Clara flow is one of the youngest lava flows along the Colorado Plateau/Basin and Range margin and is singularly important because of the insight it provides into the way in which local volcanic extrusions modify the landscape and evolve into an inverted valley. It is

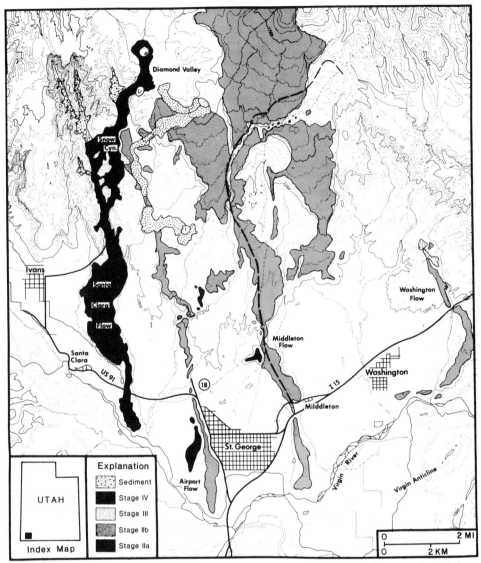

Figure 1. Map showing the major inverted valleys and cinder cones in the St. George region, Utah.

particularly important because it flowed across a variety of landforms such as open plains, deep narrow canyons, waterfalls, and broad open valleys. Each of these terrains provides a different geologic setting, illustrating different ways in which drainage systems are modified by extrusion of lava and the ways in which erosion and/or deposition will subsequently modify the flow.

Figure 2 shows the Santa Clara flow from the area where it was extruded in Diamond Valley and subsequently moved down through Snow Canyon. The flows first spread over the flat valley floor of Diamond Valley but soon found an outlet through a narrow pass in the Navajo Cliffs and flowed down the deep gorge of Snow Canyon. Here it split into several lobes before entering the main canyon. Some flows formed cascades over the Navajo Cliffs, while others moved around isolated buttes and pinnacles and blocked tributary valleys. The outline of the flow is highly irregular in this area, with peninsulas and islands of red Navajo

Sandstone protruding above the black basalt. The flow then followed the narrow deep gorge of Snow Canyon for about three miles and behaved much as intracanyon flows of the Colorado River in the Grand Canyon and the Snake River in Idaho and Washington. Here the flows are narrow (less than 700 ft [210 m] wide) and are presumably as much as 150 ft (45 m) thick (judging from the volume of lava where the flow is about 20 ft [6 m] thick and nearly 1 mi [1.2 km] wide). Upon entering the wide strike valley cut in the soft shale of the Moenave and Chinle formations, the flow spread out and is nearly one mile wide. It terminated near the Santa Clara River approximately 10 mi (16 km) from its source.

As a result of flowing over a variety of terrain types, several major types of topographic modifications are clearly preserved.

Formation of Temporary Lakes. As the flow entered the drainage system of Snow Canyon it effectively blocked many of

Figure 2. View northward toward Diamond Valley showing the inverted valley on the Airport (Stage IIb) flow (platform to the right), the recent Santa Clara flow filling Snow Canyon and the two fresh cinder cones at the source of the Santa Clara flow in Diamond Valley.

Figure 3. One of the recent cinder cones in Diamond Valley.

the tributaries and formed a number of small temporary lakes or ponds which have since been filled with silt and sand. The largest and best example of lakes formed by blocked drainage are the headwaters of Snow Canyon upstream from the area where the flow entered the main canyon. Other smaller, but more accessible, temporary lakes formed in the tributaries which enter Snow Canyon from the west (Fig. 1). All of the larger lakes have been completely filled with silt and new drainage channels have been established across the lake beds and over the lava flow where they are presently in the process of eroding channels into the lava flow.

Lava Cascades. Several spectacular lava cascades occur in the northwest corner of Snow Canyon State Park where lobes have spilled over the Navajo Cliffs in frozen lava falls some 400 ft (120 m) high. They are not readily accessible but the three largest cascades can be seen by hiking up Snow Canyon a mile north of the Snow Canyon campground.

Displaced Drainage. From the standpoint of geomorphology, one of the more significant results of the extrusion of lava flows is the displacement of drainage and the evolution of inverted valleys. The lower part of the Santa Clara flow, south of the Navajo escarpment, provides important insight concerning the processes by which drainage patterns are modified and topographic inversion is initiated. South of the Navajo escarpment the Santa Clara flow issues from Snow Canyon, where it was confined as an intracanyon flow, and spreads out across the strike valley formed in the Chinle shales and then follows the ancestral drainage of Snow Canyon until it enters Santa Clara River. As can be seen in Figure 1, two relatively large tributaries were displaced by the lava and were forced to establish a new course along the margins of the flow. These channels are now well established and focus the processes of stream erosion along the margins of the flow. The new drainage is probably developed both by extension of the channel downslope along the margins of

the flow and by headward migration of a nick point or rapids upslope from the distal margins of the flow. It is quite clear, however, that topographic inversion proceeds from the distal margin of the flow upstream because the Santa Clara flow has 30 ft (9 m) of topographic inversion where it enters the Santa Clara River and no inversion near the base of the Navajo escarpment. Indeed, a waterfall (dry) on the east flank stream marks the point where accelerated erosion is occurring and topographic inversion is taking place. Upstream from the waterfall, the east flank stream has extended itself on top of the flow near the margin.

The west flank stream is somewhat different. It drains a relatively large area on the Navajo escarpment and is blocked by the lava flow and probably at one time formed a shallow temporary lake. Sediment deposited across the flow suggests that the lake silted up and drainage then overflowed across the lava. The drainage was subsequently captured by headward erosion of the lower part of the west flank stream.

It therefore seems evident that in an arid climate the establishment of flank streams which erode the flow to produce an inverted valley is complex and involves both downslope extension of blocked tributaries and upslope migration of nick points, rapids, and waterfalls.

Cinder Cones in Diamond Valley, 10 mi (16 km) North of St. George on Utah 18. The two fresh cinder cones which mark the site of extrusion of the Santa Clara flow are among the most recent volcanic features in Utah. The cones are almost perfectly preserved, being essentially unaltered by erosion and weathering. They are so young and fresh that a soil has not been established sufficiently to support vegetation, except for local sparse sagebrush near their base. Both cones are relatively small, approximately ¼ mi (.4 km) in diameter, and are slightly over 200 ft (60 m) high. They have summit craters that are fresh and unaltered (Fig. 3). (The northern cone, unfortunately, has been quarried for road material).

Judging from the fresh, unaltered appearance, the ages of the cones and the Santa Clara flow would be roughly equivalent to Sunset Crater in Arizona, perhaps slightly more than 1 ka.

The St. George Flows (Stage II): on Utah 18 from St. George to Diamond Valley. Remnants of two inverted valleys form the black ridges just west of St. George, Utah. The highest is nearly 2 Ma and is only a small remnant of a once more extensive flow. The lower inverted valley serves as a platform on which the St. George Airport is built and is referred to as the Airport flow. It, together with the Middleton and Washington flows, is 1 Ma. Utah 18 follows the top of the Stage II Airport flow from St. George northward to Diamond Valley, a distance of approximately 10 mi (16 km). This inverted valley extends as a long, narrow, sinuous ridge generally less than ¼ mi (0.4 km) wide. All of the original margins of the flow have been destroyed by erosion, and original surface features, such as aa crust, pressure ridges, etc., have been obliterated, making the top of the flow a smooth surface. In some places the flow is dissected into small isolated buttes. In the vicinity of St. George the flow is 200 ft (60 km) above the surrounding surface where the bedrock is the relatively soft Chinle and Moenave shale and siltstone. Where it traverses the resistant Navajo Sandstone inversion is much less, about 30 ft (9 m) and for several miles it flows over the Navajo Sandstone and there is little or no inversion adjacent to the flow. At Snow Canyon inversion is 200 ft (60 m). The surface gradient is 10 ft/mi (7.9 m/km).

Isolated remnants of higher and older inverted valleys are preserved in the Vicinity of Twist Hollow and on West Black Ridge above the St. George Airport.

The Middleton Flows (gravel road from Middleton to Cottonwood Guard Station). The Middleton flows form a long, narrow, inverted valley (Fig. 4) extending from the base of the Pine Valley Mountains southward to the Virgin River, a distance of approximately 10 mi (3 km) (Fig. 1). A well-maintained gravel road follows the top of the inverted valley so there is easy access to most of the geologic features in the area. At first glance, the Middleton inverted valley appears to be similar to that capped by the Airport flow just west of St. George, but it is one of the most complex inverted valleys in the western Colorado Plateau. It is composed of multiple flow units which occur in almost every conceivable relationship with each other and with the surrounding topography. South of I-15 a single flow unit rests upon the non-resistant shales of the Chinle, Moenave, and Kayenta formations. As a result the topographic inversion is a classic, simple, narrow ridge, generally less than ⅛ mi (0.2 km) wide. North of I-15 younger flow units interbedded with fluvial sand and gravel rest upon the basal basalt, and when traced upstream they occupy a "medial" valley cut in still older flow units which merge with the basalts that form a black cap on the prominent white Navajo Sandstone platform at the base of the Pine Valley Mountains.

The complex flow units of the Middleton inverted valley are important because they indicate a series of basaltic eruptions which are closely spaced in time, extruded from the same general vent area. The source area of these flows is marked by Twin Peaks, a series of dikes and volcanic necks and remnants of old cinder cones.

The Washington Flow (Views from I-15). The Washington flow forms a prominent inverted valley called Washington

Figure 4. The Middleton inverted valley, an example of a Stage IIb flow. Note the small higher butte held up by a remnant of a Stage IIa flow.

Black Ridge, approximately 2 mi (3.2 km) east of the town of Washington. The flow extends in a southeast direction for almost 3 mi (5 km) where it encounters the hogback of the Virgin anticline. It then turns 90 degrees to the west and follows the ancestral Virgin River for more than 1.5 mi (2.4 km). The inverted valley is generally 200 ft (60 m) above the surrounding area and is breached by erosion at Grapevine Pass and at several places where it follows the ancestral Virgin River. Good cross sections can be seen in road cuts I-15. The distal end of the flow crosses the Washington fault zone and has been offset by recurrent movement, which formed several horsts and grabens. Mass movement is especially active along the lower margins where the flow overlies the Chinle Shale and the slopes of the inverted valley are covered with slump blocks and talus debris. Upstream, several small remnants of three older Stage II flows occur near the place where they were extruded, together with two small Stage III flows and their associated cinder cones.

The Washington flow is a good example of topographic inversion which takes place by headward erosion and upstream migration of water falls and rapids. The upper (northern) margins of the Washington flow show no topographic inversion whereas the downstream margin of the flow is inverted 200 ft (60 m).

REFERENCES CITED

Hamblin, W. K., 1963, Late Cenozoic basalts of the Saint George Basin, Utah: Intermountain Association of Petroleum Geologists, Guidebook to the Geology of Southwestern Utah, 25th Annual Field Conference, p. 84–89.
—— , 1970, Late Cenozoic basalt flows of the western Grand Canyon: Guidebook to the Geology of Utah, no. 23, The Western Grand Canyon District, p. 21–37.
—— , 1984, Direction of absolute movement along boundary faults of the Basin and Range-Colorado Plateau margin: Geology, v. 12, p. 116–119.
Hamblin, W. K., Damon, P. E., and Bull, W. B., 1981, Estimates of vertical crustal strain rates along the western margins of the Colorado Plateau: Geology, v. 9, p. 293–298.

Beaver Dam Mountains: The geology Rosetta Stone for southwestern Utah

James L. Baer, Geology Department, Brigham Young University, Provo, Utah 84602

LOCATION AND ACCESSIBILTY

The Beaver Dam Mountains are located in the southwest corner of Utah about 18 mi (30 km) west of St. George (Fig. 1). The range is bisected by U.S. 91 and has well-maintained, graded roads suitable for passenger cars along the eastern side of the range.

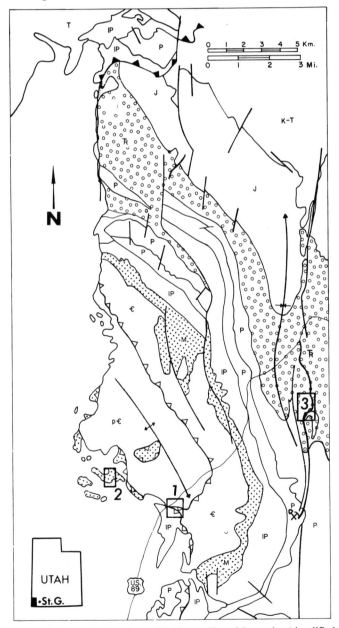

Figure 1. Generalized geology of the Beaver Dam Mountains (simplified from Hintze, 1985). Numbered boxes refer to selected sites.

SIGNIFICANCE

More than 29,500 ft (9,000 m) of strata are exposed in the Beaver Dam Mountains. These rocks range in age from Precambrian to Recent, with only the Ordovician and Silurian periods not represented; the Beaver Dam Mountains display the most complete stratigraphic section of rock in southern Utah (Fig. 2). The Beaver Dam Mountains represent the westernmost outcrops of platform stratigraphy and Colorado Plateau geology in Utah. Furthermore, Sevier folds and thrust faults, along with Tertiary extensional features, are well documented. The Red Hollow Fault along the west side of the mountains is a major structural feature and separates Colorado Plateau geology from the highly folded and faulted terranes of the Basin and Range province to the west.

SITE INFORMATION

Three sites were selected as representative of critical structural or stratigraphic elements of the range. Each selected site is accessible by passenger car.

Site 1—Castle Cliff Attenuation Area. Tertiary extensional tectonics has long been invoked to explain the exotic terranes of portions of southern Nevada, western Arizona, and southern California (Rowley and others, 1979). Only recently have investigators begun to recognize this tectonic mode as a probable cause for structures in western Utah. Recent mapping of the Beaver Dam Mountains by Hintze (1985) indicates that rocks in much of the range, principally western and southwestern portions, exhibit attenuation (stretching of the rocks), as well as detached and displaced blocks where younger rocks are over older ones. Commonly, these rocks are brecciated at a variety of scales and in certain areas are selectively attenuated. Many locations display tectonic thinning of as much as 50 percent. Because of the brecciation and the presence of many small faults, the amount of thinning can only be estimated. The general pattern of deformation is shown in Figure 4.

Site 1 is located 0.75 mi (1.1 km) east of the small cave at Castle Cliff (Fig. 1). The hillside to the southeast shows a highly attenuated section of Paleozoic rocks (Fig. 3). The prominent massive unit above midway up the hillside is the Mississippian Redwall Limestone, the same unit that makes up Castle Cliff. Below it are attenuated Cambrian Bonanza King Formation and Tapeats Quartzite, resting upon Precambrian schists and gneisses. Missing from this hillside are the Bright Angel Shale, Nopah Dolomite, and Muddy Peak Dolomite. The total section from the Cambrian through the Redwall Limestone is approximately 6,550 ft (2,000 m) thick two miles east of this site, but here near Castle Cliff the total section is only 980 to 1,250 ft (300 to 380

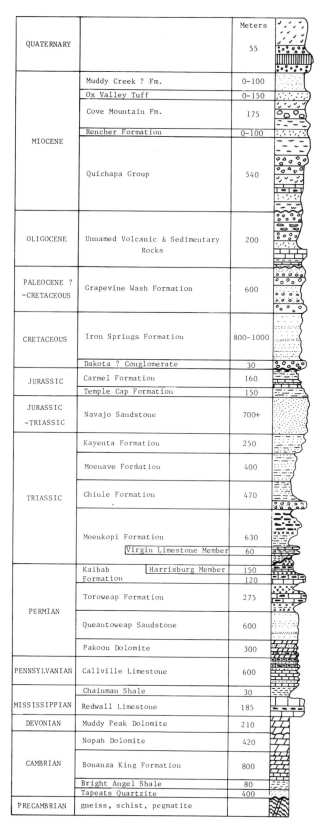

			Meters	
QUATERNARY			55	
MIOCENE	Muddy Creek ? Fm.		0-100	
	Ox Valley Tuff		0-150	
	Cove Mountain Fm.		175	
	Rencher Formation		0-100	
	Quichapa Group		540	
OLIGOCENE	Unnamed Volcanic & Sedimentary Rocks		200	
PALEOCENE ? -CRETACEOUS	Grapevine Wash Formation		600	
CRETACEOUS	Iron Springs Formation		800-1000	
	Dakota ? Conglomerate		30	
JURASSIC	Carmel Formation		160	
	Temple Cap Formation		150	
JURASSIC -TRIASSIC	Navajo Sandstone		700+	
TRIASSIC	Kayenta Formation		250	
	Moenave Formation		400	
	Chinle Formation		470	
	Moenkopi Formation		630	
		Virgin Limestone Member	60	
PERMIAN	Kaibab Formation	Harrisburg Member	150	
			120	
	Toroweap Formation		275	
	Queantoweap Sandstone		600	
	Pakoon Dolomite		300	
PENNSYLVANIAN	Callville Limestone		600	
MISSISSIPPIAN	Chainman Shale		30	
	Redwall Limestone		185	
DEVONIAN	Muddy Peak Dolomite		210	
CAMBRIAN	Nopah Dolomite		420	
	Bonanza King Formation		800	
	Bright Angel Shale		80	
	Tapeats Quartzite		400	
PRECAMBRIAN	gneiss, schist, pegmatite			

Figure 2. Generalized stratigraphy of the Beaver Dam Mountains (simplified from Hintze, 1985).

Figure 3. Site 1 view toward the south from U.S. 89, east of Castle Cliff. lPc, Pennsylvanian Callville Limestone; Mr, Mississippian Redwall Limestone; C-D, Cambrian to Devonian formations; pC, Precambrian.

m) thick. This exposure shows typical effects of attenuation upon the Paleozoic rocks on the west side of the range and is best seen up the hillside.

Site 2—Massive Slide Blocks. Site 2 is located 2.5 mi (4 km) northwest of the intersection of the main gravel road and U.S. 91 (Fig. 3). Except for occasional washouts, the gravel road is easily traversed by passenger car.

Prominent blocks of Paleozoic limestones and dolomites of Mississippian, Devonian, and Cambrian ages can be seen to the northeast of the road. Most of these blocks are both brecciated and attenuated Redwall Limestone. These blocks rest upon either Precambrian metamorphic rocks or upon the Tertiary Muddy Creek Formation.

Investigators (Cook, 1960; Jones, 1963) have different views about the manner of emplacement of these blocks. Cook (1960) interpreted the blocks as klippen of a Tertiary thrust, possibly one that occurred in late Miocene. Jones (1963) con-

Figure 4. Block of Mississippian Redwall resting upon Tertiary Muddy Creek Formation at Site 2.

cluded that the blocks are slide blocks which slide downslope into Miocene clastic sediments. This interpretation is based on their downslope position, the fact that these thin blocks could not sustain stresses associated with thrusting, and grooves on undersides of the blocks that are parallel to the inferred direction of transport. Cook (1960) concluded that the blocks came from a previously existing thrust sheet.

My interpretation differs from both of these, but agrees more closely with that of Jones (1963). Both earlier interpretations invoke thrust sheet emplacement as an integral part of the origin for the blocks. Except for the northern and northwest portions of the Beaver Dams, no evidence indicates that thrusting played a role in the structure of these mountains. Further, all the so-called "thrusts" involve young rocks upon older rocks and the directional properties of the slide blocks indicate that the blocks moved generally westward, opposite to the general regional movements of thrusts found in the region. I view the blocks as a result of extensional tectonics. These blocks could have been produced during late stages of the same tectonic regime that attenuated the rocks at Site 1.

At Site 2, the road is situated on the Tertiary Muddy Creek Formation. A large slide block of mostly Mississippian Redwall is to the northeast (Fig. 4), and another smaller block is situated just off the road to the southwest. Groove marks, brecciated nature, and contact relationships are best seen by hiking up the hill to the northeast.

Site 3—Triassic-Permian Unconformity. Site 3 is located on the Apex Mine Road 5.1 mi south of its intersection with U.S. 91. This road is a haul road for the Apex Mine and is usually well maintained (Fig. 1).

The unconformity that separates the Permian Kaibab Formation and the Triassic Moenkopi Formation (Fig. 5) is well developed along the eastern side of the range. This unconformity is particularly easy to see here because red beds of the lower red

Figure 5. View looking north from Apex Mine Road at Site 3. Wavy line marks position of the unconformity. Trlr, Lower Red Member of the Triassic Moenkopi Formation; Pkh, Harrisburg Member of the Permian Kaibab Formation.

member of the Moenkopi are present in some places and absent in others, where they fill in around paleo hills eroded on the gypsum beds of the Harrisburg Member of the Kaibab. Local relief of over 100 ft (30 m) on the unconformity is common, with some areas having nearly 590 ft (180 m) of relief. These were spectacular ancient canyons that were gradually infilled by fluvial and shallow marine sediments during Early Triassic time.

In order to better appreciate this geologic feature, after viewing it from the overlook, move northward approximately 0.25 mi where the unconformity is adjacent to the road. The Virgin Limestone Member is exposed on the east side of the road, but immediately west of the road the Harrisburg Member crops out. Locally angular debris from the eroded Harrisburg beds occurs at the unconformity.

REFERENCES CITED

Cook, E. F., 1960, Breccia blocks (Mississippian) of the Welcome Spring area, southwestern Utah: Geological Society of America Bulletin, v. 71, p. 1709–1712.

Jones, R. W., 1963, Gravity structures in the Beaver Dam Mountains, southwestern Utah: Intermountain Association of Petroleum Geologists Guidebook 12, p. 90–95.

Hintze, L. F., 1985a, Geologic map of the Shivwits and West Mountain Peak quadrangles, Washington County, Utah: U.S. Geological Survey Open-File Report 85–119.

——— 1985b, Geologic map of the Castle Cliff and Jarvis Peak quadrangles, Washington County, Utah: U.S. Geological Survey Open-File Report 85–120.

Rowley, P. D., Steven, T. A., Anderson, J. J., and Cunningham, C. G., 1979, Cenozoic stratigraphic and structural framework of southwestern Utah: U.S. Geological Survey Professional Paper 1149, 22 p.

Laramide fault blocks and forced folds of the Livermore-Bellvue area, Colorado

Vincent Matthews III, Champlin Petroleum Company, Houston, Texas 77002

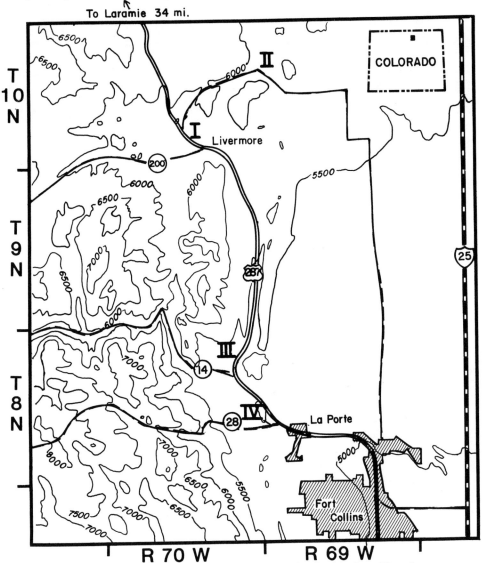

Figure 1. Index map of northeastern flank of Front Range showing location of four sites on or near U.S. 287 north of Fort Collins, Colorado.

INTRODUCTION

Location and Access. Four localities north of Fort Collins, Colorado (Fig. 1) display many of the structural features found throughout the Wyoming province of the Rocky Mountain Foreland. All of the localities can be viewed from public roads that are easily accessible to passenger cars or buses.

Significance of Locality. The northeastern flank of the Front Range displays several varieties of Laramide folds as well as critical relationships for demonstrating that the folds in the sedimentary rocks are forced folds resulting from differential up-

lift of faulted blocks of Precambrian basement. Within this small area are several beautifully exposed monoclines; differentially uplifted blocks (both tilted and untilted) of Precambrian basement; and well-displayed asymmetrical anticlines, synclines, domes, and basins. The area is also cut by a major lineament that separates two major structural styles in the Front Range and Denver Basin (Matthews, 1976). In this small area, one can see most of the crucial relationships which bear on the controversy over the nature and origin of Laramide deformation in the Rocky Mountain

Figure 2. View to north from Site I showing uplifted Precambrian block with flat surface on top (PS1). PS2 is surface on top of Precambrian block that is faulted down relative to block PS1 (Compare with Figure 3). Surface PS2 has remnant patches of Pennsylvanian strata preserved on it. GBRM is Grayback Ridge monocline. Knoll at right of photo just south of Grayback Ridge monocline is made up of horizontal, Triassic strata (Tr).

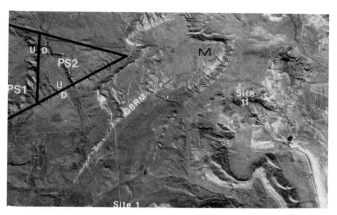

Figure 3. NASA U-2 photo showing relationships at Sites I and II. M is monocline that can be viewed from Site II. (Symbols are same as Fig. 2.)

Foreland Province. The sites are described in more detail in Matthews and others (1976), Matthews and Sherman (1976), and Matthews and Work (1978).

SITE DESCRIPTIONS

General. From Sites I and II in the northern part of the area, can be seen differentially uplifted, basement blocks that are untilted, and to the east, one can observe monoclinal folding that is of the same scale and shape as the exposed basement blocks on the west. One can also observe the basement blocks passing to the east beneath the sedimentary cover and see the relationships of deformation in the sedimentary rocks to the block edges. Sites I and II are located along a major lineament which marks the northern boundary of the Front Range uplift and the Denver basin downdrop (Matthews, 1976). South of this lineament the basement blocks are also differentially uplifted and are also tilted 15-30° toward the basin. Therefore, the corresponding folds in the overlying sedimentary rocks are asymmetrical anticlines, synclines, domes, and basins, rather than the monoclines that occur in the northern area. Sites III and IV show the tilted basement blocks and resulting folds in the overlying strata.

Site I. This site is on the east side of U.S. 287 at the intersection with a gravel road that goes north along the section line between Sec. 27 and 28, T.10N.,R.70W., 0.6 mi northwest of Livermore, Colorado (Fig. 1). The view to the north from Site I shows the character and interrelationships of the basement fault blocks and the overlying monoclinal folding (Fig. 2). In the foreground are flat-lying Permian strata which are representative of the structural attitude of most of the sedimentary rocks in the Livermore embayment. Disrupting the continuity of the horizontal beds is the Grayback Ridge monocline which is 5.3 mi (8.5 km) long. This monocline is linear, trending N47°E.

Northward of the monocline are flat-lying strata. On the far horizon can be seen two uplifted basement blocks from which most of the sedimentary strata have been eroded. Figure 3 shows the plan view of these two exposed blocks, which is the same shape as the monocline to the east that can be viewed from Site II.

Site II. This site can be reached by following the gravel road east from Site I 5.6 mi to a vantage point above the artificial reservoir north of the road. After leaving U.S. 287, the gravel road goes due north for one mile. Be sure to turn east at this point rather than following the private quarry road which extends northward from the county road. Toward the northwest is a magnificent view of a northwest-striking monocline that is not disrupted by faulting (Figs. 3 and 4). The shape of this monocline is similar to the exposed basement blocks west of here. This supports the interpretation that the monocline is a forced fold formed by the differential uplift of an underlying, rigid basement block (Matthews and Sherman, 1976).

Site III. This site is on the north side of U.S. 287 at a wide pullout 0.4 mi west of where the railroad tracks cross the highway, about 1 mi north of the junction with Colorado 20. At this

Figure 4. View northwestward from Site II of well exposed monocline. Point A is where strike of monocline changes abruptly from N25°E to N87°W. Strata at B are nearly horizontal. Strata at C are dipping 40°SE.

Figure 5. View northward from Site III showing exhumed, tilted surface (S) on Precambrian rocks (P) of the Owl Canyon block. Massive resistant strata at the crest of the hogback are gently dipping limestones of the Ingleside (I) formation. Slope-forming strata are arkoses and mudstones of the Fountain (F) formation. Ingleside strata at the base of the fault-line scarp (P) are nearly vertical.

site, one can observe the southern end of a large, well-exposed basement block that has been tilted toward the east (Fig. 5). This particular basement block (the Owl Canyon block) also illustrates one of the reasons why controversy has arisen about these structures. Here at the southern end of the block (the area that has been most intensely studied and described in the literature) the exposures are not complete. Therefore, there is considerable room for controversy about the relations of the sedimentary rocks to the underlying basement rocks. Braddock and others (1973) interpret the sedimentary rocks in the valley to the northeast to be cut by a fault. They would interpret the fold at the south end of the block to be a fold-thrust feature on the upthrown side of the fault. Matthews and Work (1978) interpret the basement fault to extend under the axis of the fold and to die out beneath the Permian Ingleside formation, which forms the first prominent hogback above the Precambrian basement. However, at the northern end of the block, the exposures are probably the best that can be found anywhere on the eastern flank of the Front Range. At this location is a continuous exposure of the critical beds directly overlying the basement. These beds clearly illustrate that the basement fault dies out into an unfaulted, continuous fold at the Ingleside level. Thus the folding in the sedimentary rocks is directly related to the differential uplift and rotation of the basement block (Matthews and Work, 1978). Unfortunately, that

locality is on private land, but it is well displayed on aerial photographs.

Site IV. This site, located at the picnic area at the northwest end of the State Fish Rearing Unit in Bellvue, is in the center of the Bellvue dome. Although Precambrian rocks are not exposed in the core of this structure (Fig. 6), the geometry of the fold indicates that it was created by an uplifted and rotated block of basement (Matthews and others, 1976). The steep west limb of

Figure 6. Aerial view of Site IV showing overall shape of Bellvue dome.

the fold can be observed about 900 ft (300 m) west of the lake. Various workers have interpreted the structural relations here in different ways. Some place a fault west of the exposed steep limb of the fold; others interpret a fault to be present under the lake, and still others question whether the sedimentary section was ever faulted through at all (Matthews and others, 1976). As in many places along the Front Range, the answers are probably not to be found at this locality because the critical relationships have been removed by erosion. Thus, one is forced to turn to other localities where the exposures are more complete, such as the north end of the Owl Canyon block.

REFERENCES

Braddock, W. A., Connor, J. J., Swann, G. A., and Wohlford, D. D., 1973, Geologic map cross sections of the Laporte quadrangle, Larimer County, Colorado: U.S. Geological Survey Open-File Report, OF 73-0030, scale 1:24,000.

Matthews, V., 1976, A tectonic model for the differing styles of deformation along the northeastern flank of the Front Range uplift: Geological Society of America Abstracts with Programs, v. 8, no. 6, p. 1000.

Matthews, V., Work, D. F., Lemasurier, W., and Stearns, D. W., 1976, Mechanism of Laramide deformation along the northeastern flank of the Front Range (Road Log), in Epis, R. C., and Weimer, R. D., eds., Studies in Colorado Field Geology: Colorado School of Mines Professional Contributions no. 8, p. 386–397.

Matthews, V., and Sherman, G. D., 1976, Origin of monoclinal folding near Livermore, Colorado; Mountain Geologist, v. 13, p. 61–66.

Matthews, V., and Work, D. F., 1978, Laramide folding associated with basement block faulting along the northeastern flank of the Front Range, Colorado, in Matthews, V., ed., Laramide Folding Associated with Basement Block Faulting in the Western United States: Geological Society of America Memoir 151, p. 101–124.

Precambrian structure, metamorphic mineral zoning, and igneous rocks in the foothills east of Estes Park, Colorado

Robert M. Hutchinson, *Department of Geology, Colorado School of Mines, Golden, Colorado 80401*
William A. Braddock, *Department of Geological Sciences, University of Colorado, Boulder, Colorado 80309*

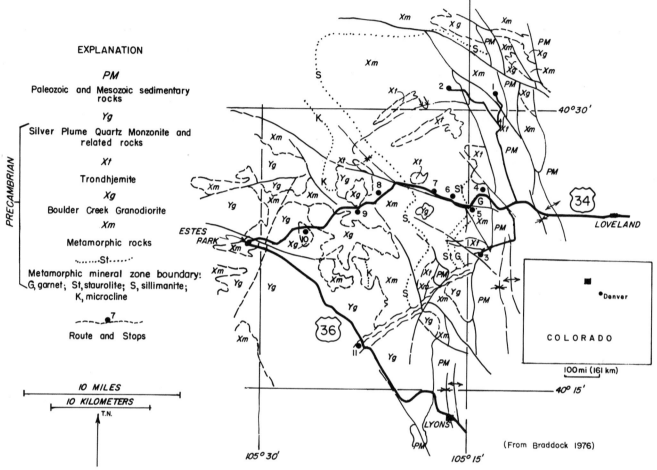

Figure 1. Map showing general locations of stops and distribution of metamorphic isograds, Big Thompson Canyon area (after Braddock, 1976a).

LOCATION

The area is located in the north-central part of Colorado and is readily accessible via U.S. 34, which roughly divides the area into north and south halves (Fig. 1). The eastern boundary of the area is 8 mi (9 km) west of Loveland, where U.S. 34 enters Big Thompson Canyon. Big Thompson power plant and Little Dam are at the immediate entrance to the canyon. Hard-bottomed dirt roads run north and south from the highway and afford ready access to the eleven stops shown on Figure 1.

SIGNIFICANCE

Within a fairly compact area of 400 mi^2 (1,040 km^2) in the foothills east of Estes Park, Colorado, there is a well documented 500-m.y. history of Precambrian sedimentation, two periods of regional metamorphism, three stages of deformation, and two major periods of plutonism extending from about 1,900 Ma to 1,390 Ma. Extensive work has been done by Braddock (1970, 1976b), Braddock, Calvert and others (1970), Braddock, Nutalaya and others (1970), Peterman and others (1968), and Abbott (1972). Information for this site description has been drawn freely from Braddock's papers. Radiometric dates for this part of the Colorado Front Range have been reported by Peterman and others (1968), Abbott (1972), and Hutchinson (1976).

In this region, a thick section of metasedimentary rocks consisting of phyllite or schist and metasandstone ranges from greenschist to upper amphibolite facies. The low- and middle-grade metamorphic rocks display evidence of three stages of deformation and two periods of regional metamorphism.

The site is significant because a geochronological sequence of structural, tectonic, metamorphic, and igneous events is readily observed in a single day's field trip. The geochronological sequence to be observed is as follows, oldest to youngest.

1. The history begins with accumulation of a very thick sequence of sediments and volcanic rocks, including pelitic shales, graywackes, rhyodacite and basalt, throughout the central Colorado Front Range starting at about 1,900 Ma (Hutchinson, 1976).

2. The first stage of deformation produced large isoclinal folds and formed a penetrative cleavage. This folding episode was accompanied by an initial regional metamorphism that produced at least greenschist assemblages.

3. The second and third stages of deformation produced crenulation cleavage and superposed folds.

4. A second period of regional metamorphism occurred after (or late in) the third stage of folding and resulted in the present areal variation of metamorphic grade. This was the Boulder Creek orogeny, 1,750 to 1,690 ± 30 Ma.

Boulder Creek Granodiorite, the first major plutonic event, was intruded as sills and phacoliths conformable with the second-stage folds, 1,700 ± 40 Ma. Some trondjemite intrusives were mostly undeformed. A northwest-trending swarm of basalt dikes was intruded after the trondjemite.

Silver Plume Quartz Monzonite, the second major plutonic

event, intruded as disconformable dikes, plugs, and partly conformable small to medium-sized stocks cutting basalt dikes and all older rocks at 1,450 ± 30 Ma. This was followed by related biotite-muscovite granite and pegmatite at 1,390 ± 30 Ma, and development of mylonite at 1,200 ± 200 Ma.

Post–Silver Plume age faults that contain mylonite were reactivated during Paleozoic, Mesozoic, and Tertiary events.

SITE INFORMATION

The geochronological sequence can be seen by one or both of two field trips. Two trips will give a more complete picture of the 500-m.y. sequence of events in the area. The least involved trip, which shows the metamorphic mineral zonation of Figure 1, can be taken by driving 8 mi (13 km) west along U.S. 34 from the entrance to Big Thompson Canyon along the Big Thompson River. The more comprehensive trip involves leaving the highway and driving the dirt roads leading north and south from U.S. 34 and making all the eleven stops located on Figure 1.

Big Thompson Canyon trip

A traverse west along Big Thompson Canyon shows a series of rocks of pelitic origin that have been metamorphosed to the amphibolite facies. The metamorphic grade increases westward,

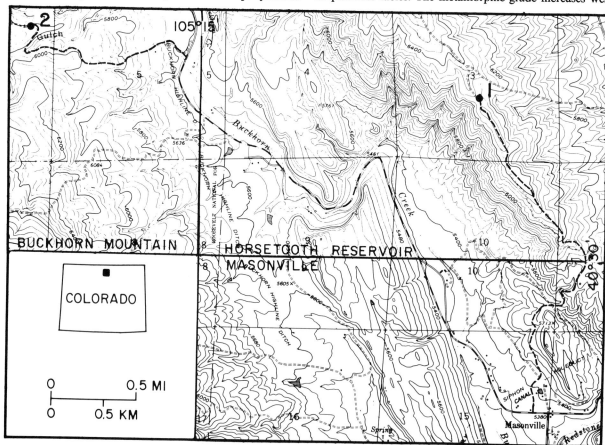

Figure 2. Map showing access to stops 1 and 2.

and 4 mi (6.5 km) beyond the town of Drake it reaches the highest grade at Stop 9. Within the map area, Braddock mapped four metamorphic mineral zone (mineral isograd) boundaries in pelitic rocks (See Fig. 1). Mineral assemblages of these zones occur at Stops 4, 5, 6, 7, 8, and 9. Petrographic analyses by Holden (personal communication, 1981) have defined additional metamorphic mineral zone (mineral isograds) reactions along the Big Thompson Canyon traverse. Reactions involved in the metamorphism, from the Big Thompson Canyon entrance to Stop 9 are as follows:

Isograd Reaction #1 Muscovite + Garnet + Chlorite → Staurolite + Biotite + H_2O

Isograd Reaction #2 Muscovite + Staurolite + Chlorite → Andalusite + Biotite + H_2O

Isograd Reaction #3 Muscovite + Staurolite → Andalusite + Biotite + Garnet + H_2O

Isograd Reaction #4 Andalusite → Sillimanite

Isograd Reaction #5 Muscovite + Quartz → Sillimanite + Microcline + H_2O

Emplacement of the Boulder Creek Granodiorite on the west side of isograd #5 caused the highest grade of metamorphism.

Areal trip north and south of U.S. 34

Visitation to Stops 1 to 11 would best be made with a copy of the road log from Braddock (1976a). A good deal more of the three stages of deformation can be seen than on only the Big Thompson Canyon traverse. Descriptions of the stops are as follows:

Stop 1. Ridge between Buckhorn and Redstone creeks (Fig. 2). Drive north from U.S. 34 for 5.6 mi (9.0 km) on paved road to Masonville. Turn left at road junction at Masonville Trading Post. After 0.1 mi (0.2 km) turn right on gravel road and drive 2.2 mi (3.5 km) to the stop.

Metasedimentary rocks here consist of thinly laminated staurolite-andalusite–bearing muscovite schist and coarse-grained to granule-sized metasandstones. Graded beds in metasandstone indicate that rocks face southwest. Weakly developed kinkfolds with axial planes striking N60°E to N80°E deflect the schistosity into steeply plunging folds having Z symmetry. Very weakly developed kinkfolds trending N20°W to N40°W deflect schistosity into steeply plunging folds having S symmetry.

Stop 2. Drive west from the Masonville Trading Post (Fig. 2) along main road for 3.9 mi (6.3 km). Turn left on private ranch road opposite yellow ranch buildings and continue 1.3 mi (2.1 km) to the stop.

In these outcrops, bedding has been disrupted to produce laminated lenticular sheets of varying composition, parallel to the schistosity and very difficult to distinguish from true bedding. Braddock has presented evidence that during the earliest period of folding, slaty cleavage developed by processes that included laminar intrusion of water-rich material across bedding (Braddock, 1976b).

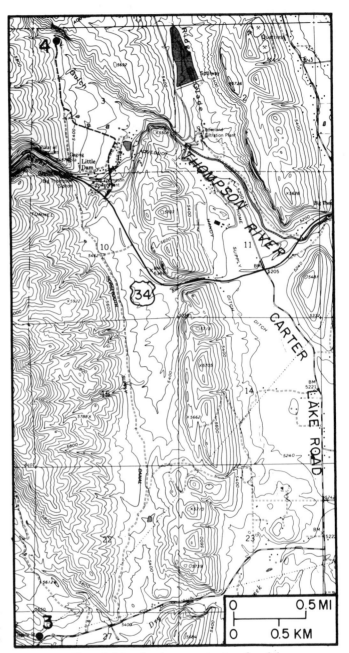

Figure 3. Map showing access to stops 3 and 4.

Stop 3. Drive south from U.S. 34 on paved road to Carter Lake. After 2.2 mi (3.5 km) continue right at junction on Carter Lake Road (Fig. 3) and drive 2.4 mi (3.8 km) to Cottonwood Creek. Stop 3 is in a quarry on the north side of Cottonwood Creek.

Muscovite-chlorite phyllite and interbedded sandstones here contain large folds of the second period of deformation. These folds have a well-developed, axial-plane crenulation-cleavage. The crenulation-cleavage plane is crinkled by an incipiently developed crenulation-cleavage of the third period of deformation.

Stop 4. Drive 1.5 mi (2.4 km) west on U.S. 34 from Carter Lake Road intersection and then turn right onto gravel road to

Sylvan Dale Rest Ranch (Fig. 3). Follow road across Thompson River for 0.3 mi (0.5 km), then turn left along barbed-wire fence. Continue to aqueduct, turn right and continue along aqueduct to stop at end of road.

The trace of the northwest-trending Green Ridge fault is visible to the east. On the ridge across the Thompson River, the fault places Precambrian rocks against Pennsylvanian Fountain and lower Permian Ingleside formations.

Metasandstones here are interbedded with biotite-muscovite-chlorite phyllite. Several metasandstone beds contain well-preserved graded units (granule-size at base, grading upward to phyllite), indicating that the rocks face north. The phyllites have strongly developed crenulation-cleavage of the third period of deformation.

Stop 5. This stop is on the south side of U.S. 34, 2.1 mi (3.4 km) west of the turnoff to Sylvan Dale Rest Ranch at the top of the narrows. Here, Thompson Canyon widens and becomes relatively straight because the canyon follows the trace of the Thompson Canyon fault. The fault passes through the topographic saddle to the east.

Metasedimentary rocks on the north side of the highway are interbedded metasandstone and biotite-chlorite-garnet-muscovite schist that contain steeply plunging folds and northwest-trending crenulation-cleavage of the third period of deformation. About 295 ft (90 m) to the east, the metamorphic rocks are cut by a thick basalt dike. The west margin of the dike is composite, containing irregular patches of quartz-rich material derived from melting of the wall rocks.

Stop 6. This stop is on the north side of U.S. 34, 1.6 mi (2.6 km) west of Stop 5.

The Thompson Canyon fault is exposed here. At the east end of the exposure, shattered staurolite-biotite-muscovite schist contains graded metasandstone. At the west end of the exposure, a north-dipping gouge zone is paralleled by a tabular body of chlorite-quartz rock that is believed to be a sheared and altered basalt dike.

Stop 7. This stop is on the north side of U.S. 34, 1.6 mi (2.6 km) west of Stop 6.

Metasandstones here contain braided pseudo-bedding believed to be a product of first-period deformation. A thick trondjemite sill contains a weakly developed foliation at a high angle to the sill contacts and about parallel to the direction of third deformation crenulation-cleavage nearby.

Stop 8. This stop is on the northwest side of U.S. 34, 4.5 mi (7.2 mi) west of Stop 7.

Crenulated biotite-sillimanite-muscovite schist here is cut by discordant bodies of pegmatite that grade into biotite-muscovite granite. The area of biotite-muscovite granite is restricted to this general vicinity and is probably a late phase of the Silver Plume intrusive complex.

Stop 9. This stop is on the south side of U.S. 34, 2.8 mi (4.5 km) west of Stop 8.

Biotite-sillimanite-microcline gneiss here contains pegmatites of probable Silver Plume age and deformed granitic veinlets that mark the first appearance of migmatite. A conformable body of Boulder Creek Granodiorite outcrops to the south and southwest of the highway.

Stop 10. This stop is on the south side of U.S. 34, 4.2 mi (6.7 km) west of Stop 9.

Medium-grained Boulder Creek Granodiorite is cut by basalt dike, and both are cut by fine-grained Silver Plume Quartz Monzonite.

Stop 11. This stop is on U.S. 36 at the junction with the gravel road to Big Elk Meadows, 11.6 mi (18.7 km) east of the junction of U.S. 34 and 36 in Estes Park.

Cataclastic rocks of the Moose Mountain shear zone are well exposed here in highway cuts. The shear zone trends N55°E, is about 2,950 to 3940 ft (900 to 1,200 m) wide, and in this area was produced by deformation of medium-grained Silver Plume Granite. The roadcuts 656 ft (200 m) north expose protomylonite, and those 656 ft (200 m) south expose protomylonite and mylonite. Similar shear zones 25 m (40 km) to the north have been dated by whole-rock Rb-Sr techniques at 1.2 Ga (Abbott, 1972).

REFERENCES CITED

Abbott, J. T., 1972, Rb-Sr study of isotopic redistribution in a Precambrian mylonite-bearing shear zone, northern Front Range, Colorado: Geological Society of America Bulletin, v. 83,, p. 487–494.

Braddock, W. A., 1970, The origin of slaty cleavage; Evidence from Precambrian rocks in Colorado: Geological Society of America Bulletin, v. 81, p. 589–600.

——, 1976a, Road log, Precambrian geology of the northern and central Front Range, Colorado; First day, northern Front Range: Professional Contributions of the Colorado School of Mines, Studies in Colorado Field Geology, no. 8, p. 1–8.

——, 1976b, Origin of slaty cleavage; Evidence from Precambrian rocks in Colorado: Professional Contributions of the Colorado School of Mines, Studies in Colorado Field Geology, no. 8, p. 8–16.

Braddock, W. A., Calvert, R. H., Gawarecki, S. J. and Nutalaya, P., 1970, Geologic map of the Masonville Quadrangle, Larimer County, Colorado: U.S. Geological Survey Quadrangle Map GQ-832, scale 1:24,000.

Braddock, W. A., Nutalaya, P., Gawarecki, S. J., and Curtin, G. C., 1970, Geologic map of the Drake Quadrangle, Larimer County, Colorado: U.S. Geological Survey Quadrangle Map GQ-829, scale 1:24,000.

Hutchinson, R. M., 1976, Precambrian geochronology of the western and central Colorado and southern Wyoming: Professional Contributions of the Colorado School of Mines, Studies in Colorado Field Geology, no. 8, p. 69–73.

Peterman, Z. E., Hedge, C. E. and Braddock, W. A., 1968, Age of Precambrian events in the northeastern Front Range, Colorado: Journal of Geophysical Research, v. 73, p. 2277–2296.

Late Cretaceous Mesaverde Group outcrops at Rifle Gap, Piceance Creek Basin, northwestern Colorado

J. C. Lorenz and A. K. Rutledge, Sandia National Laboratories, Albuquerque, New Mexico 87185

LOCATION AND ACCESS

Rifle Gap is accessible by paved road, and, except for the three southernmost flatirons, all the outcrops described here are on public land administered by the Bureau of Land Management (BLM). Permission for access to the private land is obtained from the residence just west of the southern entrance to the gap. From I-70, take the Rifle exit (Exit 90) and drive north on Colorado 13 (take Railroad Avenue through the town of Rifle) to the junction with Colorado 325. Turn right and drive about 4.1 mi (6.6 km), through Rifle Gap, to the Rifle Gap Reservoir dam. This is the stratigraphic base of the section described below.

The area of Rifle Gap is included in the adjacent corners of the Rifle, Rifle Falls, Silt, and Horse Mountain 7½-minute Quadrangles. Figure 1, a simplified topographic map, is updated to include the dam and reservoir. The approximate locations of the sandstone lenses visible from the road are plotted (Fig. 1). Though much of the section can be seen from the road, better views are afforded by scrambling up the valley slopes to view the opposite side with binoculars. The east side of the gap can be reached on foot by crossing Rifle Creek at the bridge just below the spillway to the dam.

SIGNIFICANCE

The well-exposed Mesaverde Group at Rifle Gap consists of the deposits of a major Late Cretaceous (late Campanian) progradation. The rocks are significantly younger than the type Mesaverde Formation in the Four Corners area of southwestern Colorado. Deep marine deposits intertongue with shoreface and delta plain deposits at the formation base, recording several minor transgressive and regressive episodes (Young, 1982). The onset of the Laramide Orogeny is recorded in lower and upper delta plain and fluvial deposits in the vicinity of Rifle Gap. The uppermost Mesaverde deposits record a paralic environment, deposited during the Lewis transgressive cycle.

Rocks of Late Cretaceous and Tertiary age have been rotated (to a present dip of 45 to 90 degrees along the Grand Hogback) by late Laramide uplift of the White River Plateau. About 5,000 ft (1,500 m) of the Mesaverde Formation are exposed in Rifle Gap. The vertical stratigraphic section can be traversed from older to younger rocks by traveling from Rifle Gap Reservoir south-southwest through the gap.

SITE INFORMATION

Marine/Non-marine Deposits. The valley north of the Grand Hogback is underlain primarily by the easily eroded Man-

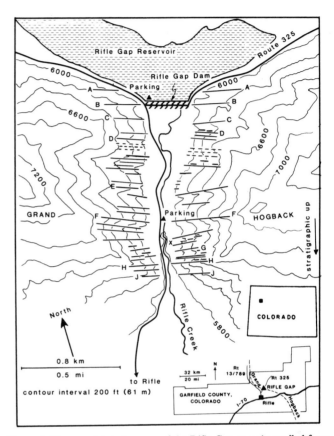

Figure 1. Topographic index map of the Rifle Gap area (compiled from the Rifle, Rifle Falls, Silt, and Horse Mountain 7½-minute quadrangles), showing approximate outcrop locations of Mesaverde sandstone lenses. Lettered lenses are discussed in the text. Dashed lines indicate smaller or obscured lenses.

cos Shale, which was deposited in the Cretaceous Western Interior Seaway. The first Mesaverde deposits at the top of the Mancos Shale may be below the present water level of the Rifle Gap Reservoir (Johnson, 1982).

The first visible units are sandy and silty beds of what may be the Corcoran Sandstone Member of the Iles Formation or Mesaverde Group (Collins, 1976). These deposits can be seen in cuts along the gravel road leading northwest from Rifle Gap Reservoir. Thin-bedded, extensive sandstones and siltstones, which usually show abundant marine trace fossils on the bedding planes, are interbedded with the gray mudrocks of the upper Mancos Shale. In the road cut, two thick units of amalgamated thinly bedded sandstones display tracks and trails and hummocky cross-stratification. A third, uppermost sandstone (Lens A, Figs. 1,

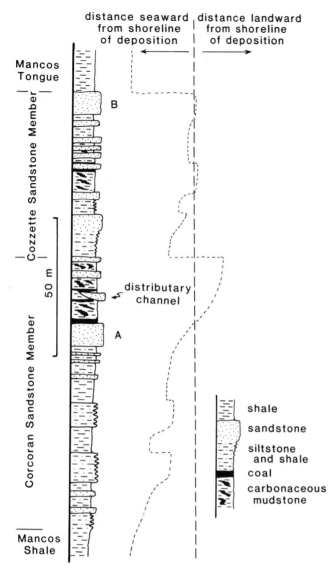

Figure 2. Measured section through the Corcoran and Cozzette Sandstone Members at Rifle Gap. Sandstones A and B are plotted on Figure 1.

Figure 3. Photograph of the Tongue of the Mancos Shale grading upward into the Rollins Sandstone (east side of Rifle Gap). Upsection is to the right.

2) contains *Ophiomorpha* burrows, as well as hummocky cross-bedding. A rooted zone is present in the top of this sandstone; it is overlain by a thin coal. The sequence is capped by a series of shaly and carbonaceous nonmarine sediments. These include a distributary channel (exposed at road level) which is cross-bedded and contains groups of climbing ripple structures, indicative of associated splay deposition.

This stratigraphic sequence records the progradation of the Cretaceous shoreline to the southeast, into the seaway (Warner, 1964). Environments of deposition, from the bottom to the top of the sequence, are interpreted as lower shoreface, upper shoreface and swash zone, and deltaic. This progradational sequence was terminated abruptly by transgression and a return to deep marine, Mancos Shale deposition. A well-burrowed sandstone provides

evidence of the transgression and marks the beginning of the next progradational sequence. This sequence, the Cozzette Sandstone Member, occurs as 260 ft (80 m) of thin sandy units interbedded with fine-grained carbonaceous sediments. These deposits may reflect erratic deposition on a sand-starved coastline. The upper sandstone of this unit (lens B, Figs. 1 and 2) is exposed directly adjacent to the abutments of the Rifle Gap Reservoir dam and is probably a shoreface deposit.

The overlying Tongue of the Mancos Shale is about 230 ft (70 m) thick (Johnson, 1982). It contains ironstone concretions and scattered selenitic gypsum. The tongue is capped by a classic coarsening-upwards shale to sandstone profile (Fig. 3). Marine trace fossils occur on the bedding planes and within the sandstones of this thinly bedded unit. The sandstones become thicker and more numerous upsection until they coalesce into the 130-ft (40-m) thick Trout Creek or Rollins Sandstone (Fig. 1, lens C). The lower half of this sandstone contains *Ophiomorpha* trace fossils and lag deposits of *Inoceramus* shell fragments. This has been interpreted as a shoreface to foreshore deposit, while the upper half of the Rollins Sandstone is suggested to be a tidal channel complex (Newman, 1982; Madden, 1985). The entire sequence above and including the Tongue of the Mancos Shale and the Rollins Sandstone records a shoreline progradation.

Delta Plain Deposits. Overlying the Rollins Sandstone is a 1,000-ft (300-m) thick series of coal-bearing lower-delta plain deposits. The road cut exposes the lower half of this unit. Most of the coals have been burned and only clinker zones remain. Notable in the road cut are a 16-ft (5-m) thick clinker zone, a 1-ft (30-cm) bentonite layer, and several small normal faults (probably associated with the uplift of the hogback). No distributary channels are exposed in the road cut, though they are poorly exposed elsewhere in the hillsides. A 65-ft (20-m) thick sandstone

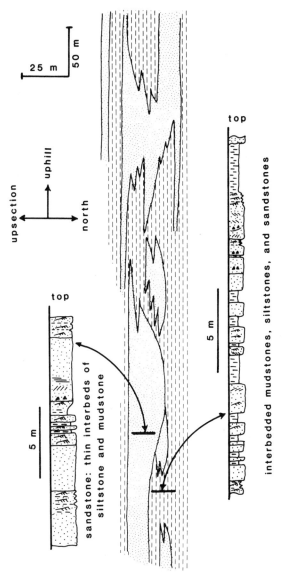

Figure 4. Schematic drawing of the mid-fluvial lens F complex (Fig. 1) on the west side of the gap. Schematic has a vertical exaggeration of 2×.

Figure 5. Photograph of lateral accretion (point bar) bedding from lens G (Fig. 1). Sandstone is about 20 ft (6 m) thick. Photograph shows uphill accretion at the top of the bed, but the unit is composite, and downhill accretion is visible in the basal part of the bed when viewed from the proper angle. Upsection is to the right.

unit, located at the upsection end of the roadcut (Lens D, Fig. 1), contains marine trace fossils. This sandstone probably represents a period of marine deposition. Most of the other deposits exposed here probably represent channel-margin, marsh, lacustrine, and other well-burrowed, brackish-water environments.

Above the coal-bearing strata there is an interval of lenticular sandstones without associated coals. Good examples can be seen on the west side of the gap in the valley and hillslopes just upsection from (south of) the red, clinkered, coal-bearing beds. Most of the lenses are limited in horizontal extent and terminate abruptly. Internal structures are obscure. In one location, visible from the road, a small (15 × 80-ft; 5 × 25-m) lens is joined to the downhill side of a more extensive sand body (Lens E, Fig. 1).

This sequence has been interpreted as an upper delta plain depositional environment which overlies the coal-bearing, lower delta plain deposits. These deltaic deposits, and the overlying fluvial sequence, continue the progradational cycle which began with the Tongue of the Mancos Shale.

Fluvial Deposits. The next 2,000 ft (600 m) of section consists of sandstones alternating with thin-bedded, gray, purple, and brown mudstones and siltstones. The major sandstone beds are usually exposed as flatirons jutting above the more easily eroded mudstones and siltstones. Lichen cover and viewing angle obscure much of the internal structure. Several of the beds (Lens G among others, Figs. 1, 5), display lateral accretion bedding typical of point bar deposits in meandering streams. On the west side of the gap, Lens F (Fig. 1), presents aspects of a vertically accreting, anastamosing, multiple-channel, fluvial system (Fig. 4). On the east side of the gap, the correlative sand beds are suggestive of lateral accretion of meander belts (Fig. 5) which, with some vertical accretion, developed a vertically stacked offlap configuration.

Root traces and subaqueous tace fossils are common in the intervals of fine-grained sediments. Mudstones may be carbonaceous. A 2.5-ft (76-cm) coal seam of limited extent (position X, Fig. 1) occurs near the middle of this section (Horn and Gere, 1954). An upright tree stump has been found in these fine-grained deposits (downhill side of Lens F, Fig. 1). Sandstones are often rippled, with mudstone ripup clasts occurring at the bases of sandstone beds. Subsets within the standstone beds show distinct, cyclic fining-upward trends (Lorenz and others, 1985).

The evidence suggests that these sediments were laid down during the continuing progradation by meandering fluvial systems in the Rifle Gap area. Most of the fluvial systems at Rifle Gap (and in the eastern part of the Piceance Creek basin) were probably small streams which originated on a coastal plain, rather than in the contemporaneous fold and thrust highlands of central Utah, or on the Uncompaghre Uplift of southwestern Colorado.

Paralic Deposits. The uppermost member of the Mesaverde Group is the Ohio Creek Member. Lorenz and Rutledge consider it to be a transitional facies between the fluvial Ohio Creek deposits to the south and west, and the Lewis Shale and Fox Hills Sandstone to the north. Its presence at Rifle Gap has been inferred on the basis of paralic trace fossils and sedimentary structures (Lorenz and Rutledge, 1985).

The Ohio Creek section is approximately 400 ft (125 m) thick, and includes several sandstone lenses and interbedded shales. The lower sandstone (Lens H, Fig. 1) of the Ohio Creek Section at Rifle Gap is a flatiron 40–50 ft (12–15 m) high, and 65–80 ft (20–25 m) thick. The most notable feature in this sand lens is the presence at its base of more than 35 fossilized logs ranging from 6–16 ft (2–5 m) in length. Most of the logs show concentrations of the marine- to brackish-water trace fossil *Teredolites* (Fig. 6). The zones of large shale ripup clasts at the base of the sandstone, the juxtaposition of sandstone which contains Teredinid-burrowed logs, and freshwater mudstones containing *Plesielliptio* sp. immediately below indicates that the base of the sandstone is an erosion surface. This surface is taken to be the contact between the fluvial Mesaverde and the Ohio Creek Member at Rifle Gap. The deposits are the products of a distribu-

Figure 6. Fossil log containing abundant *Teredolites* trace fossils (from the base of the lowest sandstone of the Ohio Creek Member).

tary or estuarine environment associated with an Ohio Creek fan delta (broad sense of usage for "fan delta").

The upper surface of this lens displays large areas of rippled and burrowed sandstone. On the west side of the gap, regularly spaced holes the size and shape of tree trunks are present. This surface is interpreted as a colonized, aggrading tidal flat (Lorenz and Rutledge, 1985).

The uppermost sandstone lens at Rifle Gap (Lens J, Fig. 1) is locally conglomeratic at the top, with a matrix of white kaolinitic sandstone. Large diagonal bedding planes within this sandstone on the west side of the gap are suggestive of a migrating (distributary) channel environment. The conglomeratic and coarse sandstones of this uppermost Ohio Creek member are probably synorogenic deposits associated with the beginning of local Laramide uplift. They represent the last stages of deposition prior to the erosion and weathering which created the overlying unconformity (Johnson and May, 1980).

REFERENCES

Collins, B. A., 1976, Coal deposits of the Carbondale, Grand Hogback, and Southern Danforth Hills coal fields, eastern Piceance basin, Colorado: Quarterly of the Colorado School of Mines, v. 71, p. 1–138.

Horn, G. H., and Gere, W. C., 1954, Geology of the Rifle Gap district, Garfield County, Colorado: U.S. Geological Survey Open-File Report, 13 p., 1 plate.

Johnson, R. C., 1982, Measured section of the upper Cretaceous Mesaverde Formation and lower part of the lower Tertiary Wasatch Formation, Rifle Gap, Garfield County, Colorado: U.S. Geological Survey Open-File Report OF 82-0590, 11 p., 1 log.

Johnson, R. C., and May, F., 1980, A study of the Cretaceous–Tertiary unconformity in the Piceance Creek basin, Colorado; The underlying Ohio Creek Formation (Upper Cretaceous) redefined as a member of the Hunter Canyon or Mesaverde Formation: U.S. Geological Survey Bulletin No. 1482-B, 26 p.

Lorenz, J. C., and Rutledge, A. K., 1985, Facies relationships and reservoir potential of the Ohio Creek interval across the Piceance Creek basin, northwestern Colorado: Oil and Gas Journal, v. 83, p. 91–96.

Lorenz, J. C., Heinze, D. M., Clark, J. C., and Searls, C. A., 1985, Determination

of widths of meander belt sandstone reservoirs from vertical downhole data, Mesaverde Group, Piceance Creek basin, Colorado: American Association of Petroleum Geologists Bulletin, v. 69, p. 710–721.

Madden, D. J., 1985, Description and origin of the lower part of the Mesaverde Group in Rifle Gap, Garfield County, Colorado: The Mountain Geologist, v. 22, p. 128–138.

Newman, K. R., 1982, Stratigraphic framework of Upper Cretaceous (Campanian) coal in western Colorado, *in* Averett, W. R., ed., Southeastern Piceance basin, western Colorado: Grand Junction Geological Society field trip guidebook, p. 61–64.

Warner, D. L., 1964, Mancos–Mesaverde (Upper Cretaceous) intertonguing relationships, southeast Piceance basin, Colorado: American Association of Petroleum Geologists Bulletin, v. 48, p. 1091–1107.

Young, R. G., 1982, Stratigraphy and petroleum geology of the Mesaverde Group, southeastern Piceance Creek basin, Colorado, *in* Averette, W. R., ed., Southeastern Piceance basin, western Colorado: Grand Junction Geological Society field trip guidebook, p.45–54.

Tertiary mineralization—Idaho Springs, Colorado

S. Budge, P. J. LeAnderson, and G. S. Holden, Department of Geology, Colorado School of Mines, Golden, Colorado 80401

INTRODUCTION

As I-70 climbs westward from Denver to the Continental Divide, it traverses the Idaho Springs, Central City, Empire, Georgetown, and Silver Plume mining districts, which lie near the northeast end of the Colorado Mineral Belt. Together these districts comprise a zone of nearly continuous mineralization extending about 16 mi (27 km) east-northeast from Silver Plume to Central City. Originally gold, then later Ag, Cu, Pb, Zn, and U were produced from the area, beginning in 1859. Most mines at present are inactive. Primary precious-metal production was dominated by gold, with some silver, from the Idaho Springs and Central City districts to the northeast and mainly by silver from the Silver Plume and Georgetown Districts to the southwest. Mineralization was associated with shallow Tertiary igneous activity in country rock composed of Precambrian (middle Proterozoic) granite and gneiss.

The Idaho Springs site is one of the few easily accessible locations where the Tertiary intrusive suite and mineralization are well exposed on the surface. The site is a road cut along the north side of I-70, just west of Idaho Springs (Fig. 1). A sketch, drawn to scale, is provided to show the features that may be observed (Fig. 2). Access to the outcrop is along a 20-ft (6-m) wide service road protected from I-70 by a guardrail.

REGIONAL GEOLOGY

At approximately 1800–1750 Ma, precursors to the composite Idaho Springs Formation (host rock for mineralization) were deposited as interlayered immature sedimentary, volcaniclastic, and felsic to mafic volcanic units. At about 1720 Ma this sequence was subjected to deformation and metamorphism to upper amphibolite grade, producing amphibolite and locally migmatitic biotite gneiss. Deformation produced early isoclinal folds with axial planar schistosity and later, large, mainly open folds, trending north to northeast. Late synkinematic granodiorite, gabbro, quartz diorite, and biotite-muscovite granite plutons (including the Boulder Creek and Mt. Evans batholiths) were intruded during this metamorphic event. Much of the area was later intruded by the nonorogenic, shallow level, Silver Plume quartz monzonite (1450 Ma) and by spatially related biotite-muscovite granite and pegmatite (1390 Ma). Good exposures of the Idaho Springs formation may be viewed along I-70 east of the town of Idaho Springs. The Silver Plume suite is well exposed on either side of the town of Silver Plume.

Silver Plume intrusive activity was followed by deformation producing the cross-cutting, northeast-trending Idaho Springs–Ralston Creek shear zone. The zone, which later localized the Colorado Mineral Belt (Tweto and Simms, 1963), extends 20 mi (32 km), to the northeast and 3 mi (2 km) to the southwest of the

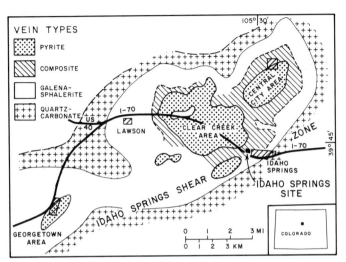

Figure 1. Map of the Idaho Springs area of the Colorado Mineral Belt showing the location of this field site and the distribution of vein types related to mineralization.

town of Idaho Springs. It is the dominant feature of a set of northwest- and northeast-trending faults, all initially formed at this time. Although evidence is sparse, these faults were probably reactivated during Paleozoic and Tertiary uplift events, which are not otherwise recorded in rocks of the area.

During the Laramide Orogeny (72–50 Ma), the region was pervasively fractured, faulted, and intruded by a suite of shallow subvolcanic plutons and dikes. Fractures are generally steeply dipping and may be classed into NNW and ENE sets. Intrusive units range from early granodiorite porphyry stocks to younger, more felsic (to granite and syenite) dikes.

Galena and sphalerite are the predominant metallic minerals in lead-zinc veins (which are present in the road cut). Pyrite is fairly abundant locally but is distinctly subordinate to galena and sphalerite. Chalcopyrite and tennantite are generally present only in small amounts. Quartz is the dominant gangue mineral, and ankerite, siderite, rhodochrosite, dolomite, barite, and fluorite may be present. The walls of lead-zinc veins usually have sharp contacts, and wallrocks are not strongly altered. Locally the veins are well-defined fissure fillings more than a foot thick, but they may also be zones of thin subparallel interconnecting veinlets and lenses. The veins are valued chiefly for their silver, gold, and lead content. The primary ores are richer in lead and silver than those of other vein types, but poorer in gold and copper.

Vein emplacement in the Central City and Idaho Springs districts began at 59 Ma (Rice and others, 1982), and was associated with early stocks and bostonite (leucocratic alkalic felsite) dikes. After emplacement of bostonite dikes and prior to emplacement of the youngest intrusion in the districts (biotite-latite porphyry), east-northeast- and northeast-trending faults devel-

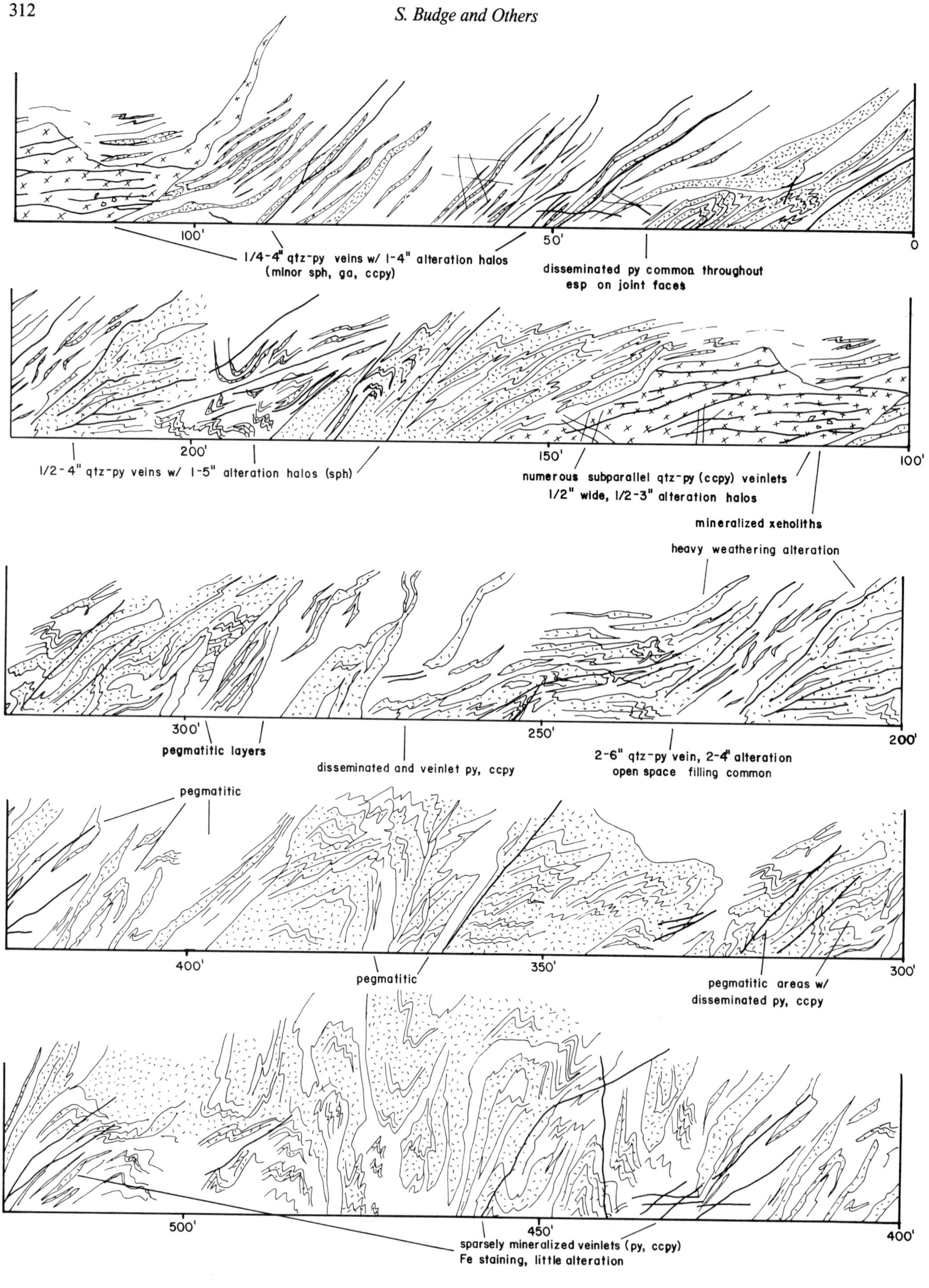

100'
50'
0
1/4-4" qtz-py veins w/ 1-4" alteration halos
(minor sph, ga, ccpy)
disseminated py common throughout
esp on joint faces
200'
150'
100'
1/2-4" qtz-py veins w/ 1-5" alteration halos (sph)
numerous subparallel qtz-py (ccpy) veinlets
1/2" wide, 1/2-3" alteration halos
mineralized xenoliths
heavy weathering alteration
300'
250'
200'
pegmatitic layers
disseminated and veinlet py, ccpy
2-6" qtz-py vein, 2-4" alteration
open space filling common
pegmatitic
400'
350'
300'
pegmatitic
pegmatitic areas w/
disseminated py, ccpy
500'
450'
400'
sparsely mineralized veinlets (py, ccpy)
Fe staining, little alteration

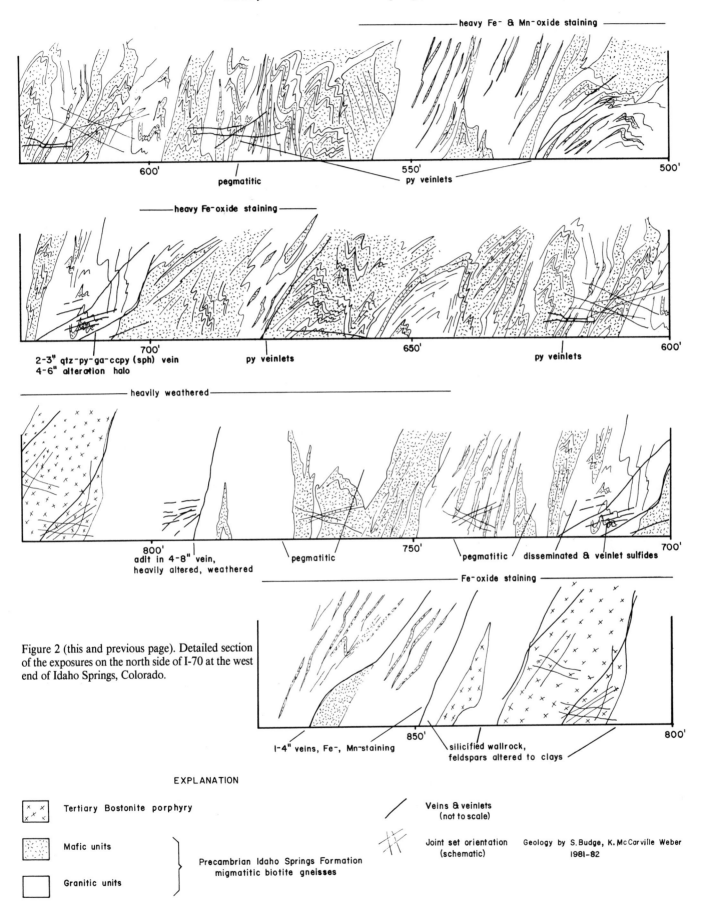

heavy Fe- & Mn-oxide staining

600'   pegmatitic   550'   py veinlets   500'

heavy Fe-oxide staining

2-3" qtz-py-ga-ccpy (sph) vein
4-6" alteration halo   py veinlets   700'   650'   py veinlets   600'

heavily weathered

800'
adit in 4-8" vein,
heavily altered, weathered   pegmatitic   750'   pegmatitic   disseminated & veinlet sulfides   700'

Fe-oxide staining

Figure 2 (this and previous page). Detailed section
of the exposures on the north side of I-70 at the west
end of Idaho Springs, Colorado.

1-4" veins, Fe-, Mn-staining   850'   silicified wallrock,
feldspars altered to clays   800'

EXPLANATION

Tertiary Bostonite porphyry

Mafic units

Granitic units   Precambrian Idaho Springs Formation
migmatitic biotite gneisses

Veins & veinlets
(not to scale)

Joint set orientation
(schematic)   Geology by S. Budge, K. McCarville Weber
1981-82

oped, followed by reactivation of the northwest-trending fault system. Mineralization of these structures was concurrent with reactivation. Some late mineralization occurred after movement, concurrent with the main mineralization (at 35 Ma) in the Georgetown district.

Mineralization in the Idaho Springs–Central City districts is vein dominated, with accompanying alteration that is generally restricted to small halos about individual veins and is not pervasive. Principal metallic vein minerals are pyrite, sphalerite, chalcopyrite, and tennantite. Native gold is sparse, but widespread. Gold tellurides, enargite, pearceite, polybasite, marcasite, wolframite, pitchblende, and coffinite occur locally. The veins are classified according to the dominant ore minerals as pyrite, pyritic-copper, pyritic lead-zinc (composite), lead zinc (galena-sphalerite), and barren quartz-carbonate types. These types are distributed in a crudely concentric zonal pattern about the districts' center between Central City and Idaho Springs (Fig. 1).

Quartz is the dominant gangue mineral; carbonate minerals are locally abundant, and fluorite and barite are sparse. The walls of some veins are gradational, marked by a zone of broken, silicified, and pyritic wallrock. Base-metal minerals form veinlets and lenses that cut quartz and pyrite and may locally cement brecciated quartz-pyrite aggregates.

SITE DESCRIPTION

The I-70 road cut is mainly in Precambrian biotite gneiss, cut by several dikes of bostonite porphyry and mineralized veins. The position of important features is indicated in Figure 2, a sketch of the road cut, beginning at the east end of the outcrop. The site is within the lead-zinc vein zone of the Central City–Idaho Springs districts and contains several galena-sphalerite-pyrite-quartz veins, which are stringers of the Stanley vein system, mined at the Stanley Mine, just south across I-70. The best vein is exposed at 110 ft (34 m), where it is associated with a bostonite porphyry dike.

REFERENCES

Harrison, J. E., and Wells, J. D., 1956, Geology and ore deposits of the Freeland-Lamartine district, Clear Creek County, Colorado: U.S. Geological Survey Bulletin 1032-B, p. 33–127.

Hawley, C. C., and Moore, F. B., 1962, Geology and ore deposits of the Lawson-Dumont–Fall River district, Clear Creek County, Colorado: U.S. Geological Survey Bulletin 1231, 92 p.

Lovering, T. S., and Goddard, E. N., 1950, Geology and ore deposits of the Front Range, Colorado, U.S. Geological Survey Professional Paper 223, 319 p.

Moench, R. H., 1964, Geology of Precambrian rocks, Idaho Springs district, Colorado: U.S. Geological Survey Bulletin 1182-A, 70 p.

Moench, R. H., and Drake, A. A., 1966, Economic geology of the Idaho Springs district, Clear Creek and Gilpin Counties, Colorado: U.S. Geological Survey Bulletin 1208.

Rice, C. M., Harmon, R. S., and Shepard, T. S., 1982, Porphyry molybdenum style mineralization near Central City, Colorado: Geological Society of America Abstracts with Programs, vol. 14, no. 7, p. 598.

Simmons, E. C., and Hedge, C. E., 1979, Minor element and Sr-isotope geochemistry of Tertiary stocks—Colorado Mineral Belt: Contributions to Mineralogy and Petrology, vol. 62, p. 379–396.

Sims, P. K., Drake, A. A., and Tooker, E. W., 1963, Economic geology of the Central City district, Gilpin County, Colorado: U.S. Geological Survey Professional Paper 359, 224 p.

Tweto, O., and Simms, P. K., 1963, Precambrian ancestry of the Colorado Mineral Belt: Geological Society of America Bulletin, vol. 74, no. 8, p. 991–1014.

Paleozoic-Mesozoic section: Red Rocks Park, I-70 road cut, and Rooney Road, Morrison area, Jefferson County, Colorado

Robert J. Weimer, and L. W. Le Roy, Department of Geology, Colorado School of Mines, Golden, Colorado 80401

INTRODUCTION

Several significant sections of Paleozoic and Mesozoic rocks occur in close proximity to one another at the foot of the Front Range west of Denver. The Paleozoic section on the east flank of the central Front Range is well exposed in Red Rocks Park. Excellent exposures of the Mesozoic section are at the I-70 Road Cut, and along Rooney Road, east of the Dakota Hogback on the southeast side of I-70.

Location and access. Red Rocks Park is located northwest of Morrison, Colorado, 10 mi (16 km) west of Denver (Fig. 1). The I-70 Road Cut is located west of Denver, 3 mi (5 km) south of Golden, Colorado, and is the point at which I-70 cuts the Dakota Hogback. Rooney Road is east of the Dakota Hogback on the southeast side of I-70. To reach all three sites, proceed west from Denver about 10 mi (16 km) on I-70. Leave I-70 at the Morrison Exit, which is well marked. A parking lot and footpath adjacent to the exit provide access to the I-70 Road Cut site. To reach the Red Rocks and Rooney Road sites, proceed south from the exit on Colorado 26. About 1 mi (1.6 km) south of I-70, signs at a marked intersection will direct one west to Red Rocks Park, or east to Rooney Road via Colorado 26 (Alameda Parkway). All three localities are on publicly owned lands and can be reached by passenger car. Adequate parking is available at each locality. Generalized lithologies and thicknesses for formations exposed in the area are summarized on Figure 2.

Significance of site. Collectively, the three localities are the best places to observe the Paleozoic and Mesozoic section along the Colorado Front Range. The most important events, recorded in the deposits of this area, are the late Paleozoic orogeny, the encroachment of the Cretaceous Western Interior Seaway, and several phases of the Laramide orogeny. A number of regional unconformities, representative of the cratonic region of North America, are also present.

No type section has been designated for the Laramide orogeny, but the Rooney Road could serve as a reference section, if one were needed.

RED ROCKS PARK

Red Rocks Park takes its name from the spectacular red monuments of the Fountain Formation of Pennsylvanian and Early Permian age. The formation is composed of red arkoses, sandy mudstones, and conglomerates, and is the most conspicuous and oldest sedimentary unit in the park. It forms the beautiful and unique setting for a 10,000-seat amphitheater. The 975-ft (300-m) formation has an outcrop width of approximately 0.6 mi

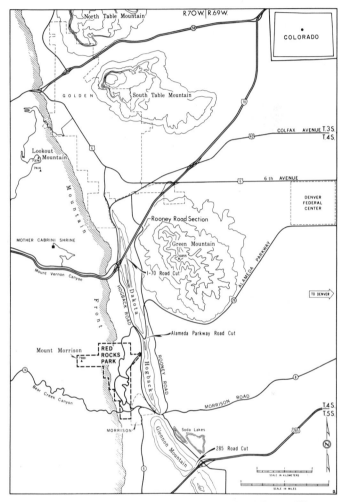

Figure 1. Location map of Red Rocks Park, I-70 Road Cut, and Rooney Road Section (from LeRoy and LeRoy, 1978).

(1 km) and a length of 1.25 mi (2 km). Structural dip is 26°E (Fig. 3). The lower half of the section is resistant to erosion and forms the scenic landscape that was the basis for establishment of the Park by the City and County of Denver.

The Fountain is an excellent example of a vast complex of shallow braided channels deposited on alluvial fans, or in small fault-controlled valleys marginal to and east of the nearby ancestral Front Range Highland. The angular nature of mineral and rock grains and clasts, the immature nature of the sediment, the trough cross-stratification (showing a dominant east transport), and occasional mud-cracked layers and root zones all support this interpretation of depositional setting.

The oldest Fountain strata were deposited on an irregular

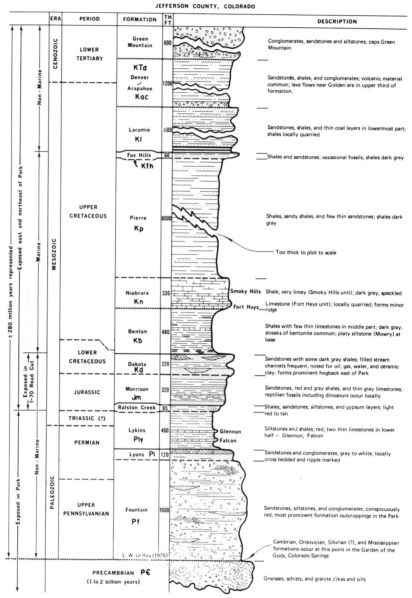

GENERALIZED STRATIGRAPHIC SECTION

Golden-Morrison Area

JEFFERSON COUNTY, COLORADO

Figure 2. Stratigraphic section in the Red Rocks Park—Rooney Road area (from LeRoy and LeRoy, 1978).

weathered surface on top of the Precambrian Idaho Springs Formation (Fig. 2), which can be observed along the access road northwest of the upper parking area at the amphitheater (Figs. 1 and 3). The Idaho Springs Formation represents the crystalline basement, and is the oldest group of rocks within the Park; estimated age is older than 1.7 Ga. The formation is composed of amphibolite and granite gneisses with minor schist layers and is resistant to erosion. The top of the Idaho Springs forms an east-dipping surface from which much of the Fountain Formation was stripped by erosion.

The upper 3 to 5 ft (1 to 1.5 m) of the Idaho Springs (observed at road level) is a weathered zone in which iron-rich silicate minerals have been oxidized to hematite. Upward along the exposed contact, this weathered zone is removed by scour at the base of a channel in the lowermost Fountain. A concentration of cobbles marks the base of this channel.

The contrast in fabric and lithology above and below the unconformity (nonconformity) is the most striking geologic change in the Park. Foliation in the gneiss generally trends east-west and dips 70° to 80°S. Several sets of closely spaced joints are obvious features; in addition, a shear zone 100 ft (30 m) north of the contact controls the position of a small east-west drainage

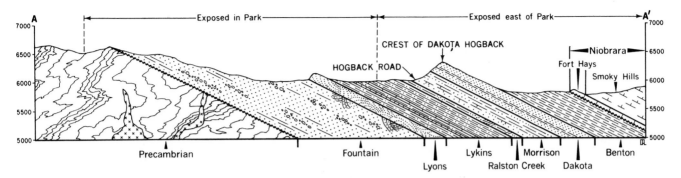

Figure 3. Southwest-northeast geologic cross section, Red Rocks Park (from LeRoy and LeRoy, 1978).

observed on the slope west of the curve in the road. Strata of the overlying Fountain Formation strike N10°W and dip 26°E. Few joints are observed in the sedimentary strata overlying the basement. Because the age of the lowermost Fountain is estimated to be approximately 300 Ma, the missing record of the unconformity represents a hiatus (time gap) of 1.4 b.y.

Two Permian formations, the Lyons and Lykins, are intermittently exposed in the eastern half of the park (Fig. 3). However, the best locality from which to view these formations is the southeast corner of the park, along the north side of Colorado 74, in the town of Morrison (Fig. 1). The overall thickness of the Lyons is 230 ft (70 m) and of the Lykins is 390 ft (120 m). The upper portion of the Lykins may be Triassic in age.

The Lyons can be subdivided into three units in the Park, all well developed at the Morrison outcrop. The lower part is light gray conglomeratic arkose composed of genetic units 3 to 10 ft (1 to 3 m) thick. A complete genetic unit begins with a scour surface overlain by cross-stratified conglomerate that grades upward into fine- to medium-grained feldspathic sandstone, occasionally capped by thin green shale beds at the top. The shale beds may contain simple sand-filled burrows, possibly made by insects. These stacked genetic units are interpreted as narrow channels of a braided stream deposit. The middle unit is fine-grained parallel- to cross-laminated feldspathic sandstone. Several green mud-cracked shales (illite), 0.5 to 2 in (1 to 5 cm) thick, are locally developed in the Morrison section. The upper unit is a mixture of the lithologies of the lower and middle units.

The Lyons is interpreted as deposits of an eastward-flowing braided stream complex on the east side of the ancestral Front Range Highland; overall setting is similar to that of the Fountain. The fine-grained, well-sorted middle Lyons, and an interval that shows the same lithology in the upper unit, is interpreted as eolian deposits blown from the wadis (arroyos) by northerly winds. The middle unit has both dune and interdune deposits: the deposits viewed at the Morrison section are largely interdune, but 0.6 mi (1 km) to the north, cross-stratified fine-grained sandstone characteristic of eolian dunes is dominant in the middle Lyons Formation. The Morrison section is a well known locality for illustrating the intertonguing of deposits of coarse-grained fluvial and fine-grained eolian environments.

The Lykins Formation consists of brick-red, thinly-bedded siltstones and shales. Two crinkly laminated carbonate layers, the Glennon and Falcon members (Fig. 2), are present in the lower 100 ft (30 m) of the Lykins. At Morrison, the Falcon Member is 60 ft (18 m) above the base of the Lykins; the Glennon Member is 110 ft (33 m) above the base. Three thin (1.5 in; 4 cm thick) carbonates are locally present in the lower 10 ft (3 m) of the Lykins at Morrison. The Lykins is interpreted as a nonmarine aqueous deposit, either related to widespread supersaline restricted marine, or to saline freshwater lake conditions. The absence of fossils prevents accurate reconstruction of the environments.

I-70 ROAD CUT

Dakota Group sandstones (Early Cretaceous) form a conspicuous hogback east of the Park (Figs. 1 and 3). West of the hogback, the nonmarine Morrison, Ralston Creek, and upper Lykins formations form a valley cut by Ralston Creek; east of the hogback marine shales of the Benton, Niobrara, and Pierre formations are eroded as a valley. Thicknesses and lithologies of these units are shown in Figure 2. The best exposures are along Alameda Parkway and at the I-70 Road Cut, 1 mi (1.6 km) to the north. The I-70 Road Cut is designated as a Point of Geologic Interest along the U.S. Interstate Highway System (Fig. 1). Parking areas and walkways have been provided for ease in viewing the stratigraphic sections on both the north and south sides of the cut. Signs with illustrations and descriptions have been placed along the exposures to guide visitors through 650 ft (200 m) of the exposed Jurassic and Lower Cretaceous section. Descriptions and interpretations of this classic section are provided by LeRoy and Weimer (1971) and Weimer and Land (1972). In addition, guides to the textures and structures of the Dakota Group at the Alameda Parkway Road Cut have been published by MacKenzie (1968) and Chamberlain (1976).

ROONEY ROAD

Exposures along Rooney Road are an important reference section for the uppermost Cretaceous strata in the Golden-Morrison area (Figs. 4, 5, and 6; Weimer, 1976; Weiner and Tillman, 1980). Approximately 1,600 ft (500 m) of strata are

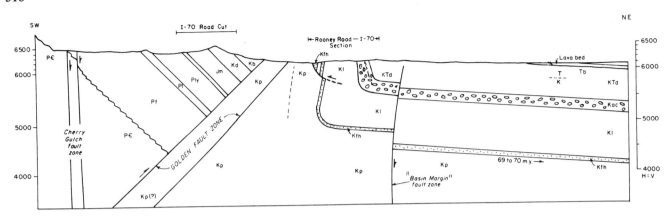

Figure 4. Structure cross section of east flank of Front Range uplift along south side of I-70 Highway. Horizontal scale = vertical scale (from Weimer, 1976).

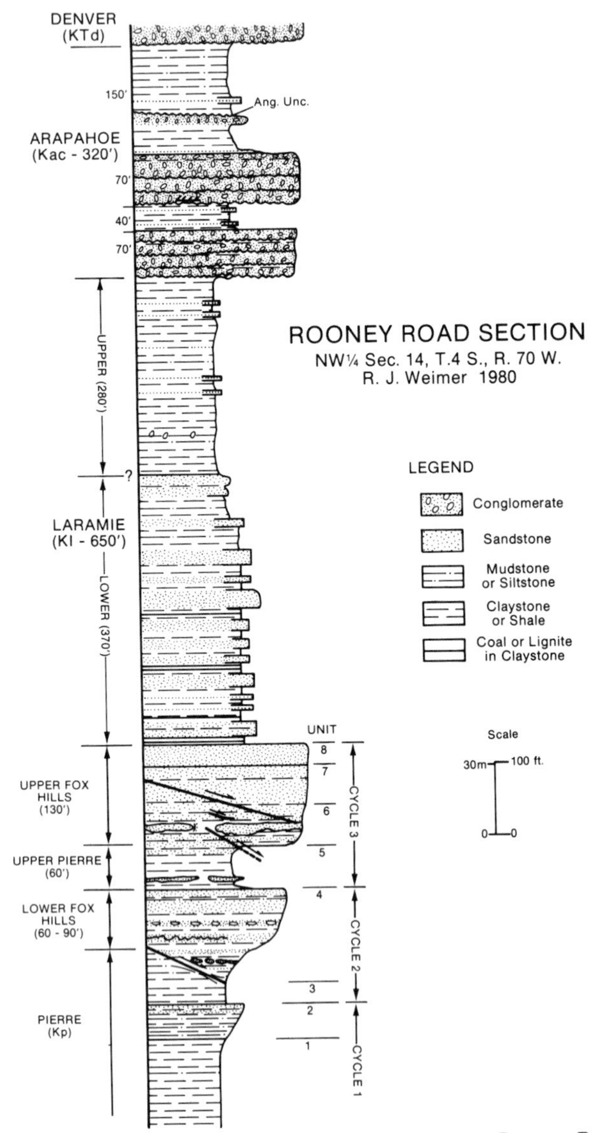

FORMATION   LITHOLOGY

ROONEY ROAD SECTION
NW¼ Sec. 14, T.4 S., R. 70 W.
R. J. Weimer 1980

LEGEND

	Conglomerate
	Sandstone
	Mudstone or Siltstone
	Claystone or Shale
	Coal or Lignite in Claystone

Scale

30m — 100 ft.

0 — 0

Figure 5. Lithologies and thicknesses of formations at Rooney Road section (from Weimer and Tillman, 1980). Formational symbols shown on Figure 2.

exposed, from the upper Pierre Shale on the west to the lowermost beds of the Denver Formation to the east. At the west end of the section the strata are overturned and dip 80°W; at the east end, strata dip 60°E (toward the Denver Basin). The overall structure on the east flank of the Front Range is shown by the structure section through the I-70 Road Cut and the Rooney Road section (Fig. 4).

The formations exposed along Rooney Road record the final regression of the Cretaceous sea from eastern Colorado and also indicate the early structural movements of the Laramide orogeny. Four phases of the Laramide orogeny are recorded: (1) uplift of the Front Range with erosion of the sedimentary cover and deposition of upper Pierre, Fox Hills, Laramie, and lower Arapahoe formations; (2) movement to develop an angular unconformity within the Arapahoe Formation with a 20° dip discordance; (3) resumption of sedimentation in the upper Arapahoe Formation, with a volcanic event in the Front Range source area recorded in the Denver Formation; and (4) deformation of these strata to steep normal or overturned dip.

Stratigraphic details of the vertical sequence, starting in the upper Pierre and extending through the Arapahoe, are illustrated in Figures 5 and 6. The sequence is interpreted as having been deposited by processes in a river-dominated delta in an unstable tectonic setting. Specific environments are prodelta (Pierre), delta front (Fox Hills), delta plain (Laramie), and alluvial channels (Arapahoe).

Fissile laminated shales and siltstones represent the main lithologies of the Pierre; the Fox Hills is made up of tan and white fine-grained sandstones. Deformational structures and growth faults suggest oversteepened depositional slopes in both formations, a result of high rates of sedimentation. Three progradational deltaic cycles can be identified in the Pierre and Fox Hills (Figs. 5 and 6).

The Laramie Formation is alternating sandstone, kaolinitic claystone, and siltstone with minor coal or carbonaceous clay (Fig. 5). The lower Laramie Formation is mined for kaolinitic clay used in the manufacture of bricks and ceramics. The upper Laramie is not mined for clay, because kaolinite content de-

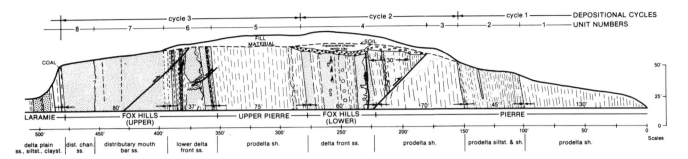

Figure 6. Sketch of lower part of Rooney Road section where upper Pierre, Fox Hills, and Laramie formations are exposed (from Weimer, 1976).

creases upwards, while content of montmorillonite and illite increases. The Laramie is interpreted as fresh-water delta plain deposits, dominated by crevasse splay sandstones, laminated lacustrine clays, oxidized swamp deposits, light gray claystones of well-drained swamps, and coals or carbonaceous claystones of poorly drained swamps. The Laramie Formation thins from 650 ft (200 m) at Rooney Road to 375 ft (115 m) at U.S. 40, 1 mi (1.6 km) to the north, with thinning occurring in both the upper and lower Laramie.

Conglomerates in the Arapahoe Formation are well developed along Rooney Road. In the lower Arapahoe Formation, two units of conglomeratic sandstone, each about 80 ft (25 m) thick, are separated by 40 ft (12 m) of mudstone that contains minor pebbly lenses (Fig. 5). The conglomerate and sandstone units are cross-stratified and contain numerous lenticular graded sub-units, each 3 to 6 ft (1 to 2 m) thick. Conglomerate at the base of each unit grades vertically to medium- to coarse-grained sandstone, in places capped by gray claystone. The Arapahoe Formation is the principal water-bearing unit in the Greater Denver area to the east.

The composition and texture of the two 80-ft (25-m) conglomerate units are different (Fig. 5). The lower unit contains 1 to 2 in (2.5 to 5 cm) chert and quartzite pebbles, with a lesser quantity of igneous and metamorphic pebbles. In contrast, the upper unit contains much more cobble-sized material, which is dominantly igneous and metamorphic in composition. The upper 115 ft (35 m) of the Arapahoe Formation consists of dark gray claystone or mudstone and thin layers of siltstone, with minor thin coarse-grained to conglomeratic sandstone. An angular unconformity with a dip discordance of 20° occurs in the upper Arapahoe (Fig. 5). The Arapahoe is a fluvial deposit dominated by channels in the lower half and floodplain deposits in the upper half.

The dozen feet (few meters) of the Denver Formation that are exposed at the east end of the Rooney Road cut, consist of greenish-brown, fine- to coarse-grained, hornblende and pyroxene andesite-rich, clayey sandstone. Local lenses of conglomeratic sandstone contain pebbles as much as 0.5 in (1 cm) in diameter. Dominant pebble lithology is andesite porphyry. Dark brown and yellow-brown sandy mudstone layers are intercalated with

the sandstones. The volcanic-rich sandstones are distinctly different from the arkosic sandstones and conglomerates and the lighter-colored mudstones of the underlying Arapahoe Formation. This abrupt lithologic change is the result of the introduction of large quantities of volcanic material into the drainage basin in the adjacent Front Range area, which was the source of the sediments in the Denver Formation. The Denver is approximately 1,000 ft (300 m) thick and is interpreted as fluvial channel and floodplain deposits.

The conspicuous lava flows that cap South and North Table mountains (Fig. 1) are present within the Denver Formation. A famous fossil locality that establishes the Cretaceous-Tertiary boundary occurs below the capping lava bed at the east end of South Table Mountain (Brown, 1943).

The youngest Tertiary formation in the area is the Green Mountain Conglomerate, which caps the high mesa (Green Mountain) east of Rooney Road (Figs. 1 and 2).

REFERENCES CITED

Brown, R. W., 1943, Cretaceous-Tertiary boundary in the Denver basin, Colorado: Geological Society of America Bulletin, v. 54, p. 65–86.

Chamberlain, C. K., 1976, Field guide to the trace fossils of the Cretaceous Dakota Hogback along Alameda Avenue, west of Denver, Colorado, *in* Epis, R. C., and Weimer, R. J., eds., Studies in Colorado field geology: Colorado School of Mines Professional Contributions 8, p. 242–250.

LeRoy, L. W., and LeRoy, D. A., 1978, Red Rocks Park: Colorado School of Mines Press, Golden, 29 p.

LeRoy, L. W., and Weimer, R. J., 1971, Geology of the Interstate 70 Road Cut, Jefferson County, Colorado: Colorado School of Mines Professional Contributions 7.

MacKenzie, D. B., 1968, Studies for students; Sedimentary features of Alameda cut, Denver, Colorado: The Mountain Geologist, v. 5, no. 6, p. 3–13.

Weimer, R. J., 1976, Cretaceous stratigraphy, tectonics and energy resources, western Denver basin, *in* Epis, R. C., and Weimer, R. J., eds., Studies in Colorado field geology: Colorado School of Mines Professional Contributions 8, p. 186–223.

Weimer, R. J., and Land, C. B., 1972, Field guide to Dakota Group (Cretaceous) stratigraphy, Golden-Morrison area: The Mountain Geologist, v. 9, nos. 2–3, p. 240–267.

Weimer, R. J., and Tillman, R. W., 1980, Tectonic influence on deltaic shoreline facies, Fox Hills Sandstone, west-central Denver basin: Colorado School of Mines Professional Contributions 8, p. 123–138.

The Black Canyon of the Gunnison, Colorado

Wallace R. Hansen, U.S. Geological Survey, Denver Federal Center, P.O. Box 25046, Denver, Colorado 80225

LOCATION AND ACCESS

The Black Canyon in southwestern Colorado is easily reached by passenger car from U.S. 50 (Fig. 1). Turn north 8 mi (13 km) east of Montrose, Colorado, to reach Black Canyon of the Gunnison National Monument in the most impressive part of the canyon. The National Monument boundary is 5 mi (8 km) from the turnoff, and the approach is very scenic. A key visitation point in the monument is Chasm View overlook. Most visitors will want to see several other points also. Some parts of the canyon are accessible only by 4-wheel-drive trails or footpaths.

SIGNIFICANCE

Aside from its stunning physiography, this world-class gorge is an excellent example of superimposed drainage in a youthful river valley deepened by rejuvenation, and its geologic history is well documented by structural and stratigraphic evidence. Although the drainage pattern was set by superposition across middle Tertiary volcanic rocks, the canyon itself is a product of continued downcutting through the crystalline basement, in response to a gradual rise of the Gunnison uplift in late Tertiary and Quaternary time.

SITE INFORMATION

At the turnoff from U.S. 50 in the valley of Cedar Creek, the road is on Mancos Shale (Upper Cretaceous) near the axis of the Montrose syncline. This broad trough separates the Gunnison uplift on the northeast from the Uncompahgre uplift on the southwest. The Gunnison uplift is the setting for the Black Canyon, and as the road climbs the steep 1,900 ft (580 m) to the rim of the canyon, some idea of the rather anomalous course of the river across the heights of the uplift is conveyed. The course is an accident of superposition; the uplift has been eroded into bold relief by the partial removal of the softer sedimentary rocks that once topped, and still flank, the hard crystalline core. Further erosion sought out large joints and faults.

About 1½ mi (2½ km) north of the turnoff, the road leaves the drainage of Cedar Creek and climbs onto the broad alluvial surface of Bostwick Park. This alluvium contains well-washed, rounded gravel derived from the volcanic rocks of the San Juan Mountains to the south and deposited by ancestral Bostwick Creek, which once flowed north into the Black Canyon. Ancestral Bostwick Creek was beheaded by Cedar Creek in late Pleistocene time, and the upper part of the fill in Bostwick Park, therefore, is all locally derived. Another remnant of the same old valley, Shinn Park, south of Cedar Creek, was beheaded farther upstream by the Uncompahgre River more than 600,000 years ago. After Shinn Park was beheaded, ancestral Bostwick Creek

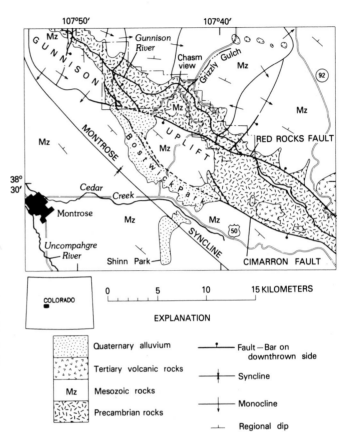

Figure 1. General structure and location map of the Black Canyon area. Dashed and dotted line is boundary of Black Canyon of the Gunnison National Monument.

no longer contributed significant drainage to the Black Canyon, but locally-derived sediment continued to accumulate in both parks up to the present time. These sediments are important in helping establish the time and rate of canyon cutting.

At the north margin of Bostwick Park, the road to the monument climbs onto the Burro Canyon Formation (Lower Cretaceous), which is faulted up against the Mancos Shale along the Cimarron fault (here, the boundary of the Gunnison uplift). Another large fault a bit farther up the road, the Red Rocks fault, separates the crystalline core of the Gunnison uplift on the north from the flanking younger sedimentary rocks on the south. The traces of the Cimarron and Red Rocks faults overlap, and as the Cimarron fault dies out, the Red Rocks fault takes up the displacement. Conversely, the Cimarron fault extends much farther southeastward.

Rock Formations. Precambrian crystalline rocks dominate the terrane in the national monument, but rocks of many ages and origins crop out along the length of the canyon (Table 1; Hansen, 1971, 1981; Hansen and Peterman, 1968; several additional ref-

TABLE 1. ROCK FORMATIONS OF THE BLACK CANYON AREA

Subdivision	Formation	Thickness (m)	Age (millions of years)[1]
Quaternary System	Alluvium, talus, landslides	≥25	≥1
Tertiary System	Hinsdale Formation	<45	18.5
	Carpenter Ridge Tuff	<70	27.5
	Fish Canyon Tuff	<90	27.8
	Sapinero Mesa Tuff	<70	28
	Dillon Mesa Tuff	<25	>28
	Blue Mesa Tuff	<80	>28
	West Elk Breccia	≥300	≥32
Cretaceous System	Mancos Shale	>680	~80
	Dakota Sandstone	<30	~105
	Burro Canyon Formation	<35	~110
Jurrassic System	Morrison Formation		~145
	Brushy Basin Member	<120	
	Salt Wash Member	35-55	
	Wanakah Formation	<80	~155
	Junction Creek Sandstone member	<30	
	Pony Express Limestone member	<2	
	Entrada Sandstone	<30	~170
Ordovician or Cambrian System	Diabase	<90	510±60
Precambrian (Proterozoic Eon)	Pegmatite	<100's	≥1700
	Curecanti Quartz Monzonite		1420±15
	Vernal Mesa Quartz Monzonite		1480±40
	Pitts Meadow Granodiorite		1730±190
	Metamorphic Rocks	>1000's	>1700

[1]Tertiary ages from Lipman and others, 1978.

```
<   up to
>   more than
≥   equal to or greater than
~   about
```

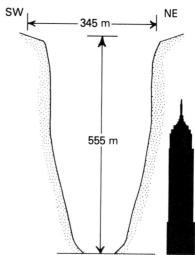

Figure 2. Profile of Black Canyon of the Gunnison at Chasm View overlook, scaled in meters. Silhouette of Empire State Building is at same scale.

erences appear in the papers cited.) Metamorphosed arenites, graywackes, and pelites, mostly of almandine-amphibolite grade, are intruded by countless deep-seated dikes, sills, and varied small plutons. Pegmatite is the most conspicuous intrusive rock, and its distribution is pervasive. In aggregate, its total volume is very large. Individual pegmatite bodies range widely in shape and size, from small, irregular vein-like networks to straight-walled concordant intrusions thousands of meters long and hundreds of meters wide; some sheets form buttress-like erosional fins in the canyon walls.

Some of the more awesome reaches of the gorge are carved from granites. At Chasm View overlook, in the most rugged part of the canyon, the walls are Vernal Mesa Quartz Monzonite. Here the canyon is more than half again as deep as it is wide, and its full dimensions are difficult for the viewer to grasp (Fig. 2). Blocks of rock on the canyon floor, looking like stepping stones from the rim, are the size of railroad freight cars. The resistance of the Vernal Mesa pluton to erosion is largely responsible for the precipitous canyon walls at Chasm View and for the steep gradient of the river below. At this point the river drops about 370 ft/mi (70 m/km). The average fall through the monument is about 94 ft/mi (18 m/km). Through the 50 mi (80 km) length of the canyon, the average fall is 42 ft/mi (8 m/km). By way of comparison, the Green River in Dinosaur National Monument—a comparable-size stream in a comparable-size canyon—averages a drop of about 12 ft/mi (2.25 m/km), and its maximum drop is near the average of that of the Gunnison through the Black Canyon.

Vernal Mesa Quartz Monzonite forms several separate intrusive bodies in the Black Canyon area, the largest of which is the pluton at Chasm View. This pluton has the form of a large, semiconcordant phacolith, moderately well foliated by protoclastic flowage. All along its outcrop, conspicuous elongate inclusions of wall rock and large phenocrysts of potassium feldspar are oriented parallel to the walls of the intrusion.

Ancestral Uncompahgre Highland. Throughout the Black Canyon, a conspicuous unconformity separates the Precambrian core of the Gunnison uplift from the overlying younger rocks. The unconformity is all the more obvious because of the marked differential resistance of the rocks above and below. Where exposures are good, moreover, as in the lower reaches of the Black Canyon, the brightly colored Mesozoic rocks contrast with the more somber crystalline rocks below. This unconformity marks the bevelled root of the ancestral Uncompahgre highland, a part of the ancestral Rocky Mountains that arose in Pennsylvanian time in western Colorado and adjacent parts of Utah and New Mexico. The ancestral Uncompahgre highland was gradually reduced to a peneplain in the latter part of the Paleozoic Era and the first half of the Mesozoic. In cross section the peneplain had the shape of a broad low arch, highest near the town of Gunnison. In the Black Canyon area the flank of the old arch sloped west about 4 ft/mi (0.75 m/km), and successively younger rocks thus overlap the unconformity toward the east. By Late Jurassic time, the last vestiges were finally buried beneath the Morrison Formation.

At Chasm View overlook and elsewhere along the south rim, the Mesozoic rocks have been stripped off the old Uncompahgran surface, but across the canyon to the north, the Mesozoic section is intact, though poorly exposed, and a thin band of pink or gray Entrada Sandstone tops the inner canyon wall. The Entrada is 65 ft (20 m) thick or more on the west flank of the Gunnison uplift, but it thins gradually to zero near the east boundary of the national monument. The entire section, from

Entrada to Dakota Sandstone, is preserved in the upper wall north of Chasm View overlook but is mostly concealed by brush and colluvium. Far downstream, in the lower reach of the canyon, the Mesozoic section is totally exposed, and the unconformity can be studied close at hand, complete with a well-preserved fossil regolith at the Entrada-Precambrian contact. Structure contouring on the unconformity (Hansen, 1971) shows that the Gunnison uplift is basically a tilted, upfaulted block, rather than a domical uplift, although it is gently flexed across its crest and is warped locally by monoclinal drag along faults.

Laramide Uplift and Subsequent Bevelling. The Gunnison uplift is basically a Laramide structure, raised along old fault-lines that certainly date back to the rise of the ancestral Rocky Mountains and probably to the Precambrian. Thus, a conspicuous WNW fracture trend, widespread in the eastern Colorado Plateau and marked by the Cimarron and Red Rocks faults, locally contains 510 ± 60 Ma diabase dikes (Hansen and Peterman, 1968). The Red Rocks fault had a pre-Laramide component of left-lateral strike slip that offset the Vernal Mesa pluton and several large diabase dikes about 3.4 mi (5½ km). Laramide displacements are largely high-angle readjustments along such preexisting fractures. The Gunnison uplift took form when renewed movements along these old faults raised the uplift above its surroundings. Erosion then reduced the heights to a nearly flat plain, and streams draining west from the newly risen Sawatch Range 75 mi (120 km) to the east flowed unhindered across the truncated crest of the Gunnison uplift. This second regionwide erosion surface is the well-known late Eocene surface of Central Colorado (Epis and Chapin, 1975), preserved in the Black Canyon area as the unconformity beneath the Oligocene volcanic sequence. Its development was an essential early step in the later cutting of the Black Canyon—the Gunnison uplift had to be truncated before drainage could be superimposed across it.

Tertiary Volcanism. From the crest of the Gunnison uplift just inside the monument boundary, remnants of the Tertiary volcanic sequence are visible several kilometers to the east, still capping the uplift on the south flank of the West Elk Mountains. The whole upstream third of the canyon retains this cap of Tertiary volcanic rocks, and travelers headed east along U.S. 50 will see the thick sections of breccia and welded tuff in the upper canyon walls.

Volcanic eruptions began in the nearby West Elk and San Juan Mountains in middle Tertiary (Oligocene) time, burying the bevelled core of the Gunnison uplift under successive blankets of volcanic debris. Eruptions of intermediate lavas and volcanic breccias (West Elk Breccia and San Juan Formation) from both volcanic centers were succeeded in late Oligocene time by more silicic pyroclastic eruptions of welded tuff from the San Juan Mountains and, finally, by localized outpourings of Miocene mafic alkalic lava—the Hinsdale Formation (Olson and others, 1968; Lipman and others, 1978). As the volcanic piles grew, the west-flowing drainage through the area became channeled between the two volcanic centers. Drainage was curtailed from time to time by volcanic outbursts, but between eruptions it left sheets

of gravel carried in from the Sawatch terrane to the east. Some of these gravels are well exposed along Colorado 92 on the northeast side of the Black Canyon. They contain scattered, well rounded granitic cobbles and other crystalline rocks that increase proportionally upward in the section. Volcanism persisted in the San Juan Mountains long after it had ended in the West Elks, as great sheets of pyroclastic tuff poured north from the San Juans across the Gunnison uplift and onto the flanks of the West Elks. Concomitantly, synclinal warping began along the axis of the present Black Canyon, probably caused by unloading and doming of the West Elk Mountains (Fig. 3). Consequently, each ash flow acquired a synclinal structure and, because of ponding, each ash flow thickened over the Black Canyon (Hansen, 1981). Warping continued until after the eruption of the Hinsdale Formation.

Erosion of the Black Canyon. The stage was now set for the cutting of the Black Canyon. Volcanic activity affecting the area ended with the eruption of the Hinsdale Formation, and the Gunnison River began to cut its canyon. When its course became fixed, the river was positioned along the axis of the late Tertiary syncline on a thick fill of volcanic rock and gravel. Fortuitously, its course also crossed the buried Gunnison uplift, so that when the uplift was finally again breached by erosion, the river could only erode downward into the Precambrian core. Had its course been a few kilometers to the south, beyond the bordering faults of the Gunnison uplift, the river would have met only soft sedimentary rocks and today, instead of the Black Canyon, there would be just another broad valley, cut in Mancos Shale.

Formation of the syncline was critical to the ultimate cutting of the Black Canyon. Because volcanic activity in the West Elk Mountains had largely ended by the onset of the ash-flow eruptions in the San Juan Mountains, erosion was unloading the West Elks all the while that the ash flows were spreading north across the Black Canyon area. Updoming of the West Elks, probably caused by unloading, reversed the dips in the distal parts of the ash flows, one after another, and produced the developing syncline along the Black Canyon axis. Without the syncline, drainage would have migrated down the depositional slope of the ash flow sheets, locating the Gunnison River far to the north and off of the Gunnison uplift, and there would be no Black Canyon.

Rate of Cutting. For a long time after volcanic activity ended, the Gunnison River must have flowed through a broad valley eroded into volcanic rocks. As early as 1.2 Ma, however, the crystalline gorge of the Black Canyon in the western part of the National Monument was already about 1,100 ft (350 m) deep, judging from the occurrence of Mesa Falls ash (Izett and Wilcox, 1982) in the alluvial fill of Bostwick Park just south of the Black Canyon. This ash occurrence is about 1,100 ft (350 m) lower than the canyon rim, and inasmuch as Bostwick Park is tributary to the canyon, the floor of the canyon must have been at least that deep. The river has since cut about 1,100 ft (350 m) more to its present grade, at an average rate of roughly 1 ft (0.3 m) per thousand years.

Bostwick Park and a few other localities downstream along

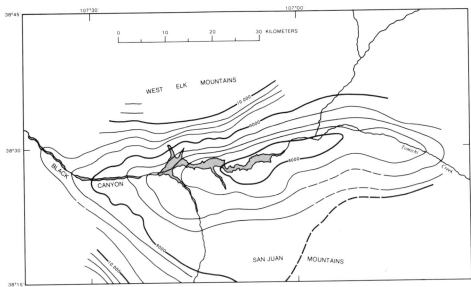

Figure 3. Syncline depicted by structure contours drawn at the base of the ash-flow tuff sequence. Dashed contours cross area of probable nondeposition. Blue Mesa reservoir at center. Contour interval 250 feet (about 78 m). From Hansen (1981), courtesy of New Mexico Geological Society.

the Black Canyon also contain the 0.62 Ma Lava Creek B ash (Izett and Wilcox, 1982). The ash deposits downstream are about 1,400 ft (425 m) above modern drainage; their position suggests an increased rate of downcutting—about 2.25 ft/thousand years (0.7 m/thousand years). Only part of the downcutting in that reach of the canyon, however, was through crystalline rock, and entrenchment through the less resistant sedimentary cap then mantling that part of the canyon may accordingly have been more rapid. These rocks still form the outer rims of the canyon downstream from the monument. We can safely assume that while the river was deeply entrenched in hard crystalline rock in the national monument 600,000 years ago, it was still flowing through a broad valley in softer Mesozoic rocks farther downstream.

At the head of the Black Canyon the entrenchment rate was much lower. The Lava Creek B ash (Izett and Wilcox, 1982) is only about 200 ft (60 m) above valley bottom near the mouth of the Lake Fork of the Gunnison and near the right (north) abutment of Blue Mesa dam, so the river there has cut only 200 ft (60 m) in 0.62 m.y. Structurally, the Gunnison uplift has a distinct northeasterly tilt, shown by the southwestward rise of the Tertiary

volcanic rocks and the subjacent Uncompahgran and late Eocene unconformities. The position of the Lava Creek B ash, combined with evidence of tilt, suggests that a continued rise of the Gunnison uplift to the southwest, lasting into Quaternary time, constrained downcutting at the head of the canyon. A lack of Quaternary terracing farther upstream near the town of Gunnison supports this view (Hansen, 1981). On the north side of the Black Canyon east of Grizzly Gulch, moreover, terrace deposits that are graded to Grizzly Gulch and contain cobbles derived from the West Elk Mountains now rise west toward Grizzly Gulch, away from their source (Hansen, 1971). Grizzly Gulch now is a headless hanging valley, beheaded by north-flowing drainage between the Gunnison uplift and the West Elk Mountains. The floor of Grizzly Gulch hangs 1,800 ft (550 m) above the Gunnison River.

Many visitors to the Black Canyon of the Gunnison wonder aloud if the river is still scouring its bed and deepening its canyon. The unleashed fury of the river at flood stage, and its ungraded long profile, both suggest that it is. On each of my many descents to the canyon floor, moreover, I have found the climb out longer, steeper, and more arduous than the time before.

REFERENCES

Epis, R. C., and Chapin, C. E., 1975, Geomorphic and tectonic implications of the post-Laramide, late Eocene erosion surface in the Southern Rocky Mountains, in Curtis, B., ed., Cenozoic history of the Southern Rocky Mountains: Geological Society of America Memoir 144, p. 45–74.

Hansen, W. R., 1971, Geologic map of the Black Canyon of the Gunnison River and vicinity, western Colorado: U.S. Geological Survey Miscellaneous Geologic Investigations Map I-584, scale 1:31,680.

——1981, Geologic and physiographic highlights of the Black Canyon of the Gunnison River and vicinity, Colorado: New Mexico Geological Society Guidebook, 32nd Field Conference, Western Slope, Colorado, p. 145–154.

Hansen, W. R., and Peterman, Z. E., 1968, Basement-rock geochronology of the

Black Canyon of the Gunnison, Colorado: U.S. Geological Survey Professional Paper 600-C, p. C80–C90.

Izett, G. A., and Wilcox, R. E., 1982, Map showing localities and inferred distributions of the Huckleberry Ridge, Mesa Falls, and Lava Creek ash beds (Pearlette family ash beds) of Pliocene and Pleistocene age in the Western United States and Southern Canada: U.S. Geological Survey Miscellaneous Investigations Series Map I-1325, scale 1:4,000,000.

Lipman, P. W., Doe, B. R., Hedge, C. E., and Steven, T. A., 1978, Petrologic evolution of the San Juan volcanic field, southwestern Colorado: Pb and Sr isotope evidence: Geological Society of America Bulletin, v. 89, p. 59–82.

Olson, J. C., Hedlund, D. C., and Hansen, W. R., 1968, Tertiary volcanic stratigraphy in the Powderhorn-Black Canyon region, Gunnison and Montrose Counties, Colorado: U.S. Geological Survey Bulletin 1251-C, p. C1–C29.

The carbonatite complex at Iron Hill, Powderhorn district, Gunnison County, Colorado

Theodore J. Armbrustmacher U.S. Geological Survey, Denver Federal Center, Denver, Colorado 80225
Spencer S. Shannon, Jr., * Los Alamos National Laboratory, Mail Stop D462, Los Alamos, New Mexico 87545

LOCATION

The carbonatite complex at Iron Hill is located southeast of the town of Powderhorn along Cebolla Creek, Deldorado Creek, and Beaver Creek in Gunnison County, Colorado. To reach the complex, travel 9 mi (14.5 km) west of Gunnison on U.S. 50 to the intersection with Colorado 149, turn south on Colorado 149 and travel 15 mi (24 km) to the intersection with Cebolla Creek road; turn southeast on Cebolla Creek road; after about 2.5 mi (4 km), the outcrops on the left are in the carbonatite stock of the

*Present address: Department of Geology, Adams State College, Alamosa, Colorado 81102

complex. Roads to the east off Cebolla Creek road along Deldorado Creek and Beaver Creek connect with a variety of dirt ranch roads that provide excellent access to the rocks of the complex. These roads are periodically maintained by Humphreys Mineral Industries, Inc., of Denver, Colorado, the property managers. Therefore, most of the complex is easily accessible by short hikes from roads navigable by most vehicles. A number of patented mining claims exist in the area, but permission to access the property has never been denied. The geologic map of the complex (Fig. 1) is located within the Rudolph Hill Quadrangle (Olson, 1974) and within the Powderhorn Quadrangle (Hedlund and Olson, 1975). Access roads are also shown on the Rudolph

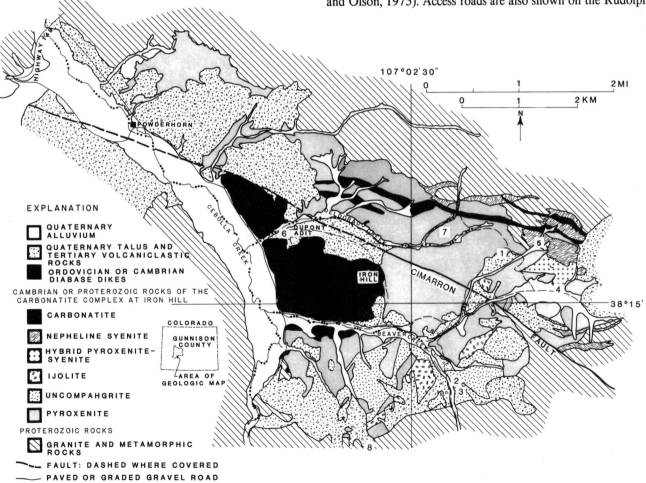

Figure 1. Geology of the carbonatite complex at Iron Hill, Colorado. Modified from Hedlund and Olson (1975) and Olson (1974).

Hill and Powderhorn 7.5-minute quadrangles (1:24,000). Some of the better localities at which to see specific rock exposures are shown on Figure 1.

SIGNIFICANCE

The carbonatite complex at Iron Hill (Fig. 1) is the best example of the carbonatite-ijolite type of subvolcanic alkaline intrusive complex in the United States. Nearly all the principal rock types developed in this association, namely carbonatite, ijolite, pyroxenite, and fenite, are well represented at Iron Hill. This complex is also the type locality of the rock uncompahgrite. Two different structural levels of the complex are exposed because the complex is offset along the Cimarron fault, which bisects it (Fig. 1). Deposits of thorium, uranium, iron, titanium, niobium, rare-earth elements, vermiculite, and manganese are spatially and genetically associated with rocks of the complex.

GEOLOGIC DESCRIPTION

The carbonatite complex of Iron Hill has been studied since 1912 (Singlewald, 1912), not only because of the uncommon lithologies but also because of the potentially important mineral deposits associated with them. Various hypotheses regarding the origin of the Iron Hill rocks have paralleled the development of ideas concerning carbonatite petrogenesis. The complex was originally described by Larsen (1942), who thought that the carbonatite was formed by assimilation of marble followed by crystal differentiation. Temple and Grogan (1965) interpreted their data to suggest that the carbonatite formed by metasomatism of preexisting pyroxenite. Nash (1972) favored formation of the carbonatite through immiscibility of magmatic liquids. Major- and minor-element analyses and Rb/Sr and Sm/Nd isotope systematics of rocks of the complex show that the carbonatite, ijolite, uncompahgrite, and pyroxenite may be comagmatic and that the rocks have been derived from alkaline magmas generated in the upper mantle (Armbrustmacher and Futa, 1985).

Prior to about 570 Ma, rocks of the complex were emplaced into Early Proterozoic Powderhorn Granite and older Proterozoic metamorphic rocks that are locally fenitized (alkali metasomatized) adjacent to the complex. Parts of the complex are covered by ash-flow tuffs, welded tuffs of Oligocene age, and colluvium and alluvium of Quaternary age. The complex consists chiefly of pyroxenite, magnetite-ilmenite-perovskite segregations, uncompahgrite, ijolite, hybrid pyroxenite-syenite rocks, nepheline syenite, and carbonatite, listed oldest to youngest. Diabase dikes of Cambrian or Ordovician age intrude rocks of the complex along the north side of the complex. Carbonatite dikes intrude all rocks of the complex except the carbonatite stock; they also intrude the Proterozoic country rocks, especially those within the fenitized aureole. Veins containing quartz, thorite, and potassic feldspar—interpreted to be genetically related to the episode of intrusion—intrude Proterozoic country rocks. All the uncompahgrite and the carbonatite stock and most of the ijolite are found on the southwest side of the Cimarron fault; the magnetite-ilmenite-perovskite segregations, the nepheline syenite, and most of the diabase dikes are found on the northeast side of the fault.

Pyroxenite, which is well exposed at locality 1 on Figure 1, is highly variable in chemical and mineralogical composition. This medium- to coarse-grained, locally pegmatitic rock contains 55–90 percent clinopyroxene, 10–15 percent magnetite and ilmenite, 5–10 percent melanite garnet, up to 5 percent fluorapatite, and 0–10 percent biotite and phlogopite. Accessory minerals are chiefly sphene, sodic amphibole, calcite, perovskite, leucoxene, pyrite, chalcopyrite, and pyrrhotite. Calcic plagioclase is absent. Vermiculite and magnetite-ilmenite-perovskite segregations are locally abundant. Much of the outcrop area of pyroxenite is altered and forms a slightly topographically low ring around the carbonatite core.

Magnetite-ilmenite-perovskite segregations, also well exposed at locality 1, consist chiefly of magnetite, ilmenite, and perovskite with minor amounts of apatite and biotite. The dark steely gray segregations become gunmetal blue in samples where perovskite becomes the dominant mineral. The segregations are commonly less than 3 ft (1 m) thick but may attain a thickness of 165 ft (50 m). The perovskite content, which may reach major proportions, makes the segregations attractive titanium-prospecting targets. Seven samples of these rocks averaged about 60 ppm uranium and 300 ppm thorium, and the anomalous radioactivity registered by them is a useful characteristic in their exploration. Outcrops of magnetite-ilmenite-perovskite segregations are too small to be shown at the scale of the geologic map (Fig. 1).

Uncompahgrite is a light-gray, medium-grained to very coarse-grained rock composed of melilite, apatite, phlogopite, melanite garnet, and perovskite. Good outcrops are found at locality 2 (Fig. 1). Late alteration of the uncompahgrite has produced a suite of minerals including vesuvianite and other unusual replacement minerals.

Ijolite is variable in grain size, usually medium grained; it has a hypidiomorphic-granular texture (Hedlund and Olson, 1975). This rock contains 30–50 percent nepheline, 30–40 percent sodic clinopyroxene, 10–30 percent melanite garnet, and minor amounts of orthoclase, magnetite, apatite, biotite, sphene, and alteration products of nepheline. Because of the high percentage of mafic minerals, the rocks tend to be fairly dark colored. Good outcrops occur at locality 3 (Fig. 1).

Hybrid pyroxenite-syenite is a bimodal rock consisting of brecciated pyroxenite with numerous, closely spaced, irregular fracture fillings and small dikes of nepheline syenite. The pyroxenite contains clinopyroxene, melanite garnet, sphene, apatite, and melilite; the syenite contains orthoclase, microperthite, sodic clinopyroxene, nepheline, and melanite garnet. Good outcrops occur at locality 4 (Fig. 1).

Nepheline syenite is medium to coarse grained, light gray, and commonly has a trachytoid texture. Alignment of minerals resembling foliation has been observed in some outcrops. The rock consists of orthoclase, microperthite, and albite, with about 10 percent interstitial sodic clinopyroxene and nepheline, and accessory melanite garnet, magnetite, sphene, biotite, apatite, and zircon. Several lines of field and geochemical evidence suggest that at least part of the nepheline syenite is rheomorphic fenite. Good outcrops occur at locality 5 (Fig. 1).

Carbonatite in the form of a stock underlies Iron Hill and the ridge toward the northeast across Deldorado Creek. It consists chiefly of light-brown to light-gray foliated to massive rauhaugite (dolomite carbonatite). The following minerals, listed approximately in the order of their frequency of occurrence, have been identified: dolomite, barite, hematite, calcite, quartz, fluorapatite, pyrochlore, pyrite, magnetite, biotite, rutile, fluorite, bastnaesite, aegirine, anatase, sphalerite, synchysite, zircon, magnesite, and manganese oxide minerals. In comparison with average igneous rocks, the carbonatite stock contains over 20 times more barium, cerium, neodymium; 15–20 times more lanthanum, niobium, phosphorus, and total rare-earth elements; and nearly 10 or more times manganese, molybdenum, and strontium (Armbrustmacher, 1980). The carbonatite stock is cut by narrow martite-fluorapatite veins and jasper veins, neither of which are shown on the geologic map (Fig. 1). Good exposures of carbonatite occur in and around the Dupont adit at locality 6 on the map.

A variety of mineral deposits of possible economic value are spatially and genetically associated with the rocks of the Iron Hill carbonatite complex. Commodities studied in the past include thorium, rare-earth elements, niobium, iron, titanium, vermiculite, and manganese. Thorium is mainly concentrated in veins and shear zones outside the complex, and rare-earth elements and niobium are mainly concentrated in carbonatite dikes and in the carbonatite stock. The magnetite-ilmenite-perovskite segregations were discussed as early as 1912 (Singewald, 1912) as a source of iron and titanium. Rose and Shannon (1960) reported an average grade of 11.7 percent iron and 6.5 percent titania and a possible tonnage in excess of 100 million short-tons (90 million metric tons) in pyroxenites containing the segregations. In the February 25, 1976, issue of the *Denver Post,* Buttes Gas and Oil Company announced results of a study that indicated 419 million tons (380 million metric tons) of reserves averaging 12 percent TiO_2 in the pyroxenites containing the segregations. Studies of the carbonatite stock show an inhomogeneous distribution of thorium with values ranging between 0.0007 percent and 0.017 percent ThO_2; the average thorium content is 0.0041 percent ThO_2 (Armbrustmacher, 1980). The carbonatite stock contains reserves totaling 29,775 tons (27,000 metric tons) of ThO_2, 9180 tons (8330 metric tons) of U_3O_8, 2,865,500 tons (2,599,000 metric tons) of total rare-earth oxides, and 412,000 tons (374,000 metric tons) of Nb_2O_5. Vermiculite deposits are found in altered pyroxenite, but quality and quantity of expansible materials are not great and data on reserves are not available. Pits dug in the vicinity of locality 7 (Fig. 1) mainly evaluated vermiculite potential. Veins of manganese-rich material occur at the south edge of the complex at locality 8 on the map, but they are poorly known.

REFERENCES

Armbrustmacher, T. J., 1980, Abundance and distribution of thorium in the carbonatite stock at Iron Hill, Powderhorn district, Gunnison County, Colorado: U.S. Geological Survey Professional Paper 1049-B, 11 p.

Armbrustmacher, T. J., and Futa, K., 1985, Petrology of alkaline rocks in the carbonatite complex at Iron Hill, Powderhorn district, Gunnison County, Colorado; New geochemical and isotopic data: Geological Society of America Abstracts with Program, v. 17, no. 3, p. 149.

Hedlund, D. C., and Olson, J. C., 1975, Geologic map of the Powderhorn quadrangle, Gunnison and Saguache Counties, Colorado: U.S. Geological Survey Geologic Quadrangle Map GQ-1178, scale 1:24,000.

Larsen, E. S., Jr., 1942, Alkalic rocks of Iron Hill, Gunnison County, Colorado: U.S. Geological Survey Professional Paper 197-A, p. 1–64.

Nash, W. P., 1972, Mineralogy and petrology of the Iron Hill carbonatite complex, Colorado: Geological Society of America Bulletin, v. 83, p. 1361–1382.

Olson, J. C., 1974, Geologic map of the Rudolph Hill quadrangle, Gunnison, Hinsdale, and Saguache Counties, Colorado: U.S. Geological Survey Geologic Quadrangle Map GQ-1177, scale 1:24,000.

Rose, C. K., and Shannon, S. S., Jr., 1960, Cebolla Creek titaniferous iron deposits, Gunnison County, Colorado: U.S. Bureau of Mines Report of Investigations 5679, 30 p.

Singewald, J. R., Jr., 1912, The iron ore deposits of the Cebolla district, Gunnison, County, Colorado: Economic Geology, v. 7, p. 560–573.

Temple, A. K., and Grogan, R. M., 1965, Carbonatite and related alkalic rocks at Powderhorn, Colorado: Economic Geology, v. 60, p. 672–692.

The Florissant Fossil Beds National Monument, Teller County, Colorado

R. M. Hutchinson and K. E. Kolm, Department of Geological Engineering, Colorado School of Mines, Golden, Colorado 80401

LOCATION

The Florissant National Monument is easily accessible by taking U.S. 24 west from Colorado Springs to the small town of Florissant, 35 mi (56 km) away (Fig. 1). At the town center, turn south toward Cripple Creek on the unpaved Teller County Road No. 1. The national monument is 0.5 mi (0.8 km) from the town of Florissant. The entire national monument area can be seen in half a day's time.

SIGNIFICANCE

The Florissant flora is the only one now known from the central Rocky Mountain region that is intermediate in age between the middle Eocene Green River, Wyoming flora and the late Pliocene flora of Creede, Colorado and the High Plains. Thus it furnishes critical evidence regarding environmental conditions in the area during middle Tertiary time. This evidence is important in the interpretation of biologic and climatic sequences found in the succession of rich Tertiary floras in the High Plains and Rocky Mountains. The Florissant Oligocene fossil beds form the most extensive fossil record of its type in the world. The fossil-bearing shales have yielded more than 80,000 specimens that contain more than 1,100 species of insects, including almost all the fossil butterflies of the New World, more than 140 plant species, and several species of fish, birds, and small mammals. The rare quality of Florissant is the delicacy with which thousands of fragile insects, tree foliage, and other forms of life—completely absent or extremely rare in most paleontological sites—have been preserved in stone.

SITE INFORMATION

The Tertiary deposits of the Florissant valley are largely of volcanic origin. These were deposited on the eroded surface of the 1.02 Ga Pikes Peak granite (Hutchinson, 1976) and consist of lava flows, massive pumiceous tuffs, river gravels, agglomerates, and finely laminated, fossiliferous paper shales of lacustrine origin. The lake-bed shales form the most prominent outcrops, but they constitute only a minor part (16%) of the total thickness. As emphasized by MacGinitie (1953, p. 4), (1) the beds, instead of forming a single unit, comprise a complex and varied series of sediments and volcanics that can easily be divided into at least five members; (2) these beds span lower and middle Oligocene time; (3) the plant-bearing member makes up less than a third of the total thickness, and it alone is of lacustrine origin with the remaining beds being mudflows and reworked, river-deposited tuffs; and (4) the present outline of the beds is due to complex faulting and subsequent erosion, and does not represent, in any sense, the old lake margin.

The succession of the Tertiary deposits is as follows, top to bottom, with lithologic divisions numbered from base to top:

	~thickness	
	ft	(m)
7. Trachyandesite	0–538	(0–164)
Hiatus		
6. Basic breccia with augite andesite (Thirty-nine Mile volcanics)	108–1,076	(32.8–328)
Hiatus		
5. Pumiceous andesite tuffs, shales, and agglomerates, and volcanic river gravels	323	(98.4)
Hiatus		
4. Rhyolitic tuff	±215	(±65.6)
3. Lake shales and associated volcanic sediments	±538	(±164)
2. Bedded andesite tuffs	±591	(±180)
1. Basal water-laid pebbly arkose ...	±108	(±32.8)
Total	3,388	(1,032.8)

The Florissant basin is irregularly sickle-shaped, the arc being about 7.2 mi (11.6 km) long and 2.0 mi (3.2 km) wide, and concave to the west. The Florissant beds are preserved in two synclinal structures whose axes intersect near Florissant at an angle of approximately 50°. The axis of the west basin, between Florissant and Lake George, trends N55°W and that of the basin extending southward is about N5°W. These are not simple synclinal structures. The structural trends and fault pattern in the Florissant area correspond in direction to those occurring throughout the central Colorado Rockies. In the Florissant basin the structural lines fall roughly into three groups having directions of N50°–55°W, N–S to N60°E, and N80°W to E–W. The northerly and northeasterly faults have been offset by those trending northwesterly. The varying and erratic course of the contact between the Tertiary sediments and the Pikes Peak granite appears to be largely controlled by these systems of intersecting faults. In some areas the soft and incompetent Tertiary beds, where intersected by high angle faults, have been moderately folded. The time of major deformation was later than the extrusion of the Thirtynine Mile volcanics to the west (early Oligocene, Stark and others, 1949, p. 103) and earlier than the eruption of the trachyte (late Oligocene, Stark and others, 1949,

p. 66, 109). It was most probably in the later Oligocene. According to MacGinitie (1953), all the evidence considered points to the conclusion that the Florissant flora is between early and middle Oligocene in age. It is MacGinitie's considered opinion that the age is best described as being lower Oligocene (MacGinitie, 1953, p. 75).

SUMMARY OF ENVIRONMENTAL CONDITIONS INDICATED BY THE FLORA AND SEDIMENTS

1. Climate. Warm temperate with relatively warm winters and hot summers; the average annual temperature not less than 65°F (18°C); the average temperature of the warmest month approximately 80°F (27°C). Rainfall not adequate to support a true forest except along the streams; rainfall concentrated in the summer half-year, with a large part of the precipitation in the late spring and early summer; annual rainfall approximately 20 in (51 cm). According to the symbols of the Koppen climatic classification, the indicated climate would be close to the boundary between Bshw and Cwa. Similar climates are found at present between latitudes 20° and 30° in northeastern Mexico in the northern Sierra Madre, in northern Argentina just east of the Andes, in northeastern Australia, in southeastern Africa, and in northwestern India.

2. Vegetation. Rich mesic forest along the streams and lake shores; scrub forest and grass on the higher ground.

3. Physiography. The basin of deposition a region of moderate elevation, not exceeding 3,000 ft (910 m) and a very moderate relief; the streams with wide straths and low interfluvs; higher terrain immediately to the west with volcanic mountains reaching elevations of 7,900 ft (2,400 m) or more.

SELECTED REFERENCES

Cockerell, T.D.A., 1906, The fossil flora and fauna of the Florissant shales: University of Colorado Studies, v. 3, p. 157–176.

—— , 1908, The fossil flora of Florissant, Colorado: American Museum of Natural History Bulletin, v. 24, p. 71–110.

Hutchinson, R. M., 1976, Granite-tectonics of Pikes Peak batholith, Colorado: Professional Contributions of Colorado School of Mines, Studies in Field Geology, no. 8, p. 32–44.

Kirchner, W.C.G., 1898, Contributions to the fossil flora of Florissant, Colorado: St. Louis, Transactions of the Academy of Science, v. 8, p. 161–188.

Knowlton, F. H., 1916, A review of the fossil plants in the U.S. National Museum from Florissant lake beds at Florissant, Colorado: Proceedings of the U.S. National Museum, v. 51, p. 183–197.

—— , 1923, Fossil plants from the Tertiary lake beds of south-central Colorado: U.S. Geological Survey Professional Paper 131, p. 183–197.

Lesquereux, L., 1883, Contributions to the fossil flora of the Western Territories, III, The Cretaceous and Tertiary floras: Report U.S. Geological Survey Territory, v. 8, 283 p.

MacGinitie, H. D., 1953, Fossil plants of the Florissant beds, Colorado: Washington, D.C., Carnegie Institution of Washington Publication 599, p. 1–198.

Stark, J. T., and others, 1949, Geology and origin of South Park, Colorado: Geological Society of America Memoir 33, 188 p.

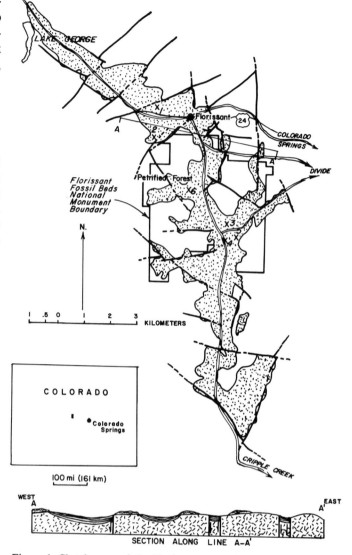

Figure 1. Sketch map of the Florissant area, showing outline of the Tertiary sediments and the most prominent fault lines. Figures refer to the more extensively worked fossil localities: 1. Denver Museum of Natural History, 1915 or 1916; 2, 3, 4. University of California localities 3731, 3732, 3733, respectively, 1936–1937; 5. The "Princeton locality" 1880; 6. Scudder's original locality, 1879.

Granite-tectonics of the Pikes Peak intrusive center of Pikes Peak composite batholith, Colorado

Robert M. Hutchinson, Department of Geology, Colorado School of Mines, Golden, Colorado 80401

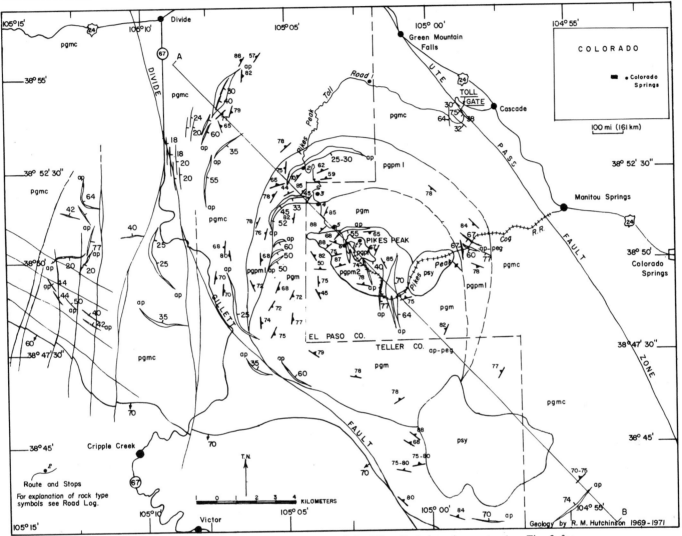

Figure 1. Granite tectonics and geologic map of the Pikes Peak intrusive center (see Fig. 3 for explanation).

LOCATION

The site lies within Pike National Forest and does not require permission to enter. The only formal requirement is the toll charged at the entrance to the Pikes Peak toll road, which goes all the way to the summit of the peak, elevation 14,110 ft (4,300 m). The hard-bottom dirt road is kept in excellent condition. Two-wheel drive vehicles can make the climb, but vehicles with carburetors tuned to low elevations may have trouble reaching the summit. Depending on snow conditions, the toll road is generally open from May 1 to October 31. The Pikes Peak toll road is reached by driving west on U.S. 24 from Colorado Springs to the small town of Cascade, 12 mi (19 km) away (Figs. 1 and 2). A

leisurely trip of a single day should be sufficient to observe the geological relationships of the intrusive center.

SIGNIFICANCE

The Pikes Peak intrusive center is the most accessible and best exposed of the three intrusive centers that make up Pikes Peak batholith. The batholith is composite, with three separate but texturally intergrading zoned intrusive centers: (1) Buffalo Park, (2) Lost Park, and (3) Pikes Peak (Fig. 3). Differentiation Index-oxide plots and computerized petrofabric plots of 21,126 granite-tectonic features indicate magmatism dominant with crys-

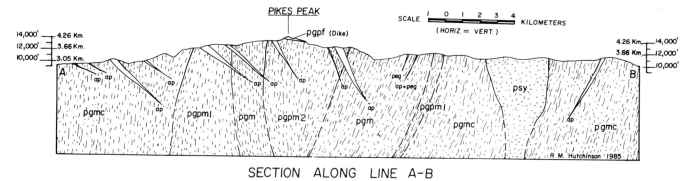

Figure 2. Geologic cross section along A-B, Fig. 1 of the Pikes Peak intrusive center, Colorado.

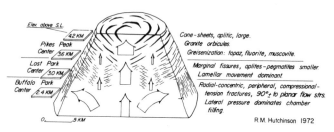

Figure 3. Granite tectonics vs. emplacement depth of intrusive centers, Pikes Peak composite batholith (a composite model).

tallization proceeding from rim to core for each of the three intrusive centers.

Each intrusive center is exposed at different but overlapping altitudes: proximate elevations are Buffalo Park intrusive center, 11,400 to 11,700 ft (3,500 to 3,600 m); Lost Park intrusive center, 9,800 to 12,300 ft (3,000 to 3,800 m); and Pikes Peak intrusive center, 9,800 to 14,110 ft (3,000 to 4,300 m). The intrusive centers have an average diameter of 12 to 16 mi (20 to 25 km). The number and degree of development of individual granite-tectonic structural features within each center vary as a function of elevation of exposure and original depth at which these features were formed within each intrusive center. A composite interpretive working model of this dominantly epizonal batholithic emplacement has been prepared using field relationships between type of granite-tectonic structural features present and the structural elevation or level at which these are developed in each of the three intrusive centers (Fig. 3). The Pikes Peak area provides a unique opportunity to readily observe the structural features and relationships of the intrusive center, coupled with superb scenery and an almost total exposure of the rock formation.

SITE INFORMATION

The Pikes Peak intrusive center exhibits structural relationships between primary flow structures, primary fracture systems, small to large aplite and/or pegmatite ring dikelike (cone sheet) bodies within an epizonal-type intrusive center. There is almost complete exposure of rock formations above an elevation of 9,800 ft (3,000 m) above sea level.

The granite of the intrusive center is composed mainly of

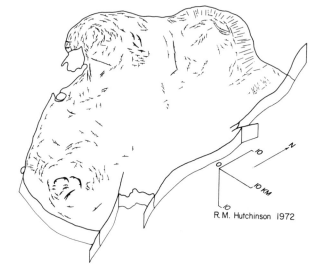

Figure 4. Planar flow structures and intrusive centers of Pikes Peak composite batholith, Colorado.

hornblende-bearing biotite peraluminous granite. Textural variations include equigranular to inequigranular seriate to hiatal, medium to coarse and very coarse granite, slightly to strongly porphyritic medium-grained types and porphyritic microphaneritic dikelike apophyses into metamorphic wall rocks. Several textural varieties and gradations of aplite, aplogranite, porphyritic-aplitic, and porphyritic-aplogranitic rock also occur as intrabatholithic dikes and marginal fissures, in some places associated with pegmatites.

The intrusive center has an equigranular to mostly inequigranular, seriate, medium- to coarse-grained outer border. Hypidiomorphic-granular, medium- to coarse-grained rocks occupy intermediate central intrusive center positions. Porphyritic medium- to coarse-grained textures occur in the core. Unlike the other two intrusive centers of the Pikes Peak batholith, Pikes Peak intrusive center is capped by 200- to 300-ft (60 to 90 m) thick dike of porphyritic very fine to fine-grained fluorite-bearing peraluminous granite dipping 35–45 degrees south (Figs. 1 and 2). The a–c joints (longitudinal) and a–b joints (normal to strike of planar flow structure) of the outer zone have a well-developed, girdle-polar distribution. Aplite and/or pegmatite cone sheets

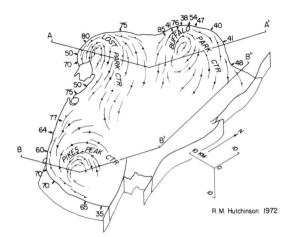

Figure 5. Aplite-pegmatite orientations in intrusive centers of Pikes Peak composite batholith, Colorado.

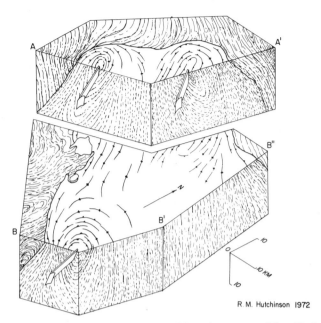

Figure 6. Magma movement linears of intrusive centers, Pikes Peak composite batholith, Colorado.

have an intermediate central, moderate, girdlelike distribution, indicative of an upward surge of magma during formation of the Pikes Peak intrusive center. Rock zones, progressing inward and including the core, show increasingly weaker girdle-polar patterns of concentrations. Aplite and/or pegmatite orientations, progressing inward develop increasingly steeper dips in both small and large dikelike and cone sheet bodies.

The Pikes Peak intrusive center is dominated by (1) a strongly developed inward-dipping ring-dike-like complex of porphyritic-aplite, porphyritic-aplogranite, and aplite dikes sub-parallel to rock zonation (Figs. 1, 2, and 4), and (2) incipient greisenitization of central rock zones via topazitization, fluoritization, and more than normal sericitization of plagioclase. The ring dikes (cone sheets) vary from small sizes up to 0.25 × 1.5 mi (0.4 × 2.5 km) and dip 20–60 degrees inward. Orbicular granite structures (new North American locality), averaging 4 to 5 in (10 to 13 cm) in diameter, occur sparingly and have aplitic shells with radial biotite-rich pegmatitelike cores. The granite orbicules were formed when slowly rising intrabatholithic pegmatite fluids were trapped in their present position as the magma stiffened and crystallized (Hutchinson, 1972).

A steeply inward-dipping funnel-shaped structural pattern that plunges 65–80 degrees in a N.40–50 degrees W. direction is defined by the granite-tectonic features displayed in the Pikes Peak intrusive center. These are the concentric planar and linear flow structures, primary fracture systems, spatial distribution and general inward dip of the aplite and/or pegmatite-filled cone sheets, and concentric pattern of the textural rock zonation (Figs. 5 and 6). Magma rose at an angle of 65–80 degrees in a S.40–50 degrees E. direction. This appears to be true for the other two intrusive centers as well. Magma moved upward along each of the three "magma-movement linears," partly coalesced, and continued to enlarge in an upward and eastward direction. Each intrusive center was completed, and the resulting larger, composite Pikes Peak batholith in its present form was formed in its structural entirety.

The composite working model for the entire Pikes Peak batholith shows relationships between granite-tectonic style vs. depth zone of emplacement (Fig. 3). During emplacement, each intrusive center developed a characteristic and slightly varying array of granite-tectonic structural features. Some of these features are best developed and most readily accessible in the Pikes Peak intrusive center. Additional important references on Pike's Peak batholith are Hutchinson (1960, 1976) and Barker and others (1976).

SITE DESCRIPTION AND ROAD LOG

Mi (km)

0.0 (0.0). Toll gate to Pikes Peak highway. For the next 5.4 mi (8.6 km) the road will be in medium- to coarse-grained peraluminous granite (pgmc), varyingly equigranular to inequigranular, seriate. Narrow pegmatite and aplite dikes strike northwesterly and dip 32–64 degrees southwest toward the Pikes Peak intrusive center.

5.4 (8.6) Stop 1. Crystal Lake and dam. Road cut on north side of road is typical holocrystalline, equigranular to mostly inequigranular, seriate, medium- to coarse-grained, hornblende-bearing biotite peraluminous granite (pgmc). The granite is a moderate reddish-orange (10R 6/6), is massive with no apparent planar flow structure visible on the outcrop, and typifies the outer zone of the intrusive center.

5.4–11.0 (8.6–17.6). Road continues in pgmc that is mostly covered by vegetation and granite gruss.

11.0–12.4 (17.6–19.8). At approximately 11.0 (17.6) the road crosses a covered contact between the medium- to coarse-grained granite (pgmc) and pophyritic medium-grained, hornblende-bearing peraluminous granite with coarse phenocrysts (pgpml).

12.4 (19.8) Stop 2. Glen Cove. Steep cliffs 500 ft (154 m) to the west are part of a ring-dike (cone sheet) of aplite to porphyritic aplite. The dike is concave to the southeast, up to 2.4 mi (4 km) in length, and 325 to 500 ft (100 to 500 m) thick. Strike of the dike trends generally N50–75E and dips 25–30 degrees SE (Figs. 1 and 2). Immediately east of the Glen Cove buildings are massive jointed cliffs of porphyritic medium-grained, hornblende-bearing biotite peraluminous granite with coarse phenocrysts (pgpml). Planar flow structure has a general trend of N38–40E and dips 78 degrees NW. Visible in the steep cliffs due south are several wide, intensive shear zones with near vertical dips. The fault zones trend N to NNW. Age of the faulting is early Miocene; it caused fragmentation of the late Eocene surface throughout the region. Regional block faulting of basin-and-range style in Miocene and Pliocene time resulted in offsets of 5,000 to possibly 39,000 ft (1,500 to 12,000 m) (Epis and others, 1976).

12.4–15.7 (19.8–25.1). Throughout this distance, with a series of switchbacks, the road cuts through porphyritic medium-grained, hornblende-bearing biotite peraluminous granite (pgpml), moderate reddish-brown (10R 6/6) to pale reddish-brown (10R 5/4). Planar flow structures strike generally N40–80E and dip 59–70 degrees NW.

15.7 (25.1) Stop 3. Top of the switchback turns. Exposures of the porphyritic medium-grained peraluminous grained granite (pgpml) are east and west of the road. Good exposures are several hundred feet west of the road. Planar flow structures strike N52–60E west of the road and N60–72E east of the road. Dips vary from 75–80 degrees NW.

15.7–16.3 (25.1–26.1). Outcrops of porphyritic medium-grained, hornblende-bearing biotite peraluminous granite (pgpml). Planar flow structures are N45–50E, 80–85 degrees NW dips.

16.3 (25.1) Stop 4. The Bottomless Pit. View to east of city of Colorado Springs. Contact between pgpml and holocrystalline, equigranlar, medium-grained granite (pgm) strikes almost east-west through bend of the road at this stop. Planar flow structure in pgm strikes N50E and dips 80 degrees NW. Contact is gradational over several hundred feet. A large xenolith of metamorphic rocks is present along the edge of the cliff to the north of the stop and strikes N54E and dips 36 degrees SE. Some granite orbicules can be seen if you walk north along the cliff several hundred ft and look down along the edge of the cliff.

16.3–18.2 (25.1–29.1). Outcrops of the equigranular medium-grained granite (pgm).

18.2 (29.1) Stop 5. Three different lithologic units are in the general area of this stop. You need to walk north, east, south, and west from this stop to observe the field relationships. Medium-grained, equigranular granite is to the north (pgm); porphyritic medium-grained granite with a range of coarse to very coarse phenocrysts (pgpm2) is to the south; and along the contact of these two units is a relatively short, stubby ring-dike (cone sheet)

of porphpyritic aplite (ap). Outcrops of the dike are well exposed to the east at the edge of the cliff. The ring-dike is concave to the southeast and has a general strike of N75E and dips SE 62 degrees.

18.2–18.6 (29.1–29.8). Road cuts through the pgpm2 unit.

18.6 (29.8) Stop 6. Contact between porphyritic medium-grained granite with coarse to very coarse phenocrysts (pgpm2) and a porphyritic fine-grained, fluorite-bearing peraluminous granite (pgpf). Exposures of this contact occur along the east side of the highway. Trend of the planar flow structure in the pgpm2 is N80–82E with dips 75–87 degrees NW. Contact with the dike rock is sharp. Footwall of the dike dips 35–45 degrees SE (Figs. 1 and 2).

18.6–19.7 (29.8–31.5). Road crosses through dike rock that caps the peak.

19.7 (31.5) Stop 7. Summit of Pikes Peak. Detached blocks of moderate reddish-brown (10R 4/6) porphyritic fluorite-bearing, fine-grained peraluminous granite (pgpf) lie about over the entire summit. Immediately east of the Summit House the contact between the dike rock (pgpf) and the pgpm2 unit can be seen. Planar structure of the pgpm2 strikes N-S to N20E and dips 74–85 degrees NW. With its fine-grained texture, the dike rock has resisted erosion to a greater extent than the coarser-grained granites and has helped preserve the elevation of Pikes Peak.

The other two intrusive centers of Pikes Peak batholith are partly visible in the distance to the north and northwest. Buffalo Park intrusive center is 37 mi (60 km) in a N10–20W direction. Lost Park intrusive center is 30 mi (48 km) in a N40W direction. On a clear day the interior of Lost Park intrusive center is visible with its deep-cut valley flanked by 300 to 500 ft (100 to 150 m) cliffs of heavily jointed, coarse to very-coarse, hornblende-bearing biotite peraluminous granite (pgmc).

REFERENCES

Barker, F., Hedge, C. E., Millard, H. T., Jr., and O'Neil, J. R., 1976, Pikes Peak batholith; Geochemistry of some minor elements and isotopes, and implications for magma genesis: Professional Contributions of Colorado School of Mines, no. 8, Studies in Field Geology, p. 44–56.

Epis, R. C., Scott, G. R., Taylor, R. B., and Chapin, C. E., 1976, Cenozoic volcanism, tectonic, and geomorphic features of central Colorado: Professional Contributions of Colorado School of Mines, no. 8, Studies in Field Geology, p. 323–338.

Hutchinson, R. M., 1960, Petrotectonics and petrochemistry of late Precambrian batholiths of central Texas and the north end of Pikes Peak batholith, Colorado: International Geological Congress, XXI Session, Part IV, The Granite-Gneiss Problem, p. 96–107.

—— , 1972, Pikes Peak batholith and Precambrian basement rocks of the central Colorado Front Range; Their 700-million year history: International Geological Congress, Section I, Precambrian Geology, p. 210–212.

—— , 1976, Granite-tectonics of Pikes Peak batholith, Colorado: Professional Contributions of Colorado School of Mines, no. 8, Studies in Field Geology, p. 32–44.

—— , 1976, Precambrian geochronology of western and central Colorado and southern Wyoming: Professional Contributions of Colorado School of Mines, no. 8, Studies in Field Geology, p. 73–78.

The Garden of the Gods and basal Phanerozoic nonconformity in and near Colorado Springs, Colorado

Jeffrey B. Noblett, Andrew S. Cohen, Eric M. Leonard, Bruce M. Loeffler, and Debra A. Gevirtzman, Colorado Colloege, Colorado Springs, Colorado 80903

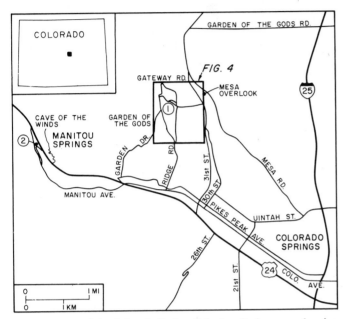

Figure 1. Map of the Colorado Springs/Manitou Springs area showing locations of Stops 1 and 2 and of the Mesa overlook.

LOCATION AND ACCESS

The Garden of the Gods, a city park within the Colorado Springs city limits, is located on the eastern flank of the Colorado Front Range (Fig. 1). It can be reached by following the Garden of the Gods Road (exit 146 from I-25) west 2.4 mi (3.8 km) until it curves to the south and becomes 30th Street. Continue south 1.5 mi (2.4 km) to an unnamed road marked by a small sign (Garden of the Gods) and turn right (west) into the park. Mesa Road, which veers uphill from 30th Street 0.8 mi (1.3 km) north of the turnoff to the park, crests at a spectacular overview of the park and surrounding areas. Several U.S. Geological Survey 7½-minute quadrangle maps cover this region. The Garden of the Gods lies mostly within the Pikeview and Cascade Quadrangles; with its southern part in the Colorado Springs and Manitou Springs Quadrangles. The two stops described here are all on public land and can be reached by passenger car.

SIGNIFICANCE OF THE SITES

Precambrian metamorphic and granitic rocks (1.75 Ga and younger), and also rocks from every geologic period except the Silurian, are exposed within the city limits of Colorado Springs (Grose, 1960; Scott and Webus, 1973). The rock record includes evidence of a variety of sedimentary environments, involving two marine transgressive cycles. Three periods of Phanerozoic uplift

are also recorded. A simplified cross section from Pikes Peak across I-25 shows the major units and structures in Colorado Springs (Fig. 2). The Garden of the Gods is an unusually well-exposed display of late Paleozoic and Mesozoic stratigraphy characteristic of the Colorado Piedmont, and of structures associated with Laramide orogenesis. The rocks were deposited in sedimentary environments, including alluvial fan, lacustrine, aeolian, and shallow marine. The primary structure is that of a faulted monocline (Rampart Range fault) with numerous smaller reverse faults, strike-slip faults, and complicated fracture systems. The Manitou Springs locality (Stop 2) is one of the more accessible and well-exposed views of the Precambrian/Phanerozoic contact near the axis of the Transcontinental Arch. The sedimentary sequence of beach sandstone, tidal flat dolomite, and marine limestone is a spectacular one-stop cross section of the early Paleozoic marine transgression. These rocks are cut by a strand of the Laramide-age Ute Pass fault. Noblett (1984) discusses six additional stops of interest in the Colorado Springs area and presents a brief summary of the historical geology.

STOP 1: THE GARDEN OF THE GODS

Exposed Rock Units. The units in the park include the Pennsylvanian Fountain Formation through the Cretaceous Pierre Shale (Fig. 2). Because so many schools use the park as a mapping problem, we do not here present a finished geologic map, but reference the more interesting features to a topographic map (Fig. 3). The units are generally younger to the east, but have been offset by extensive faulting. The oldest unit in the park is the Pennsylvanian (post Morrowan) Fountain Formation, which is composed of conglomerates, arkosic sandstones, siltstones, and shales deposited under alluvial fan and fan delta conditions. Lithic fragments are common; fluvial structures, indicative of deposition under braided stream conditions, are ubiquitous. Such structures are particularly apparent at a vertical outcrop just west of the Gateway Rocks (Fig. 3a). Elsewhere in the park, extensively reworked, planar and low-angle cross-bedded sands suggest deposition in beach and shoreface environments, whereas marine fossils occur in some shaly facies to the northwest. The Fountain Formation is up to 3,000 ft (900 m) thick in the Manitou Springs Embayment to the west, but wedges out rapidly to the east, grading from nonmarine alluvial fan deposition to a coastal marine, fan delta facies. This formation is the primary evidence for the beginning of the orogenic event that created the ancestral Rockies in Late Mississippian–Pennsylvanian times.

The Lyons Formation (Permian) lies in conformable and occasionally interfingering contact with the Fountain Formation. In the Garden of the Gods, the Lyons Formation consists of three

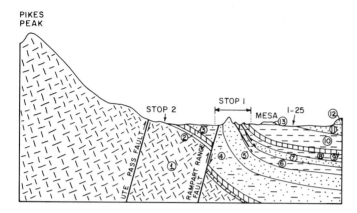

Figure 2. Schematic west-east cross section from Pikes Peak across I-25 showing relative locations of Stops 1 and 2. 1, Pikes Peak Granite; 2, Upper Cambrian Sawatch Sandstone and Peerless Dolomite; 3, Ordovician Manitou Limestone and Harding Sandstone, Devonian-Mississippian Williams Canyon and Hardscrabble limestones; 4, Pennsylvanian Fountain Formation; 5, Permian Lyons Sandstone; 6, Triassic Lykins, Jurassic Morrison, and Cretaceous Purgatoire formations; 7 to 11 are Cretaceous: 7, Dakota Formation; 8, Benton Group; 9, Niobrara Formation; 10, Pierre Shale; 11, Fox Hills and Laramie formations; 12, Tertiary Dawson Arkose; 13, Quaternary gravels.

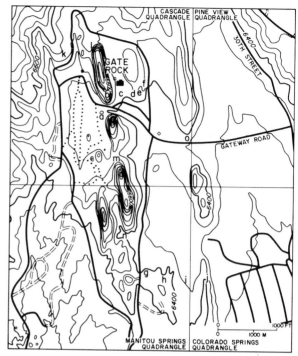

Figure 3. Topographic map of the Garden of the Gods. Contour interval is 40 ft. a, interfingering Fountain and Lyons formations with four facing indicators; b, lower Lyons; c, middle Lyons; d, upper Lyons; e, Lykins Formation; f, Morrison Formation; g, Purgatoire Formation; h, Dakota Formation; i, Benton Group; j, Fort Hays Formation; k, breccia zone in Rampart faults; l, m, and n, northern, middle and southern reverse faults; o, strike-slip fault.

members: a lower and upper hogback-forming sandstone (the big ridges in the park), and a middle valley-forming arkosic conglomerate, which is nearly identical to some of the conglomeratic facies of the Fountain Formation (Fig. 3b, 3c, 3d). In this locality, the two sandstones are distinguishable by color; the lower sandstone is red, and the upper is white. Both sandstones are fine-grained, rounded, moderately well-sorted quartz arenites, with occasional potassium feldspar. The most notable features in outcrop are thick (up to 33 ft; 10 m) high-angle cross-beds. Good exposures of these structures are visible on the south side of the Gateway Rocks, and in Bear Creek Park to the south. Depositional environment was apparently dune fields, which were locally reworked by intermittent streams, as suggested by occasional small-scale low-angle cross-beds and planar beds found at the base of the upper Lyons. The middle, Fountain-like unit indicates that there may have been continuous dynamic interaction between dune and stream environments, possibly as a result of episodic tectonism in the ancestral Rockies through much of Lyons time.

The valley-forming Lykins Formation disconformably overlies the upper Lyons unit (Fig. 3e); it is probably Permo-Triassic in age, and consists of unfossiliferous red shales, siltstones, sandstones, and stromatolitic dolomites. Siltstones are thin bedded to very thin bedded. Stromatolitic bedforms generally show positive relief, and are best seen near Bear Creek Park. The unit shows local mudcracks, which suggest periodic desiccation. Possible depositional environments include both sabkha and playa lake settings.

The Upper Jurassic (Kimmeridgian?) Morrison Formation lies in disconformable contact with the Lykins Formation (Fig. 3f). It consists mostly of variegated pastel purple and green

shales and siltstones, with local limestones, chert pebble and bone fragment conglomerates, sandstones, and bedded gypsum. The gypsum unit (up to 100 ft; 30 m thick) is placed at the base of the formation. A meter or two above the gypsum is a thin (few cm) but prominent limestone. This is mostly micritic, with local oolitic, pisolitic, and stromatolitic lenses. The formation was deposited in a mosaic of fluvial, alluvuial plain, and lacustrine environments.

In the Colorado Springs region, the Purgatoire Formation separates the Morrison and Dakota Formations (Fig. 3g). The Purgatoire is a Lower Cretaceous formation, and contains two members. The lower Aptian member (Lytle) is a coarse, poorly cemented, white conglomeratic sandstone, with a basal chert pebble conglomerate. Thin sets of planar, low-angle cross-beds indicate deposition under beach conditions. The upper member, the Glencairn Shale (Aptian-Albian) is a black shale of probable lagoonal origin. Both members of the formation are best exposed in Bear Creek Park, and along Colorado 24 west of 31st Street.

The Dakota Formation (Fig. 3h) in the Colorado Springs region is probably Albian-Cenomanian in age. Dakota sands consistently are organic-rich quartz and quartzofeldspathic wackes. Trace fossils, small channels, and herringbone cross-stratification are occasionally found, while dinosaur footprints occur in a small outcrop just south of Colorado Springs. The Dakota Formation

Figure 4. View, to south, of reverse fault at locality "1" of Figure 3; Lyons faulted up over Fountain. Note bending of the Fountain layers.

at the Garden of the Gods appears to have been deposited under alternating tidal flat-lagoonal conditions. Lateral facies thicknesses vary considerably throughout the park. At the north end of the Garden of the Gods, the Dakota forms the typical lichen-covered, pink, fine-grained sandstone hogback that is prominent all along the Front Range. Near the Visitor's Center in the south part of the park, the formation consists of three thinner ridges (8 to 10 ft; 2 to 3 m thick) separated by 33 ft (10 m) of siltstone and shale beds. Through most of the park, the Dakota Formation is present only as tiny outcrops. While an east-dipping reverse fault could explain the missing section and the overall thinning of units to the north, a lateral facies change cannot be ruled out.

The Late Cretaceous Benton Group (Fig. 3i) consists of three shallow-marine formations: the Graneros Shale, the Greenhorn Limestone, and the Carlile Shale (Scott, 1969). These form a valley just east of the Dakota sandstones. The Cenomanian Graneros Formation is a black marine shale. The Cenomanian-Turonian Greenhorn Formation is a marine shale that contains thin beds of lime mudstones and is exposed on the small hill just east of the Dakota ridge at the north end of the park. The Turonian Carlile Formation is primarily a black shale, but the uppermost portion, the Codell Sandstone Member, is a thin (8 to 10 ft; 2 to 3 m) well-sorted, subrounded, calcareous quartz arenite containing trace amounts of petroleum. All of the Benton Group is fossiliferous; sharks teeth, ammonites, and pelecypods (especially *Inoceramus* sp.) may all be found. Numerous thin bentonitic layers serve as useful marker beds.

The Cretaceous Niobrara Formation is divided into two members. The Fort Hays Limestone (Coniacian) forms a promi-

nent white hogback (Fig. 3j) together with the Codell Sandstone. Limestone units in the Fort Hays Member are interbedded with thin shale units; all are fossiliferous, with a fauna similar to that found in the Benton Group. The upper member, the Smoky Hills Chalk (Coniacian-Santonian) is exposed on the eastern edge of the hogback as a yellow fossiliferous chalk. The Smoky Hills Member underlies much of the valley east of the Garden of the Gods.

The thickest unit in the Colorado Springs area is the Pierre Shale (Campanian). It is a black marine shale that contains thin layers of organic-rich feldspathic wackes, lithic wackes (with abundant detrital chert and amphiboles), and limestone. The Pierre Shale is fossiliferous, containing pelecypods, foraminifera, and ammonoids (*Baculites* sp.). The Pierre Formation is very widespread and thick (to 5,000 ft; 1,500 m); it is a shallow marine deposit laid down under conditions that frequently were anoxic. A small exposure of Pierre Shale is visible at the intersection of 30th Street and the east entrance to the park; a better view is obtained under the mesa on Uintah Street west of I-25, where the Pierre forms an angular unconformity with the overlying Mesa gravels.

Structures. The hogback ridges that dominate the scenery in the Garden of the Gods are the vertical to overturned limb of a faulted monocline of Laramide age. The major fault associated with the monocline, the Rampart Range fault, passes immediately west of the large Lyons and Fountain hogbacks, and trends north-northwest. The fault plane dips to the west; its relatively straight trace suggests that the dip is fairly high angle. In the park, the fault separates low-dip Pennsylvanian and older sediments in the hanging wall from the vertically dipping younger footwall sediments; farther north, the fault has superposed Precambrian rocks of the Rampart Range over Paleozoic and Mesozoic sediments. At the south end of the park, most of the fault displacement has occurred on a single plane, which dies out in a monocline farther south. The width of the fault zone increases northward until, at the north end of the park (Fig. 3k), the fault zone may be as much as 330 ft (100 m) wide, and structure becomes unclear.

At least three eastward-dipping reverse faults cut across the hogbacks and offset hogback-forming units. Most spectacular is the northernmost of these (Fig. 3l), which is seen at the north end of North Gateway Rock (Fig. 4). A second fault, passing south of South Gateway Rock, may be seen from the road east of the hogbacks (Fig. 3m). This fault truncates upper (white) Lyons and Morrison ridges immediately north of the Gateway Road, and juxtaposes the continuation of these two units against lower (red) Lyons at the south end of South Gateway Rock. A third east-dipping reverse fault occurs farther to the south (Fig. 3n), but is less conspicuous from the road. It is probable that one or more of these reverse faults has eliminated the Dakota ridge from the park.

STOP 2: THE NONCONFORMITY, WEST END OF MANITOU SPRINGS

To reach the second stop at the western edge of Manitou

Springs, proceed south from the Garden of the Gods to Colorado 24 (Fig. 1), go west on Colorado 24 to the exit opposite the Cave of the Winds. Turn left (south) off the highway until the road curves sharply, at the Ute Trail stone marker. Several cars can be parked in the pullout by the trail mark. Start by looking at the wall of rocks to the north, from a point west of the marker (Fig. 5). The Pikes Peak Granite is the lowest rock exposed to the west. The nonconformity is almost planar here, and is overlain by three east-dipping formations: the basal Sawatch Sandstone, the dark red and green Peerless Dolomite, and the thin-bedded white Manitou Limestone (best seen in fallen blocks). Some deformation that is present may be related to the Ute Pass fault, which cuts through the valley to the south.

The Pikes Peak Granite (1030 Ma) is exposed along the Front Range and Rampart Range from about the latitude of Castle Rock (40 mi; 65 km north of Colorado Springs), through the southern limits of Colorado Springs. Aeromagnetic mapping suggests that the batholith continues subsurface to the east for another 50 mi (80 km); the granite is visible in outcrop for 40 mi (65 km) west. Total surface exposure is some 1,500 mi^2 (3,900 km^2). The Pikes Peak Batholith is a complex anorogenic granite, with rocks ranging in composition from gabbro to syenite to fayalite granite, along with the more typical granites and granodiorites (see articles in Epis and Weimer, 1976). Textures may vary from inequigranular to porphyritic, fine-grained equigranular, pegmatitic, and microlitic. The granite as exposed here is a coarse-grained, red, hypidiomorphic-granular rock composed of potassium feldspar, quartz, and biotite. The bulk of the batholith is similar in composition to the rock in this location.

The Pikes Peak Granite is directly overlain by the Late Cambrian (Dresbachian) Sawatch Sandstone, a 14 ft (4.5 m) thick unit of well-sorted quartz-quartzofeldspathic arenite. No metamorphism is present at the basal contact; this clearly shows that the contact is not intrusive, but is a nonconformity. At this locality, much of the Sawatch Formation consists of graded units with 1.5 to 3 ft (0.5 to 1 m) thick cycles of rounded to subrounded quartz pebble conglomerates, which fine upwards into massive or trough cross-bedded sands, and finally grade into planar-wavy laminated sands and silty sands. The upper surfaces of subcycles are often scoured and may be bioturbated. Minor glauconite occurs in the upper part of the formation. The glauconite, the well-sorted sands, small areal extent, and the lack of fluvial features all suggest that the Sawatch Sandstone is a shallow subtidal, transgressive sand. The graded cycles and hummocky cross-beds (best seen in nearby Williams Canyon) are probably the result of frequent storm activity.

The Sawatch Formation grades conformably into the overlying Late Cambrian (Franconian) Peerless Formation. The formation boundary is marked by the appearance of dolomite crystals; the Peerless Formation at this location is 51 ft (15.5 m) thick, and strikingly colored in red and green. This is a texturally complex unit that contains deeply embayed, moderately sorted, angular clastic grains of quartz, feldspars, and chert, together with rounded glauconite grains, in a coarsely crystalline dolomite ma-

Figure 5. View of the four units at the Precambrian nonconformity. a, Pikes Peak Granite; b, Sawatch Sandstone; c, Peerless Formation; and d, Manitou Limestone. Contacts are shown by short black lines.

trix. At its base, the Peerless Formation is medium to coarse sand, and shows low-angle planar and trough cross-bedding; individual cross-bed sets are up to 4 in (10 cm) thick and 28 in (70 cm) wide. Higher in the formation, the large-scale cross-beds become thin-bedded to massive, and commonly are burrowed. Bedding surfaces may be scoured. The depositional environment is thought to have been a shallow subtidal setting, with the transition from large-scale cross-beds to planar-bedded, finer sands representing continuing deposition in progressively deeper water.

The thin-bedded Early Ordovician (Canadian) Manitou Limestone here overlies the Peerless Formation in minor disconformity. It appears high above the old quarry at this stop, and may best be seen by examining fallen blocks, or by walking downhill to the bridge over Fountain Creek, and then west along the creek to a small cave. The Manitou Limestone is a thin-bedded lime wackestone, which contains abundant skeletal and intraclastic grains in a lime mud matrix. Locally the formation may be dolomitic, cherty, shaly, or oolitic. Wackestone beds are normally separated by wavy, very thin-bedded lime mudstone partings. Numerous gastropods, ostracodes, nautiloids, and horizontal feeding traces are present as fossils. This shallow intertidal-subtidal limestone completes a marine transgressive cycle at this outcrop.

REFERENCES CITED

Epis, R., and Weimer, R., eds., 1976, Studies in Colorado field geology: Golden, Colorado: Professional Contributions of Colorado School of Mines, no. 8, 522 p.

Grose, L. T., 1960, Geologic formations and structure of Colorado Springs area, Colorado, in Weimer, R., and Haun, J., eds., Guide to the geology of Colorado: Denver, Colorado, Rocky Mountain Association of Geologists, p. 188–194.

Noblett, J., 1984, Introduction to the geology of the Colorado Springs region: Colorado Springs, Colorado, Pikes Peak Lithography, 38 p.

Scott, G., 1969, General and engineering geology of the northern part of Pueblo, Colorado: U.S. Geological Survey Bulletin 1262, 124 p.

Scott, G., and Wobus, R., 1973, Geologic map of Colorado Springs area: U.S. Geological Survey Map MF-482, scale 1:62,500.

Paradox Valley, Colorado; A collapsed salt anticline

William L. Chenoweth, Consulting Geologist, Grand Junction, Colorado 81506

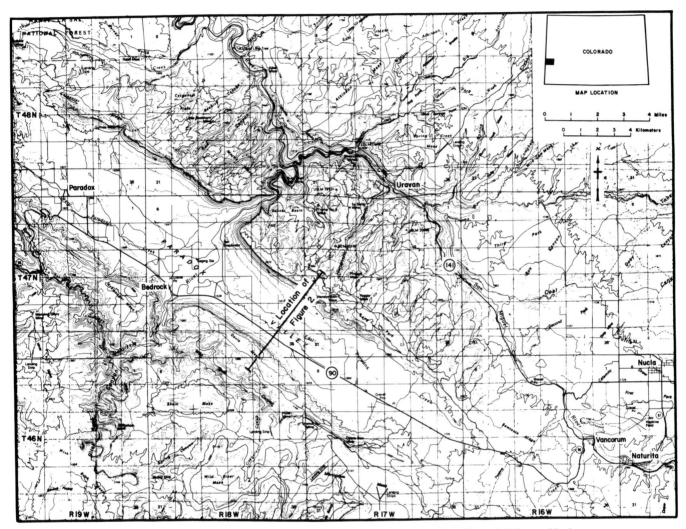

Figure 1. Location map of Paradox Valley, Colorado. Reduced from U.S. Geological Survey Nucla Quadrangle, 1:100,000 scale; contour interval 50 m.

LOCATION AND ACCESS

Paradox Valley is located in west-central Colorado, in western Montrose County (Fig. 1). The valley is approximately 23 mi (37 km) long and averages 3 mi (5 km) in width. It occupies a large part of T.46N.,R.17W.;T.47N.,R.18 and 19W.; and T.48N.,R.19W. Colorado 90 traverses the valley for nearly its full length. This highway begins at Vancorum, which is on Colorado 141 about 2 mi (3 km) west of Naturita, Colorado. Naturita, the largest settlement in the area, is about 100 mi (161 km) south of Grand Junction, Colorado, on Colorado 141. After leaving the northwest end of Paradox Valley, Colorado 90 enters Utah as Utah 46; it joins U.S. 191 some 22 mi (35.4 km) south of Moab,

Utah. Paradox and Bedrock, Colorado, are the only settlements in the valley.

County roads (marked with signs) provide access to the rims of the valley as well as the canyon of the Dolores River on both sides of the valley. Molenaar and others (1981, p. 7–9) give a detailed road log of the valley.

The majority of land in the Paradox Valley area is in the public domain, administered by the Bureau of Land Management. West of the Dolores River, the valley floor is nearly all private land, owned by ranchers in Paradox. About 15 percent of the valley floor east of the river is privately owned. The valley rims are all public domain, with the exception of numerous patented mining claims.

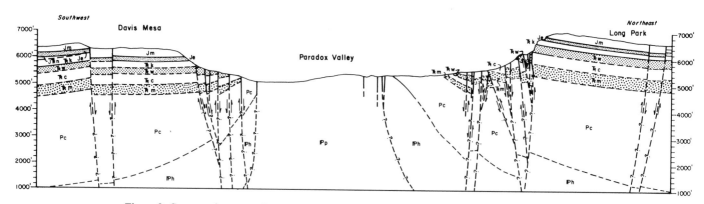

Figure 2. Cross section across Paradox Valley. Location shown on Figure 1. Ⲣp, Paradox Formation; Ⲣh, Honaker Trail Formation; Pc, Cutler Formation; Ⱦm, Moenkopi Formation; Ⱦc, Chinle Formation; Ⱦw, Wingate Sandstone; Ⱦk, Kayenta Formation; J Ⱦn, Navajo Sandstone; Je, Entrada Sandstone; Jm, Morrison Formation; Modified from Cater (1955b).

SIGNIFICANCE

Paradox Valley is the largest of several salt anticlines in the Paradox basin of southwestern Colorado and southeastern Utah. The valley is the type area of the Paradox Formation of Middle Pennsylvanian age. The valley is an erosional artifact due to solution of the Paradox salt core and subsequent collapse of the anticline. Diapirism of the salt core during the interval from the Late Pennsylvanian through the Late Jurassic is indicated by the stratigraphy of the beds on the flanks of the valley. Carnotite deposits in the Morrison Formation, located on the rim of the valley, have been mined for radium, vanadium, and uranium since 1910.

REGIONAL SETTING

Paradox Valley was named by Henry Gannett in 1875. Gannett, a topographer with the U.S. Geological and Geographical Survey of the Territories (Hayden Survey), noted the paradox in that the valley "was crossed at right angles by the Dolores" (Gannett, 1877, p. 343). The first geologic map with a description of the valley was published by A. C. Peale (1878), a geologist with the Hayden Survey.

The regional geologic setting of Paradox Valley is shown on a map by Williams (1964). Paradox Valley and vicinity have been mapped at the scale of 1:24,000 by the U.S. Geological Survey. These quadrangle maps (Cater, 1954, 1955a, 1955b; Cater and others, 1955; Shoemaker, 1956; Withington, 1955) show various details of the stratigraphic relations and the structure in the valley, which may be difficult for the first-time visitor to observe.

Paradox Valley is the largest of a series of breached anticlinal valleys in the northeastern part of the Paradox basin. The rim of the valley consists of sandstones of Jurassic age, and the valley floor is composed of diapiric Paradox evaporitic rocks of Middle Pennsylvanian age (Fig. 2). The diapiric rocks form structures commonly referred to as salt anticlines of the Paradox basin.

The Paradox basin is about 200 mi (330 km) long and 80 mi (130 km) wide, at its widest point. It was formed in Pennsylvanian time along a dominant set of northwest-trending faults of pre-Pennsylvanian age (Baars and Stevenson, 1981). As the basin subsided, the ancestral Uncompahgre highland was elevated along the northeasternmost of these faults. All of the paralleling faults are downthrown in the northeast direction; the "basement" was lowered stepwise to form the basin. Thus, the deepest part of the basin is along its northeast margin, adjacent to the present-day Uncompahgre Plateau. Salt of the Paradox Formation was deposited as the basin was repeatedly lowered; the thickest deposition occurred in the northeastern part of the basin.

The pre-Pennsylvanian fault that underlies Paradox Valley trends to the northwest into Utah. The same salt core that formed the Paradox Valley is also responsible for the formation of the Castle Valley anticline. The North Mountain intrusive group of the La Sal Mountains, an intrusion of early Miocene age, separates these two collapsed salt anticlines.

The interbedded salt members of the Paradox Formation began to rise diapirically shortly after their deposition, in response to movement along the pre-Pennsylvanian faults. The rising salt core of the Paradox Valley anticline profoundly affected the thicknesses of formations deposited in post-Paradox and pre-Morrison times (see Table 1).

Subsidence of the Paradox basin and simultaneous uplift of the ancestral Uncompahgre highland continued during deposition of the Honaker Trail and Cutler Formations (Pennsylvanian/Permian). As the salt core of the Paradox anticline began to rise diapirically, the Honaker Trail beds were lifted and beveled in response. The Honaker Trail Formation is thinner along the flanks of the present valley than in the adjacent synclines due to this erosion along the anticline. The added weight of the Cutler beds during deposition in the synclines enhanced the movement of the evaporites and squeezed the salt upward and through the overlying Cutler sediments (Cater, 1970, p. 52).

Crustal movements associated with the uplift of the ancestral

TABLE 1. GENERALIZED SECTION OF PALEOZOIC AND MESOZOIC FORMATIONS
EXPOSED IN PARADOX VALLEY, COLORADO

System	Formation	Thickness (feet)	Character and distribution
Cretaceous	Mancos Shale	50+	Dark-gray fissile marine shale. Basal beds exposed in center of Coke Oven syncline.
	Dakota Sandstone	150–195	Gray and brown partly conglomeratic sandstone with interbedded carbonaceous shale, all nonmarine, locally coal-bearing. Caps high mesas on each side of valley, also exposed in Coke Oven syncline.
	Burro Canyon Formation	110–200	Light-colored sandstone and conglomerate with interbedded green and purplish shale, nonmarine. Caps mesas on each side of valley.
Jurassic	Morrison Formation	290–420	Brushy Basin Member: varicolored bentonitic mudstone, some sandstone lenses, nonmarine. Forms slopes below Burro Canyon.
		280–350	Salt Wash Member: light-colored lenticular sandstone interbedded with dominantly red mudstone, contains uranium-vanadium deposits, nonmarine. Forms bench on rim of valley.
		60–110	Tidwell Unit: thin-bedded red, gray, green, and brown sandy shale and mudstone, some sandstone lenses, nonmarine. Forms slope below Salt Wash. Previously mapped as Summerville Formation, contains units of the Wanakah Formation (O'Sullivan, 1984).
	Entrada Sandstone	0–150	Slick Rock Member: Light-colored fine- to medium-grained massive and crossbedded sandstone, eolian. Forms a vertical cliff on rim of valley.
		0–80	Dewey Bridge Member: Reddish-brown fine-grained sandstone and siltstone, nonmarine. Forms a cliff below Slick Rock.
	Navajo Sandstone	0–190	Light-colored fine-grained massive and crossbedded sandstone, eolian. Forms a vertical cliff. Present only in the northwest end of the valley.
	Kayenta Formation	0–240	Irregularly bedded red, buff, gray and lavender fine- to coarse-grained sandstone, siltstone and shale, nonmarine. Forms a bench on top of the Wingate.
	Wingate Sandstone	0–100	Reddish-brown fine-grained massive and crossbedded sandstone, predominantly eolian. Forms a vertical cliff on walls of valley.
Triassic	Chinle Formation	0–500	Reddish-brown or varicolored mudstone, siltstone, and sandstone, in part bentonitic, lenses of quartz-pebble conglomerate and grit at base, nonmarine. Forms a steep slope on walls of valley.
	Moenkopi Formation	30–500	Chocolate-brown, reddish-brown, and purple shale, mudstone, and sandstone, in part arkosic and conglomeratic, local gypsum beds near base, marine and marginal marine. Forms a slope in the lower portion of the northeast wall of the valley.
Permian	Cutler Formation	600–3500	Maroon, red, mottled light-red, and purple conglomerate, arkose and arkosic sandstone, thin beds of sandy mudstone, locally reddish-gray marine limestone in lower part, predominately nonmarine. Forms lower slope on the northeast wall of the valley.
Pennsylvanian	Honaker Trail Formation	450–1500	Gray fossiliferous limestone with thin beds of shale, minor arkose, predominately marine. Exposed in the northeast floor of the valley.
	Paradox Formation	2000+	Carbonaceous shale, sandstone, limestone, gypsum and salt, marine. Exposed in floor of valley.

Uncompahgre highland had ceased by the end of Cutler time; however, minor upwelling of evaporites continued through the Triassic and into the Jurassic. As a result, successively older formations dip more steeply on the flanks of the anticline. In general, younger formations were deposited across the upturned and truncated edges of older formations. This process of upturning, erosion, and deposition of overlapping younger formations continued until the time of deposition of the Morrison Formation, when the salt supply from the adjacent synclines had been largely exhausted. Actual wedgeouts and unconformable contacts between beds, which are not normally subjacent, are best seen on the southwest valley wall, southeast of the Dolores River; thinning of the formations can be seen on the northeast wall of the valley.

It appears that the evaporites were probably stable throughout the deposition of Cretaceous sediments. By the end of the deposition of the Mesaverde rocks of Late Cretaceous age, the Paradox salt core had been covered by some 5,000 ft (1,500 m) of sediments, and an anticline had been formed along the old salt structure. The collapse of the crest of the anticline began as gra-

bens were downdropped as much as several hundred feet (few hundred meters). These grabens may have been the result of cessation of the lateral stress that had caused the folding. The grabens remained structurally inactive until the general uplift of the Colorado Plateau began in middle and late Tertiary times.

The uplift of the Colorado Plateau rejuvenated the streams and increased the rate of groundwater circulation. Collapse of the Paradox Valley anticline probably began at the point where the antecedent Dolores River eroded a channel across the anticlinal crest. The canyon cut by the Dolores River exposed the salt to rapid solution and removal, and collapse of the anticlinal crest began. Ancestral East and West Paradox creeks removed material from both the core and from overlying beds during the process of headward erosion. Cater (1970, p. 71) has estimated that the upper surface of the salt core of the anticline was at least 3,000 ft (900 m) higher than the present valley floor.

The Continental Oil Scorup #1 well, which was drilled in 1958 about 1.5 mi (2.5 km) north of Bedrock, penetrated the base of the salt at 14,670 ft (4,471 m) and bottomed in Mississippian rocks at 15,000 ft (4,572 m) (Molenaar and others, 1981,

p. 9). This is the thickest salt section ever penetrated by drilling in the Paradox basin.

STRUCTURAL UNITS

Paradox Valley contains several distinct structural units, which developed as a result of the collapse of the salt anticline. A basinlike downwarp known as the Coke Oven syncline, at the southeast end of the valley, appears to be a result of removal of salt by pressure-induced flowage. The Dry Creek anticline at the southeast end of the valley, on the southwest valley flank, is a result of the draping of sediments over the faulted margin of the Paradox Valley anticline during its collapse.

A central unit occupies most of Paradox Valley where the anticlinal crest was downfaulted. Numerous, closely spaced faults on either side of the valley divide the rocks into long, linear ridges that trend parallel to the valley axis. Most of these ridges have been successively downdropped toward the valley, but a few have been squeezed upward (Fig. 2). The central area retains a salt core in which a number of cupolas are observed.

The northwest unit, which Williams (1964) called the Willow Basin syncline, is a collapsed basin formed by both downsagging and downfaulting. The Mesozoic rocks preserved in this area can easily be viewed from Colorado 90 south of Paradox, Colorado.

ECONOMIC GEOLOGY

The carnotite deposits in the Salt Wash Member of the Morrison Formation of southwestern Colorado have been an important source of radium, vanadium, and uranium since the early 1900s. Two important mining areas are located in the east-central portion of Paradox Valley. The Long Park area is on the north rim, and the Jo Dandy area is on the south side. These deposit clusters are part of a mining region known as the Uravan mineral belt (Fischer and Hilpert, 1952).

The majority of the ore bodies occur in the uppermost sandstone lenses of the Salt Wash Member. A single ore body may range in size from a single fossil log to a mass of impregnated sandstone nearly 100 ft (30 m) wide and 500 ft (150 m) long. Ore thicknesses range from less than 1 ft to over 20 ft (0.3 to 6 m). Ore bodies can occur in clusters elongated parallel to the sedimentary trend of the sandstone host. In unoxidized ore bodies, the principal uranium minerals are uraninite and coffinite. The main uranium minerals in oxidized ores are tyuyamunite and metatyuyamunite, which are more abundant than carnotite.

In past years, brine wells near the Dolores River in the center of the valley have provided salt for livestock as well as uranium-vanadium processing at Uravan, Colorado. As much as 205,000 tons of salt annually enter the Dolores where it crosses Paradox Valley.

REFERENCES CITED

Baars, D. L., and Stevenson, G. M., 1981, Tectonic evolution of the Paradox Basin, Utah and Colorado, *in* Geology of the Paradox Basin: Rocky Mountain Association of Geologists, Guidebook 1981 Field Conference, p. 23–31.

Cater, F. W., Jr., 1954, Geology of the Bull Canyon Quadrangle, Colorado: U.S. Geological Survey Geologic Quadrangle Map GQ-33, scale 1:24,000.

——— , 1955a, Geology of the Naturita NW Quadrangle, Colorado: U.S. Geological Survey Geologic Quadrangle Map GQ-65, scale 1:24,000.

——— , 1955b, Geology of the Davis Mesa Quadrangle, Colorado: U.S. Geological Survey Geologic Quadrangle Map GQ-71, scale 1:24,000.

——— , 1970, Geology of the salt anticline region in southwestern Colorado, with a section on stratigraphy by F. W. Cater and L. C. Craig: U.S. Geological Survey Professional Paper 637, 80 p.

Cater, F. W., Jr., Butler, A. P., Jr., and McKay, E. J., 1955, Geology of the Uravan Quadrangle, Colorado: U.S. Geological Survey Geologic Quadrangle Map GQ-78, scale 1:24,000.

Fischer, R. P., and Hilpert, L. S., 1952, Geology of the Uravan mineral belt: U.S. Geological Survey Bulletin 988-A, p. 1–13.

Gannett, H., 1877, Topographical report on the Grand River District *in* Hayden, F. V., Report of progress of the exploration for the year 1875: U.S. Geological and Geographical Survey of the Territories Ninth Annual Report, p. 337–350.

Molenaar, C. M., Craig, L. C., Chenoweth, W. L., and Campbell, J. A., 1981, First day, road log from Grand Junction to Whitewater, Unaweep Canyon, Uravan, Paradox Valley, La Sal, Arches National Park, and return to Grand Junction via Crescent Junction, Utah, *in* Western slope Colorado, western Colorado and eastern Utah: New Mexico Geological Society, Guidebook Thirty-Second Field Conference, October 8–10, 1981, p. 1–15.

O'Sullivan, R. O., 1984, Stratigraphic sections of middle Jurassic San Rafael Group and related rocks from Dewey Bridge, Utah, to Uravan, Colorado: U.S. Geological Survey Oil and Gas Investigations Chart OC—124.

Peale, A. C., 1878, Geological report on the Grand River District, *in* Hayden, F. V., Report of progress of the exploration for the year 1876: U.S. Geological and Geographical Survey of the Territories Tenth Annual Report, p. 163–185.

Shoemaker, E. M., 1956, Geology of the Roc Creek Quadrangle, Colorado: U.S. Geological Survey Geological Quadrangle Map GQ-83, scale 1:24,000.

Williams, P. L., 1964, Geology, structure, and uranium deposits of the Moab Quadrangle, Colorado and Utah: U.S. Geological Survey Miscellaneous Investigations Series Map I-360, scale 1:250,000.

Withington, C. F., 1955, Geology of the Paradox Quadrangle, Colorado: U.S. Geological Survey Geologic Quadrangle Map GQ-72, scale 1:24,000.

Grenadier fault block, Coalbank to Molas Passes, southwest Colorado

D. L. Baars, 29056 Histead Drive, Evergreen, Colorado 80439
J. A. Ellingson, Geology Department, Ft. Lewis College, Durango, Colorado 81301
R. W. Spoelhof, Pennzoil Exploration and Production Company, P.O. Drawer 1139, Denver, Colorado 80201

LOCATION AND ACCESS

The Grenadier fault block is a paleotectonic complex exposed in the heart of the San Juan Mountains of southwestern Colorado. Structural and stratigraphic complexities are well exposed by the effects of Pleistocene glaciation and Recent erosion at and above timberline in a rugged alpine setting (Fig. 1). U.S. 550 traverses the region between Durango and Silverton, affording excellent access. The most critical exposures lie within T.40N.,R.7,8W., near the intersection of 37°45′N. and 107°45′W. in LaPlata and San Juan Counties. Elevations range between 9,000 and 14,000 ft (3,300-5,133 m), effectively limiting field observations to mid-summer months. The area lies within the San Juan National Forest; access is therefore unrestricted. However, due to the rough terrain, appropriate safety precautions should be observed.

SIGNIFICANCE

There is an intimate relationship throughout the Colorado Plateau and southern Rocky Mountains between structure and stratigraphy. Although peculiar distributions of rock units in the vicinity of various structures have been mapped and measured for generations, it was not until the San Juan Mountains were studied in detail that the true nature of events was realized. The core of the San Juan dome is an open textbook of the many interrelationships of structure and sedimentation. The growth history of the many faults can be documented here by a detailed study of the stratigraphic relationships. It was in the San Juans that it was first realized that the structural features of the Colorado Plateau originated through repeated rejuvenation of basement structures, and the infamous Laramide orogeny served only to uplift and expose the true sequence of tectonic events that shaped the land. Here we can see proof positive that the present-day structure has resulted from cumulative events that date back perhaps 1,600 m.y., and we can document rejuvenation of basement structure throughout Phanerozoic time.

PRECAMBRIAN ROCKS

The Precambrian rocks in and immediately adjacent to the Grenadier fault block in the Coal Bank Pass–Molas Pass area are the Twilight Gneiss (1,780 Ma), the Ten Mile Granite (1,720 Ma), the Whitehead Granite (1,677 Ma), and the Uncompahgre Formation (older than 1,460 Ma; Barker, 1969).

The Twilight Gneiss crops out in the Animas Valley from Electra Lake on the south to the Coal Bank–Snowdon Peak area where it is in fault contact with the Uncompahgre Formation. Near Electra Lake, the Twilight Gneiss is trondhjemitic and intrusive into an amphibolite unit. To the north, the gneiss appears to be largely surface accumulations of bimodal volcanics with minor sediments (Baars and Ellingson, 1984).

Following the formation of the Twilight igneous and associated sedimentary rocks, the region was intruded and deformed by the Boulder Creek orogeny (Barker, 1969; Hutchinson, 1976). During this event, the Twilight rocks were metamorphosed to amphibolite facies.

Two post-tectonic plutons were emplaced in the western Needles Mountains during the interval 1,720 to 1,677 Ma. Bakers Bridge Granite is located between Rockwood and Bakers Bridge, and the Ten Mile Granite is located just south of Elk Park and Snowdon Peak. The Ten Mile Granite is composed of microcline, plagioclase (An20-8), quartz, biotite with minor chlorite, sericite, and epidote. Unlike the Bakers Bridge Granite, the structure is moderately gneissic. Many granitic dikes cut the Ten Mile Granite.

The granitic dikes in the Ten Mile are very similar to the small stocks and many dikes in Whitehead Gulch just north of Elk Park. Collectively, these intrusive bodies were called the Whitehead Granite by Cross and others (1905). Barker (1969), on the basis of age and bulk composition, tied these rocks with the Ten Mile Granite and had the name Whitehead Granite officially removed.

Following the Boulder Creek orogeny and post-tectonic intrusion and erosion, a younger Proterozoic sequence was deposited; it included, in ascending order, the Vallecito Formation, the Irving Formation, and the metaconglomerate of Middle Mountain (Baars and Ellingson, 1984). The allochthonous Uncompahgre Formation has the youngest stratigraphic position in the layered sequence and is the only formation of this group that crops out in the western Needles Mountains.

The Uncompahgre Formation crops out in an arc extending for 21.75 mi (35 km) from Lime Creek to Emerald Lake. This formation is a bedded sequence, at least 8,000 ft (2,438 m) thick, of white to lavender sandy and pebbly quartzite interlayered with gray slate, phyllite, and minor schist. The quartzites are composed mostly of sand and grit-sized quartz grains. Occasional argillaceous layers and pebbly layers accentuate the bedding and define graded-bedding and cross-bedding.

In the western Needles, the Uncompahgre is covered by

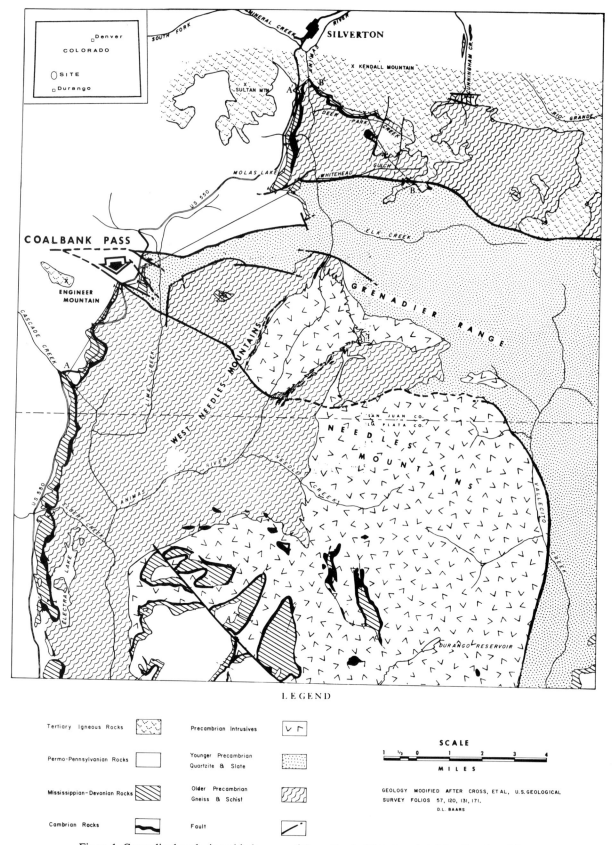

Figure 1. Generalized geologic and index map of Grenadier fault block and vicinity, San Juan Mountains, southwestern Colorado.

Paleozoic sediments to the west, is in contact with the Twilight Gneiss to the south along the Coal Bank fault, and is in contact with Paleozoic and Proterozoic rocks to the north along various faults, the main one being the Molas-Beartown. Volcanics from the Silverton caldera lap onto the north edge of the Uncompahgre block north of Elk Creek.

The second Proterozoic metamorphic event in the Needles Mountains corresponds to the regional Silver Plume orogeny and thermal event (Hutchinson, 1976). The allochthonous Uncompahgre Formation was metamorphosed during the Silver Plume event at its present site. In general, the grade of metamorphism increases to the southeast, never exceeding upper greenschist facies except next to younger plutons.

PALEOTECTONIC SETTING

The Grenadier fault block is an exposed segment of the regional Olympic-Wichita Lineament, which extends from Oklahoma northwestward through Utah as a linear swarm of basement wrench faults. The gross structure "kinks" sharply westward in the eastern Grenadier Range and strikes nearly east-west in the vicinity of Coal Bank and Molas passes, where the bounding faults are vertical, dextral strike-slip faults (Baars and Ellingson, 1984). Quartzites of the Uncompahgre Formation within the fault block are thrust strongly southward within the range as a result of the "kinking" of the wrench fault block. The gross structural features were frozen in situ by the intrusion of the Eolus pluton ca 1,460 Ma. These and many related structural features were in place when the Late Cambrian seas encroached upon the site of the San Juan Mountains (Baars and See, 1968).

CAMBRIAN ROCKS

There was considerable topography on the top of the Precambrian terrain when the Ignacio Formation was deposited in latest Cambrian time (Fig. 2). The Grenadier fault block was well above sea level and supplied coarse clastics along its flanks at both Coal Bank and Molas faults. Large angular boulder conglomerates composed entirely of Uncompahgre quartzite flank the faults at both Coal Bank Pass and Molas Lake. On both flanks, the conglomerates grade very rapidly to quartzose sandstones away from the uplift, and to sandy siltstones, mudstones, and even some thin interbedded dolomites within a mile of the faults. The fine-grained clastics contain oboloid (inarticulate) brachiopods of latest Cambrian age between Coal Bank Pass and Cascade Creek to the south. Thus, the Grenadier fault block was conspicuously high and a prominent local source of clastics in Late Cambrian time.

ORDOVICIAN AND SILURIAN TIME

No rocks of Ordovician or Silurian age are known to be present in the San Juan Mountains or Colorado Plateau, due to nondeposition or pre-Late Devonian erosion.

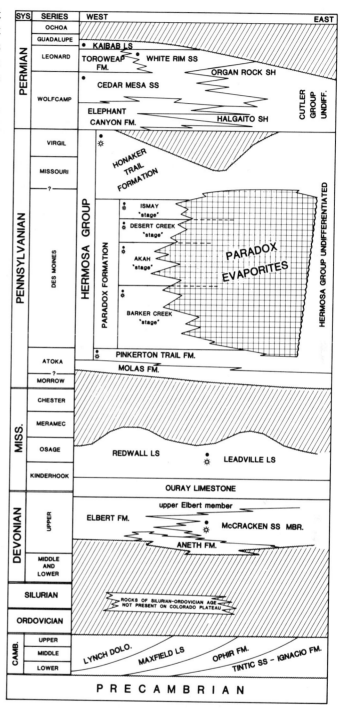

Figure 2. Paleozoic stratigraphic column for the San Juan Mountains and Paradox basin (after Baars and Stevenson, 1982.)

DEVONIAN SYSTEM

The Devonian System is represented in the San Juan Mountains by the Late Devonian Elbert Formation and the overlying Ouray Limestone. No Early or Middle Devonian strata have been found in the San Juans or Colorado Plateau. The Elbert

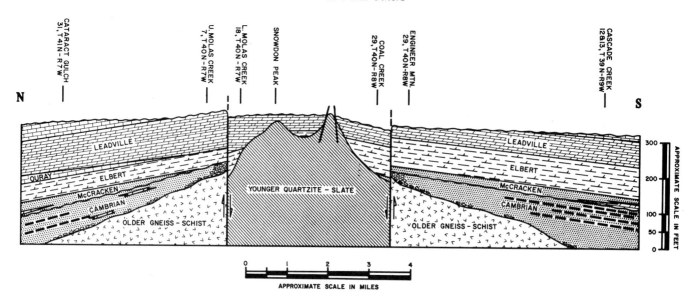

Figure 3. Grenadier Highland at the close of Mississippian time, San Juan Mountains, Colorado

Formation is subdivided into a lower McCracken Sandstone Member and an upper unnamed member.

The McCracken is a white or purplish mottled quartzite that lies paraconformably on the Ignacio. Where both the McCracken and Ignacio occur together in their sandstone facies, it is nearly impossible to distinguish them and find the disconformity that spans some 125 million years. The McCracken abuts the bounding faults of the Grenadier fault block, and is not present on top of the basement block. However, like the Ignacio, McCracken quartzites grade to shales and sandy dolomite within a mile of both bounding faults. Obviously, the Grenadier block was again (or was still) positive in McCracken time (Fig. 3).

The upper member of the Elbert Formation is a variable mixture of green and red mudstone, siltstone, and shaly dolomite. Invariably it is a slope-former and is poorly exposed. Because it contains stromatolitic zones, mud cracks, salt casts, and ripple marks, it is considered to be intertidal in origin in the San Juan Mountains (Baars, 1966). The upper member thins dramatically over the Grenadier fault block and buried all but the highest paleotopography. The Elbert Formation is dated as Late Devonian (Frasnian) on the basis of the fish plates (*Bothriolepis coloradensis*) it contains.

Gradationally overlying the Elbert Formation is the Ouray Limestone, a typically brown, micritic limestone. On both flanks of the Grenadier fault block it has been thoroughly dolomitized and is indistinguishable from the overlying lower member of the Leadville Formation. From Cascade Creek to Coal Bank Pass, the Ouray contains supratidal stromatolites, indicating the structural flank was high and the waters extremely shallow. North of the Coal Bank fault on the present-day downthrown side and on the Grenadier block, the Ouray is again a limestone with a normal marine fauna, suggesting that the fault had reversed its sense of throw since Elbert time. The Ouray is generally considered to

be Fammenian (latest Devonian) in age, although there is still some doubt because of mixed opinions on its contained fauna (Baars and Ellingson, 1984).

MISSISSIPPIAN ROCKS

The Leadville Formation is a massive carbonate unit that forms prominent cliffs and benches wherever it is exposed. The formation is informally subdivided into a lower, generally dolomitic, member, and an upper, generally limestone, member, which are separated by a regional disconformity. The formation is thin or missing along the flanks of the Grenadier fault block, due to Late Mississippian or Early Pennsylvanian weathering and erosion, and is missing on top of the central block. Crinoidal bioherms of the upper member are partially preserved in isolated localities along the flanks of the paleostructure, and may be seen between Cascade Creek and Coal Bank Pass in highly weathered remnants and at Molas Lake in "paleokarst towers." It would appear that favorable sites for shallow-water carbonate bank development were present on the paleostructural flanks, but were largely destroyed by erosion as the Grenadier block became once again a positive feature in Late Mississippian time.

PENNSYLVANIAN SYSTEM

Evidence for continuous maintenance of local Lower Pennsylvanian topographic/bathymetric differences and, hence, of paleostructure, is contained in the stratigraphy of the Molas, Pinkerton Trail, and Honaker Trail Formations of southwestern Colorado.

Molas Formation. The middle or upper Atokan Molas Formation (Merrill and Winar, 1958) in the San Juan Mountains is a regolithic residue of the underlying Leadville and Ouray carbonates. The formation has been divided into lower, essentially in situ oxidized residue of the carbonates; a middle, nonmarine but reworked red siltstone unit; and upper red, fossiliferous, well-stratified, marine members (Merrill and Winar, 1958).

The distribution of the various members of the Molas gives the oldest clue to the presence of Pennsylvanian tectonics between the locations of Silverton and Coal Bank Pass. The residual lower member and the marine upper member are recognized only in the vicinity of the type section at Molas Lake. Likewise, the transported middle member is thickest in the same area, indicating that the type section was located in a Pennsylvanian low. Specifically, the area north of a fault near Andrews Lake contains lower member residue that was not removed by erosion during middle member time; it also contains marine strata deposited by upper member seas that only entered that same area. Perpetuations of the low throughout Molas time suggests active down-to-the-north faulting.

Exposures of the residual lower member are best viewed in and around the karst tower of Leadville-Ouray near the Molas Lake parking area. The middle member is poorly exposed, but the sparsely fossiliferous upper member is exposed in road cuts between Molas Lake and Sultan Creek.

Pinkerton Trail Formation. The early(?) Desmoinesian Pinkerton Trail Formation in the subject area is typical of the unit elsewhere in the Paradox basin (Spoelhof, 1976). It consists of interbedded shale and partly dolomitized limestone and occasional medium- to very coarse-grained sandstone beds. The carbonates and shales are exceedingly fossiliferous. The biota includes small spired gastropods, bellerophontid gastropods, pelecypods, brachiopods, crinoidal debris, small foraminifers, fusulinids, rugose corals, *Chaetetes*, bryozoans, phylloid algae, and *Komia*. The stratification, lithology, and fossil assemblage indicate deposition in a relatively open shallow-marine environment.

The Pinkerton Trail does not appear to be present in the steep slopes flanking Engineer Mountain at Coal Bank Pass. It is recognized 20 mi (32 km) to the south at the type section in Hermosa Mountain, as well as in the vicinity of Molas Lake. No evidence of removal of the unit by erosion south of the Snowdon fault was found. It is concluded that the unit was not deposited in that area because of the presence of a paleostructure that resulted in continued differential subsidence in the Molas Lake area (and, presumably in the area of Hermosa Mountain).

Well-exposed road cuts of fossiliferous carbonates and dark shales of the Pinkerton Trail Formation are located just west of Molas Lake along U.S. 550.

Honaker Trail Formation. The Honaker Trail Formation in the area consists of approximately 600 ft (800 m) of "interbedded coarse-grained arkosic sandstone, siltstone, and shale, and 10 to 15 percent carbonate" that is age equivalent to Paradox Formation halite deposits in the Paradox basin to the northwest. In outcrop, the entire sequence is composed of numerous cycles of deposition, as is typical of the Desmoinesian elsewhere. Here, complete cycles consist in ascending order of: (1) Limestone: gray, clean, fossiliferous, relatively well stratified. Interpreted as

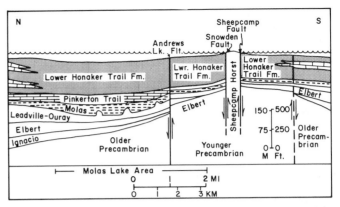

Figure 4. Paleostructure in early Honaker Trail (lower Desmoinesian) time. (After Spoelhof, 1976).

an open-marine deposit. (2) Shale: gray, sparsely fossiliferous. Interpreted as prodeltaic. (3) Sandstone: cross-stratified, convoluted, and ripple marked. Interpreted as delta fringe deposits. (4) Sandstone: coarse-grained, trough cross stratified, log bearing. Interpreted as distributary channels. (5) Shale: red below, grading to green above. Interpreted as channel fill and delta plain deposits. change in color taken to indicate increasing water depth at the end of the cycle. (6) Limestone: basal deposits of succeeding cycle.

The cycles of deposition are well exposed in road cuts along U.S. 550 just north of Coal Bank Pass. They are not, however, especially complete there. The section is dominated by siltstone and shale, with only a few well-developed channel sands. The major channel-sand packages are better developed a few kilometers to the north.

The apparent concentration of channels to the area of Molas Lake is taken as further evidence of continued fault movement, which maintained a low near Molas Lake into which channels were funneled (Fig. 4). The area of Coal Bank Pass was high and received only channel-margin sediments.

The interpreted history of paleo fault movement extends well into Desmoinesian time in this area. It was maintained at least until deposition of widespread carbonates in middle Honaker Trail time. Until then, the structural and stratigraphic relationships were a continuation of that established in early Honaker Trail time.

CUTLER FORMATION

Reddish brown arkosic sandstone, siltstone, and mudstone of the Cutler Formation overlie the Honaker Trail Formation with apparent disconformity. A basal conglomerate consisting of cobbles of sedimentary rocks derived from all underlying formations in exposures high on Engineer Mountain marks the erosional contact. Age of the formation in the San Juan Mountains has not been determined, as the red beds overlie rocks of Middle Pennsylvanian age, and underlie Triassic strata, with no known contained fossils. The Coal Bank Pass fault dies out into Cutler exposures, but heavy cover and foliage preclude determination of the exact relationships in that area north of Engineer Mountain. The fault may be terminating laterally *or* upward stratigraphically. Recent erosion has stripped the Cutler and younger rocks from the crest of the San Juan dome, and local evaluation of Permian and younger paleostructure is impossible (Baars and Ellingson, 1984).

REFERENCES CITED

Baars, D. L., 1966, Pre-Pennsylvanian paleotectonics–key to basin evaluation and petroleum occurrences in Paradox basin: American Association of Petroleum Geologists Bulletin, v. 50, p. 2082–2111.

Baars, D. L., and Ellingson, J. A., 1984, Geology of the western San Juan Mountains, in Brew, D. C., ed., Field Trip Guidebook: 37th Annual Meeting, Rocky Mountain Section, Geological Society of America, p. 1–45.

Baars, D. L., and See, P. D., 1968, Pre-Pennsylvanian stratigraphy and paleotectonics of the San Juan Mountains, southwestern Colorado: Geological Society of America Bulletin, v. 79, p. 333–350.

Baars, D. L., and Stevenson, G. M., 1982, Subtle stratigraphic traps in Paleozoic rocks of Paradox basin: *in* Halbouty, M. T., ed., Deliberate search for the subtle trap: American Association of Petroleum Geologists Memoir 32, p. 131–158.

Barker, F., 1969, Precambrian geology of the Needle Mountains, southwestern Colorado: U.S. Geological Survey Professional Paper 644A, 35 p.

Cross, W., Howe, E., Irving, J. D., and Emmons, W., 1905, Description of the Needle Mountains Quadrangle, Colorado: U.S. Geological Survey Atlas, Folio 131, 14 p.

Hutchinson, R. M., 1976, Precambrian geochronology of western and central Colorado and southern Wyoming, in Epis, R. C., and Weimer, R. J., eds., Studies in Colorado Field Geology: Professional Contributions of Colorado School of Mines, no. 8, p. 73–77.

Merrill, W. M., and Winar, R. M., 1958, Molas and associated formations in San Juan basin–Needle Mountains area, southwestern Colorado: American Association of Petroleum Geologists Bulletin, v. 42, p. 2107–2132.

Spoelhof, R. W., 1976, Pennsylvanian stratigraphy and paleotectonics of the western San Juan Mountains, southwestern Colorado, in Epis, R. C., and Weimer, R. J., eds., Studies in Colorado Field Geology: Professional Contributions of Colorado School of Mines, no. 8, p. 159–179.

The geology of Summer Coon volcano near Del Norte, Colorado

Jeffrey B. Noblett and Bruce M. Loeffler, Department of Geology, Colorado College, Colorado Springs, Colorado 80903

LOCATION AND ACCESS

The Summer Coon volcano is located about 6 mi (9.6 km) north of Del Norte, Colorado, on the western edge of the San Luis Valley (Fig. 1). The intrusive core and northern half of the volcano can be found on the Twin Mountains and Twin Mountains SE 7½-minute quadrangles. The southern part of the volcano lies in the Indian Head and Del Norte 7½-minute quadrangles. To reach the area of the volcanic core, take Colorado 112 from Del Norte to County Road 33; travel north on 33 about 6.2 mi (9.9 km); turn left on Road A32 (Forest Service road 660), marked by a small sign to "Natural Arch," and drive into the volcano.

The volcano lies almost entirely within the Rio Grande National Forest and is accessible by dirt roads, which are usually graded for passenger cars. Some areas, especially those that are near the Natural Arch and off the main road, should only be attempted during periods of clement weather or with four-wheel drive. The main road lies close to some of the radiating dikes and to good outcrops of extrusive material; it passes within 0.5 mi (0.8 km) of the heart of the intrusive complex. The round-trip time (from Del Norte) is a half-day minimum.

SIGNIFICANCE OF THE SITE

The Summer Coon volcanic complex is an Oligocene stratovolcano that has been eroded down to its base, revealing a variety of extrusive materials, hundreds of nearly perfect radiating dikes, and an intrusive core. Rock composition ranges from early mafic through middle silicic to late intermediate character. The rocks of the complex are included in the Conejos Formation, a sequence of Oligocene volcanic rocks that was extruded onto the eastern San Juan volcanic field prior to the eruption of the ash-flow tuff units that now comprise much of the San Juan Mountains. Summer Coon is one of the best-preserved relics of that earlier sequence. The unusual petrogenesis of mafic to silicic to intermediate rocks led Zielinski and Lipman (1976) to study the trace element history of this volcano. As a result of their study, they were able to hypothesize a mantle origin for the continental-interior andesite.

GENERAL DESCRIPTION

The Summer Coon volcano is part of the San Juan volcanic field, which covers an area of about 5,000 mi^2 (13,000 km^2) in southwestern Colorado. The field was developed on a platform composed of Precambrian rock overlain by a thin veneer of Phanerozoic sedimentary strata. Most of the sedimentary rocks were removed by erosion, following the uplift of the Uncompahgre highland in the late Paleozoic or during the Laramide orogeny.

The beginning of activity in the San Juan volcanic field was

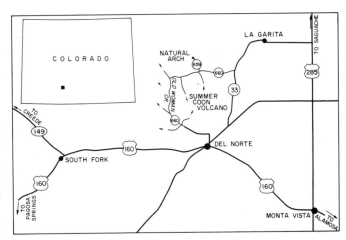

Figure 1. Map showing location of Summer Coon volcano (northwest of Alamosa) and surrounding areas.

marked by the eruption of numerous volcanos over this surface in the early Oligocene (Lipman, 1968). Eruptive units included intermediate lavas and breccias of the San Juan, Lake Fork, and West Elk formations in the western San Juans and of the Conejos Formation (with the Summer Coon volcano) in the eastern San Juans. Over a dozen calderas were active between 30 and 22 Ma (Lipman and others, 1970). Silicic ash-flows from these calderas covered most of the present San Juan area. It is possible that this volcanic field was at one time continuous across Colorado to the Thirty-Nine Mile volcanic field in South Park. A gravity anomaly suggests that the entire San Juan field may be underlain by a single large batholith.

The Rio Grande Rift began to develop in what is now the San Luis Valley at least 26 Ma, as evidenced by an early basalt flow from the Hinsdale Formation, which intruded alluvial and volcaniclastic sediments of the Los Pinos Formation. An upper limit for the beginning of rifting (at about 31 Ma) is established at Summer Coon volcano itself. The dips of extrusive material are asymmetric and change from 25–35° on the eastern side to dips of 5–15° on the western side. Assuming that the distribution of lava was initially symmetrical, the eastward tilting of the volcano (due probably to development of the Rift) occurred after the volcano was formed. Two dates (32.4 Ma and 34.7 Ma) have been determined from silicic dikes in the volcano (Lipman, 1976).

Erosion has uncovered the former stratovolcano down to its base, revealing a complete basal section of this approximately 8- to 10-mi (13- to 16-km) diameter cone. The rocks present within the volcano may be classified by structural occurrence or by chemical composition. The volcano is composed of three main structural types: (1) extrusive flows and breccias, (2) evenly distributed radial dikes, and (3) a circular intrusive complex in the center of the volcano. Due to the lack of cross-cutting relationships, correlation among these units is based largely on chemical composition as well as on observed or inferred field occurrence. The exposed rocks are classified into three main compositional types (Table 1): early mafic, middle silicic, and late intermediate compositions (Lipman, 1968; Mertzman, 1971a). Geologic maps may be found in Lipman (1976) and in Mertzman (1971a).

An obvious feature of the Summer Coon volcano is the nearly perfect radial pattern of dikes. Lipman (1968) interpreted these dikes as being the result of (1) small-scale doming and (2) radial fracturing, the result of doming due to a rising intrusion. Pre-existing horizontal stress in the area was apparently isotropic; the dikes lack any prevailing orientation. This supports the hypothesis that the volcano predates the north-south opening of the Rio Grande Rift.

There is a distinct lack of faulting in the volcano. Loken (1982) suggested that this may be the reason for the lack of mineralization in the intrusive core, which is barren, despite the existence of zones of hydrothermal alteration (which are a typical association of ore deposits).

IGNEOUS ROCKS

The volcanistratigraphy of the Summer Coon volcano includes the early mafic unit (2800 to 3100 ft; 850 to 950 m thick), which is almost entirely a poorly stratified breccia (see Fig. 2). The infrequent presence of spindle-shaped bombs suggests that this unit originated as an explosion breccia near a vent (Lipman, 1968; Mertzman, 1971a). Presence of a few scoriaceous beds supports this hypothesis. Lava flows grade into the breccia; these are petrographically identical to the breccia. The flows are thicker (to 20 ft; 6 m) and more abundant near the edges of the volcano. This circumstance may have resulted from autobrecciation of thin flows, with less-viscous portions flowing downhill while the angular blocks of breccia were left behind (Lipman, 1968).

The mafic dikes, though numerous, are easily overlooked in the field; they are typically only 1 to 2 ft (0.5 m) wide and rarely stand more than 1 ft (0.3 m) above the ground surface. These dikes are usually less than 500 ft (150 m) in length. None of the published maps attempt to show all of these dikes, though Lipman (1976) mapped many of them. Dikes range in composition from alkali olivine basalt to trachyandesite; clinopyroxene and plagioclase are the dominant phases.

Most of the mafic rocks have been classified as olivine andesites of basaltic appearance (Zielinski and Lipman, 1976). Color varies from dark gray to dark red where oxidized. These rocks contain about 30% phenocrysts (clinopyroxene, calcic andesine, and olivine) in a groundmass of plagioclase and glass, with small amounts of olivine and clinopyroxene. SiO_2 content varies from about 51 to 56 wt%.

The middle unit of the volcanic sequence is the most silicic; SiO_2 content is around 70 wt%. This unit is divided into two parts: a lower rhyodacite member and an upper rhyolite member (Mertzman, 1971a) (Table 1). The lower rhyodacite member is made up of dikes and a few lava flows that have a total thickness

of about 100 ft (30 m). Flows of this member are restricted to the southern side of the volcano. The dikes of this unit trend both southeast and southwest from the central intrusive complex. Dikes are up to 50 ft (15 m) thick and may rise 150 ft (45 m) above the ground surface and extend for 2 to 3 mi (3 to 5 km). The dike rocks are rich in phenocrysts of sodic andesine and biotite; they have been classed as quartz latite (Lipman, 1968).

The upper rhyolite member also is composed of both extrusive units and dikes. Flows may reach 330 ft (100 m) in thickness in the southwest part of the volcano. Units may show well-developed flow lamination or may be brecciated. The dikes trend northeast-southwest and may be traced for nearly 6 mi (9.6 km) across the base of the cone; they outcrop within 75 ft (23 m) of equivalent intrusive rocks in the core. A resistivity map (Mertzman, 1971a) suggests that the dike crosscuts the core.

The rhyolite unit is light tan and, in comparison with the rhyodacite, is poor in phenocrysts. The groundmass has devitrified to alkali feldspar and silica.

The late Intermediate unit is divided into two subunits: the lower pyroclastic member and the upper andesite member. Rocks of this unit have a silica content between 57 and 67 wt% and range in composition from quartz latite to rhyodacite. The lower pyroclastic member, concentrated on the southeast flank of the volcano, is a breccia of quartz latitic composition and has a minimum thickness of 600 ft (180 m); neither the top nor the bottom of this subunit is exposed. The breccia blocks are about 2 ft (0.6 m) in diameter. The large quantity of matrix material present, in addition to the lack of fiamme, led Mertzman (1971a) to hypothesize a laharic origin for this member.

The upper andesite member consists of flows, the aggregate thickness of which may reach 5,000 ft (1,500 m). Individual flows vary from 50 to 300 ft (15 to 90 m) in thickness. Basal and upper portions of the flows are brecciated, whereas the central portions are massive. The rocks are reddish-brown and average 10–20 percent phenocrysts. Plagioclase (mostly andesine) is the dominant phase. Other phases present include abundant hornblende, biotite, minor clinopyroxene, and sparse orthopyroxene and olivine.

These rocks may have been erupted from local fissures as lava lakes (Mertzman, 1971a).

Dikes of this unit radiate in all directions from the central core. Dikes may reach 200 ft (160 m) in height; most are 25 to 50 ft (8 to 15 m) wide and 2 to 4 mi (3 to 5 km) in length. Phenocrystic hornblende characterizes the dikes, although plagioclase, biotite, and augite are also common. The Natural Arch has been weathered through one of these dikes (Fig. 2).

TABLE 1. STRATIGRAPHY OF THE SUMMER COON VOLCANO

Conejos Formation	Extrusive	Intrusive
Late	Upper Andesite Lower Pyroclastic	Augite Monzonite (minor)
Middle	Upper Rhyolite Lower Rhyodacite	Upper Breccia Lower Granodiorite
Early	Mafic (andesitic)	Mafic

Figure 2. Natural Arch, weathered through a late andesitic dike that in turn cuts early mafic breccia.

The central intrusive complex appears as a group of low hills running north-northwest in the center of the volcano (Fig. 3). The hills are surrounded by an approximately circular, alluvium-filled valley about 2 mi (3.2 km) in diameter. All of the volcanogenic units occur within this complex.

The early mafic unit (intrusive) occurs in a variety of lithologic types, classified on the basis of composition (SiO_2 less than 58% by weight) and spatial distribution. Common phenocrysts include plagioclase, two pyroxenes, and hornblende. The groundmass is typically holocrystalline and contains abundant plagioclase, with clinopyroxene, quartz, hornblende, rare alkali feldspar, and opaques. Plagioclase composition ranges from An_{34} to An_{58}. The unit appears to be the result of a series of small intrusions that cooled at various rates.

The middle silicic unit (intrusive) consists of two subunits: a lower granodiorite porphyry member and an upper breccia member (Table 1). The lower granodiorite crops out as two distinct bodies, each about 0.5 mi (0.8 km) wide and separated by the upper breccia member. The granodiorite is the analogue of the lower rhyodacite dike member, chemically and mineralogically. It is an altered light-gray or buff porphyritic rock that contains phenocrysts of sodic oligoclase, minor biotite and hornblende, and some opaques. The groundmass consists of alkali feldspar, plagioclase, and quartz.

The lower granodiorite member becomes finer-grained near the contact with the early mafic intrusive unit; therefore, Mertzman (1971a) concluded that the granodiorite is the younger unit. However, since several early mafic dikes cut the granodiorite, the emplacement events of these two units may have overlapped.

The upper breccia member cuts between the two bodies of lower granodiorite and lies on strike with the major upper rhyolite dike. This member appears to be a poorly sorted light tan to yellow tuff breccia, which is fragmental and intensely altered. The presence of pumiceous blocks is indicated by slight changes in texture and color. Phenocrysts cannot be identified. The ground-

mass is mostly cryptocrystalline quartz with some vitric fragments; feldspars have altered to clay. The rock contains about 10% void space by volume, either as vesicles or as weathered-out phenocrysts.

Intrusive rocks of the late intermediate sequence occur only in a single minor pipe, about 20 ft (6 m) in diameter. This pipe is located on the southeast side of the hill, in Sec.30,T.41N.,R.6E., which is underlain by the early mafic intrusive unit (b, Fig. 3). The rock of this unit has been described as an augite monzonite porphyry (Mertzman, 1972); it can only be correlated with the late intermediate dike and flow units on the basis of chemistry and mineralogy. This single pipe could not have been the sole vent for the large volume of late flow material; other sources must exist. One such source may be Indian Head (see "Selected Stops") on the south flank of the volcano.

All of the units present in the Summer Coon complex show some degree of hydrothermal alteration. This alteration shows the classic concentric pattern, proceeding from propylitic alteration at the lithologic boundary between the early mafic and middle silic units outward through argillic and quartz-sericite zones. Despite numerous indications of mineralization, the Summer Coon complex probably does not host any economic deposits (Loken, 1982).

An interesting petrogenetic pattern is observed in the volcanic sequence exposed at Summer Coon. The rocks in this area were not extruded and deposited in order of increasing silica content. The earliest rocks in the sequence are basaltic/andesitic; the most likely model for origin of the primary magma is the partial melting at depth of a garnet-bearing, eclogitic equivalent of crustal material. Such a source might have evolved from a subducted slab of lithosphere or from conversion of mafic lower crustal material (from a depth of 25 to 30 mi; 40 to 50 km) to plagioclase-depleted garnet granulite. This supposition is supported by available Sr^{87}/Sr^{86} ratios and Pb-isotope and trace element data (Zielinski and Lipman, 1976).

The rocks in the later silicic sequence become more mafic with time (rhyodacite-rhyolite to andesite). This petrogenetic evolution may be explained by low-pressure fractional crystallization in an andesitic magma chamber. Vertical migration (settling or flotation) of phenocrysts might produce a stratified body that could then be progressively tapped to yield the observed sequence. The tapping of such a chamber at various levels can also explain some supposedly contradictory temporal relationships, in which middle rhyodacite dikes are seen cutting "late" andesite flows. This model is also supported by trace element data (Zielinski and Lipman, 1976).

SELECTED STOPS

Summer Coon can be explored with the assistance of either of the published maps and with Mertzman's road log (Mertzman, 1971b). A few of the readily accessible and interesting locations will be mentioned here.

Stop 1. Stop at the junction of County Road 33 and Road A32 (Forest Service road 660), about 9.5 mi (15.2 km) from Del

Figure 3. View to the southeast from northwest corner of intersection of Forest Roads 660 and 659 (Stop 3). Labeled points are (a) the middle silicic intrusive hills, (b) early mafic intrusive, (c) lower rhyodacite dikes cutting early mafic member, (d) upper andesite dike, and (e) early mafic extrusives.

Norte. The high ridge east of the County Road is vitrophyric Carpenter Ridge Tuff. The "elephant rocks" west of the road are outcrops of the slightly older Fish Canyon Tuff (27.8 Ma).

Stop 2. Continue west on the Forest Service road about 0.9 mi (1.4 km) from the junction. Several hundred yards to the north is a small hill with a well-exposed outcrop of early mafic extrusives. A basaltic andesite flow, which has east-dipping platy joints, overlies explosion breccia material.

Stop 3. Continue on Forest Service road 660 to the junction with the side road (659.1, or Road 35C) which leads to the Natural Arch. At this point you are 13.6 mi (21.8 km) from Del Norte, or 4.1 mi (6.5 km) from Stop 1; you are now within the intrusive complex (Fig. 3). Rocks on the hill northeast of the junction are part of the early mafic unit. The high hills to the southeast are made of middle silicic rocks. A hike to the top of these hills is recommended as the best way to view the alteration zones, the intrusive units, and the spectacular radiating pattern of the dikes. The main middle silicic–lower rhyolite dike cuts through these hills and can be traced across the road about 0.5 mi (0.8 km) east of the junction. A late intermediate dike crops out northwest of the road junction and trends north-northwest. A group of middle silicic–upper rhyodacite dikes crops out about 1.5 mi (2.4 km) south of the junction, along the four-wheel drive road.

Stop 4. About 0.4 mi (0.6 km) north of the junction in Stop 3, near the crest of a small hill, is a sharp but subtle contact between the early mafic breccia and early mafic intrusives. A small, typical early mafic dike cuts the breccia. The tall dikes west of the road belong to the late intermediate unit.

Stop 5. Continue 1.7 mi (2.7 km) north from the junction of Forest Service road 660 with Road 35C (Stop 3). This is the Natural Arch, a fine example of a late intermediate dike that cuts early mafic breccia (Fig. 2). This has been described as a porphyritic quartz latite (Mertzman, 1971a, b). The view from inside the arch includes the Sangre de Cristo Mountains and Great Sand Dunes National Monument to the east, across the San Luis Valley. Most of the Summer Coon volcano, including the intrusive complex, is visible to the east and south.

Stop 6. Return to Stop 3 and continue west on the main Forest Service road through the volcano. At about 22 mi (35 km) from Del Norte (including the 3.3-mi—5.3-km—round trip to Natural Arch), there is a large rhyodacite dike on the left. At 23.4 mi (37.5 km), a rhyolite flow occurs on the left. The next hill south (on the eastern side of the road) at the county line contains more rhyolite flows, overlain by late andesite lavas. The low hills visible to the east and north for the next few miles are also late andesites. Turn onto the paved road (a junction) at 26.8 mi (42.9 km). The outcrops immediately east of this junction are part of the lower pyroclastic member of the late intermediate flows. At mile 27.8 (44.5 km), Indian Head juts up immediately to the north; this may have been a vent for late intermediate material. Turn right across the canal at the junction with County Road 15 (29.8 mi; 47.8 km), and rejoin Colorado 112 at 30.6 mi (49 km).

REFERENCES

Lipman, P. W., 1968, Geology of the Summer Coon volcanic center, eastern San Juan Mountains, Colorado: Colorado School of Mines Quarterly, v. 63, p. 211–236.

—— , 1976, Geologic map of the Del Norte area, eastern San Juan Mountains, Colorado: U.S. Geological Survey Miscellaneous Investigations Map I-952.

Lipman, P. W., Steven, T. A., and Mehnert, H. H., 1970, Volcanic history of the San Juan Mountains, Colorado, as indicated by potassium-argon dating: Geological Society of America Bulletin, v. 81, p. 2329–2352.

Loken, T., 1982, Hydrothermal alteration and oil show at the Summer Coon intrusive center, Saguache County, Colorado [M.S. thesis]: Corvallis, Oregon State University, 91 p.

Mertzman, S. A., Jr., 1971a, The Summer Coon volcano, eastern San Juan Mountains, Colorado: New Mexico Geological Society Guide, 22nd Field Conference, p. 265–273, map and chemical analyses in pocket.

—— , 1971b, Supplemental roadlog no. 2, Del Norte to Summer Coon volcanic area and return: New Mexico Geological Society Guide, 22nd Field Conference, p. 73–75.

—— , 1972, The geology and petrology of the Summer Coon volcano, Colorado [Ph.D. thesis]: Cleveland, Ohio, Case Western Reserve University, 253 p.

Zielinski, R. A., and Lipman, P. W., 1976, Trace element variations at Summer Coon volcano, San Juan Mountains, Colorado, and the origin of continental-interior andesite: Geological Society of America Bulletin, v. 87, p. 1477–1485.

Recurrent Quaternary normal faulting at Major Creek, Colorado: An example of youthful tectonism on the eastern boundary of the Rio Grande Rift Zone

James P. McCalpin, Department of Geology, Utah State University, Logan, Utah 84322

LOCATION AND ACCESS

A well-preserved fault scarp resulting from recurrent Quaternary normal faulting occurs at the western edge of the northern Sangre de Cristo Mountains of south-central Colorado, at the mouth of west-draining Major Creek (Fig. 1). The site is approximately 56 mi (90 km) north-northeast of Alamosa, Colorado, and may be approached from Colorado 17 between Alamosa and Poncha Springs, Colorado. Turn east off of Colorado 17, 50 mi (80 km) north of Alamosa, opposite the junction with U.S. 285, onto a dirt road that leads due east for 6 mi (10 km) across the valley floor toward Valley View Hot Springs. Instead of turning off to the hot springs, bear right and follow the road as it turns south to parallel the range front. The road will continue south past the mouth of Garner Creek [1 mi (1.7 km) south of the hot springs turnoff] to the upper part of the Major Creek alluvial fan [2 mi (3.2 km) south of the hot springs turnoff]. Approximately 0.25 mi (400 m) north of the crossing of Major Creek (Fig. 2) turn onto a dirt driveway leading due east to the head of the fan where the fault scarps occur. Entry through the locked gate will require permission from the landowner on the north side of the fanhead, Dr. Ben Eismann, Dept. of Surgery, University of Colorado Health Science Center, Denver, CO 80262. The fault scarps on the south side of the road (including the 1980 trench site, Profile 35 on Fig. 2) are owned by Mr. Stan Pavlin, 4004 Hillside Dr., Pueblo, CO 81008; permission is also necessary for access to this property.

SIGNIFICANCE OF THE SITE

The Major Creek fanhead displays one of the best multiple-event fault scarps within the Rio Grande Rift Zone. Quaternary alluvial fan deposits of five ages are offset by a single strand of the Sangre de Cristo Fault Zone. Detailed scarp profiling, soil description, and scarp trenching yield a coherent picture of multiple surface-faulting earthquakes (as many as 13) in roughly the last 400,000 years, with displacements of 5.2 ft to 7.2 ft (1.6 m to 2.2 m) per event. The last such event occurred about 7,660 ± 140 B.P. The timing and style of faulting are important for two reasons: (1) they indicate the probable way in which the structural relief of the Rio Grande Rift has developed by numerous small-scale earthquake displacements, and (2) they identify the Sangre de Cristo Fault as one of Colorado's few active faults, based on both Holocene activity and recurrent late Pleistocene activity. The significance of this locality in relation to regional tectonics is described in recent reports on Quaternary faulting for Colorado (Kirkham and Rogers, 1981; Colman, 1985) and for the northern Rio Grande Rift Zone (McCalpin, 1983; Colman and others, 1985).

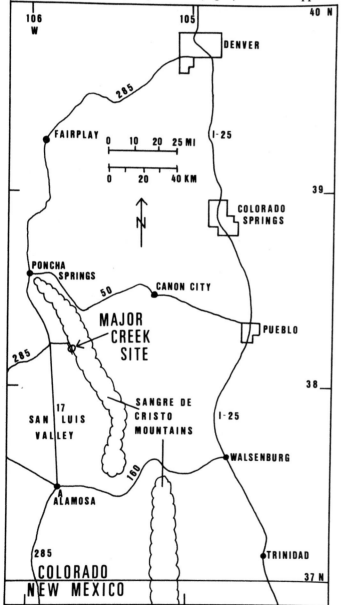

Figure 1. Regional location map for the Major Creek site. Detailed access directions given in text refer to this figure and Figure 2.

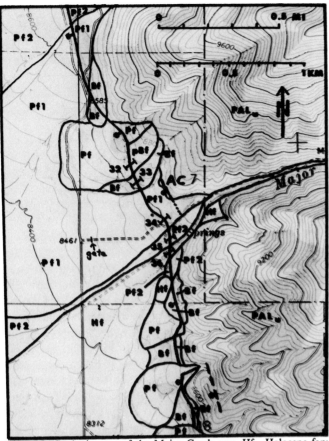

Figure 2. Geologic map of the Major Creek area. Hf—Holocene fan; Pfl—early Pinedale fan; Pf2—mid-late Pinedale fan; Pf—Pinedale fan, undivided; Bf—Bull Lake fan; pBf—pre-Bull Lake fan; Palu—Paleozoic bedrock, undivided. Fault scarp is a heavy line with bar and ball on downthrown side; subsidiary(?) fault segments to E are dashed lines. The numbered barred lines across the scarp show profile locations of Figure 3. Base map: Valley View Hot Springs 7½′ quadrangle: contour interval = 40 ft (12.19 m).

SITE INFORMATION

The heads of many alluvial fans on the west side of the Sangre de Cristo Mountains are offset by fault scarps, but the scarps at Major Creek show clearly the geomorphic relations with a sequence of Quaternary terraces. A set of five climatically induced fan terraces is inset into the head of the Major Creek alluvial fan (Fig. 2). Although the range-bounding Sangre de Cristo Fault offsets all five terraces, some terraces are present both above *and* below the scarp. Such geometry indicates that the terraces are not purely strath terraces cut into the upthrown block (as described by Soule, 1978), because strath terraces would not continue below the scarp. Instead the terraces owe their origin to changes in stream regimen, presumably accompanying climatic change, and resemble non-faulted fanhead terraces common elsewhere along the range front and elsewhere in Colorado.

The fault scarp is simple in geometry as defined by Slemmons (1957, p. 367), meaning that a single high-angle break offsets surfaces that preserve their original depositional gradient. The scarp becomes progressively higher as it offsets older fan deposits, indicating recurrent movement contemporaneous with episodes of fan formation. Deposits of five ages (Fig. 2) offset at the fanhead are correlated to glacial episodes in the Sangre de Cristo Mountains by McCalpin (1983, p. 9–37), based on relative-dating criteria and by comparison to nearby fans that can be traced directly to late Pleistocene moraines. Figure 3 shows scarp profiles across each faulted unit; the age of the unit and the vertical displacement it has undergone are given in Table 1. The number of faulting events may be estimated by assuming that the smallest scarp (#36, 5.2 ft—1.6 m—throw) represents a single event or alternatively, that the second smallest (#35, 12.5 ft—3.8 m—throw) represents two events. The number of inferred fault events with displacements of 5.25–7.25 ft (1.6–2.2 m) ranges from 5 since Pinedale time to 13 since mid-Pleistocene time

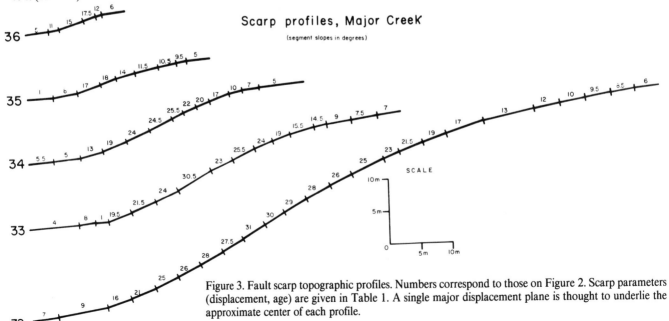

Scarp profiles, Major Creek

(segment slopes in degrees)

Figure 3. Fault scarp topographic profiles. Numbers correspond to those on Figure 2. Scarp parameters (displacement, age) are given in Table 1. A single major displacement plane is thought to underlie the approximate center of each profile.

TABLE 1. FAULT SCARP PARAMETERS, MAJOR CREEK SITE

Profile Number[1]	Deposit Offset[2]	Total Vertical Fault Displacement[3]		Number of Fault Events[4]	Deposit Age (ka)[5]	Recurrence Interval (ka)[6]
		ft	m			
36	Hfl	4.6	1.4	1	8	8.0
35	Pf2	12.5	3.8	2	13	5.0
34	Pfl	29.2	8.9	5	25	4.0
33	Bf	44.3	13.5	7	150	62.5
32	pBf	76.7	23.4	13	400	41.6

[1]Corresponds to numbers on Figure 2.
[2]Hfl, early Holocene; Pf2, mid-late Pinedale; Pfl, early Pinedale; Bf, Bull Lake; pBf, pre-Bull Lake.
[3]Measured by graphical projection of upper over lower surface, assuming a 70° fault dip.
[4]Calculated by dividing an average 1.6-2.2 m displacement per event into the total vertical fault displacement, rounded to the nearest even number.
[5]Ages from local C14 dates, and via correlation to dated glacial deposits elsewhere in the western U.S. (see McCalpin, 1983, Fig. 32 and Table 10.
[6]Calculated by dividing the number of fault events (column 4) into the length of time in which they occurred (difference between ages in column 5); Example: 2 events (7 minus 5) occurred between 150 ka and 25 ka, so 125 ka divided by 2 events = 62.5 ka per event.

(Table 1, column 4). To confirm all assumptions, the scarp at surface profile #35 was trenched; the log is presented as Figure 4.

Study of the trench confirmed that two faulting events had occurred since deposition of unit Pf 2. The first event offset lenticularly bedded, sandy to cobbly late Pinedale alluvium and created a topographic trough in which clayey, organic-rich sag pond deposits had begun accumulating by at least 10,400 ± 240 B.P. The base of sag deposits was not exposed. Faulting is therefore roughly bracketed between the formation of the mid-late Pinedale terrace (approximately 15,000 B.P.) and 10,400 ± 240

B.P. A second fault event severely deformed this soft muck and created a free face that dumped 5.9 ft (1.8 m) of coarse colluvium onto the sag deposits. The uppermost sag deposit carried an organic A horizon that dated at 7,660 ± 140 B.P.—presumably no younger carbon was added after burial. The mean residence time of organic material in the A horizon may be hundreds or thousands of years, so the date of latest faulting may be somewhat younger than 7,660 B.P. Fault scarp profiling of the younger scarps (#35, #36) does not yield a comparably young date in relation to other published scarp height-slope angle data (e.g.,

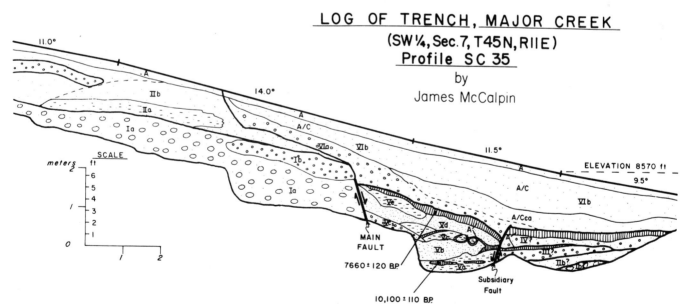

Figure 4. A portion of the log of a trench across profile 35 in mid-late Pinedale alluvium. Heavy lines bound six depositional units: I through IV—mid-late Pinedale alluvium; V—early Holocene tectonic colluvial wedge; VI—mid-Holocene tectonic colluvial wedge. Facies of major units are shown by thinner lines and lower case letters, soil horizons developed on deposits by upper case letters like A/C. Two colluvium-producing fault events are inferred.

Bucknam and Anderson, 1979), because fine sands in which the scarps are developed are highly erodable.

Faulting recurrence intervals for several time periods were calculated based on scarp data. Resulting recurrence intervals seem to be considerably shorter since the Early Pinedale (Table 1, column 6). This conclusion, also reached for most other faulted fanheads along the Sangre de Cristo Fault, has two possible causes: (1) seismicity has rapidly increased within the last 25,000 years on this fault, or (2) not all earlier faulting events resulted in rupture on this fault trace. If (2) is true, as we examine longer pieces of Quaternary history, we may miss more and more fault events that occurred on other less visible traces.

Possible evidence to favor the second hypothesis is found immediately southeast of Profile 33. The base of the range front approximately 425 ft (130 m) east of the scarp here is very steep and linear. A 165-ft (50-m) high outcrop on this face exposes hydrothermally altered Paleozoic carbonate rock now altered to variegated clays (marked AC on Fig. 2). This alteration may be the result of shearing and fluid migration along a second, less active normal fault that parallels the measured scarp. However, no Quaternary deposits are offset by a projection of this suspected fault across the narrow Major Creek valley.

Larger features of the fault-generated range front are also visible on the approach to Major Creek. Well-developed facets truncate ridge lines above the fault zone, as described for other areas by Wallace (1978). Between narrow faceted ridges, elongate drainage basins extend perpendicularly to the range crest. Faceted spurs exhibiting multiple benches and steps are best displayed in the 12.5-mi (20-km) long range segment stretching from Major Creek south to San Isabel Creek. Steps sloping valleyward at 20° to 34° are separated by lower-angled ridge crests with 7° to 15° slopes. Four persistent facet sets have crests at approximately 10,825 ft (3,300 m), 9,512 ft (2,900 m), 9,180 ft (2,800 m), and 8,760 ft (2,670 m). The higher, larger facets are severely gullied, but the two lower sets are inset within the larger set and exhibit less dissection. The lowest facet set includes numerous small, very steep, ungullied planar facets that rise directly above fault scarps in Quaternary deposits. This geometry suggests that periodic rapid uplift of the mountain block has alternated with tectonic quiescence and range-front parallel retreat within the late Cenozoic. Recurrence interval data in Table 1 also show that variable activity occurs within shorter time spans and suggest that intervals between individual earthquakes or swarms of earthquakes have varied widely in the late Cenozoic.

REFERENCES

Bucknam, R. C., and Anderson, R. E., 1979, Estimation of fault scarp ages from a scarp-height-slope-angle relationship: Geology, v. 7, p. 11–14.

Colman, S. M., 1985, Map showing tectonic features of late Cenozoic origin in Colorado: U.S. Geological Survey Map I-1556, scale 1:1,000,000.

Colman, S. M., McCalpin, J., Ostenaa, D. A., and Kirkham, R. M., 1985, Map showing upper Cenozoic rocks and deposits and Quaternary faults, Rio Grande rift, south central Colorado: U.S. Geological Survey Map I-1594, scale 1:125,000.

Kirkham, R. M., and Rogers, W. P., 1981, Earthquake potential in Colorado; A preliminary assessment: Colorado Geological Survey Bulletin 43, 171 p.

McCalpin, J., 1983, Quaternary geology and neotectonics of the west flank of the northern Sangre de Cristo Mountains, south central Colorado: Colorado School of Mines Quarterly, v. 77, no. 3, 97 p.

Slemmons, D. B., 1957, Geological effects of the Dixie Valley–Fairview Peak, Nevada, earthquakes of December 16, 1954: Bulletin of Seismological Society of America, v. 47, no. 4, p. 353–375.

Soule, C. H., 1978, Tectonic geomorphology of the Big Chino Fault, Yavapai County, Arizona [M.S. thesis]: Tucson, University of Arizona, 114 p.

Wallace, R. E., 1978, Geometry and rates of change of fault-generated range fronts, north-central Nevada: U.S. Geological Survey, Journal of Research, v. 6, no. 5, p. 637–650.

Alteration zones related to igneous activity, Spanish Peaks area, Las Animas and Huerfano counties, Colorado

Robert M. Hutchinson, *Department of Geology, Colorado School of Mines, Golden, Colorado 80401*
J. D. Vine, *21736 Panorama Drive, Golden, Colorado 80401*

LOCATION

The Spanish Peaks area is composed of two prominent peaks and associated dike swarms about 19 to 22 mi (30 to 35 km) southwest of Walsenburg in south-central Colorado (Fig. 1). Although the entire area is surrounded by roads, the Apishapa Pass road on the south flank of the high peaks is open only in the summer time. From Walsenburg, the north slope is accessible by driving west on U.S. 160 about 11 mi (18 km) to Colorado 111. Continue 5 mi (8 km) west and south on Colorado 111 to the town of La Veta. Take a country road from the southeast edge of town for about 7 mi (11 km) south and southeast past twin reservoirs to Wahatoya Canyon and the Forest Service Wahatoya trail head at an elevation of about 8,400 ft (2,560 m). The main trail leading south goes to the Bulls Eye Mine, elevation 11,200 ft (3,414 m), on the north slope of West Spanish Peak, elevation 13,626 ft (4,153 m).

Access to the south flank of the Spanish Peaks is from Apishapa Pass, which can be reached by continuing from La Veta south on Colorado 111 about 16 mi (26 km) to Cucharas Pass. Turn left onto the Forest Service road to the northeast and drive about 6 mi (9.6 km) to the trail head at Apishapa Pass (closed in the winter), elevation 11,248 ft (3,428 m). From the pass, a trail leads to timberline on West Spanish Peak at an elevation of about 11,800 ft (about 3,600 m). The climb from there to the top of West Spanish Peak is over rocks and scree but does not require technical equipment. The Forest Service road east from Apishapa Pass continues past many beautiful rock exposures and vistas to Aguilar on I-25, some 17 mi (27.4 km) south of Walsenburg. Although it is possible to see much beautiful scenery in a half day by driving the main roads surrounding the area, one or more days are recommended to hike the trails and study the several zones of alteration.

SIGNIFICANCE

The Spanish Peaks have long been regarded as a classic area in which to study a great variety of igneous intrusions and dikes of post-Laramide (mid-Tertiary) age. More recently the area has also been recognized for its pervasive alteration of the intruded sedimentary rocks. East Spanish Peak and West Spanish Peak form twin conical summits that rise 6,500 and 7,500 ft (2,000 and 2,300 m) above the adjacent high plains, like sentinals guarding the approach to the Sangre de Cristo Range to the west. Although the summit of East Spanish Peak is formed from an igneous stock, the summit of West Spanish Peak is composed mostly of altered Tertiary sedimentary rocks. It also represents

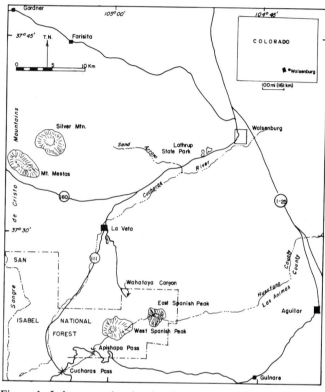

Figure 1. Index map showing location of the Spanish Peaks area, Colorado.

the focal center for a conspicuous swarm of radial dikes as well as the higher grade zones of alteration.

SITE INFORMATION

Petrology and Structure of the Igneous Rocks. The sedimentary rocks enclosing the two Tertiary stocks of the Spanish Peaks area have been invaded by a prominent swarm of dikes radiating outward from each. The eastern stock of granite porphyry with a central core of somewhat later granodiorite porphyries has slightly domed the surrounding rocks. The later, somewhat smaller western stock of pyroxene "syenodiorite" (monzonite) has invaded a syncline without doming the sedimentary rocks and has metamorphosed them for at least 900 ft (274 m) outward from the contact (Fig. 2).

Most, but not all, of the 500 or more radial dikes converge on West Spanish Peak stock, the swarm occupying an ellipse extending largely eastward from the focal area. None of these dikes come in contact with the stock and only a few cut its

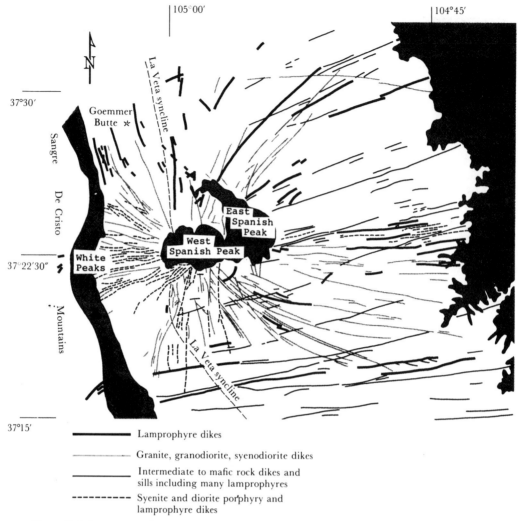

Figure 2. Dike swarms of lamprophyres and associated rocks surrounding the Spanish Peaks stocks, Colorado. After Johnson (1961, plate 1; 1968).

metamorphosed aureole, suggesting that the dike magmas are younger and unrelated to the stock magmas.

Petrographically, the radial dikes range from basalt and gabbro lamporphyre to microgranite and granite porphyry. The rock types of the dikes are basalt and gabbro lamprophypre; feldspathoidal diorite lamprophyre, diorite porphyry, diorite lamprophyre, and microdiorite; syenodiorite porphyry and microsyenodiorite; syenite porphyry, syenite lamprophyre, and microsyenite; granodiorite porphyry; and granite porphyry and microgranite.

Gabbro lamprophyre and basalt dikes are scattered throughout the dike swarm. Most of them are less than 3 mi (4.83 km) long and are relatively far from the West Spanish Peak stock. These dikes do not cut any other types of dikes but are cut by syenodiorite porphyry, granodiorite porphyry, and granite porphyry dikes.

The general sequence of magmatic intrusion of the various igneous features was determined in the field (Johnson, 1961,

1964, 1968) by structural relationships of the various types of igneous rocks. Intersections of the dikes are comparatively common. Generally, the order of intrusion does not seem to have been from mafic to silicic. Instead, it seems to have been almost the reverse: the oldest intrusion is probably the granite porphyry of the East Spanish Peak stock, and mafic dikes and sills of the parallel dike system are among the youngest. However, the order of intrusion of the radial dikes seems to have been from mafic to silicic, the dikes range from gabbro lamprophyre and basalt to granite porphyry and microgranite. Johnson's (1961) proposed sequence of magmatic intrusion is summarized as follows:

(1) granite porphyry stock of East Spanish Peak
(2) granite porphyry stock of the White Peaks
(3) granodiorite porphyry stock of East Spanish Peak
(4) syenodiorite stock of West Spanish Peak
(5) basalt sills of West Spanish Peak
(6) radial dike swarm of West Spanish Peak

(A) gabbro lamprophyre and basalt dikes
(B) feldspathoidal diorite lamprophyre dikes
(C) diorite porphyry, diorite lamprophyre, and microdiorite dikes
(D) syenodiorite porphyry and microsyenodiorite dikes (two or three separate phases)
(E) syenite porphyry, syenite lamprophyre, and microsyenite dikes (two separate phases)
(F) granodiorite porphyry dikes
(G) granite porphyry and microgranite dikes
(7) later parallel dike swarm and related sills
(A) Olivine gabbro lamprophyre; gabbro lamprophyre, and basalt dikes, dikes and sills; feldspathoidal diorite and syenite lamprophyre, diorite lamprophyre, syenodiorite porphyry and microsyenodiorite, and syenite porphyry and lamprophyre dikes (sequence indeterminate)
(8) microsyenodiorite dike and sill
(9) latite plug of Goemmer Butte

Knopf (1936) and earlier workers surmised that the lamprophyres and other dikes all originated in the Spanish Peaks stocks. Johnson (1961) pointed out, however, that there is no reason to believe that the intrusion of the stocks should result in radial fractures; that the dikes do not converge on a single source; that some dikes are exposed discotinuously along their lengths, indicating that the magma rose from below; and that no dikes contact the West Spanish Peak stock at the surface, and the lamprophyres are rare among the sills and metamorphosed rocks surrounding the stock—also indicating that the magma came from below. Johnson proposed that regional joints are related to orogenic stresses of varying magnitude and direction, resulting in a complex and random pattern. The radial dike pattern would result from selective injection of low-viscosity magmas into those joints that are normal to nearly domical equipotential magma-pressure surfaces.

Alteration of the Sedimentary Rocks. Sedimentary rocks of early Tertiary age that are exposed in the area peripheral to the Spanish Peaks have been pervasively altered over a broad area. These rocks, including the Poison Canyon Formation of Paleocene age and the Cucharas Formation of Eocene age, have been intruded by the stocks of East Spanish Peak and cut by numerous dikes, including the conspicuous radial dike swarm, the focal center of which is West Spanish Peak. A concentric arrangement of alteration zones also centers on West Spanish Peak (Fig. 3). Characteristic minerals of these zones, away from the center are: amphibole → prehnite-pumpellyite → authigenic plagioclase → laumontite → nonlaumontite zeolite → and weakly altered zone (Utada and Vine, 1985). The amphibole zone contains hornblende or actinolite in addition to authigenic plagioclase in hornfels and is present in an aureole less than 0.6 mi (1 kmF) wide where it forms a ridge above timberline on the south slope of West Spanish Peak. Prehnite and pumpellyite were found only locally in a zone peripheral to the amphibole zone. The authigenic feldspar zone is more widely distributed in an aureole about

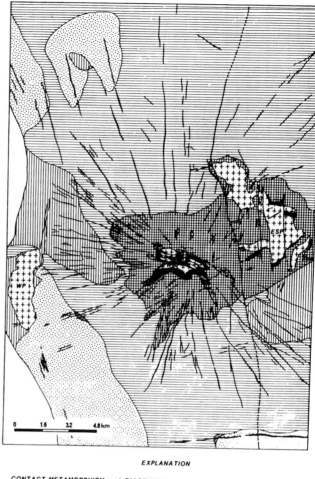

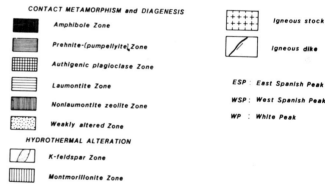

EXPLANATION

CONTACT METAMORPHISM and DIAGENESIS

Amphibole Zone

Prehnite-(pumpellyite) Zone

Authigenic plagioclase Zone

Laumontite Zone

Nonlaumontite zeolite Zone

Weakly altered Zone

HYDROTHERMAL ALTERATION

K-feldspar Zone

Montmorillonite Zone

Igneous stock

Igneous dike

ESP : East Spanish Peak
WSP : West Spanish Peak
WP : White Peak

Figure 3. Distribution of the alteration zones in the Spanish Peaks region. After Utada and Vine (1983, fig. 5).

2.5 mi (4 km) wide surrounding both East and West Spanish Peaks, where it is transitional with the surrounding laumontite zone. The laumontite and nonlaumontite zeolite zones are also transitional and widely distributed from the vicinity of Gulnare on the Apishapa Pass road, some 14 mi (22 km) southeast of West Spanish Peak to the vicinity of Gardner, more than 28 mi (45 km) to the northwest.

Excellent exposures of laumontite-bearing feldspathic sandstones and conglomerates are present intermittently along the Apishapa Pass road west of Gulnare to the vicinity of Apishapa Pass. One of the best natural exposures of altered conglomerate is in a cliff about 2,000 ft (610 m) to the northwest and 600 ft (200 m) above the Forest Service picnic area on Apishapa Creek. Additional exposures of laumontitic arkose are present along U.S. 160 west of Walsenburg in the vicinity of Lathrop State Park and north of the highway along the tributaries of Sand Arroyo, which crosses the highway about 6 mi (10 km) west of Walsenburg (see Vine, 1974).

Other types of alteration occur locally. Wairakite has been observed in the authigenic plagioclase and amphibole zones where it replaces plagioclase. It also occurs in hornfels as spherical nodules up to several cm in diameter, especially in the upper part of Wahatoya Canyon. Analcime occurs locally in the laumontite and lower grade zones of alteration. In the cliff exposures above the Apishapa Creek picnic ground, analcime occurs as 0.4- to 0.8-in (1- to 2-cm) thick leached selvages adjacent to less altered reddish-brown mudstone. K-feldspar alteration forms a selvage as much as 1.6 ft (0.5 m) thick along the contact zone of dike rocks and is especially conspicuous in the laumontite zone of alteration because it forms a hard ledge that is more resistant to weathering than the laumontitic sandstone and locally is more resistant than the mafic dike rock that it encloses. K-feldspar is also associated with metal-bearing veins at the Bulls Eye mine on the north slope of West Spanish Peak. Authigenic epidote is abundant locally in the authigenic plagioclase zone and forms hard ledges on the south flank of West Spanish Peak. Authigenic montmorillonite is widespread in the laumontite and lower grade zones of alteration. In areas north of U.S. 160, along Sand Arroyo, sheetlike beds of sandstone in the Poison Canyon Formation that are as much as 25 ft (7.6 m) thick display laumontite alteration in the upper part, authigenic montmorillonite alteration in the middle, and nonlaumontite zeolite such as stilbite in a thin zone at the base. The laumontite alteration zone weathers into characteristic massively rounded shapes that are more resistant to weathering than the underlying montmorillonite-cemented sandstone. This gives the distinctive character to the topography in this area. Calcite cement is abundant locally along the bluffs adjacent to the Cucharas River near La Veta. This may represent a late stage of alteration superimposed on laumontitic sandstone, perhaps by weathering. In the Cucharas Formation north of the Spanish Peaks, pink sandstone tongues that are lenticular in cross section and surrounded by mudstone contain a laumontite cement throughout. Although alteration of various kinds extends for many miles along the trough of the La Veta syncline, it fades out along the steeply tilted rocks that forms the east flank of the Sangre de Cristo range.

Halos of alteration are also recognized in areas peripheral to intrusive rocks other than the Spanish Peaks. For example, there is an extensive zone of laumontite alteration in the red arkosic sandstones of the Sangre de Cristo Formation of Pennsylvanian and Permian age where these are exposed in road cuts of U.S. 160 along the flank of Mount Mestas and in the vicinity of North La Veta Pass. This alteration is probably centered on the felsite intrusion of Mount Mestas. Alteration zones of a grade higher than laumontite also occur peripheral to Silver Mountain (Dike Mountain), about 5 mi (8 km) north of U.S. 160, but this area has not been studied or mapped in the same detail as the Spanish Peaks.

Because laumontite is a calcium zeolite, $Ca(Al_2Si_4O_{12})$ $4H_2O$, and is the most widespread alteration mineral replacing K-feldspar and micaceous minerals, the amount of calcium in the sandstone has apparently increased at the expense of the alkalis, potassium, and sodium. Laumontite is known to form by hydrothermal alteration and is locally associated with hot springs (Sharp, 1970). Thus, the entire northern part of the Raton basin may have been heated and altered by calcium-rich hydrothermal solutions associated with the intrusions at the Spanish Peaks and adjacent areas. These intrusions had to penetrate a sequence of Paleozoic limestones, including those in the Minturn Formation of Pennsylvanian age. If the limestones provided a source for the calcium, what happened to the CO_2? The presence of a CO_2 gas field in the Sheep Mountain intrusive area north of Mount Mestas may be one answer.

REFERENCES

Johnson, R. B., 1961, Patterns and origin of radial dike swarms associated with West Spanish Peak and Dike Mountain, south-central Colorado: Geological Society of America Bulletin, v. 72, p. 579–590.
—— , 1964, Walsen composite dike near Walsenburg, Colorado: U.S. Geological Survey Professional Paper 501-B, p. B69–B73.
—— , 1968, Geology of igneous rocks of the Spanish Peaks region, Colorado: U.S. Geological Survey Professional Paper 594-G, p. G1–G47.
Knopf, A., 1936, Igneous geology of the Spanish Peaks region, Colorado: Geological Society of America Bulletin, v. 47, p. 1727–1784.

Sharp, W. N., 1970, Extensive zeolitization associated with hot springs in central Colorado: U.S. Geological Survey Professional Paper 700-B, p. B14–B20.
Utada, M., and Vine, J. D., 1985, Zonal distribution of zeolites and authigenic plagioclase, Spanish Peaks region, southern Colorado: Conference proceedings Sixth International Zeolite Conference, Reno, Nevada, 10–15 July, 1983, p. 604–615.
Vine, J. D., 1974, Geologic map and cross sections of the La Veta Pass, La Veta and Ritter Arroyo Quadrangles, Huerfano and Costilla Counties, Colorado: U.S. Geological Survey Map, MI-833.

Monument Valley, Arizona and Utah

Ronald C. Blakey, *Geology Department, Box 6030, Northern Arizona University, Flagstaff, Arizona 86011*
Donald L. Baars, *29056 Histead Drive, Evergreen, Colorado 80439*

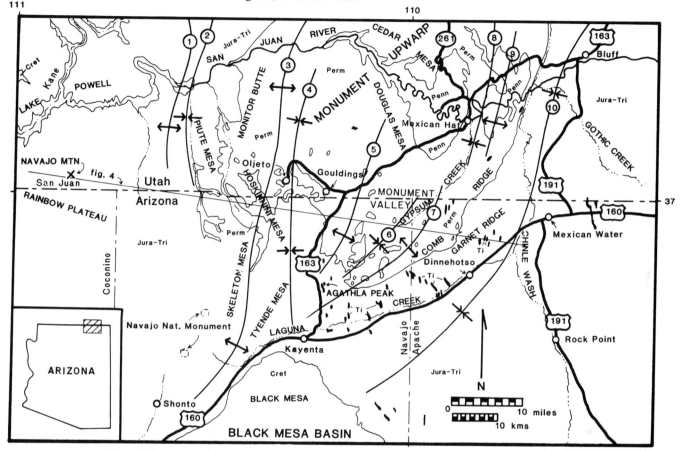

Figure 1. Maps of Monument Valley and vicinity showing structure and simplified geology. Key to numbered structural features: (1) Balanced Rock anticline, (2) Nokai syncline, (3) Organ Rock anticline, (4) Oljeto syncline, (5) Douglas Mesa arch, (6) Mitten Butte syncline, (7) Cedar Mesa anticline, (8) Mexican Hat syncline, (9) Raplee anticline, (10) Tyende syncline.

LOCATION AND ACCESS

Standing astride the Utah-Arizona border with bronzed shafts of sandstone, the buttes and pinnacles of Monument Valley rise above the high desert floor of the central Colorado Plateau. The valley is tucked into the apex of the southern end of the Monument Upwarp, a large asymmetrical anticline bounded on its southern and eastern margins by the Comb Ridge monocline (Fig. 1). As recently as 35 years ago, no paved roads served the remote region; today three U.S. highways (U.S. 160 running east-west, and U.S. 163 and U.S. 191 running north-south), provide access to the region. But the best way to see Monument Valley is via the back roads that provide access to some spectacular backcountry.

Monument Valley is located within the Navajo Indian Reservation. Travel on paved and main dirt roads is generally unrestricted. However, travel in some areas is restricted and a permit is required for rock collecting or extensive geological

study. Inquiries should be addressed: Navajo Nation, Window Rock, Arizona 86515.

Monument Valley and vicinity is covered by geologic maps of various scales: 1:62,500 (Witkind and Thaden, 1963), 1:96,000 (Baker, 1936), 1:250,000 (U.S.G.S. I-series, 345, 629, 744, 1003), and 1:500,000 (Geological Maps of Utah and Arizona). Witkind and Thaden (1963) have provided the most comprehensive geological report of the region.

SIGNIFICANCE

The Monument Valley region is not just a region of spectacular beauty, but also a significant geologic area. Included are key regional upper Paleozoic and Mesozoic stratigraphic sections, excellent examples of ancient marine and continental depositional environments (Fig. 2), key upper Cenozoic deposits, varied structural features, unusual igneous features, and classic examples of plateau geomorphology. In spite of several recent studies on some

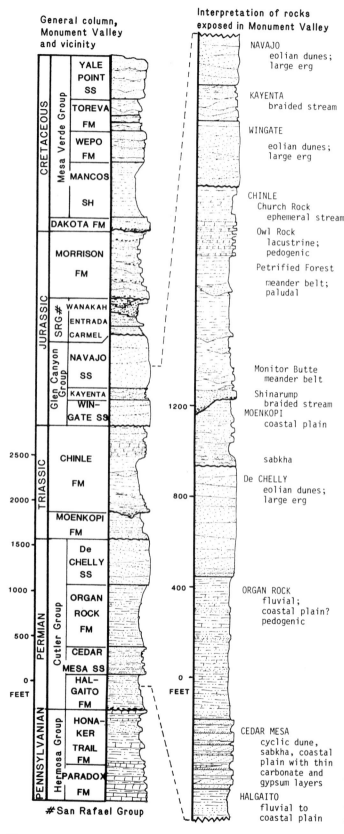

Figure 2. Stratigraphic column of sedimentary rocks exposed in and around Monument Valley.

of the above topics, the area is far from well-known geologically, and modern comprehensive studies remain to be done.

PALEOZOIC STRATIGRAPHY

Pennsylvanian System. The oldest rocks exposed in the Monument Valley region are marine strata of Middle and Late Pennsylvanian age (Fig. 2), beautifully displayed in the canyons of the San Juan River to the north. There rocks of the Paradox and Honaker Trail formations of the Hermosa Group form the canyon walls. The lower and oldest of the two formations, the Paradox, is seen in its typical evaporitic facies in Soda Basin, carved from the heart of the Raplee anticline. Equivalent limestone facies are to be seen in the Goosenecks, entrenched meanders cut into the Cedar Mesa anticline. Fusilinids date these cyclic deposits as middle Desmoinesian (Middle Pennsylvanian) in age.

The Honaker Trail Formation forms the ledge and cliff upper canyon walls. The formation, named for exposures along the trail west of the Goosenecks, is entirely marine and composed of alternating cyclic deposits of limestone, sandstone, and shale. Marine fossils are ubiquitous, dating the Honaker Trail as latest Desmoinesian in the lower part and Missourian (early Late Pennsylvanian) in the upper three quarters of the section. Resistant limestones of the upper Honaker Trail Formation cap much of the barren landscape surrounding the canyons.

Permian Halgaito Shale. The lowermost reddish brown siltstone-mudstone formation above the Late Pennsylvanian disconformity is the Halgaito Shale, named for a spring in the Mexican Hat syncline. It is probably a low coastal plain to intertidal deposit, adjacent to marine environments of the Pakoon-Elephant Canyon basins to the west. The slope-forming Halgaito forms the badlands terrane in the Mexican Hat syncline and the basal slopes of the monuments in the Valley of the Gods north of Mexican Hat, Utah. Local sand and gravel filled channels contain bone fragments and teeth of fresh water sharks, crossopterygian fishes, and lungfish, and of amphibians of both labyrinthodont and lepospondylous varieties. The formation becomes thicker and coarser toward the east and grades into undifferentiated Cutler arkoses in Colorado (Baars, 1962). It interfingers with marine fossiliferous limestones and clastics of the Elephant Canyon Formation (Wolfcampian) in Cataract Canyon to the north.

Permian Cedar Mesa Sandstone. Light colored massively cross-stratified sandstones exposed capping the high, broad Cedar Mesa north of the Goosenecks comprise the Cedar Mesa Sandstone. It is conformable with both the underlying Halgaito Shale and the overlying Organ Rock Shale, and forms impressive massive cliffs between the two red bed units. The Cedar Mesa is composed of very fine- to fine-grained, calcite-cemented quartzose sands that contain significant quantities of sand-size skeletal fragments. Massive cross-stratification typifies the Cedar Mesa, but horizontal bedding planes and red mudstone interbeds are common.

Some 800 ft (240 m) of Cedar Mesa Sandstone of the type

Figure 3. Typical scenery of Monument Valley.

section grade rapidly eastward to gypsiferous pink mudstones, siltstones, and sandstones along the Comb Ridge monocline. This evaporative "lagoonal" facies is widespread in the subsurface in the Four Corners region (Baars, 1962). The top of the formation forms the floor of Monument Valley within the belt of rapid facies change. This impressive facies change follows the contours of the east flank of the Monument Upwarp northward into the Needles District of Canyonlands National Park, suggesting a paleotectonic control of the eastern margin of the sandstone facies in late Wolfcampian time.

Permian Organ Rock Shale. Another sequence of reddish brown mudstone and siltstone, the Organ Rock Shale, overlies the Cedar Mesa Sandstone throughout the Monument Upwarp. The thin bedded slope-forming formation forms the lower slopes of the buttes and mesas in Monument Valley. Like the Halgaito below, the Organ Rock grades eastward into undifferentiated Cutler arkosic clastics in Colorado, and correlates directly with the Hermit Shale of the Grand Canyon. Plant remains found in the Hermit indicate an early Leonardian age. The formation was deposited on coastal lowlands. The formation varies from 600 to 700 ft (180 to 210 m) thick in the Monument Valley area.

Permian De Chelly Sandstone. The magnificent buttes, spires, and cliffs in Monument Valley have been carved by erosion from the De Chelly Sandstone of Leonardian age (Fig. 3). The scenically impressive formation is composed of fine-grained, quartzose sandstone that derives its reddish color from orange-red hematitic coatings on its individual grains. Vertebrate trackways on lee slopes of the dune faces are fairly common in the De Chelly, both on the Defiance Uplift and in the Monument Valley area.

The De Chelly Sandstone is between 300 and 400 ft (90 and 120 m) thick in Monument Valley, but thins northward and pinches out along a line roughly paralleling the San Juan River. It thickens southward into the Black Mesa Basin to a maximum

thickness exceeding 1,000 ft (300 m). It is 825 ft (251 m) thick at its type section in Canyon De Chelly.

MESOZOIC STRATIGRAPHY

Triassic Moenkopi Formation. The Moenkopi Formation forms a thin redbed sequence that caps many of the buttes and mesas of Monument Valley. Three members are recognized in the northwestern portion of the region. The basal Hoskinnini Member, type section on Hoskinnini Mesa, comprises chiefly silty, very poorly sorted sandstone that displays poorly developed ripple lamination and local large-scale contorted bedding. The Torrey Member consists of ledge-forming very fine-grained sandstone and slope-forming siltstone; the Moody Canyon Member is mainly slope-forming siltstone and mudstone. The formation is less than 300 ft (100 m) in thickness and thins southeastward across the area. Locally it is absent where erosion associated with paleovalleys at the base of the overlying Chinle Formation has scoured into Permian rocks.

Triassic Chinle Formation. The Chinle Formation comprises more than 1,000 ft (300 m) of heterogeneous assemblages of bentonitic mudstone, siltstone, sandstone, conglomerate, and cherty micritic limestone (Fig. 2). The formation forms broad benches and badlands throughout the area. The Shinarump Member consists of tan, cross-stratified sandstone and conglomerate made up of amalgamated sandstone bodies forming broad sheets or filling deep paleovalleys (Blakey and Gubitosa, 1983). The Monitor Butte Member is chiefly lenticular sandstone and interbedded bentonitic mudstone, and the Petrified Forest Member is mostly bentonitic mudstone with local lenticular sandstone bodies (Stewart and others, 1972). Bentonic material sharply decreases upwards in the Chinle; the Owl Rock Member is chiefly interbedded mudstone, siltstone, and micritic limestone,

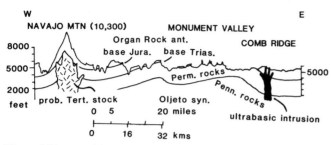

Figure 4. Topographic and structural cross section of Monument Valley region oriented E-W along the Utah-Arizona border.

and the Church Rock Member consists of red sandstone, siltstone, and mudstone.

Jurassic Wingate Sandstone. Towering, vertical, un-broken cliffs of cross-stratified quartz sandstone characterize the Wingate Sandstone across the central Colorado Plateau. The 400 ft-high (120 m) Wingate cliffs rim Monument Valley to the south, west, and east and can easily be mistaken for the Permian De Chelly Sandstone. Large-scale, high-angle cross-stratified sandstone dominates the formation but some plane-bedded, ripple-laminated, and small-scale cross-stratified units are present.

Jurassic Kayenta Formation. The Kayenta Formation (type section at Kayenta, Arizona) caps the Wingate cliffs and forms a ledge-forming sequence up to 160 ft (50 m) thick. Amalgamated lenticular, cross-stratified, and plane-bedded sandstone and minor intercalated mudstone dominate the Kayenta in the Monument Valley area.

Jurassic Navajo Sandstone. The Navajo Sandstone consists of approximately 600 ft (180 m) of large-scale, cross-stratified, texturally and mineralogically mature quartz sandstone. The Navajo is similar to both the DeChelly and Wingate sand-stones and is most easily distinguished from them by stratigraphic position.

Younger Jurassic and Cretaceous Rocks. The southern Monument Upwarp is bounded by basins and sags in which Jurassic and Cretaceous rocks are exposed. The San Rafael Group comprises the Carmel Formation and Entrada Sandstone. The Entrada is overlain by and intertongues with the Wanakah Formation (Summerville of previous usage).

The Morrison Formation overlies the Wanakah Formation throughout much of the Four Corners region. South of Kayenta in Black Mesa Basin, Cretaceous rocks consisting of the Dakota Formation, Mancos Shale, Toreva Formation, Wepo Formation,

and Yale Point Sandstone form the youngest lithified sedimentary strata in the region.

CENOZOIC ROCKS

Igneous Rocks. Igneous rocks in the Monument Valley region consist of (1) lamprophyric intrusive breccia, (2) associated lamprophyric dikes, and (3) ultrabasic rubble pipes (Witkind and Thaden, 1963). The intrusive breccia formed as volcanic necks and now stand as impressive monoliths as much as 1,400 ft (420 m) tall. The ultrabasic rubble pipes contain serpentine and garnet and are similar in composition and structure to the South African diamond pipes.

Quaternary Sediments. Quaternary deposits of Monument Valley consist of fluvial, eolian, and colluvial deposits. Fluvial deposits are generally referenced to archeological sites and include those that formed before, during, and since occupation by several groups of Indians (Witkind and Thaden, 1963). Eolian deposits include both active and stabilized dunes.

STRUCTURE AND GEOMORPHOLOGY

The Monument Valley area is dominated by the southward-plunging Monument Upwarp and associated Comb Ridge monocline but also includes several other important structured features (Figs. 1, 4). Balanced Rock, Organ Rock, and Raplee anticlines are uplifts with up to several thousand ft (m) of structured relief that strongly influence local geomorphology. Associated downwarps include the Copper Canyon, Nokai, Oljeto, Mitten Butte, and Mexican Hat synclines. South of Monument Valley lies Black Mesa Basin, a major structural downwarp on the Colorado Plateau.

Since the Late Cretaceous, and probably since the middle Cenozoic, the Colorado Plateau in the Monument Valley region has been uplifted more than 7,000 ft (2,100 m). Major rivers including the San Juan and Colorado have incised into flat-lying and folded sedimentary rocks and their tributaries have worked to keep pace in downcutting and have created the fantastic Mesa, Butte, and "cockscomb" topography of the central Colorado Plateau. Thus the color and form of the sedimentary and igneous rocks, the structure of the region coupled with plateau uplift, and the forces of erosion in the high desert have combined to form some of the spectacular topography and scenery of the Colorado Plateau.

REFERENCES CITED

Baars, D. L., 1962, Permian system of the Colorado Plateau: American Association of Petroleum Geologists Bulletin, v. 46, p. 149–218.

Baker, A. A., 1936, Geology of the Monument Valley–Navajo Mountain Region, San Juan County Utah: U.S. Geological Survey Bulletin 865, 106 p.

Blakey, R. C., and Gubitosa, R., 1983, Late Triassic paleogeography and depositional history of the Chinle Formation, southern Utah and northern Arizona, *in* Reynolds, M. W., and Dolly, E. D., eds., Mesozoic paleogeography of west-central U.S.: Rocky Mountain Section, Society of Economic Paleon-

tologists and Mineralogists, p. 57–76.

Stewart, J. H., Poole, F. G., and Wilson, R. F., 1972, Stratigraphy and origin of the Chinle Formation and related Upper Triassic strata in the Colorado Plateau region: U.S. Geological Survey Professional Paper 690, 336 p.

Witkind, I. J., and Thaden, R. E., 1963, Geology and uranium-vanadium deposits of the Monument Valley area, Apache and Navajo counties, Arizona: U.S. Geological Survey Bulletin 1103, 171 p.

The mouth of the Grand Canyon and edge of the Colorado Plateau in the Upper Lake Mead area, Arizona

Ivo Lucchitta, U.S. Geological Survey, 2255 North Gemini Drive, Flagstaff, Arizona 86001

LOCATION AND ACCESSIBILITY

The Pierce Ferry area in northern Arizona is easily reached via the paved all-weather Dolan Springs–Meadview road (Fig. 1). The road is poorly maintained, however, and stretches of the pavement have disappeared entirely. The area is also subject to flash floods during the summer thunderstorm season. Meadview, 12 mi (19 km) south of Pierce Ferry, offers modest facilities that include a gas station, country store, motel, and restaurant. A National Park Service ranger station is nearby.

Even though the upper Lake Mead area as a whole offers interesting geology and breathtaking views, the best place to visit is the northern end of Grapevine Mesa, which affords a truly remarkable overview of much geologic interest and great beauty. This viewpoint, 2.5 mi (4 km) southwest of Pierce Ferry and about 1,700 ft (520 m) higher, is reached by traveling north from Meadview on the paved road to Sand Cove. About 5 mi (8 km) north of Meadview, the paved road abruptly drops off the mesa through a narrow slot cut into the resistant limestone that caps the mesa. Just before the slot are a gate and a dirt road to the right (east). This well-graded dirt road is easily passable in dry weather, but not in wet, and leads to the viewpoint at the north end of Grapevine Mesa. Near the viewpoint, on National Park Service land, is a gravel landing strip used routinely by light aircraft of the general public.

The paved road near Meadview affords views of the imposing Grand Wash Cliffs to the east, and of the south Virgin Mountains, culminating in Gold Butte and Bonelli Peak, to the west. Spectacular glimpses of Lake Mead far below in Greggs Basin are obtained where the road skirts the western edge of Grapevine Mesa.

SIGNIFICANCE

The upper Lake Mead area near Pierce Ferry is critically situated to provide information on three separate, but interrelated and important, geologic topics: the transition from the little-deformed Colorado Plateau to the highly deformed Basin and Range province; the stratigraphy and depositional environment of a typical interior-basin deposit, exemplified by the Miocene Muddy Creek Formation; and the time and manner of development of the Colorado River and its Grand Canyon. These topics are discussed individually.

GEOLOGIC INFORMATION

The information presented here is condensed largely from a Ph.D. thesis (Lucchitta, 1966) whose purpose was to explore the topics outlined above. Pertinent published articles that include more extensive reference lists are Blair, 1978; Hunt, 1969; Longwell, 1936; Lovejoy, 1980; Lucchitta, 1972, 1979; and Young and Brennan, 1974.

The stratigraphic nomenclature used in this paper for Cenozoic rocks is that of Longwell (1936), Lucchitta (1966, 1972,

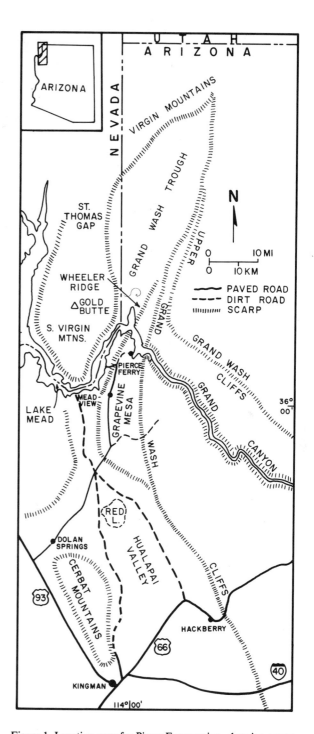

Figure 1. Location map for Pierce Ferry region, showing access.

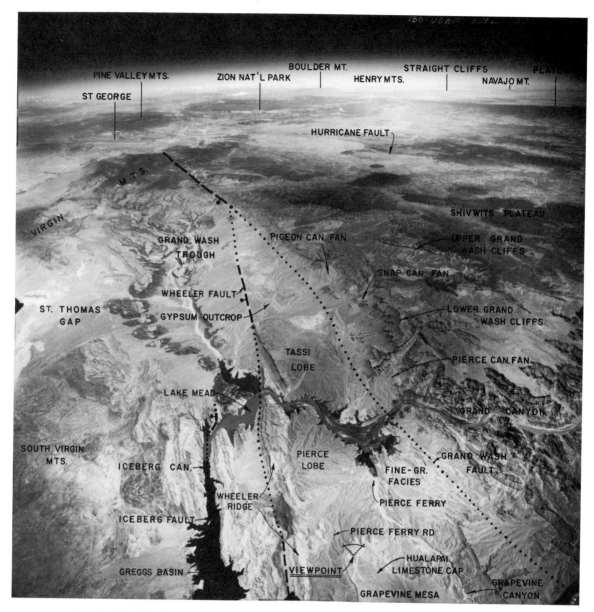

Figure 2. High-altitude aerial view of the Pierce Ferry area, Grand Wash trough, and adjacent Colorado Plateau, looking north-northwest. U.S. Air Force U-2 photograph.

1979), and Blair (1978) rather than that proposed by Bohannon (1984). Bohannon does not regard the basin fill in the Grand Wash (called the red sandstone unit by him) as the Muddy Creek Formation, and he views the Hualapai Limestone above the Muddy Creek as a separate formation.

SETTING

The Pierce Ferry area is in the Grand Wash trough, a structural depression that is but one of a continuous system of such basin-range depressions along the western margin of the Colorado Plateau in Arizona (Fig. 2). The floor of the trough is at an altitude of about 1,200 ft (366 m) along the Colorado River, and rises gradually toward the north. About 50 mi (80 km) north of Pierce Ferry the trough ends because the Virgin Mountains,

which are high both topographically and structurally, abut the Colorado Plateau directly. South of Pierce Ferry the trough is occupied by Grapevine Mesa, at an altitude of about 3,000 ft (900 m). This mesa owes its altitude to the presence of the resistant Hualapai Limestone of late Miocene age, the uppermost unit of the Muddy Creek Formation. South of Grapevine Mesa is the Hualapai Valley, which is lower topographically and structurally than the mesa, from which it is separated by a northwest-striking fault.

The various parts of the Grand Wash structural depression are filled to a great depth by the middle and upper Miocene Muddy Creek Formation, an interior-basin deposit that predates establishment of the Colorado River system. East of the depression are the north-trending Grand Wash Cliffs, an imposing fault-line scarp ranging in height from somewhat more than 2,000 ft to about 5,500 ft (600 to 1,700 m). These cliffs are the result of movement on the Grand Wash high-angle normal fault, whose western side has been downthrown. The cliffs mark the western edge of the little-deformed Colorado Plateau, which is part of the North-American craton.

The cliffs expose the entire Paleozoic section. South of the Colorado River, where they are lower in both altitude and relief than north of the river, the cliffs expose Cambrian, Devonian, and Mississippian rocks. North of the river, the cliffs are in two parts, the lower of which exposes Cambrian through Permian rocks, and the upper the Permian Hermit, Coconino, Toroweap, and Kaibab formations. Late Tertiary basalt caps the upper cliffs in places. The bench separating the upper and lower cliffs is formed by erosion and retreat of the Hermit Shale, which is more easily erodable than units above and below it.

The Grand Wash Cliffs are near the hinge line separating shelf-sequence rocks to the east from geosynclinal-sequence rocks to the west. Therefore, the rocks that form the cliffs show strong affinities to the familiar stratigraphic section of the Grand Canyon, from which, however, they differ by the thickening of individual units, the appearance of new units, and the increase in carbonate rocks at the expense of clastic ones.

The western edge of the Grand Wash trough is formed by the South Virgin Mountains, separated from the Virgin Mountains to the north by a structural and topographic sag known as St. Thomas gap. In these ranges, Precambrian crystalline rocks occur at heights much greater than they do on the adjacent Colorado Plateau, even after the basin-range rifting to which they have been subjected. Near Lake Mead, the Grand Wash trough is corrugated by fault-bounded north-trending ridges composed of Paleozoic rocks that are tilted steeply to the east. The most easterly, prominent, and laterally persistent of these is Wheeler ridge, bounded on its west side by the Wheeler fault.

TRANSITION BETWEEN THE COLORADO PLATEAU AND THE BASIN AND RANGE PROVINCE

Colorado Plateau

The view northward from Grapevine Mesa displays a remarkably sharp transition from the little-deformed rocks of the Colorado Plateau on the east to the strongly rotated and deformed ridges of the Basin and Range province on the west. In this region, the rocks that form the Colorado Plateau dip 1° to 2° northeast. The dip steepens to 4° to 5° near the western edge of the Plateau at the Grand Wash Cliffs. Folds are few and consist chiefly of small monoclinal flexures, some of which pass along strike into faults. Major faults are high-angle normal faults with the west side down. Strike lengths are tens to hundreds of miles (km). Minor faults are also high-angle normal faults, but their strike lengths are miles (km) to tens of miles (km) and their sense of offset is not systematic.

The Grand Wash high-angle normal fault (Fig. 2) is the master structural feature of the area. Extending from southwest Utah into west-central Arizona, it marks the boundary between the Colorado Plateau and the Basin and Range province. In northernmost Arizona, offset is several tends to a few hundred ft (m) at most (unpublished maps by Lucchitta, Bohannon, and Beard; and Lucchitta, Beard, and Rieck). In the Pierce Ferry area, estimated offset is 10,000 to 16,000 ft (3,000 to 5,000 m) on geometric grounds (Lucchitta, 1966). In Hualapai Valley at Red Lake, offset is estimated at a maximum 15,000 to 20,000 ft (4,500 to 6,000 m; Lucchitta, 1966). Farther south, near Hackberry, Arizona, the throw decreases but is at least 1,000 ft (300 m; Young and Brennan, 1974). In the Pierce Ferry area, the downthrown block west of the Grand Wash fault has been rotated strongly, as evidenced by eastward dips of 30° to 40° on Wheeler ridge. In northernmost Arizona, rotation of the downthrown block along the fault has been minor, in consonance with the small displacement. South of Lake Mead, documentable rotation is 10° to 20° eastward near Red Lake and near Hackberry.

Basin and Range Province. Wheeler and Iceberg Faults (Figs. 2 and 3)

Wheeler fault is a high angle down-to-the west normal fault that strikes north-northeast and dips west about 60°. The fault is about 40 mi (65 km) long. Offset is at least 5,000 ft (1,500 m) and possibly as much as 11,000 ft (3,500 m; Lucchitta, 1966). To the north the fault merges with the Grand Wash fault; the southern termination is uncertain. The Wheeler fault block, west of Wheeler fault, locally dips eastward as much as 70°.

Iceberg fault is exposed two miles (3 km) to the west of Wheeler fault, whose trace it parallels. Only about one mile (1.6 km) of the trace of Iceberg fault is exposed now, directly north of Iceberg Canyon. According to Longwell (1936), the fault was well exposed in Iceberg Canyon before the filling of Lake Mead, and it dipped 5° to 15° west, was curviplanar, and concave upward. Beds in the Iceberg block, west of Iceberg fault, dip eastward 70° or more. Offset is about 4,000 ft (1,200 m). Iceberg and Wheeler faults, and the associated rotated fault blocks, strongly suggest structural relations typical of the "listric" faults that are so common along the lower Colorado River. Such faults merge or terminate downward in subhorizontal detachment faults, giving rise to the "thin-skin tectonics" of Anderson (1971). Thus, a detachment surface and the associated features of thin-skin tectonics may be present in the Pierce Ferry area, only a few miles (km) west of the Colorado Plateau.

Age of Deformation

Along most of its length, the Grand Wash fault has not moved since deposition of the exposed part of the middle and upper Miocene Muddy Creek Formation (Longwell, 1936; Lucchitta, 1966). Muddy Creek beds are offset, however, north of the junction of the Grand Wash and Wheeler faults, as are upper Miocene and Pliocene (?) lavas. Rotation of the block between the Grand Wash and Wheeler faults occurred entirely after deposition of the middle Miocene Horse Spring Formation because small patches of this unit on Wheeler Ridge are in structural conformity with underlying rotated Paleozoic rocks near the Colorado River and also near the northern end of the ridge.

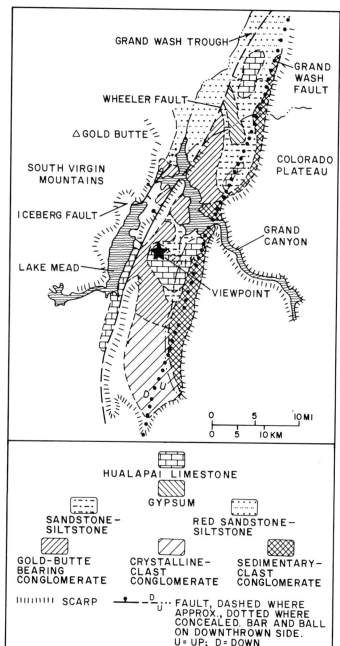

Figure 3. Sketch map showing distribution of facies of the Muddy Creek Formation, an interior-basin deposit of Miocene age. From Lucchitta, 1966.

Wheeler fault was active while the Hualapai Limestone was being deposited, and moved an additional 1,000 ft (300 m) after deposition of the Hualapai. Movement on Iceberg fault can only be dated as having occurred in the interval between Permian and Quaternary. According to Lucchitta (1979), the top of the Muddy Creek basin fill was near sea level at 5.5 Ma and has been uplifted about 3,000 ft (880 m) since that time, partly by faulting, partly by warping. The uplift has affected both the upper Lake Mead area and the adjacent Colorado Plateau.

Summary

1. The transition from stable Colorado Plateau to highly deformed Basin and Range is remarkably abrupt and well exposed in the Pierce Ferry area.

2. The structural styles differ greatly: subhorizontal rocks, widely spaced high-angle major faults, and little rotation on the Plateau; highly tilted rocks, strong rotation, closely spaced curviplanar faults in the Basin and Range.

3. Thin-skin tectonism, marked by "listric" faults, rotated blocks, and subhorizontal detachment faults, probably occurred within a few miles (km) of the Colorado Plateau. This type of tectonism started after deposition of the Horse Spring Formation (12 to 20 Ma) and had ceased by the time the Muddy Creek Formation was being deposited, between 17 and 8 Ma and more likely between 12 and 5.5 Ma (see below).

4. Faulting and regional upwarping have affected the Plateau and adjacent Basin and Range terrane along the lower Colorado River since the end of the Miocene.

CHARACTERISTICS AND DEPOSITIONAL ENVIRONMENT OF THE MUDDY CREEK FORMATION—A TYPICAL INTERIOR-BASIN DEPOSIT

The Grand Wash trough near upper Lake Mead provides a rare opportunity to view in three dimensions a classic and little-deformed interior-basin deposit, the Muddy Creek Formation. Dissection by the Colorado River has provided nearly 2,000 ft (600 m) of vertical exposure. The distribution of Muddy Creek facies is shown in Figure 3.

Paleogeography

The Muddy Creek Formation was deposited in an asymmetrical basin whose axis trended north-northeast and was near the eastern margin of the basin. The floor of the basin sloped gently from the north as well as from the south toward a low point located approximately where Pierce Ferry and the north edge of Grapevine Mesa are now. The surface of Grapevine Mesa is near the original (now uplifted) stratigraphic top of the basin fill in this low spot. Filling of the trough was dominantly from the west, as indicated by the predominance of igneous and metamorphic debris that includes clasts of the Gold Butte Granite of Longwell (1936), a distinctive and easily identifiable coarsely porphyritic rapakivi granite. This granite crops out extensively in the south Virgin Mountains 4 to 12 mi (6 to 19 km) west of Wheeler Ridge. Thirty-foot (9-m) boulders of the granite are present at Wheeler Ridge, and 20-ft (6-m) boulders at the foot of the lower Grand Wash Cliffs, a minimum transport distance of 12 mi (19.3 km) from the nearest possible source area. Two prominent Muddy Creek fan lobes are visible to the north from Grapevine Mesa: one, the Pierce lobe, is south of the Colorado River and west of Pierce Ferry; the other, Tassi lobe, is north of the Colorado (Fig. 2). Both lobes are east of Wheeler Ridge. During development of the fans, the country west of Wheeler Ridge consisted of a pediment cut on bedrock and was a zone of transport rather than deposition. This ended late in Muddy Creek time when movement on the Wheeler fault isolated the fans from their source areas and created a separate basin of deposition in what is now the Greggs Basin area. Influx of material into the Grand Wash trough from the Grand Wash Cliffs to the east was minimal, partly because streams on the cliffs' face drain small areas, and partly because the carbonate rocks that form most of the cliffs yield relatively little debris. Substantial fan lobes are present only at the foot of the Grand Wash Cliffs at Pierce, Snap, and Pigeon canyons (Fig. 2). The canyons that existed at these localities in Muddy Creek time were as deep and narrow as the present ones, but shorter and steeper.

Facies

All the fans are composed of poorly sorted breccia and conglomerate that contain chiefly angular to subangular clasts in pell-mell arrangement and with a high matrix-to-clast ratio. The tops of many depositional units show evidence of reworking by flowing water. These features, which suggest deposition by debris flows, can be studied conveniently in a prominent outcrop on the east side of the Pierce Ferry road about 2.5 mi (4 km) north of the Sand Cove turnoff. The areas between fan lobes are underlain by a fine-grained facies composed of sandstone and mudstone, and by a chemical-precipitate facies composed of fresh-water limestone and dolomite, and gypsum. Silicic airfall tuffs are common, many with delicate glass shards and bubbles still preserved. Twenty-three individual tuff layers are exposed about 1.5 mi (2.5 km) southeast of Pierce Ferry. Transition from the fan material to the fine-grained facies occurs partly through progressive decrease in grain size, partly through abrupt interfingering. Rocks of the fine-grained facies are very well bedded and well sorted. Deposition in quiet water is indicated by even bedding, tuffs, evaporite, and carbonates. Many of these features are exposed near Pierce Ferry.

The fine-grained facies were deposited in intermittent playas and lakes. Initially, these playas and lakes were small and restricted to the lowest parts of the basin because the influx of clastics was high. As the basin filled and relief waned, playas and lakes occupied progressively greater areas. Finally, a lake occupied much of the basin in the Pierce Ferry area. The Hualapai Limestone was deposited in the lakes and eventually transgressed widely over other lithologies. These relations are well displayed on the north face of Grapevine Mesa. The transgressive stacking of limestone, fine-grained facies, and fanglomerate can be observed in road cuts and natural exposures on the east side of the Pierce Ferry road where it starts dropping off the north end of Grapevine Mesa, just north of the turnoff to the viewpoint. The Hualapai Limestone therefore is not a sheet-like deposit that was laid down in a restricted interval at the end of Muddy Creek time, even though limestone deposition was most widespread at that time. Instead, it was deposited throughout the time represented by the 1,700 ft (500 m) of Muddy Creek section exposed in the Pierce Ferry area. The walls of Grapevine Canyon (Fig. 2), cut into Grapevine Mesa, are composed entirely of Hualapai Limestone.

Age

The Muddy Creek Formation is younger than the 17–18

Ma Peach Springs Tuff (Young and Brennan, 1974) and the 12–20 Ma Horse Spring Formation, which predate development of the basin where the formation was deposited. The Muddy Creek includes basalts dated 5 to 6 Ma (Anderson, 1978; Damon and others, 1978) and 10.9 Ma (Blair, 1978) and tuffs about 8 Ma (Blair, 1978; R. G. Bohannon, oral communication, 1982; Bohannon, 1984). The formation is older than establishment of through-flowing drainage, probably about 5.5 Ma.

Summary

(1) The axis of the basin was near its eastern boundary. (2) The basin was filled predominantly with granitic and metamorphic material of westerly derivation. (3) The lowest part of the basin, near Pierce Ferry, was occupied by playas and lakes that became larger as filling of the basin progressed. (4) Canyons in the Grand Wash Cliffs were few, but as deep as the present ones, though shorter and steeper. These canyons shed debris actively but in small quantities into the trough. (5) The configuration of the basin was disturbed late in Muddy Creek time when movement on Wheeler fault produced a subsidiary basin west of Wheeler Ridge in the present Greggs Basin area. This interrupted influx of debris from the west into the Pierce Ferry area. (6) Facies distribution, lithologic characteristics, and bedrock highs indicate a closed basin of interior deposition. (7) Muddy Creek deposition probably began after 12 Ma and had ended by about 5.5 Ma.

INCEPTION OF THE COLORADO RIVER AND DEVELOPMENT OF THE WESTERN GRAND CANYON

Hypotheses

The Pierce Ferry area lies athwart the mouth of the Grand Canyon. Consequently, it provides important restrictions on interpretations of the history of the Colorado River and the development of its Grand Canyon. Longwell (1936) and Blackwelder (1934) long ago indicated that the presence of interior-basin deposits across the mouth of the Grand Canyon in the Pierce Ferry area effectively precluded the existence of the Colorado River in its present course in Muddy Creek time. Others, however, do not agree with this view. Contrary hypotheses fall into two main groups:

1. The Grand Canyon existed in Muddy Creek time. The Colorado River emptied into the Grand Wash trough and ultimately exited by a course different from the present one (Lovejoy, 1980). Or: the Grand Canyon was a dry valley carrying neither water nor sediment (D. Elston, oral comm., 1983).

2. The Colorado River existed in Muddy Creek time, but flowed elsewhere. Eventually, the river was ponded near the western margin of the Colorado Plateau, whence it drained by subterranean piping to form springs and a lake in the Grand Wash area. The river eventually established its course in post-Muddy Creek time (Hunt, 1969). Much information from the Pierce Ferry area bears on these hypotheses.

Information Bearing on Hypotheses of Group 1

There are no known deposits or structures in the Muddy

Creek Formation that point to the existence of the Colorado River during Muddy Creek time. On the contrary, internal fabrics, facies distribution, as well as facies composition and relations all point to a closed basin of interior deposition. It is unlikely that Colorado River sediments were deposited and then dispersed by wave action or currents, given the abundant evidence for quiet-water deposition in the Muddy Creek Formation. Even the oldest preserved Colorado River gravels, some at altitudes of hundreds of meters above present grade, are immediately recognizable because of their coarse size (pebble-and-cobble gravel), excellent rounding, and exotic, far-traveled lithologies.

The distribution of facies in the Muddy Creek Formation shows no evidence for an inlet or outlet, as would be required had the Colorado River existed. Specifically, such facies do not occur either at the mouth of the Grand Canyon, or at the south end of Grapevine Mesa, Lovejoy's postulated outlet (1980). The latter locality is at an altitude of about 4,000 ft (1,200 m), or 1,000 ft (300 m) higher than the top of the basin fill near Pierce Ferry, and is underlain exclusively by locally-derived fanglomerate and bedrock.

There is little reason to suppose that a drainage system nearly as large as the present Colorado River would not have supplied sediment when drainage systems such as Pierce, Snap, and Pigeon canyons, which are orders of magnitude smaller and only a few miles (km) from the Grand Canyon, supplied sediment in abundance. Nor can the absence of river sediments be attributed to a reduced base flow. The western United States is characterized by drainage systems (washes) that are dry most of the time, yet flood on occasion. It is during such floods that the streams transport sediment in great quantities. The amount of material transported by a drainage system is not closely related to its base flow.

A remnant of the Muddy Creek fan issuing from Pierce Canyon is present directly south of the mouth of the Grand Canyon. This material could not have been deposited in that position had the Grand Canyon existed at the time.

Information Bearing on Hypotheses of Group 2

The evidence provided by the Muddy Creek Formation for conditions of interior drainage rather than through-flowing drainage near the mouth of the Grand Canyon cannot be bypassed by postulating another course for the Colorado River because locally derived interior-basin deposits of Miocene age are ubiquitous in the lower Colorado river region.

There is no evidence for springs feeding Hualapai Lake in the Grand Wash trough. The Hualapai Limestone was deposited during much of Muddy Creek time in scattered topographically low spots often subject to playa conditions. Only at the end of Muddy Creek time was limestone deposition widespread, when it occurred not only in the Grand Wash trough but also tens of miles (km) away and perhaps in separate basins.

Preferred Hypothesis

The preferred hypothesis offered here is that summarized from data presented by Lucchitta (1966, 1972, 1979) and McKee and others (1967). The Muddy Creek Formation of the Grand Wash through shows that no Colorado River existed at the mouth of the Grand Canyon when the Muddy Creek Formation was being laid down. The youngest radiometric dates obtained from rocks correlative with the Muddy Creek Formation at the Grand Wash trough range from 5 to 8 Ma. The oldest date on rocks that reflect the existence of the Colorado River is the 5.3 Ma date obtained from the Bouse Formation (Damon and others, 1978), an estuarine deposit that crops out widely along the lower Colorado River. The conclusion is that the lower Colorado River came into being after the end of Muddy Creek deposition and after opening of the Gulf of California in latest Miocene time. The lower Colorado worked its way onto the Colorado Pleateau by headward erosion, capturing an ancestral upper Colorado River probably in the stretch between the Kaibab Plateau and the mouth of the Grand Canyon. In the process, the canyon as we know it today was formed.

REFERENCES

Anderson, R. E., 1971, Thin-skin distension in Tertiary rocks of southeastern Nevada: Geological Society of America Bulletin, v. 82, p. 43–58.
——1978, Geologic map of the Black Canyon 15-minute quadrangle, Mohave County, Arizona and Clark County, Nevada: U.S. Geological Survey Quadrangle Map GQ-1394.
Blackwelder, E., 1934, Origin of the Colorado River: Geological Society of America Bulletin, v. 45, p. 551–566.
Blair, W. N., 1978, Gulf of California in Lake Mead of Arizona and Nevada during late Miocene time: Bulletin of the American Association of Petroleum Geologists, v. 62, p. 1159–1170.
Bohannon, R. G., 1984, Nonmarine sedimentary rocks of Tertiary age in the Lake Mead region, southeastern Nevada and northwestern Arizona: U.S. Geological Survey Professional Paper 1259, 72 p.
Damon, P. E., Shafiquallah, M., and Scarborough, R. B., 1978, Revised chronology for critical stages in the evolution of the lower Colorado River: Geological Society of America Abstracts with Program, v. 10, no. 3, p. 101.
Hunt, C. B., 1969, Geologic history of the Colorado River, in The Colorado River region and John Wesley Powell: U.S. Geological Survey Professional Paper 669-C, p. 59–130.

Longwell, C. R., 1936, Geology of the Boulder Reservoir floor, Arizona-Nevada: Geological Society of America Bulletin, v. 47, p. 1393–1476.
Lovejoy, E.M.P., 1980, The Muddy Creek Formation at the Colorado River in Grand Wash—The dilemma of the immovable object: Arizona Geological Society Digest, v. 12, p. 177–192.
Lucchitta, I., 1966, Cenozoic geology of the upper Lake Mead area adjacent to the Grand Wash Cliffs, Arizona [Ph.D. Thesis]: Pennsylvania State University, 218 p.
——1972, Early history of the Colorado River in the Basin and Range Province: Geological Society of America Bulletin, v. 83, p. 1933–1948.
——1979, Late Cenozoic uplift of the southwestern Colorado Plateau and adjacent lower Colorado River region: Tectonophysics, v. 61, p. 63–95.
McKee, E. D., Wilson, R. F., Breed, W. J., and Breed, C. S., 1967, Evolution of the Colorado River in Arizona: Museum of Northern Arizona Bulletin, v. 44, 68 p.
Young, R. A., and Brennan, W. J., 1974, The Peach Springs Tuff—Its bearing on structural evolution of the Colorado Plateau and development of Cenozoic drainage in Mohave County, Arizona: Geological Society of America Bulletin, v. 85, p. 83–90.

Geology along the South Kaibab Trail, eastern Grand Canyon, Arizona

Stanley S. Beus, Northern Arizona University, Flagstaff, Arizona 86001

LOCATION AND ACCESS

The South Kaibab Trail head is at the South Rim of the Grand Canyon about 2.25 mi (3.5 km) due east of the National Park headquarters and visitors center at Grand Canyon Village. Convenient access is provided by the paved south entrance road, east Rim Drive Road and Yaki Point Road leading east and north from the village, a driving distance of 3.5 mi (5.75 km) from the Park Headquarters (Fig. 1). Grand Canyon Village at the South Rim is accessible from the south by US 180 and Arizona 64 from Flagstaff (80 mi; 129 km) or by Arizona 64 from just east of Williams (60 mi; 95 km), or from the east via Arizona 64 (57 mi; 92 km) leading west from Cameron.

Detailed geology is shown on the geologic map of the Eastern Grand Canyon, 1:62,500 (Huntoon and others, 1986). A larger scale (1:24,000) topographic, shaded relief map is also available (Washburn, 1978), a part of which is used as Figure 1.

Overnight motel accommodations and campsites are available at the Grand Canyon Village but generally must be reserved in advance during the major tourist season, April through October. A full day is recommended for a one way traverse of the South Kaibab Trail. Overnight accommodations at the bottom are available at Phantom Ranch or at Bright Angel Creek campground (Park Service permit required), but also generally require reservations several months in advance. The return trip to the rim can be made back up to the South Kaibab Trail (not recommended in summer; 6.5 mi; 10.5 km and no water available!!) or by the longer, but gentler route along the river trail 2 mi (3.2 km) and up the Bright Angel Trail 7 mi (11.3 km) where water is available at three points. Caution: carry adequate water and avoid hiking in the heat of midsummer.

SIGNIFICANCE

The eastern part of the Grand Canyon is one of the most scenically spectacular, and geologically instructive parts of the canyon. Perhaps at no other single locality in North America are so many events over such a long interval of earth history displayed in one view. From near Desert Tower along the South Rim of the Canyon can be seen the entire Paleozoic and most of the Precambrian rock record for the southern Colorado Plateau, in addition to a substantial part of the Mesozoic stratigraphic section and Cenozoic volcanic features.

The South Kaibab Trail traverses the entire Paleozoic section and much of the Precambrian rock record from rim to river in a distance of 6.5 mi (10.5 km), and through a vertical distance of nearly 5,000 ft (1,500 m). It crosses 18 formations beginning with the Permian Kaibab Formation at the rim and ending in the early Proterozoic Vishnu Group at the river (Figs. 1, 2).

A complex system of high-angle faults is well displayed in the Cambrian and Precambrian rocks near the mouth of Bright Angel Creek and records up to six different episodes of faulting.

Bright Angel Canyon follows a remarkably straight-line trend for some 8 mi (13 km) upstream to the northeast from its mouth owing to alignment long the Bright Angel fault. The fault strikes about N35E and dips 75 to 88 degrees to the southwest. It is cut by a number of northwest-trending faults near the mouth of Bright Angel Creek. Two of these, the Tipoff and Cremation faults (Fig. 3), account for the downdropped Tipoff graben, which inserts Unkar Group strata into the Vishnu Group complex along the South Kaibab Trail.

The several episodes of vertical displacement along the Bright Angel fault result in up to 200 ft (61 m) of stratigraphic separation of the Paleozoic strata in a normal sense, down to the southeast, and a net offset of the Precambrian Unkar Group strata of up to 1,000 ft (305 m) in a reverse sense, down to the northwest (Huntoon and Sears, 1975). A 3.5 mi (5.6 km) walk up the North Kaibab Trail following Bright Angel Creek affords a splendid view of drag and overturned beds expressed in the Bass Limestone along the fault.

Finally, the landforms of the canyon are a profound testament to the power of running water and its allies to sculpture bedrock. Early studies of these landscapes, beginning in 1869 (1875) with the epic journey of John Wesley Powell and extending well into this century, gave rise to fundamental concepts in geology, including base level of erosion, antecedent and superimposed streams, and the power of rivers to cut their canyons.

LOCALITY 83: PALEOZOIC ROCKS

Cambrian Tonto Group. The three formations of Cambrian age in eastern Grand Canyon are, in ascending order, the Tapeats Sandstone, Bright Angel Shale, and Muav Limestone. Their aggregate thickness is nearly 1,000 ft (300 m) along the South Kaibab Trail. The Tapeats Sandstone is a ledge- or cliff-forming unit of gray to red-brown sandstone that is coarse to medium-grained and locally conglomeratic at the base. The Tapeats rests on the Precambrian rocks beneath with marked relief of up to several hundred feet and locally pinches out entirely against resistant hills of Shinumo Quartzite. Medium-scale wedge-planar and tabular-planar cross-beds are common in the Tapeats. Asymmetrical ripples and small-scale trough cross-beds occur locally. Trilobite crawling traces, *Cruziana* sp., and small U-burrows are common trace fossils in the upper Tapeats.

The Tapeats Sandstone grades upward into the Bright Angel Shale through a transition zone of alternating coarse sandstone and light brown shaly mudstone. The Bright Angel Shale consists

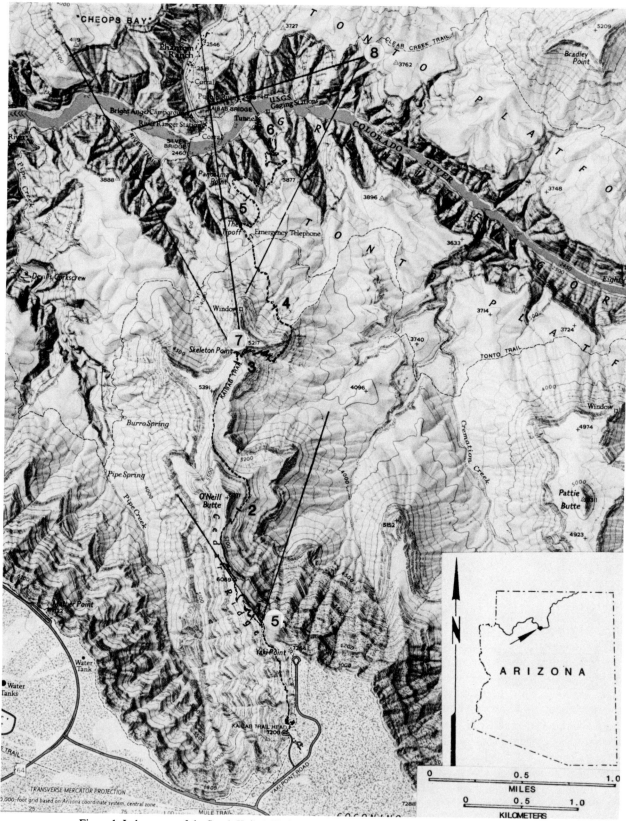

Figure 1. Index maps of the South Kaibab Trail. Numbers indicate mileposts. Numbers 5, 7, and 8 on white circles locate viewpoints for Figures 5, 7, and 8, respectively. Map from Washburn, 1978, used with permission of the National Geographic Society.

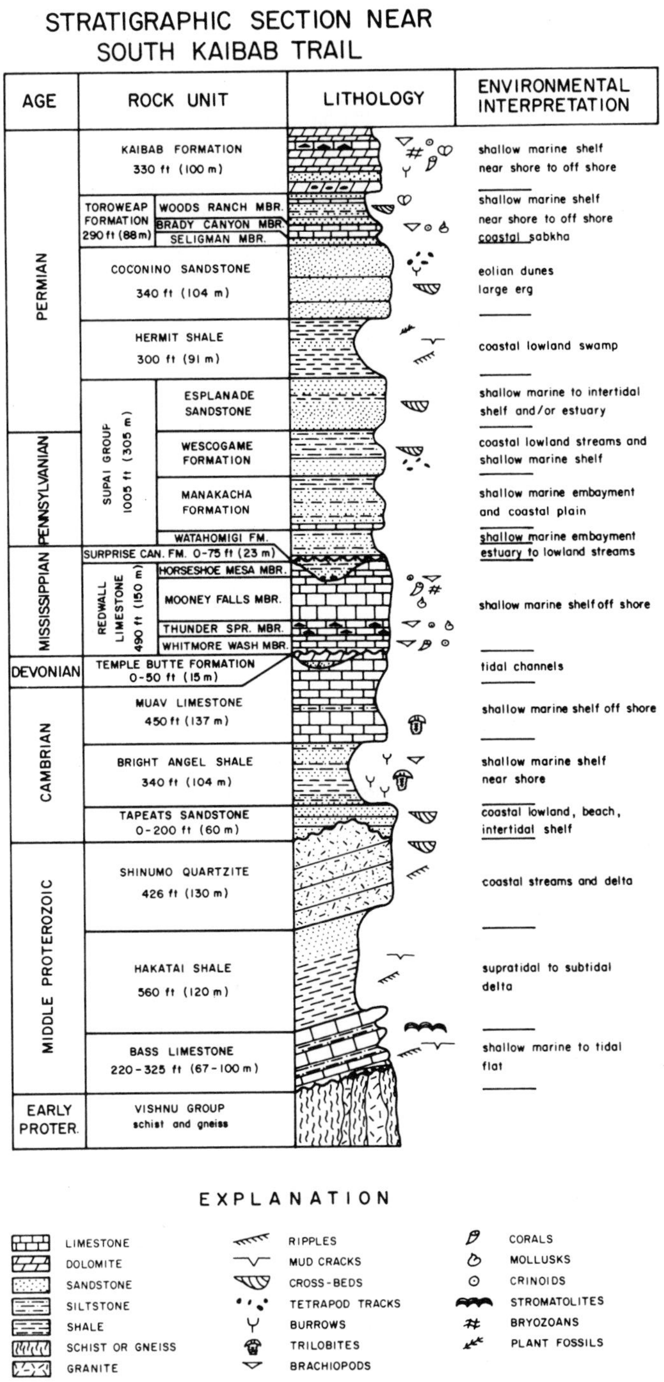

STRATIGRAPHIC SECTION NEAR SOUTH KAIBAB TRAIL

AGE	ROCK UNIT	LITHOLOGY	ENVIRONMENTAL INTERPRETATION
PERMIAN	KAIBAB FORMATION 330 ft (100 m)		shallow marine shelf near shore to off shore
	TOROWEAP FORMATION 290 ft (88 m) — WOODS RANCH MBR. / BRADY CANYON MBR. / SELIGMAN MBR.		shallow marine shelf near shore to off shore coastal sabkha
	COCONINO SANDSTONE 340 ft (104 m)		eolian dunes large erg
	HERMIT SHALE 300 ft (91 m)		coastal lowland swamp
PENNSYLVANIAN	SUPAI GROUP 1005 ft (305 m) — ESPLANADE SANDSTONE		shallow marine to intertidal shelf and/or estuary
	WESCOGAME FORMATION		coastal lowland streams and shallow marine shelf
	MANAKACHA FORMATION		shallow marine embayment and coastal plain
MISSISSIPPIAN	WATAHOMIGI FM. / SURPRISE CAN. FM. 0-75 ft (23 m)		shallow marine embayment estuary to lowland streams
	REDWALL LIMESTONE 490 ft (150 m) — HORSESHOE MESA MBR. / MOONEY FALLS MBR. / THUNDER SPR. MBR. / WHITMORE WASH MBR.		shallow marine shelf off shore
DEVONIAN	TEMPLE BUTTE FORMATION 0-50 ft (15 m)		tidal channels
CAMBRIAN	MUAV LIMESTONE 450 ft (137 m)		shallow marine shelf off shore
	BRIGHT ANGEL SHALE 340 ft (104 m)		shallow marine shelf near shore
	TAPEATS SANDSTONE 0-200 ft (60 m)		coastal lowland, beach, intertidal shelf
MIDDLE PROTEROZOIC	SHINUMO QUARTZITE 426 ft (130 m)		coastal streams and delta
	HAKATAI SHALE 560 ft (120 m)		supratidal to subtidal delta
	BASS LIMESTONE 220-325 ft (67-100 m)		shallow marine to tidal flat
EARLY PROTER.	VISHNU GROUP schist and gneiss		

EXPLANATION

LIMESTONE	RIPPLES	CORALS
DOLOMITE	MUD CRACKS	MOLLUSKS
SANDSTONE	CROSS-BEDS	CRINOIDS
SILTSTONE	TETRAPOD TRACKS	STROMATOLITES
SHALE	BURROWS	BRYOZOANS
SCHIST OR GNEISS	TRILOBITES	PLANT FOSSILS
GRANITE	BRACHIOPODS	

Figure 2. Stratigraphic section along the South Kaibab Trail, eastern Grand Canyon.

of predominant dark red-brown to gray-green sandstone, locally conglomeratic, and subordinate micaceous green shale and shaly mudstone. The sandstone is commonly glauconitic. Sedimentary structures include common trough and less abundant tabular-planar cross-beds. Trace fossils are ubiquitous and strikingly displayed in the Bright Angel Shale. Martin (1985) recognized 25 different ichnospecies of traces. Body fossils include trilobites and inarticulate brachiopods.

The Muav Limestone, the uppermost unit of the Tonto Group, overlies and intertongues with the Bright Angel Shale. It is mainly thick-bedded, mottled gray limestone in western Grand Canyon, where McKee and Resser (1945) recognized seven members, most bounded by a thin basal conglomerate. These members thin eastward and grade into light brown or tan dolomite and dolomitic sandstone beds, which intertongue with the Bright Angel Shale. Along the South Kaibab Trail, the Muav is mostly thin-bedded, ledge-forming, mottled gray limestone of the upper two members. Fossils in the Muav include trilobites, inarticulate brachiopods, and algal structures referred to *Girvanella*. McKee and Resser (1945, Fig. 1) have demonstrated that the entire Tonto Group is time transgressive, becoming younger from west to east through Grand Canyon. In the eastern Grand Canyon the Tapeats Sandstone is essentially Early Cambrian, and the Tapeats and Bright Angel are Middle Cambrian in age.

An undifferentiated Cambrian dolomite sequence from 0 to a few yards (meters) thick occurs conformably above the Muav in eastern Grand Canyon, becoming thicker to the west, but as yet has no formal name status.

Devonian Temple Butte Formation. Rocks of Ordovician and Silurian age are missing in the Grand Canyon. The Late Devonian Temple Butte Formation unconformably overlies the Muav and crops out as scattered lens-shaped channel fill deposits up to 75 ft (23 m) thick in eastern Grand Canyon and upstream in Marble Gorge. The Temple Butte consists of pale red to purple dolomite that is commonly gnarly bedded and locally silty and sandy. A small patch only a few yards (meters) thick is visible along the South Kaibab Trail. In central and western Grand Canyon the Temple Butte is a continuous layer of gray to pale red-brown dolomite and is 400 ft (125 m) thick at the mouth of Grand Canyon. Rare fossil fish and conodonts from the Temple Butte in central Grand Canyon indicate a Late Devonian (Frasnian) age.

Mississippian Redwall Limestone. The Redwall Limestone typically forms a 500-ft (150-m) sheer, red-stained cliff about midway up the Canyon wall. Four members of the Redwall are recognized by McKee and Gutschick (1969) throughout the Grand Canyon.

The lowermost Whitmore Wash Member is a fine-grained, thin- to thick-bedded, oolitic and fossiliferous dolomite in eastern Grand Canyon. It forms the lower fifth of the Redwall cliff and rests unconformably on the Muav or Temple Butte.

The Thunder Springs Member is the most prominent unit in the Redwall. It is 90 ft (27 m) thick along the South Kaibab Trail and is characterized by a distinct banded appearance, owing to

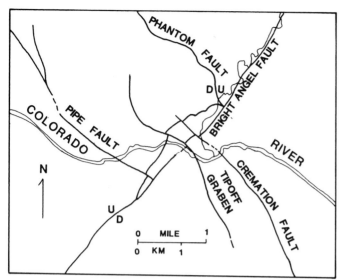

Figure 3. Major fault structures near mouth of Bright Angel Canyon (from Huntoon and Sears, 1975).

the alternation of gray, thin-bedded, crinoidal limestone and thin beds or lenses of light- to dark-gray chert. The Whitmore Wash and Thunder Springs together are interpreted as the record of a major marine transgression-regression cycle (McKee and Gutschick, 1969).

The Mooney Falls Member is the thickest unit in the Redwall and makes up 245 ft (75 m) or more than half the Redwall cliff. It is mainly thick-bedded limestone containing oolites, pelloids, and a variety of skeletal fragments dominated by crinoid plates.

The uppermost Horseshoe Mesa Member is the thinnest and least extensive unit of the Redwall. It is about 70 ft (20 m) thick along the trail and consists of light gray, thin-bedded limestone and minor chert. This unit is considered the record of a slow marine regression following a major and prolonged transgression recorded by the underlying Mooney Falls Member.

The Redwall Limestone is fossiliferous throughout and is dated as Kinderhookian through Meramecian by index brachiopod, foraminiferid, and coral fossils (Mckee and Gutschick, 1969).

Late Mississippian Surprise Canyon Formation. The Surprise Canyon Formation (Billingsley and Beus, 1985) is exposed as isolated lens-shaped outcrops of strata filling old channels cut into the top of the Redwall Limestone. In central and western Grand Canyon, channel-fill beds of limestone, sandstone, and local conglomerate contain a variety of both plant and animal fossils indicative of a Late Mississippian (Chesterian) age. In eastern Grand Canyon, outcrops are smaller, less abundant, and composed of up to 65 ft (20 m) of red-brown mudstone, not easily distinguishable from redbeds of the overlying Supai Group.

No outcrops occur on the South Kaibab Trail, but a small red-brown patch is visible at the base of Pattie Butte to the east (see trail guide, mile 2.5).

Supai Group (Pennsylvanian-Permian). The red sandstone and siltstone beds of the Supai Group form 1,000 ft (300 m) of canyon wall. McKee (1982) defined four formations in the Supai Group, each recording a major marine transgression and marked by a limestone pebble conglomerate at the base. The lowermost, Watahomigi Formation, is 158 ft (48 m) thick along the trail. It is predominantly a slope-forming, pale red siltstone and claystone but includes a 12-ft (3.6-m) and 35-ft (10-m) cliff-forming brownish gray limestone in the lower part, as well as minor sandstone.

The Manakacha Formation overlies the Watahomigi with a basal limestone-pebble conglomerate. It includes a lower cliff unit of alternating limestone ledges and sandstone or mudstone slopes. The upper half is mainly slope-forming mudstone and very fine-grained sandstone.

The Wescogame Formation is predominantly fine-grained, red-brown sandstone and mudstone. It crops out as a series of cliffs and slopes, including a prominent 60-ft (18-m) cliff of brown, cross-bedded sandstone near the base.

The uppermost Esplanade Sandstone is the most resistant unit of the Supai Group. The lower 70 ft (21 m) is slope-forming claystone, but the upper 200 ft (61 m) crops out as nearly vertical cliffs or ledges of fine-grained sandstone. Large-scale, tabular-planar or wedge-planar cross-beds are common and dip mainly to the southeast.

Marine invertebrate fossils, including diagnostic foraminifera, are abundant in the Supai Group in western Grand Canyon and decrease in abundance eastward. Index fossils indicate a Pennsylvanian age for the lower three formations and an Early Permian (Wolfcampian) age for the Esplanade Sandstone (McKee, 1982). Fragments of plant fossils occur locally throughout the Supai, and vertebrate trackways are present in the upper two formations in eastern Grand Canyon.

Hermit Shale. The Hermit Shale is predominantly a deep red-brown siltstone and consistently forms a slope between more resistant sandstones above and below. It has yielded an extensive flora of "upper Lower Permian age" (White, 1929, p. 38). The Hermit is about 300 ft (91 m) thick at the South Kaibab Trail but increases to more than 1,000 ft (300 m) in western Grand Canyon.

Coconino Sandstone. The Coconino Sandstone forms a bold 300-ft (91-m) light yellow cliff, generally clear of vegetation, near the top of the Canyon wall. It consists of well-sorted subrounded quartz sand grains and displays prominent large-scale wedge-planar eolian cross-beds having a consistent south to southeast dip (McKee, 1934). The Coconino thins to about 65 ft (20) in western Grand Canyon and to about 60 ft (18 m) in Marble Canyon to the east. Tetrapod vertebrate trackways are common on some bedding planes in the Coconino, as are insect trails and burrows.

Permian Toroweap and Kaibab formations. McKee

Figure 4. View to the east from Mather Point of Permian Strata at the South Rim. K, Kaibab Formation; T, Toroweap Formation; C, Coconino Sandstone; H, Hermit Shale; E, Esplanade Sandstone.

(1938) recognized the lower strata of the Kaibab Formation as a separate unit, the Toroweap Formation, and demonstrated that the two units record two major marine transgression/regression cycles across northern Arizona during late Early Permian (Leonardian) time. The Toroweap is divided into three members (Sorauf, 1962). In eastern Grand Canyon the lower Seligman member is a 40-ft (12-m) slope forming sandstone very similar to the underlying Coconino Sandstone except for the thin, flat beds in contrast to the prominent cross-bedding of the Coconino. The middle Brady Canyon member is a prominent 60-ft (18-m) cliff-forming, grayish brown limestone or dolomite. The upper Woods Ranch member is commonly exposed as a tree-covered slope of sandstone, siltstone, and minor limestone and gypsum forming the upper two-thirds of the Toroweap.

The Kaibab Formation forms a 330-ft (100-m) cliff of the canyon wall and is almost everywhere the rimrock of Grand Canyon. The lower Fossil Mountain member is nearly 300 ft (91 m) thick along the South Kaibab Trail and is mostly sandy dolomitic limestone with common chert nodules. The lower boundary is conveniently placed at the lowest cherty carbonate above light gray or yellowish sandstone of the Toroweap Formation. The upper Harrisburg member is about 35 ft (10 m) thick and consists of sandstone and dolomite marked at the base by a light gray nodular to bedded chert unit up to 8 ft (2.5 m) thick. Both the Kaibab and Toroweap contain a variety of invertebrate marine fossils including brachiopods, bryozoans, crinoids, sponges, and locally abundant mollusks.

TRAIL LOG, PALEOZOIC ROCKS ALONG THE SOUTH KAIBAB TRAIL

Mile 0.0. Begin descent of Kaibab Trail through limestone ledges of the Kaibab Formation. The Kaibab Formation forms a steep 200 to 300-ft (60 to 90-m) cliff at the canyon rim (Fig. 4).

Third switchback corner: note brown ellipsoid or irregular chert nodules parallel to bedding in the Kaibab.

Fifth switchback corner: fossil productid brachiopods and sponges abundant in cherty dolomite beds of the Fossil Mountain member of the Kaibab.

Sign: Kaibab-Toroweap contact. The Toroweap has three distinct members: the upper Woods Ranch member forms a slope; the middle Brady Canyon member is a resistant ledge-forming limestone 70 ft (21 m) thick; and the lower Seligman member is a cross-bedded to flat-bedded quartz sandstone 40 ft (12 m) thick.

Huge slump blocks of Kaibab Limestone lie scattered on the Toroweap slope. Specimens of brachiopods, sponges, and trilobites locally occur on weathered surfaces of the slump blocks.

Mile 0.5. Cross-bedded quartz sandstone beside trail is in the Woods Ranch Member of the Toroweap. Sixty-foot (18 m) vertical cliff of gray-brown limestone is the Brady Canyon member of the Toroweap.

Sign: Toroweap-Coconino contact; clearly marked by color boundary between red-brown flat-bedded sandstone of the Toroweap and the cross-bedded sandstone of the Coconino, which

Figure 5. View northwest of the Supai Group traversed by South Kaibab Trail. E, Esplanade Sandstone at O'Neill Butte; We, Wescogame Formation; M, Manakacha Formation; W, Watahomigi Formation; R, Redwall Limestone.

Figure 6. Air view to west of South Kaibab Trail route through Redwall Limestone (R), Muav Limestone (M), and Bright Angel Shale (B).

forms a massive 340-ft (104-m) buff-yellow cliff of canyon wall (Fig. 4).

View point just beyond corner of tenth switchback affords panoramic view of trail ahead and entire Paleozoic section down to the Cambrian beds along the Tonto Platform at the top of the Inner Gorge. Figure 5 was photographed here.

Cross-beds up to 30 ft (9 m) high in Coconino Sandstone are mainly planar-tabular or planar-wedge shape with predominantly southerly dips. A few bedding planes display gentle broad ripple marks and, in other localities of the Grand Canyon, vertebrate trackways.

Mile 1.0. Sign: Coconino-Hermit contact; actual contact largely obscured. The Hermit Shale forms a smooth slope on deep red siltstone and shale beneath cliffs of Coconino and is about 300 ft (90 m) thick.

Cedar Ridge. Good place for rest stop where trail enters juniper-covered platform area near base of Hermit Shale. View of plant fossils under glass-covered display case about 150 ft (45 m) to left of trail. Mud-crack casts, rain prints, fossil ferns, conifers, and insects occur in this part of the Hermit.

Hermit-Supai contact about where trail begins descent from Cedar Ridge. The Supai Group is more than 1,000 ft (300 m) of dark red to orange-brown sandstone, siltstone, and limestone of Pennsylvanian and Permian age. The system boundary is approximately one-quarter the distance down from the top of the Supai in the slope beneath the massive upper cliff of Esplanade Sandstone (Fig. 5). Irregular limestone pebble conglomerate marks Permian-Pennsylvanian boundary and contact between Esplanade Sandstone and Wescogame Formation.

Mile 1.5. Trail on saddle between Cedar Ridge and O'Neill Butte in Wescogame Formation.

Mile 2.0. Pennsylvanian reptile or amphibian trackway in

sandstone beds to left side and about 3 or 4 ft (0.9 to 1.2 m) above trail are in the lower Wescogame Formation of the Supai Group. Wescogame-Manakacha boundary is near the base of the cliff in which these tracks occur.

Trail on ridge crest.

Mile 2.5. Trail in lower Watahomigi Formation crosses some blue-gray finely crystalline limestone. Pennsylvanian fusulinids and brachiopods occur in similar limestones as far east as Marble Canyon.

East of the trail can be seen the Tapeats Sandstone onlapping against a promontory of Zoroaster Granite (Precambrian) in the Inner Gorge across the Colorado River.

Just beyond hill 5391 the trail comes on to the light gray limestone of the upper Redwall Limestone. Thin lenses of dark red siltstone of Surprise Canyon Formation occur at this stratigraphic position at the base of Pattie Butte, 2 mi (3.3 km) to the east, but are not present here.

Mile 3.0. Trail begins descent through sheer 500-ft (150-m) cliffs of Redwall Limestone of Mississippian age (Fig. 6).

Third switchback corner in Redwall below Skeleton Point. Good place for a rest stop to view northwest across canyon to "Cheops Bay" on Tonto Platform across the river. The Cambrian Tapeats Sandstone beds (red-brown rounded ledges), in the flat east and below Cheops Pyramid, pinch out against promontories of Precambrian Shinumo Quartzite on both sides of a mile-wide Tapeats outcrop. Gray-green patches of Bright Angel Shale rest on Tapeats Sandstone in the center, but on Shinumo Quartzite around the margins of this ancient Cambrian bay (Fig. 7).

Solution cave in Redwall Limestone to the right and above the trail; white minerals filling vugs in limestone at trail level are calcite and barite.

Mile 3.5. Chert and crinoidal limestone abundant in the lower Redwall strata.

Devonian Temple Butte Formation crops out as a purple-gray dolomite in the cliff just northeast of the last switchback corner in the Redwall.

Sign: Temple Butte Formation. The Temple Butte forms thin lenses of channel-fill deposits between the Redwall and the Cambrian Muav Limestone.

Muav Limestone forms irregular ledges below massive Redwall cliffs. Blue-gray limestone of the Upper Muav grades downward into green sandy limestone and sandstone beds rich in glauconite; horizontal trace fossil burrows and trails are abundant on the underside of some beds. The lower boundary of the Muav is placed at the base of the lowest resistant ledge-forming bed above the slope-forming Bright Angel Shale.

Mile 4.0. Sign: Bright Angel Shale. Trail traverses smooth slope eroded on soft green shale above the Tonto Platform. Window in Redwall Limestone on skyline to the west.

Mile 4.5. Junction with Tonto Trail, which crosses the Kaibab Trail east-west. In the bottom of the gully to the left of the trail, about 50 yds (45 m) past the outhouses, is a large (4 by 4 ft; 1.2 by 1.2 m) buff-colored sandstone block of Bright Angel Shale exhibiting crawling and resting traces presumably made by trilobites as well as numerous other trace fossils.

LOCALITY 84: PROTEROZOIC ROCKS
Early Proterozoic Rocks

Dark green mica and quartz-feldspathic schist of the Vishnu Group is exposed in most of the inner gorge of Grand Canyon along the lower Kaibab Trail. Foliation in the schist is at or near vertical and has a generally northeast strike. Bands of light gray to pink felsic gneiss are included in the Vishnu in lower Bright Angel Creek and are locally elastically deformed into boudinage structures and migmatite. Degree of metamorphism ranges from greenschist to upper amphibolite facies. In some areas of eastern Grand Canyon, Zoroaster Gneiss occurs as foliated granitic plutons, dikes, or sills emplaced in the Vishnu, probably during stages of early metamorphism. In addition, pegmatite/aplite dikes are intruded into the gneiss and schist with crosscutting relationships. The Vishnu records metamorphism of original sedimentary and volcanic rocks as part of a major orogeny about 1,700 Ma (Pasteels and Silver, 1965).

Middle Proterozoic Rocks

Bass Limestone. The Grand Canyon Supergroup exposed in eastern Grand Canyon includes more than 13,000 ft (4,000 m) of sedimentary and volcanic rocks. Only the lower quarter of this section, the lower part of the Unkar Group, is exposed in the Kaibab Trail area, where it is preserved within the inner gorge in narrow blocks of prominently layered strata down-faulted or inset into the Vishnu Group basement complex.

The basal unit of the Unkar group and of the Grand Canyon Supergroup is the Bass Limestone, which is predominantly dolomite with subordinate arkose, sandy dolomite, shale, argillite, and locally conglomerate. Carbonates in the Bass contain several varieties of stromatolites, including "biscuit shaped," matte, and

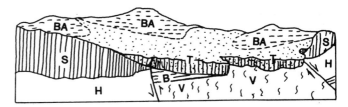

Figure 7. Sketch of "Cheops Bay" area as viewed from near Skeleton Point, in Redwall Limestone along South Kaibab Trail. BA, Cambrian Bright Angel Shale; T, Cambrian Tapeats Sandstone; S, Precambrian Shinumo Quartzite; H, Hakatai Shale; B, Bass Limestone; V, Vishnu Group.

rare biohermic forms. Ripple marks, mud-crack casts and intraformationally collapsed breccia in the upper Bass suggest subareal exposure and the former presence of evaporites.

Hakatai Shale. The Bass is overlain conformably by the bright orange-red Hakatai Shale. The Hakatai forms relatively smooth slopes in contrast to the more resistant formations above and below. It is nearly 1,000 ft (300 m) thick at the type section but thins eastward to about 560 ft (170 m) at the South Kaibab Trail. Three subdivisions are recognized—a basal red mudstone, a middle orange mudstone, and an upper purple sandstone. The abundant ripple marks and mudcrack casts, plus local raindrop impressions and salt crystal pseudomorphs, suggest deposition in a near-shore and locally subareally exposed environment.

Shinumo Quartzite. The Shinumo is a thick succession of red and purple quartz and feldspathic sandstone that unconformably overlies the Hakatai Shale. Sand grains are commonly coated with hematite and set in siliceous cement. Small-scale trough and tabular-planar cross-beds are common throughout the formation, and ripple marks occur locally. The Shinumo forms resistant cliffs and is more than 1,300 ft (400 m) thick in eastern Grand Canyon. Only the lower third is exposed in the South Kaibab Trail area.

Additional units of the Unkar Group above the Shinumo include the Dox Formation (3,100 ft; 950 m) and Cardenas Lavas (about 1,000 ft; 300 m), extensively exposed in easternmost Grand Canyon some 15 mi (24 km) upstream of the Kaibab Trail. The Cardenas Lavas have yielded a Rb-Sr isochron date of about 1.1 b.y. (McKee and Noble, 1976), which provides a minimum age for the Unkar Group.

CONTINUATION OF TRAIL LOG, PROTEROZOIC ROCKS ALONG THE KAIBAB TRAIL

The Tipoff. Trail begins descent into Inner Gorge through thin beds of Tapeats Sandstone, which is only a few tens of feet thick here where it is deposited on an erosion surface of considerable relief (100 ft; about 30 m) cut on Precambrian Shinumo Quartzite.

Switchback corner at Cambrian-Precambrian boundary affords first full view of the Inner Gorge (Fig. 8); buttress unconformity where Tapeats Sandstone buries old cliff of Shinumo

Quartzite above the trail. Blocks of Shinumo are incorporated in the Tapeats.

Sign: Shinumo Quartzite; purple-gray quartz sandstones display small-scale cross-beds and current ripples.

Fault brings Hakatai Shale up against Shinumo Quartzite; the nearly vertical fault surface forms a sheer cliff below the trail to the northwest.

Mile 5.0. Trail in bright orange Hakatai Shale passes around massive outlier of Tapeats Sandstone (landslide block?) resting on Hakatai. Ripple marks and mud cracks are common in the Hakatai Shale.

Panorama Point (elevation 3,600 ft; 1,100 m) in the Hakatai Shale affords a spectacular view of the Inner Gorge and Bright Angel Creek delta.

Sign: Bass Limestone (actually dolomite and argillite): exposures to left of trail display thin algal laminae.

Buddha Temple to the north and Brahma Temple to the northeast are impressive slender spires of Coconino and Toroweap strata on the skyline across the gorge.

Mile 5.5. Sign: Vishnu Schist; trail traverses greenish black hornblende mica schist having vertical foliation and locally injected by granite dikes rich in pink feldspar.

Mile 6.0. Trail crosses sloping surface on tightly cemented boulder talus in small steep drainage.

Junction with river trail on left; suspension bridge and tunnel visible below.

Tunnel entrance to suspension bridge; trail crosses Colorado River at mile 87.4 below Lee's Ferry; cable car for river gaging station just upstream.

Mile 6.5. Trail leads downstream on right bank past ruins of 12th century Indian dwellings on left.

Bright Angel Creek was named by John Wesley Powell (1875) when he camped here in August 1869. Powell's journal reads: "August 16.—We must dry our rations again to-day and make oars. The Colorado is never a clear stream, but for the past three or four days it has been raining much of the time, and the floods poured over the walls have brought down great quantities of mud, making it exceedingly turbid now. The little affluent which we have discovered here is a clear, beautiful creek . . . We have named one stream, away above, in honor of the great chief of the 'Bad Angels,' and as this is in beautiful contrast to that, we conclude to name it 'Bright angel.'"

Bridge over Bright Angel Creek on left; cross bridge to enter Bright Angel campground just upstream or stay on right (east) side of creek for trail to Phanton Ranch (0.5 mi; 0.8 km upstream).

REFERENCES CITED

Billingsley, G. H., and Beus, S. S., 1985, The Surprise Canyon Formation; An Upper Mississippian and Lower Pennsylvanian (?) rock unit in the Grand Canyon of Arizona: U.S. Geological Survey Bulletin 1605-A, p. 27–33.

Huntoon, P. W., and Sears, J. W., 1975, Bright Angel and Eminence faults, eastern Grand Canyon, Arizona: Geological Society of America Bulletin, v. 86, p. 465–472.

Figure 8. View west of inner gorge and lower part of the South Kaibab Trail. Precambrian rock units downfaulted in the Tipoff graben are: S, Shinumo Quartzite; H, Hakatai Shale; B, Bass Limestone; and V, Vishnu Group. Kaibab Trail crosses river at Kaibab (nearest) bridge. Silver bridge 0.5 mi (0.8 km) downstream carries water pipeline bringing water from near the head of Bright Angel Creek (Roaring Springs) to the South Rim.

Huntoon, P. W., Billingsley, G. H., Breed, W. J., Sears, J. W., Ford, T. D., Clark, M. D., Babcock, R. S., and Brown, E. H., 1986, Geologic map of the eastern part of Grand Canyon, Arizona: Grand Canyon Natural History Association, scale 1:62,500.

Martin, D., 1985, Depositional systems and ichnology of the Bright Angel Shale (Cambrian), eastern Grand Canyon, Arizona [M.S. thesis]: Flagstaff, Northern Arizona University, 365 p.

McKee, E. D., 1934, The Coconino Sandstone; Its history and origin: Carnegie Institution of Washington Publication 440, Contributions to Paleontology, p. 77–115.

——, 1938, Environment and history of the Toroweap and Kaibab formations in northern Arizona and southern Utah: Carnegie Institution of Washington Publication 492, 268 p.

——, 1982, The Supai Group of Grand Canyon: U.S. Geological Survey Professional Paper 1173, 504 p.

McKee, E. D., and Resser, C. E., 1945, Cambrian history of the Grand Canyon region: Carnegie Institution of Washington Publication 563, 232 p.

McKee, E. D., and Gutschick, R. C., 1969, History of the Redwall Limestone: Geological Society of America Memoir 114, 726 p.

McKee, E. H., and Noble, D. C., 1976, Age of the Cardenas Lava, Grand Canyon, Arizona: Geological Society of America Bulletin, v. 87, p. 1188–1190.

Pasteels, P., and Silver, L. T., 1965, Geochronological investigations in the crystalline rocks of the Grand Canyon [abs.]: Geological Society of America Special Paper 87, p. 124.

Powell, J. W., 1875, Exploration of the Colorado River of the West and its tributaries explored in 1869–1872: Smithsonian Institution, 291 p.

Sorauf, J. E., 1962, Structural geology and stratigraphy of the Whitemore area, Mohave County, Arizona [Ph.D. thesis]: Lawrence, Kansas University, 361 p.

Washburn, B., 1978, The heart of Grand Canyon (color-shaded topographic map): Washington, D.C., National Geographic Society, scale, 1:24,000.

White, D., 1929, Flora of the Hermit Shale, Grand Canyon, Arizona: Carnegie Institute of Washington Publication 405, 221 p.

Geology of the Gray Mountain area, Arizona

Charles W. Barnes, Department of Geology, Northern Arizona University, Flagstaff, Arizona 86011

LOCATION

Gray Mountain is a topographic prominence that occupies most of the Coconino Point 15-minute Quadrangle, Coconino County, in northern Arizona (Fig. 1). That quadrangle is delimited by 111° 30′ to 111° 45′ west longitude, and 35° 45′ to 36° 00′ north latitude. The total structure extends westerly into the adjacent Grandview Point 15-minute Quadrangle (which includes a very small part of the eastern Grand Canyon), and extends south into the adjacent SP Crater 15-minute Quadrangle.

ACCESSIBILITY

Road Access

Access to the eastern part of the area is by a series of dirt roads that trend west from U.S. 89A North; these roads trend westerly from the village of Gray Mountain, about 45 mi (72 km) north of Flagstaff, Arizona, on U.S. Highway 89A North. The "main" road westerly is the Gray Mountain Truck Trail (so labelled on the Coconino Point quadrangle). Following this road to its end takes one generally westerly and northwesterly across all of Gray Mountain, where the road intersects Arizona 64.

Access to the northern area is by numerous dirt roads leading from Arizona 64. Arizona 64 runs essentially due west from its intersection with U.S. 89N, near Cameron, Arizona. Arizona 64 provides excellent all-weather access to the northern reaches of Gray Mountain, and allows closeup views of the strongly deformed norther flanks of Gray Mountain, and spectacular views of the Little Colorado River valley. Continuing to drive westward on Arizona 64 takes one to the east entrance of the Grand Canyon National Park, where the structures that form a part of Gray Mountain are superbly exposed in three dimensions (Reches, 1978).

Dirt roads are passable by four wheel drive vehicles, ordinary pickup trucks, and passenger vehicles with sufficient clearance. One should inquire locally about access in bad weather.

Vegetation and Exposure

The area's vegetation (within the lower zones around 5,000 ft [1,500 m]) is characterized by sparse short grasses and low blackbrush. The top of Gray Mountain reaches elevations around 7,000 ft (2,100 m); vegetation on top includes some pinyon and ponderosa pine. Access by foot is easy and exposure quality is superb throughout the whole area. There are numerous smaller hiking trails, and cross-country hiking is easy.

Jurisdictional Concerns

The majority of the Gray Mountain area is within the jurisdiction of the Navajo Tribe, and forms a part of the Navajo Reservation. Hogans and small homes dot the entire area, which is a small part of the Navajo Reservation. Casual access does not require permission, nor do day visits with classes. The Indians for whom this area is home are generally friendly, but do not appreciate being disturbed; courtesy and prudence suggest that you avoid their homesites, which are very thinly scattered throughout the area. The land is open range; watch for livestock. The land to both the south and to the northwest of the Coconino Point Quadrangle is within the Coconino National Forest; no permits are needed for work within these areas.

Camping

There are *no* camping facilities within the area. There is a primitive (dry!) USFS campsite right off of Arizona 64 on the extreme western edge of the Coconino Point Quadrangle; recommended camping is around Flagstaff (about 60 mi [96 km] south on U.S. 89A), or at Desert View in the eastern Grand Canyon (about 20–40 mi [32–64 km] west of Gray Mountain on Arizona 64). For camping within the Grand Canyon, reservations are required; contact Campground Reservations Office, Grand Canyon National Park, Grand Canyon, AZ 86023, or phone 602-638-7761. Student groups, for which a Grand Canyon stay is part of a class experience, may use the Group Campground at Mather Campground (602-638-7851) without payment of fees, if arranged by letter in advance. From Mather (a nice campground!) to Gray Mountain is about 50–60 mi (80–96 km).

SIGNIFICANCE

Gray Mountain is a topographic prominence that has resulted from polyphase uplift of intermittently active basement blocks; the surface expression (Billingsley and others, 1985) is a series of complexly interfingering monoclines, whose junctions include a large scale right-oblique-down hinge fault, thrust faults, graben fields, high-angle reverse faults, and a very large-scale graben. Subsidiary to these structures are two poorly-defined mesodermal(?) circular gliding structures, whose intersection has led to a superbly exposed epidermal slip-sheet structure.

Excellent exposures provide superb examples of the interplay between forced folding, hinge faulting, oblique-slip faulting, and graben development at numerous scales. Equally spectacular exposures of fixed hinge fold development and slip-sheet epidermal gliding provide another focus. T-shaped monoclinal fold intersections, the transition from a hinge fault into a very large graben, overturned folds, flatirons, neotectonism, breccia pipes, Plio-Pleistocene basaltic volcanism, multiple graded pediment surfaces, barbed drainage patterns, collapse structures, and classic Grand Canyon stratigraphy provide other points of interest. Just to the northwest of the area, within the eastern Grand Canyon,

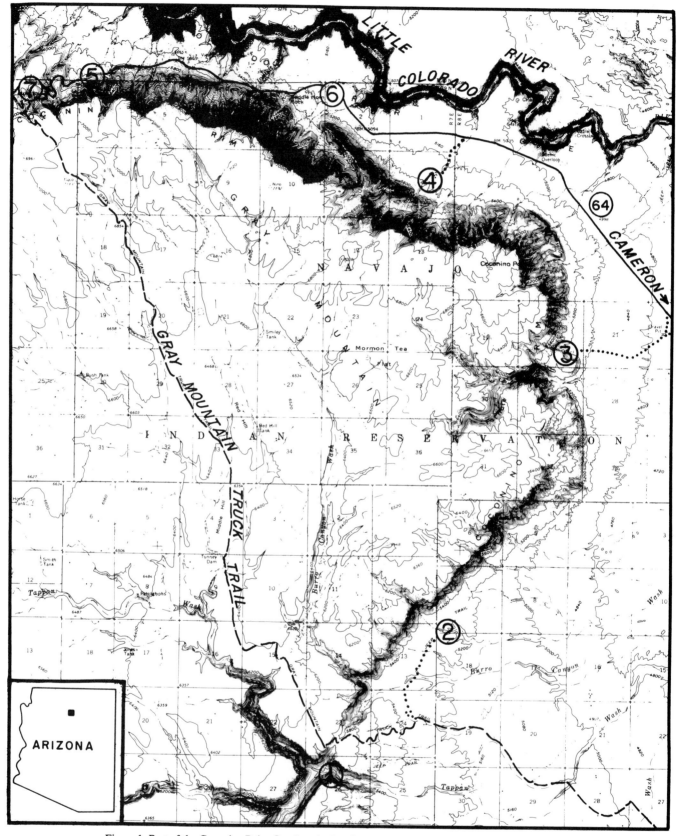

Figure 1. Part of the Coconino Point Quadrangle showing access to the localities discussed in the text.

one gains three-dimensional views of the structures that, by extension, might reasonably underlie the Gray Mountain area (Reches, 1978).

SITE INFORMATION

Geomorphology

The Coconino Point Quadrangle encompasses about 250 mi^2 (650 km^2) of the southern Colorado Plateau. The quadrangle includes two geomorphic subprovinces: the Little Colorado River valley, and the Coconino Plateau–Gray Mountain area. Elevations range from 3,800 ft (1,150 m) at the bottom of the Little Colorado river gorge to 7,200 ft (2,200 m) at the northern rim of Gray Mountain. Rocks exposed within the map area include those from the Permian to Plio-Pleistocene. The Kaibab Formation of Permian age forms the resistant erosional surface that defines the land surface over most of this area. This stripped surface is the fundamental unit of topography.

Abandoned and barbed drainage systems in the southeast quadrant of the map are probably of late Tertiary age, as are three separate pediment levels. All of these surfaces and drainages are graded to the Little Colorado River, which is a subsequent drainage with respect to the Gray Mountain uplift. Younger drainage systems have eroded the weak Triassic sedimentary rocks and are superimposed on the resistant Kaibab surface; in areas of uplift, these drainages are incised.

Structural Geology

The major topographic uplift within this quadrangle is locally termed Gray Mountain. This generally high area is the result of the varying uplift of five monoclines (Fig. 2), which join one another.

To the west, the Grandview monocline forms a low, prominent north-facing scarp, which trends westerly and then northwesterly into the Grand Canyon, where its three-dimensional structure can be readily seen within the Sinking Ship area, about 7 mi (11 km) west of Desert View. The Grandview monocline trends eastward into the Coconino Point Quadrangle. The northerly-trending East Kaibab monocline joins the Grandview monocline, with the trace of the fold axes making an approximate 60° acute angle where they intersect. This intersection zone is structurally complex and is the site of a trapezoidal strong negative gravity anomaly suggesting isostatic decompensation along a probable downdropped basement block.

Both the Grandview and East Kaibab monoclines are paralleled by high-angle reverse faults, which generally diminish by a few hundred ft (m) the stratigraphic offset produced by monoclinal folding. The trend of the East Kaibab monocline, southeast of its intersection with the Grandview, is a series of grabens, collectively termed the Burro Canyon Fault System, which trend southeasterly, and almost join the Additional Hill monocline. The Burro Canyon Fault System subdivides the Gray Mountain mass

(Fig. 3) into a topographically higher northeasterly trapezoidal block, which displays distinctive isosceles-like linear patterns, twin, intersecting radial joint patterns, faint evidence of mesodermal(?) gliding, and a southwesterly lower block, having generally homoclinal dip to the southwest.

East of the Grandview–East Kaibab monoclinal junction, the continuation of the Grandview structure is known as the Coconino Point monocline. The Coconino Point monocline is the largest of the five monoclines (Fig. 2) within the quadrangle.

The stratigraphic throw increases easterly along the Coconino Point fold axes, reaching a maximum of approximately 2,000 ft (610 m) at Coconino Point, the northeastern "corner" of Gray Mountain (cf. Coconino Point 15-minute Quadrangle). From Coconino Point, both to the west and to the south, the stratigraphic throw steadily diminishes. At Coconino Point, the offset is taken up wholly by folding; bedding dip is near-vertical to overturned in some locales. Dips remain quite high over the whole course of the Coconino Point structure, as forced folding is the dominant process that produced the stratigraphic offset.

Paralleling the Coconino Point fold axes is a high-angle reverse fault, which serves to slightly diminish the total throw on the structure. Fault exposures are seen in a series of aligned saddles, gouge zones, and slickensided zones of abrupt stratigraphic contact; zones of overturned bedding parallel the fault throughout its course.

To the southeast, this rim fault loses throw, and then scissors into a hinge fault as the Black Point (Fig. 2) monoclinal axis interferes with the Coconino Point structure. Paralleling part of this hinge fault is a superbly exposed thrust fault (cf. Billingsley and others, 1985), necessary to accommodate the room problem as Black Point and Coconino Point structures interfere. Still further southwest on Gray Mountain, the hinge fault, at its fulcrum, loses throw and then passes into the Mesa Butte Graben (Barnes, 1974), which parallels one of the major dislocation zones in the basement of northern Arizona, the Mesa Butte lineament (Shoemaker and others, 1974). The monoclinal folding, characteristic of the Coconino Point area, is interrupted by northeasterly faulting having quite a diversity of styles. The Mesa Butte trend defines the southeast "edge" of Gray Mountain.

This regmental fault system exhibits both positive magnetic anomalies, and an anomalous regional geomagnetic variation field (Towle, 1984), with a regional telluric current polarized parallel the orientation of the Mesa Butte lineament. The high conductivity, polarization, and magnetic anomaly all suggest that shallow (<6 mi [10 km]) Plio-Pleistocene volcanism within this area is only dormant.

A series of grabens radiate out from the northern limbs and the northeastern corner of Gray Mountain; these grabens offset the stripped Kaibab surface, with throws generally less than 50 ft (15 m). They all lose throw as they approach the steep northern middle limb of the Coconino Point monocline, and hinge down to zero throw within a few hundred ft (m) of the steeply-dipping north face of Gray Mountain.

The generally radial map pattern of these grabens suggests

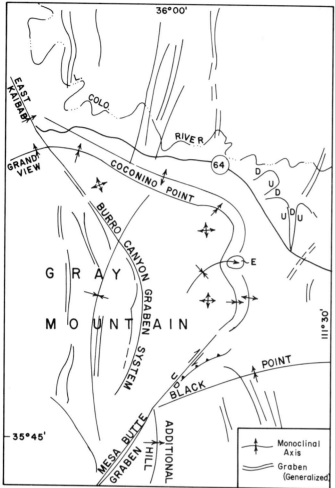

Figure 3. Photo of Gray Mountain area for comparison with Fig. 2. Photo courtesy of U.S. Air Force, Mission 374, 06 Sept. 68, Photo # 033V.

Figure 2. Sketch map of Gray Mountain area; major structures shown with conventional symbols. *E* = epidermal slip sheet. For complete geologic map, see Billingsley and others, 1985.

that they have formed parallel to maximum compressive-stress trajectories; the pattern reminds one of the classic studies of an indentor impinging on an elastic plate (Jaeger and Cook, 1976). If this mechanical model is accepted, that suggests that the Gray Mountain mass has shifted relatively north-northeasterly with respect to the generally flat-lying Permian units incised by the Little Colorado River. The Mesa Butte structural zone is the accommodating tear. Further evidence of the generally sinistral-normal character of the motion is seen in poorly exposed slickensides in the fault zone north of the Mesa Butte Graben. These slickensides rake 70°S35°W, and document the predicted movement pattern.

Further evidence of the recency of motion along the Mesa Butte tear is obtained from analysis of the surface offset by throw along the radial graben system. In one locality, the grabens break Tappan Wash basalts that have flowed north from Mesa Butte; these basalts yield radiometric ages (Moore and others, 1974) of less than 520 ka. The grabens also break terrace gravels along the Little Colorado that contained bones of *Elephas*

columbi (Columbian mammoth); these terrace gravels approximate 500 ka (Reiche, 1937). Thus one may assume that a generally northeasterly shift of the Gray Mountain mass continues into modern times; Arizona's largest earthquake occurred in the Gray Mountain area in the 1920s (Sturgul and Irwin, 1971).

Still more evidence of the overall movement pattern is gained by studying the shear joint sets prominently displayed in the flat-lying Kaibab limestone near the Little Colorado River. The bisectrix of the dihedral angle of these shear joints is precisely perpendicular to the Coconino Point monoclinal fold axes, with the dihedral angle opening up to the north, away from the source of stress, in the manner suggested by Muehlberger (1961).

Interestingly, the same movement pattern is repeated in parallel along the Doney Cliffs structure within the Wupatki National Monument to the east (**obtain formal permission** from Monument rangers **prior to** entering Doney Cliffs area), where the Black Point monocline provides a perfectly congruent example of the same movement pattern, but on a much smaller scale. Along the southwestern part of the Doney Cliffs, the sinistral normal motion provides elegant examples of small right-stepping

Figure 4. View looking northeast. Permian Kaibab Formation on skyline defines an open cylindrical fold on the upper limb of the Coconino Point monocline.

Figure 5. View looking southerly from atop epidermal slip sheet. Permian Kaibab limestone layers define a fixed-hinge flat-slab fold, broken by rim reverse fault.

en echelon folds that join segments of the bounding faults. The Doney Cliffs area, within the western part of the Wupatki National Monument is a particularly instructive place to bring classes. The evidence of sinistral normal motion on the fault is everywhere, including well-exposed slickensides, right-stepping en echelon folds, and fault splinters of the Triassic Moenkopi still hanging on the hanging wall block.

Tectonics

On a regional scale, the Doney Cliffs linear, and the parallel Mesa Butte and Bright Angel linears to the northwest all have documented sinistral normal motion (Sears, 1973), as do the Shylock and Chaparral fault zones in central Arizona (Shoemaker and others, 1974). If one *assumes* (currently this is undocumented) sinistral shift on the Sinyala Fault System, which parallels the Bright Angel System in the western Grand Canyon (Shoemaker and others, 1974), these linears define a northeasterly-trending set of fault systems that, *in toto,* lead to northeasterly to southwesterly sinistral shift in a "card-deck" model that covers most of north-central Arizona. The observations of Towle (1984, p. 225) that the regional telluric current is polarized parallel to the Mesa Butte, Bright Angel, and Sinyala fault systems over a regional conductive fabric, suggests that the Sinyala, Bright Angel, Mesa Butte, and, by extension, Doney Cliff fault systems are fundamental, deep crustal/upper mantle features, capable of allowing the shift proposed.

Recommended Field Trip Stops

Following are several particularly instructive localities (cf. Barnes and others, 1974; also Billingsley and others, 1985).

1. NW¼Sec.26,T.28N.,R.7E. Drive to this area on Gray

Mountain Truck Trail. Road ascends dip slope of Additional Hill monocline, with cross faults, and includes fine views of the Black Point monocline to the east, with numerous grabens perpendicular to the fold axes of Black Point. Road takes one to the zone where the hinge fault along the southwest zone of Gray Mountain loses all its throw, and scissors into the Mesa Butte Graben. Spectacular views, including an overview of all the structures on the southwest flank of Gray Mountain (Fig. 4).

2. NW¼Sec.18,T.28N.,R.8E. Well-exposed collapse structure, defined by vertically-dipping Triassic Moenkopi siltstones and shales, with Z-shaped folds having vertical fold axes. Bounding fault well-exposed, as is the chert breccia that commonly forms the uppermost part of the Permian Kaibab Formation within the Gray Mountain area. Just to the northwest of this area are excellent exposures of a low-angle thrust, which brings Kaibab Formation out over Moenkopi siltstones and shales. Walking out this thrust to its northern terminus yields well-exposed structural relations too complicated to display on the published map (Billingsley and others, 1985) for the Gray Mountain area.

3. SE¼,Sec.20,T.29N.,R.8E. Look at *north* side of this prominent butte; superb exposures of an epidermal slip sheet, complete with exposures of flat-lying Kaibab slip-sheet on top of vertical Kaibab and Moenkopi. Excellent locale to have students sketch the geologic picture and try to figure it all out. Climb to top of the butte and walk to southwest edge; superb exposure of fixed-hinge flat-slab folding to the southwest (Fig. 5). Also excellent exposures of flatirons, high-angle reverse rim fault, including multiple repetition of section just north of the butte. Just north of Section 20 (in unsurveyed Section 17) are spectacular exposures of vertical faults. Both this area and previously described area are reachable by rather rough, dirt roads—high clearance vehicles REQUIRED.

4. S½,Sec.12,T.29N.,R.7E. Fairly rough dirt road southerly

to hogans; then walk south-southwesterly up major drainage. Spectacular view of near-vertical Permian section, heavily slickensided. To the south, one crosses one splinter of the rim fault, and abruptly enters the orangish-red sandstones of the uppermost Supai Formation (locally termed the Schnebly Hill member). Near the hogan, there are excellent exposures of the Chinle Formation, including the Shinarump conglomerate member, and the "purple platy beds," which are probably equivalent to the Mossback Formation in Utah.

5. NW¼,Sec.6,T.29N.,R.7E. Drive up on the old highway alignment. Outstanding exposures on ripple-marked Moenkopi siltstones. Dips are extremely steep, as these exposures are on the middle limb of a strongly deformed monocline. Excellent exposures of rim fault, including overturned beds and gouge zones, hillside creep over Moenkopi, and view to west along Grandview monocline and view to northwest along East Kaibab monocline. Spectacular view to north over Little Colorado River valley.

6. NW¼,Sec.2,T.29N.,R.7E. Drive on old highway alignment. Excellent exposures of grabens, including hinge characteristic of southern terminus. Drag-fold very well-exposed in wash immediately south of old highway. Aragonitic material in fault zone; hinging to zero extremely well-exposed.

7. Unsurveyed Section 1, T.29N., R.6E. Drive Gray Mountain Truck Trail to top of mountain. Road crosses superb exposures of rim fault, overturned beds, and gouge zones. Walking east-northeast along the monoclinal axis brings one across grabens associated with the intersection of East Kaibab and Grandview monoclines. Intersection zone is in the NE¼,Sec.6, T.29N.,R.7E. If one takes the Gray Mountain Truck Trail all the way to the top of Gray Mountain, the Burro Canyon Graben system can easily be seen. Exposure at Coco Benchmark (elev. 7,113 ft [2,168 m] extreme west edge, Sec.7,T.29N.,R.7E.), is fairly typical. Views to north are incredible. On a clear day, one can trace the East Kaibab monocline north into Utah.

REFERENCES CITED

Barnes, C. W., 1974, Interference and gravity tectonics in the Gray Mountain area, Arizona, *in* Karlstrom, T.N.V., Swann, G. A., and Eastwood, R. L., eds., Geology of Northern Arizona: Geological Society of America, Rocky Mountain Section Meeting, p. 442–453.

Barnes, C. W., Eastwood, R., and Beus, S., 1974, Geologic resume and field guide, north-central Arizona, *in* Karlstrom, T.N.V., Swann, G. A., and Eastwood, R. L., eds., Geology of Northern Arizona: Geological Society of America, Rocky Mountain Section Meeting, p. 423–441.

Billingsley, G., Barnes, C. W., and Ulrich, G., 1985, Geologic map of Grandview and Coconico Point Quadrangles, Coconino County, Arizona: U.S. Geological Survey Map MI-1644.

Jaeger, J. C., and Cook, N.G.W., 1976, Fundamentals of rock mechanics (2nd edition): London, Chapman, and Hall, 585 p.

Moore, R. B., Wolfe, E. W., and Ulrich, G. E., 1974, Geology of the eastern and northern parts of the San Francisco volcanic field, Arizona, *in* Karlstrom, T.N.V., Swann, G. A., and Eastwood, R. L., eds., Geology of Northern Arizona: Geological Society of America, Rocky Mountain Section Meeting, p. 465–494.

Muehlberger, W. R., 1961, Conjugate joint sets of small dihedral angle: Journal of Geology, v. 69, no. 2, p. 211–219.

Reches, Z., 1978, Development of monoclines, Part 1, Structure of the Palisades Creek branch of the East Kaibab monocline, Grand Canyon, Arizona, *in* Matthews, V., ed., Laramide folding associated with basement block faulting in the western United States: Geological Society of America Memoir 151, p. 235–272.

Reiche, P., 1937, Quaternary deformation in the Cameron district of the Plateau province: American Journal of Science, 5th Ser., v. 34, p. 128–138.

Sears, J. W., 1973, Structural geology of the Precambrian Grand Canyon Series, Arizona [M.S. thesis]: University of Wyoming, 100 p.

Shoemaker, E. M., Squires, R. L., and Abrams, M. J., 1974, The Bright Angel and Mesa Butte fault systems of northern Arizona, *in* Karlstrom, T.N.V., Swann, G. A., and Eastwood, R. L., eds., Geology of Northern Arizona: Geological Society of America, Rocky Mountain Section Meeting, p. 355–392.

Sturgul, J. R., and Irwin, T. D., 1971, Earthquake history of Arizona and New Mexico, 1850–1966: Arizona Geological Society Digest, v. 9, p. 1–22.

Towle, J. N., 1984, The anomalous geomagnetic variation field and geoelectrical structure associated with the Mesa Butte fault system, Arizona: Geological Society of America Bulletin, v. 95, p. 221–225.

SP Mountain cinder cone and lava flow, northern Arizona

G. E. Ulrich, *U.S. Geological Survey, 2255 N. Gemini Road, Flagstaff, Arizona 86001*

LOCATION

SP Mountain is a prominent cinder cone in the northern part of the San Francisco volcanic field near the southern margin of the Colorado Plateau. It is easily visible from U.S. 89 among a cluster of older cinder cones. The area is most readily reached by driving north from Flagstaff about 34 mi (55 km) or south from Cameron 16 mi (26 km) to a graded ranch road that leads westward 5.2 mi (8.4 km). The turnoff is 2.1 mi (3.4 km) north of the entrance to Wupatki National Monument and is marked only by a stop sign. The site is accessible to all vehicles. Ownership of most of the land is divided between the CO Bar Livestock Company and the state of Arizona in a checkerboard pattern of one-mile sections; permission to visit and collect samples is not required (1984). The northern tip of the SP lava flow extends into the Navajo Indian Reservation (Fig. 1).

SIGNIFICANCE

SP Mountain and its associated lava flow (Fig. 2) are excellent examples of volcanic landforms formed by basaltic andesites. Their uneroded appearance and easy access make them among the most readily observable volcanic features of their kind anywhere. The cone's sharp-rimmed profile, radial symmetry, and steep flanks mark it as the youngest volcano in the immediate area; its thick flow is a classic example of block lava as described by Macdonald (1972, p. 91–93).

DESCRIPTION

The geology of the SP area and the eastern two thirds of the late Miocene to Holocene San Francisco volcanic field is illustrated by Ulrich and others (1984). The SP cinder cone is 820 ft (250 m) high and approximately 3,940 ft (1200 m) in diameter at its base. Its summit crater is 1,300 ft (400 m) across and about 400 ft (120 m) deep (Fig. 3). The crater's resistant circular rim of agglutinate and the cone's symmetrical shape as seen from all directions suggested to early cowboys the shape of a chamber pot, the abbreviation for which (in the vernacular) became its accepted geographic name. The cone is built upon its vent, which in the early stages of eruption was also the source of the lava flow that extends 4.3 mi (7 km) to the north. The lava flow overlies older basaltic flows, derived from other vents in the area, as well as outcrops of the Permian Kaibab Formation (Fig. 2). It follows the regional slope to the north, spreading laterally over north-south-trending fault scarps that displace several of the underlying units. The flow is 820 ft (250 m) wide near SP Crater where it is crossed by the ranch road and 2.4 mi (3.9 km) wide at its widest point. The thickness of the flow is 50 ft (15 m) near the vent and 180 ft (55 m) near the terminus. A similar thickness (180 ft) was measured by core drilling on the upthrown side of a

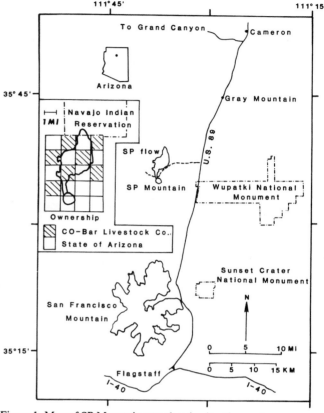

Figure 1. Map of SP Mountain area showing location and access route. Enlarged inset shows ownership of land.

fault scarp buried by the flow near its eastern margin; hence the flow is probably even thicker toward its center, over the downthrown side. The blocks on the flow top and margins of the flow appear randomly arranged to an observer on the ground (Fig. 4); however, in aerial view, distinctive patterns of cross-flow pressure ridges, flow-parallel crevasses, and flow-lobe margins are readily apparent (Fig. 2; see also radar images in Schaber and others, 1980).

AGE

The age of SP Mountain was first discussed by Johnson (1907) who reported it to be the youngest volcano in the San Francisco volcanic field. Later comparative studies (Hodges, 1962; Colton, 1967) concluded that it is older than Sunset Crater, 17 mi (27 km) to the southeast, dated by dendrochronologic methods at less than 1,000 years. More recently, a K-Ar study by Baksi (1974) reported a preferred age for the SP flow of $71 \pm 4 \times 10^3$ yrs (corrected for new decay constants).

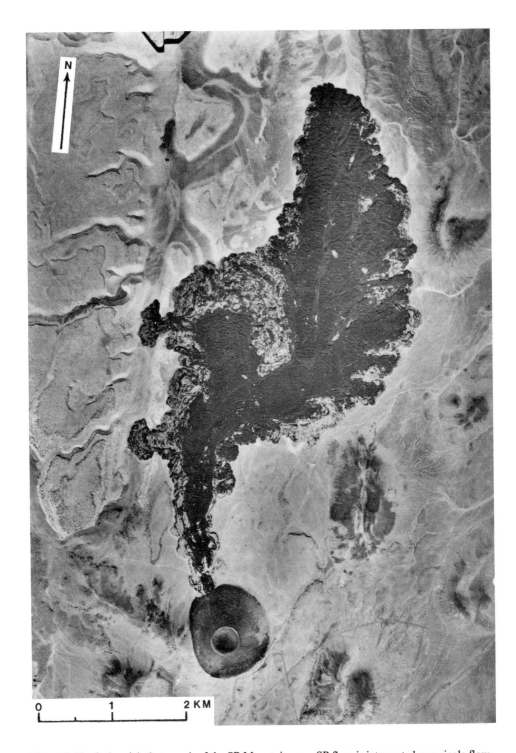

Figure 2. Vertical aerial photograph of the SP Mountain area. SP flow is interpreted as a single flow. Lighter colored parts of flow are characterized by smaller rock fragments, eolian sand and silt, and sparse grass cover. The flow overlies older basaltic flows and the Permian Kaibab Formation (dissected light-gray areas). It overran fault scarps of a horst-and-graben area on its east and west margins. Note the older meandering valley constrained by graben walls and subsequently filled by a pre-SP basalt flow.

Figure 3. SP Mountain and lava flow. Oblique aerial view toward the northeast.

Figure 4. Typical blocky surface of the SP lava flow showing both the rubble of polyhedral blocks and a polygonal pavement described by Hodges (1962). Blocks are approximately 10 cm to one meter in maximum dimension.

COMPOSITION

The basaltic andesite of SP Mountain and its flow are typical of other basaltic andesites in the San Francisco volcanic field. Spatter from the cone contains phenocrysts of clinopyroxene, olivine, and embayed and sieved plagioclase in a hypocrystalline groundmass of the same minerals plus opaque oxide. The flow is similar but contains, in addition, sparse orthopyroxene and embayed quartz. The interior of the flow as seen in drill cores is more crystalline than samples of the flow surface and vent spatter. The chemical composition of the flow appears somewhat differentiated with respect to that of the cone (table 1), suggesting that the magma body may have resided in the crust long enough to undergo fractionation (or contamination?) prior to eruption.[1] These chemical relations, combined with the physical evidence that the base of the cone overlies the lava flow and is not deformed by it (Fig. 2), indicate that the more differentiated magma was extruded first from the top of the reservoir, forming the SP flow. The less differentiated magma then built the cinder cone of SP Mountain.

---

[1]Although 54 analyses of the flow are averaged and compared with one analysis of the cone, the only constituents which overlap in value are Al_2O_3, Fe_2O_3, FeO, MnO, and volatiles.

TABLE 1. CHEMICAL COMPOSITION* OF
SP MOUNTAIN AND SP FLOW

	SP Mountain	SP Flow**
SiO_2	54.78	56.80
Al_2O_3	15.15	14.90
Fe_2O_3	2.32	2.50
FeO	4.91	4.10
MgO	6.04	5.30
CaO	8.40	7.60
Na_2O	2.86	3.70
K_2O	1.81	2.40
TiO_2	1.17	0.95
P_2O_5	0.74	0.48
MnO	0.13	0.13
H_2O	0.80	1.04
CO_2	0.10	<0.05
Total	99.21	99.90

*Analysts:
  SP Mountain - R.L. Swenson
  SP Flow - P. Elmore, L. Artis, H. Smith,
    J. Kelsey, S. Botts, G. Chloe, and
    J. Glenn

**Average of 54 analyses from core samples
  from 55-m-thick flow.

REFERENCES CITED

Baksi, A. K., 1974, K-Ar study of the S.P. flow: Canadian Journal of Earth Science, v. 11, p. 1350–1356.

Colton, H. S., 1967, The basaltic cinder cones and lava flows of the San Francisco Mountain volcanic field: Museum of Northern Arizona Bulletin 10, revised, 58 p.

Hodges, C. A., 1962, Comparative study of S.P. and Sunset Craters and associated lava flows: Plateau, v. 35, no. 1, p. 15–35.

Johnson, D. W., 1907, A recent volcano in the San Francisco Mountain region, Arizona: Geographic Society of Philadelphia Bulletin, v. 5, no. 3, 6 p.

Macdonald, G. A., 1972, Volcanoes: Englewood Cliffs, New Jersey, Prentice-Hall, 510 p.

Schaber, G. G., Elachi, Charles, and Farr, T. G., 1980, Remote sensing data of SP Mountain and SP lava flow in north-central Arizona: Remote Sensing of Environment, v. 9, p. 149–170.

Ulrich, G. E., Billingsley, G. H., Hereford, Richard, Wolfe, E. W., Nealey, L. D., and Sutton, R. L., 1984, Map showing geology, structure, and uranium deposits of the Flagstaff 1° × 2° quadrangle, Arizona: U.S. Geological Survey Miscellaneous Investigations Map I-1446, scale 1:250,000.

San Francisco Mountain: A late Cenozoic composite volcano in northern Arizona

Richard F. Holm, Department of Geology, Northern Arizona University, Flagstaff, Arizona 86011

LOCATION AND ACCESS

San Francisco Mountain, about 6 mi (10 km) north of Flagstaff, Arizona (Fig. 1), is a composite volcano in the San Francisco volcanic field of northern Arizona; the volcano is situated on the southern Colorado Plateau about 31 mi (50 km) north of the escarpment that forms its topographic edge. Flagstaff, the major city in the volcanic field, is at the junction of I-40, U.S. 180, and U.S. 89, the latter two highways passing San Francisco Mountain on its west and east sides, respectively. Access to the mountain is by unpaved U.S. Forest Service roads, most of which are passable by low-clearance vehicles; a few primitive roads require 4-wheel-drive vehicles.

A ski resort on the west side of the mountain normally operates its chairlift during the summer months, providing easy access to the 11,600 ft (3,512 m) level on Agassiz Peak. Hiking restrictions are in effect on Agassiz and Humphreys Peaks, and the designation of the upper part of San Francisco Mountain as the Kachina Peaks Wilderness Area imposes wilderness area restrictions. The Inner Basin is reached by foot or with a vehicle permit from the U.S. Forest Service. The Elden Ranger District office in Flagstaff should be consulted in planning a back-country trip.

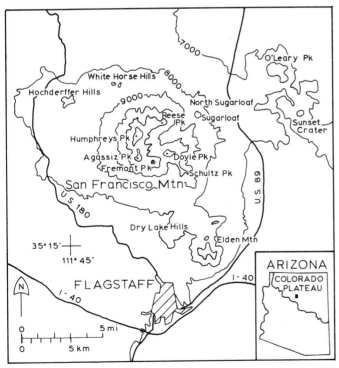

Figure 1. Index map of San Francisco Mountain. Contours are in feet above mean sea level.

Topographic maps available from the U.S. Geological Survey include the Humphreys Peak and Sunset Crater West quadrangles of the 7½-minute series (1:24,000), the San Francisco Mountain map at 1:50,000, and the Flagstaff, Arizona map of the 30- × 60-minute series (1:100,000); a planimetric map of the Coconino National Forest (scale ½-inch = 1 mile) is available from the U.S. Forest Service.

SIGNIFICANCE

Rising to an elevation of 12,633 ft (3,850 m), San Francisco Mountain is the most prominent geomorphic and volcanic structure in the San Francisco volcanic field, a late Miocene to Holocene volcanic province approximately 1,935 mi^2 (5,000 km^2) in area overlying eroded strata of Permian and locally Triassic age typical of the Colorado Plateau in northern Arizona. The volcanic field is dominated by basalt lava flows; more than 600 vents, basaltic to rhyolitic in composition, have been identified. Basaltic scoria cones are scattered throughout the field, whereas the intermediate-to-silicic rocks are localized in a few structurally controlled centers (Luedke and Smith, 1978). San Francisco Mountain, the only composite volcano in the southwestern part of the Colorado Plateau, displays well-preserved examples of a variety of features of volcanism and glaciation; a complete spectrum of lithologies from low-silica andesite to alkali rhyolite is present as lava flows, lava domes, pyroclastic deposits, and hypabyssal plutons. A large valley that breaches the core and northeast flank of the mountain provides exceptional exposures of the volcano's internal structure and stratigraphy, and affords relatively easy access to its interior.

VOLCANIC GEOLOGY

Morphology. In most profiles, San Francisco Mountain presents smooth, concave-up slopes that rise more than 3,900 ft (1,200 m) above the Colorado Plateau to culminate in several peaks over 11,100 ft (3,400 m) in elevation (Fig. 2). The individual peaks are erosional remnants of a compound composite volcano that predates the Inner Basin, a large bowl-shaped valley (caldera) that occupies the central part of the mountain. Nearly encircling the Inner Basin, the peaks form a horseshoe-shaped ridge open to the northeast where the Interior Valley breaches the mountain's flank. In plan, the volcano is slightly elliptical along a northeast-southwest axis parallel to the valley. The generally smooth slopes are interrupted locally where silicic domes were erupted low on the mountain's flanks; these include North Sugarloaf, Sugarloaf, and Schultz Peak. Centers of silicic domes peripheral to San Francisco Mountain are Hochderffer Hills, White Horse Hills, O'Leary Peak, Elden Mountain, and Dry Lake Hills.

Figure 2. Photograph of San Francisco Mountain viewed toward the southwest from O'Leary Peak. Humphreys Peak is the highest peak on the right, Agassiz Peak is in the middle, and Fremont Peak is on the left. The Interior Valley, Inner Basin, and Core Ridge are behind Sugarloaf, a small rhyolite dome in the middle distance with excavation scars on its lower slopes. North Sugarloaf underlies the treeless slope to the right of Sugarloaf.

San Francisco Mountain has the appearance of being extensively dissected, but calculations based on geologic mapping indicate that only about 7 percent of the original volume of the volcano has been removed. Most of the displaced material constitutes nine large debris fans that were deposited radially about the base of the volcano and contribute significantly to its classic profile. Fluvial erosion has cut deep ravines on the mountain's slopes, but the amount of material eroded is a very small part of the original mass (Robinson, 1913). Steep slopes and cliffs in the Inner Basin resulted largely from glacial erosion, and well-preserved cirques and moraines, along with other glacial features, are evident in the Inner Basin, Interior Valley, and in local areas on the outer slopes (Updike, 1977).

Occupying the southwest part of the Inner Basin is Core Ridge, together with a smaller ridge on its southeast side, trending northeast parallel to the Interior Valley. These ridges are formed by erosion-resistant hypabyssal plutons of the volcano's central conduit system.

Structure. San Francisco Mountain was constructed on a platform of late Miocene, and possibly younger, basalt lava flows that unconformably cover Paleozoic and Triassic strata. Basalt lavas and pyroclastic deposits from peripheral vents that were active during and after the growth of San Francisco Mountain locally interfinger with and overlie lavas and deposits erupted from the central volcano. Regional structures of the Colorado Plateau on northeast, north-south, and northwest trends controlled the locations and orientations of some vents, dikes, and peripheral silicic centers, and may even have influenced the location of San Francisco Mountain; the regional structures possibly contributed in the development of the Interior Valley as well.

Complexly related lava domes, radially outward-dipping lava-tuff sequences, and hypabyssal plutons comprise the com-

posite volcano. Dacite and rhyolite domes that were buried by younger lavas and tuffs are exposed on the south and northeast sides of the Inner Basin, and partly buried dacite domes occur on the outer slopes of Humphreys, Fremont, and Doyle Peaks. A water well intersects a deeply buried rhyolite and obsidian dome on the lower west flank of Humphreys Peak. Stratified lavas and pyroclastic deposits of andesite and dacite form an impressive 3,300-ft (1,000-m)-thick section in the north wall of the Inner Basin below Humphreys Peak, and similar deposits form the upper slopes of the other major peaks. The absence of young lavas on Agassiz Peak imply that it is the eroded remnant of an early stratovolcano cone that blocked the southwestward flow of younger dacite and andesite lavas capping all of the other peaks and presumably erupted from central vents on a late stratovolcano cone constructed to the northeast of Agassiz.

Core Ridge and the adjacent ridge to the southeast consist of lava flows, tuffs, tuff breccias, flow breccias, intrusion breccias, breccia pipes, agglomerates, agglutinates, and small plutons that were deposited in and intruded into the central vent and conduit system of the volcano. Intracone dikes of andesite and dacite in the walls of the Inner Basin are arranged radially, converging on Core Ridge. Tuffisite dikes on the north side of the Inner Basin, a large, complex quartz monzodiorite dike along and near the crest of Core Ridge, and a large andesite dike in the valley between the two core ridges strike northeast parallel to the Interior Valley and appear to be related to regional structures beneath the volcano. Plugs of pyroxene diorite underlie the central part of the smaller core ridge.

Petrology. Volcanic rocks of San Francisco Mountain and its five peripheral silicic centers constitute a coherent and continuous suite of lithologies from low-silica andesite to alkali rhyolite (comendite); the suite forms a compositional continuum with hawaiite and transitional- to alkali-basalt lavas of the surrounding volcanic field (Fig. 3). The compositions of the rocks define a sodium-rich, mildly alkaline series. Correlations of phenocrysts with geochemical variation diagrams suggest that the lavas are related by processes of magmatic differentiation; this conclusion is supported by trace element studies (Wenrich-Verbeek, 1979), but field and petrographic observations demonstrate that crustal contamination and magma mixing also were involved in their origin, although only to minor extents. It is unlikely, however, that all of the lavas are comagmatic from a single magma chamber.

In San Francisco Mountain, the andesites are dark- to medium-gray lavas containing phenocrysts of plagioclase, augite, hypersthene, and olivine; pigeonite occurs locally and hornblende appears in medium- to high-silica andesites. The dacites are dark to light gray or pink, depending on the content of glass and degree of oxidation; commonly they carry phenocrysts of plagioclase, hypersthene, hornblende, biotite, and, in the low-silica dacites, augite. Some dacites, however, lack hydrous minerals, a few contain phenocrysts of quartz, and some have microphenocrysts of olivine. Rhyolites are light gray to light bluish-gray; locally they have lenses and bands of black obsidian. Characteristic pheno-

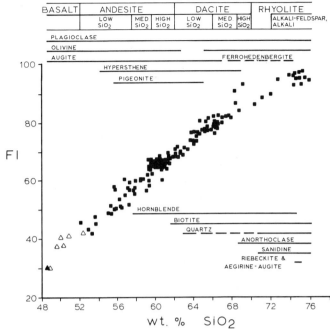

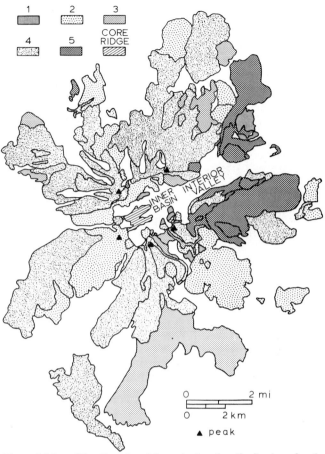

Figure 3. Silica vs. fractionation index (FI=Q + or + ab + ac + ns) for 169 chemical analyses of rocks from San Francisco Mountain, its peripheral silicic centers (squares), and seven basalts from the San Francisco volcanic field (triangles); the Bonito lava flow at Sunset Crater is identified with a filled triangle. Included are the ranges of phenocrysts and microphenocrysts relative to silica, and the petrographic classification of the rocks; dash lines indicate sparse occurrence.

Figure 4. Map of San Francisco Mountain showing distribution of rocks erupted during 5 stages of growth. Names of peaks are shown on Figure 1.

crysts and microphenocrysts in alkali-feldspar rhyolite are anorthoclase, quartz, sanidine, oligoclase-albite, biotite, hornblende, and locally olivine; alkali rhyolite, on the other hand, contains quartz, sanidine, aegirine-augite, riebeckite, and aenigmatite. Magnetite occurs throughout the suite as microphenocrysts.

Volcanic Development. The growth and development of San Francisco Mountain occurred during the late Pliocene and Pleistocene by the eruption of approximately 26 mi³ (110 km³) of lava and pyroclastic material. Estimated volumes of the lithologic types in the central volcano are andesite, 22 mi³ (93 km³), dacite, 3 mi³ (13 km³), and rhyolite, 0.24 mi³ (1 km³); erosion has removed 0.72 mi³ (3 km³) of unclassified rocks and 1.2 mi³ (5 km³) of identifiable units. The volume of andesite probably is overestimated and that of dacite and rhyolite underestimated because the buried portions of the volcano that could not be interpreted to be dacite or rhyolite were assumed to be andesite; nevertheless, the calculations indicate that San Francisco Mountain is an andesitic volcano. The explosive index of the volcano is about 20, suggesting that most of the magmas were relatively dry.

Reconstruction of the volcano's history is possible based on geologic mapping and K-Ar ages by P. E. Damon (written communication, 1977) and E. H. McKee (written communication, 1973). Five stages of growth that are characterized by distinctive styles of volcanism and lithologies are illustrated in Figure 4. Early in stage 1, lava domes of dacite (North Sugarloaf) and

rhyolite (Inner Basin), 2.78 ± 0.13 Ma and 1.82 ± 0.16 Ma, respectively, were extruded after which an andesitic stratovolcano (about 12 mi³) that rose over 5,600 ft (1,700 m) in height was constructed. The lavas and pyroclastic deposits of this early cone are exposed low only on the north and south sides of the Inner Basin, and at the north end of Schultz Peak where they were uplifted during stage 2. Stage 2 began with extrusion of rhyolite and dacite lava domes and flows; resumption of andesitic stratovolcano construction in stage 2 increased the volume to 23 mi³ (95 km³) and the height to about 6,900 ft (2,100 m). Agassiz Peak is an eroded remnant of the stage 2 cone, and an excellent section through its flank is well exposed in the steep slopes and cliffs on the north side of the Inner Basin; the entire section below Humphreys Peak has normal magnetic polarity and has been assigned to the Brunhes chronozone (T. Onstott, oral communication). In stage 3, the structure of the andesitic stratovolcano was disrupted, but 1 mi³ (4 km³) was added to its volume by a plinian eruption of rhyolite to dacite pumice and the emplacement of dacite lava domes on the northeast flank at Reese Peak and on the south flank between Doyle and Fremont Peaks. Stratovolcano construction during stage 4 shifted northeast relative to the stage 2 cone, and increased the volcano's height to 8,200 ft

(2,500 m) and its volume to about 26 mi^3 (108 km^3). Central eruptions produced stage 4 lavas of dacite and andesite that flowed down all flanks of the cone except toward the southwest where they were blocked by the stage 2 cone of Agassiz Peak. A flank eruption on the west slope of Agassiz Peak early in stage 4 extruded andesite lava similar in composition to lavas of stage 2. Parasitic eruptions of andesite and dacite lava flows on the northeast side during stage 5 added 0.5 mi^3 (2 km^3), but may have initiated events that led to the development of the Interior Valley that breaches the volcano's flank. Sugarloaf rhyolite dome, extruded 0.22 ± 0.02 m.y. ago in the mouth of the Interior Valley, concluded stage 5 and was the last known eruptive activity on San Francisco Mountain.

Origin of the Inner Basin and Interior Valley. The origin of the Inner Basin and Interior Valley has attracted interest for over 70 years (Robinson, 1913), and several processes have been suggested to account for their existence; these include: 1) erosion, 2) explosion, 3) collapse, with downward displacement, and 4) collapse, with outward displacement. Although all four mechanisms may have contributed, the latter is most consistent with available information. The distribution of stage 4 lavas indicates that San Francisco Mountain must have had a complete, if not perfectly symmetrical, cone by 0.43 ± 0.03 m.y. ago, the age of the youngest andesite lava flow on top of Humphreys Peak. The eruption of Sugarloaf dome in the mouth of the Interior Valley implies that it, as well as the Inner Basin, was in existence by 0.22 ± 0.02 m.y. ago, restricting their period of development to approximately 210,000 years.

Lavas and pyroclastic deposits that once formed the upper portion of the San Francisco Mountain composite volcano now reside in debris fans about its base (Fig. 5). The estimated volume of 1.1 mi^3 (4.4 km^3) in nine debris fans that were deposited distally and on the lower flanks of San Francisco Mountain plus 0.8 mi^3 (3.4 km^3) of debris buried by younger volcanic deposits northeast of the mountain compares very favorably with the 2 mi^3 (8.0 km^3) volume of the Inner Basin and Interior Valley (0.8 mi^3) and the restored cone (1.2 mi^3).

San Francisco Mountain still had a complete stratovolcano cone after early stage 5 parasitic eruptions of dacite on the northeast flank. Subsequent failure and collapse may have occurred after the cone became gravitationally unstable as a result of struc-

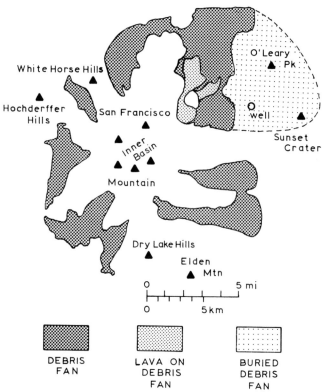

Figure 5. Map showing the distribution of debris fans around San Francisco Mountain; uncertain boundary is dashed. The water well that penetrates the buried debris fan is the Sunset Crater No. 2. (A-23-8) 21 aab 3.

tural displacements due to magma intrusion and extrusion or tectonic faults in the volcano related to regional northeast-trending fractures in the platform below. Explosive eruptions or even an earthquake may have caused the unstable cone to collapse into debris avalanches and debris flows that formed the fans around the mountain's base. More than one collapse event is indicated by cuttings from a water well that penetrates a buried debris fan on Sunset Crater National Monument (Fig. 5); however, a late stage 5 dacite lava flow that partially covers the debris fan on the northeast side of the mountain implies that the Inner Basin originated in a fairly short period of time. Final sculpting of the volcano has been by glacial and ongoing fluvial processes.

REFERENCES CITED

Luedke, R. G., and Smith, R. L., 1978, Map showing distribution, composition, and age of late Cenozoic volcanic centers in Arizona and New Mexico: U.S. Geological Survey, Map I-1091-A, scale 1:1,000,000.

Robinson, H. H., 1913, The San Franciscan volcanic field, Arizona: U.S. Geological Survey, Professional Paper 76, 213 p.

Updike, R. G., 1977, The Geology of the San Francisco Peaks, Arizona [Ph.D. thesis]: Tempe, Arizona, Arizona State University, 423 p.

Wenrich-Verbeek, K. J., 1979, The petrogenesis and trace-element geochemistry of intermediate lavas from Humphreys Peak, San Francisco volcanic field, Arizona: Tectonophysics, v. 61, p. 103–129.

Holocene scoria cone and lava flows
at Sunset Crater, northern Arizona

Richard F. Holm, Department of Geology, Northern Arizona University, Flagstaff, Arizona 86011
Richard B. Moore, U.S. Geological Survey, Hawaiian Volcano Observatory, Hawaii 96718

LOCATION AND ACCESS

Sunset Crater, about 15 mi (25 km) north of Flagstaff, Arizona (Fig. 1), is in the eastern part of the San Francisco volcanic field, on the southern margin of the Colorado Plateau. Access to the Sunset Crater National Monument is by Forest Road 545, an all-weather road off of U.S. 89. Included within the boundaries of the monument are the scoria cone of Sunset Crater and most of one of its associated lava flows, the Bonito flow. Another lava flow associated with Sunset Crater, the Kana-a flow, occurs in the Coconino National Forest east of the monument. Hiking is prohibited on Sunset Crater scoria cone, and collection of specimens is not permitted within the boundaries of the national monument.

Topographic maps published by the U.S. Geological Survey that cover the monument and surrounding area are the Sunset Crater East, Sunset Crater West, O'Leary Peak, and Strawberry Crater quadrangles of the 7½-minute series. The geology of the area is shown by Moore and Wolfe (1976).

SIGNIFICANCE

Sunset Crater and its associated lava flows, together with nearby small scoria and agglutinate cones, are the youngest volcanic features in the San Francisco volcanic field, a late Miocene to Holocene volcanic province in northern Arizona. Although Sunset Crater is only one of more than 550 basaltic vents in the volcanic field, the colorful, nearly symmetrical cone and its basalt lava flows are distinctive because they are virtually untouched by weathering and erosion. Many primary flow structures are well preserved and displayed on the Bonito lava flow. Studies in volcanology, archeology, dendrochronology, and paleomagnetism provide an exceptionally detailed documentation of the history of the Sunset Crater eruption and its effects on the indigenous Indian population of northern Arizona (Pilles, 1979). The volcano was originally named Sunset Mountain by Major J. W. Powell when he was director of the U.S. Geological Survey, in reference to the colorful appearance of the top, as though it was bathed in the rays of the setting sun (Colton, 1945).

VOLCANIC GEOLOGY

Introduction

Volcanic deposits formed during the Sunset Crater eruption include the scoria cone of Sunset Crater, two basalt lava flows that extruded from its base, three rows of small scoria and agglu-

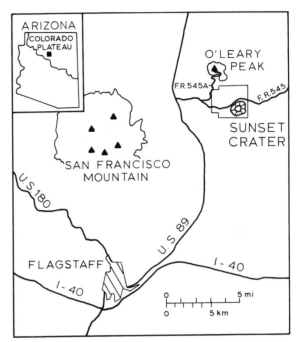

Figure 1. Index map showing location of Sunset Crater and access roads. *Triangles* are major peaks.

tinate cones east-southeast of Sunset Crater, a basalt lava flow from vent 512 about 6.2 mi (10 km) east-southeast of Sunset Crater, and a tephra blanket that originally covered over 800 mi[2] (2,080 km[2]) (Colton, 1932; Moore and Wolfe, 1976; Fig. 2). The vents lie in a zone that trends S 60° E from Sunset Crater and probably were controlled by a major fracture in the Paleozoic and Precambrian rocks beneath the volcanic field. The lava flows and essential pyroclasts are all similar in chemistry and petrography (Moore, 1974), having formed from alkali basalt magma (ne = 1.5%) that contained microphenocrysts of plagioclase (An_{70}), olivine, augite, and magnetite, and sparse phenocrysts of olivine (Fo_{83}).

Sunset Crater Scoria Cone

Sunset Crater is a nearly symmetrical cone, composed dominantly of black scoria, that is about 1,000 ft (300 m) in height and 1 mi (1.6 km) in base diameter. The symmetry of the cone is lowered only by the west-of-center location of the main crater and a gentle recess in the east side of the cone above the presumed extrusion point of the Kana-a flow. Older volcanoes that were partially buried by Sunset Crater form large rounded mounds at the base on the northeast, southeast, and southwest sides.

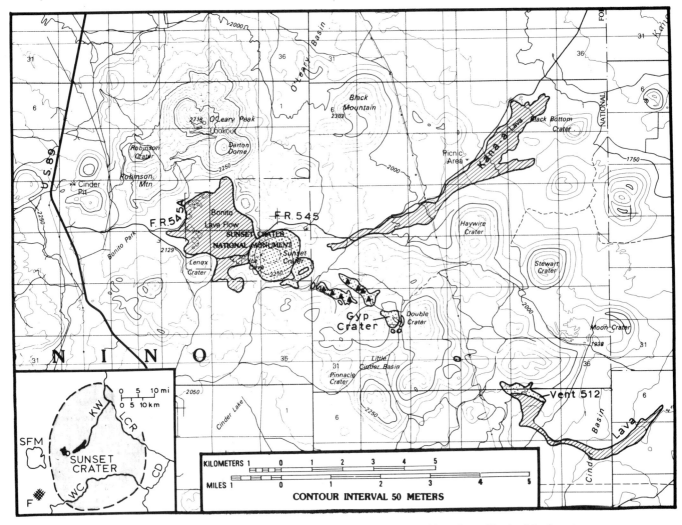

Figure 2. Map showing distribution of vent deposits (*dot pattern*) and lava flows (*lines*) of the Sunset Crater eruption (from Moore and Wolfe, 1976); some of the small scoria cones southeast of Sunset Crater are shown with *filled triangles*. Small scale map shows original distribution (*dashed line*) of the Sunset Crater tephra blanket (from Colton, 1932); *F,* Flagstaff; *SFM,* San Francisco Mountain; *WC,* Walnut Creek; *KW,* Kana-a Wash; *LCR,* Little Colorado River; *CD,* Canyon Diablo.

At the summit of Sunset Crater a rim encloses the main crater, which is more than 400 ft (120 m) deep, and a small crater on the east side of the larger that is about 160 ft (50 m) deep. The rim is highest on its eastern side, which probably is a reflection of the prevailing wind direction during the eruption, but prominent notches in the rim on its eastern and western sides above the extrusion points of the Kana-a and Bonito flows suggest that disruption of the original shape of the rim occurred by partial collapse events. Planar bedded deposits of red scoria on the crater rim were cemented by sublimation of silica, gypsum, and iron oxide from fumarolic vapors, which give the summit its distinctive red, pink, and yellow colors. The cemented beds on the rim dip outward around the main crater, but display inward dips around the smaller crater. Welded spatter and agglutinate deposits occur only in a few loose blocks up to 3 ft (1 m) in diameter in local areas on the upper surface of the cone. Unconsolidated

black lapilli and bombs overlie the red summit scoria around the small crater and on the outer north slope of the main crater. Particles ranging in size from fine ash to spheroidal bombs up to 3 ft (1 m) in diameter mantle the slopes of the cone. Although granulometric data have not been obtained, the most common pyroclast size appears to be fine lapilli, but ash is also conspicuous in its abundance.

The Sunset Crater eruption produced a mantling blanket of black scoriaceous lapilli and ash, originally covering more than 800 mi² (2,100 km²), that extended in a broad ellipse generally to the east of the volcano (Colton, 1932; Fig. 2). Amos and others (1981) noted the large size of this deposit compared to those from other Strombolian events and suggested that eruption columns may have been at least several hundred ft (m) high. Fluvial and eolian erosion has removed and locally redeposited much of the tephra, and today it covers an area of only about 122 mi² (315

Figure 3. Photograph of the Bonito lava flow; view is to the west from the summit of Sunset Crater. The early scoria-mantled part of the flow has a smoother, lighter surface, whereas the later scoria-free margin is dark. In the foreground are a pit crater and several hornitos. The large mounds in the middle of the flow are agglutinate deposits rafted from an early cone of Sunset Crater. Scoria-covered islands of the early Bonito flow are surrounded by scoria-free zones in the upper right corner.

km^2; Moore, 1974). Near Sunset Crater the tephra blanket is several decimeters to over a meter thick, and it covers parts of the Bonito and Kana-a flows as well as agglutinate deposits to the east-southeast at Gyp Crater and vent 512.

Lava Flows From Sunset Crater

Bonito Flow. The Bonito lava flow was extruded at 2 or more points in a zone along the western to northwestern base of Sunset Crater. The lava spread northwesterly and ponded in an intercone basin bounded by Sunset Crater and older volcanic deposits. Low areas between older cones were inundated, and on the south side of the basin the lava overtopped a low divide and flowed into the crater of an older, extensively eroded scoria cone. Exposures around the periphery of the flow demonstrate that it is on top of black scoria of the Sunset Crater tephra blanket. The lava flow covers an area of 1.79 mi^2 (4.63 km^2) to a depth of around 6 ft (2 m) locally along its margin to perhaps more than 100 ft (30 m) in the center (Moore, 1974).

Many flow units and a wide variety of flow structures create a highly complicated appearance of the Bonito flow. Although the flow is dominantly slab pahoehoe, massive, ropy, festoon, and shelly surface structures occur locally, and aa clinkers are common around the flow's margin. Chains of hornitos, 3 ft to 50 ft (1 m to 15 m) high, and a few tumuli dot the top of the flow, presumably above lava tubes, and an impressive pit crater 320 ft (100 m) long is almost 50 ft (15 m) deep (Fig. 3).

The most conspicuous features of the Bonito lava flow are mounds of crudely stratified spatter, welded spatter, agglutinate, and rootless flows on the top of the flow (Fig. 3), and large fissures that break its crust. The mounds of welded pyroclasts are strung out like beads in a chain in a north-northwesterly direction for about 3,700 ft (1,130 m) from the base of Sunset Crater; the mounds appear to be near-vent deposits from an earlier, and now buried, cone of Sunset Crater that were rafted away when the Bonito flow was extruded from the base and caused partial collapse of the earlier cone. The fissures, up to 2,900 ft (880 m) long and 24 ft (7 m) deep occur in most areas of the flow, although they are most numerous near and in the margins; many of the fissures contain elongated squeeze-ups marked with vertical grooves, but some are gaping, open cracks with steep walls. The fissures were opened as the flow's molten interior continued to spread outward, and many are located in subsided areas where the interior lava drained away.

Most of the central and southeastern part of the Bonito flow is mantled with black scoria about 1 ft (several decimeters) to 3 ft (a meter) or more thick; locally, on the tops of mounds that may be small tumuli or lava blisters, fumarolic vapors emitted from the flow's interior oxidized the scoria to reddish colors. The scoria mantle is probably an air-fall blanket from Sunset Crater. Flow units exposed in the pit crater, fissures, and tumuli in the southeastern part of the flow commonly contain thin interbeds of scoria, which demonstrate concurrent extrusion of the early portions of the Bonito flow and eruption of scoria from Sunset Crater's vent.

Surrounding the scoria-mantled part of the Bonito flow on all sides, except the southeast where it emerges from Sunset Crater, is a scoria-free zone that extends to the edge of the flow (Fig. 3). In the north-central part of the flow, scoria-free zones surround large islands of scoria-mantled lava. The distribution of the scoria-mantled and scoria-free zones (Moore and Wolfe, 1976) appears to have resulted from air-fall mantling by scoria of the initially extruded part of the flow after which the scoria eruptive phase ended, or subsided; the lava extrusion continued, or resumed with a new pulse, to advance the flow beyond the scoria-mantled zone, and even broke up and floated away parts of the scoria-mantled crust like pack ice.

Kana-a Flow. The Kana-a flow was extruded from the eastern base of Sunset Crater and flowed as a stream of lava northeasterly down Kana-a Wash for about 4 mi (6.4 km) to where it bifurcated around an older scoria cone; one branch continued down Kana-a Wash for 1 mi (1.6 km) and the other followed another wash northeasterly for 2 mi (3.2 km). Most of the lava flow is mantled with black scoriaceous ash and lapilli of the Sunset Crater tephra blanket, which increases in thickness toward the cone. Within 1 mi (1.6 km) of Sunset Crater the scoria is thick enough to completely bury the flow, but it can be followed as a low scoria-covered mound to within 0.6 mi (1 km) of the base of the cone. Although the lava flow cannot be followed continuously to its presumed eruption point at the eastern base of Sunset Crater, it is likely that the gentle indentation of the eastern slope of the cone resulted from its disruption on extrusion of the lava (Hodges, 1960). A small area of scoria-free lava near the distal end of the Kana-a flow apparently was extruded from a lava tube after the main scoria eruption of Sunset Crater and flowed on top of the scoria-mantled lava. Although there is no readily identified extrusion point of the scoria-free flow, several hornitos built on tumuli upstream on the scoria-mantled part of the flow imply the existence of a lava tube whose lower end may have been blocked to force the breakthrough of the lava to the surface of the flow.

The Kana-a lava flow is largely slab pahoehoe containing many large (to several cm), round to stretched, smooth-walled vesicles. At the distal end of the flow, the edge is steep, 12–20 ft (4–6 m) high, and typically covered with aa clinkers. The scoria-free part is 3–20 ft (1–6 m) thick and also is highly vesicular slab pahoehoe with local transition to aa.

Vents Southeast of Sunset Crater

The eruption of Sunset Crater occurred at the northwestern end of a linear zone of related vents that trends S 60° E for 6.2 mi (10 km) from a spatter rampart at the northwest base of Sunset Crater to a complex of spatter ramparts at vent 512. Midway along this zone a shorter row of vents branches off to trend S 45° E for about 0.9 mi (1.4 km). Similar lithologies of the deposits and relatively close timing of the eruptions indicate that the same magma, or closely related magmas, supplied all of the vents. Nearly perfect alignment of the vents is strong evidence that the

magma followed a major crustal fracture on its way to the surface. The shorter line of vents appears to have been fed by magma from the major fracture that was injected into an intersecting fracture.

The spatter rampart at the northwest base of Sunset Crater trends S 56° E, which together with its position on the linear zone of vents strongly suggests that it was built by lava erupting from the crustal fracture and not just from a crack in the crust of the Bonito lava flow. East-southeast of Sunset Crater the volcanic zone is marked by at least 9 small scoria cones, a few ft (m) to 164 ft (50 m) high, so closely spaced that their deposits coalesce. Each cone is composed of red scoria that overlies the black tephra blanket of Sunset Crater. Spatter is absent from these cones, and they are generally unconsolidated; typically they are capped by several inches (cm) of scoria cemented by vapor-phase minerals deposited from post-eruption fumaroles. The row of vents that branch off to trend S 45° E contains 14 small scoria cones, about 10 ft to 115 ft (3 m to 35 m) high, which coalesce along the zone. All of the cones consist of red scoria that overlies the black tephra blanket; in some cones the scoria is unconsolidated, in others a cap of cemented scoria is up to 1 ft (30 cm) thick, and a few contain spheroidal bombs 4 to 12 in (10 to 30 cm) in diameter.

Gyp Crater, about at the midpoint of the volcanic zone, is strongly elongated along the S 60° E trend; the crater is only about 1,100 ft (335 m) long and 600 ft (180 m) wide, but it is more than 200 ft (60 m) deep. The early stage of the eruption produced thick deposits of red scoriaceous lapilli and bombs along the northeast and southwest sides of the vent, after which the eruption style changed and spatter, welded spatter, and agglutinate were deposited in layers that mantle the inner and outer slopes of the crater. Most of the deposits are red, but the top layer on the southeast side is a black agglutinate, which grades northward along the inner east side of the crater into a black, densely welded, rootless flow about 20 in (50 cm) thick. The crater's rim and inner and outer slopes are covered by red scoriaceous lapilli, which are generally unconsolidated, but locally on the crest of the rim they are cemented by vapor-phase minerals into layers about 8 in (20 cm) thick that are maroon, red, and yellow. The mantling scoria may be part of the Sunset Crater tephra blanket that was laid down while the Gyp Crater deposits were still hot and emitting oxidizing fumarolic vapors.

Vent 512 is a complex zone nearly 1 mi (1.6 km) long in which several dozen individual vents produced spatter ramparts 3–30 ft (1–10 m) high and 30–600 ft (10–200 m) long, all generally elongated along the S 60° E trend; a few, more circular, spatter cones are 10–30 ft (3–10 m) high. The deposits consist of very fresh-appearing spatter, welded spatter, and agglutinate with cow dung bombs up to 2 ft (60 cm) in diameter; locally, the agglutinate deposits were mobilized and formed densely welded rootless flows. Black scoria of the Sunset Crater tephra blanket covers the area of vent 512; scoria that landed on the spatter ramparts and cones was oxidized red by fumarolic vapors escaping from the still-hot welded pyroclastic deposits. Basalt lava extruded from vent 512 flowed nearly 4 mi (6.4 km) to the east.

HISTORY OF THE ERUPTION

The Sunset Crater eruption began between the growing seasons of A.D. 1064 and 1065 (Smiley, 1958) and continued, probably intermittently, for about 150 years (Shoemaker and Champion, 1977). During this time about 0.07 mi³ (0.3 km³) of magma was erupted from the vent area at Sunset Crater. Approximately three-fourths of the magma erupted explosively as scoria and was deposited in nearly equal amounts in the scoria cone of Sunset Crater and the tephra blanket; about one-fourth of the magma was extruded as lava, with nearly three-fourths of the lava contributed to the Bonito flow.

Eruptions began as lava reached the surface through a 6.2-mi-long (10 km) dike that tapped a source of primitive basalt magma in the upper mantle (Brookins and Moore, 1975). It is likely that concurrent eruptions occurred at several vents along the dike, and eruptive activity at vent 512 and the rows of small cones east-southeast of Sunset Crater probably resembled the well-known "curtain of fire" eruption style on Hawaii. Most of the lava was erupted from vents associated with Sunset Crater, and it is at this long-lived vent area that a complex sequence of eruptive episodes took place.

On the basis of field relationships and volcanological interpretations, six eruptive episodes are identifiable at the Sunset Crater vent area; these episodes have been correlated with the time scale by dendrochronology (Smiley, 1958, dates in parentheses) and by paleomagnetism (D. E. Champion, written comm., 1985; Shoemaker and Champion, 1977, dates in brackets).

1. Deposition of an extensive tephra blanket of black scoria (began A.D. 1064–1065) [continued to A.D. 1090 or later]; construction of early Sunset Crater with unconsolidated scoria.

2. Extrusion of early Kana-a flow [A.D. 1064]; disruption of the east flank of Sunset Crater, with unconsolidated scoria collapsing onto the Kana-a flow; probable extrusion of the vent 512 lava flow.

3. Continued, or resumed, construction of Sunset Crater with deposition of agglutinate layers; tephra deposition on top of early Kana-a flow and the vent 512 flow.

4. Extrusion of early Bonito flow [A.D. 1180] contemporaneously with scoria deposition on flow units; Bonito flow disrupts west flank of Sunset Crater and rafts away collapsed agglutinate layers.

5. Continued, or resumed, construction of Sunset Crater with deposition of unconsolidated black scoria; tephra deposition on top of Kana-a and Bonito lava flows; latest deposits around the summit are of red scoria; cementation of crater rim scoria deposits by vapor-phase minerals.

6. Extrusion of late Bonito flow and late Kana-a flow; small eruption of black lapilli and bombs results in minor collapse of the east rim of the crater and formation of the small crater on the east side of the summit; no, or only minor, deposition of tephra beyond the cone.

Eruptions at vent 512 [A.D. 1100] and at Gyp Crater [A.D. 1100] must have occurred prior to the fourth eruptive episode because deposits at both vents are covered with the upper part of the tephra blanket; deposition of agglutinates at both vents may have been roughly contemporaneous with the third episode, and fumarolic vapors from the vents oxidized the mantling black scoria from either the third or fifth episodes, or both. The rows of small red scoria cones on top of the black scoria tephra blanket east-southeast of Sunset Crater erupted during or after the sixth episode.

REFERENCES CITED

Amos, R. C., Self, S., and Crowe, B., 1981, Pyroclastic activity of Sunset Crater: Evidence for a large magnitude, high dispersal Strombolian eruption: EOS, Transactions, American Geophysical Union, v. 62, no. 45, p. 1085.

Brookins, D. G., and Moore, R. B., 1975, Sr isotope initial ratios from the San Francisco volcanic field, Arizona: Isochron/West, no. 12, p. 1–2.

Colton, H. S., 1932, Sunset Crater: The effect of a volcanic eruption on an ancient pueblo people: The Geographical Review, v. 22, no. 4, p. 582–590.

Colton, H. S., 1945, Sunset Crater: Plateau, v. 18, no. 1, p. 7–14.

Hodges, C. A., 1960, Comparative study of S. P. and Sunset Craters and associated lava flows, San Francisco volcanic field, Arizona [M.S. thesis]: Madison, Wisconsin, University of Wisconsin, 129 p.

Moore, R. B., 1974, Geology, petrology and geochemistry of the eastern San Francisco volcanic field, Arizona [Ph.D. thesis]: Albuquerque, New Mexico, University of New Mexico, 360 p.

Moore, R. B., and Wolfe, E. W., 1976, Geologic map of the eastern San Francisco volcanic field, Arizona: U.S. Geological Survey Map I-953, scale 1:50,000.

Pilles, P. J., 1979, Sunset Crater and the Sinagua: A new interpretation, *in* Sheets, P. D., and Grayson, D. K., eds., Volcanic activity and human ecology: Academic Press, p. 459–485.

Shoemaker, E. M., and Champion, D. E., 1977, Eruption history of Sunset Crater, Arizona: Investigator's Annual Report, Wupatki–Sunset Crater National Monument, Arizona. (unpublished).

Smiley, T. L., 1958, The geology and dating of Sunset Crater, Flagstaff, Arizona: New Mexico Geological Society, Ninth field conference guidebook, Socorro, New Mexico, p. 186–190.

Meteor Crater, Arizona

Eugene M. Shoemaker, U.S. Geological Survey, Flagstaff, Arizona 86001

LOCATION

The location of Meteor Crater is shown on nearly all highway maps of Arizona. It lies 6 mi (9.7 km) south of I-40 between Flagstaff and Winslow in northern Arizona; the turnoff from I-40, 34 mi (55 km) east of Flagstaff, is well marked by signs along the highway. Access to the crater is by a paved road that leads directly to a visitor center and museum on the rim of the crater. Qualified scientists may obtain permission to hike to various parts of the crater by writing or calling in advance to Meteor Crater Enterprises, 121 East Birch Avenue, Flagstaff, Arizona 86001.

SIGNIFICANCE

Meteor Crater was the first recognized impact crater on Earth; it remains the largest known crater with associated meteorites. It is the most thoroughly investigated and one of the least eroded and best exposed impact craters in the world. Study of Meteor Crater provided clues for the recognition of impact craters on other solid bodies in the solar system, as well as on Earth, and stimulated the search for other terrestrial impact structures.

STRUCTURAL AND STRATIGRAPHIC SETTING

Meteor Crater lies in the Canyon Diablo region of the southern part of the Colorado Plateau. In the vicinity of the crater, the surface of the plateau has low relief and is underlain by nearly flat-lying beds of Permian and Triassic age. The crater lies near the anticlinal bend of a gentle monoclinal fold, a type of structure characteristic of this region. The strata are broken by wide-spaced, northwest-trending normal faults, which generally are many miles (kilometers) in length but have displacements of only a few feet (meters) to about 100 ft (30 m).

Rocks exposed at Meteor Crater range from the Coconino Sandstone of Permian age to the Moenkopi Formation of Triassic age. Drill holes in and around the crater have intersected red beds in the upper part of the Supai Formation of Pennsylvanian and Permian age, which conformably underlies the Coconino. The Coconino Sandstone consists of about 700 to 800 ft (210 to 240 m) of fine-grained, saccharoidal, white, cross-bedded, quartzose sandstone of eolian origin. Only the upper half of the formation is exposed at the crater. The Coconino is overlain conformably by a unit 10 ft (3 m) thick of white to yellowish- to reddish-brown, calcareous, medium- to coarse-grained sandstone interbedded with dolomite, which is referred here to the Toroweap Formation of Permian age.

The Kaibab Formation of Permian age, which rests conformably on the Toroweap, includes 260 to 265 ft (79-81 m) of fossiliferous marine sandy dolomite, dolomitic limestone, and minor calcareous sandstone. Three informal members are recognized (McKee, 1938). The lower two members, the gamma and beta members, are chiefly massive, dense dolomite; the upper or alpha member is composed of well-bedded dolomite and limestone with several sandstone interbeds. The Kaibab is exposed along the steep upper part of the crater wall.

Beds of the Moenkopi Formation (McKee, 1954) of Triassic age form a thin patchy veneer resting disconformably on the Kaibab in the vicinity of the crater. Two members of the Moenkopi are present. A 7- to 20-ft (2- to 6-m) bed of pale reddish brown, very-fine-grained sandstone (McKee's lower massive sandstone), which lies virtually at the base, constitutes the Wupatki Member. Above the Wupatki are dark, reddish brown, fissile siltstone beds of the Moqui Member. About 7 to 30 ft (2 to 10 m) of Moenkopi strata are exposed in the wall of the crater.

QUATERNARY STRATIGRAPHY AND STRUCTURE OF THE CRATER

Meteor Crater is a bowl-shaped depression 600 ft (180 m) deep and about 0.75 mi (1.2 km) in diameter encompassed by a ridge or rim that rises 100 to 200 ft (30 to 60 m) above the surrounding plain. The rim is underlain by a complex sequence of Quaternary debris and alluvium resting on deformed and uplifted Moenkopi and Kaibab strata (Figs. 1 and 2).

The debris units on the crater rim consist of angular fragments ranging from splinters less than 1μ in size to blocks up to 100 ft (30 m) long. Because of the striking lithologic contrast between the formations from which the debris is derived, it is easy to distinguish and map units or layers in the debris by the lithic composition and stratigraphic source of the component fragments.

The stratigraphically lowest debris unit of the rim is composed almost entirely of fragments derived from the Moenkopi Formation. Within the crater this unit rests on the edge of upturned Moenkopi beds (Fig. 2) or very locally grades into the Moenkopi Formation; away from the crater wall the debris rests on the eroded surface of the Moenkopi. A unit composed of Kaibab debris rests on the Moenkopi debris. The contact is sharp where exposed within the crater, but at distances of 0.5 mi (0.8 km) from the crater, there is slight mixing of fragments at the contact. Patches of a third debris unit, composed of sandstone fragments from the Coconino and Toroweap, rest with sharp contact on the Kaibab debris. No fragments from the Supai are represented in any of the debris.

The bedrock stratigraphy is crudely preserved, inverted, in the debris units. Not only is the gross stratigraphy preserved, but even the relative position of fragments from different beds tends to be preserved. Thus, most sandstone fragments from the basal sandstone bed of the Moenkopi occur near the top of the Moen-

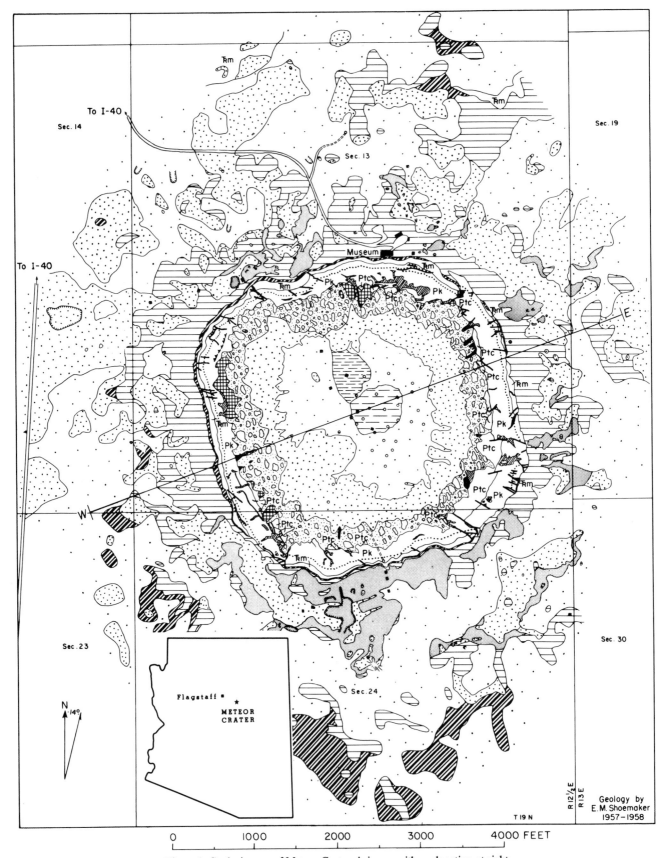

Figure 1. Geologic map of Meteor Crater, Arizona, with explanation at right.

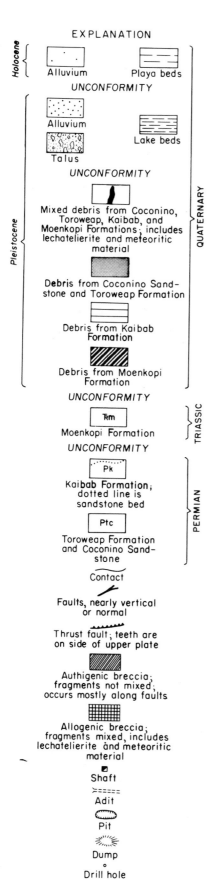

Holocene

Alluvium   Playa beds

UNCONFORMITY

Alluvium   Lake beds

Talus

UNCONFORMITY

Mixed debris from Coconino,
Toroweap, Kaibab, and
Moenkopi Formations; includes
lechatelierite and meteoritic
material

Debris from Coconino Sandstone and Toroweap Formation

Debris from Kaibab
Formation

Debris from Moenkopi
Formation

UNCONFORMITY

Ɫm
Moenkopi Formation

UNCONFORMITY

Pk
Kaibab Formation;
dotted line is
sandstone bed

Ptc
Toroweap Formation
and Coconino Sand-
stone

Contact

Faults, nearly vertical
or normal

Thrust fault; teeth are
on side of upper plate

Authigenic breccia;
fragments not mixed;
occurs mostly along faults

Allogenic breccia;
fragments mixed, includes
lechatelierite and meteoritic
material

Shaft

Adit

Pit

Dump

Drill hole

Pleistocene

QUATERNARY

TRIASSIC

PERMIAN

kopi debris unit, fragments from the alpha member of the Kaibab occur at the base of the Kaibab debris unit, and brown sandstone fragments from the Toroweap occur just above the Kaibab debris unit.

Pleistocene and Holocene alluvium rests unconformably on all the debris units, as well as on bedrock. The Pleistocene alluvium forms a series of small, partly dissected pediments extending out from the crater rim and also occurs as isolated patches of pediment or terrace deposits on the interstream divides. It is correlated on the basis of well-developed pedocal paleosols with the Pleistocene Jeddito Formation of Hack (1942, p. 48–54) in the Hopi Buttes region, some 50 mi (80 km) to the northeast. Holocene alluvium blankets about half the area within the first 0.5 mi (0.8 km of the crater and extends along the floors of minor stream courses (Fig. 2). It includes modern alluvium and correlatives of the Holocene Tsegi and Naha formations of Hack (1942) in the Hopi Buttes region.

Both the Pleistocene and the Holocene alluvium are composed of material derived from all formations represented in the debris and also contain meteorite fragments, lechatelierite (silica glass derived from the Coconino Sandstone), other kinds of fused rock (Nininger, 1956), and less strongly shocked rocks containing coesite and stishovite (Chao and others, 1960, 1962; Kieffer, 1971). Oxidized meteoritic material and fragments of relatively strongly shocked Coconino Sandstone are locally abundant in the Pleistocene alluvium where it occurs fairly high on the crater rim. Sparse unoxidized meteoritic material occurs in two principal forms: (a) large crystalline fragments composed mainly of two nickel-iron minerals, kamacite and taenite; and (b) minute spherical particles of nickel-iron. The bulk of the meteoritic material distributed about the crater apparently is in the form of small particles. The total quantity of fine-grained meteoritic debris about the crater, which occurs in the Pleistocene and Holocene alluvium and also as lag and dispersed in colluvium, has been estimated by Rinehart (1958) as about 12,000 tons.

Low on the crater wall, the bedrock generally dips gently outward. The dips generally are steeper close to the contact with the debris on the rim, and beds are overturned along various stretches totaling about one-third the perimeter of the crater. Along the north and east walls of the crater, the Moenkopi locally can be seen to be folded back on itself, the upper limb of the fold consisting of a flap that has been rotated in places more than 180° away from the crater (Fig. 2). At one place in the southeast corner of the crater, the flap grades outward into disaggregated debris, but in most places there is a distinct break between the debris and the coherent flap.

Rocks now represented by the debris of the rim have been peeled back from the area of the crater somewhat like the petals of a flower. The axial plane of the fold in three dimensions is a flat cone, with apex downward and concentric with the crater, that intersects the crater wall. If eroded parts of the wall were restored, more overturned beds would be exposed.

The upturned and overturned strata are broken or torn with scissors-type displacement along a number of small, nearly verti-

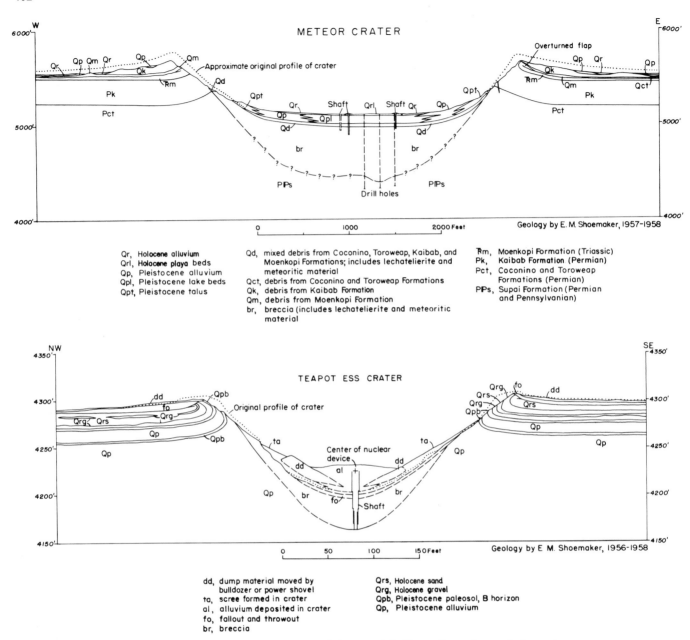

Figure 2. Cross sections of Meteor Crater, Arizona, and the Teapot Ess nuclear explosion crater.

cal faults. A majority of these tears are parallel with a northwesterly regional joint set, and a subordinate number are parallel with a northeasterly set. Regional jointing has controlled the shape of the crater, which is somewhat squarish in outline (Fig. 1); the diagonals of the "square" coincide with the trend of the two main sets of joints. The largest tears occur in the "corners" of the crater. In the northeast corner of the crater a torn end of the overturned flap on the east wall forms a projection suspended in debris. A few normal faults, concentric with the crater wall and along which displacement is down toward the crater, occur on the southwest side. Small thrust faults occur on the north and west sides of the crater; relative displacement of the lower plate is invariably away from the center of the crater. Crushed rock (authigenic breccia) is locally present along all types of faults.

The floor of the crater is underlain by Quaternary surficial deposits, debris, and breccia. Pleistocene talus mantles the lowest parts of the crater walls and grades into and is overlain by Pleistocene alluvium along the floor. The Pleistocene alluvium, in turn, interfingers with a series of lake beds about 100 ft (30 m) thick toward the center of the crater. Up to 6 ft (1.8 m) of Holocene alluvium and playa beds rest unconformably on the Pleistocene. Where exposed in shafts, the lowermost Pleistocene lake beds contain chunks of pumiceous, frothy lechatelierite.

A layer of mixed debris underlies the Pleistocene talus and

lake beds and rests on breccia and on bedrock. This layer is composed of fragments derived from all formations intersected by the crater and includes much strongly shocked rock and oxidized meteoritic material. Material from the different stratigraphic horizons is thoroughly mixed. Where intersected by a shaft in the crater floor, the mixed debris is about 35 ft (10.5 m) thick and almost perfectly massive, but it exhibits a distinct grading, from coarse to fine, from base to top. The average grain size, about 2 cm, is much less than in the debris units of the rim or in the underlying breccia; the coarsest fragments at the base of the mixed debris rarely exceed 1 ft (0.3 m) in diameter. Evidently this unit was formed by fallout of material thrown to considerable height. It has not been recognized outside the crater and apparently has been entirely eroded away. However, its constituents have been partly redeposited in the Pleistocene and Holocene alluvium on the crater rim.

Where exposed at the surface, the breccia underlying the mixed debris is composed chiefly of large blocks of Kaibab, but the breccia exposed in shafts under the central crater floor is made up chiefly of shattered and twisted blocks of Coconino. Extensive drilling conducted by Barringer (1906) and his associates (Tilghman, 1906) has shown that, at a depth of 300 to 650 ft (100 to 200 m), much finely crushed sandstone and some fused and other strongly shocked rock and meteoritic material are present. Some drill cuttings from about 600 ft (180 m) depth contain fairly abundant meteoritic material, chiefly in the form of fine metallic spherules dispersed in glass. Cores of ordinary siltstone and sandstone of the Supai were obtained at depths of 700 ft (210 m) and deeper. The lateral dimensions of the breccia are not known because the drilling was concentrated in the center of the crater. The age of the breccia, and hence the crater, has been determined by thermoluminescence techniques as 49,000 ± 3,000 years (Sutton, 1985).

MECHANISM OF CRATER FORMATION

Most of the major structural features of Meteor Crater are reproduced in a crater in the alluvium of Yucca Flat, Nevada, formed by the underground explosion of a nuclear device. The Teapot Ess crater (Fig. 2), about 300 ft (90 m) across and originally about 100 ft (30 m) deep, was produced in 1955 by a 1.2-kiloton device detonated at a depth of 67 ft (20 m) below the surface. Beds of alluvium exposed in the rim are peeled back in an overturned syncline, just as the bedrock is peeled back at Meteor Crater. The upper limb of the fold is overlain by and locally passes outward into debris that roughly preserves, inverted, the original alluvial stratigraphy. Shock-formed glass and other strongly shocked materials, some containing coesite, are present in the uppermost part of the debris. A thin layer of debris formed by fallout or fall-back is also present in the crater. The floor and lower walls of the crater are underlain by a thick lens of breccia containing mixed fragments of shock-compressed alluvium and dispersed glass.

A crater that has the structure of Meteor Crater or Teapot Ess crater is formed by propagation of a shock wave either from the penetration path of a high velocity projectile or from an explosion originating at moderate depth beneath the surface (Shoemaker, 1960, 1963). In the case of impact, a strong shock races ahead of the projectile into the target rocks, and another engulfs the projectile. At projectile speeds corresponding to encounter speeds of asteroids with the Earth (typically in the range 9 to 19 mi/sec; 15 to 30 km/sec), initial shock pressures generally exceed the dynamic yield strengths of rocks and meteorites by several orders of magnitude. The target rocks engulfed by shock are initially accelerated more or less radially away from the penetration path of the projectile. An expanding cavity is formed by the divergent flow of the target material. Near the cavity wall, the direction of flow is approximately tangent to the wall, owing to upward acceleration in an expanding rarefaction between the shock front and the wall. Part of the target rock and much of the projectile flow up the wall and out of the growing cavity.

As the shock wave expands, pressures at the shock front drop rapidly. A limit for displacement of the target rocks is reached where the stresses in the shock wave drop below the dynamic yield strength of the rocks. This limit defines the outer boundary of deformation and uplift of bedrock beneath the crater rim. The position of the final wall of a structurally simple crater like Meteor Crater or Teapot Ess is located approximately where the outward flow along the wall of the growing cavity is stopped by the combined effects of gravity and the shear strength of the target material. At Meteor Crater, a sheath of fragmented rock that stopped flowing up the cavity wall collapsed back toward the center of the crater to form the thick breccia lens beneath the crater floor.

The kinetic energy of the projectile required to form a crater the size of Meteor Crater can be estimated from computer simulation of the impact process (Roddy and others, 1980) to be about 15 megatons TNT equivalent. This energy is equal to that of a spherical body of meteoritic iron about 130 ft (40 m) in diameter travelling at a speed of 12 mi/sec (20 km/sec; the rms encounter speed of Earth-crossing asteroids). Somewhat greater energy was required if the projectile struck at an oblique angle, as suggested by the presence of faults with underthrust displacement on the north and west walls of Meteor Crater.

REFERENCES CITED

Barringer, D. M., 1906, Coon Mountain and its crater: Proceedings of the Academy of Natural Sciences of Philadelphia, v. 57, p. 861–886.

Chao, E.C.T., Shoemaker, E. M., and Madsen, B. M., 1960, First natural occurrence of coesite: Science, v. 132, p. 220–222.

Chao, E.C.T., Fahey, J. J., Littler, J., and Milton, D. J., 1962, Stishovite, SiO$_2$, a very high pressure new mineral from Meteor Crater, Arizona: Journal of Geophysical Research, v. 67, p. 419–421.

Hack, J. T., 1942, The changing physical environment of the Hopi Indians: Peabody Museum of Natural History Papers, v. 35, p. 3–85.

Kieffer, S. W., 1971, Shock metamorphism of the Coconino Sandstone at Meteor Crater, Arizona: Journal of Geophysical Research, v. 76, p. 5449–5473.

McKee, E. D., 1938, The environment and history of the Toroweap and Kaibab formations of northern Arizona and southern Utah: Carnegie Institution of Washington Publication 492, 221 p.

——— , 1954, Stratigraphy and history of the Moenkopi Formation of Triassic age: Geological Society of America Memoir 61, 133 p.

Nininger, H. H., 1956, Arizona's meteorite crater: Denver, Colorado, World Press, 232 p.

Rinehart, J. S., 1958, Distribution of meteoritic debris about the Arizona meteorite crater: Smithsonian Institution Contributions to Astrophysics, v. 2, p. 145–160.

Roddy, D. J., Schuster, S. H., Kreyenhagen, K. N., and Orphal, D. L., 1980, Computer code simulations of the formation of Meteor Crater, Arizona; Calculations MC-1 and MC-2: Proceedings of the Eleventh Lunar and Planetary Science Conference, New York, Pergamon Press, p. 2275–2308.

Shoemaker, E. M., 1960, Penetration mechanics of high velocity meteorites, illustrated by Meteor Crater, Arizona: International Geological Congress, 21st, Copenhagen, pt. 8, p. 418–434.

——— , 1963, Impact mechanics at Meteor Crater, Arzona, *in* Middlehurst, B., and Kuiper, G. P., eds., The moon, meteorites, and comets; The Solar System, Volume 4: Chicago, Illinois, University of Chicago Press, p. 301–336.

Sutton, S. R., 1985, Thermoluminescence measurements on shock-metamorphosed sandstone and dolomite from Meteor Crater, Arizona—2. Thermoluminescence age of Meteor Crater: Journal of Geophysical Research, v. 90, p. 3690–3700.

Tilghman, B. C., 1906, Coon Butte, Arizona: Proceedings of the Academy of Natural of Sciences of Philadelphia, v. 57, p. 887–914.

Petrified Forest National Park, Arizona

S. R. Ash, Department of Geology, Weber State College, Ogden, Utah 84408

LOCATION

Petrified Forest National Park is located near the southern margin of the Colorado Plateau in the Painted Desert area of east-central Arizona (Fig. 1). The park extends about 26 mi (42 km) north and south. The north entrance is about 68 mi (109 km) west of Gallup, New Mexico, on I-40 and the south entrance is on U.S. 180 about 17 mi (27 km) east of Holbrook, Arizona. Visitors traveling west can leave I-40 at exit 311 and enter the park at the north entrance. After entering the park they may follow the scenic park road southward to the southern exit, a distance of about 27 mi (43 km). They then can travel westward on U.S. 180 to Holbrook where they rejoin I-40. Visitors traveling east may leave I-40 at exit 285 at Holbrook and drive east on U.S. 180 to the south entrance of the park, a distance of about 18 mi (29 km). Frequent pullouts and short side roads are present throughout the park. Much of the park is accessible to all types of vehicles, although their use is restricted to established roads. There are also two large, undeveloped, wilderness areas in the park. The removal of petrified wood or any other natural, archeological, or historical object from the park is strictly prohibited.

SIGNIFICANCE

Petrified Forest National Park is one of the world's greatest storehouses of terrestrial Upper Triassic plant and animal fossils. It contains not only some of the largest known and most famous deposits of colorful petrified wood but an extraordinary number and variety of other Upper Triassic plant and animal fossils as well. The park also includes well-developed badlands, which make it an unexcelled laboratory for study of the sedimentology and stratigraphy of a highly varied Upper Triassic fluvial/lacustrine complex.

DESCRIPTION

Petrified Forest National Park is an area of spectacular desert scenery that includes badlands, buttes, mesas, and low plateaus carved out of the Chinle Formation. Large quantities of petrified wood and other fossils that occur in the formation have been and are being exposed by erosional processes. The erosion also has produced exceptional three-dimensional exposures of the Chinle Formation, which facilitate observations and interpretations of the many sedimentary features present in the deposit.

The petrified wood is concentrated principally at four discrete sites in the park called "fossil forests." They are geographically isolated from each other, and each has been given a distinctive name based on the character of the wood it contains (Fig. 1). From north to south they are Black Forest (so called because the wood is usually black in color), Jasper Forest (where the wood is reddish in color), Crystal Forest (here the wood often

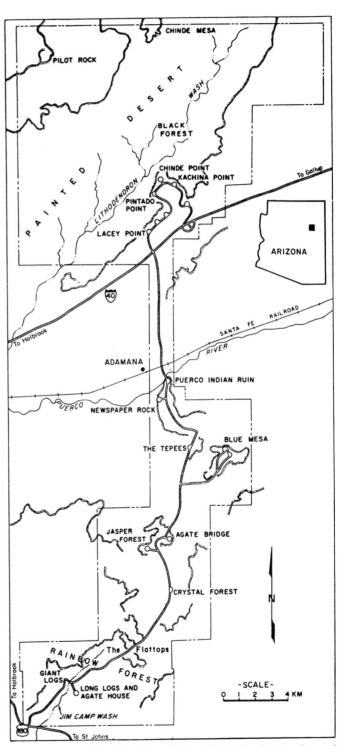

Figure 1. Map of Petrified Forest National Park showing location and access routes.

contains crystal filled cavities), and Rainbow Forest (the wood is highly colored in this area). Rainbow Forest has been subdivided into two sections: the Long Logs section to the east and the Giant Logs section to the west.

GEOLOGY

The geology of Petrified Forest National Park is shown on the map of central Apache County prepared by Akers (1964). More recently it has been discussed briefly by Billingsley (in Colbert and Johnson, 1985). As those authors indicate, the most widespread pre-Pleistocene formation exposed in the park is the fossil-bearing Chinle Formation of Late Triassic age. The only other pre-Pleistocene unit exposed in the park is the Pliocene Bidahochi Formation, which occurs in a few small areas in the park. In addition, there are deposits of Quaternary alluvium in the valley of the Puerco River and in some of the larger washes in the park. Thin deposits of colluvium are present on the tops of some of the broad mesas and ridges, and sand dunes occur in a few areas.

Chinle Formation. The Chinle Formation is probably the most widely exposed terrestrial Late Triassic Formation in the world (Gillette and others, in Nations and others, 1986). It com-prises multicolored bentonitic mudstone, siltstone, sandstone, conglomerate, and silty limestone. The formation is present throughout most of the Colorado Plateau and is exceptionally well exposed in northeastern Arizona. It was deposited in a broad basin by westward and northwestward flowing streams, as well as in lakes (Stewart and others, 1972; Blakey and Gubitosa, 1983). The basin was flanked on the east and southeast by the Ancestral Rocky Mountains and the Mogollon Highlands, respectively, and on the west and southwest by a complex volcanic arc. Generally it has been thought that most of the sediments were derived from volcanic and sedimentary sources to the south (e.g., Stewart and others, 1972), but Blakey and Middleton (in Nations and others, 1986) suggest that at least some of the volcanic debris found in the Chinle was derived from the volcanic arc that flanked the Chinle basin to the west and southwest.

In the Petrified Forest area the Chinle is about 932 ft (284 m) thick and disconformably overlies the Early–Middle(?) Triassic Moenkopi Formation (Fig. 2). The upper contact is not present but elsewhere the Chinle is separated from overlying strata by an unconformity. In Petrified Forest National Park the following members are recognized: Petrified Forest Member (lower part) at the base, Sonsela Sandstone Bed, Petrified Forest Member (upper part), and Owl Rock Member (at the top).

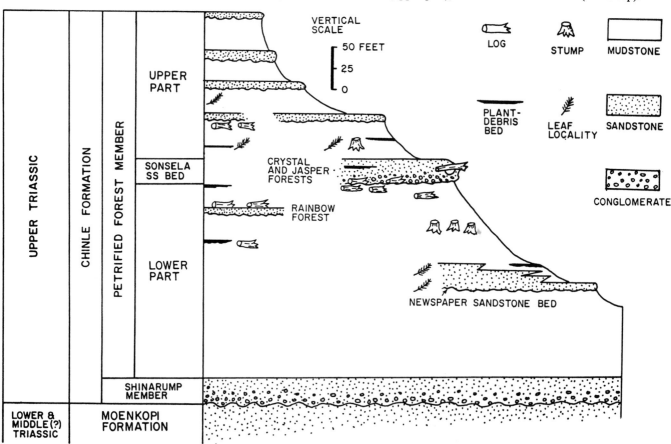

Figure 2. Stratigraphic section of the Chinle Formation in the southern part of Petrified Forest National Park showing the location of the fossil forests and other plant localities. The section profile runs from The Flattops on the left (south) to Puerco Indian Ruin on the right (north).

The regional dip of the Chinle Formation in the Petrified Forest is to the north at a few degrees. Consequently, the oldest part of the Chinle is exposed in the southern part of the park, and the youngest is in the northern part. A few broad folds occur locally in the park but no significant faulting is evident.

Petrified Forest Member. The Petrified Forest Member is widely exposed and it is not surprising that the park is the type locality for the member (Stewart and others, 1972). The member consists principally of blue, green, and gray bentonitic mudstone and siltstone and several beds of sandstone, but is reddish in color near the top. At many localities in the southern part of the Colorado Plateau, including the Petrified Forest, the member is divided into upper and lower parts by a distinctive bed of conglomeratic sandstone, the Sonsela Sandstone Bed (Stewart and others, 1972).

In the Petrified Forest the lower part of the member is exposed from the Puerco River (Fig. 1) to the southern boundary of the park. The base of the unit is not exposed in the park, but the log of a well at Adamana about 1.3 mi (2.0 km) west of Puerco Indian Ruin shows that about 100 ft (30 m) of the unit are present in the subsurface. About 170 ft (52 m) are exposed in the park so that the total thickness of the lower Petrified Forest Member is about 270 ft (82 m).

One particularly prominent unit in the lower part of the member is the Newspaper Sandstone Bed. It consists of brown weathering, hard, gray cross-bedded and ripple-laminated sandstone that is composed of very fine-grained to fine-grained quartz sand with small amounts of silt and clay. The unit is strongly crossed-bedded with micro–cross laminations. This bed has a maximum thickness of about 30 ft (9 m) along the west-facing escarpment near the type locality at Newspaper Rock. East of this locality the bed divides into several tongues that thin and disappear fairly abruptly within a few miles. The bed has not been recognized outside of the park. Horseshoe crab trackways are the only fossils known in this unit.

Some of the best and most typical exposures of the lower part of the member are present near the two steep hills 2.5 mi (4 km) south of the Puerco River, which are called the Tepees. The Tepees are eroded from blue and red mudstone-dominated facies in the lower part of the Petrified Forest Member of the Chinle Formation. In the vicinity of the Tepees, sequences of peppermint-striped beds of mudstones fill scours, and in turn, the tops of the beds are eroded and overlain by mudstones of different color and texture. These sequences occupy low-lying areas adjacent to levees or channels and may represent flood-plain deposits. Well-developed paleosols have been recognized in some of the mudstones here (Blakey and Middleton, in Nations and others, 1986).

The lower part of the Petrified Forest Member is very fossiliferous, and plant fossils occur almost everywhere the unit is exposed. Rainbow Forest occurs in the unit in addition to several smaller unnamed concentrations of logs and stumps. Many of the logs in Crystal Forest occur at the top of the unit just below the Sonsela Sandstone Bed. The lower part of the Petrified Forest

Member also contains the richest deposits of leaves known in the park and a number of plant debris beds (Daugherty, 1941; Ash, 1974; Gillette and others, in Nations and others, 1986). Palynomorphs are very common in this part of the Chinle (e.g., see Litwin, 1984) as are vertebrate fossils (Long, in Colbert and Johnson, 1985; Colbert, in Colbert and Johnson, 1985).

Sonsela Sandstone Bed. A generally thick bed of light grayish, cliff-forming conglomeratic sandstone occurs about 130 ft (40 m) above the Newspaper Rock Bed in the central part of the park. This unit is usually correlated with the Sonsela Sandstone Bed as it seems to occupy the same stratigraphic position as the Sonsela at its type locality about 80 mi (129 km) northeast of the park (Billingsley, in Colbert and Johnson, 1985). The Sonsela ranges up to 30 ft (9 m) thick in the park. It rests disconformably on the lower part of the Petrified Forest Member with a sharp, somewhat irregular contact; the upper contact is generally gradational but is fairly sharp in places. The bed is present between Blue Mesa and the Flattops but does not seem to occur in the area of Rainbow Forest. It has tentatively been recognized in the Painted Desert Section of the park (Billingsley, in Colbert and Johnson, 1985).

The Sonsela Sandstone Bed caps many of the mesas and buttes in the central part of the park. Typical exposures of the bed are on Blue Mesa. Here the unconformable contact between the Sonsela and the underlying beds is visible at many places and shows several feet of relief. In this area the Sonsela is about 28 ft (8.5 m) thick and consists of light grayish, cross-bedded, conglomeratic sandstone. The sandstone ranges from very fine to very coarse grained and from well sorted to poorly sorted. The sand grains are mainly quartz and chert and are weakly to strongly cemented with calcite. Gray mudstone pebbles are common in the unit, particularly near the base, and laminations of gravel composed predominantly of quartzite and chert commonly occur along bedding planes and in lenses. Some of the chert pebbles contain fusulinids, brachiopods, bryozoa, and other marine fossils and were derived from late Paleozoic strata in the region. Cross-bedding in the Sonsela consists primarily of the trough type, but some planar is also present. Studies of this cross-bedding indicate that the predominate direction of stream flow was to the northeast (average N57°E). The Sonsela apparently was deposited in a braided stream system. According to Blakey and Middleton (in Nations and others, 1986), this is a change in fluvial style from that found in the lower part of the Petrified Forest Member and reflects either a major change in gradient or increased rates of basin subsidence.

Much of the petrified wood in the park occurs in the Sonsela Sandstone Bed, including the logs in the concentration on Blue Mesa and in Jasper Forest together with many of the logs in Crystal Forest. Leaf and vertebrate fossils occur sparingly in this unit (Gillette and others, in Nations and others, 1986).

The upper part of the Petrified Forest Member rests on the Sonsela Sandstone Bed. It is most widely exposed in the Painted Desert section of the park. Smaller exposures occur in the southern part of the Rainbow Forest section, particularly in the area of the Flattops, and along the southwestern boundary of the park where it forms a prominent escarpment. The unit is nearly 512 ft (156 m) thick and is composed principally of grayish red, reddish brown, and reddish purple mudstone and siltstone. The strata are more flat-bedded than those in the lower part and they can be traced farther laterally.

Some easily accessible exposures are on the Flattops, two small but prominent mesas in the southern part of the park. In this area the basal beds of the unit are greenish gray in color but they grade upward into more typical reddish beds. Several thin beds of sandstone are present at several levels on the sides of the Flattops. Some of the best exposed channel and near-channel deposits within the park reportedly occur in this area (Blakey and Middleton, in Nations and others, 1986). Those authors suggest that the lateral accretion sets and heterolithic lithologies found in some of the sandstone beds here indicate that they were deposited in point bars. Trough and planar-tabular cross-stratification and convoluted bedding and other sedimentary features present in other beds support the idea that they were deposited in channels characterized by in-channel dune migration and fluctuating discharges. Levee deposits of sandstone and mudstone occur adjacent to the channel sands.

The upper part of the Petrified Forest Member contains a distinctive bed of pale pink to light purple tuffaceous sandstone called the Black Forest Bed. This unit forms a distinctive light-colored band on the tops of the low mesas in the central part of the Painted Desert section of the park. The upper part of the Petrified Forest Member is thought to have been deposited in channels and on associated flood plains as well as in lakes.

Plant fossils are locally abundant in the unit and include the Black Forest, a few leaf localities, and several plant debris beds. Vertebrate fossils and invertebrate fossils are locally abundant in this unit (Long, personal communication, 1986).

Owl Rock Member. The Owl Rock Member is exposed only on Chinde Mesa in the north-central part of the Painted Desert section of the park. It is about 120 ft (36 m) thick and consists mainly of pale red siltstone interbedded with light greenish gray, nodular limestone. The member was deposited primarily in lakes (Stewart and others, 1972). The base of the unit intertongues with the upper part of the Petrified Forest Member. The upper contact is not exposed in the park.

Age and correlation. Recent evaluation of the plant megafossils, palynomorphs, and vertebrate fossils in the Chinle Formation indicates that the lower part of the Petrified Forest Member and the Sonsela Sandstone Bed are late Carnian and that the upper part of the Petrified Forest Member is early Norian (Gillette and others, in Nations and others, 1986). The age of the Owl Rock Member is uncertain at this time, but it is assumed that it is probably late Norian.

Bidahochi Formation. The Bidahochi Formation of Pliocene age is represented in the park by a few thin deposits of pale yellow to pink silty mudstone and siltstone and some basalt flows. It rests disconformably on the Chinle Formation. Extensive exposures of the basalt underlie Pilot Rock in the northwestern corner of the park and they are also present at many places along the rim drive in the northern part of the park. The formation was deposited in a large fresh water lake and associated fluvial environments about 3 to 6 Ma (Billingsley, in Colbert and Johnson, 1986).

PALEONTOLOGY

The fossils in the Chinle Formation in the Petrified Forest indicate that this area was populated by a large variety of plants and animals during the Late Triassic. They indicate that the environment of the Petrified Forest and vicinity probably was warm and moist without strongly marked seasons.

Fossil forests. Most of the fossil forests consist of long lengths of large, uncompressed logs together with smaller fragments derived from the logs. Many of the logs have been only partially exposed by erosion, and since the formation is relatively flat lying, most of these in situ logs are still relatively horizontal. Some, however, clearly are no longer in place and have shifted and broken since exposure.

The preservation of the logs in the park varies greatly. Some have been totally replaced whereas others have been permineralized. In every specimen that has been examined, the principal mineral involved is quartz containing varying amounts of iron, manganese, and a few other elements (Sigleo, 1979).

The majority of the logs in the park are the remains of fairly large, straight, unbranched trunks 12 to 28 in (30 to 70 cm) or more in diameter. They range up to about 203 ft (62 m) in length. The topmost parts of the trunks are nearly always missing, but the butt ends with stumps of the roots are often present. In most specimens, 4 to 6 roots typically surround a larger, thick central tap root. The limbs are spirally arranged all along the trunk. Although many of the limbs were as much as 8 in (20 cm) in diameter, only rarely do any of them now extend out from the trunk for more than a few inches. Generally they are represented by depressions in the trunk or are nearly flush with the surface of the logs. Bark has been discovered on only one of the logs. Some of the logs have cavities of various sizes and shapes, and fragments are missing from the sides of some of them. These structures are filled with sediment which resembles that in which the logs are buried. Apparently they represent injuries that occurred to the logs prior to burial, possibly while the trees were alive.

Sorting apparently occurred during transportation of the logs, as most are the remains of fairly large trunks 12 in (30 cm) or more in diameter. Rarely are smaller trunks found in the park, and isolated limbs seem to be almost entirely absent. This, together with the general absence of bark and foliar material with the trunks, seems to indicate that the logs were transported into the area. During transportation the trunks were worn, and the smaller, more fragile bits and pieces were broken off and disappeared; only the larger fragments were deposited in the park and fossilized.

In a few places there are short lengths of logs in the park that are more or less perpendicular to the dip of the bedding in the

Chinle. Investigation (Gottesfeld, in Breed and Breed, 1972) has shown that at least some of these are stumps that are in the position of growth and have not been disturbed significantly since the trees died. They occur at several places in the park, but the only petrified forests that contain any such in-situ stumps are in the Black Forest in the northern part of the park.

Other plant fossils. Other plant fossils (up to 200 genera) include leaves and reproductive structures of both spore- and seed-bearing plants and spores and pollen grains. These fossils indicate that most major groups of land plants except the angiosperms, or flowering plants, were present in the area during the Late Triassic.

The remains of the primitive plants called horsetails are quite abundant in the park. These plants had slender, hollow, joined green stems like those of the living members of the group, but were much larger. The smaller horsetails in the park are referred to *Equisetites,* and the larger forms are assigned to *Neocalamites.* A large variety of ferns occur in the Petrified Forest. Most of them had pinnately compound leaves (also called fronds) in which the pinnae were arranged along opposite sides of a stem. Some of the leaves, such as *Cynepteris* and *Wingatea,* were large and filmy. Others, such as *Phlebopteris* and *Clathropteris,* had robust, palmately compound leaves with pinnae radiating outward from a common center at the end of a stem. The aerial stem of a small tree fern, *Itopsidema,* has been described from the park. This plant would have looked, superficially at least, like a miniature palm with a crown of filmy leaves, although it was actually closely related to some of the tree ferns that now grow in the humid tropics.

Conifer fossils are abundant in the Petrified Forest. Most of the petrified wood is similar to the wood that was described from here many years ago as *Araucarioxylon arizonicum* (see Daugherty, 1941; Ash, 1974). This wood is similar to that of the living *Araucaria* of the southern hemisphere, but that is not sufficient evidence to assign it to that family. Also, the trunks have non-*Araucaria* features so they should be merely considered the remains of an extinct conifer of some type. Other coniferous fossils include leafy shoots and twigs (*Pagiophyllum* and *Brachyphyllum*), and several undescribed cones, seeds, and pollen grains. The park also contains the wood of two other possible coniferous taxa, *Schilderia* and *Woodworthia.*

Petrified stems of two genera of cycads have been described from the park. One of them, *Lyssoxylon,* had a columnar stem that may have stood 3 to 4 ft (1 to 1.2 m) tall. The other, *Charmorgia,* had a short, globular stem about 1 ft (30 cm) in diameter. A crown of coarse leaves was present at the apex of both types of stems, and the exteriors were covered with broken leaf bases. These plants are closely related to the living cycads of the tropical parts of the southern continents.

Coarse leaves that look something like the leaves of cycads also occur in the park. However, these leaves, which are called *Zamites,* came from a group of extinct plants sometimes called the Cycadeoids. The leaves have a row of stiff oblong leaflets arranged on each side of a narrow stem. Although the stems of

the plants that bore these leaves is not known, they may have been tall and columnar.

Invertebrate fossils. Coquinas of clam and snail shells occur at several places in the park, particularly in the upper part of the Petrified Forest Member in the Painted Desert section. Trails made by horseshoe crabs have been described from a locality near the Tepees (see Breed, in Breed and Breed, 1972). The remains of clam shrimps commonly occur in the park, and a well-preserved specimen of the oldest North American nonmarine crayfish has recently been discovered there. Insects apparently were fairly common in the area during Late Triassic time as their remains and evidences of their activities are fairly common here. These fossils include compressions of beetles and cockroaches in addition to tunnels and galleries that have been excavated into the logs. Some of the fern leaves have been nibbled by insects.

Vertebrates. A large variety of freshwater fish, amphibians, and reptiles lived in the park area during the Late Triassic (Colbert, in Colbert and Johnson, 1985; Ash, 1986). The fish ranged in size from small ones like *Semionotus,* which was only 3 to 5 in (8 to 13 cm) long, to the giant *Chinlea,* which may have reached a length of 5 ft (1.5 m) and a weight of as much as 150 lb (68 kg). Other fish found in the park include a lungfish, *Ceratodus,* and two types of freshwater sharks, *Lissodus* and *Xenocanthus.*

The most common amphibian found in the park is *Metoposaurus.* This animal, which looked somewhat like a large modern salamander, was however about 10 ft (3 m) long and weighed about 1,000 lb (450 kg). It had short, weak legs, a large flat head, and a thick, heavy body. *Metoposaurus* was one of the last of a primitive group of amphibians that had been dominant during the Paleozoic.

The reptiles are much better represented in the park than are the amphibians. The most abundant of them is the crocodilelike reptile called *Rutiodon* (also called *Phytosaurus*). These average about 17 ft (5 m) in length, but the remains of one that was at least 30 ft (9 m) long have been found in the park. *Rutiodon* had a long narrow snout, and its teeth were large and sharp. The thick, heavy body and the long, narrow tail of the reptile were covered with heavy, bony plates. Its teeth indicate that *Rutiodon* fed on fish and any other small animals that came near the water sources. Several other large armored crocodilelike reptiles, including *Typothorax, Desmatosuchus,* and *Calyptosuchus* were also present.

The remains of one of the earliest known dinosaurs, *Coelophysis,* have been found in the park at several localities. This dinosaur was only about 8 ft (2.4 m) long with a long, slender neck and long, slender tail. Its body was narrow, and the head was rather small. Sharply pointed teeth indicate that it was predator.

In nearby areas the Chinle Formation has yielded the remains of several other types of reptiles that probably also frequented the park area, although their remains have not yet been found there. They include the remains of distant relatives of the lizards, true lizards, flying reptiles, and others that resembled living "horned toads." Another is the giant mammallike reptile *Placerias* that had a bulky body and a huge head that bore two strong tusks.

REFERENCES CITED

Akers, J. P., 1964, Geology and ground water in the central part of Apache County, Arizona: U.S. Geological Survey Water-Supply Paper 1771, 107 p.

Ash, S. R., ed., 1974, Guidebook to the Devonian, Permian, and Triassic plant localities of east-central Arizona: Paleobotany Section of the Botanical Society of America, 63 p.

—— , 1986, Petrified Forest; The story behind the scenery: Petrified Forest National Park, Arizona, Petrified Forest Museum Association, 48 p.

Blakey, R. C., and Gubitosa, R., 1983, Late Triassic paleogeography and depositional history of the Chinle Formation, southern Utah and northern Arizona, *in* Reynolds, M. W., and Dolly, E. D., eds., Mesozoic paleogeography of west-central United States, Rocky Mountain Paleogeography Symposium 2: Rocky Mountain Section, Society of Economic Paleontologists and Mineralogists, p. 57–76.

Breed, W. J., and Breed, C. S., eds., 1972, Investigations in the Triassic Chinle Formation: Museum of Northern Arizona Bulletin 47, 103 p.

Colbert, E. H., and Johnson, R. R., eds., 1985, The Petrified Forest through the ages, 75th Anniversary Symposium, November 7, 1981: Museum of Northern Arizona Bulletin 54, 91 p.

Daugherty, L. H., 1941, The Upper Triassic flora of Arizona: Carnegie Institution of Washington Publication 526, 108 p.

Nations, J. D., Conway, C. M., and Swann, G. A., eds., 1986, Geology of central and northern Arizona: Geological Society of America, Rocky Mountain Section, Field Trip Guidebook, 180 p.

Litwin, R. J., 1984, Fertile organs and in situ spores of ferns from the Late Triassic Chinle Formation of Arizona and New Mexico, with discussions of associated dispersed spores: Review of Paleobotany and Palynology, v. 42, p. 101–146.

Sigleo, A. C., 1979, Geochemistry of silicified wood and associated sediments, Petrified Forest National Park, Arizona: Chemical Geology, v. 26, p. 151–163.

Stewart, J. H., Poole, F. G., Wilson, R. F., Cadigan, R. A., Thordarson, W., and Albee, H. F., 1972, Stratigraphy and origin of the Chinle Formation and related Upper Triassic strata in the Colorado Plateau region: U.S. Geological Survey Professional Paper 690, 336 p.

Ship Rock, New Mexico: The vent of a violent volcanic eruption

Paul T. Delaney, U.S. Geological Survey, Flagstaff, Arizona 86001

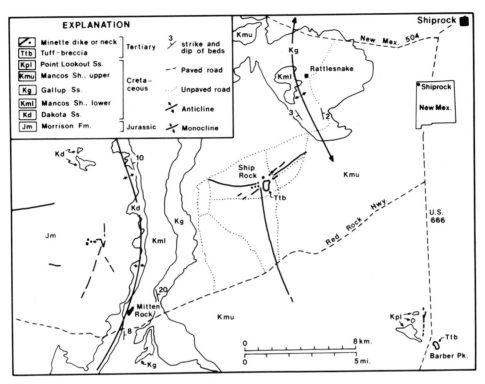

Figure 1. Simplified geologic map showing location of Ship Rock.

LOCATION

Ship Rock, the erosional remnant of an Oligocene volcanic vent, is a prominent feature of the skyline in northwest New Mexico. It rises 1,700 ft (500 m) above a surrounding plain, 24 mi (38 km) south-southeast of the Four Corners and 10 mi (17 km) southwest of the town of Shiprock.

ACCESSIBILITY

Ship Rock can be reached only via unimproved dirt roads. It is on the Navajo Indian Reservation, and permission to visit should be obtained from the Navajo Nation, Parks Department, Window Rock, Arizona. The most direct access road extends northward from the Red Rock Highway along the east side of the south dike (Fig. 1); this road is often muddy during the rainy season. Alternatively, one can take the gravel road from New Mexico 504 past Rattlesnake, turning onto a dirt road leading to the east side of Ship Rock.

SIGNIFICANCE

Ship Rock is perhaps the most spectacular among the exhumed diatremes of the Four Corners region and provides excel-lent exposures of the vent facies of a violent volcanic eruption. Dikes and necks surrounding Ship Rock, typical of others in the field, resulted from a style of magma ascent more quiescent than that of the diatremes.

REGIONAL STRUCTURE

Ship Rock is part of the Navajo volcanic field, which extends along the Defiance and Comb Ridge monoclinal uplifts that parallel the New Mexico–Arizona border northward into southern Utah. Although the Navajo magmas probably ascended along deep fractures associated with these uplifts, the vents are not located along exposed faults.

Ship Rock is on the Four Corners platform, a Laramide (Late Cretaceous to early Tertiary) tectonic element of the Colorado Plateau province intermediate in structural height between the San Juan basin to the east and the Defiance uplift to the west. The eastern boundary of the platform, 16 mi (25 km) east of Ship Rock, is the northeast-trending Hogback monocline, which has about 3,300 ft (1,000 m) of structural relief. The western boundary, about 5 mi (8 km) west of Ship Rock, begins at the Mitten Rock monocline, which has about 1,650 ft (500 m) of structural relief. Farther west in the Defiance uplift, the total structural relief is about 3,300 ft (1,000 m) relative to Ship Rock. Between the monoclines, the sedimentary strata are flat lying or gently folded, with maximum dips of about 5°.

STRATIGRAPHY

The sedimentary sequence of the Four Corners platform is about 3,300 ft (1,000 m) thick near Ship Rock and is underlain by Precambrian crystalline rocks. The sedimentary rocks range from Cambrian to Eocene in age. The plain that surrounds Ship Rock consists of the upper part of the Mancos Shale, which is separated from the lower part by a tongue of the Gallup Sandstone. Both units are Late Cretaceous, marine, tan argillaceous shale and siltstone to fine-grained sandstone. Some beds are as thick as 5 ft (1.5 m), but most are thinner than 2 in (50 mm).

The Ship Rock eruption probably formed a crater at the ancient surface, and so its present height above the surrounding plain (1,650 ft—500 m) provides a minimum estimate of the amount of post-eruptive erosion. The regional stratigraphy indicates that erosion removed less than 8,200 ft (2,500 m) of overburden (Delaney and Pollard, 1981). More likely, about 3,300 ft (1,000 m) or less was removed.

IGNEOUS ROCKS

Rocks of the Navajo volcanic field were erupted during Oligocene time, between 19 and 30 Ma (Roden and others, 1979; A. W. Laughlin, personal communication, 1985). Subsequently, the Colorado Plateau was uplifted relative to the surrounding provinces, and much cover was removed. Surficial volcanic deposits, such as lava flows, tuffs, agglomerates, and airfall material, are preserved only in the Chuska Mountains, south and southwest of Ship Rock. Presently, igneous rocks occur as dikes, necks, and diatremes, with little or no direct evidence that these structures lead upward to vents. However, several of the minette diatremes, including Ship Rock, contain airfall deposits.

The Navajo volcanic rocks are potassic lamprophyres. The dominant rock-type is minette, characterized by phenocrysts of phlogopitic biotite, diopside or diopsidic augite, and, in some cases, olivine set in a groundmass of phlogopitic biotite, diopside, alkali feldspar (commonly sanidine, less commonly orthoclase), and apatite (Williams, 1936). Diopside, alkali feldspar, and biotite are typically present in the proportions 2:2:1. Minettes are magnesium rich and aluminum poor; the lack of aluminum accounts for the absence of plagioclase. In some of the larger volcanic centers, a felsic minette with 52–60 wt % SiO_2 occurs as a differentiate of the more common minette that has 48–52 wt % SiO_2 (Roden, 1981). Monchiquite, olivine leucitite, and vogesite are present but are quantitatively subordinate to minette.

Many of the Navajo magmas were rich in volatiles during ascent and eruption. Vent facies of the diatremes are typically dominated by tuff-breccia derived from the gas-rich minette and comminuted wallrock. Diatremes composed of breccias with kimberlitic affinities are also present within the field. The most spectacular examples are at Buell Park and along the Comb

Figure 2. Ship Rock from the east. Note the en echelon character of the dike in foreground. Courtesy of D. L. Baars.

monocline. The volatile phases of kimberlites are thought to have been derived from the upper mantle. Vesicular minette is relatively uncommon, and minette clasts in breccias are typically dense, angular fragments. These characteristics are taken as evidence that the minette diatremes result from violent interaction between groundwater and volatile-poor magma and are thus phreatomagmatic deposits (Roden, 1981).

Upper mantle and crustal xenoliths are present at many localities. Common xenoliths are granite, gneiss, schist, and diorite. Of particular interest are peridotite xenoliths, commonly spinel lherzolite, and rarely garnet lherzolite, as well as eclogite and websterite xenoliths.

SITE GUIDE

Rattlesnake Anticline

From the intersection of New Mexico 504 and U.S. 666 at Shiprock, New Mexico, proceed west 5 mi (8 km) and turn south on a gravel road to the townsite of Rattlesnake, located on the crest of the Rattlesnake anticline. This structure is the trap for the second oil find in the San Juan Basin, which in 1924 was the highest gravity crude oil ever discovered. It made a millionaire of the man who purchased the 4000-acre (1620-ha) lease for only $1,000.

After the road turns south-southeast from Rattlesnake, it becomes dirt and is unmaintained. After 2 mi (3.3 km) turn west-southwest and proceed down the limb of the anticline, driving on beds of the Gallup Sandstone. This drive provides a good view of Ship Rock (Fig. 2). The massive buff-colored unit is mainly composed of tuff-breccia. The saucer-shaped, inward-dipping stratification near the top probably reflects settling of material that fell back into the vent during the waning stages of eruption.

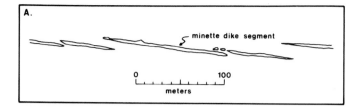

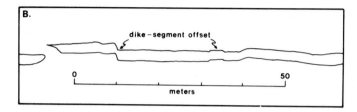

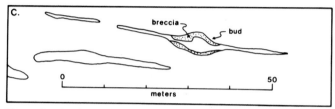

Figure 3. (A) A portion of the northeast dike, showing en echelon arrangement of segments. (B) Dike-segment offsets. (C) Bud and related breccia.

Near the vent walls, this stratification is locally as steep as 80 degrees. The tuff-breccia is cut by irregularly shaped minette dikes. The major dikes and four necks are visible from this vantage. One neck is high on the north flank of Ship Rock but separated from it by sedimentary rocks. In the background, the Red Rock monocline is visible beyond the west Ship Rock dike. The Chuska Mountains are southwest of Ship Rock; the laccolithic Carrizo Mountains are to the west.

East Side of Ship Rock

Continue on the dirt road for 1 mi (1.6 km) from the last turnoff, turning south-southwest near a small wash at the base of the anticline. Drive 1.1 mi (1.8 km), past a windmill and cattle tank, until the road crosses the northeast Ship Rock dike. Continue for 1.5 mi (2.5 km) to the southwest along the road that parallels the dike.

The dikes and necks on the east side of Ship Rock were the subject of a study of emplacement processes by Delaney and Pollard (1981). The outcrop length of the northeast dike is 9,528 ft (2,904 m), it has an average thickness of 7.5 ft (2.3 m), and it consists of 35 discrete segments separated by weakly deformed Mancos Shale. The segments are approximately en echelon (Figs. 2, 3a); the strike of individual segments differs by as much as 14 degrees from the overall N. 65°E. strike of the dike. Sites where segments coalesced during dike growth are marked by abrupt steps, or offsets, of the contact (Fig. 3b). Along most of the dike, wallrocks would fit neatly back together if the minette were removed.

The process of the dike emplacement was modeled as a magma-filled, segmented hydraulic fracture that propagated in response to a magmatic pressure about 20 bars greater than the horizontal regional stress. The host rock responded as an elastic medium during emplacement, except between closely spaced segments where inelastic deformation was appreciable. Joints in the host rocks adjacent to the dike are commonly parallel to the dike contacts. Joints of this orientation are generally absent at distances greater than about 50 ft (15 m) from the dike.

Along 30% of the outcrop length of the northeast dike, wallrock has been fractured, comminuted, and locally mixed with the minette. Ten percent of the dike is occupied by breccias consisting of wallrock material. Where breccia is present, the dike is commonly anomalously thick—up to 24 ft (7.2 m)—and such areas are called "buds." Elsewhere, the buds lack breccia, which presumably was carried away by the flowing magma. Examples of buds can be found 1,380 ft (420 m), 3,230 ft (1,000 m) (Fig. 3c), and 4,265 ft (1,300 m) (Fig. 4) from the southwest end of the dike.

Assuming that viscous magma flowed vertically along the entire length of the dike at the same time, Delaney and Pollard (1981) estimated that one segment, located 3,000 ft (900 m) from the southwest end of the dike, provided more than 20% of the total delivery of magma but was responsible for less than 4% of the heat loss. This segment is about 4% of the dike length and twice the mean thickness. The rates of fluid flow are sensitive to dike thickness, whereas rates of heat loss are not.

The three necks on the east side of Ship Rock are approximately circular to oval in map view, with diameters or major axes of about 100 ft (30 m). Each neck contains abundant breccia. Some of the most interesting breccia is on the south side of the northwest neck, where crystalline and sedimentary clasts are intermixed with the minette and finely comminuted sedimentary material. Each neck intersects a dike. A dike is visible on the northeast and southwest sides of the northeast neck. Although not continuously exposed, this same dike is visible on the northeast side of the northwest neck. This is probably the same dike exposed through the colluvium beneath the neck on the north flank of Ship Rock.

On the southwest side of the southern neck, breccia separates the minette of the neck from that of the dike (Fig. 5). Minette fragments are present in the breccia, which has a groundmass that consists primarily of sedimentary material. Locally, fractures in the adjacent wallrocks are filled with breccia and calcite. Elsewhere, the minette near the neck margin is cut by calcite-filled fractures. From the crosscutting relations and the presence of the breccia, it appears that the neck grew from the dike by stoping and removal of wallrock. If so, each widened portion of the northeast dike, discussed above, represents an early stage in the growth of a volcanic neck. Magma must have continued flowing in the southern neck after magma in the adjacent dike solidified. About 700 times more magma could have flowed through the neck than the dike from which it grew (Delaney and Pollard, 1982).

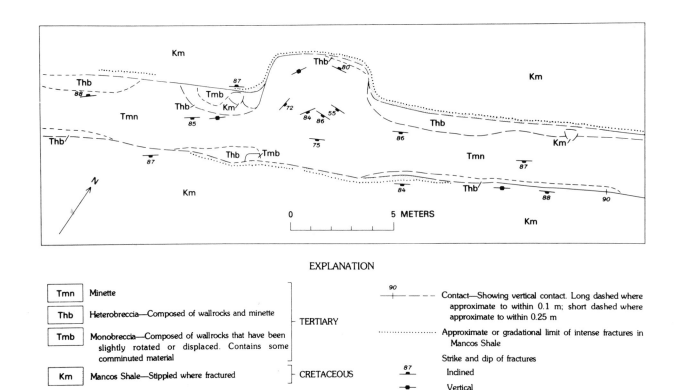

EXPLANATION

Tmn	Minette	
Thb	Heterobreccia—Composed of wallrocks and minette	TERTIARY
Tmb	Monobreccia—Composed of wallrocks that have been slightly rotated or displaced. Contains some comminuted material	
Km	Mancos Shale—Stippled where fractured	CRETACEOUS

90
——┼—— — — — Contact—Showing vertical contact. Long dashed where approximate to within 0.1 m; short dashed where approximate to within 0.25 m

················· Approximate or gradational limit of intense fractures in Mancos Shale

Strike and dip of fractures
—•—   Indined
—■—   Vertical

Figure 4. Map of bud on northeast dike. Location is 4265 ft (1300 m) from southwest end of dike.

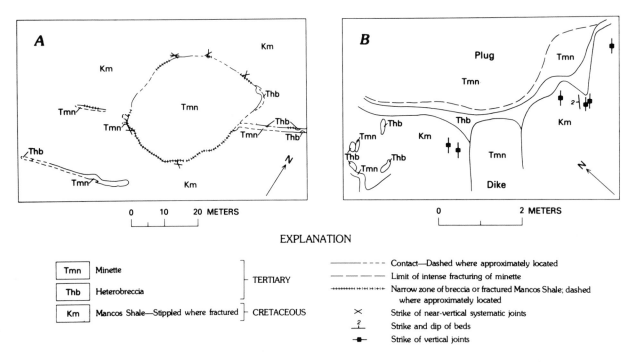

EXPLANATION

Tmn	Minette	TERTIARY
Thb	Heterobreccia	
Km	Mancos Shale—Stippled where fractured	CRETACEOUS

———— — — — Contact—Dashed where approximately located

— — — — — Limit of intense fracturing of minette

++++++ ++ ++++ Narrow zone of breccia or fractured Mancos Shale; dashed where approximately located

×   Strike of near-vertical systematic joints

2
—┼—   Strike and dip of beds

—■—   Strike of vertical joints

Figure 5. Maps of (A) southern neck and dike and (B) dike–joint–host rock relations on southwest side of neck, east side of Ship Rock.

The inferred sequence of events—dike intrusion, stoping of wallrocks along a portion of the dike to form buds, and localization of magma flow in a neck widened from the dike—corresponds to that observed during many basaltic fissure eruptions. For example, at Kilauea Volcano in Hawaii, when a dike intersects the earth's surface, the initial eruption is characterized by a "curtain of fire." Such an eruption generally lasts less than a day, before effusion of magma localizes to a single site where a cone or shield develops. By analogy, the necks on the east side of Ship Rock may have led upward to cinder cones that are now eroded away.

Ship Rock

Good exposures of tuff-breccia can be found on the east side of Ship Rock. Here, near the margin of the vent, blocks slumped from the vent walls and were disaggregated and mixed with the ascending minette tuff. Some of these blocks are greater than 33 ft (10 m) in their greatest dimension. Locally, breccia is composed almost entirely of sedimentary material, and only rare fragments of minette or flakes of biotite indicate that the sedimentary rocks were completely disaggregated. High on Ship Rock the tuff-breccia appears to be fine grained and of airfall origin. The vent apparently was enlarged by slumping while intermixed minette tuff and comminuted host-rock material streamed up the center of the conduit. The contact between the vent and wallrock is best exposed on the north side of Ship Rock, 492 ft (150 m) above its base. Here, in-place Point Lookout Sandstone of the Mesaverde Group is sharply truncated against tuff-breccia. Lack of deformation and thermal alteration of wallrock is a characteristic of the numerous basaltic necks and diatremes of the Colorado plateau (e.g., Williams, 1936). Stringers of minette have sharp contacts against the tuff-breccia and appear to have been emplaced after the tuff-breccia had settled into its present position.

The sequence of an initial violent eruption of a gas-rich magma (as recorded by the tuff-breccia), followed by a quiescent upwelling of gas-poor magma (as recorded by the irregularly shaped minette dikes), is typical of many of the diatremes on the Colorado Plateau.

West Side of Ship Rock

A dirt road skirts the base of Ship Rock, crossing the 5.5-mi (9-km) long south dike, one of the largest in the Navajo volcanic field. The north end of the south dike terminates about 165 ft (50 m) west of Ship Rock. The west dike, which is 2.5 mi (4 km) long, terminates north of Ship Rock. Thus, the dikes are not strictly radial about Ship Rock at their present level of exposure. Their outcrop pattern probably does not result from the stresses induced by the emplacement of the volcanic vent. The absence of a consistent strike among the 29 minette dikes on the Four Corners platform (see O'Sullivan and Beikman, 1963) provides evidence that the dikes did not propagate perpendicular to a single direction of least compressive horizontal stress. Rather, position and strikes of the dikes may reflect the distribution and orientation of fractures in the crystalline basement, only 1.6 mi (1 km) beneath the surface.

Also note the three aligned breccia bodies on the west side of Ship Rock. These bodies are on strike with the northeast dike and may connect with it at depth. The most easterly of the breccia bodies is composed entirely of Mancos Shale that has been only slightly broken and displaced.

Returning to the Highway

Return to the road that leads south along the east side of the southern dike. Barber Peak, Bennett Peak, and Ford Butte are visible to the southeast. These vents, which are comparable to Ship Rock, are easily accessible from U.S. Highway 666 and are not so precipitous. Mitten Rock and "The Thumb" are easily accessible to the west via the Red Rock Highway. These necks resemble those on the east side of Ship Rock but contain abundant mafic xenoliths. Surficial deposits of an explosive vent similar to Ship Rock can be inspected at Wahington Pass, 40 mi (65 km) south of Ship Rock. The geologic map compiled by O'Sullivan and Beikman (1963) serves as an excellent guide to locate other Navajo volcanic rocks in the Four Corners area. Akers and others (1971) provide brief descriptions of all Cenozoic igneous-rock localities on the Navajo Reservation.

SELECTED REFERENCES

Akers, J. P., Shorty, J. C., and Stevens, P. R., 1971, Hydrogeology of the Cenozoic igneous rocks, Navajo and Hopi Indian Reservations, Arizona, New Mexico, and Utah: U.S. Geological Survey Professional Paper 521-D, 18 p.

Delaney, P. T., and Pollard, D. D., 1981, Deformation of host rocks and flow of magma during growth of minette dikes and breccia-bearing intrusions near Ship Rock, New Mexico: U.S. Geological Survey Professional Paper 1202, 61 p.

—— 1982, Solidification of magma during flow in a dike: American Journal of Science, v. 282, p. 856–885.

O'Sullivan, R. B., and Beikman, H. M., compilers, 1963, Geology, structure, and uranium deposits of the Shiprock quadrangle, New Mexico and Arizona: U.S. Geological Survey Map I-345, 1:250,000.

Roden, M. F., 1981, Origin of coexisting minette and ultramafic breccia, Navajo volcanic field: Contributions to Mineralogy and Petrology, v. 77, p. 195–206.

Roden, M. F., Smith, D., and McDowell, F. W., 1979, Age and extent of potassic volcanism on the Colorado Plateau: Earth and Planetary Science Letters, v. 43, p. 279–284.

Williams, H., 1936, Pliocene volcanoes of the Navajo-Hopi country: Geological Society of America Bulletin, v. 47, p. 111–171.

Upper Cretaceous-Paleocene sequence, northwestern New Mexico

Barry S. Kues and Spencer G. Lucas, Department of Geology, University of New Mexico, Albuquerque, New Mexico 87131

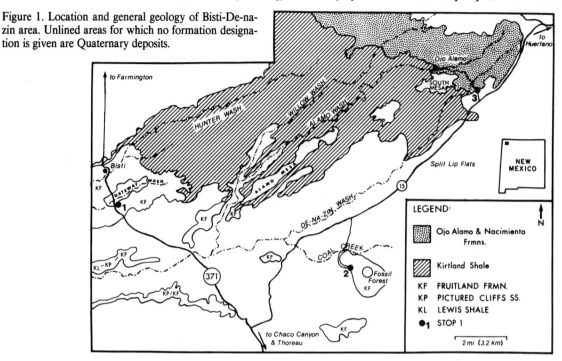

Figure 1. Location and general geology of Bisti-De-na-zin area. Unlined areas for which no formation designation is given are Quaternary deposits.

LOCATION AND ACCESSIBILITY

The area discussed here is in San Juan County, New Mexico, approximately 30 mi (48 km) south of Farmington and 20 mi (32 km) northwest of Chaco Canyon National Monument (Fig. 1). The small settlement of Bisti is most conveniently reached by traveling south on New Mexico 371 from Farmington or north on New Mexico 57 and New Mexico 371 from Thoreau, which is about 73 mi (117 km) east of Gallup on I-40. Both of these routes are paved to within a few miles of Bisti. From New Mexico 44 at Huerfano an improved dirt road (County Road 15) runs southwest about 25 mi (40 km) to intersect New Mexico 371 about 9 mi (14 km) south of Bisti.

All of the stops indicated on Figure 1 are accessible by two-wheel drive vehicles, but many of the side roads into the badlands and arroyo areas require four-wheel drive. Viewing of areas off the main roads is best done on foot. The nearest gasoline, food, and water are in Farmington and several stores along New Mexico 44. Most of the area under consideration is within recently established federal wilderness areas, managed by the U.S. Bureau of Land Management (BLM). A few sections are owned by the State of New Mexico or the Navajo Nation, but no part of the area is within the Navajo Reservation. Permission is not required to examine exposures on federal or state land (a permit from the BLM is required to collect fossils), but permission should be obtained from the few local Navajo landowners before entering their private land.

SIGNIFICANCE

Study of the Late Cretaceous-early Tertiary sequence in the

San Juan Basin has contributed significantly to our understanding of the depositional patterns and processes associated with periodic transgressions and regressions along the western shoreline of the Late Cretaceous midcontinent epeiric sea (e.g., Sears and others, 1941) and to our knowledge of Late Cretaceous, Paleocene, and Eocene terrestrial faunas and floras. Indeed, the long succession of superposed Late Cretaceous to Eocene vertebrate faunas present in the San Juan Basin is absolutely unique; no sedimentary sequence as long, complete, and richly fossiliferous spanning the Mesozoic-Cenozoic boundary exists in a single basin anywhere else on earth.

The uppermost Cretaceous delta and flood plain deposits of the Fruitland and Kirtland formations (Fig. 2) have a varied and distinctive biota of nearly 200 plant, invertebrate, and vertebrate species, together with economically important coal deposits. The Cretaceous-Tertiary boundary is at or very near the base of the overlying Ojo Alamo Sandstone, and magnificent mammalian faunas of early Paleocene age (Puercan and Torrejonian land-mammal "ages") occur within the Nacimiento Formation (Fig. 2). Recognition of the pre-Eocene but post-Cretaceous aspect of these mammals, beginning with the work of E. D. Cope in the 1880s, played an important role in the acceptance of the Paleocene as a distinct epoch of the Tertiary Period (Matthew, 1937). These formations crop out extensively in the San Juan Basin, but the sequence is not exposed in its entirety at a single location. We have, therefore, selected a restricted area of exposures in the Bisti and De-na-zin Wash areas, observable in three stops, to illustrate

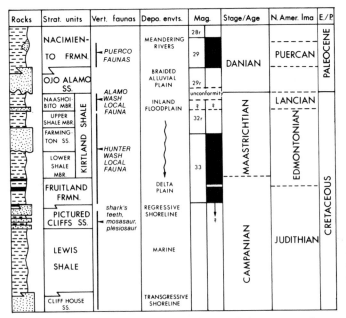

Figure 2. Generalized stratigraphic section, and data on vertebrate faunas, depositional environments, magnetostratigraphic units, stages/ages, and North American land mammal "ages" for Bisti-De-na-zin area.

Figure 3. Upper delta-plain facies of Fruitland Formation; mesa to left capped by delta top sandstone marking uppermost level of Fruitland of some workers. About 2 mi (3 km) east of Stop 1, Bisti badlands.

the major parts of this sequence that span the Cretaceous-Tertiary boundary.

GEOLOGY AND PALEONTOLOGY

The geology and paleontology of the Bisti-De-na-zin area are discussed at each of the three designated stops (Fig. 1).

Stop 1—Bisti Badlands. This stop is along New Mexico 371, 0.6 mi (1 km) south of the Gateway coal mine and 1.6 mi (2.5 km) South of Bisti, on south side of Gateway Wash, NE¼SW¼ Sec.5,T.23N.,R.13W. From this point an area of highly dissected, nearly barren badlands that developed in the Fruitland Formation and lower part of the Kirtland Shale is exposed mainly to the north and east along Hunter, Willow, Gateway, and Alamo Washes. The stratigraphic nomenclature of these units was first established by Bauer (1916) and more recently studied by Fassett and Hinds (1971). The arroyos, hills, ridges, and grotesquely eroded pinnacles that constitute the present topography of the area began forming very recently, some 2,800 to 5,600 ka (Wells and others, 1983).

The lower and middle parts of the Fruitland near the road include distributary channel sandstones, carbonaceous overbank and interdistributary shales, and a few thin coal beds that represent a swampy, in part brackish, delta-plain environment that during the late Campanian prograded to the northeast over the regressing Pictured Cliffs shoreline (Fassett and Hinds, 1971). A short distance to the north and east one moves upsection through middle and upper delta-plain facies of the Fruitland (Fig. 3), to a locally continuous sheet-sand unit that marks the delta top and the boundary between the Fruitland and Kirtland formations in this area (best seen near the top of low cliffs just north of Bisti). Above this, the lower shale member of the Kirtland is characterized by lighter gray, less carbonaceous shales, less numerous and

more vaguely-bounded channel sandstones, and an absence of significant coal beds, suggestive of a well-drained flood plain environment farther inland from the shoreline than the deltaic facies of the Fruitland.

The Fruitland and lower Kirtland formations, especially middle and upper parts of the Fruitland, contain a rich and diverse biota consisting of nearly 200 reported species of plants, invertebrates, and vertebrates of late Campanian age (Lucas, 1981; Tidwell and others, 1981). A majority of these have been recorded from exposures within the Bisti area. Approximately one-fourth of the taxa have not been reported outside of the San Juan Basin, and composition of the biota differs significantly from contemporaneous biotas in the northern Rocky Mountains. In the Bisti badlands, plants are represented by petrified logs (mostly conifers), palm leaf impressions, and beds of carbonized leaves, primarily of ferns and a large variety of angiosperm shrubs and trees. A single locality in the SE¼Sec.32,T.24N.,R.13W. yielded 30 plant species.

Unionid bivalves and nonmarine gastropods, the most common invertebrates, are present locally in great abundance, especially in the upper Fruitland. Among vertebrates, turtle remains are virtually ubiquitous, and crocodilian teeth and scutes (*Brachychampsa, Leidyosuchus*) are common. Dinosaurs are represented by isolated bones, articulated skeletal elements, and at least two skulls discovered in the Bisti area. Hadrosaurs (*Kritosaurus, Parasaurolophus*) predominate, with lesser numbers of ceratopsians (*Pentaceratops*), coelurosaurs, carnosaurs (*Albertosaurus*), ankylosaurs, and dromaeosaurs. Several significant microvertebrate localities have been sampled in the area between Hunter and Gateway Washes (Clemens, 1973); holostean and teleost fish (*Amia, Lepisosteus, Paralbula*), chondrichthyan (*Myledaphus* and several small sharks), crocodilian, and, less commonly, lizard and amphibian teeth are the major elements in these assemblages. In addition, about 10 species of mammals, including several new taxa, have been reported. Clemens (1973) emphasized the unique composition of the Fruitland-lower Kirtland vertebrate fauna and termed it the Hunters Wash local fauna. Lucas (1981) comprehensively summarized the composition and relationships of the Fruitland-Kirtland fauna in the Bisti and nearby areas.

Two small coal strip-mining operations are active on state land in this area; the Gateway mine is visible east of New Mexico 371 just north of this stop. About 4,000 acres north, east, and south of this mine have been established as the Bisti Wilderness Area.

One to two miles south of this stop, along the north and south side of De-na-zin Wash, marine units underlie the Fruitland. These include, in ascending order, the shoreline sandstones of the Cliff House Sandstone (more extensively exposed in the cliffs of Chaco Canyon National Monument); the prodeltaic, dark gray Lewis Shale; and the distal-bar, distributary mouth bar, distributary channel, and beach-front facies of the Pictured Cliffs Sandstone, immediately below the Fruitland. The Cliff House and lower Lewis beds here represent nearly the maximum extent of the last transgression of the Midcontinent sea across northwestern New Mexico; the upper Lewis and Pictured Cliffs were deposited during the final regression of this sea, and, in part, are laterally equivalent to the nonmarine delta-plain deposits of the Fruitland Formation.

Stop 2—"Fossil Forest". Travel 5.7 mi (9 km) northeast of intersection of New Mexico 371 and County Road 15 on New Mexico 371, then 2.9 mi (4.6 km) south on unimproved dirt road across Split Lip Flats to end of road; walk east along fence 0.2 mi (0.3 km), to SE¼Sec.14, and N½Sec.23,T.23N.,R.12W. Erosion of stream channel deposits in the upper part of the Fruitland Formation has exposed scores of petrified stumps in their original growth positions, together with some coniferous logs up to 50 ft long—the remnant of part of a Late Cretaceous delta-plain forest. In this area, the rich biota also includes leaf assemblages, nonmarine molluscs, turtles, crocodilians, dinosaurs, and mammals. Several partially articulated dinosaur skeletons, mostly of hadrosaurs, are known from the vicinity, and at least two localities have yielded mammalian teeth, jaw fragments, and postcranial remains. The great number of *in situ* stumps and diverse associated fauna make this area unique in the San Juan Basin and possibly in the North American Upper Cretaceous. The biota preserved here, thus, offers an unparalleled opportunity to study in detail the structure and paleoecology of a Late Cretaceous delta-plain forest. Congress has established the "Fossil Forest" as an area to be preserved in its natural state.

Stop 3—Cretaceous-Paleocene Boundary, Alamo and De-na-zin Washes. Take small dirt road leading to south rim of De-na-zin Wash, about 6.5 mi (10.5 km) northeast of turnoff to "Fossil Forest" from County Road 15, and 11.0 mi (18 km) southwest of intersection of County Road 15 with New Mexico 44 at Huerfano, center NE¼,Sec.16,T.24N.,R.11W. This road quickly becomes a jeep trail as it descends into De-na-zin Wash and ultimately crosses Alamo Wash near Ojo Alamo Spring. From this vantage point on the southern side of De-na-zin Wash

Figure 4. View to northwest across De-na-zin Wash towards South Mesa, from near Stop 3, showing uppermost Cretaceous to lower Paleocene sequence. K_{us} = upper shale member of Kirtland; K_n = Naashoibito Member of Kirtland; P_o = Ojo Alamo Sandstone; P_n = Nacimiento Formation.

("Barrel Springs Arroyo" of some early workers) the channel sandstones and conglomerates of the Ojo Alamo Sandstone may be observed. The Cretaceous-Tertiary boundary is placed at the base of (or possibly within) the Ojo Alamo in this part of the San Juan Basin. However, except for petrified logs, wood fragments, pollen, and occasional pieces of reworked dinosaur bone, the Ojo Alamo is unfossiliferous. The predominantly gray to red and purple strata immediately underlying the Ojo Alamo to the north and northwest represent the uppermost, or Naashoibito Member of the Kirtland, and the dark gray, green, and maroon-striped units above the Ojo Alamo to the northeast belong to the early Paleocene (Puercan) part of the Nacimiento Formation (Fig. 4).

The early literature on this area refers to the dinosaur-bearing "Ojo Alamo beds," but Baltz and others (1966) restricted the term Ojo Alamo to the sandstones unconformably overlying the uppermost dinosaur-bearing units, and named the thin, finer-grained unit beneath it the Naashoibito Member of the Kirtland Formation. Dinosaur bones and teeth indicating a fauna of middle to late Maastrichtian ("Lancian") age are present in the Naashoibito to within 10 ft (3 m) of the base of the Ojo Alamo. This fauna (Alamo Wash local fauna) includes some animals not present lower in the Kirtland or in the Fruitland, such as the dinosaurs *Alamosaurus, Torosaurus* and *Tyrannosaurus*? (Lehman, 1981), and represents a community that was not only somewhat younger than those of the Fruitland and lower Kirtland, but which also lived in a considerably more inland environment (Lucas, 1981). Microvertebrate localities are rare in the Naashoibito and little is known about the smaller elements of the fauna; likewise, leaf assemblages and nonmarine molluscs are negligible constituents of the Naashoibito biota.

The basal Nacimiento Formation, exposed to the east along the heads of Alamo and De-na-zin Washes (Fig. 5), contains a large variety of early Paleocene mammals and represents one of two principal reference localities for faunas of the Puercan land-mammal "age" (the other being "Mammelon Hill," along Tsosie Wash some 17 mi to the southeast). Puercan mammals from the San Juan Basin were first studied by Cope in the 1880s and monographed by Matthew (1937). At least 40 species and nearly 20 genera of mammals have been established on the basis of material from the San Juan Basin Puercan, and some of these have never been reported outside of the basin. The Nacimiento exposures along the heads of Alamo and De-na-zin Washes continue to yield numerous specimens of Puercan mammals.

The magnetic polarity zonation of Upper Cretaceous and Paleocene strata in the San Juan Basin presented by Lindsay and others (1981) relied principally on sections in the drainages of Alamo and De-na-zin Washes. These workers identified a normal polarity magnetozone in the lower, Puercan, mammal-bearing interval of the Nacimiento Formation as chron 28, and therefore placed the Cretaceous-Tertiary boundary in the San Juan Basin in the underlying zone of reversed polarity ("chron 28r"), which encompasses the lower part of the Ojo Alamo Sandstone. This placement of the Cretaceous-Tertiary boundary at a time younger than this boundary (and concomitant marine extinctions) in the pelagic marine limestones at Gubbio, Italy, led Lindsay and others (1981) to conclude that the Late Cretaceous marine and terrestrial extinctions were diachronous. However, biostratigraphic evidence that counters this interpretation (Lucas and Schoch, 1982) has led to a general acceptance that the Puercan mammal-bearing interval of the Nacimiento Formation corresponds to chron 29, and that the Cretaceous-Tertiary boundary here is in chron 29r, as it is at Gubbio. The search for anomalously high iridium concentrations at the Cretaceous-Tertiary boundary in the San Juan Basin (Orth and others, 1982), undertaken near De-na-zin Wash, only produced a small iridium spike that is thought to be due to geochemical enrichment processes.

Some 24,000 acres of the Alamo and De-na-zin Wash drainage are within the De-na-zin Wilderness Area.

The badlands visible at Stop 3 are essentially continuous with those of the Bisti area about 14 mi (23 km) to the west, and in their entirety these exposures display in excellent detail a sequence of marine and nonmarine sediments and biotas from Campanian to early Paleocene age. Elsewhere in the San Juan Basin (Torreon Wash, Kutz Canyon) the upper part of the Nacimiento bears late early Paleocene (Torrejonian) mammals, the Animas Formation in the northern part of the basin contains the typical late Paleocene (Tiffanian) mammal fauna, and the San Jose Formation along the east side of the basin yields diverse early Eocene (Wasatchian) vertebrate faunas.

REFERENCES CITED

Baltz, E. H., Jr., Ash, S. R., and Anderson, R. Y., 1966, History of nomenclature and stratigraphy of rocks adjacent to the Cretaceous-Tertiary boundary, western San Juan Basin, New Mexico: U.S. Geological Survey Professional Paper 524-D, 23 p.

Figure 5. Exposure of basal Nacimiento Formation, De-na-zin Wash drainage to northeast of Stop 3, showing "double red" from which many Puercan mammal remains have been collected. The bar is about 15 ft (4.6 m) long.

Bauer, C. M., 1916, Contributions to the geology and paleontology of San Juan County, New Mexico, 1. Stratigraphy of a part of the Chaco River valley: U.S. Geological Survey Professional Paper 98-P, p. 271–278.

Clemens, W. A., Jr., 1973, The role of fossil vertebrates in interpretation of Late Cretaceous stratigraphy of the San Juan Basin, New Mexico: Four Corners Geological Society Memoir, 1973, p. 154–167.

Fassett, J. C., and Hinds, J. S., 1971, Geology and fuel resources of the Fruitland Formation and Kirtland Shale of the San Juan Basin, New Mexico and Colorado: U.S. Geological Survey Professional Paper 676, 76 p.

Lehman, T. M., 1981, The Alamo Wash local fauna: a new look at the old Ojo Alamo fauna, *in* Lucas, S. G., Rigby, J. K. Jr., and Kues, B. S., eds., Advances in San Juan Basin Paleontology: Albuquerque, University of New Mexico Press, p. 189–221.

Lindsay, E. H., Butler, R. F., and Johnson, N. M., 1981, Magnetic polarity zonation and biostratigraphy of Late Cretaceous and Paleocene continental deposits, San Juan Basin, New Mexico: American Journal of Science, v. 281, p. 390–435.

Lucas, S. G., 1981, Dinosaur communities of the San Juan Basin: a case for lateral variations in the composition of Late Cretaceous dinosaur communities, *in* Lucas, S. G., Rigby, J. K., Jr., and Kues, B. S., eds., Advances in San Juan Basin Paleontology: Albuquerque, University of New Mexico Press, p. 337–393.

Lucas, S. G., and Schoch, R. M., 1982, Discussion, magnetic polarity zonation and biostratigraphy of Late Cretaceous and Paleocene continental deposits, San Juan Basin, New Mexico: American Journal of Science, v. 282, p. 920–927.

Matthew, W. D., 1937, Paleocene faunas of the San Juan Basin, New Mexico: American Philosophical Society Transactions, n.s., v. 30, 510 p.

Orth, C. J., Gilmore, J. S., Knight, J. D., Pillmore, C. L., Tschudy, R. H., and Fassett, J. E., 1982, Iridium abundance measurements across the Cretaceous-Tertiary boundary in the San Juan and Raton Basins of northern New Mexico: Geological Society of America Special Paper 190, p. 423–433.

Sears, J. D., Hunt, C. B., and Hendricks, T. A., 1941, Transgressive and regressive Cretaceous deposits in southern San Juan Basin, New Mexico: U.S. Geological Survey Professional Paper 193-F, p. 101–121.

Tidwell, W. D., Ash, S. R., and Parker, L. R., 1981, Cretaceous and Tertiary floras of the San Juan Basin, *in* Lucas, S. G., Rigby, J. K., Jr., and Kues, B. S., eds., Advances in San Juan Basin Paleontology: Albuquerque, University of New Mexico Press, p. 307–332.

Wells, S. G., Bullard, T. F., Smith, L. N., and Gardner, T. W., 1983, Chronology, rates, and magnitudes of late Quaternary landscape changes in the southeastern Colorado Plateau: American Geomorphological Field Group, 1983 Conference Field Trip Guidebook, p. 177–186.

Capulin Mountain Volcano and the Raton-Clayton Volcanic Field, northeastern New Mexico

John C. Stormer, Jr., Department of Geology and Geophysics, Rice University, P.O. Box 1892, Houston, Texas 77251

LOCATION

The Raton-Clayton Volcanic Field is located on the High Plains of northeastern New Mexico. The field extends for more than 85 mi (137 km) from near Trinidad, Colorado, southeastward through Raton, New Mexico, to Clayton, New Mexico. U.S. 64/87 runs along the axis of the field, and most of the important features can easily be seen from this highway. This is one of the best routes from Texas and Oklahoma to Denver and other cities in Colorado. Capulin Mountain and a few other points can easily be visited while traveling this route without significantly extending the trip. All of the locations described here are on paved highways, accessible with almost any type of vehicle. Most of the land in this area is private or leased state land so that permission must be obtained for collecting and travel off the highways.

Capulin Mountain, a large cinder cone with associated lava flows, is near the center of the volcanic field. The Capulin Mountain National Monument encompasses the cone and 775 surrounding acres (314 ha). The entrance is on New Mexico 325, 3 mi (4.8 km) north of the town of Capulin which is on U.S. 64/87, 54 mi (87 km) west of Clayton and 30 mi (48 km) east of Raton. The monument is open all year from 8:00 a.m. to 4:30 p.m. (later during the summer). There is a road spiraling up the cone to the rim of the crater about 1,000 ft (300 m) above the base. The trail around the rim, 1 mi (1.6 km) in length, provides excellent panoramic views of the whole volcanic field and nearby physiographic features. Collecting is not permitted in the monument, and the surrounding land is private. However, fresh specimens of the lava can be collected from a road cut located 1.9 mi (3 km) east of the town of Capulin on the north side of U.S. 64/87.

An attractive alternative route between Raton and Capulin Mountain is New Mexico 72, a paved road that runs over the top of the high Johnson Mesa to Folsom where it intersects New Mexico 325 about 6 mi (10 km) north of the entrance to Capulin Mountain National Monument. The features to be seen along these routes are described in the road logs of the New Mexico Bureau of Mines publication *Scenic Trips to the Geologic Past No. 7: High Plains.* (At the time of writing, 1985, this publication was out of print, but a revised version was to be made available).

SIGNIFICANCE

The Raton-Clayton field is part of a larger province related to the Rio Grande Rift (see Fig. 1A). The Taos Plateau, Ocate (Cimarron), and Raton-Clayton fields nearly coincide in age, and there are many parallels in types of magma erupted. The rocks illustrate many of the petrogenetic problems of current interest,

such as the nature of mantle processes giving rise to continental tholeiites, transitional basalts, and associated alkalic magmas; the role of crustal contamination and assimilation in producing silicic and intermediate magmas; and the relationship of magma generation to extensional tectonics.

Many of the physiographic features of active volcanic regions are preserved because of the relatively recent age of the volcanism (<8000 years for Capulin Mountain and the relatively dry climate. Capulin Mountain is perhaps the most accessible well-preserved example of a basaltic cinder cone in North America. The cone is nearly symmetrical and little modified by erosion. The cuts along the road to the crater rim provide excellent exposures of near-vent pyroclastic deposits, and the rim trail provides the equivalent of an aerial view of the morphology of the flows from the Capulin vent and nearby spatter cones and fissure vents.

PETROLOGY AND DISTRIBUTION OF ROCK TYPES

General Statement. The Raton-Clayton Volcanic Field contains a very wide range of volcanic rock types. Baldwin and Muehlberger (1959) provided the best overview of distribution and stratigraphy of volcanic units. Stormer (1972) presented a study of the mineralogy, geochemistry, and petrology of the area, and Phelps and others (1983) discussed the petrogenesis of the feldspasthoidal volcanic rocks. Figure 1 shows the approximate distribution of the volcanic rock types and the locations discussed in this paper. A schematic representation of the ages, volumes, and lithologic relationships is given in Figure 2.

Many of the most abundant rock types in this field are commonly found throughout the western United States (Lipman and Mehnert, 1975). In the Raton-Clayton field, there is a general correspondence of rock type with the formally designated volcanic formations (i.e., Clayton Basalt, Red Mt. Dacite). However, several distinct magma types may occur in the same formation, and a particular magma type may occur in more than one formation. The discussion below will focus on magma types rather than stratigraphically defined formations.

Low-K Olivine Basalt. The most abundant rocks are "transitional," mildly alkalic, olivine basalts that are analogous to continental olivine tholeiites. These lavas have silica contents near 50%, low potassium and incompatible trace element contents, and relatively flat rare earth element patterns. The rocks have small olivine phenocrysts in a finely crystalline groundmass with a distinctive diktytaxitic texture. These were thin, wide spreading, and apparently very fluid flows that now cap the mesas and form

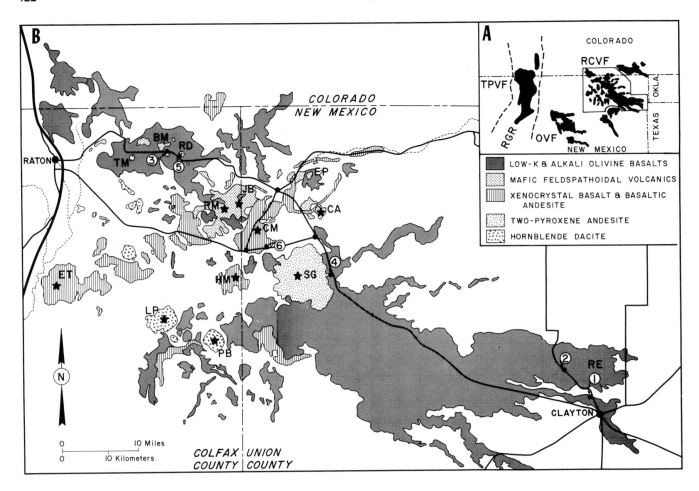

Figure 1. A. Location of the Raton-Clayton Volcanic Field (RCVF) with respect to the Ocate Volcanic Field (OVF), Taos Plateau Volcanic Field (TPVF), and the Rio Grande rift (RGR). The outlined area is shown in B. B. Distribution of volcanic rock types in the Raton-Clayton Volcanic Field (modified from Stormer, 1972). The numbered locations refer to suggested stops (see text). The following volcanic features are identified: BM—Bellisle Mt., CA—Carr Mt., CM—Capulin Mt., EP—Emery Peak, ET—Eagle Tail Mt., HM—Horseshoe Mt., JB—Jose Butte, LP—Laughlin Peak, PB—Palo Blanco, RE—Rabbit Ears Mt., RD— Red Mt., RM—Robinson Mt., SG—Sierra Grande, TM—Towndrow Mt.

the greater part of the Raton Basalt and the Clayton Basalt formations.

The lavas capping the high mesas can be seen at almost any point from Clayton to north of Trinidad, Colorado. This basalt is well exposed in the road cut on the edge of the Clayton mesa on New Mexico 370 about 1 mi (1.6 km) north of the intersection with U.S. 64/87 at the eastern edge of the town of Clayton (location 1, Fig. 1B). (This is the road to Clayton Lakes State Park.)

Alkali Olivine Basalt. In the Raton Basalt and Clayton Basalt formations, alkali olivine basalts occur interspersed with and overlying the low-potassium olivine basalts. Phenocrysts of olivine and clinopyroxene are common in these basalts, whereas plagioclase is rare. The groundmass is generally aphyric. These lavas contain 3–10% normative *ne* and are enriched in incompatible elements.

Rabbit Ears Volcano, 6 mi (10 km) NW of Clayton, and many of the other more eroded cones and shields seen between Clayton and Capulin are of this type. This rock type is exposed in road cuts on the flank of Rabbit Ears Mountain, on New Mexico 370 about 7.5 to 8.5 road mi (12 to 14 km) NW of its intersection with U.S. 64/87 (location 2, Fig. 1B).

Mafic Feldspathoidal Volcanics. These rocks are basanites and nephelinites. Abundant phenocrysts of clinopyroxene, olivine, and occasional hauyne occur in a fine grained groundmass containing nepheline and, rarely, mellilite. These rocks are strongly enriched in incompatible elements and contain up to 25% normative *ne*. These occur as the uppermost parts of the Clayton Basalt formation, and one is included in the Capulin Basalt formation.

The feldspathoidal volcanic rocks occur as heavily eroded vents in an area centered around Capulin Mountain. The most highly alkaline rocks can be found at Carr Mountain (formerly called Gaylor Mountain) 10 mi (16 km) ENE of Capulin Mountain, but access to this vent is difficult. The most convenient location at which to see these rocks is along New Mexico 72 on top of Johnson Mesa where the road crosses the south flank of Belisle, 1 mi (1.6 km) east of an old stone church, about 18 mi (29 km) east of Raton or the same distance west of Folsom (location 3, Fig. 1B). Blocks of the flows from Belisle Mountain can be found along the roadside, and some of these contain occasional small blue hauyne phenocrysts.

Two-pyroxene Andesite. These rocks are characterized by large Mg-rich orthopyroxene phenocrysts, some of which have cores of hypersthene. Smaller clinopyroxene and plagioclase phenocrysts are set in a matrix that is often glass. The two-pyroxene andesites have silica contents of about 60% and are moderately enriched in incompatible elements.

These andesites form the shield volcano of Sierra Grande, the largest volcanic edifice in the region. This large volcano is located southeast of Capulin Mountain and can be seen from U.S. 64/87 most of the way from Clayton. About 4.2 mi (6.8 km) southeast of the town of Des Moines, or about 40 mi (64 km) west of Clayton, U.S. 64/87 cuts a flow from Sierra Grande. The flow is exposed in low cuts on the southwest side of the highway (location 4, Fig. 1B).

Hornblende Dacites. Phenocrysts of hornblende and plagioclase (about An$_{35}$) are characteristic of the hornblende dacites, although biotite, apatite, and quartz are also present as phenocrysts at some localities. The groundmass contains feldspar microlites and glass and has a trachytic texture. The dacites have silica contents of about 65–70% and form the domes, plugs, and thick flows that compose the Red Mt. Dacite formation.

The dacites are not exposed along any of the major roads. However, New Mexico 72 skirts within a few hundred yards of the south and west flanks of Red Mountains on Johnson Mesa, 20 mi (32 km) east of Raton and 17 mi (27 km) west of Folsom (location 5, Fig. 1B). Permission to visit the outcrops should be obtained from Mr. John Floyd (ranch house directly south of the mountain).

Xenocrystal Silicic Alkalic Basalts and Basaltic Andesites. These are the most recent lavas erupted in the volcanic field; some are possibly as young as 6,000 years (Baldwin and Muehlberger, 1959). These basalts and basaltic andesites (50–55% silica) contain abundant olivine phenocrysts in an aphyric groundmass. Resorbed and fritted, reversely zoned plagioclase, and large quartz grains are found in variable abundance (common in the flow from Capulin Mountain). These form cinder cones and short blocky flows of scoria. Capulin Mountain is one of the vents, and the Capulin Basalt is almost completely composed of this rock type. A few of the youngest vents of the Clayton Basalt erupted rock of this type.

These basalts and basaltic andesites were erupted from a number of vents in the Capulin-Folsom area and to the south and west. The largest and most westerly of these vents is Eagle Tail Mountain, a very prominent volcano south of Raton. The best

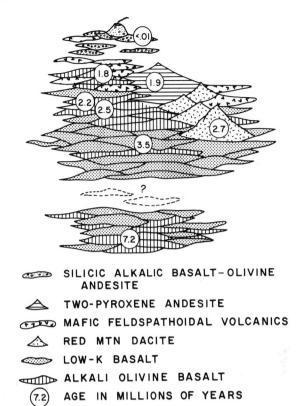

SILICIC ALKALIC BASALT–OLIVINE ANDESITE

TWO-PYROXENE ANDESITE

MAFIC FELDSPATHOIDAL VOLCANICS

RED MTN DACITE

LOW-K BASALT

ALKALI OLIVINE BASALT

(7.2) AGE IN MILLIONS OF YEARS

Figure 2. Age and stratigraphic relationships of rock types in the Raton-Clayton Volcanic Field (schematic). Circled numbers refer to absolute ages obtained on samples of the indicated rock type.

locality for collecting is the flow from Capulin Mountain on the north side of U.S. 64/87, 1.9 mi (3 km) east of the town of Capulin (location 6, Fig. 1B).

More recent data and theory require some modification of the conclusions reached by Stormer (1972) regarding processes necessary for the generation of such a diverse suite of volcanic rocks. The voluminous low-K olivine basalts and the alkali olivine basalts were probably derived by variable degrees of partial melting in the upper mantle. Some variation may be due to differences in the depths at which the melt segregated. A small amount of contamination with lower crustal material (with unradiogenic isotope ratios) is also probable.

For the development of xenocrystal basalts and two-pyroxene andesites, recent unpublished modeling and oxygen isotopic data suggest the importance of direct assimilation of lower crustal material or, perhaps more likely, efficient mixing of anatectic crustal melts with mantle-derived magmas similar to the low-K olivine basalts. Advances in understanding of isotopic systematics of lower crustal materials now permit such an interpretation. (Compare with interpretation for the Ocate and Taos Plateau volcanic fields, Nielson and Dungan, 1985, and Dungan and others, 1981).

The mafic feldspathoidal lavas have been interpreted by Phelps and others (1983) as the products of partial melting of a recently metasomatically enriched mantle source (presumably at greater depths than the preceding types). The metasomatizing fluid must have been CO_2-rich.

The hornblende dacites seem to be independently derived partial melts of lower crustal material. They are not a part of the mixing series discussed above. Some contribution of fractionated mantle derived magma is possible.

VOLCANIC FEATURES VISIBLE FROM CAPULIN MOUNTAIN

Because it is centrally located in the volcanic field, the summit of Capulin Mountain is an ideal place to get an overview of the important features in the field. Since most recent activity took place in the immediate vicinity of Capulin Mountain, it is also a convenient place to see volcanic landforms. The road approaching the entrance to Capulin Mountain National Monument from either direction runs on top of flows from the Capulin vent. Squeeze-ups and pressure ridges can be seen standing above the general surface of the blocky flow. Within the monument, past the visitors center, the road turns sharp left (north) and passes a small valley (picnic area) that was formed by the natural levees of the flow that went south of the mountain. The road then makes a sharp right turn and begins spiraling up the mountain to a parking lot on the west rim. The area from which the Capulin flows issued is at the west base of the cone, directly below the parking lot; flows went both north and south from this point. Looking west and northwest from the parking lot, one can see the basalt-capped surface of the high mesas rising from the low cliffs about 2 mi (3.2 km) away up toward the northwest where they are over 2,500 ft (760 m) above the Canadian River at Raton. The basalts that cap these mesas flowed down broad valleys. They then protected the former valley floor while erosion proceeded elsewhere. In the next episode of volcanism, the flows occupied slightly lower valleys. Repetition of this cycle has produced a "reverse" stratigraphy in which the basalts capping the highest mesas are the oldest. The most recent (Capulin) flows are essentially at the present erosional level.

On top of these mesas are two of the mafic feldspathoidal vents, Jose Butte [azimuth N40°W, 4 mi (6.4 km) and Robinson Mountain (N70°W, 5 mi; 8 km). Between these, farther in the distance, can be seen the conical, hornblende dacite domes, Red Mountain (N50°W, 13 mi; 20.9 km) and Towndrow Mountain (N62°W, 18 mi; 29 km). To the southwest are a number of basalt-capped mesas and two large hornblende dacite volcanos,

Laughlin Peak (S50°W, 16 mi; 25.7 km) and Palo Blanco (S25°W, 15 mi; 24 km). Horseshoe Mountain, a treeless cone in front of Palo Blanco (S25°W, 7 mi; 1.3 km) is another of the cinder cones of xenocrystal basaltic andesite of the Capulin Basalt formation.

Taking the crater rim trail counterclockwise provides the best views of the Capulin flows to the south. To the south and southeast the flows spread out over relatively flat terrain in a lobate pattern. The concentric pattern of pressure ridges on the surface of the flow can easily be seen. Sierra Grande (summit S45°E, 8 mi; 12.9 km) can be seen on the other side of U.S. 64/87. This shield of two-pyroxene andesite flows is about 7 mi (11.3 km) across and almost 2,000 ft (610 m) above the surrounding land surface.

To the east of Capulin Mountain there are spectacular views out across the high plains. On a clear day, Rabbit Ears Mountain (S68°E, 44 mi; 70.8 km) north of Clayton can be seen across the northeastern shoulder of Sierra Grande. To the northeast, Carr Mountain (N75°E, 9 mi; 14.5 km) a low eroded cone, is the most extreme in composition of the mafic feldspathoidal centers (<36% silica; hauyne phenocrysts are abundant, and the groundmass contains predominantly nepheline with some mellilite). Emery Peak (N45°E, 9 mi; 14.5 km) is another of the feldspathoidal vents with lavas that are less extreme in composition.

In the foreground to the north and northeast are a number of vents younger than Capulin Mountain. The stratigraphy of these lavas was worked out in detail by Baldwin and Muehlberger (1959), and they give a detailed description of the features in this area. The most distinctive feature is the small cinder cone named Baby Capulin (N40°E, 3 mi; 4.8 km). Baby Capulin erupted lavas similar to those of Capulin Mountains but with few of the distinctive quartz and feldspar xenocrysts. The lava from Baby Capulin flowed almost 20 mi (32 km) down the Dry Cimarron River to the northeast beyond Emery Peak. Twin Mountain is an elongated cinder cone 2 mi (3 km) east of Baby Capulin. It has been extensively mined for cinders and not much of its original form is visible. Beyond it to the east is a line of four low vents, the Purvine Hills. Twin Mountain and the Purvine Hills were probably produced by essentially contemporaneous fissure eruptions and are the latest volcanic activity in the area. The eroded pyroclastic cone of Mud Hill and the great wall lies between Capulin and Baby Capulin. This is an older vent of mafic feldspathoidal magma that is partially covered by the Capulin age lavas.

REFERENCES

Baldwin, B. and Muehlberger, W. R., 1959, Geologic studies of Union County, New Mexico: New Mexico Bureau of Mines and Mineral Resources Bulletin 63, 171 p.

Dungan, M. A., and others, 1981, Continental rift volcanism, in Basaltic volcanism study project, basaltic volcanism on the terrestrial planets: New York, Pergamon Press, p. 108–131.

Lipman, P. W., and Mehnert, H. H., 1975, Late Cenozoic basaltic volcanism and development of the Rio Grande Depression in the southern Rocky Mountains: Geological Society of America Memoir 144, p. 199–153.

Nielson, R. L., and Dungan, M. A., 1985, The petrology and geochemistry of the Ocate volcanic field, north-central New Mexico: Geological Society of America Bulletin, v. 96, p. 296–312.

Phelps, D. W., Gust, D. A., and Wooden, J. L., 1983, Petrogenesis of the mafic feldspathoidal lavas of the Raton-Clayton volcanic field, New Mexico: Contributions to Mineralogy and Petrology, v. 84, p. 182–190.

Stormer, J. C., Jr., 1972, Mineralogy and petrology of the Raton-Clayton volcanic field, northeastern New Mexico: Geological Society of America Bulletin, v. 83, p. 3299–3322.

The Valles Caldera, Jemez Mountains, New Mexico

Kenneth J. De Nault, Department of Earth Science, University of Northern Iowa, Cedar Falls, Iowa 50614

LOCATION AND ACCESS

The Valles Caldera is in Sandoval County, New Mexico (Fig. 1). The caldera, subsequent resurgent dome, later intracaldera domes and flows, and present solfatara activity can be seen from New Mexico 4. Most of the caldera floor itself is owned or controlled by the Baca Ranch and is not accessible. In winter, New Mexico 4 is intermittently closed through the caldera due to ice or snow.

GEOLOGIC SIGNIFICANCE

The Valles Caldera is one of the best-preserved Quaternary caldera complexes in the world. It is the type example of both the Valles type and resurgent type calderas. From the Valle Grande overlook, a vantage point located between the ring fracture zone and the present topographic wall of the caldera, one can observe a portion of the caldera floor, Redondo Peak (a resurgent dome), rhyolite domes that erupted along the ring fracture system, and a portion of the present topographic wall of the caldera. On the drive to the overlook are exposures of Bandelier Tuff, a thick sequence of ash-flow and air-fall deposits erupted during formation of the Toledo and Valles calderas. Beyond the overlook are exposures of postcaldera lava flows, pyroclastic deposits, and San Diego Canyon, a deep gorge formed by draining a large lake that once filled the caldera. Geophysical data suggest that magma may still be present under the caldera, and the area has been the focus of major geothermal exploration.

GEOLOGIC SETTING

The Valles Caldera sits astride the intersection of the Pajarito fault, the western bounding fault of the Rio Grande Rift, and the Jemez volcanic zone, a major northeast-southwest alignment of volcanic centers. Both the Jemez volcanic zone and the Rio Grande Rift have had active volcanism for at least the last 10 m.y. The intersection of these tectonic zones concentrated mantle-derived basalt, which in turn increased crustal preheating and melting and allowed a large volume of magma to accumulate and differentiate underneath the Jemez Mountains. Without this tectonic focusing of basalt into a restricted zone of the crust, the Jemez Mountains locus would probably be just another field of cinder cones (Smith and Luedke, 1984, p. 54).

GEOLOGIC HISTORY

The volcanic history of the Jemez Mountains can be divided into three episodes. The first episode began about 10 Ma with the eruption of the Keres Group, a series of volcanic rocks ranging in composition from basalts to rhyolites, but which is dominated by

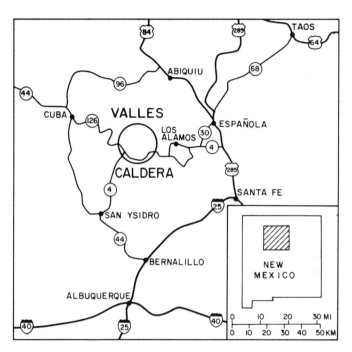

Figure 1. Location and access map of the Valles Caldera.

andesites of the Palizo Canyon Formation (Fig. 2). The volume of erupted material probably exceeded 240 mi^3 (1,000 km^3). The second episode began about 6.8 Ma, lasted until about 2.0 Ma, and produced about 100 mi^3 (450 km^3) of volcanic rocks, predominantly latites (the Tschicoma Formation). Flows and pyroclastic deposits from these two eruptive episodes formed a large volcanic pile. The third episode began about 1.4 Ma when a major Plinian-type eruption expelled about 12 mi^3 (50 km^3) of pumice and ash, causing the top of the Jemez Mountains to collapse into the underlying evacuated magma chamber to form the Toledo Caldera. Ash-flow and air-fall deposits from this eruption formed the Otowi Member of the Bandelier Tuff. The Toledo eruption was followed 300,000 years later by a smaller-sized eruption that destroyed most of the Toledo Caldera, formed the Valles Caldera, and deposited the Tshirege Member of the Bandelier Tuff. Deposits from these two colossal eruptions blanketed the surrounding region with thick ash-flow sheets and air-fall deposits (Fig. 3).

Of great interest are events surrounding formation and later modification of the Valles Caldera. The following description is taken predominantly from Smith and Bailey (1966; 1968) and Bailey and Smith (1978). Following formation of the Toledo Caldera, the Jemez volcanic pile was regionally domed and placed in tension. A ring-fracture system formed. Either before or during creation of the ring-fracture system, a single vent opened to produce a Plinian-type eruption. The erupted pumice settled to form the Tsankawi Pumice Bed. The ring-fracture system then became the primary locus for rapid eruption of over 8 mi³ (33 km³) of pumice and ash. Rapid evacuation of the magma chamber caused sagging or partial foundering of the caldera block along the ring-fracture system. The down-dropped block wedged shut the ring-fracture conduits, bringing the eruption to a temporary halt. Sporadic eruptions continued until the ring-fracture system reopened to erupt more than 4 mi³ (17 km³) of magma. The caldera block finally collapsed into the magma chamber to form the Valles Caldera.

Following collapse, a large lake formed in the centrally drained basin, and the topographic expression of the caldera expanded as the steep original walls weathered and eroded. Alluvium, talus, landslide, and lacustrine deposits interbedded with some pyroclastic material accumulated to over 2,000 ft (600 m) on the caldera floor. Rhyolite erupted in the center of the caldera.

Less than 100,000 years after caldera collapse, the center of the caldera block was uplifted and domed to form Redondo Peak. The uplifted mass stretched and broke, forming a northeast-trending graben. Along these tensional fractures, rhyolitic lavas erupted to fill the cracks and form small domes. Uplift of the caldera floor caused the intracaldera lake to overflow and rapidly erode San Diego Canyon. Following formation of the resurgent dome (Redondo Peak) and its accompanying lavas, rhyolitic lava erupted along the old ring-fracture system to form a series of discontinuous domes. Accompanying dome emplacement was explosive activity that deposited intracaldera ash-flow and air-fall sediment. The last eruption from this series of vents occurred less than 100,000 years ago. Presently solfataras and hot springs are active along the western margin of the caldera and along the Jemez Fault.

VISITING THE CALDERA

There is an excellent geologic map of the Caldera by Smith and others (1970) and two good road guides: Bailey and Smith (1978) and Goff and Bolivar (1983). Many of the descriptions in this chapter are taken or modified from Bailey and Smith (1978).

Santa Fe to the intersection of New Mexico 4 and Alternate New Mexico 4. Traveling north from Santa Fe on U.S. 84 and U.S. 285 and then west on New Mexico 4, one crosses the Santa Fe Group, a thick, late Tertiary sequence of alluvium with minor volcanic ash that was deposited in the Rio Grande Rift. On the western skyline are the Jemez Mountains. The highest point is Redondo Peak. The high plateau is the Pajarito Plateau, the eroded surface of the Bandelier Tuff. About 1 mi (1.6 km) after

GROUP	FORMATION	MEMBER
TEWA GROUP	VALLES RHYOLITES	Banco Bonito Member El Cajete Member <0.1(?) Ma Battleship Rock Member Valle Grande Member 1.1 - 0.4 Ma Redondo Creek Member Deer Canyon Member
	BANDELIER TUFF	Tshirege Member 1.1 Ma (Includes Tsankawi Pumice Bed) Otowi Member 1.4 - 1.5 Ma (Includes Guaje Pumice Bed)
POLVADERA GROUP 9.8 - 2.0 Ma		
KERES GROUP >13 - 6.2 Ma		

Figure 2. Stratigraphic nomenclature and general chronologic relations of volcanic and associated volcaniclastic rocks of the Jemez Mountains. Ages show range of K-Ar dates available and do not necessarily indicate maximum and minimum ages of rock units (modified from Goff and Bolivar, 1983).

crossing the Rio Grande at Otowi bridge, one can observe the Quaternary Puye Formation unconformably overlying the Santa Fe Group. The Puye Formation is composed of coarse fluvial debris eroded from the Jemez Mountains. The basal gravels, Precambrian igneous and metamorphic rocks eroded from the core of the Jemez Mountains, grade upward into pyroclastic and laharic deposits derived and associated with eruption of the Tschicoma Formation. Overlying these fluvial deposits are lacustrine beds composed of clay and volcanic ash deposited in a lake that formed when a basalt flow dammed the ancestral Rio Grande.

About 3.5 mi (5.6 km) from Otowi Bridge, New Mexico 4 enters the Puye Formation and begins to climb up the side of Los Alamos Canyon. Rounding the bluff one observes the tongue of a 2.4-Ma basalt flow that was injected into sand and gravel beds. Around the bend to the right the basalt becomes a pillowed palagonite breccia. Farther around the bend the flow contains less palagonite and larger pillows. The flow finally becomes massive, broken only by large explosion pipes. Across the canyon the basalt flow forms columnar-jointed cliffs. The vent for the flow is exposed in the canyon.

A thin soil layer on the basalt flow is covered by the Guaje Pumice Bed, the basal Plinian air-fall deposit of the Otowi Member of the Bandelier Tuff. The bed is a good example of a typical "popcorn sized" Plinian air-fall deposit with excellent sorting and an absence of ash. About 165 ft (50 m) of nonwelded Otowi ash flows underlie the slopes and are exposed in gullies at the base of the cliffs. Though nonwelded, vaporphase crystallization can be found in pumice fragments. Three ft (1 m) of the Tsankawi Pumice Bed of the Tshirege Member of the Bandelier Tuff occurs at the base of the cliffs and 165 ft (50 m) of partly welded Tshirege ash flows occurs above. In the upper 100 ft (30 m) of columnar-jointed tuff, at least eight distinct flow units separated by sandy partings and pumice concentrations are discernible.

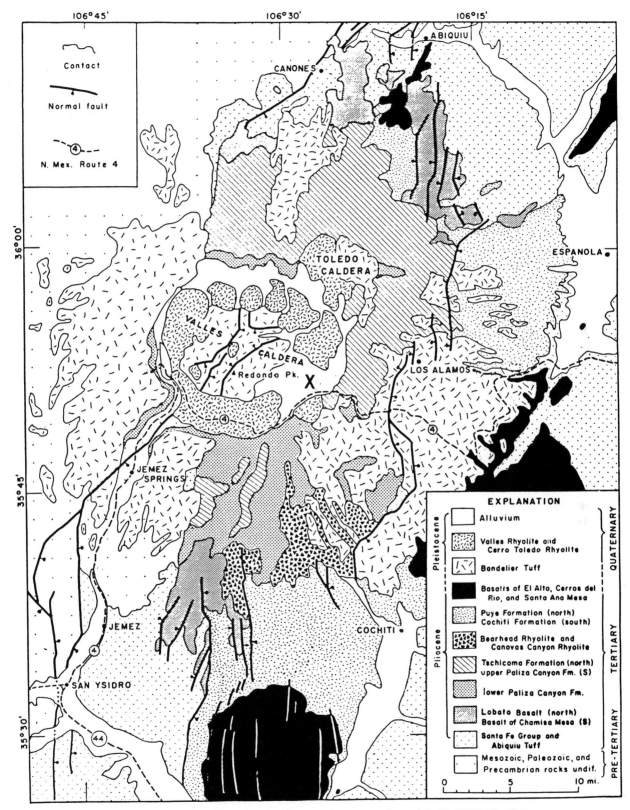

Figure 3. Generalized geologic map of the Jemez Mountains (from Bailey and Smith, 1978). The Valle Grande overlook is marked by an X.

The intersection of New Mexico 4 (also marked "Truck Route") and Alternate New Mexico 4 is about 1 mi (1.6 km) ahead. To see Los Alamos, continue on Alternate New Mexico 4; otherwise, continue on New Mexico 4 to White Rock and Bandelier National Monument.

Intersection of New Mexico 4 and Alternate New Mexico 4 to Valle Grande overlook.

If you drive to the caldera via Los Alamos, Alternate New Mexico 4 becomes Trinity Drive. Continue north on Trinity Drive through Los Alamos until it dead ends at Diamond Drive. Turn left (south) on Diamond Drive and after three blocks turn right (west) onto West Jemez Road. West Jemez Road is the continuation of Alternate New Mexico 4.

The route via New Mexico 4 gives an opportunity to view the valley of the Rio Grande and visit Bandelier National Monument. In about 1.2 mi (1.9 km), where the truck route to Los Alamos turns right, continue straight on New Mexico 4 to White Rock. Exposed 2.3 mi (3.7 km) past White Rock is a small mound of buff, nonwelded ash-flow tuff (Otowi Member) overlain by a 3-ft-thick (1 m) gray pumice fall (Tsankawi Pumice Bed) and white nonwelded ash-flow tuff (Tshirege Member). Ancient indurated and zeolitized groundwater tables are preserved in the Tshirege Member.

New Mexico 4 rejoins Alternate New Mexico 4 past the entrance to Bandelier National Monument. From this intersection the highway climbs the escarpment of the Pajarito fault. The fault extends about 30 mi (50 km) along the east side of the Jemez Mountains and vertically offsets the Bandelier Tuff 300 to 500 ft (100 to 150 m). In roadcuts on the right, several gouge zones separate large blocks of densely welded Bandelier Tuff that have been steeply tilted eastward. To the left and below is the Pajarito Plateau, a gently eastward-sloping surface formed by ash-flow deposits of the Bandelier Tuff. The ash flows filled preexisting canyons and blanketed the southwest segment of the Española Basin.

New Mexico 4 climbs up to the rim of the caldera and then descends into the caldera. The Valle Grande overlook is marked by a historic marker at a large pullout on the right (west).

Valle Grande overlook.

In the foreground is the Valle Grande, a portion of the present topographic floor of the caldera. The topographic rim of the caldera begins on the far left (southwest), continues behind the observer, and extends to the grass-covered peaks of Cerro Grande and Pajarito Peak on the far right. The high flat-topped mountain to the west is Redondo Peak, elevation 11,350 ft (3,460 m), summit of the resurgent dome. The high knob just to the north (Redondito) and the lower ridge extending northeast (Redondo Extension) are also structural elements of the resurgent dome. All are underlain by densely welded Bandelier Tuff, which dips generally southeastward toward the observer. The tuff has been uplifted as much as 3,280 ft (1,000 m) above the caldera rim and possibly as much as 5,920 ft (1,500 m) above its postcollapse position within the caldera. Postcaldera domes of the Valle Grande Member that erupted along the ring-fracture system include 0.49 Ma South Mountain

(just south of Redondo Peak); 0.50 Ma Cerro La Jara (the small treed knob immediately to the east of South Mountain); and the heavily logged mountains in the middle distance to the north.

Valle Grande overlook to Banco Bonito glass flow.

About 9.6 mi (15.5 km) down New Mexico 4 is a road cut exposing filled valleys eroded into 0.49 Ma South Mountain Rhyolite. The topography developed on the South Mountain Rhyolite was mantled by air fall and filled with intracaldera ash-flow deposits of the El Cajete Member and then overridden by vitrophyric blocks of the Banco Bonito Member glass flow. The Banco Bonito glass flow is the youngest recorded volcanism in the caldera. The exposure gives a graphic display of the mantling of topography by Plinian air-fall deposits. Ash flows are seen to be confined to the swales and pinch out laterally against valley walls. Surge deposits can be found at the base of the ash flows.

Banco Bonito glass flow to Redondo Peak overlook and Fenton Hill.

Six mi (10 km) ahead is the junction of New Mexico 4 and 126. To obtain a view of Redondo Peak, the Banco Bonito glass flow, and the Fenton Hill Hot Dry Rock drilling site, turn right onto New Mexico 126. Continue for 4.4 mi (7.0 km) and turn left on the unmarked side road. Proceed to the parking area and walk another 130 ft (40 m) to the overlook. Fenton Hill is 0.5 mi (0.8 km) farther on New Mexico 126.

The high domical mountain due east is Rendondo Peak. The nearer and lower irregularly crested ridge to the left of Redondo Peak is Redondo Border, which forms the western half of the resurgent dome. The valley between the two is a northeast-trending medial graben which separates the two halves of the dome. Recent drilling for geothermal energy inside the graben, along with gravity data, indicate that fault blocks at depth are relatively flat lying. These data pose problems for the accepted model of resurgence proposed by Smith and Bailey (1968).

In the foreground to the east-southeast is the west moat of the caldera. The southern part of this moat is occupied by the Banco Bonito glass flow. The flow can be seen heading from El Cajete Crater down the valley toward San Diego Canyon.

San Diego Canyon to San Ysidro.

Return to the intersection of New Mexico 126 and 4, turn right and descend into San Diego Canyon. San Diego Canyon generally follows the trace of the Jemez fault. The river originated when water filling the caldera overflowed due to the rise of Redondo Peak. Ahead 1.5 mi (2.4 km) the frozen remnants of the Banco Bonito glass flow are preserved where it cascaded over the east wall of the canyon in what must have been an impressive lava fall.

Ahead 1.5 mi (2.4 km) is Battleship Rock, a spectacular outcrop of columnar-jointed welded tuff formed by a series of postcaldera small-volume ash flows that issued from a vent near El Cajete Crater. The ash-flow deposits preserved at Battleship Rock initially extended a considerable distance down San Diego Canyon and filled it to a depth of about 330 ft (100 m). Subsequent erosion has removed all but the outcrops in the Battleship Rock area and one or two small remnants down-canyon. Battleship Rock is the filling of a narrow vertical-walled gorge. The curved columnar jointing in the lower part of Battleship Rock is a consequence of cooling against the former gorge walls. Subsequent erosion has removed the adjacent less resistant sedimentary rocks and left the more resistant welded tuff standing as a promontory—an interesting example of inverted topography.

The tuff at Battleship Rock is about 260 ft (80 m) thick and contains two main flow units that constitute a single cooling unit. The tuff is entirely vitric from bottom to top. The basal 50 ft (15 m) is composed of poorly consolidated pumiceous tuff breccia, which becomes increasingly compacted upward and grades into partly welded tuff. The tuff becomes gradually less welded and passes again into unconsolidated pumaceous tuff breccia about 210 ft (65 m) above the base. Surprisingly, the contact between the two main flow units is in the middle of the most densely welded part, about 115 ft (35 m) from the base, indicating that the two units were emplaced in very rapid succession. The contact is marked by a 12-in-thick (30 cm) zone of large, flattened, vitrophyric pumice blocks that accumulated at the top of the lower flow unit as a result of their buoyant upward concentration during flow. Two or three similar (but nonwelded) coarse pumice concentrations occur in the uppermost 50 ft (15 m) of unconsolidated tuff, suggesting the presence of two or three additional thin flow units near the top of the section.

Those interested in collecting samples of the various zones within a welded tuff should note that Battleship Rock provided many of the samples illustrated in Smith's 1960 classic work on welding in ash flows. The climb up the back of the outcrop is somewhat arduous, and the steep slopes are slippery and treacherous. Care should be used.

Continuing down San Diego Canyon, New Mexico 4 generally follows the Jemez fault zone. Down canyon is Soda Dam, a travertine deposit precipitated from carbonate rich thermal waters discharged from the Jemez fault zone. Remnants of older travertine dams enclosing Pleistocene stream deposits occur as much as 100 ft (30 m) high on the canyon walls, indicating that springs have been active for a long time during downcutting of the canyon.

While traveling down the canyon, one is treated to spectacular exposures of Bandelier Tuff, which attains a maximum thickness of 1,000 ft (300 m). The contact between the Otowi and Tshirege members is placed immediately below the orange zone about 280 ft (85 m) above the base.

Continue on New Mexico 4 to San Ysidro and the junction with New Mexico 4. Turn left to return to I-25.

REFERENCES CITED

Bailey, R. A., and Smith, R. L., 1978, Volcanic geology of the Jemez Mountains, New Mexico, *in* Hawley, J. W., ed., Guidebook to the Rio Grande rift in New Mexico and Colorado: New Mexico Bureau of Mines and Mineral Resources Circular 163, p. 184–196.

Goff, F. E., and Bolivar, S. L., 1983, Field trip guide to the Valles Caldera and its geothermal systems: Los Alamos National Laboratory, 53 p.

Smith, R. L., 1960, Zones and zonal variations in welded ash flows: U.S. Geological Survey Professional Paper 354-F, p. 149–159.

Smith, R. L., and Bailey, R. A., 1966, The Bandelier Tuff; A study of ash-flow eruption cycles from zoned magma chambers: Bulletin Volcanologique, v. 29, p. 83–104.

——, 1968, Resurgent cauldrons: Geological Society of America Memoir 166, p. 613–662.

Smith, R. L., and Luedke, R. G., 1984, Potentially active volcanic lineaments and loci in western conterminous United States, *in* Explosive volcanism; Inception, evolution, and hazards: National Academy of Sciences, p. 47–66.

Smith, R. L., Bailey, R. A., and Ross, C. S., 1970, Geologic map of the Jemez Mountains, New Mexico: U.S. Geological Survey Map I-571, scale 1:250,000.

Quaternary basalt fields of west-central New Mexico: McCartys pahoehoe flow, Zuni Canyon aa flow, Zuni Ice Cave, Bandera Crater, and Zuni Salt Lake maar

Wolfgang E. Elston, Department of Geology, University of New Mexico, Albuquerque, New Mexico 87131
Kenneth H. Wohletz, ESS1, Los Alamos National Laboratory, Los Alamos, New Mexico 87545

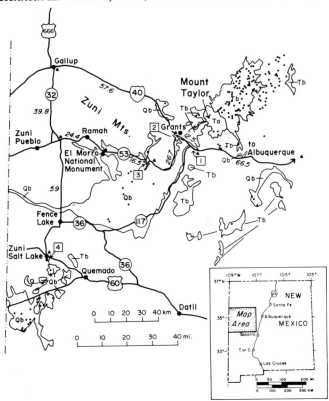

Figure 1. Mount Taylor–Zuni–Zuni Salt Lake volcanic field, west-central New Mexico. Ta, Pliocene andesite, dacite, and trachyte; Tb, Pliocene basalt; Qb, Quaternary basalt. Numbers in squares indicate sites described in this paper. 1, McCartys flow; 2, Zuni Canyon flow; 3, Zuni Ice Cave and Bandera Crater; 4, Zuni Salt Lake. Distances are in miles.

INTRODUCTION

The U.S. Geological Survey was only five years old when its director, John Wesley Powell, approved Captain Clarence Dutton's plan for one field season on Mount Taylor and the Zuni Plateau (Fig. 1). Dutton's description of west-central New Mexico, published in 1885 as part of the U.S. Geological Survey's 6th Annual Report, has become a geologic classic, not only for its lucid literary style but also because it laid the foundation of basic geologic concepts. In his account of erosion and uplift, the reader can discern the germ of ideas that would lead Dutton within four years, to the principle of isostasy. In a discussion of mountain building, he contrasted compressional fold belts, such as the Alps and Appalachians, with mountains of western interior North America, in which vertical uplift and rifting had been the princi-

pal forces. Finally, he pondered the significance of the great volcanic fields that nearly encircle the Colorado Plateau.

In the Mount Taylor–Zuni area, Dutton recorded a history of volcanism from late Tertiary to late prehistoric time. Basalt predominates; the youngest lavas issued from cones that are still well preserved, and the lava flowed down modern valleys. Older lavas cap mesas (in general, the higher the mesa, the older the cap; see Fig. 2), and many of their vents have been reduced to volcanic necks. There are dozens of small basalt volcanoes in various stages of preservation as well as one large central volcano (Mt. Taylor, elevation 11,389 ft; 3472 m) of more varied composition, including alkali basalt, andesite, dacite, trachyte, and rhyolite. Radiometric and archaeological dates now bracket volcanism between 4.4 Ma (Pliocene) and AD 700, with a probable peak between 3.0 and 2.5 Ma (Lipman and Mehnert, 1980; Nichols, 1946). All Pliocene basalts are alkalic; Quaternary basalts include both tholeiite and alkali basalt.

This guide describes three of the youngest, best preserved, and most accessible volcanic features of the Mount Taylor–Zuni field: (1) the contrast between the Holocene Zuni Canyon aa and McCartys pahoehoe flows, (2) Bandera crater, a Holocene cinder cone of alkali basalt, with bombs cored by ultramafic nodules, and a pahoehoe flow honeycombed by lava tubes and tunnels, and (3) Zuni Salt Lake, a late Pleistocene maar crater and cinder-cone cluster flooded with salt water.

In recent decades, all three localities have been examined closely by students of comparative planetology. The late Gerard Kuiper, Chief Scientist of Ranger (the first unmanned mission to the moon), was much impressed with similarities between collapse depressions on the McCartys flow and small craters in the dark lunar maria ("seas") and between collapsed lava tunnels of the Bandera area and sinuous rilles of the moon. The Hadley Rille, a steep-sided meandering valley, was visited by Apollo 15 in 1971. Eugene M. Shoemaker, founder and first chief of the U.S. Geological Survey Branch of Astrogeologic Studies, compared lunar craters with the Zuni Salt Lake maar and Meteor Crater, Arizona.

The area holds great beauty and historical interest. Visitors driving along New Mexico 53 should leave time to see El Morro National Monument (where Spanish conquistadores and American pioneers carved their names into a great cliff of Jurassic Zuni Sandstone) and Zuni Pueblo, already ancient when first seen by Spaniards searching for the Seven Cities of Cibola, three generations before the Pilgrims landed at Plymouth Rock.

Guidebooks 10 (1959) and 18 (1967) of the New Mexico

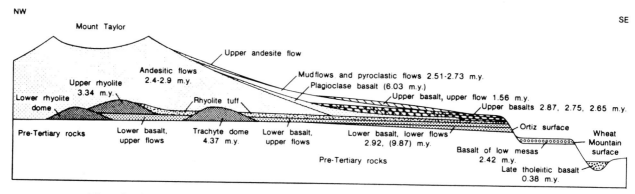

Figure 2. Diagrammatic representation of age relations between volcanic units dated by K-Ar, south side of Mount Taylor. Unreliable ages are in parentheses. From Lipman and Mehnert (1980).

Geological Society are compendia of the geology, anthropology, and history of the region; both contain detailed geologic road logs. The Society's 1:1,000,000 Highway Geologic Map (1982) provides a wealth of information. For the casual tourist, *Southern Zuni Mountains* by R. W. Foster (Scenic Trips to the Geologic Past No. 4, New Mexico Bureau of Mines and Mineral Resources, 1971) is recommended for the stretch between Grants and Gallup via New Mexico 53 and 32. These publications can be obtained from the New Mexico Bureau of Mines and Mineral Resources, Campus Station, Socorro, NM 87801, (505) 835-5410.

AA AND PAHOEHOE: ZUNI CANYON AND McCARTYS BASALT FLOWS

Clarence Dutton, who had studied volcanoes of the Cascade Range and Hawaii before coming to New Mexico, found Mount Taylor uninteresting; a modern geologist might wonder how so large a volume of intermediate-composition magma originated 600 mi (1,000 km) from a plate margin and 28 m.y. after subduction had ceased. Dutton was more impressed with the similarities between the basalt flows of New Mexico and Hawaii. The McCartys flow, 12 mi (20 km) east of Grants, New Mexico, is typical *pahoehoe* (a term coined by Dutton); its ropy surface formed as slowly moving viscous lava congealed. A typical *aa* flow can be seen at San Rafael, along New Mexico 53, 1.2 mi (2 km) south of Grants, where the Zuni Canyon alkali basalt flow debouched from a narrow canyon. Its rubbly top formed when the crust of the rapidly moving flow broke up. In Zuni Canyon, most of the flow is covered, but remnants can be seen as a "bath-tub ring" on the canyon walls.

The primary features of the McCartys pahoehoe flow can be seen in rest areas on both sides of I-40, about 66.5 mi (107 km) west of the intersection of I-40 and I-25 in Albuquerque (Fig. 1). The rest areas are in the terminal segment of the flow, an olivine tholeiite that flowed from a small cone about 25 mi (40 km) south of I-40. The main part of the flow can be seen along New Mexico 117. Just south of I-40, a tongue flowed through a narrow constriction and then turned sharply east for its last 6 mi (10

km), following the valley of the Rio San José. The river was partly dammed and the area is swampy, even though precipitation is only 10 in. (25 cm) per year. Indian potsherds of Pueblo I period, AD 700–900, have been found buried beneath 4 ft (1.2 m) of the youngest valley fill, 17 mi (27 km) east of the terminus of the McCartys flow. Nichols (1946) correlated this valley fill with material beneath the McCartys flow.

The primary features of the McCartys flow were described in numerous publications by R. C. Nichols, especially Nichols (1946). In addition to the ropy surfaces characteristic of pahoehoe lavas, the flow is characterized by pressure ridges, collapse depressions, and minor features such as spatter cones, squeeze-ups, cavities, grooved lava, tree molds, cracks, and banded lava. Longitudinal and transverse pressure ridges abound in the last mile (1.6 km) above the terminus, which includes the rest areas. They surround a general collapse area (Fig. 3) formerly a dome 27 ft (8 m) high (Nichols, 1946). After a crust 6–9 ft (2–3 m) thick had formed, lava drained from the dome, and pressure ridges formed by lateral compression, sliding, and hydrostatic pressure. The next 2 mi (3.2 km) above the terminus are characterized by numerous collapse depressions, formed when liquid lava drained out from lava tubes beneath a solidified crust. The

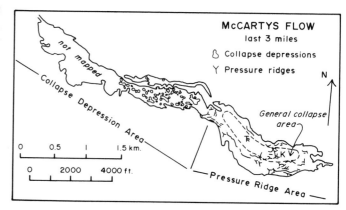

Figure 3. Distribution of collapse depressions and pressure ridges, terminus of McCartys flow. From Nichols (1946).

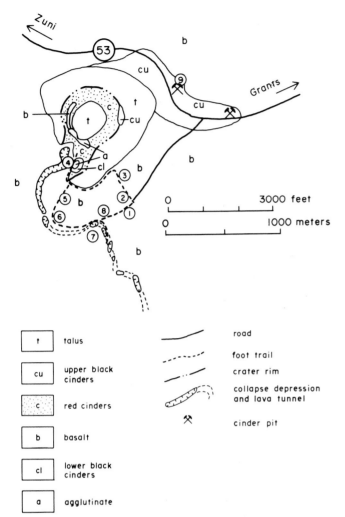

Figure 4. Geologic sketch map of Bandera Crater and Zuni Ice Cave, adapted from J. D. Causey, unpublished M.S. thesis, University of New Mexico (1971). 1, parking area; 2, tree molds; 3, hornito; 4, (heading into the crater breach) spatter (agglutinate) rampart on right, drained lava lake and head of channel-tunnel system on left; 5, aa; 6, pahoehoe with ropy surface, pressure ridges, lava tubes, collapse depressions; 7, ice cave.

surface of the flow is highly irregular; ridges and valleys tend to be parallel to flow direction.

The McCartys flow is an olivine tholeiite (XRF analysis, in weight percent, by J. Renault, New Mexico Bureau of Mines and Mineral Resources: $SiO_2 = 49.93$, $TiO_2 = 1.38$, $Al_2O_3 = 16.62$, $Fe_2O_3 = 1.54$, $FeO = 9.25$, $MnO = 0.17$, $MgO = 8.45$, $CaO = 8.90$, $Na_2O = 2.89$, $K_2O = 0.75$, $P_2O_5 = 0.25$). It has phenocrysts of olivine and ophitic clinopyroxene to 2 mm and plagioclase to 1 mm in a matrix of clinopyroxene, plagioclase, and opaque oxides. Embayed quartz grains may be xenocrysts.

BANDERA CRATER AND ZUNI ICE CAVE

Travelers crossing western New Mexico on a hot summer day might be surprised to learn that billboards advertising a perpetual ice cave refer to a real geologic phenomenon, not a tourist trap. Zuni Ice Cave is the most accessible of numerous lava tunnels in the Zuni volcanic field; several of them are partly filled with ice. Alkali basalt lava that once flowed through the tunnel originated at Bandera Crater, one of the youngest cinder cones in the southwestern United States (Fig. 4). The rim of Bandera Crater stands 430 ft (130 m) above the surrounding country; the central crater is 650 ft (200 m) deep. Bandera Crater and Zuni Ice Cave have long been owned and maintained by the David Candelaria family (phone 505/783-4303) and are open daily from 8:00 a.m. to one-half hour before sunset. They can be reached via a short (0.6 mi or 1.0 km) side road off the south side of New Mexico 53, 25.2 mi (40.3 km) from its intersection with I-40 in Grants (Fig. 1). In 1985, admission was $3.50 for adults and $1.75 for children aged 5 to 11 years. Bandera Crater straddles the Continental Divide and is surrounded by ponderosa forests at an elevation of 7,900 ft (2400 m). In winter, cold weather and snow are common. Trails rise about 120 ft (35 m) over 0.5 mi (0.8 km) between the Candelaria trading post and a breach on the southwest side of the crater; after that, they are downhill or level.

There are 74 vents in the Zuni volcanic field, most of them aligned along NE structural trend, parallel to a positive gravity anomaly. The field has erupted between 15 and 30 mi^3 (64 and 123 km^3) of basalt (Ander et al., 1981). Bandera Crater is superimposed on an earlier and smaller crater on its south side. According to an unpublished study by J. D. Causey (University of New Mexico), it probably formed by the following sequence of events: (1) tephra eruption from the small crater; (2) agglutinate eruption from the small crater; (3) major tephra eruption from Bandera Crater; (4) lava eruption in Bandera Crater; (5) eruption of red cinders to form the present cone, cinders have partly fused inclusions of Permian sedimentary rocks; (6) quiescence; and (7) explosive eruption to form a deep crater, ejection of black cinders and alkali basalt bombs with ultramafic inclusions. The general sequence of tephra-lava-tephra is common to many cones in the field. The lava flow that developed during stage 4 breached the cone and covered about 40 mi^2 (100 km^2). Lava ponded just outside the breach and drained through a lava tunnel, 17.9 mi (28.6 km) long. As the tunnel has partly collapsed, much of its course is now marked by depressions and short channels. Ice accumulated in part of the tunnel where air circulated in winter but was stagnant in the summer, because of differences in the angle of incident sunlight (Fig. 5). The ice is layered and colored green by algae. Ice caves exist only in cool climates; at Bluewater, New Mexico (the weather station nearest the Zuni Ice Cave), the mean annual temperature is 47.6°F (8.7°C). The temperature in the ice cave is 31°F (−0.5°C) throughout the year.

The lava flow of stage 4 is an alkali basalt with phenocrysts of olivine (to 3 mm) and plagioclase laths (1 mm). From the degree of weathering, an age of about 10,000 years can be estimated. Along its axis it is typical pahoehoe with the same primary features as the McCartys flow, but it turns into aa along its flanks and blocky lava near its terminus. The black cinders from stage 8

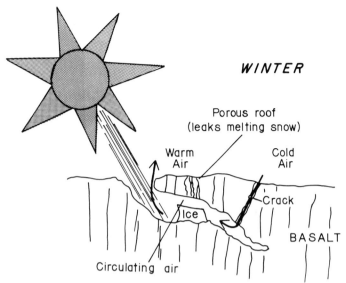

Figure 5. Winter air circulation in Zuni Ice Cave. In summer, the sun's rays are steeper and do not reach the bottom of the cave, where cold air stagnates. Air circulation was determined through smoke experiments by Harrington (1940).

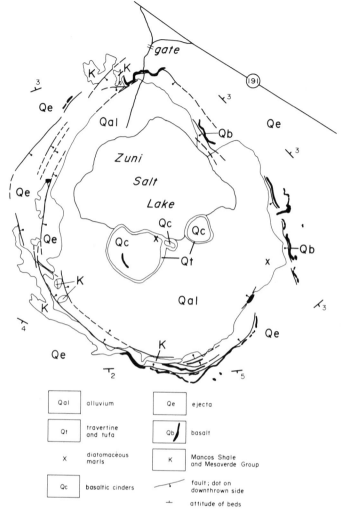

Figure 6. Geologic map of Zuni Salt Lake, from Bradbury (1967) and Cummings (1968).

could be much younger. A Pueblo I stone axe (AD 700–900) was found in a cinder pit by a bulldozer operator, who thought he had uncovered it from beneath the cinders. The possibility that it had fallen to the floor of the pit from above the cinders cannot be ruled out. The state of erosion of Bandera Crater is about the same as at Sunset Crater, Arizona, dated at AD 1065 by the tree-ring method. Bombs found in cinder pits along NM Highway 53 are cored by inclusions of olivine-red spinel lherzolite and pyroxene-green spinel lherzolite, as well as sandstone, shale, anorthoclase crystals, and jasper (Laughlin et al., 1971). The rim around a lherzolite inclusion is a nepheline-normative alkali basalt (analysis by K. Aoki in weight percent: SiO_2 = 44.47, TiO_2 = 3.04, Al_2O_3 = 15.22, Fe_2O_3 = 4.39, FeO = 8.42, Mn = 0.42, MgO = 9.30, CaO = 8.80, Na_2O = 3.38, K_2O = 1.60, H_2O^+ = 0.28, H_2O^- = 0.08, P_2O_5 = 0.58). Although the last-stage eruption of Bandera Crater may be close in age to the McCartys flow, the products are chemically quite different.

ZUNI SALT LAKE

In 1598 Don Juan Oñate came to Cibola in search of gold but settled for "unas salinas famosas mejores que Christianos han discobierto" (one of the most noted and best salt pans which Christians have discovered). During hot and dry weather, salt precipitates from Zuni Salt Lake (mean depth 2.8 ft—70 cm), which occupies the northern half of a maar crater, about 1.0 mi (1.6 km) in diameter (Fig. 6). Its maximum salinity reaches 35% by weight, about ten times that of sea water. Actually, there are two salt lakes; a second small saline pool occupies the crater of the largest of three cinder cones that rise from the center of the maar. The archeology, ethnology, limnology, biology, volcanol-

ogy, and economic geology of Zuni Salt Lake were described by Bradbury (1967); a 1:6,000 geologic map was published by Cummings (1968).

Zuni Salt Lake occupies the best preserved and northernmost member of a NNE-trending belt of maars and cinder cones, about 35 mi (60 km) long and probably controlled by a fracture zone. Elevation of the lake is about 6,215 ft (1895 m). In the surrounding country, outcrops of shale and sandstone of the Upper Cretaceous Mancos and Mesaverde Groups are covered by sparse vegetation.

In general, maar craters form by phreatomagmatic or Surtseyan processes, i.e., repeated steam explosions resulting from interactions between ascending magma and surface water or shallow groundwater. Zuni Salt Lake documents a history of decreasing magma-water interactions. An early Surtseyan stage of nearly continuous phreatomagmatic surges gave way to a Strombolian stage of intermittent explosions (Fig. 7) and a final stage of cinder eruptions, to form the central cones. The particles of Surtseyan

ZUNI SALT LAKE

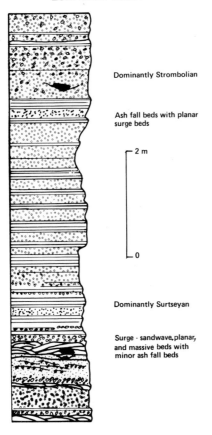

Dominantly Strombolian

Ash fall beds with planar surge beds

2 m

0

Dominantly Surtseyan

Surge - sandwave, planar, and massive beds with minor ash fall beds

Figure 7. Stratigraphic column of upper third of crater-rim beds of Zuni Salt Lake maar, showing transition from dominantly Surtseyan (phreatomagmatic) eruptions of ash, emplaced in pyroclastic surges, to Strombolian eruptions of ash, lapilli, and scoria falls.

deposits, the coarser particles of Strombolian fall deposits are better sorted, angular, glassy (sideromelane), and vesicular. Fine ash was deposited downwind as much as 5 mi (8 km) from Zuni Salt Lake. Large (to 12 in—30 cm) clasts of basalt and pre-volcanic sedimentary rocks were ejected ballistically all during the eruption.

The cause for the decrease in water-magma interaction may have been a basalt ring dike that was emplaced during the eruption. In places, it penetrated to the surface to form spatter and short flows. The ring dike may have progressively sealed rising magma from contact with groundwater.

Subsidence of the Zuni Salt Lake maar occurred during the eruption, as is shown by numerous small faults. At the end of the eruption, the rim of the maar crater stood at an elevation of about 6,500 ft (2,000 m), 100 ft (30 m) above the surrounding country. Its floor was about 400 ft (120 m) below the rim. According to Bradbury (1967), the climate was more humid than at present, and the crater filled to the 6,270-ft (1,910-m) contour with a brackish lake in which diatoms, ostracodes, gastropods, and algae lived. P. E. Damon (University of Arizona) dated ^{14}C from the charophyte *Chara* at 22,900 ±1400 years. The basin partly filled with tufa, marl, sand, gravel, and, as the climate became drier, salt and gypsum. The lake shrank to one-third of its former size, and its brines now only support bacteria, algae, the brine shrimp *Artemia salina,* and the shore fly *Hydropyrus hians.* Salinity varies with evaporation and influx of fresh water; an analysis made in 1935 reported (in ppm) Ca = 345, Mg = 2255, Na = 75,000, K = 498, $CO_3^=$ = 46, $(HCO_3)^-$ = 235, $SO_4^=$ = 14,650, Cl^- = 113,100, Br^- = 35, total dissolved solids 206,306, specific gravity at 20°C = 1.158. In the small pool in the cinder cone, total dissolved solids were 99,935 ppm (ocean water contains about 35,000 ppm, dissolved solids and the Great Salt Lake 263,000 ppm). All investigators have concluded that Permian evaporite deposits are the source of salt. The fault that controls the Zuni Salt Lake maar may have acted as a conduit for brines.

Indians have obtained salt from Zuni Salt Lake for at least 1000 years. In modern times salt was harvested for cattle feed, water softeners, ice cream makers, metallurgical processes, and highway salting. At present the plant is abandoned. The road leading from the crater rim to the lake is barred for vehicles, but visitors can go on foot. At times of low water, salt-encrusted tumbleweeds, grasshoppers, etc., can be collected along the lakeshore.

surge deposits were propelled and rapidly transported along the ground in a cushion of turbulent steam. Consequently, they are abraded, fine-grained, and moderately to poorly sorted. During the early high-energy Surtseyan stage, sand-wave and massive surge beds built up a tuff ring. The ash particles are strongly altered to palagonite; accretionary lapilli and armored mudballs are common. During transition to the Strombolian stage, weakening surges deposited their load in planar beds; coarser ash-and-lapilli fall deposits became abundant and mantled the earlier tuff ring. In contrast to the abraded and altered particles of surge

REFERENCES CITED

Ander, M. E., Heiken, G., Eichelberger, J., Laughlin, A. W., and Huestis, S., 1981, Geologic and geophysical investigations of the Zuni-Bandera volcanic field, New Mexico: Los Alamos National Laboratory Report LA-8827-MS, 39 p.

Bradbury, J. P., 1967, Origin, paleolimnology and limnology of Zuni Salt Lake maar, west-central New Mexico [Ph.D. thesis]: Albuquerque, University of New Mexico, 247 p.

Cummings, D., 1968, Geologic map of the Zuni Salt Lake volcanic crater, Catron County, New Mexico: U.S. Geological Survey Miscellaneous Geologic Investigations Map I-544, scale 1:6,000.

Harrington, E. R., 1940, Desert ice box: New Mexico Magazine, July 1940, p. 14–15, 45.

W. E. Elston and K. H. Wohletz

Laughlin, A. W., Brookins, D. G., Kudo, A. M., and Causey, J. D., 1971, Chemical and strontium isotopic investigations of ultramafic inclusions and basalt, Bandera Crater, New Mexico: Geochimica et Cosmochimica Acta, v. 35, p. 107–113.

Lipman, P. W., and Mehnert, H. H., 1980, Potassium-argon ages from the Mount Taylor volcanic field, New Mexico: U.S. Geological Survey Professional Paper 1124-B, p. B1–B8.

Nichols, R. C., 1946, McCartys basalt flow, Valencia County, New Mexico: Geological Society of America Bulletin, v. 57, p. 1049–1086.

ACKNOWLEDGMENT

Elston's work was funded by NASA grant NGR 32-004-062.

Precambrian–Upper Paleozoic geology along I-40 east of Albuquerque, New Mexico

James R. Connolly, Department of Geology and Meteoritics, University of New Mexico, Albuquerque, New Mexico 87131
Barry S. Kues, Department of Geology, University of New Mexico, Albuquerque, New Mexico 87131

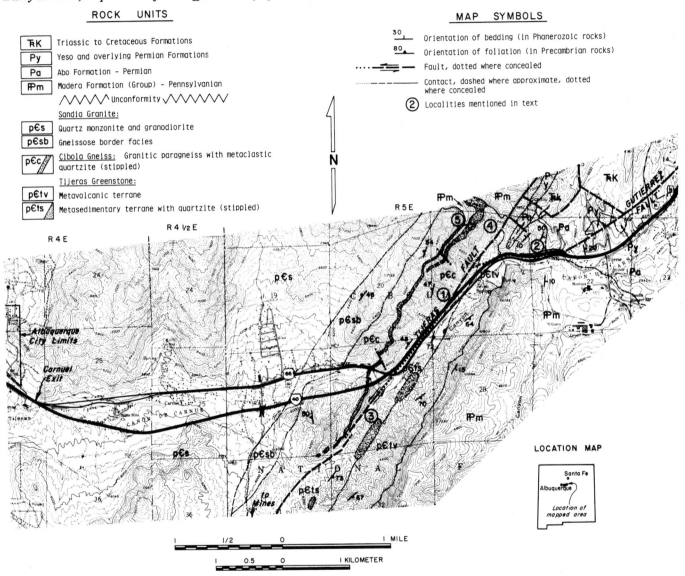

Figure 1. Generalized bedrock geologic map along Interstate 40, east of Albuquerque, New Mexico. Geology generalized from Connolly (1982) and Kelley and Northrop (1975). Topographic base from USGS Tijeras and Sedillo 7½′ quadrangles.

LOCATION AND ACCESSIBILITY

I-40 follows Tijeras Canyon through the Precambrian core and overlying Pennsylvanian–Lower Permian sequence of the south end of the Sandia Mountains, beginning immediately east of the Albuquerque city limits. Geologic features of this area are well displayed in road cuts and exposures near I-40 and are summarized in this text for an 8.6-mi (13.8-km) interval of I-40,

beginning at the eastbound Carnuel (locally pronounced "carn-way") exit (Fig. 1). This section of I-40 was constructed with wide shoulders, and there are numerous places where, with reasonable care, it is safe to stop along the highway to examine road cuts. Site descriptions, below, progress from west to east (from Precambrian to Permian), but the traveler should be aware that most of the road cuts referred to are on the north side of the interstate highway, along the west-bound lane. To examine them closely, if proceeding eastward, it is necessary to carefully cross the interstate or to drive to a nearby exit and backtrack. Short hikes off the highway, particularly in Precambrian terranes, can be most informative, and old U.S. 66, which parallels I-40 between Carnuel and Tijeras, provides the safest access for a "hands-on" look at the Precambrian geology. Recent geologic maps of the area considered here were published by Kelley and Northrop (1975); Myers and McKay (1976) and Connolly (1982).

SIGNIFICANCE

This short, easy drive traverses the Precambrian granitic core of the Sandia Mountains. They are a classic east-tilted rotational fault block, uplifted since Miocene time, that forms part of the eastern margin of the subsiding Rio Grande rift. Precambrian rocks include an exposure of rare orbicular rock in the granite, a debatably "gradational" intrusive contact between granite and granitic paragneiss, and two contrasting multiply deformed Precambrian metamorphic terranes that are juxtaposed across a major northeast-striking fault that is part of a structural zone 62 mi (100 km) or more long. A major regional unconformity, representing over a billion years of geologic time, separates Precambrian and Paleozoic rocks. Excellent exposures of a variety of late Paleozoic lithologies and faunas typical of central New Mexico occur above the unconformity. In addition, an unusually well exposed section of Pennsylvanian cyclothemic sedimentation is present, and the area has historical importance as the location of some of the earliest described and studied Pennsylvanian fossils in the western United States.

GEOLOGY

The geology of Tijeras Canyon site is discussed below as a series of stops and checkpoints along I-40, beginning at the Carnuel Exit (No. 170), just east of the Albuquerque city limits, and continuing to the Zuzax Exit (No. 178), 8.6 mi (13.8 km) to the east. (Checkpoints are given in miles.) This interval of highway, with its numerous well-exposed road cuts, provides an unusually good view of both the Precambrian and overlying late Paleozoic units that compose most of the Sandia and Manzano Mountains.

0.0. Carnuel Exit (No. 170). Good exposures of Sandia Granite in road cuts along preceding 0.5 mi (0.8 km); spheroidally weathered exposures of Sandia Granite continue for the next 2.5 mi (4 km).

1.5. Stop 1. Sandia Granite in road cuts and for next 0.4 mi (0.6 km); it holds up overpass at Mile 1.7. "Granite" in this area is typically quartz monzonite containing large (0.4–1.6 in. or 1–4 cm) microcline megacrysts. U-Pb zircon and Rb-Sr whole rock dates indicate a crystallization age of 1.44 ± 0.04 b.y. for Sandia Granite (Brookins and Majumdar, 1984). Along its southeast margin, Sandia Granite intrudes granitic paragneiss (Fig. 1), the Cibola Gneiss of Kelley and Northrop (1975). A gradational contact between gneiss and granite has been produced by a com-

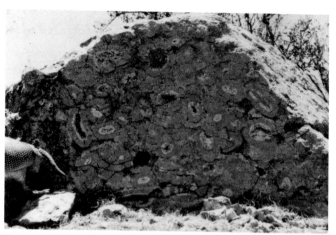

Figure 2. Orbicular rock outcrop in Tijeras Canyon (photograph courtesy of K. A. Affholter).

bination of cataclastic deformation in border facies granite and gneiss, recrystallization, intrusion of aplite and pegmatite dikes, and overprinting of a potassic/sericitic alteration halo in granite and gneiss. Connolly (1982) argued in favor of a metasomatic origin for the microcline megacrysts in the Cibola Gneiss, but megacrysts are rare in paragneiss, and cataclasis in megacryst-bearing rocks commonly obscures original textures (R. H. Vernon, pers. comm., 1985). Barium zoning studies of megacryst-bearing granitic rocks in northern New Mexico suggest an igneous origin for the megacrysts, but comparable studies have not been done in Sandia Granite.

2.3. Stop 2. Cibola Gneiss in road cuts on right and continuing to mile 2.5. North of this stop, the area just west of the granite-gneiss contact contains one of three known exposures of orbicular granitic rock (Fig. 2) in the Sandia Mountains. The orbicules, typically 2.8–4 in. (7–10 cm) across, are set in an aplite-pegmatite granitic matrix and have magmatic or metamorphic cores surrounded by a white plagioclase shell and one or two salmon-colored microcline shells (Affholter and Lambert, 1982). A metasomatic origin for the Sandia Mountains orbicular rocks has been proposed, but recent crystallographic and geochemical studies point to an igneous origin for this unusual rock.

2.7–3.0. Stop 3. Cibola Gneiss in large road cut on right. Excluding the gneissose border facies probably best considered as part of the Sandia Granite, the Cibola Gneiss is chiefly metaarkose with a prominent resistant ridge of gray metaclastic quartzite (visible in these road cuts). Foliation in gneiss varies from weak to strongly gneissose, with cataclastic textures prominent in thin section. Both large- and small-scale structures indicate the gneiss/quartzite terrane has been isoclinally folded about gently northeast-plunging F_1 axes and refolded tightly about steeply northwest-plunging F_2 axes prior to granite intrusion. Andalusite, muscovite, and quartz are dominant in folded metapelitic lenses in quartzite suggesting low-pressure amphibolite facies metamorphism during F_1–F_2 folding; late post-kinematic sillimanite related to granite intrusion is present locally (Connolly, 1982).

Figure 3. Close-up view of Tijeras fault in roadcut on I-40 at Stop 4. Highly fractured Cibola Gneiss (Gn) and colluvium derived from gneiss (Qc) are juxtaposed with sheared and fractured phyllitic greenstone (Gr). Arrows show probable sense of most recent movement only, which has juxtaposed greenstone and colluvium.

Rb-Sr whole rock isochron ages suggest Cibola Gneiss formation occurred about 1.6 b.y. ago (Brookins and Majumdar, 1984). Reddish Cibola Gneiss, in eastern road cuts, is cut by small, dark-green Tertiary lamprophyre dikes.

3.8. Stop 4 (Loc. 1, Fig. 1). Tijeras fault is well exposed in road cut on north side of highway. The fault here is approximately vertical, strikes N40°E, and juxtaposes Cibola Gneiss and Quaternary colluvium on the northwest with Tijeras Greenstone on the southeast (Fig. 3), indicating that at least part of the Tijeras fault must still be considered active. Intense shearing related to the Tijeras fault has affected most bedrock exposures in road cuts here, and a short hike up canyons to the northwest is required to observe intact gneiss. The Tijeras fault is part of a system of faults and intrusive centers that extend at least 62 mi (100 km) northeast from Albuquerque and have a complex history of activity from Precambrian through Quaternary time. En echelon aplites, pegmatites, and quartz veins in Tijeras Greenstone adjacent to the fault (Loc. 3, Fig. 1) and shear fractures in Cibola gneiss offsetting pinnate quartz veins (Loc. 4, Fig. 1) together suggest left-lateral strike slip movement on the Tijeras fault roughly contemporaneous with granitic magmatism. Facies changes in Pennsylvanian strata across the fault near Tijeras suggest dip-slip movement (down on the north) in late Paleozoic time, and offset structures in the same area and farther north suggest late Cenozoic left-slip and dip-slip movement.

The Tijeras Greenstone, exposed between Stop 4 and the

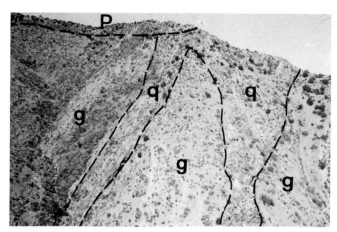

Figure 4. Structural closure in isoclinally folded quartzite (q) in Cibola Gneiss (g), with overlapping upper Paleozoic strata (P). View is northeast toward Loc. 4 on Fig. 1, near Stop 5.

prominent late Paleozoic limestone ledges to the southeast, is one of several Proterozoic greenstone terranes in New Mexico. Metavolcanic rocks, chiefly tholeiitic metabasalts and minor felsic metavolcanics, are dominant and underlie the steep slopes below the Paleozoic strata. The low hills southwest of the arroyo are underlain by a thin slice of metasedimentary phyllites, schists, and quartzites adjacent to the Tijeras fault (Fig. 1). The Tijeras Greenstone was isoclinally folded about southeast plunging F_1 axes and tightly refolded about sub-parallel F_2 axes, resulting in an internal fold geometry distinct from F_1-F_2 folds in Cibola gneiss. Metamorphic mineral assemblages in phyllites and metabasalts indicate low-pressure amphibolite facies prograde metamorphism synkinematic with F_1 and F_2 (at about 550°C and 2 kb water pressure) followed by local retrograde metamorphism at greenschist facies conditions (Connolly, 1982).

4.2. Stop 5. Mile marker 173; good road-cut exposure of somewhat sheared Tijeras Greenstone. Large-scale fold closure in quartzite of Cibola Gneiss (Fig. 4; Loc. 5, Fig. 1) is northwest of this outcrop.

Poorly exposed gray to maroon shales and gray limestone ledges of Madera Formation above Tijeras Greenstone in the eastern part of the road cut. To the south, across Tijeras Arroyo, steeply dipping beds of the Madera Formation are visible in the hills. This area is of historical importance, for in 1853 Jules Marcou, the first geologist to study the geology of New Mexico, visited these outcrops. Some of the fossil invertebrates he collected were later described as new species, among the first to be established on the basis of New Mexico specimens.

4.6. High road cut exposes highly deformed and faulted Madera limestones, shales, and minor sandstones. Deformation here is probably related to passive faulting and detachment of Paleozoic cover above the basement-rooted Tijeras fault.

4.8. Stop 6 (Loc. 2, Fig. 1). Steeply dipping, cyclically deposited units of the Madera Formation exposed in road cuts along the interstate highway and in a high road cut along the frontage road, above, immediately to the north (Fig. 5). The Pennsylvanian sequence in the Sandia and Manzano Mountains disconfor-

Figure 5. Road cut in upper part of Madera Formation at Stop 6. Lower part of cut is along I-40; upper part is along a frontage road just north of I-40. Numbers refer to the upper limestones of each of the five cycles exposed here. Mollusc-dominated marine faunas are present just beneath limestones 3 and 4.

mably overlies Precambrian units in most places (remnants of Mississippian rocks are also present locally) and reaches a thickness of nearly 2,000 ft (610 m). It represents deposition from Atokan through Virgilian time. Regionally the sequence consists of a basal, relatively thin, predominantly clastic unit (Sandia Formation), overlain by the very thick Madera Formation (Kelley and Northrop, 1975). The lower part of the Madera is mainly composed of thick, massive, gray, marine limestones, with an increasing proportion of gray, black, green, and red shales and mudstones and brown to red sandstones in the upper half, representing a wide variety of marine and nonmarine depositional environments. The variability of the upper Madera rocks is visible at this stop and in road cuts along I-40 to the east. The lithologies, age, and relationships with the underlying Precambrian displayed by the Madera in Tijeras Canyon are typical of the Pennsylvanian over most of central and northern New Mexico, although different lithostratigraphic names are used elsewhere.

Deposition of these Pennsylvanian units occurred in subsiding, generally north-south trending shallow elongate basins adjacent to intermittently rising uplifts. In the Sandia–Manzano Mountains area, most nearshore, shallow marine deposition occurred in relatively unstable, restricted belts between basin and uplift, resulting in abrupt facies changes, especially in the upper part of the Madera Formation. Near the end of the Pennsylvan-

ian, renewed uplift to the north increased the volume of terrigenous sediments shed into the basins. Alternating red shales and sandstones and marine limestones characterize the uppermost part of the Madera Group, with the overlying Lower Permian Abo Formation documenting the final inundation of marine environments in this area by southward prograding alluvial plain sediments.

5.2–5.5. Stop 7. Exit 175 to Tijeras and Cedar Crest via New Mexico 14. The Ideal Cement Plant is quarrying Madera Group limestones in the distance to the south. A long road cut through the brick-red sandstones and mudstones of the Lower Permian Abo Formation is exposed at this interchange; some minor faulting is visible in the cut. Although no fossils have been reported in the Abo Formation of this area, elsewhere in central New Mexico (for example, at the south end of the Manzano Mountains and in the Nacimiento uplift northwest of the Sandias) this formation has yielded a moderately diverse Wolfcampian (Early Permian) flora and occasional reptile and amphibian remains, including some complete skeletons.

New Mexico 14 runs southward from Tijeras through the Madera Group of the Manzano Mountains for many miles, and road cuts through all levels of the Madera are plentiful. An especially interesting marginal marine to nonmarine facies of dark shales is exposed in the Kinney clay pit, 8 mi (13 km) south of Tijeras. The Virgilian biota here, not yet well studied, includes numerous plants and nonmarine bivalves, together with rare fish and amphibian skeletons, crustaceans, eurypterids, and insects (Kelley and Northrop, 1975).

6.4. Stop 8. A medium brown sandstone unit (Meseta Blanca Member of the Yeso Formation, which overlies the Abo) is faulted down into the Abo Formation along this long road cut. Farther to the east, gypsiferous members of the Yeso predominate, and the formation gradually becomes a normal marine unit with abundant Leonardian invertebrate fossils, toward the south.

7.0. Ridge extending away from the highway to the left is capped by massive brown sandstones of the Upper Cretaceous Mesaverde Formation, underlain by slope- and valley-forming Mancos Shale. The Gutierrez fault, running just north of the interstate highway, has brought a large wedge of Triassic to Cretaceous sediments into contact with the late Paleozoic units exposed in road cuts in this area.

8.3. Mile marker 177; brick red Abo formation again in road cuts.

8.6. Zuzax Exit (No. 178); Abo in road cut to left.

REFERENCES

Affholter, K. A., and Lambert, E. E., 1982, Newly described occurrences of orbicular rock in Precambrian granite, Sandia and Zuni Mountains, New Mexico: New Mexico Geological Society Guidebook 33, p. 225–232.

Brookins, D. G., and Majumdar, A., 1984, Geochronologic study of Precambrian rocks of the Sandia Mountains, New Mexico: Geological Society of America Abstracts with Programs, v. 16, p. 217.

Connolly, J. R., 1982, Structure and metamorphism in the Precambrian Cibola gneiss and Tijeras greenstone, Bernalillo County, New Mexico: New Mexico

Geological Society Guidebook 33, p. 197–202.

Kelley, V. C., and Northrop, S. A., 1975, Geology of Sandia Mountains and vicinity, New Mexico: New Mexico Bureau of Mines and Mineral Resources, Memoir 29, 135 p.

Myers, D. A., and McKay, E. J., 1976, Geologic map of the north end of the Manzano Mountains, Tijeras and Sedillo quadrangles, Bernalillo County, New Mexico: U.S. Geological Survey Miscellaneous Geological Investigations Map I-968, scale 1:24,000.

The Emory resurgent ash-flow tuff (ignimbrite) cauldron of Oligocene age, Black Range, southwestern New Mexico

Wolfgang E. Elston, Department of Geology, University of New Mexico, Albuquerque, New Mexico 87131
William R. Seager, Department of Earth Sciences, New Mexico State University, Las Cruces, New Mexico 88003
Richard J. Abitz, Department of Geology, University of New Mexico Albuquerque, New Mexico 87131

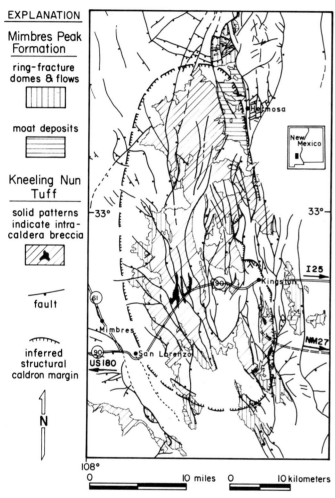

Figure 1. Geologic sketch map of the Emory cauldron. Note access roads.

LOCATION

The Emory cauldron is traversed by New Mexico 90 about 35 mi (55 km) east of Silver City, in southwestern New Mexico.

SIGNIFICANCE

The Emory cauldron (Fig. 1), one of the world's largest extinct volcanoes, is a fine example of an eroded Valles-type resurgent ash-flow tuff (ignimbrite) caldera. Since its eruption about 34 Ma, uplift and erosion have exposed its internal struc-

ture, including the septum between caldera floor and underlying pluton. It is typical of hundreds of known or suspected mid-Tertiary cauldrons in the southwestern United States and western Mexico, which represent links between granitic batholiths and relatively uneroded Pleistocene rhyolite calderas, such as those in the Jemez Mountains (New Mexico), Yellowstone National Park (Wyoming), and Long Valley (California). The main-stage collapse of an ignimbrite caldera is either a single catastrophic event or several closely spaced events, although resurgence and lesser eruptions may continue for over a million years.

INTRODUCTION

The Emory cauldron erupted during early stages of the Oligocene "extensional orogeny" that may have doubled the width of the Basin and Range province. During early (probably ductile) stages of extension, the lower crust partly melted, and a granitic magma body, tens of miles in diameter, came so close to the surface that its roof ruptured, causing a catastrophic release of hot gas, magma foam particles (pumice and shards), crystals, and pieces of wall rock. Its roof subsided, and much of the erupted material ponded in the resulting caldera, where viscous pumice and shards were compacted and welded into a plug of solid rock more than 0.6 mi (1 km) thick. The rest spilled out of the caldera, rapidly moved over the ground as ash flows, and welded into an outflow sheet up to 500 ft (150 m) thick, covering thousands of square miles. Elutriated tephra probably spread beyond the edge of the continent. The present limits of the welded outflow sheet are about 20 mi (30 km) from the caldera rim, but the original extent must have been much greater. We conservatively estimate the original volume of Kneeling Nun Tuff, the principal product of the Emory cauldron, at about 360 mi^3 (1,500 km^3) and the total volume of erupted rocks at about 480 mi^3 (2,000 km^3). The devastation must have equaled that of a vast thermonuclear battlefield. At Mount St. Helens the eruption of <1 km^3 of juvenile magma devastages 230 mi^2 (600 km^2) on May 18, 1980.

The present interpretation of the Emory cauldron follows Elston and others (1975), who also summarized earlier work. Only the southern end of the Emory cauldron has been mapped in detail (Kuellmer, 1954; Hedlund, 1977; Seager and others, 1982). We are currently investigating the northern end, previously known only from reconnaissance mapping (Ericksen and others, 1970). New Mexico Geological Society Special Publication 7 (Seager and others, 1978), contains a detailed road log across the Emory cauldron. The Society also publishes a very handy 1:1,000,000 New Mexico Highway Geologic Map. These and most other publications cited in this text can be ordered from

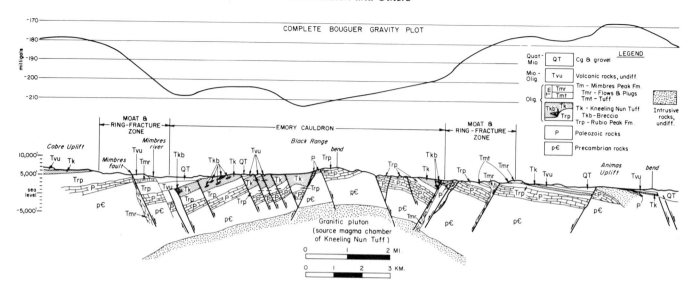

Figure 2. Geologic section and gravity profile through the Emory cauldron, approximately at the latitude of New Mexico 90. No vertical exaggeration.

GEOLOGIC DEVELOPMENT

Late Eocene to early Oligocene volcanism began with eruption of andesite and latite of the Rubio Peak Formation, about 38–36 Ma. By the time the Kneeling Nun Tuff erupted, the source volcanoes of the Rubio Peak Formation had been eroded to stumps and the intervening low areas filled with tuffaceous sandstone, sandy tuff, and ash-flow tuff beds of the Sugarlump Formation. Little is known about the sources of these units, consequently it has not been established whether eruptions of Kneeling Nun Tuff had precursors.

K-Ar dates of Kneeling Nun Tuff cluster around 34 Ma During its eruption, unstable caldera walls collapsed, and blocks of wallrock, from inches to hundreds of feet long, were engulfed. The caldera became enlarged until it measured 34 × 16 mi (55 × 25 km), elongated N-S. Current work is partly designed to show whether the Emory is a single elongated cauldron or a dumbbell-shaped double cauldron, as suggested by aeromagnetic and gravity patterns. No plinian airfall has been described, but probable surge deposits are known from the base of the outflow sheet, about 9 mi (15 km) west of the structural caldera wall (Seager and others, 1978). The southern end of the outflow sheet is normally zoned, with at least two major members. According to an unpublished study by D. L. Giles (University of New Mexico), SiO_2 content ranges from about 73 percent near the base to 66 percent 250 ft (75 m) above the base; crystals are abundant (30–50 percent) and up to 3 mm in diameter. Near the base, quartz, plagioclase, and sanidine occur in nearly equal amounts; higher in the section, plagioclase >sanidine >quartz. Biotite is the ubiquitous mafic phase; hornblende and augite occur in traces.

After an unknown interval, the caldera fill was domed dur-

ing resurgence; quaquaversal dips were about 10 degrees. At the northern end of the cauldron, the porphyritic andesite of Curtis Canyon erupted in the ring-fracture zone; it may represent a defluidized residue of Kneeling Nun magma. Following doming, there was a period of quiescence, perhaps lasting hundreds of thousands of years, in which the caldera was further enlarged by erosion to the present topographic caldera wall (Fig. 2). Kneeling Nun Tuff was stripped off the topographic wall and from the moat between the topographic wall and the resurgent dome (Fig. 3); it is now missing in a zone as much as 5 mi (8 km) wide.

Quiescence ended when a fresh pulse of siliceous lava reopened the ring fractures of the Emory cauldron. Several concentric zones of discontinuous fractures had formed during cauldron subsidence; some were located in the moat and others on the flanks of the volcano. These now became conduits for pyroclastic flows and lavas of the Mimbres Peak Formation. The pyroclastic flows erupted first and partly filled the moat with pumice beds ("moat deposits"); smaller deposits occur on the flank and lie unconformably on the Kneeling Nun outflow sheet. They were partly covered by lava flows from ring-fracture domes. Typical lavas of the Mimbres Peak Formation are high-silica rhyolite ($SiO_2 \approx 76$ percent), flow banded, and low in crystal content (1–10 percent, usually ≤5%). Crystals are quartz, sanidine, plagioclase, and biotite in variable ratios.

Hot-spring activity continued long after volcanic eruptions of the Emory cauldron had ceased. It accounts for pervasive alteration of moat deposits of the Mimbres Peak Formation to clays and zeolites and for lesser degrees of alteration of other units. Hot-spring activity may also account for mineralization in small mining districts, arranged in a crude zonal pattern around the Emory cauldron: Carpenter (Zn-Pb skarns); Tierra Blanca, Kingston, and Hermosa (Pb-Ag limestone replacement deposits); Lake Valley and Georgetown (oxidized Ag or Ag and Mn ores in limestone); and Northern Cooks Range (fluorspar). However,

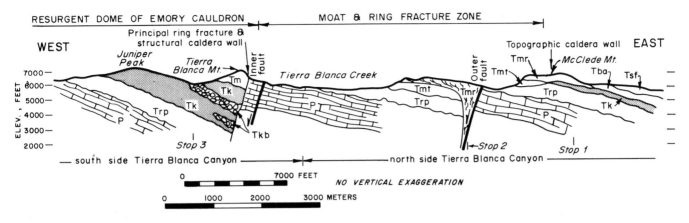

Figure 3. Geologic section through the eastern margin of the Emory cauldron in Tierra Blanca Canyon, showing positions of Stops 1, 2, and 3 in the road log. No vertical exaggeration. Symbols are the same as in Figure 2.

none of these deposits have been dated, and no direct evidence links them to the Emory cauldron.

After the Emory cauldron had become extinct, it was deeply and irregularly eroded and partly buried by volcanic rocks from other centers, especially Caballo Blanco Tuff, a coarse-grained crystal-rich ignimbrite (29.8 Ma), Bear Springs basaltic andesite (28.1 Ma), sediments, and a thick and varied sequence of aphyric and porphyritic lavas, domes, lahars, and tuffs of intermediate to silicic composition. Subsequently, it was subjected to Basin and Range faulting, and the down-faulted sides were partly buried by sand and gravel of the Miocene to Pleistocene Santa Fe and Gila Groups

STRUCTURE AND METAMORPHISM

The Emory cauldron collapsed along normal faults that symmetrically dip inward (Fig. 2). During resurgence, an apical horst developed on the dome, in contrast to the more usual apical graben of resurgent cauldrons and other domal uplifts. The horst accounts for the exposures of disturbed rocks in the septum between caldera floor and underlying pluton. In spite of cauldron subsidence and resurgence, the rocks of the septum have kept their stratigraphic integrity. On the east side of the resurgent dome they have been intensely shattered and hydrothermally altered; on the west side they have been metamorphosed. Mississippian Lake Valley limestone, for example, has been recrystallized to coarse marble in which chert nodules have reaction rims of wollastonite. Skarn assemblages (garnet + epidote + tremolite-actinolite + calcite + clinozoisite or garnet + diopside + wollastonite + calcite) developed along metasomatic veinlets. Maggiore (1981) used phase relations and overburden thicknesses to estimate T = 420–620°C; P = 0.5–1.0 kb.

During Basin and Range faulting, movement on reactivated cauldron faults was in the same sense as movement during collapse and resurgence. About 75 percent of all fault movements

and of all present dips can be attributed to this period. The flanks of the resurgent dome now dip about 40°; overlying post-cauldron rocks dip 30°.

RELATIONSHIP TO PLUTON

The relationship of the Emory cauldron to the underlying pluton can be inferred from several lines of evidence. On its east flank, intensely altered quartz monzonitic sills, laccoliths, and plugs intrude rocks as young as the Rubio Peak Formation and have yielded a zircon fission-track age of 34 Ma (Maggiore, 1981), indistinguishable from the age of Kneeling Nun Tuff. The quartz monzonite may well be cogenetic with Kneeling Nun Tuff, but alteration makes chemical affinities difficult to document. Even if fission tracks were annealed during activity of the Emory cauldron, intrusion into the Rubio Peak Formation sets an upper age limit of about 38 Ma.

A variety of rhyolitic plugs intrude Kneeling Nun Tuff, and all observers have interpreted them as cogenetic. Kuellmer (1954) described gradations from fine-grained to granitoid textures; one coarsely porphyritic plug hosts high-temperature sanidine pegmatites. A megabreccia zone on the west flank of the cauldron contains clasts of a granite unknown from outcrops but possibly cognate.

A negative Bouguer anomaly coincides with the Emory cauldron; Figure 2 illustrates the anomaly in profile. Elsewhere in the region; positive anomalies characterize outcrops of Paleozoic and Precambrian rocks uplifted into the midst of thick Tertiary volcanic rocks. The negative anomaly associated with pre-Tertiary rocks in the heart of the Emory cauldron suggests that they are a rootless septum between the caldera floor and a shallow low-density pluton. This interpretation is confirmed by the evidence for thermal metamorphism.

ROAD GUIDE

The following summary of the detailed illustrated road log

by Seager and others (1978) guides the visitor from I-25 westward across the Emory cauldron; road-log distances are in miles.

0.0. Junction of I-25 and New Mexico 90, about 15 mi (24 km) south of Truth or Consequences. Head west on New Mexico 90.

12.9. Outflow sheet of Kneeling Nun Tuff.

17.5. Hillsboro, junction of New Mexico 90 and New Mexico 27. The next 34.5 mi (55.5 km) of the road log are a detour to see relations illustrated in Fig. 3 and then return to Hillsboro. The detour involves driving on an unpaved but well-graded Forest Road suitable for passenger cars except for the last 2 mi (3 km), which should only be attempted with a pickup truck or 4WD vehicle. IF TIME DOES NOT PERMIT THE DETOUR, CONTINUE THIS LOG FROM MILE 52.0. Turn south on New Mexico 27.

25.4. Turn right (west) on Forest Road 552, up Tierra Blanca Canyon.

28.2. Contact between basaltic andesite and outflow facies of Kneeling Nun Tuff, about 400 ft (120 m) thick.

29.5. (Fig. 3, Stop 1) Climb low hill south of road and face north to see topographic caldera wall on the face of McClede Mountain. Two cooling units of Kneeling Nun Tuff outflow facies pinch out beneath the unconformity at the base of the Mimbres Peak Formation (ring-fracture lavas and moat tuffs). Outcrops near the road are andesite of the Rubio Peak Formation, underlain (near mile 30.8) by Pennsylvanian limestone.

31.1. For the next mile, road traverses moat deposits and ring-fracture lavas of the Mimbres Peak Formation. The moat deposits are zeolitized bedded pyroclastic flows (vitrified to perlite near contacts with lavas), with abundant angular to subrounded clasts of pre-cauldron rocks, to about 4 in (10 cm). There also are rounded clasts of Kneeling Nun Tuff, sparse but conspicuous by their size to 3 ft (1 m), even though no *in situ* Kneeling Nun Tuff remains in this moat zone (Fig. 3). Lavas are crystal-poor flow-banded rhyolite that emerged from at least two vents and now overlie moat tuff. For details of this complex area, see Seager and others (1978).

32.7. Road junction, turn right. THE ROAD BEYOND THIS POINT IS NOT SUITABLE FOR PASSENGER CARS! If you continue, please visit Wittenberry Ranch on left fork and inform the rancher of your presence. Drive or walk 0.1 mi (0.16 km) up a rutted rise to a gate. View *east* is to pyroclastic moat deposits capped by a rhyolite flow. The flow was fed by a conspicuous funnel-shaped intrusion on the north side of Tierra Blanca Canyon. To the *west*, massive outcrops on barren slopes in middle distance are cauldron facies of Kneeling Nun Tuff, beyond the structural caldera wall. Between here and Stop 3, the road mainly traverses Paleozoic sedimentary rocks (Fig. 3). Mine dumps from the Tierra Blanca mining district (lead-silver mantos) mark the contact between Fusselman Dolomite (Silurian) and Percha Shale (Devonian); higher cliffs are Lake Valley Limestone (Mississippian).

34.3. Road forks, stay left.

34.5. Fault, interpreted as structural caldera wall. Note that bedded Paleozoic rocks end abruptly halfway across the face of Tierra Blanca Peak, to the south (Fig. 3).

34.8. End of road (Fig. 3, Stop 3). Cauldron facies of Kneeling Nun Tuff. In foreground, andesite and green volcaniclastic sandstone has variable rock types and dips, suggesting it is a collapse breccia slumped off caldera wall and engulfed by Kneeling Nun Tuff. Eutaxitic foliation of tuff dips 30 to 40°E.; exposed thickness of tuff is 2,600 to 3,000 ft (800 to 900 m). RETURN TO HILLSBORO.

52.0. Hillsboro. Junction of New Mexico 90 and New Mexico 27. Turn left (west) on New Mexico 90. Road follows Percha Creek. For the next 4 mi (6.5 km), it passes a variety of post-cauldron volcanic rocks and lake sediments, 29 to 28 Ma, and Pennsylvanian limestone. Note unmetamorphosed nature of limestone.

56.1. Cross poorly exposed outer fault of ring-fracture zone; post-cauldron andesitic lavas and breccias faulted against Pennsylvanian limestone.

56.5. For next 1.3 mi (2 km), ring-fracture lava and moat tuff of the Mimbres Peak Formation (see mile 31.1); beyond are andesite of Rubio Peak Formation and unmetamorphosed Paleozoic sedimentary rocks. As in Tierra Blanca Canyon, Kneeling Nun Tuff is absent in this moat zone.

59.2. Cross strand of fault that forms structural caldera wall. Between here and Mimbres fault (mile 88.3) all pre–Kneeling Nun rocks (especially carbonates) are badly altered, fractured, and metamorphosed; they are interpreted as a septum between caldera floor and pluton. The rocks are Bliss Sandstone (Cambrian), El Paso Limestone and Montoya Group (Ordovician), Fusselman Dolomite (Silurian), Percha Shale (Devonian), Lake Valley Limestone (Mississippian), Sandia Formation and Madera Limestone (Pennsylvanian), Abo Sandstone (red beds, Permian), Beartooth Sandstone (Cretaceous), and Rubio Peak Formation (andesite and latite, late Eocene to early Oligocene). The comspicuous black and fissile Percha Shale makes a good marker between lower and upper Paleozoic rocks.

On the east side of the Black Range and across Emory Pass, New Mexico 90 runs below the caldera floor, therefore no Kneeling Nun Tuff cauldron facies is exposed until mile 74.3.

60.9. Main strand of fault zone of eastern structural cauldron margin; Abo and Rubio Peak Formations faulted down against lower Paleozoic rocks.

62.3. For the next 3.4 mi (5.5 km), numerous sills and plugs of altered quartz monzonite. Brown color is from oxidation of pyrite. This road may be part of the pluton beneath Emory cauldron. WINDING ROAD!

64.9. Emory Pass, elevation 8,228 ft (2,508 m), turn right to view area. View east is toward Kingston, Hillsboro, and the eastern cauldron margin in Tierra Blanca Canyon (miles 29.5 to 34.8). Fault blocks on eastern margin of Emory cauldron dip east, toward Rio Grande rift. Cliffs of Kneeling Nun Tuff above and north of view area. At altitudes above 9,000 ft (2,750 m), they are at the summit of the resurgent arch. North of New

Mexico 90, the arch nearly extends across the full width of the Black Range. To the south, the arch is replaced by the apical horst, in which precauldron rocks are exposed.

71.5. For next 0.9 mi (1.5 km), Mississippian and Pennsylvanian limestone have been metamorphosed to coarse marble. Note wollastonite rims on chert nodules and small veinlets of green garnet (zoned andradite-grossularite).

72.4. Western margin of apical horst. Coarse-grained rhyolite porphyry intrusion, about 34 Ma and probably cogenetic with Kneeling Nun Tuff, for next 0.1 mi (0.16 km); Rubio Peak Formation beyond.

74.3. Kneeling Nun–Rubio Peak contact. For the next 7.9 mi (12.7 km), cauldron-fill facies of Kneeling Nun Tuff is virtually the only rock in sight.

75.4. Devil's Backbone; excellent roadcuts in closely jointed cauldron facies of Kneeling Nun Tuff. Slow cooling has permitted partial submicroscopic exsolution of albite from sanidine (cryptoperthite with silky luster).

76.6. Several north-striking megabreccia zones over next 2.2 mi (3.5 km).

77.8. Excellent exposures of megabreccia. Most clasts are andesite of Rubio Peak Formation, some are of Paleozoic rocks. Cognate(?) granite clasts are common a bit farther down the road. Tuff matrix is locally glassy or altered. The megabreccia has

been variously interpreted as a vent or collapse breccia; it could be both, as vents tend to be controlled by marginal ring fractures.

79.3. Kneeling Nun Lookout; view across cauldron facies to outflow sheet in far distance, across ring-fracture zone in Mimbres Valley (Fig. 2). Kneeling Nun Tuff is named after a column in the outflow sheet, here visible in the distance. It overlooks the Chino mine at Santa Rita.

82.2. For next 0.4 mi (0.6 km), fault zone brings Kneeling Nun Tuff in contact with post-cauldron rocks (Caballo Blanco Tuff, Bear Springs basaltic andesite, Gila Conglomerate).

86.6. Cross Mimbres River at San Lorenzo.

86.9. Junction of New Mexico 90 and New Mexico 61. Continue on New Mexico 90.

88.3. Cross Mimbres fault zone, which brings Precambrian greenstone against Miocene Gila Conglomerate (strand along highway brings El Paso Limestone against Gila Conglomerate). This fault was active before, during, and after activity of the Emory Cauldron; here it was part of the ring-fracture zone and was reactivated by Basin and Range faulting. To see details of zoned outflow sheet (including possible base-surge deposits) in Lucky Bill Canyon, near Vanadium, New Mexico, see Seager and others (1978) for directions. However, the exposures are on private property. Permission should be obtained in advance from the Resident Geologist, Chino Mines Division, Kennecott Copper Corp., Hurley, NM 88043, (505) 537-3381, ext. 557.

REFERENCES CITED

Elston, W. E., Seager, W. R., and Clemons, R. E., 1975, Emory cauldron, Black Range, New Mexico: Source of the Kneeling Nun Tuff: New Mexico Geological Society Guidebook of Las Cruces County, 26th Field Conference, p. 283–292.

Ericksen, G. E., Wedow, Helmuth, Jr., Eaton, G. P., and Leland, G. R., 1970, Mineral resources of the Black Range Primitive Area, Grant, Sierra, and Catron Counties, New Mexico: U.S. Geological Survey Bulletin 1319-E, 162 p.

Hedlund, D. L., 1977, Geologic map of the Hillsboro and San Lorenzo quadrangles, Sierra and Grant Counties, New Mexico: U.S. Geological Survey Map MF-900A, 1:48,000.

Kuellmer, F. J., 1954, Geologic section of the Black Range at Kingston, New Mexico: New Mexico Bureau of Mines and Mineral Resources Bulletin 33, 100 p.

Maggiore, P., 1981, Deformation and metamorphism in the floor of a major ash-flow tuff cauldron, The Emory cauldron, Grant and Sierra Counties, New Mexico [M.S. thesis]: Albuquerque, University of New Mexico, 133 p.

Seager, W. R., Clemons, R. E., and Elston, W. E., 1978, Road log from Truth or Consequences to Silver City via Hillsboro and Tierra Blanca Canyon, *in* Chapin, C. E., and Elston, W. E., eds., Field guide to selected cauldrons and ore deposits of the Datil-Mogollon volcanic field, New Mexico: New Mexico Geological Society Special Publication 7, p. 33–49.

Seager, W. R., Clemons, R. E., Hawley, J. W., and Kelley, R. E., 1982, Geology of northwest part of Las Cruces 1° × 2° sheet, New Mexico: New Mexico Bureau of Mines and Mineral Resources Geologic Map 53.

ACKNOWLEDGMENTS

Seager wishes to thank G. R. Keller and P. H. Daggett for providing him with gravity data, including those used for the profile in Figure 2. His research was supported by the New Mexico Bureau of Mines and Mineral Resources, and the encouragement of the director, F. E. Kottlowski, is gratefully acknowledged. The work of Elston and Abitz was supported by NSF grant EAR 83-06397.

Paleozoic reef complexes of the Sacramento Mountains, New Mexico

Arthur L. Bowsher, Yates Petroleum Corporation, Artesia, New Mexico 88210

INTRODUCTION

Reef complexes of Mississippian, Pennsylvanian, and Early Permian age are conspicuous in the western escarpment of the Sacramento Mountains of southeastern New Mexico (Fig. 1). The coincidence of reef trends and the fault zone at the western front of the Sacramento Mountains appears to be significant. The basement fault zone on the western edge of the Sacramento block created platform margins along which Paleozoic shelf-edge reefs accumulated. Facies in Late Paleozoic formations were strongly controlled by structural trends in the Precambrian basement rocks. The Sacramento escarpment lies parallel to and adjacent to the western margin of the "Pedernal Landmass" (Pray, 1961) that trends northerly, as do the Sacramento Mountains.

The La Luz Anticline (Pray, 1961) in the northern part of the Sacramento Mountains plunges to the north (Fig. 1). Beds on the west limb dip 8 to 12 degrees. Numerous deep canyons off this west flank expose thick sequences including nearly all the Paleozoic formations. Exposures perpendicular to the reef-trends that are available in these canyons (Fig. 2) are unique. Facies

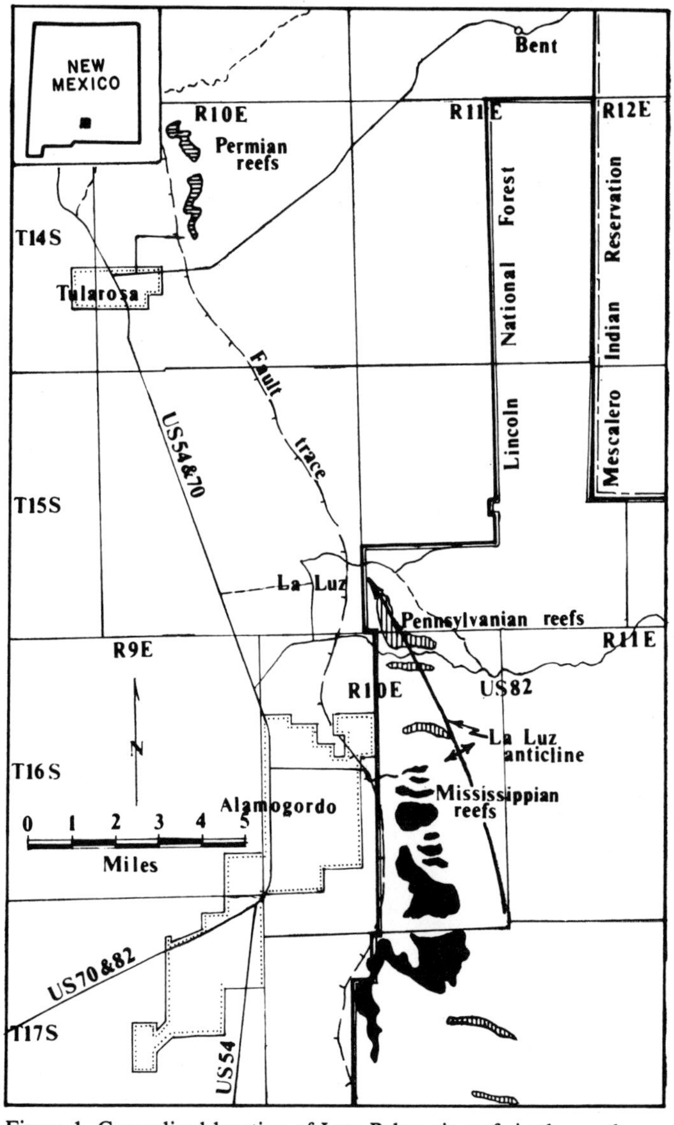

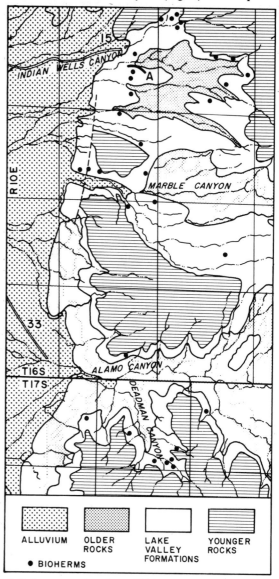

Figure 2. Index map to locations of bioherms in Mississippian strata in the northern part of the Sacramento Mountains, New Mexico. The center of the area covered by the map lies approximately 3 miles east of Alamogordo, Otero County, New Mexico (modified from Bowsher, 1948, Fig. 2).

Figure 1. Generalized location of Late Paleozoic reefs in the northern part of the Sacramento Mountains.

Figure 3. Panoramic view of the Mississippian reef complex, toward the south, along the south wall of Indian Wells Canyon; locality A of Figure 2, E½ of Sec.15,T.16S.,R.10E. Forereef rocks dip to the right and lagoonal strata are to the left. The basal ledge of the Andrecito Member is clearly evident in the small gully in the center of the photograph.

changes in the reef complexes are readily available to the field geologist. This area of the Sacramento Mountains is of extreme importance to students of carbonate rocks because of the excellent, numerous, and distinct exposures of the reef complexes.

MISSISSIPPIAN REEFS

Mississippian bioherms of the Lake Valley Formation (Laudon and Bowsher, 1949) are numerous throughout the northern part of the Sacramento Mountains (Fig. 2). Excellent examples can be seen in Indian Wells Canyon by going east on Indian Wells Road from its intersection with Pennsylvania Avenue (U.S. 54) toward the International Space Hall of Fame (Bowsher, 1986, p. 1). The little-used trail to Indian Wells Canyon leaves the macadam surface on the right, just before the rise leading to the International Space Hall of Fame. The trail runs along the east side of an old retaining dike for nearly a quarter of a mile before it turns eastward, up the alluvial fan, to Indian Wells Canyon parking area. The trail is extremely rough and bouldery; a four-wheel-drive vehicle is advised. Exposures in Indian Wells Canyon are

on U.S. Forest Service Land. Permission is not required for entry. Trails also lead from Alamogordo into Marble and Alamo Canyons where other reefs are abundant in the Mississippian.

The most instructive exposure for studying Mississippian reefs is at locality A (Fig. 2), E½ of Sec.15,T.16S.,R.10E., Otero Co., which furnishes a cross section at right angles to the reef trend. The exposure (Fig. 3 and 4) extends from forereef facies on the west (right), through compound bioherms (Bowsher, 1948) of the reef facies and eastward into the back reef or lagoonal facies. The exposure is less than 1,400 ft (425 m) long. All facies can be examined firsthand, but a bit of climbing is required. This is, undoubtedly, one of the most completely exposed cross sections of a Mississippian reef in North America. The canyon and adjacent canyons are excavated completely through the Lake Valley Formation. The reef complex (Figs. 1 and 2) is thin and narrow in the north but widens and thickens southward. The isolated, large bioherms to the south are related to the overall pattern of reef growth.

PENNSYLVANIAN REEFS

The Holder Formation reef complex (Late Pennsylvanian) was first described by Plumley and Graves (1953). The phylloid algal mounds of this complex have been the object of study by numerous geologists (Pray and others, 1977; Wilson, 1967, 1975). However, these mounds are only a part of a much larger reef complex (Fig. 5). The distribution of the Holder reefs is shown in Figure 1. Figures 6 and 7 illustrate parts of the reef complex.

Relationships of facies of the Pennsylvanian reef complex can be readily examined on hillsides above and in roadcuts along U.S. 82 between Alamogordo and Cloudcroft in the northern part of T.16S., R.10E. (Fig. 1). Parking is available along U.S. 82 near the exposures of interest. Cyclic Pennsylvanian sequences are well developed in the roadcuts. Strata here are awash with numbers assigned the beds by the many geologists who have studied these exposures.

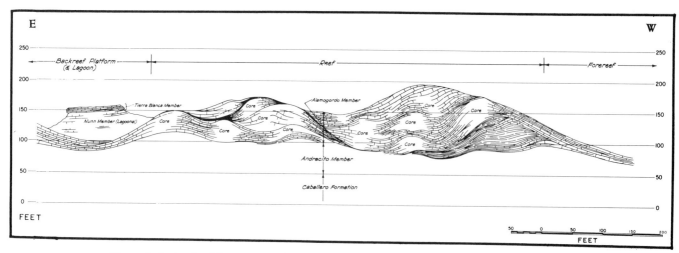

Figure 4. A sketch of the structure of the reef shown in Figure 3; Lake Valley Fm. (Miss.), Indian Wells Canyon.

W

E

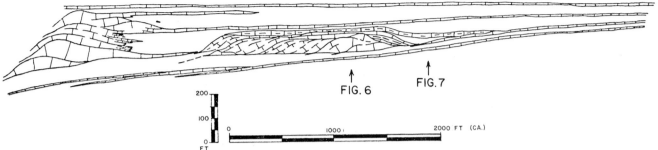

Figure 5. Reconstruction of the Holder Formation reef complex, showing approximate relative locations for Figures 6 and 7.

You are advised to watch the traffic closely; the road is paved, steep, and heavily traveled.

Excellent examples of lagoonal strata also can be examined to the north in La Luz Canyon; reached by going north through the village of La Luz. The road is paved.

The western margin of the reef complex along U.S. 82 is incomplete because of erosion. The limestones equivalent to the phylloid algal masses thin to the east and are onlapped by limestone, shale, and sandstone of the lagoon. These, in turn, pass farther east into red beds of shale, sandstone, and conglomerate. The cyclic lagoonal strata were studied by Cline (1959).

PERMIAN REEFS

Early Permian reefs occur in the Laborcito Formation northeast of Tularosa (Fig. 1). These reefs trend northerly and are transected by arroyos running west, but the bioherms are incompletely exhumed.

The most instructive locality in the Laborcito Formation reef complex is reached by traveling seven blocks east on U.S. 70 from the junction of U.S. 70 and 54 in the north edge of Tularosa. Turn left (north) and travel 0.8 mi (1.3 km) to a section corner. Turn right and travel east 0.9 mi (1.4 km). There is a place to park here, about 100 yards south of the Julius Martinez home. Walk east and north about 0.4 mi (0.6 km) into the first canyon north of the Julius Martinez home to examine the Early Permian Reefs (Fig. 8). Mr. Martinez has never objected to our going up the slope, but may do so if visitors are discourteous.

At this locality remnants of interreef and lagoonal beds remain so that one can examine the beds laterally equivalent to the phylloid algal mounds and the lagoonal strata laid down near the reef mounds (Fig. 9). Thick fusuline beds in the lagoon pinch out on the back slope of the mound. Patches of crinoidal debris are found atop the mounds. The fusuline facies grades lagoonward (east) into a gastropod facies with sponges and trilobites that

Figure 6. The lower phylloid bank of the Holder Formation in Dry Canyon, just above U.S. 82.

Figure 7. Cyclic lagoonal strata of the Holder Formation along U.S. 82, just east of the algal mounds. Rocks are partly equivalent to the phylloid algal bank of Figure 6, and partly correlate to younger bioherms of the Holder Formation lying west from the algal mounds (Fig. 5).

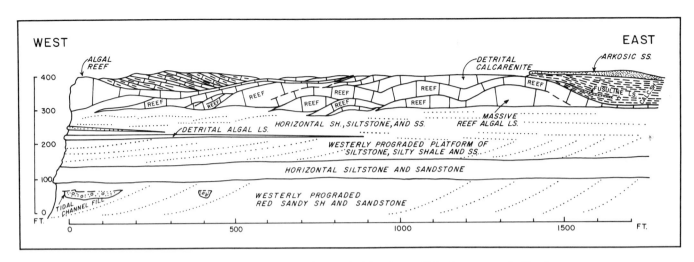

Figure 8. Cross section of the Laborcito reef and underlying foreset siliciclastic strata; north side of canyon above the Julius Martinez home, east of Tularosa (SW¼ of Sec.16,T.14S.,R.10E., Otero Co.).

passes farther east into a gray calcareous shale with abundant productids and a few pelecypods. Deposition in the lagoon closed out with shale containing only pelecypods. These strata represent excellent examples of Permian lithologic and faunal facies.

Exposures beneath the phylloid algal mounds in the SE¼ of the NW¼ of Sec.16,T.14S.,R.10E., northeast of Tularosa in an arroyo that cuts the reef, clearly indicate that the reef began growth upon a westward-prograding siliciclastic platform. It is proposed that the Recent fault that lies west of the exposures reflects later movement of the mountain block. The progradation of the Abo-like siliciclastic sediments of the Laborcito Formation spread westward over the fault and into the Orogrande Basin (Pray, 1961). The reef mounds are embedded within the siliciclastic sequences and appear to lie just above a hiatus on the prograding shelf.

SUMMARY

The reef complexes of the Mississippian, Pennsylvanian, and

Permian in the northern Sacramento Mountains are unusually well developed and exhumed. They have been the target of much study. Early work by Laudon and Bowsher (1949), Pray (1961), Otte (1959), and Plumley and Graves (1953) was followed by a series of papers published by the Roswell Geological Society and the West Texas Geological Society. These reports furnish details of the complexes and excellent road logs of the area.

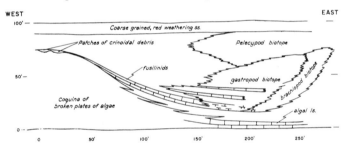

Figure 9. Generalized sketch of the east end of the Laborcito Formation reef showing facies zonation in the lagoonal beds that onlap the bioherm on the lagoonal side.

REFERENCES CITED

Bowsher, A. L., 1948, Mississippian bioherms in the northern part of the Sacramento Mountains: The Compass of Sigma Gamma Epsilon, v. 25, no. 2, p. 21–29.

Bowsher, A. L., 1986, Roadlog from Alamogordo to Indian Wells Canyon, *in* Ahlen, J. and Hansen, M., Southwest Section of AAPG Transactions and Guidebook of 1986 Convention, Ruidoso, New Mexico, Sponsored by the Roswell Geological Society: Socorro, New Mexico Bureau of Mines and Mineral Resources.

Cline, L. M., 1959, Preliminary studies of the cyclical sedimentation and paleontology of the upper Virgil strata of the La Luz area, Sacramento Mountains, New Mexico: Roswell Geological Society and Permian Basin Section, Economic Paleontologists and Mineralogists, Guidebook of the Sacramento Mountains, p. 172–185.

Laudon, L. R., and Bowsher, A. L., 1949, Mississippian formations of southwestern New Mexico: Geological Society of America, Bulletin, v. 60, p. 1–88.

Otte, C., 1954, Late Pennsylvanian and Early Permian stratigraphy of the north-

ern Sacramento Mountains, Otero County, New Mexico: State Bureau of Mines and Mineral Resources, Bulletin 50, 111 p.

Plumley, W. J., and Graves, R. W., 1953, Virgilian reefs of the Sacramento Mountains: Journal of Geology, v. 61, p. 1–16.

Pray, L. C., 1961, Geology of the Sacramento Mountains Escarpment, Otero County, New Mexico: New Mexico Bureau of Mines and Mineral Resources, Bulletin 35, 144 p.

Pray, L. C., Wilson, J. L., and Toomey, D. F., 1977, Geology of the Sacramento Mountains, Otero County, New Mexico: West Texas Geological Society, Field Trip Guidebook, October 21-22, 1977, Publication No. 1977-68, 230 p.

Wilson, J. L., 1967, Cyclic and reciprocal sedimentation in Virgilian strata of southern New Mexico: Geological Society of America Bulletin, v. 78, p. 805–818.

——1975, Carbonate facies in geologic history: New York, Springer-Verlag, p. 156–160, 182–184, and 214–216.

White Sands National Monument, New Mexico

David V. LeMone, Department of Geological Sciences, University of Texas at El Paso, El Paso, Texas 79968

LOCATION

White Sands National Monument is located in west-central Otero and east-central Dona Ana Counties in south-central New Mexico (Fig. 1). It is within and bounded by the White Sands Missile Range. The monument has a visitor's center with the usual facilities. It is open year-round with the single exception of Christmas Day. Summer hours (Memorial Day to Labor Day) are from 8 A.M. to 10 P.M. Winter hours are from 8:30 A.M. to sunset. Periodic trips to Lake Lucero led by monument personnel are planned for every other month. These trips are subject to change depending on weather conditions and the firing schedule at White Sands Missile Range. Information concerning these special trips may be obtained by phoning (505) 437-1058.

ACCESS

White Sands is easily accessible from any direction on U.S. 70-82 between Las Cruces and Alamogordo. However, this highway may be closed infrequently (seldom exceeding 1 to 1½ hours) for missile testing on the White Sands Missile Range.

SIGNIFICANCE

White Sands National Monument is the most extensive and best-documented field of gypsum dunes in North America. Relationships of the dunes to the source of the gypsum in Lake Lucero have been clearly demonstrated. The national monument, established in 1933, includes some of the most spectacular dunes in the U.S. It includes 230 mi^2 (596 km^2) or 147,200 acres (59,595 ha) of territory, of which only a small portion can be seen in the 16-mi (26-km) drive through the dunes.

SITE DESCRIPTION

The centerpiece for any serious examination of the White Sands remains the four classic papers by McKee and his associates. The first paper by McKee (1966) carefully treated the internal structures of the dunes, comparing them with dunes of Libya (Fezzan), Saudi Arabia (south of Zalim), and central-southern Colorado (Great Sand Dunes of the San Luis Valley). McKee and Douglass (1971), in their paper on the growth and movement of dunes, examined dunes in the dynamic system based on long-term measurements. McKee, Douglass, and Rittenhouse (1971) documented lee-side accumulation and deformation of these structures in barchan, transverse, and parabolic dunes in the eastern part of the field. Comparative laboratory studies and the effects of moisture were also discussed.

The study by McKee and Moiola (1975) established the areal extent of the dune field (278 mi^2 [720 km^2] or 177,920

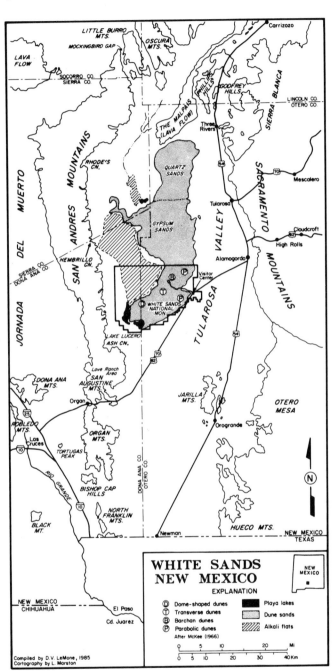

Figure 1. Location map of White Sands, New Mexico.

acres [72,032 ha]) and downwind genesis of dune type. These authors developed a theory for the mechanism of sand sea growth, based on core and trench data from this and earlier studies. Simpson and Loope (1985) have tested this model at White Sands with a detailed trench study and have proposed an alternate model based on amalgamated interdune deposits. Their paper contains an excellent reference list of the more pertinent recent theoretical papers. Allmendinger (1971) contributed a detailed study of the Lake Lucero source area and proposed a groundwater model.

The most northern part of the dune field grades into yellowish, wind-deposited, quartz sands (Fig. 1). Talmadge and Wooten estimated in 1937 that the monument proper contained approximately 4.5 billion tons of 96 percent pure gypsum. These gypsum sands support a specialized flora and fauna. The visitor's center at the monument has an area of properly labeled plants for examination and comparison. Technical and popular texts on the flora and fauna are also available. It is illegal to remove flora, fauna, or sand from the monument without the specific permission of the Park Service.

The dunes have accumulated in a large graben that extends southward to the Texas border and northward to the Malpais basalt flows in the vicinity of Carrizozo in central New Mexico. The graben (Tularosa Basin) is bounded on the west by west-dipping rocks of the San Andres Mountains and those of the more complex, cauldron-influenced, Organ Mountains south of Organ Pass on U.S. 70-82 (Fig. 1). The Sacramento Mountains, with drainage influenced by fairly recent, alluvial fan-cutting fault scarps, form the eastern boundary.

The most logical source for the gypsum of the White Sands seems to be from leached late Paleozoic formations (Fig. 2) (Panther Seep, Yeso, and San Andres Formations) and groundwater-leached Pleistocene evaporite deposits. Early and middle Paleozoic and Mesozoic sedimentary sources seem very unlikely.

Gypsum of the Panther Seep (Late Pennsylvanian, Virgilian) of the San Andres Mountains may have contributed minor amounts by leaching. The quantity of gypsum decreases in Panther Seep rocks from 176.5 ft (54 m) at Rhodes Canyon in the north to 9.5 ft (2.9 m) at Ash Canyon in the south (Fig. 1). A similar condition exists in the Middle Permian (Leonardian) Yeso Formation in the range where the total thickness of gypsum beds decreases southward from 635 ft (194 m) north of Rhodes Canyon to 178 ft (54 m) at Hembrillo Canyon. Gypsum is absent south of the Love Ranch area (Kottlowski and others, 1955). Contorted, folded, and slumped beds described in their report may be, in part, collapse and solution breccias. The authors suggested that the thinning of the entire formation is due to a loss of gypsum beds. Yeso (Spanish for gypsum) source areas are possible from the north (Oscura Mountains) and east (Sacramento Mountains) also.

The overlying San Andres Formation (Middle Permian, Leonardian-Guadalupian, and Guadalupian) probably was not a major source in the San Andres. However, San Andres outcrops in the Oscura Mountains (north) and the Sacramento Mountains (east), may have been significant contributors. In summary, the best immediate source area for gypsum seems to be from the Yeso Formation of the San Andres Mountains.

The graben was the site of a large Pleistocene lake (Lake Otero), probably better described as playa-like or a salina(s).

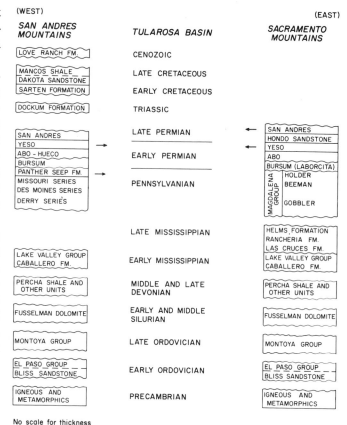

No scale for thickness
→ Potential gypsum source

Figure 2. Potential gypsum source beds, White Sands, New Mexico.

Reported terraces near Alamogordo need to be validated to determine if they are fault scarps or not. Lake Lucero (Fig. 1) is a remnant of Lake Otero. As Lake Otero shrank in size to Lake Lucero, massive amounts of selenite gypsum crystals formed as a result of evaporation of the sulfate-bearing waters. These selenite crystals are the source of the White Sands. Crystals precipitating in a clay matrix will displace the matrix by the force of crystallization, whereas those in sand form gypsum rosettes. The dune grains, in part, are fragments spalled from the crystals, largely through mechanical weathering. The white color is the result of microfractures formed on the grains during the process of saltation.

Studies by Allmendinger (1971) indicated that the gypsum sand source has been deflated away from the Lake Lucero area to the dune field. He recognized a western selenite zone that extends to 30 ft (9 m) above the playa surface with the base of the unit near, but below, the surface. Selenite crystals at the top of the unit are more abundant, larger (up to 4 ft [1.2 m]), have sharp crystal boundaries, distinct cleavage surfaces, and tend to be dark brown to golden yellow. Size of crystals decreases downward to an average of 10 to 20 mm. These crystals are frequently solution-pitted by fresh water and are gray to colorless. The environment of deposition of the selenite deposits has been interpreted to be mud flats forming at a late stage of Lake Otero between the coarse-grained alluvium derived from the San Andres and the fine-grained, saline, lacustrine deposits to the east.

The process of deflation carries the clay and silt-size particles off in suspension, leaving a lag of small crystals to be saltated away to the dune field. Allmendinger (1971) also has noted a crust that forms during times of playa filling by surface water. No gypsum salts precipitate prior to groundwater infiltration. As the mud of the playa floor dries, the thin, very fine grained, efflorescent crust thickens with precipitates from waters rising by capillary action during the process of evaporation—a sort of evaporite pumping action. The crust becomes puffy due to extension from crystallization. It then breaks down to a fine powder that is carried off in suspension in white clouds that may rise thousands of feet above the playa floor. Displacive action of crystallization apparently also brings up the small (10-20 mm) clear gypsum crystals, which then may be saltated to the dune fields. The time of origin of these crystals is unknown and may be penecontemporaneous with crust formation.

Surface waters apparently contribute little dissolved salt to Lake Lucero. Groundwater brines leached from late Paleozoic and basinal Pleistocene evaporitic sources around the basin margin may be important and continuing sources. Deflation of the original selenite crystal unit, combined with crystals precipitated from evaporite-leaching groundwater brines, must provide the necessary source for the sand.

McKee's (1966) initial and classic monograph on White Sands described the basic dune types (barchan, parabolic, dome-shaped, and transverse) at the monument. Internal analyses of the dunes were done along trenches dug through the dunes with a bulldozer, parallel to and at right angles to the wind direction. The vertical walls were trimmed and smoothed with machetes, and 5-ft (1.5-m) vertical and horizontal blue markers were sprayed on for precise positioning. The entire cut was then recorded on graph paper utilizing a Brunton compass for angular measurements. Selected features were photographed and the old-style Latex peels (up to 6 ft [1.8 m] × 8 ft [2.4 m]) were made.

McKee determined that sets of cross-stratification were mostly medium- to large-scale, with nearly all laminae dipping downwind at 30 to 34°, which exceeds the normal values of quartz sand. He found that cross-stratification was dominantly in tabular planar sets with sets being thicker at the base and thinner at the crest of a dune. Simple (nonerosion) tabular form sets were less common, wedge planar forms were scarce, and trough-type sets were rare. Bounding surfaces of the planar forms were typically either virtually horizontal or dipping at a moderate to high angle.

The gypsum dune sand is mostly medium-grained, tends to be tabular in shape, ranges from angular to subrounded, and has fair to good sorting. Variations occur in relationship to distance from the source area at Lake Lucero. The development of dunes, which rise to 65 ft (20 m), involves primary deposition by settling from suspension, saltation, and creep; redeposition by avalanching of the crest; and penecontemporaneous erosion. In active areas the interdune flat areas occupy between half and three-quarters of the area. Exposed interdune areas frequently show exhumed, eroded cross-beds, especially in the western area.

The results of the study by McKee and Douglass (1971) established approximate annual rates of movement. Embryonic dome-shaped dunes that are near the source, under constant, unobstructed wind, rarely exceed 18 ft (5.5 m) high and move 24 to 38 ft (7.3 to 11.6 m). V- and U-shaped, blowout, parabolic dunes, partially anchored by vegetation, move only 2 to 8 ft (0.6 to 2.4 m). Barchans move 6 to 13 ft (1.8 to 4 m); transverse dunes move 4 to 12 ft (1.2 to 3.7 m). Dunes of the eastern margin, farthest from the source, show no to a maximum of 5 ft (1.5 m) of movement.

Winds blow almost constantly from the southwest. Wind velocity of 17 mph (15 knots) is necessary to move the sand. Spring sandstorms (especially March and April) are the most effective for movement of the sand. Gusts of up to 55 mph are not uncommon at this time. Rate of sand movement varies with the cube of the velocity (e.g., 24 mph = 1, 36 mph = 3.5, and 55 mph = 11) (McKee and Douglass, 1971). Exceptions to this wind regime may occur in winter when the cold "northers" blow in the opposite direction. The effect of this wind is apparently negligible, with the exception that it is the cause of the infrequently preserved counterdunes.

Field observations using magnetite marker laminae indicate that moisture is a critical factor in dune crest avalanching and the formation of dune structures. Four sand-moisture relationships are recognized. They are: dry sand, wet sand, sand with wet crust, and saturated sand. These relationships give rise to the nine documented structural types recognized by McKee and others (1971), which are rotated plates and blocks, stairstep and monoclinal folds, stretched laminae, warped and gentle folds, drag folds and flames, high-angle asymmetrical folds, overthrusts and overturned folds, breakapart laminae, and fade-out laminae.

These features of crestal lee side avalanching are the result of sand flow (grains move independently) and slump (general movement of a cohesive mass). Slump structures are characteristically sharp and distinct while sand-flow structures tend to be more indistinct with distance of the sand flow travel. Dune structures at the top of the dune tend to be tensional while those at the base are compressional. Typically, near the base of the dunes, avalanche sands intertongue with deposits of well-stratified sands moved into the area by crosswinds.

In 1968 Stokes (see McKee and Moiola, 1975) developed a model to explain the multiple parallel-truncation bedding planes that he observed in the Entrada, Navajo, Wingate, DeChelly, and Coconino aeolian sandstones. The model proposed that a sand sheet was deposited, followed by a rise in the water table, and then a period of deflation down to the wet sand with a repetition of the process. The problem recognized by McKee and Moiola was that the water table rises under the mounds and dunes of sand. They proposed an alternate model in which the dunes climb as they migrate and preserve both dune and interdune sediments. Simpson and Loope (1985) pointed out that this concept, which has been refined and quantified, today represents "the single most useful concept for interpretation of vertical sequences of aeolian strata." They wrote that it requires no change in the energy or material of the system in the stacking of dune and interdune deposits. They refer to the McKee-Moiola model as being an autocyclic system (integral part of the sedimentary model). These authors tested the model at White Sands by examining a series of hand-dug trenches through the interdune deposits.

The expected lateral tops of the dune pinchouts were not observed. Utilizing two of the trenches dug parallel to prevailing wind direction (barchanoid ridge 118 ft [36 m] and transverse dune 75 ft [23 m], depth ranged from 1.6 to 2.3 ft [0.5 to 0.7 m]), two interdune facies (horizontally laminated and cross-bedded gypsum) were revealed. The horizontally laminated facies contained irregular dark and light laminations, parallel to the present interdune surface, that can be traced for distances of 3 to 26 ft (1 to 8 m). Thin beds of well-sorted, structureless, and cross-bedded sands were included and enclosed. The dark organic laminations were algal in origin. Algal layers were selectively deformed by salt ridge growth (expansion by crystallization). Simpson and Loope also documented excellently preserved examples of irregular sand mounds and shadow dunes formed by vegetation. These are observable in processes of formation today at White Sands. The second facies (cross-bedded gypsum) is conspicuously cross-bedded and has a lower tangential contact; typically, it has an eroded, corrugated upper contact of ridges (caused by early diagenetic cementation) and less resistant swales. The swales are commonly marked by a lag of siliciclastic granules. Observed thicknesses of these products of differential wind erosion range from 2 to 12 in (5 to 30 cm).

The amalgamated interdune deposit model of Simpson and Loope (1985) is in response to an allocyclic mechanism (originating outside the sedimentary model). They attributed their allocyclic model to fluctuations in sand supply, wind energy, or groundwater level, which in turn is controlled by climate. The estimated age of 12,000 to 24,000 years for the dune field would provide ample opportunity to develop the conditions necessary to periodically shut off the sand supply. Times of no or severely limited sand supply provide amalgamated deposits, whereas times of ample supply would provide the climbing dunes and interdunes of the McKee-Moiola model. It is significant to note that, without the black organic algal laminae, the amalgamated interdune deposit model would be extremely difficult to recognize.

The contribution of the classic core papers by McKee and McKee and others have set the standards for aeolian studies for the past nearly two decades. They form the basis for procedures of investigation and an integral part of our modern theoretical concepts in ancient and extant dune deposits as discussed by Simpson and Loope (1985). The rare opportunity to see this unusual accumulation of gypsum dunes in a setting that is relatively unspoiled, scenically attractive, active, and extant makes a visual and photographic delight for not only the professional geologist but also the nonprofessional tourist.

REFERENCES CITED

Allmendinger, R. J., 1971, Hydrologic control over the origin of gypsum at Lake Lucero, White Sands National Monument, New Mexico [M.S. thesis]: Socorro, New Mexico Institute of Mining and Technology, 85 p.

Kottlowski, F. E., Flower, R. H., Thompson, M. L., and Foster, R. W., 1955, Stratigraphic studies of the San Andres Mountains, New Mexico: New Mexico Bureau of Mines and Mineral Resources Memoir, 132 p.

McKee, E. D., 1966, Structures of dunes at White Sands National Monument, New Mexico (and a comparison with structures of dunes from other selected areas): Sedimentology, v. 7, special issue, 69 p.

McKee, E. D., and Douglass, J. R., 1971, Growth and movement of dunes at White Sands National Monument, New Mexico: U.S. Geological Survey Professional Paper 750-D, p. D108–D114.

McKee, E. D., and Moiola, R. J., 1975, Geometry and growth of White Sands dune field, New Mexico: Journal of Research, U.S. Geological Survey, v. 3, p. 59–66.

McKee, E. D., Douglass, J. R., and Rittenhouse, S., 1971, Deformation of lee-side laminae in eolian dunes: Geological Society of America Bulletin, v. 82, p. 359–378.

Simpson, E. L., and Loope, D. B., 1985, Amalgamated interdune deposits, White Sands, New Mexico: Journal of Sedimentary Petrology, v. 55, p. 361–365.

Banded Castile evaporites, Delaware basin, New Mexico

Roger Y. Anderson, *Department of Geology, University of New Mexico, Albuquerque, New Mexico 87131*
Douglas W. Kirkland, *Mobil Research and Development Corporation, Dallas, Texas 75381*

LOCATION AND ACCESS

This site is located along U.S. 62, 0.75 mi (1.2 km) north of the Texas–New Mexico state line and 12 mi (20 km) southwest of Carlsbad Caverns (Fig. 1). This is the main two-lane highway between Carlsbad, New Mexico, and El Paso, Texas. The shoulder of the highway at the outcrop is wide enough to accommodate a single row of cars. The geographic relationship of the outcrop to the Gypsum Plain, a vast expanse (1000 sq mi—2600 sq km) of outcropping gypsum, and to other physiographic features in the vicinity is shown in Figure 1.

SIGNIFICANCE

The famous "state-line outcrop" (Fig. 2) has been visited by many geologists because of excellent exposures of remarkable laminations in the Upper Permian (Ochoan) Castile Formation. The seasonal and annual laminations (varves) have been correlated over more than 60 mi (100 km) in the Delaware basin. Each lamination pair has been measured and assembled into a long (200,000-year) paleoclimatic record that contains evidence for the Milankovitch orbital climatic perturbations. Laminations exposed in the outcrop are sporadically contorted into beautifully complex microfolds that have attracted rock collectors for years and that may have inspired native pottery and weaving design. This brief account describes these outcrop features and refers to their larger significance within the Delaware basin and in geological and climatic processes.

DESCRIPTION OF OUTCROP AND FEATURES OF INTEREST

Good exposures of laminated Castile gypsum occur along both sides of the road cut, which is more than 300 ft (100 m) long. The main features of interest, in addition to the laminations themselves, include microfolded laminations that occur in pods on both sides of the road cut, beds of nodular gypsum within the banded gypsum, a prominent thick limestone bed near the top of the exposure on the north side of the highway, and thin beds of dissolution breccia near the middle of the outcrop on the north side.

Castile Laminations

The banded or laminated character of the Castile section is the result of the seasonal precipitation of calcium carbonate and calcium sulfate from evaporating sea water. The Delaware basin in Permian time was near the equator and near the western coast of the Gondwana supercontinent. This position brought hot

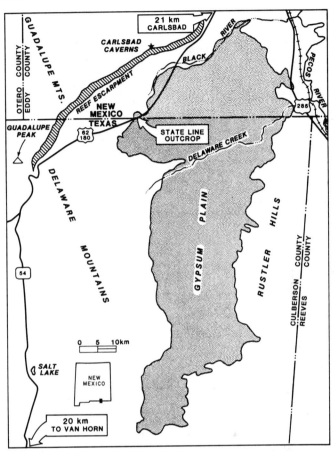

Figure 1. Location of state-line outcrop in relationship to the Gypsum Plain and to other physiographic features in the vicinity.

Figure 2. State-line outcrop of Castile Formation looking southwest toward New Mexico–Texas state line.

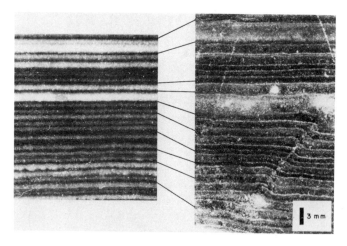

Figure 3. Correlative slabs of anhydrite (left) and gypsum showing expansion of laminae due to hydration.

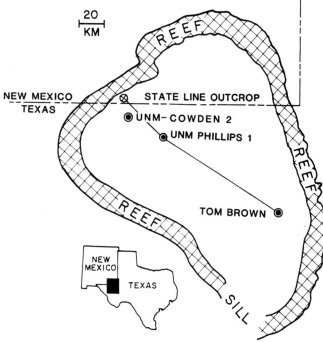

Figure 5. Location of state-line outcrop in relationship to the Permian (Guadalupian) Capitan Reef (position inferred in part), location of UNM-Cowden 2, and line of section of Figure 8.

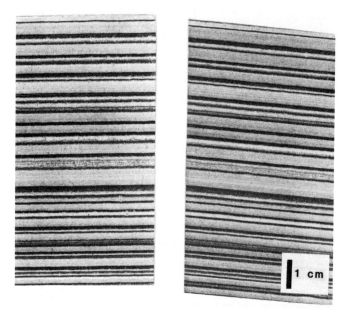

Figure 4. Correlative core slabs of Castile Formation; UNM-Cowden 2 (left) is separated from the UNM-Phillips 1 by 20 mi (32 km). See Figure 5 for locations of boreholes with respect to state-line outcrop.

desert conditions with high rates of evaporation, and the laminae precipitated as sea water that flowed into the restricted basin became concentrated. The thin dark laminae are composed of calcium carbonate stained with organic matter (~1.5 percent). The thicker white laminae are composed of gypsum near the surface and anhydrite at depth. Correlation of calcite-anhydrite laminations in cores at depth with the same laminations near the surface where the anhydrite has been converted to gypsum shows a volume increase of 44 percent resulting from hydration (Fig. 3). Many fractures evident at the outcrop may be a consequence of

this expansion. The character of the laminations has been described in detail by Anderson and Kirkland (1966) and Anderson et al. (1972).

Seasonal changes in temperature, evaporation, and ionic concentration are believed to have controlled the timing of precipitation of calcium carbonate and calcium sulfate. Thicknesses of gypsum (anhydrite) laminae are more uniform over the basin than are thicknesses of calcite laminae. Gypsum precipitation, therefore, may have been more constant throughout the year. A thin calcite layer, however, probably accumulated almost every year to form the annual cycles or couplets of carbonate-sulfate deposition (varves). Rarely, thick (up to 4 in.—10 cm—or more) bands of gypsum occur within the sequence; these are probably a result either of recrystallization, which forced the calcite out, or of short periods of greatly diminished calcite precipitation. Some thick bands show thin "ghost" laminae.

One feature that suggests that the laminations are truly varves is their remarkable lateral persistence. Figure 4 shows a typical correlation; the locations of the boreholes are shown in Figure 5. Each lamina seen in outcrop can be correlated precisely with the same lamina in deep cores from more than 60 mi (100 km) away on the opposite side of the basin. This means that each layer was precipitated almost instantaneously over the entire basin and that the entire basin must have responded everywhere to climatic forcing that was larger than the basin. Most likely, this

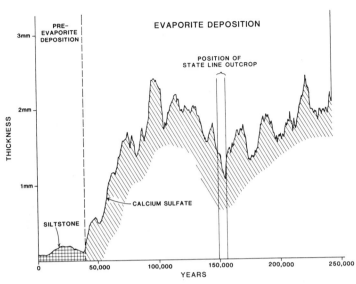

Figure 6. Smoothed plot of Castile varve thickness. Note the location of the Castile state-line outcrop in the varve time series. Note also the long-term oscillation at about 100,000 years and 10–11 shorter oscillations with a period near 20,000 years that may be related to precession of the equinox.

Figure 7. Nodular gypsum in state-line outcrop (lower one-third of outcrop, southern side); reorganization of anhydrite (gypsum) into nodules is responsible for forming some of the thick gypsum layers.

climatic forcing was dominated by strong seasonal temperature and evaporation changes, just as it is today.

A Long Paleoclimatic Record

Each pair of calcite-anhydrite laminae in cores from the Castile was counted and measured to reconstruct the longest continuous annual climatic record known to date (Anderson, 1982; Anderson, 1984). The compilation shows that the climate of that day fluctuated in a pattern of 100,000 years, which probably represents changes in the eccentricity of the earth's orbit. Varves also change in a pattern of 20,000 years, which most likely reflects changes in solar insolation that accompany precession of the equinox. The entire outcrop includes only one minimum of a 20,000-year oscillation (Fig. 6).

Salinity of the water in the Delaware basin fluctuated with a period of about 2500 years (see thickness variations in Fig. 8). A cycle began with a freshening event that reduced sulfate deposition and varve thickness. Salinity gradually increased to a maximum and culminated in thick sulfate varves that developed nodular structure after deposition. Some of these nodular zones can be seen in the outcrop (Fig. 7). Occasionally, salinity increased during these oscillations to the point that halite was deposited (Halite I, Halite II, etc., Fig. 8).

In the upper half of the Castile, salinity fluctuations became stronger as the basin filled with sediment, and freshening events had greater impact on the smaller volume of water. The thick limestone bed near the top of the outcrop on the north side of the highway represents one such freshening event. This limestone bed

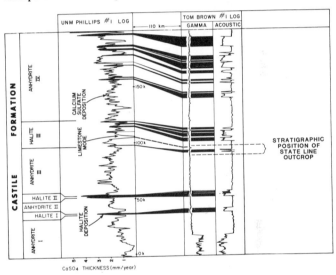

Figure 8. Plot of gamma-ray and acoustic log showing location of state-line outcrop in the Castile time series. The gamma-ray "kick" near 100,000 years can be traced to the thick limestone bed near the top of the outcrop on the north side of highway. Note also that the amplitude of the salinity oscillation (varve thickness) increases as the basin becomes shallower and that the 2500-year pattern determines the duration of halite deposition. See Figure 5 for location of boreholes.

is responsible for the differential erosion that formed the ridge in which the state-line outcrop occurs; the bed can be traced for several miles at the surface. Sulfate deposition was almost completely shut off during the freshening.

The thick limestone bed that is exposed in the outcrop produces a sharp kick on a gamma-ray log, and as a result, it can be traced over almost the entire Delaware basin. The position of the outcrop within the Halite III Member of the Castile Formation

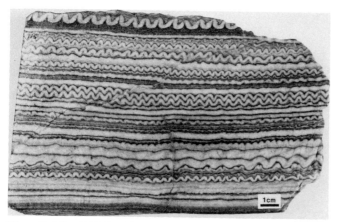

Figure 9. Microfolded calcite-gypsum laminae from state-line outcrop; note interlayering of several styles of folding.

Figure 10. "Pod" of microfolds developed in axis of larger megafold in state-line outcrop; the strike of both fold axes is parallel to Tertiary regional faulting to the west.

can be found by locating the gamma ray kick on the log in Figure 8. A bed of halite dissolution breccia that represents a salinity-concentration event can be found (with difficulty) about midway in the outcrop on the northern side. The dissolution breccia shows that the halite, now preserved in the deeply buried eastern side of the basin, once extended nearly to the western margin. This breccia has been traced through varve correlation to the base of Halite III (Anderson and others, 1978).

CASTILE MICROFOLDING

Contorted and microfolded laminations (Fig. 9) occur in isolated "pods" at various places along both sides of the outcrop. Before the microfolds were studied in detail, speculations about their origin included submarine slumping triggered by earthquakes, ripple marks, compression associated with the volume increase from anhydrite to gypsum, and crystal growth between confining layers. We were able to show that the microfolds were generated during tectonic compression, probably sometime in the early Cenozoic, long after deposition and consolidation (Kirkland and Anderson, 1970). The pods of microfolds occur within the

axial zones of subtle "megafolds" (Fig. 10). The strike of megafold and microfold axes, both about N 30°–50° W, is parallel to the trend of the major belt of Tertiary faulting west of the erosional edge of the Castile Formation. We correlated folded with unfolded laminae and compared thicknesses of laminae that were folded with those that were not folded. The gypsum or anhydrite laminae that were thick before compression generally did not fold, but became even thicker after compression. The thinner laminae are usually the ones that buckled to produce the microfolds. The more ductile calcite laminae have variable thickness and shape within the folded zones. It is the interlayering of laminae of differing thickness and ductility that is responsible for the complex style of microfolding in the outcrop (Figs. 9 and 10). Certain zones within the Castile have the proper mix of lamina thickness and composition to promote microfolding, and these zones tend to be intermittently microfolded over much of the basin.

SELECTED REFERENCES

Anderson, R. Y., 1982, Long geoclimatic record from the Permian: Journal of Geophysical Research, v. 87, no. C9, p. 7285–7294.
——1984, Orbital forcing of evaporite sedimentation, *in* Berger, A. L., (ed.), Milankovitch and climate, Part 1: Amsterdam, D. Reidel Publishing Co., Amsterdam, p. 147–162.
Anderson, R. Y., Dean, W. E., Jr., Kirkland, D. W., and Snider, H. I., 1972, Permian Castile varved evaporite sequence, West Texas and New Mexico: Geological Society of America Bulletin, v. 77, p. 241–256.

Anderson, R. Y., and Kirkland, D. W., 1966, Intrabasin varve correlation: Geological Society of America Bulletin, v. 83, p. 59–86.
Anderson, R. Y., Kietzke, K. K., and Rhodes, D. J., 1978, Development of dissolution breccias, northern Delaware basin, New Mexico and Texas: New Mexico Bureau of Mines and Mineral Resources, Circular 159, p. 47–52.
Kirkland, D. W., and Anderson, R. Y., 1970, Microfolding in the Castile and Todilto evaporites, Texas and New Mexico: Geological Society of America Bulletin, v. 81, p. 3259–3282.

Index

[Italic page numbers indicate major references]

Typeset by WESType Publishing Services, Inc., Boulder, Colorado
Printed in U.S.A. by Malloy Lithographing, Inc., Ann Arbor, Michigan